Managing Postharvest Quality and Losses in Horticultural Crops

THE EDITORS

Dr. K.L. Chadha is currently the President of Indian Academy of Horticultural Sciences (erstwhile The Horticultural Society of India), New Delhi. He was also Adjunct Professor (Hort.) at ICAR-IARI, New Delhi. He received his B.Sc. (Agri) and M.Sc. (Hort.) degrees from Punjab University, and Ph.D. from Indian Agricultural Research Institute (IARI), New Delhi.

He has served in various capacities at IARI, Punjab Agricultural University, Ludhiana Central Mango Research Station, Lucknow. He was the Director, IIHR, Bangalore, Horticulture Commissioner and Executive Director, National Horticulture Board, Govt. of India; Deputy Director General (Hort.), ICAR and National Professor (Hort.) at IARI.

He has authored/edited nearly 4 dozen books and bulletins including the series "Advances in Horticulture" published in 13 volumes covering 9,410 pages. He has been associated with several scientific societies as their Chief Patron, President or Vice President. He has served as consultant to several national and international organisations.

Dr. Chadha is the recipient of several awards for his outstanding contributions to R & D in horticulture. He is a Fellow of NAAS, New Delhi and ISHS, Belgium and D.Sc. (h.c.) from three State Agricultural Universities. Some important awards won by him include Borlaug Award, Om Prakash Bhasin Award, Dr. H.M. Marigowda Award, S.K. Mitra Memorial Award, S.S. Ranade Memorial Life Time Achievement Award, HSI-Shivashakthi Life-Time Achievement Award, B.P. Pal Memorial Award, National Agriculture Leadership Award and Krishi Shiromanni Samman. Recently in the year 2021 he was honoured with IFFCO FAI lifetime achievement award for his outstanding contribution on R&D in Agriculture.

Dr K.L. Chadha has steered policy planning, research and development of horticulture in India for almost five decades. For his yeomen's service Dr Chadha was conferred the prestigious Padma Shri by Her Excellency, the President of India.

Professor R.K. Pal is the Editor-in-Chief of Indian Journal of Horticulture published by Indian Academy of Horticultural Sciences, New Delhi. He is also chairman of the Scientific Panel on fruits and vegetables of FSSAI, Min of Health & Family Welfare, Govt. of India. He served as Director of the ICAR-National Research Centre on Pomegranate at Solapur, Maharashtra, India; Resident Advisor of ACARE in Myanmar; Professor and Head of the Division of Food science and PHT at ICAR-IARI, New Delhi. He was National Consultant of FAO and Technical expert of SAARC. He obtained his M.Sc. and Ph.D. degrees form ICAR-IARI, New Delhi and undergone post-doctoral training on PHT at the University of Arkansas, Fayetteville, USA. As FAO Fellow, he completed the International Course on Quality Management at IAC-Wageningen, the Netherlands, and received a certificate of achievement with honour for securing the first rank. He is a Fellow of NAAS, IAHS, Society for Advancement of Research on Pomegranate (SARP) and recipient of many awards, honors & recognition from India and abroad. He has more than 150 research/ technical publications in Indian and international journals. He has patented technologies on process for extraction of virgin pomegranate seed oil with retention of bioactive compounds having anti-cancer properties; natural food colour and bio-formulation. Under his leadership the first bio-fortified pomegranate cultivar Solapur Lal having very high content of Zn and Fe in arils has been developed. The Indian Academy of Horticultural Sciences has conferred him IAHS-Shivashakti lifetime achievement award in 2021 for his outstanding contribution in the field of Horticulture & PHT.

Managing Postharvest Quality and Losses in Horticultural Crops

2nd Revised and Enlarged Edition

Editors

K.L. Chadha

President, Indian Academy of Horticultural Sciences,
F-1 Block, NASC Complex, DPS Marg, Todapur,
New Delhi – 110 012, India

and

R.K. Pal

Former Director, ICAR-National Research Centre on Pomegranate
NH-65, Solapur-Pune Highway, Kegaon,
Solapur – 413 255, Maharashtra, India

2024

Daya Publishing House®

A Division of

Astral International Pvt. Ltd.

New Delhi – 110 002

First Edition, 2015
2nd Revised and Enlarged Edition, 2024

ISBN 978-93-5461-960-1 (HB)

Publisher's Note:

Every possible effort has been made to ensure that the information contained in this book is accurate at the time of going to press, and the publisher and author cannot accept responsibility for any errors or omissions, however caused. No responsibility for loss or damage occasioned to any person acting, or refraining from action, as a result of the material in this publication can be accepted by the editor, the publisher or the author. The Publisher is not associated with any product or vendor mentioned in the book. The contents of this work are intended to further general scientific research, understanding and discussion only. Readers should consult with a specialist where appropriate.

Every effort has been made to trace the owners of copyright material used in this book, if any. The author and the publisher will be grateful for any omission brought to their notice for acknowledgement in the future editions of the book.

Published by : **Daya Publishing House®**
A Division of
Astral International Pvt. Ltd.
– ISO 9001:2015 Certified Company –
4736/23, Ansari Road, Darya Ganj
New Delhi-110 002
Ph. 011-43549197, 23278134
E-mail: info@astralint.com
Website: www.astralint.com

Printed at : **Replika Press Pvt. Ltd.**

Prof. M. S. Swaminathan
Founder Chairman, M S Swaminathan Research Foundation
Third Cross Street, Taramani Institutional Area
Chennai – 600 113 (India)
Email: swami@mssrf.res.in

Foreword

During 1981 the Government of India had constituted a committee under my chairmanship to suggest a scientific approach to postharvest management of perishables in the country. Among various recommendations made, the Committee suggested the establishment of a National Horticulture Board to lay emphasis on programmes related to Postharvest Management. The Committee also identified a group of 12 horticultural crops covering fruits, vegetables and spices for postharvest research apart from several other new crops that were gaining commercial importance. Among fruits apple, mango, banana, citrus, grape, guava, papaya, pineapple, pomegranate and sapota were identified. While among vegetables and spice crops commercial crops like onion, potato, pepper/chilli, turmeric and ginger were included. The National Horticulture Board of the Ministry of Agriculture with its headquarters at Gurgaon is an important outcome of the recommendations.

Even though research on Postharvest Management has been underway in India at the Central Food Technological Research Institute (CFTRI), Mysore, it was not adequate to cover all major crops and aspects. Thus ICAR as a nodal organization implemented a project on Postharvest Management under PL-480 scheme and subsequently set up the Central Institute for Postharvest Engineering and Technology at Ludhiana. In addition, Defence Food Research Laboratory (DFRL), Indian Institute of Crop Processing Technologies (IICPT) Thanjavur, Indian Institute of Technology (IIT) Kharagpur and Mumbai, Jhadavpur University, Kolkata, Bhaba Atomic Research Centre (BARC), Mumbai and many other private funded Institutions have been giving emphasis on these aspects. Of late, new institutions *viz.* National Institute of Food Technology Entrepreneurship and Management (NIFTEM) at Kondili, Haryana and National Agri-Food Biotechnology Institute (NABI), Mohali, Punjab have also been established in view of the emergent need for combating the huge postharvest losses of perishable commodities taking place in India. As a result several technologies have been developed on postharvest management of perishable horticultural crops by these institutions. However, most of these technologies have not been transferred to the end users in an appropriate manner due to non-availability of recommendations in an implementable form. Further, all these recommendations are not available at one place and are mostly concealed in research papers or reports which are neither available nor comprehensive.

I am happy that Dr. K.L. Chadha, former Deputy Director General (Horticulture), ICAR and Dr. R.K. Pal, Director, ICAR-National Research Centre on Pomegranate, Solapur, Maharashtra have made an excellent effort in consolidating the scattered information available from various source, compiled it and edited it in a lucid and understandable format in this book titled *"Managing Postharvest Quality and Losses in Horticultural Crops"* with conclusive recommendations for the benefit of different stake holders. I congratulate the editors for this effort at a time when the horticulture production has overtaken the food crop production and its postharvest management warrants the use of scientific technologies.

I hope this book will be of useful to all engaged in wholesale, retail and export trade of horticultural crops as well as researchers and students. It is also hoped that researchers will make further efforts to plug the gaps in technologies wherever these exist. At the same time, we should continue to make effort to bridge the gap between scientific know-how and field level do-how.

M.S. Swaminathan

Preface to Second Revised Edition

The first edition of the book entitled "Managing Postharvest Quality and Losses in Horticultural Crops" was published in the year 2015 in three volumes covering General Issues, Postharvest Management of Fruit Crops and Postharvest Management of Vegetables, Flowers and Plantation Crops. There were overwhelming responses from various users of this book due to many commercially relevant and practical information that were readily available as handy tool in this book. However, during the last 5 years many changes have happened with respect to introduction of new varieties, of technologies, marketing strategies and export protocol. The revised edition covers all the relevant topics as per the latest R and D findings. Apart from this one new chapter on postharvest management of seed spices was included.

In view of ease of accessibility to all concerned, the revised edition has been published combining all the three volumes in one.

The editors express their thanks and cooperation to all the contributors of the chapters for updating the revised edition with latest data, technologies and other relevant information.

K.L. Chadha

R.K. Pal

Preface

India is the second largest producer of fruits and vegetables in the world but it encounters huge postharvest losses almost equivalent to the total production of horticulture crops of the United Kingdom. Reduction in postharvest losses of horticultural crops is less expensive than bringing equivalent increase in production. Postharvest research in India was started at CFTRI, Mysore. However, with the increase in production and productivity of horticultural crops in India, ICAR launched a PL-480 scheme with the assistance of the United States of America to undertake research on postharvest technology of horticultural crops at IARI, New Delhi as lead centre. Later on an All India Coordinated Research Project on Postharvest Technology of Horticultural Crops was launched with a few centres at ICAR institutes and state agricultural universities. During the last three decades or so, lot of information on several aspects of postharvest management of horticultural crops has been generated in India. The horticulture industry in India has grown leaps and bounds with the increase in area under cultivation, production and export. However, scientific postharvest management and leading adoption of international standards for export quality of horticultural produce from India with respect to combating various tariff and non-tariff barriers is still in its infancy. This book has compiled information keeping in view the needs of horticultural entreprenerus who would like to take-up domestic and export trade of fruits and vegetables on scientific basis and do not have access to this vital information.

The book is divided into three parts comprising of 1) General Issues, 2) Postharvest Management of Fruit Crops and 3) Postharvest Management of Vegetables, Flowers and Plantation Crops. Eminent research workers who have been engaged in the area of Postharvest Management in the NARS have contributed chapters in this publication. Efforts have been made to incorporate information on status of cultivation, cultivars, association between pre-harvest crop management practices and quality after harvest, pack-house operations, storage and marketing for domestic and export trade. The book is also suitably illustrated with colour photographs, wherever necessary. Important references to each subject have been cited for detailed reading. The authors acknowledge the compilation of chapters using scattered information that is available in public domain in the form of research papers and in relevant websites.

The editors are grateful to all the contributors for their cooperation, support and timely submission of their manuscripts for bringing out this publication. The editors also express thanks to Mrs. Mandira Pal for her active support in correcting proofs of the book. The cooperation and support of the staff of the Horticultural Society of India especially Mr. Santosh Kr. Atre is duly acknowledged. The editors are extremely grateful to Prof. M.S. Swaminathan for agreeing to write the Foreword of the book.

It is hoped that this publication will be of immense value to all the stakeholders *viz.* horticultural entrepreneurs, retailers, wholesalers and exporters engaged in the business of supply chain of perishable horticultural produce, research scholars engaged in development of postharvest management protocols, and others in view of the objectives for which it has been written, *i.e.,* reducing postharvest losses and delivering quality produce both in domestic and international market.

K.L. Chadha

R.K. Pal

Contents

FRUIT CROPS

VEGETABLES, FLOWERS, PLANTATION CROPS AND SEED SPICES

List of Contributors

A.K. Misra Former Project Coordinator (Fruits)
ICAR-Central Institute for Subtropical Horticulture,
Rehmankhera, P.O. Kakori, Lucknow – 226 101, Uttar Pradesh, India
E-mail: misra_a_k@yahoo.co.in

A.K. Trivedi Principal Scientist
ICAR-Central Institute for Subtropical Horticulture,
Rehmankhera, P.O. Kakori, Lucknow – 226 101, Uttar Pradesh, India
E-mail: ajayakumartrivedi@gmail.com

Abraham Verghese Former Director
ICAR-National Bureau of Agricultural Insect Resources,
Bengaluru – 560 024, Karnataka, India and
Former Head, Entomology, ICAR-Indian Institute of Horticultural Research,
Bengaluru – 560 089, Karnataka, India
E-mail: abraham.avergis@gmail.com

Ahammed Shabeer T.P. Senior Scientist
ICAR-National Research Centre for Grapes,
P.O. No.3, Manjari Farm, Pune – 412 307, Maharashtra, India
E-mail: shabsnrcg@gmail.com

Ajay Kumar Sharma Principal Scientist
ICAR-National Research Centre for Grapes,
P.B. No. 3, Manjari Farm, Pune – 412 307, Maharashtra, India
E-mails: sharmadrajay@gmail.com

Alemwati Pongener Scientist
ICAR-Indian Agricultural Research Institute,
Gogamukh – 787 034, Dhemaji, Assam, India
E-mail: alemwati@gmail.com

Ananta Saikia Professor
Department of Horticulture,
Assam Agricultural University, Jorhat – 785 013, Assam, India
E-mail: anantasaikia@aau.ac.in

Ashok K. Senapati Assistant Professor
Department of Postharvest Technology, ASPEE College of Horticulture and Forestry,
Navsari Agricultural University, Navsari – 396 450, Gujarat, India
E-mail: aksphtc@nau.in

B.R. Jayanthi Mala Scientist, Division of Entomology, ICAR-Indian Institute of Horticultural Research, Bengaluru – 560 089, Karnataka, India
E-mail: entjaya@gmail.com

B.V.C. Mahajan Professor, Punjab Horticultural Postharvest Technology Centre, Punjab Agricultural University, Ludhiana – 141 004, Punjab, India
E-mail: bvc_mahajan@rediffmail.com

Bidyut C. Deka Assam Agricultural University, Jorhat – 785 013, Assam, India
E-mail: bidyutdeka@yahoo.com

C.K. Narayana Principal Scientist, Division of Postharvest Technology and Agricultural Engineering, ICAR-Indian Institute of Horticultural Research, Hessaraghatta, Bengaluru – 560 089, Karnataka, India
E-mail: narayana.ck@icar.gov.in

D. Balasubramanian Principal Scientist
ICAR-Directorate of Cashew Research, Puttur – 574 202, Dakshina Kannada, Karnataka, India
E-mail: bavika13@email.com

D.V. Sudhakar Rao Principal Scientist, ICAR-Indian Institute of Horticultural Research, Hessaraghatta, Bengaluru – 560 089, Karnataka, India
E-mail: sudhadvrao@gmail.com

Desh Beer Singh Former-Director,
ICAR-Central Institute of Temperate Horticulture, Old Air Field, Rangreth, Srinagar – 5, Jammu & Kashmir, India
E-mail: deshbsingh@yahoo.co.in

Dev Raj Professor and Head
Department of Postharvest Technology, ASPEE College of Horticulture and Forestry, Navsari Agricultural University, Navsari – 396 450, Gujarat, India
E-mail: drdpandir@yahoo.co.in

Giridhar Kalidasu AICRP on Spices, Dr. Y.S.R. Horticultural University, Horticultural Research Station, Lam, Guntur – 522 034, India
E-mail: gkalidasu@yahoo.com

Gopal Lal Member (Ag.), Cauvery Water Management Authority, New Delhi, India and Former Director, ICAR-NRC on Seed Spices, Ajmer, Rajasthan, India
E-mail: glal67@yahoo.co.in

H.P. Sharma Joint Director, NHRDF National Horticultural Research and Development Foundation, New Delhi – 110058, India
E-mail: hariprakash.sharma@gmail.com

Jyotsana Sharma Principal Scientist (Plant Pathology), ICAR-National Research Centre on Pomegranate, NH-65, Solapur-Pune Highway, Kegaon, Solapur – 413 255, Maharashtra, India
E-mail: jyotisharma128@yahoo.com

K. Dhinesh Babu Principal Scientist
ICAR-National Research Centre on Pomegranate, NH-65, Solapur-Pune Highway, Kegaon, Solapur – 413 255, Maharashtra, India
E-mail: ckdhinesh@gmail.com

K. Usha Kumari Assistant Professor(Hort), College of Horticulture, Dr.YSRHU,Venkata Ramanna Gudem – 534 101, Andhra Pradesh, India
E-mail: ushahrs@gmail.com

K.E. Lawande Ex-Vice Chancellor, Dr. Balasaheb Sawant Konkan Krishi Vidyapeeth, Tal. Dapoli, Dist. Ratnagiri – 415 712, Maharashtra, India
E-mail: kelawande2011@gmail.com;

K.K. Paramanick
Principal Scientist and Head ICAR-IARI
RS Amartara Cottage, 4, Cart Rd, Near R.T.O,
Shimla – 1710 04, Himachal Pradesh, India
E-mail: kallolpramanick@gmail.com

K.L. Chadha
President, Indian Academy of Horticultural Sciences,
F-1 Block, NASC Complex, DPS Marg, Todapur,
New Delhi – 110 012, India
E-mail: klchadha@gmail.com

Kaushik Banerjee
National Fellow and Principal Scientist
ICAR-National Research Centre for Grapes,
P.O. No.3, Manjari Farm, Pune – 412 307, Maharashtra, India
E-mail: kbgrape@yahoo.com

Kushal Singh
Senior Plant Physiologist
Department of Floriculture and Landscaping,
Punjab Agricultural University, Ludhiana – 141 004, Punjab, India
E-mail: kushal.flori@pau.edu;

L.C. De
Principal Scientist
ICAR-National Research Centre for Orchids,
Pakyong – 737 106, Sikkim, India
E-mail: lcde_horti@rediffmail.com;

M.A. Rashmi
Senior Technical Officer (Entomology), Regional Plant Quarantine Station, Directorate of Plant Protection Quarantine and Storage (DPPQS), Bengaluru – 560 024, Karnataka, India
E-mail: rashmigowda.ento@gmail.com

M.S. Ladaniya
Former Director, ICAR-Central Citrus Research Institute
Amaravati Road, Shankar Nagar Post Office, Kachimet, Nagpur – 440 010, Maharashtra, India
E-mail: msladaniya@gmail.com

Manjunatha N
Senior Scientist (Plant Pathology), ICAR-National Research Centre on Pomegranate,
NH-65, Solapur-Pune Highway, Kegaon, Solapur – 413 255, Maharashtra, India
E-mail: manju.ars@rediffmail.com

N. Pathak
Sikkim Himalayan Orchids Ltd., Assam Lingzey – 737 135, East Sikkim, India
Present address: Kangra, Himachal Pradesh
E-mail: narender.pathak@gmail.com

Nazeer Ahmed
Former-Vice Chancellor
Sher-e-Kashmir University of Agricultural Sciences and Technology,
Srinagar, J&K, India
E-mail: dnak59@rediffmail.com

Neelima Garg
Former Principal Scientist
ICAR-Central Institute for Subtropical Horticulture,
Rehmankhera, P.O. Kakori, Lucknow – 226 101, Uttar Pradesh, India
E-mail: neelimagargg@gmail.com

Nilesh Gaikwad
Senior Scientist
ICAR-National Research Centre on Pomegranate,
NH-65, Solapur-Pune Highway, Kegaon, Solapur – 413 255, Maharashtra, India
E-mail: nileshgaikwad98@gmail.com

P. Suresh Kumar
Principal Scientist, ICAR-National Research Centre for Banana,
Tiruchirappalli – 620 102, Tamil Nadu, India
E-mail: psureshars@gmail.com

P.C. Tripathi Principal Scientist, Indian Institute of Horticultural Research, Bangalore – 560 089, Karnataka, India
E-mail: prakaashtripathii2000@yahoo.co.in

P.G. Adsule Former Director
ICAR-National Research Centre for Grapes,
P.B. No. 3, Manjari Farm, Pune – 412 307, Maharashtra, India
E-mails: adsulepg@gmail.com

P.L. Saroj Former Director
ICAR-Directorate of Cashew Research,
Puttur – 574 202, Dakshina Kannada, Karnataka, India
E-mail: plsaroj@yahoo.co.in

R. Ezekiel Former NHB Chair Professor, National Institute of Food Technology Entrepreneurship and Management,
Kundli – 131 028, Haryana, India and
Former National Coordinator, National Agricultural Innovation Project, ICAR, New Delhi – 110 012, India
E-mail: rajarathnamezekiel@yahoo.co.in

R.D. Rawal Former Head, Division of Plant Pathology, ICAR-Indian Institute of Horticultural Research,
Hessaraghatta, Bengaluru – 560 089, Karnataka, India
E-mail: rawal.rd@gmail.com

R.G. Somkuwar Principal Scientist
ICAR-National Research Centre for Grapes,
P.B. No. 3, Manjari Farm, Pune – 412 307, Maharashtra, India
E-mails: rgsgrapes@gmail.com

R.K. Pal Former Director
ICAR-National Research Centre on Pomegranate,
NH-65, Solapur-Pune Highway, Kegaon,
Solapur – 413 255, Maharashtra, India
E-mail: rkrishnapal@gmail.com

R.P. Gupta Ex-Director, NHRDF
Chief Consultant MIDH, Dept. of Agriculture Co-operation and Farmers Welfare, Ministry of Agriculture and Farmers Welfare, Govt. of India, New Delhi
E-mail: rpgupta5@yahoo.com

R.R. Sharma Principal Scientist, Division of Food Science and Postharvest Technology,
ICAR-Indian Agricultural Research Institute, New Delhi – 110 012, India
E-mail: rrs_fht@rediffmail.com

R.T. Patil Former Director, ICAR-Central Institute of Postharvest Engineering & Technology,
P.O. PAU, Ludhiana – 141 004, Punjab, India
E-mail: ramabhau@gmail.com

Ram Asrey Principal Scientist, Division of Food Science & Postharvest Technology,
ICAR-Indian Agricultural Research Institute, New Delhi – 110 012, India
E-mail: ramu_211@yahoo.com

Ramesh Kumar Former Director, Directorate of Floricultural Research,
ICAR-Indian Agricultural Research Institute, New Delhi – 110 012, India
E-mail: directordfr@gmail.com

S. Arulraj Former Director
ICAR-Directorate of Oil Palm Research, Pedavegi – 534 450, Andhra Pradesh, India
E-mail: aruldopr@gmail.com

S. Vijay Rakesh Reddy Scientist, ICAR-Indian Institute of Horticultural Research, Hessaraghatta, Bengaluru – 560 089, Karnataka, India
E-mail: rakesh.reddy@icar.gov.in

S.C. Mandhar Emeritus Scientist, ICAR-Indian Institute of Horticultural Research, Hessaraghatta, Bengaluru – 560 089, Karnataka, India
E-mail: suresh.mandhar@gmail.com

S.K. Jawandha Department of Fruit Science, PAU, Ludhiana – 141 004, Punjab, India
E-mail: skjawandha@pau.edu

S.K. Purbey Principal Scientist
ICAR-National Research Centre on Litchi,
Mushahari, P.O. Ramna, Muzaffarpur – 842 002, Bihar, India
E-mail: skpurbey_nrcl@yahoo.com

Sanjib Sharma Assistant Professor
Department of Horticulture,
Assam Agricultural University, Jorhat – 785 013, Assam, India
E-mail: sanjib.sharma@aau.ac.in

Shiv Lal Senior Scientist (Hort.),
ICAR-National Research Centre, Ajmer, Rajasthan, India

Soumitra Goswami Assistant Professor,
Assam Agricultural University,
Jorhat – 785 013, Assam, India

Sunil Pareek Associate Professor, Department of Agriculture and Environmental Science, NIFTEM, Plot No. 97, Sector 56, HSIIDC Industrial Estate, Kundli, District-Sonepat – 131 028, Haryana, India
E-mail: sunil.pareek@niftem.ac.in

Surya Kumari Surepeddi Principal Scientist (Horticulture)
AICRP on Spices, Dr. Y.S.R. Horticultural University,
Horticultural Research Station, Lam, Guntur – 522 034, India;
E-mail: gkalidasu@yahoo.com

T.L. Preethi Assistant Professor
KVK, Tiruvallur – 602 025,
Tamil Nadu Agricultural University, Tamil Nadu, India
E-mail: preethi.chari@rediffmail.com

V. Ponnuswamy Former Dean
Tamil Nadu Agricultural University,
Periyakulam – 641 020, Tamil Nadu, India

Vishal Nath Former Director
ICAR-National Research Centre on Litchi,
Mushahari, P.O. Ramna, Muzaffarpur – 842 002, Bihar, India

W.S. Dhillon Former Assistant Director General (Horticulture),
ICAR, New Delhi – 110 012, India
E-mail: wasakhasingh@yahoo.com

General Issues

Chapter 1

Postharvest Management: An Overview

K.L. Chadha

President, Indian Academy of Horticultural Sciences, F-1 Block, NASC Complex, DPS Marg, Todapur, New Delhi – 110 012, India

E-mail: klchadha@gmail.com

R.K. Pal

Former Director, ICAR-National Research Centre on Pomegranate, NH-65, Solapur-Pune Highway, Kegaon, Solapur – 413 255, Maharashtra, India

E-mail: rkrishnapal@gmail.com

1.0 Introduction

The horticultural crops comprise of a wide range of fruits, vegetables, spices, flowers, mushroom, bamboo, plantation crops and non food crops like medicinal and aromatic plants. Three decades back the role of these crops was realized in India through cropping diversification as a remunerative, viable and sustainable alternative to rice-wheat cropping system and their comparative advantage in livelihood security. As a result, policy attention to horticulture has been given since VII Five Year Plan. Emphasis on horticulture crops has resulted in higher growth rate of 5 per cent compared to 2.5 per cent from agriculture. Horticulture also contributes 30 per cent of GDP compared to 17 per cent from agriculture. It also enjoys 52 per cent share of exports from 15 per cent cultivated area. As a result of this, the area, production and productivity of horticultural crops has increased significantly. Production of horticultural crops in India increased from 94.07 million tonnes in 1991-92 to 334.60 million tonnes in 2021-22. It is now fully recognized that in order to achieve a growth rate of 4 per cent in agriculture, horticulture has to grow at a much faster and higher rate.

1.1 Nutritional Benefits

Fruits and vegetables are vital to improve the nutritional quality of our diet: together these supply not only vitamins and minerals but also the much needed roughage; at the same time these are low on calorie count. Fruits and vegetables are sometimes called as powerhouse of antioxidants having anti-ageing and anti cancer properties. Vitamin C, E, carotenoids and mineral selenium are powerful antioxidants provided by fruits and vegetables. Fruits and vegetables also help to maintain a healthy gastrointestinal, tract skin, and a better vision. Some of the potential sources of antioxidants and their health benefits are mentioned in Table 1.

Because of high productivity per unit area compared to cereal crops, fruits and vegetables provide much better economic returns when properly managed besides serving as a good source of foreign exchange earnings. Cultivation of fruits and vegetables provides maximum potential for generating employment in the countryside.

1.2 Area and Production

Of the available 194 million hectares of land for agriculture (total cropped area), the area under fruits and vegetables is about 8 per cent. The diverse agroclimatic zones available in our country make it possible to grow almost all kinds of fruits and vegetables in India (be it tropical, sub-tropical or temperate). State wise area and production of fruits and vegetables in India during 2021-22 is presented in Table 2.

Table 1. Health Benefits of Antioxidants and their Food Sources

Antioxidant	*Health Benefits*	*Food Sources*
β-carotene	Keeps skin healthy, helps prevent night blindness and infections, promotes growth and bone development.	Red, yellow-orange, and leafy green vegetables and fruits, including carrot, apricot, cantaloupe, pepper, tomato, spinach, broccoli, sweet potato, and pumpkin.
Selenium	Helps maintain healthy hair and nails, enhances immunity, works with vitamin E to protect cells from damage. Reduces the risk of cancer, particularly lung, prostate, and colorectal.	Garlic, seeds and Brazil nuts.
Vitamin C	Destroys free radicals inside and outside cells. Helps in the formation of connective tissue, the healing of wounds, and iron absorption; also helps to prevent bruising and keep gums healthy. May reduce risk of cataracts, heart disease, and cancer.	Pepper, tomato, citrus fruits and juices, berries, broccoli, spinach, cabbage, potato, mango and papaya.
Vitamin E	Acts as the protector of essential fats in cell membranes and red blood cells. Reduces risk of cancer, heart disease, and other age- associated diseases.	Nuts, seeds, avocado and salad dressings.

Source: The American Dietetic Association and WebMD.

India is the second largest producer of fruits in the world (102.03 million tonnes) with a production share of approximately 11.56 per cent as per the final estimate of Min of Agriculture and FW 2021. During the same period banana recorded the highest production of 32.59 million tonnes, followed by mango (20.26 million tonnes), citrus (14.56 million tonnes), papaya (5.78 million tonnes), guava (4.36 million tonnes), grape (3.18 million tonnes), apple (2.81 million tonnes), pineapple (1.73 million tonnes) and sapota (0.91 million tonnes). Among the vegetables potato occupied the first rank with production of 48.56 million tonnes followed by tomato (21.17 million tonnes) and onion (26.09 million tonnes). India produced 199.88 million tonnes of vegetables during 2020-21which is 14.0 per cent of the global production. The productivity of grape (21.2 t/ha), pea (10.29 t/ha) and cassava (35.0 t/ha) in India has been reported to be the highest in the world. India is considered as "the land of spices" for being the largest producer (81.24 lakh tonnes/annum) and exporter of spices in the world. In a short span of time, India has also made significant advancement in production of cut flowers and ornamental plants. India produced 2.09 million tonnes of loose flowers and 7.9 lakh tonnes of cut flowers (Table 2). Production share of major fruits and vegetables in India is presented in Figures 1 and 2.

The export of grapes from India to different countries showed a sight decline may be due to the Covid-19 pandemic. Country wise export of grapes from India during the period from 2017-2020 is presented in Table 3. However, the import of fresh grapes to India during the same period exhibited an increasing trend (Table 4).

Table 3. Export of Grapes from India to different Countries

Country	*2017-18*		*2018-19*		*2019-20*	
	Qty in MT	*₹ Lacs*	*Qty in MT*	*₹ Lacs*	*Qty in MT*	*₹ Lacs*
Netherland	58,456.97	64,753.35	72,262.13	79,485.81	56,265.55	78,672.33
Russia	27,434.39	27,734.44	30,293.17	31,069.34	23,736.47	26,558.79
UK	18,594.24	22,193.56	17,596.82	21,319.13	14,484.40	20,544.30
Bangladesh	5,073.32	1,437.91	46,419.00	20,216.38	32,742.75	13,784.62
Germany	16,449.51	18,091.76	15,739.17	17,777.97	8,718.95	13,669.04
UAE	13,574.90	10,971.59	2,772.08	11,210.09	12,128.27	11,235.27
Saudi Arab	9,482.96	7,116.90	9,203.74	7,520.10	8,709.26	7,729.59
Thailand	5,043.03	6,554.06	3,919.42	5,574.21	3,569.99	5,885.86
Oman	2,799.36	2,328.50	4,338.30	3,991.64	3,833.54	3,528.65

https://agriexchange.apeda.gov.in/indexp/Product_description_32headChart.aspx?gcode=0205

Table 2. Area and production of horticultural crops in India 2021-22

Sl. No.	States/Uts	FRUITS		VEGETABLES		PLANTATION		AROMATICS & MEDICINAL		FLOWERS			SPICES		HONEY	TOTAL	
		A	P	A	P	A	P	A	P	A	P		A	P	P	A	P
											LOOSE	CUT					
1	Andhra Pradesh	757.24	18010.39	234.36	6434.93	348.50	1275.87	4.20	14.35	17.51	279.85	0.00	201.79	796.34	1.89	1563.59	26813.62
2	Arunachal Pradesh	48.14	125.84	2.62	17.41	2.67	11.76	0.24	0.16	0.00	0.00	0.00	12.39	37.69	0.13	66.06	193.00
3	Assam	159.98	2504.19	302.68	3669.26	89.29	159.71	4.62	0.18	5.31	35.58	57.80	100.49	310.51	1.35	662.36	6738.58
4	Bihar	363.77	4986.74	903.47	17767.84	12.16	54.06	4.85	22.49	1.21	10.95	0.19	9.21	14.64	16.00	1294.67	22872.91
5	Chhatisgarh	221.93	2471.09	490.06	6887.65	33.90	30.75	5.15	32.55	11.19	52.68	178.05	12.51	17.52	0.95	774.75	9671.25
6	Gujarat	426.11	8243.67	785.80	15687.11	34.70	153.96	0.00	0.00	19.99	189.34	0.00	665.10	992.14	0.75	1931.68	25266.96
7	Haryana	82.00	1398.97	372.37	6295.61	0.00	0.00	0.30	0.34	2.69	24.96	1.14	12.01	61.22	4.80	469.37	7787.04
8	Himachal Pradesh	234.78	796.79	91.99	1875.24	0.00	0.00	1.12	0.91	0.37	1.30	10.45	7.84	19.16	5.85	336.10	2709.69
9	J&K	335.02	2237.87	60.12	1338.27	0.00	0.00	3.83	0.01	0.10	0.21	1.02	5.04	1.14	2.30	404.12	3580.82
10	Jharkhand	105.64	1293.55	311.72	3773.67	15.60	6.40	0.00	0.00	1.20	4.81	3.81	0.00	0.00	1.65	434.16	5083.88
11	Karnataka	452.74	8549.03	500.93	8099.77	1364.21	5392.67	2.25	18.28	37.68	332.52	185.10	368.76	745.42	2.30	2726.58	23325.09
12	Kerala	313.91	1939.45	24.45	544.95	984.90	3498.92	0.01	0.00	0.26	0.12	1.73	159.50	173.09	2.00	1483.03	6160.28
13	Madhya Pradesh	348.13	6141.15	1031.06	20591.22	0.00	0.00	41.45	105.39	30.07	340.01	10.87	715.67	3602.43	2.60	2166.37	30793.67
14	Maharashtra	840.86	12296.11	1180.43	16783.40	224.40	356.64	0.84	1.53	11.30	64.63	0.35	91.99	494.61	1.85	2349.81	29999.10
15	Manipur	41.62	463.86	32.90	339.80	0.90	0.30	0.04	0.12	0.07	0.01	0.17	9.05	58.90	0.40	84.58	863.55
16	Meghalaya	37.05	376.02	49.42	519.41	26.96	34.09	0.00	0.00	12.47	0.00	0.36	15.72	76.54	0.25	141.63	1006.67
17	Mizoram	65.97	345.36	40.67	224.60	14.15	14.20	0.77	0.78	0.08	0.00	0.80	27.75	101.38	0.25	149.38	687.36
18	Nagaland	34.01	316.59	41.09	457.92	2.85	8.29	0.11	0.86	0.05	0.00	0.22	12.03	57.74	0.65	90.14	842.26
19	Odisha	366.64	2782.05	674.97	9523.71	266.74	394.62	1.92	0.61	5.61	31.57	33.56	147.88	295.21	1.65	1463.76	13062.98
20	Punjab	102.75	2202.10	290.30	5758.47	0.00	0.00	13.37	3.02	2.18	13.70	0.00	30.97	112.47	17.00	439.56	8106.77
21	Rajasthan	69.89	938.75	195.46	2247.83	0.00	0.00	408.96	235.10	3.66	8.31	0.00	972.55	1163.92	11.20	1650.52	4605.11
22	Sikkim	20.16	50.76	9.64	65.45	0.00	0.00	0.00	0.00	0.24	16.50	0.09	38.41	89.72	0.45	68.45	222.97
23	Tamil Nadu	320.10	5823.88	339.76	9074.90	657.24	3844.95	20.44	312.91	42.92	515.71	5.54	106.64	182.30	2.20	1487.10	19762.39
24	Telangana	177.81	2309.96	75.71	1682.28	0.79	6.06	0.22	3.50	3.04	44.42	0.00	141.62	793.62	0.85	399.19	4840.69
25	Tripura	55.45	571.43	48.57	848.06	15.27	36.62	0.00	0.00	0.00	0.00	0.00	6.74	28.56	0.20	126.03	1484.87
26	Uttar Pradesh	505.13	11259.24	1324.91	29584.06	0.00	0.00	135.04	13.53	22.76	49.66	69.73	394.81	269.91	22.50	2382.64	41268.63
27	Uttarakhand	181.07	648.90	99.84	1043.75	0.00	0.00	0.00	0.00	4.75	3.02	11.25	14.65	65.99	2.70	300.30	1775.62
28	West Bengal	283.56	3717.98	1511.19	28229.16	55.70	313.51	0.00	0.00	29.73	74.53	218.71	61.17	250.67	20.00	1941.35	32824.55
29	Others	15.87	121.86	38.59	516.78	119.71	258.49	0.00	0.00	0.18	0.77	0.02	1.69	2.74	0.29	176.04	900.94
	All India Total	6967.32	102923.58	11065.06	199882.49	4270.64	15851.87	649.71	766.60	266.62	2095.16	790.96	4343.98	10815.60	125.01	27563.33	333251.25

Source: https://agricoop.nic.in/sites/default/files/2021-22%20%28First%20Advance%20Estimates%29%20%281%29_0.pdf A=Area in '000 Ha; P= Production in '000 MT

Table 4. Import of fresh grapes in India

Year	*Quantity (MT)*	*Value (₹ Lacs)*
2019-20	7,583.77	8,582.14
2018-19	7,006.13	8,615.67
2017-18	5,626.96	7,352.65

https://agriexchange.apeda.gov.in/importtoindia/Product_description_32head.aspx?gcode=0205

1.3 Postharvest Losses

Although India has touched new heights in production of fruits and vegetables comparable to Brazil and China, the production is still insufficient to meet the basic nutritional requirements of the growing population. Part of this problem could be overcome by salvaging a fair proportion of huge postharvest losses

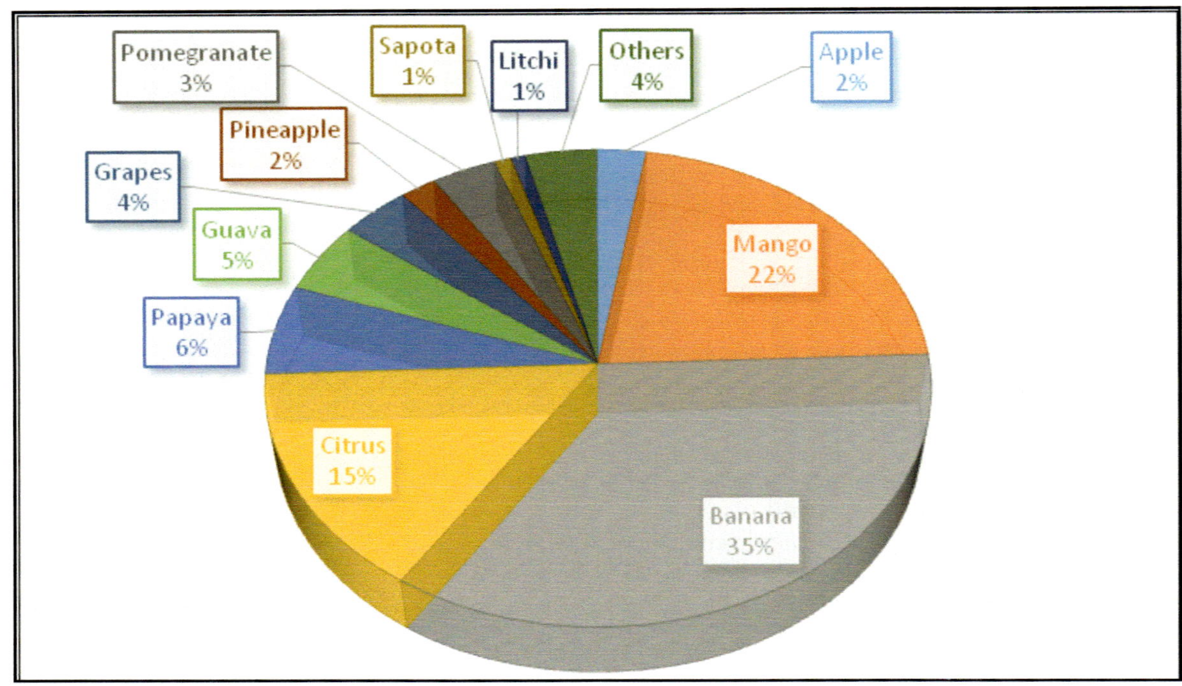

Source. **https://agricoop.nic.in/sites/default/files/2021-22%20%28First%20Advance%20Estimates%29%20%281%29_0.pdf**

Figure 1. Production Share of Major Fruits in India (2022)

Source. **https://agricoop.nic.in/sites/default/files/2021-22%20%28First%20Advance%20Estimates%29%20%281%29_0.pdf**

Figure 2. Production Share of Major Vegetables in India (2022)

of these perishables which are equivalent to total annual production of these commodities by United Kingdom. These losses occur at several stages in the postharvest management chain *viz.* harvesting, grading, packaging, transport, storage and marketing. Hence, it is better to pay more attention in managing postharvest losses besides increasing production and productivity of horticultural crops.

In India the estimated postharvest losses of fruit and vegetables have been reported to be within the range of 6-36 per cent depending upon the commodity whereas these range between 16-50 per cent in different countries of Asia-Pacific region. Such losses in developing countries have been reported to be around 32 per cent. Postharvest losses can be divided into two categories *e.g.*, quantitative and qualitative. Qualitative losses include deterioration in texture, flavour, aroma, volatile compounds, nutritional value and cosmetic look. However, quantitative losses are valued at several thousand crores of rupees. It has been estimated that every one per cent reduction in postharvest loss in India can save approximately 5 million tonnes of fruits and vegetables.

When we aim to reduce such quantitative and qualitative postharvest losses of fruits and vegetables, we must consider both the biological and environmental factors involved in the deterioration of the produce after harvest and before reaching consumers. Basically postharvest management attempts to slow down the undesirable changes that the commodity may undergo in its postharvest life.

1.4 Factors Affecting Postharvest Life

As almost all edible plant parts are living tissue, there are irreversible postharvest physiological changes leading to senescence/cell death. However management of postharvest processes depending on the plant part consumed demands different postharvest treatments/storage requirements. Additionally all fresh horticultural commodities being high in water content face desiccation causing wilting/shriveling and corresponding postharvest loss. Like all other living tissues, horticultural produce also falls prey to pests and diseases, both before and after harvesting.

Biotic factors which are responsible for postharvest degradation could be mainly identified as: stage of harvest maturity, rates of respiration, transpiration, and ethylene production, vulnerability to ethylene, changes in chemical composition, colour and flavour, tendency of sprouting, physiological reaction towards low temperature storage conditions besides pathological infections.

Respiratory breakdown leads to both quantitative and qualitative losses resulting in weight loss and food value. Moreover, senescence is also accelerated leading to shortening of storage life. Thus crops with low rate of respiration like nuts, apple, onion, garlic, potato, *etc.* degenerate slowly while others namely, broccoli, green onion, cut flowers, *etc.* are highly perishable due to very high respiratory rate. Such direct relationship can also be established between rate of ethylene production and ripening of climacteric fruits *e.g.* apple, banana, guava, tomato and sapota. Non-climacteric fruits *e.g.* citrus, grape, litchi, pineapple, pomegranate, strawberry, sweet pepper, *etc.* liberate ethylene at a very low rate. Therefore, these two types of fruits require different approach to postharvest management. Change in colour during postharvest life could be a desired trait in fruits but not in case of vegetables.

Temperature, relative humidity, gaseous combination of the storage environment, light, type of packaging material, mode of transport, *etc.* are some other important abiotic factors influencing postharvest life of horticultural produce. Bruising due to mechanical injury is a great cause of concern as it allows microbes to get an entry in the fleshy part. To minimise losses caused by pathogens maintenance of hygiene always remains a priority during postharvest operations.

Appropriate postharvest management is essential not only for controlling various physiological processes *viz.* respiration, transpiration and other metabolic activities but also for quality assurance, prolonging market life, checking market gluts and for increasing financial gains. Postharvest behavior of perishable horticultural crops depends further on species, cultivar, harvest maturity, postharvest treatments and storage conditions. Critical gaps observed in postharvest management in India are poor farm practices, improper harvesting, improper ripening practices, poor postharvest handling, poor infrastructure and logistic facility and ill equipped marketing system.

1.5 Research and Development Infrastructure

Realizing the importance of postharvest management practices in managing Postharvest quality and reducing postharvest losses, a beginning was made by Indian Council of Agricultural Research (ICAR), New Delhi during fourth five-year plan. The All India Coordinated Fruit and Vegetables Improvement Projects were started in 1970-71 by the Indian Council of Agricultural Research. The broad objectives of such projects were development of fruit and vegetable cultivars for higher yields, identification of varieties for disease resistance, better shelf life and to some extent suitability for processing. A large number of research centres were established all over the country under the above project to study scientific methods of reducing wastage during the process of handling, transport, storage and processing for ensuring full utilization of horticultural produce.

A study group under the chairmanship of Dr. M.S. Swaminathan, the then Director General, ICAR recommended setting up of a coordinated research programme on postharvest technology of fruits, vegetables and flowers in the country. The programme came into operation during August, 1978 through PL-480 grant made available under the Agricultural Special Foreign Currency Programme of the United States Department of Agriculture with Dr. J.C. Anand as the first Project Coordinator. The primary objective of the project was to reduce postharvest losses of fruits and vegetables during harvesting, farm handling, storage, marketing, transportation and processing. A Coordinated Project was started with four centers located at Indian Agricultural Research Institute (IARI), New Delhi, Indian Institute of Horticulture Research (IIHR), Bangalore, Central Institute of Horticulture for Northern Plains (CIHNP), currently known as Central Institute of Subtropical Horticulture (CISH), Lucknow and Dr. Y.S. Parmar University of Horticulture and Forestry (YSPUHF), Solan. Considering importance and impact of this project on Indian agri-economy, several more centres were later added in different states during subsequent plans.

During the 7th plan another collaborative project, the Indo-USAID sub-project on postharvest technology of fruits and vegetables, was approved under the Indo-US sub-commission on Agriculture which remained operational at four centres *viz.* IARI, New Delhi, IIHR, Bangalore, CIHNP, Lucknow and NRCC, Nagpur till 1991. In this project, seven economically important horticultural crops *viz.* mango, banana, citrus, guava, potato, onion and tomato were identified. The major objective of the coordinated projects during the period from 1971 to 1991 was to minimize postharvest loss. Subsequently research on postharvest management was started at various State Agricultural Universities in the country. A Central Institute of Postharvest Engineering and Technology was also setup by ICAR at Ludhiana with a regional station at Abohar to intensify research on this aspect. Several other old and new Institutes namely CFTRI, Mysore, Regional Research Laboratories (CSIR), National Agril. Food Biotechnology Institute, Mohali under (DBT), Bhabha Atomic Research Centre, Mumbai, Indian Institute of Crop Processing Technology, Thanjavur and National Institute of Food Technology Entrepreneurship and Management, Sonipat, Haryana, Defence Food Research Laboratory, Mysore are now involved in R&D Programmes on various agri/hort crops.

Indian agriculture scenario was becoming brighter during this period, farmers were reaping the benefits of Green Revolution and suddenly the country had more agricultural produce to manage. To handle the surplus production, stakeholders realized need of investment in storage infrastructures, and postharvest processing plants. Ministry of Food Processing Industries was set up in 1988. To promote export and to deal with related policy matters Agricultural and Processed Food Products Export Development Authority (APEDA) was established by the Government of India under the Ministry of Commerce through an Act passed by the Parliament in December, 1985.

To take care of the horticulture sector a National Horticulture Board (NHB) was set up by the Government of India in 1984 as an autonomous society. Though NHB started promoting commercial horticulture in selected agro-climatic zones, it soon integrated the idea of development of postharvest management infrastructure with area expansion projects. Development of energy efficient cold infrastructure for fresh horticultural produce is also a priority research area identified by NHB.

1.6 Managing Postharvest Losses

According to FAO, nearly one-third of food produced for human consumption (approximately 1.3 billion tonnes per year) is either lost or wasted globally along the food chain from production to consumption. A High Level Panel of Experts on Food Security and Nutrition (HLPE) was created in 2010 to provide UN's Committee on World Food Security (CFS) with evidence based and policy oriented analysis about the causes of food losses and waste (FLW) and to suggest actions to reduce them in order to improve food and nutrition security and the sustainability of food systems. The aim of the HLPE report, given the diversity of contexts, was to help all concerned actors to reduce food losses and waste by identifying the causes and potential solutions that may be implemented, alone or in a coordinated way, by the relevant actors in the food system, including the public and private sectors, civil society, individual producers, wholesalers, retailers and consumers. Successful reduction of food losses and waste will save resources and has the potential to improve food security and nutrition, goals shared with the Zero Hunger Challenge and the post-2015 sustainable development agend.

HLPE defines food losses and waste (FLW) as "a decrease, at all stages of the food chain from harvest to consumption, in mass, of food that was originally intended for human consumption, regardless of the cause". FLW represents at the same time a waste of resources, as well as an environmental issue by itself, as for example food-related waste, as part of urban total waste, that has a significant greenhouse gas (GHG) footprint. FLW also tends to become a symbol of inefficiency, unfairness and unsustainability of food systems with economic, environmental and social concerns. Evidently HLPE report mainly discussed the postharvest loss of food. Causes of FLW were discussed in the report under three headings: micro, meso and macro and these causes are often interrelated.

1.6.1 Micro Level

Micro-level causes can be found all along the food chain, and are the direct, immediate reasons for FLW taking place at a certain point of the chain, resulting from actions (or non-action) of individuals potentially leading to FLW. The major causes are:

i) Poor harvest scheduling and timing; rough and careless handling of the produce.

ii) Inadequate or lack of storage conditions and poor temperature management all along the food chain.

iii) Transport can be a major cause of FLW by introducing a time span between production and consumption and by bringing additional risks of mechanical and heat injury. Time spent because of transport can also lead to decrease of nutritional value.

iv) Conditions within the retail outlet (temperature, relative humidity, lighting, gas composition, *etc.*) and handling practices have an effect on quality, shelf-life and acceptability of the product.

Micro-level solutions at harvest and postharvest stages involve improved practices, adoption of technical innovations, investments, or a combination of these. When appropriately applied, good agricultural practices at the primary stage of production as well as good manufacturing practices and good hygienic practices during food processing can protect food from contamination or damage. A key intervention all along food chains is to improve storage conditions. Various solutions have been already successfully implemented in many places.

1.6.2 Meso and Macro Levels

Very often causes of FLW are found at "higher" meso- and macro-levels. These include:

i) At meso-level, the lack of equipment and/or of good practices, inadequate organization, coordination and communication between food chain actors *e.g.* transformation that renders the product useless at a later stage of the chain, *etc.*, inadequate infrastructure, maladapted economic conditions along the food chain (product unmarketable, *etc.*)

ii) More macro-level, systemic causes include the absence of a good, enabling environment to support coordination between actors, investment and improvement of practices.

There is no doubt that preharvest conditions and actions in the field can indirectly lead to losses at later stages in the chain, as production and agronomic practices very often influence quality at harvest, suitability for transport and shipping, storage stability and shelf-life after harvest. Nature of storage losses of some commercially important fruits is presented in Table 5.

Table 5. Nature of Storage Losses in some Commercially Important Fruits

Losses	*Crops*
Chilling Injury	Banana, citrus, guava, mango, papaya, pomegranate, sapota
Rind (also stem end rind) breakdown	Citrus
Loss of colour	Litchi
Desiccation and Decay	Litchi, pomegranate
Internal breakdown	Pineapple
Crown dehydration	Pomegranate

2.0 Postharvest Management

There are several method and approaches for reducing losses of horticultural produce. These could be broadly categorized under the following heads:

2.1 Maturity and Harvesting

The issues related to maturity of horticultural produce begin with standardization of optimum maturity indices that ensure proper size, color, good edible quality and better shelf life. Fruits and Vegetables are eaten at various stages of development *e.g.* seed (fresh and sprouted), leaves, flowers, whole fruits and nuts. Delay in harvesting and over maturity reduces the storage quality. A lot of research has been carried out on various types of maturity indices; however several maturity standards or guidelines prepared by Bureau of Indian Standards (BIS) are not adopted due to lack of awareness. Based on respiratory pattern, the fruits are classified into two major groups *viz.* climacteric fruits *e.g.* apple, banana, guava, tomato, sapota and non-climacteric fruits *e.g.* citrus, grapes, litchi, sweet pepper, pineapple, pomegranate, strawberry *etc.* Therefore, harvest maturity of these two distinct groups of fruits differs and must be taken into account for judicious postharvest management practices. A wide variety of maturity indices have been used in different crops *viz.* formation of abscission layer in melon, heat units in grape and sweet peas; leaf changes in root, bulb and tuber crops; number of days from flowering in banana, papaya, jackfruit; oil content in avocado; skin color in apple, tomato, peach, chilli, pepper; specific gravity in mango and TSS/acid ratio in grape. However, there is an urgent need for evolving commercially feasible maturity indices for all major crops.

Harvesting of horticultural crops plays a very important role on their postharvest quality. Improper harvesting makes the commodity unattractive. Harvest injuries serve entry points for microorganisms that result in early rotting. A number of fruit harvesters have been developed by various institutes *viz.* Dapoli Mango harvester, IIHR mango harvester, IARI mango harvester, CISH mango harvester, *etc.* which are commercially used in many places. Some of the physical and mechanical methods used for harvesting of major fruits crops of India are: use of sharp sickle in banana, scissor in grape; shaking of limbs in case of ber, pulling the fruit with a jerk in case of pineapple and use of clippers in litchi, kinnow *etc.*

2.2 Physico-chemical Treatments

Different pre-treatments before harvest or at postharvest stages have been standardized to keep the produce fresh for long time or to enhance their shelf life and retain their quality. For example, potatoes and onions are sprayed with maleic hydrazide (2000-2500 ppm) 2- weeks before harvest or treated with gamma irradiation (0.15 kGy) after harvest in order to inhibit sprouting during storage. This technology is now commonly used in commercial storage of potatoes and onions. Irradiation treatment has been quite promising and an effective technology which is currently available to overcome quarantine restrictions in international trade. Pretreatment of apple before storage with ethanol vapours is quite common. Vapour heat treatment (VHT) and hot water treatment (HWT) or hot air treatment are very simple and have become popular pre-treatments before storage or transportation of fruits, especially mango and citrus.

2.3 Precooling

The first and foremost important practice of handling after harvest is pre-cooling. The primary objective of pre-cooling is to remove the field heat. Pre-cooling is the process of "rapid" removal of field heat and is usually practiced for fresh fruits, vegetables and flowers immediately after harvest but before shipment or storage or processing. The latest developments in the area of pre-cooling include air handling system for forced air cooling, use of ozonated water in hydrocooling and for washing of fruits and vegetables. The primary advantages of pre-cooling are: (a) removal of field heat/respiratory heat, (b) inhibition of the growth of decay causing organisms, (c) restriction of the enzyme activity, (d) reduction of water loss, (e) reduction in rate of respiration and C_2H_4 liberation and (f) rapid wound healing. The production and action of ethylene from harvested fruits, vegetables and flowers are temperature dependent. Harvested produce kept at 25 °C with 30 per cent RH shows a tendency of 36 times more water loss as compared to that stored at 0 °C with 90 per cent RH. Hence, pre-cooling serves as an essential practice in any successful cool chain management of horticultural produce. The speed of cooling depends upon the following factors: (i) accessibility of product to the refrigerating medium. (ii) difference in temperature between the product and refrigerating medium. (iii) velocity of refrigerating medium and (iv) type of cooling medium. In order to predict the end point of pre-cooling it is essential to know the "half-cooling time". Half-cooling time is the time required to reduce the temperature difference between commodities and the coolant by one-half. This is independent of the initial temperature of the commodity. The shape, density, and surface area/volume ratio of the fruit or vegetable and the type of packaging are important factors of determining the appropriate type of cooling medium, *i.e.* air, water or ice. Some products do not tolerate exposure to water. In general, the greater the amount of packaging, the slower the cooling rate. If the packaging is exposed to water, packaging in paper-based material is not recommended unless it is waxed or coated with polymer as lamination.

The two most important factors in pre-cooling are temperature and time, *i.e.* fruit or vegetable must be cooled in the shortest time possible. Product cooling follows a logarithmic function, with initial rapid cooling followed by a slower rate. Since it is difficult to remove all the field heat, pre-cooling to 7/8th of the recommended storage temperature is advisable. The product can then be placed in storage where the remaining 1/8th of the heat can be gradually removed with less energy cost. If several products are being handled, then a system must be selected that is compatible to all of them. Highly perishable products, which have a high respiratory rate need fast cooling and require high refrigeration capacity and fast pre-cooling methods. There are basically five methods used for horticultural commodities. These are (i) Room/air cooling. (ii) Water/hydro- cooling (iii) Forced air-cooling (iv) Vacuum cooling and (v) Package icing.

2.3.1 Room Cooling

It is the practice of storing bulk or packaged produce in a refrigerated room for an indefinite period. By the processes of conduction and convection, the heat is gradually removed from the produce. Because the outside of the containers cool more rapidly than the inside, the cooling is uneven and slow. The cooling rate of room cooling may be sufficient for items such as potatoes, cabbage, or root crops, however, it has been proven unsatisfactory for highly perishable items such as strawberries, blueberries, and snap beans.

2.3.2 Hydro-Cooling

Cooling freshly harvested produce by flooding, immersing, or spraying with large quantities of cold water is done in hydrocooling. Compared to other cooling methods, hydrocooling is fast and generally thorough. Hydrocooling is limited to those produce items that will tolerate liquid water (except onion and garlic).

2.3.3 Package Icing

It is cooling fresh produce by the addition of crushed ice or a slurry of crushed ice and water over the top of a load or to each individual package. This method is limited to the produce that is not harmed by contact with ice. This is commercially used in many cut flowers.

2.3.4 Forced-Air Cooling

Forced-air cooling is a commonly used cooling method that utilizes specially constructed portable fans to draw chilled air horizontally through pallets or stacks of packaged produce. A properly designed forced-air cooling system is fast, energy efficient and relatively inexpensive. It may be utilized on most types of produce.

2.3.5 Vacuum Cooling

When warm produce is placed inside a closed container and reduced to a partial vacuum, a small portion of the water in the produce evaporates, causing a cooling effect. Vacuum cooling is fast and may be utilized very effectively on packaged produce. It is generally more effective on items with a large surface area to volume ratio such as lettuce and various greens.

2.4 Sorting and Grading

Sorting and grading of fruits and vegetables based on size, shape, color, texture, maturity, flavor and aroma is essential for better marketing. Several fruit and vegetable graders have been developed at IARI – New Delhi, IIHR – Bangalore, PKDV – Akola, JNKVV – Jabalpur and GAU – Junagarh, Gujarat. Apart from this several machineries are available for grading like ultrasonic devices, colour sorters, machine vision technology *etc.*

2.5 Ripening

Artificial fruit ripening is another important issue which needs immediate attention particularly with respect to food safety. Rule 44-AA of the Prevention of Food Adulteration (PFA) Rules, 1955 prohibits the use of acetylene gas commonly known as carbide gas produced from calcium carbide for artificial ripening of fruits. It states that "*No person shall sell or offer or expose for sale or have in his premises for the purpose of sale under any description, fruits which have been artificially ripened by use of acetylene gas, commonly known as carbide gas*". Calcium Carbide is used in some countries as source of acetylene gas, which is an artificial ripening agent. However, acetylene is not nearly as effective for ripening as is ethylene, and acetylene is not a natural plant hormone like ethylene. Also, calcium carbide may contain traces of arsenic and phosphorus, both of which are highly toxic to humans. The mango cultivars ripened with calcium carbide had significantly low carotenoid content and less firmness when compared to naturally ripened fruits (Table 6).

Table 6. Effect of Ripening with Calcium Carbide on Quality of Mango Fruits

Cultivar	*Treatments*	*Sugar/ Acid Ratio*	*Total Carotenoids µg/100g pulp*	*Firm-ness (lbs)*
Alphonso	Natural ripening	82.70	16,450	8.8
	Calcium carbide	43..20	12,800	4.4
Baneshan	Natural ripening	83.75	3,540	7.9
	Calcium carbide	35.90	2300	6.2
Langra	Natural ripening	118.90	4,687	12.8
	Calcium carbide	36.50	2,375	7.0
Pairi	Natural ripening	51.20	11,250	11.6
	Calcium carbide	37.00	8,450	4.8
Totapuri	Natural ripening	87.50	3,500	9.7
	Calcium carbide	45.00	2,250	6.6

Source: Shantha Krishnamurthy and Rao, 1981.

The following acute (short-term) health effects may occur immediately or shortly after exposure to calcium carbide

i) Contact with calcium carbide can irritate the skin causing a rash, redness and burning feeling

ii) Contact can also severely irritate and burn the eyes with possible permanent damage (corneal opacities)

iii) Exposure can irritate the mouth, nose and throat.

iv) Inhaling calcium carbide can irritate the lungs causing cough and/or shortness of breath. Higher exposures may cause a build-up of fluid in the lungs (pulmonary edema), a medical emergency, with severe shortness of breath.

v) It is flammable and reactive and thus a dangerous fire and explosion hazard.

vi) When exposed to water or moisture it forms flammable acetylene gas.

The following chronic (long-term) health effects may also occur some time after exposure to calcium carbide and can last for months or years.

i) Cancer hazard

ii) Reproductive hazard

Use of calcium carbide though banned in India has ben reported to be in use in many crops in different countries. It is however, not clear whether the use of calcium carbide is legal in these countries or not. However, several countries use calcium carbide for ripening of banana, mango, citrus, tomato, plums and peaches (Table 7).

Table 7. Countries where the Fruits Continue to be Ripened Using Calcium Carbide

Fruits	*Countries*
Banana	Australia, Egypt, India, Philippines, South Africa, Sudan, Taiwan, U.S.A, Yemen
Citrus	Australia, Philippines, South Africa
Mango	Brazil, Costa Rica, India, Malaysia, Philippines, Senegal, South Africa
Peach	South Africa
Plum	South Africa
Tomato	Australia, Morocco, Philippines, U.S.A

Source: Rahman *et al.*, 2008.

2.6 Packaging

Packaging of fresh fruits and vegetables is one of the most important steps in the long and complicated journey from grower to consumer. Bags, crates, hampers, baskets, cartons, bulk bins, and palletized containers are convenient for handling, transporting, and marketing fresh produce. Although the industry generally agrees that container standardization is one way to reduce cost, the trend in recent years has moved toward a wider range of package sizes to accommodate the diverse needs of wholesalers, consumers, food service buyers, and processing operations. Packing and packaging materials contribute a significant cost to the produce. It is therefore important that packers, shippers, buyers, and consumers have a clear understanding of the wide range of packaging options available. A significant percentage of buyer and consumer complaints may be traced to container failure because of poor design or inappropriate selection and use. A properly designed produce container should contain, protect, and identify the produce, satisfying everyone from grower to consumer.

The most important considerations in packaging are:

2.6.1 Recyclability

Many developed countries and export markets have waste disposal restrictions for packaging materials. In the near future, almost all produce packaging may have to be recyclable or biodegradable, or both.

2.6.2 Sales Appeal

Packaging plays a vital role in attracting the consumers. High quality graphics are increasingly becoming popular to boost sales appeal. Multi-color printing, distinctive lettering, and logos are now common. Modern produce packaging can be custom engineered for each commodity to extend shelf life and reduce waste.

2.6.3 Containment

The package must enclose the produce in convenient units for handling and distribution. Optimization of space inside the package to accommodate produce is very important. Small produce items that are spherical or oblong (such as potatoes, onions, and apples) may be packaged efficiently utilizing a variety of different package shapes and sizes. However, many produce items such as asparagus, berries, or soft fruits may require containers specially designed for that item. Packages of produce commonly handled by hand are usually limited to 25 kg. Bulk packages that are moved by fork lifts may weigh as much as 600 kg.

2.6.4 Protection from Mechanical Damage

The package must protect the produce from mechanical damage and poor environmental conditions during handling and distribution. To buyers, torn, dented, or collapsed produce packages usually indicate lack of care in handling the contents. Produce containers must be sturdy enough to resist damage during handling, storage, and transportation to market. Because almost all produce packages are palletized, produce containers should have sufficient stacking strength to resist crushing in a low temperature, high humidity environment. Although the cost of packaging materials has escalated sharply in recent years, poor quality, lightweight containers that are easily damaged by handling or moisture are no longer tolerated by packers or buyers. Produce destined for export markets requires containers to be extra sturdy. Air-freighted produce may require special packing, with respect to package sizes, and insulation. Marketers who export fresh produce should consult with freight companies about any special packaging requirements. APEDA of Government of India may also be able to provide specific packaging information.

Damage resulting from poor environmental control during handling and transit is one of the leading causes of rejected produce and low buyer and consumer satisfaction. Each fresh fruit and vegetable commodity has its own requirements for temperature, humidity, and environmental gas composition. Containers should help to maintain an optimum environment for the longest shelf life of produce. This may include special materials to slow down the loss of water from the produce, insulation materials to keep out the heat, or engineered plastic liners that maintain a favorable mix of oxygen and carbon dioxide.

2.6.5 Information about Produce

The package must identify and provide useful information about the produce. It is customary (and may be required in some cases) to provide information such as the produce name, brand, size, grade, variety, net weight, count, grower, shipper, and country of origin. It is also becoming common to give nutritional information, recipes, and other useful information on the packages directed specifically at the consumer. In consumer marketing, package appearance has also become an important part of point of sale displays. Universal Product Codes (UPC or bar codes) may be included as part of labeling. The UPCs used in the food industry consist of a ten-digit machine readable code. The first five digits are a number assigned to the specific producer (packer or shipper) and the second five digits represent specific product information such as type of produce and size of package. Although no price information is included, UPCs are used more and more by packers, shippers, buyers, and retailers as a fast and convenient method of inventory control and cost accounting. Efficient use of UPCs requires coordination with everyone who handles the package.

2.7 Packaging Materials

2.7.1 Wooden Pallets

Pallets literally form the base on which most fresh produce is delivered to the consumer. Pallets were first used during World War II as an efficient way to move goods. Standard size pallets make efficient use of truck and van space and can accommodate heavier loads and more stress than lighter single-use pallets. Substantial wooden pallet bins of milled timber or plywood are primarily used to move produce from the field or orchard to the packing house. Although the height may vary, the length and width is generally the same as a standard pallet (120 cm by 100 cm). More efficient double-wide pallet bins (120 cm by 200 cm) are becoming more common in some produce operations. Most pallet bins are locally made; therefore it is very important that they be consistent from lot to lot in materials, construction and especially size. For example small differences in overall dimensions of pallet bin can add up to big problems when several hundred are stacked together for cooling, ventilation, or storage. It is also important that stress points be adequately reinforced. When properly protected from the weather, pallets bins may have a useful life of 10 years or more. Wooden crates, once extensively used for apple, mango, stone fruit and potatoes have been almost totally replaced by other types of containers.

2.7.2 Corrugated Fiber Boxes (CFB)

Corrugated fiber board (often called cardboard or pasteboard) is manufactured in many different styles and forms. Due to its relativity low cost and versatility, it is the dominant produce container material. Corrugated fiberboard is made from three or more layers of paperboard. Double-faced corrugated fiberboard is the predominant one. It is produced by sandwiching a layer of corrugated paperboard between an inner and outer liner (facing) of paper-board. Unless the CFB boxes are specially treated, moisture absorbed from the surrounding air and the contents can reduce the strength of the container by as much as 75 percent. New anti-moisture coatings with plastic substantially reduce the effects of moisture.

There are numerous types of corrugated fiberboard containers. The two types most used in the industry are the one piece, regular slotted container (RSC) and the two piece, full telescoping container (FTC). The RSC is the most popular because it is simple and economical. However, it has relatively low stacking strength and therefore must be used with produce, such as potatoes, that can carry some of the stacking load. The FTC (with one container inside another), is used when greater stacking strength and resistance to bulging is required. A third type of container is the Bliss box, which is constructed from three separate pieces of corrugated fiberboard. The Bliss box is used when maximum stacking strength is required. The bottoms and tops of all three types of containers may be closed by glue, staples, or interlocking slots.

2.7.3 Shrink Wrapping

One of the newest trends in produce packaging is the shrink wrapping of individual produce items. Shrink wrapping has been used successfully to package, apple, onion, sweet corn, cucumber and a variety of tropical fruits and vegetables. Shrink wrapping with an engineered plastic wrap can reduce shrinkage, protect the produce from disease, reduce mechanical damage and provide a good surface for stick-on labels.

3.0 Spoilage of Produce

Different types of micro-organisms can cause spoilage in storage houses. However, bacteria and fungi are most common. Bacteria are generally propagated by direct contact between perishable goods or with contaminated surfaces, or else with the water used during treatment before or after storage. Some of these bacteria can be pathogenic for human as well, such as *Listeria monocytogenes,* sometimes found on certain vegetables. Vegetables are mostly spoiled by bacteria due to their high pH (4.5-7.0) whereas many fruits particularly the acidic fruits (pH < 4.5) inhibit the growth of bacteria but encourage fungal spoilage. A relative humidity of less than 94 to 95 per cent is considered low enough for growth of bacteria. Majority of the soft rot causing bacteria have their optimum temperature for 30 °C. The following are the major bacterial species that infect the fruits and vegetables (Table 8).

Table 8. Major Bacterial Diseases of Horticultural Produce

Commodity	*Causal Organism*
Apple and pear	*Pseudomonas* spp.
Cabbage, lettuce	*Pseudomonas cicorii*
Cauliflower, cabbage, tomato, bean, radish	*Xanthomonas* spp.
Mango, plum, cherry, peach	*Xanthomonas campestris*
Onion	*Pseudomonas cepacia*
Pineapple, papaya, guava	*Ewrinia* spp.
Potato	*Erwinia caratovora*
	Bacillus subtilis
	Clostridium sp.

3.1 Postharvest Diseases

3.1.1 Bacterial Diseases

Chlorine compounds added to water as chorine gas or as hypochlorite solution could be used to kill bacteria rapidly. In commercial practice 2 per cent sodium hypochlorite or 50 to 100 mg of chlorine or 1-1.5 g of chlorine dioixde per litre of water are frequently used to control the bacteria. Low temperature storage is considered to be the best method for controlling spoilage due to bacterial diseases.

3.1.2 Fungal Diseases

Several fungal rots cause severe postharvest losses to fruits and vegetables by latent or quiescent infection. Some of the examples of quiescent infections are mentioned in Table 9.

Control of quiescent infection in fruits can be achieved by

i) Strict sanitation in pack house to reduce conidial rot

Table 9. Major Fungal Organisms Affecting Fruits in Quiescent Infection

Commodity	*Causal Organism*
Apple	*Gloeosporium perennans, Diplodia perniciosa, Nectria galligena*
Citrus	*Colletotrichum gloeosporioides*
Mango	*Colletotrichum gloeosporioides*
Strawberry	*Botrytis cinerea*

ii) Avoidance of injury

iii) Postharvest dips with permitted systemic fungicides *viz.* Prochloraj carbendazim *etc.*

iv) Pre-cooling or cool chain management

Two major diseases of mango observed during postharvest stage are: a) anthracnose: caused by *Glomerella cingulata* (conidial stage *Colletotrichum gloeosporioides*) and the stem end rot of mango associated with *Botryodiplodia theobromae* or *Aspergillus niger*. Among the other pathogens *Penicillium* spp. is responsible for causing diseases to many fruits and vegetables *e.g.* apples, citrus *etc.* Finger rot of banana caused by *Botryodiplodia theobromae.*

Major fungal organisms mostly associated with postharvest diseases of vegetables are:

i) *Fusarium* spp. causing rots in round melon, bean, cowpea, chilli, tomato, cauliflower, potato, cucumber, pointed gourd and pumpkin.

ii) *Geotrichum* spp. which causes spoilage in tomato, pointed gourd, watermelon and muskmelon attack okra and cucurbits.

The following are the time temperature combinations for control of the fungal diseases in fruit and vegetables (Table 10).

Table 10. Time-Temperature Combinations of Hot Water Treatment for Control of Fungal Diseases in Fruit Crops

Organism	*Temperature (°C)*	*Time/Duration (min)*
Bortyodiolodia, Gloeosporium, Colletotrichum, Rhizopus	43-49 (Papaya)	20
Colletotrichum, Diplodia, Aspergillus, Bortyodiolodia, Rhizopus	47-55 (Mango)	10-20
Monilia, Rhizopus	49-84 (Peach)	1.5-3.5
Alternaria, Geotrichum	48-57 (Tomato)	5-10

In the recent times consumers are very much concerned about residual toxicity of pesticides on fruits and vegetables. Several bio-control formulations available in the market which could be successfully used in postharvest management of horticultural produce are listed in Table 11.

Table 11. Biological Control of Postharvest Diseases

Crops	*Name of Disease*	*Biocontrol Agent*
Apricot, avocado, cherry, citrus, litchi, Strawberry	Brown rot, Grey mold, Green Mold, Stem end rot, Alternaria rot	*Bacillus subtilis*
Mango	Anthracnose, steam end rot	*Bacillus lichenifomis*
Citrus, mango, strawberry	Green mold, Stem-end rot, Gray mold	*Tricoderma viride*
Apple, citrus, peach	Blue mold, Green and blue mold and Gray mold, Brown rot	*Pseudomonas syringae*
Banana, citrus, papaya	Penicillium rots and Crown rot, Anthracnose	*Candida oleophila*
Citrus, peach	Green and blue mold, Blue mold, Rhizopus rot, Sour rot	*Debaryomyces hansenii*

3.2 Insect Pests

A large number of insect pests adversely affect the postharvest quality of horticulture produce. These include fruit fly, mite damage in citrus, thrip infection in banana, mealy bug in pineapple *etc.* Adoption of strict regulation is required to eliminate the risk of entry of new pests. Although fumigants such as ethylene dibromide (EDB) and methyl bromide (MB) are often used for disinfestation in many countries these are slowly being phased out due to their carcinogenic effects. Quarantine treatments for fruit fly is mandatory except in European Union and Canada for fruits like mango, papaya, melons and many other fruits and vegetables. Hence there is need for chemical free produce for its acceptability in the international market. Recent developments on irradiation and insecticidal controlled atmosphere are some of the technologies for pest management strategies of fruits and vegetables.

Use of irradiation in postharvest management is limited to inhibition of sprouting in potato, onion, garlic, and ginger, reduction in pathogen inoculum in spices and disinfestation of insects in dry fruits. However, it has also been used as quarantine treatment for export of mango from India to United States. Technologies of application of irradiation on fresh pre-packed foods have been standardized by BARC, Mumbai. Several researches have proved that irradiation helps delaying ripening and senescence and controlling fugal decay. More than 21 horticultural products are now being treated with gamma irradiation about 45 countries in the world use this technology. Studies on wholesomeness, microbial safety, nutritional adequacy, mutagenicity and toxicity have shown that this technology is safe.

Vapour Heat Treatment (VHT) has been developed in Japan in the year 1993 for destruction of fruit flies in green pepper. This technology has now been emphasized to disinfest insects in several kinds of fruits exported from Thailand, Philippines, Taiwan, Hawaii, Fiji, Mexico, Australia, Brazil, and Columbia. This treatment eliminates the use of chemicals for disinfestation of fruits flies. VHT has been widely accepted in most of the countries where imports of fruit are banned due to chemical disinfestation of fruit flies. In India the facilities of VHT or hot water treatment of mango are available and standardization of VHT for different fruits needs to be done.

Several chemical growth regulators are being used for enhancing shelf life and delaying ripening of fruits and vegetables. Some examples of chemicals and edible surface coating materials used in fruits and vegetables for regulation of postharvest behaviour are mentioned in Table 12.

Table 12. Important Edible Surface Coating Materials

Coating Material	*Fruits*
Semperfresh	Apple, Banana, Guava
Brilloshine	Apple, Avocado, Citrus fruits
Citrashine, Nu-coatflo	Citrus fruits
Tal prolong, Applewax	Apple
Nutri-save	Apple (Golden Delicious)
Vapor gard	Mango
Ban-seel	Banana and Plantains
Chitosan	Raspberry, Strawberry

In recent times several new molecules have been used to retain the postharvest quality of fruits. These are Methyl Jasmonate (MeJA) for alleviation of chilling injury, *viz.* loquat, peach, pomegranate, strawberry. Similarly, inducing resistance for postharvest diseases and checking browning in cut fruit salicylic acid (SA) has been found to be useful, in case of banana, cherry, kiwi fruit, mango, and pomegranate. Use of nitric oxide (NO) has been found successful in peach for delay in fruit softening, reduction in chilling injury, microbial infection and extension of postharvest life (Table 13).

4.0 Storage

4.1 Controlled (CA)/Modified Atmosphere Storage (MAS)

Postharvest losses of fruits and vegetables during storage occur primarily due to chilling injury in case of banana, citrus, guava, mango, papaya, pomegranate, sapota: rind breakdown in case of citrus; loss of peel color in litchi, desiccation and decay in litchi and pomegranate; internal breakdown and crown dehydration in pineapple, pomegranate *etc.* The effect of gas on harvested fruits is known since ancient times. In earlier days fruits were taken to temples for improving ripening. The scientific explanation of this phenomenon could be attributed to the volatiles released on burning of incense containing hydrocarbon gases causing ripening of fruits. Earliest scientifically documented evidence on postharvest physiology was recorded by J.E. Bernard in 1819 in France who noticed that harvested fruits absorb O_2 and liberate CO_2. Atmosphere devoid of O_2 caused no ripening in peach, prunes and apricot for several days but ripening continued when they were

Table 13. Use of Chemicals in Postharvest Management

Chemicals	*Crops*	*Effect*
Calcium Chloride	Grape	Improves shelf life
	Pineapple	Better ripening and overall quality
CIPC	Potato	Inhibits sprouting
Ethephon	Grape	Hastens ripening
Ethylene	Tomato	Hastens ripening
GA	Guava, Mandarin and Mango	Delays ripening
Guar gum	Pomegranate	Reduces weight loss
Methyl Jasmonate	Loquat, peach, pomegranate, strawberry	Checks postharvest decay, alleviates chilling injury, delays fruit ripening and improves sensory apeal
MH	Onion, Potato	Checks sprouting
N-Benzyladanine	Leafy vegetables	Improves shelf life
Nitric Oxide (NO)	Peach	Delays fruit softening, chilling injury, microbial infection and extends postharvest life.
Polyamines	Pomegranate	Alleviates chilling injury, reduces respiratory rates and retards colour chang
Salicylic Acid (SA)	Banana, cherry, kiwi, mango, pomegranate	Increases resistance to diseases, delays ripening and inhibits browning in cut fruits
Skin coatings with waxes	Citrus, Mango	Reduces respiratory rate and moisture loss
Smoking	Banana	Enhances ripening

placed back in air. A commercial cold storage built in 1856 by B. Nice used ice to maintain a temperature of 1 °C. After a decade he experimented with modifying the cold store with CO_2 and O_2 gases inside. In 1907 Foulton J. observed the increase in fruit damage by large accumulation of CO_2 in the storage atmosphere. In 1915 R.W.Thatcher on their experiments with apples in sealed boxes containing various levels of gases concluded that CO_2 greatly inhibited ripening. The first scientific evidence on the effect of CO_2 on respiratory rates was established by Kidd and West in 1917. In 1927 they carried out comprehensive studies on the effect of CO_2 and O_2 on storage of fruits and vegetables. In the early 1940 the term 'gas storage' was replaced by Controlled Atmosphere storage.

Atmosphere at ambient conditions comprises of 78.08 per cent N_2, 20.98 per cent O_2 and 0.03 per cent CO_2 under normal conditions. Any deviation from this normal atmosphere composition *e.g.* elevated level of CO_2 reduced level of O_2, N_2 or any other combination is known as '**Modified Atmosphere**'. When this deviated normal atmosphere is precisely kept under control then it is termed as "**Controlled Atmosphere**". This control can be done in package (Controlled Atmosphere Packaging) or in the storage chamber (CA-storage). Similar is the case for Modified Atmosphere Storage (MA-storage) and Modified Atmosphere Packaging (MAP). Generally, O_2 below 8 per cent and CO_2 above 1 per cent are used in CA-storage. Atmospheric modification is a supplementary practice to temperature management in preserving quality and safety of fresh fruits, vegetables, ornamentals and their products throughout postharvest handling. Essentiality of CA/MA technology should be justified only if:

i) the commodities are having high market value
ii) it significantly enhances storage life
iii) it retains significantly better quality
iv) it fetches better price compared to conventional cool stored produce.

Retardation of ripening, reduction in decay, prevention of specific disorders and maintenance of product texture are some of the potential advantages of CA/MA storage. However, initiation or aggravation of certain physiological disorders *viz.* black heart in potato, brown stain in lettuce, brown heart in apples and pears *etc.* take place in CA-storage if appropriate gaseous regimes are not maintained. A concentration of less than 2 per cent O_2 or more than 5 per cent CO_2 in the storage/package atmosphere results in irregular ripening of banana, tomato and pear fruits. Too low O_2 or too high CO_2 can increase the susceptibility to decay causing organisms. Off-flavour development in fruits, stimulation of sprouting and retardation of periderm formation in potatoes are some of the ill effects of improper CA-condition. On the other hand, a very high concentration of CO_2 (up to 30 per cent) was found to be extremely beneficial in checking the *Botrytis* rot in strawberry fruits.

Controlled atmosphere storage and preservation of packaged horticultural produce is a widely utilized technology for fresh fruits, vegetables and flowers. The utilization of inert gases or vacuum can allow unique applications that control microorganisms as well as maintain product colour and freshness. It should be kept in mind that CA and MA are the supplementary practices to low temperature storage. The quality retention is mainly due to reduction in the respiratory and metabolic activities and check in ethylene liberation by fresh horticultural produce during storage.

4.1.1 Biological Basis

Exposure of fresh horticultural crops to low O_2 and/or elevated CO_2 atmospheres within the range tolerated by each commodity reduces their respiration and ethylene production rates. However, outside this range respiration and ethylene production rates can be stimulated indicating a stress response. This stress can contribute to incidence of physiological disorders and increased susceptibility to decay. Elevated CO_2-induced stresses are additive to and sometimes synergistic with stresses caused by low O_2; physical or chemical injuries; and exposure to temperatures, RH, and/or ethylene concentrations outside the optimum range for the commodity.

The shift from aerobic to anaerobic respiration depends on fruit maturity and ripeness stage (gas diffusion characteristics), temperature, and duration of exposure to stress-inducing concentrations of O_2 and/or CO_2. Up to a point, fruits and vegetables are able to recover from the detrimental effects of low O_2 and/or high CO_2 stresses (fermentative metabolism) and resume normal respiratory metabolism upon transfer to air. Plant tissues have the capacity for recovery from the stresses caused by brief exposure to fungistatic CO_2 (> 10 per cent CO_2). Climacteric fruits are less tolerant and have lower capacity for recovery following exposure to reduced O_2 and/or elevated CO_2 levels than non-climacteric fruits. The speed and extent of recovery depend upon duration and levels of stresses, stage of maturity and metabolically driven cellular repair mechanism.

Elevated-CO_2 atmospheres inhibit activity of ACC synthase (key regulatory enzyme of ethylene biosynthesis), while ACC oxidase activity is stimulated at low CO_2 and inhibited at high CO_2 concentrations and/or low O_2 levels. Thus elevated CO_2 atmospheres inhibit ethylene action. Optimum modified atmospheric compositions retard chlorophyll loss (green color), biosynthesis of carotenoids (yellow and orange colours), anthocyanins (red and blue colors), biosynthesis and oxidation of phenolic compounds (brown color). Controlled atmospheres slow down the activity of cell wall degrading enzymes involved in softening and enzymes involved in lignification leading to toughening of vegetables. Low O_2 and/or high CO_2 atmospheres influence flavor by reducing loss of acidity, starch to sugar conversion, sugar inter-conversions, and biosynthesis of flavor volatiles.

Severe stress CA conditions decrease cytoplasmic pH and ATP levels, and reduce pyruvate dehydrogenase activity while pyruvate decarboxylase, alcohol dehydrogenase, and lactate dehydrogenase are induced or activated. This causes accumulation of acetaldehyde, ethanol, ethyl acetate, and/or lactate, which may be detrimental to the commodities if they are exposed to stress CA conditions beyond their tolerance. Specific responses to CA depend upon cultivar, maturity and ripeness stage, storage temperature and duration, and in some cases, ethylene concentrations.

Nitrogen is an inert component of CA. Replacing N_2 with argon or helium may increase diffusivity of O_2, CO_2 and C_2H_4, but they have no direct effect on plant tissues and are more expensive than N_2 as a CA component. Super-atmospheric levels of O_2 up to about 80 per cent may accelerate ethylene-induced de-greening of non-climacteric commodities and ripening of climacteric fruits, respiration and ethylene production rates, and incidence of some physiological disorders (such as scald on apples and russet spotting on lettuce). At levels above 80 per cent of O_2 some commodities and postharvest pathogens suffer from O_2 toxicity. Use of super-atmospheric O_2 levels in CA will likely be limited to situations in which they reduce the negative effects of fungistatic or elevated CO_2 atmospheres on commodities that are sensitive to CO_2-injury.

4.1.2 Commercial Application

Controlled atmosphere (CA) storage systems are used commercially for the long-term storage of fresh horticultural crops. Commercial use of CA storage is maximum in apples and pears worldwide. Recent research has shown the potential advantages of this method in short-term (a few days) and medium-term (a few weeks) storage of certain types of produce. Optimizing storage conditions requires facilities that allow the temperature and composition of gases in storage rooms to be controlled precisely. Each product reacts in different ways to different concentrations of gases.

There are various types of CA systems used in commercial storage. These include generating nitrogen by separation from compressed air using molecular sieve beds or membrane systems. Various types of CA storage system include ultra low O_2 (1.0 to 1.5 per cent) storage, low ethylene ($< 1\ \mu L\ L^{-1}$) CA storage; rapid CA storage (rapid establishment of optimal levels of O_2 and CO_2), and programmed or sequential CA storage (*e.g.*, storage in 1 per cent O_2 for 2 to 6 weeks followed by storage in 2 to 3 per cent O_2 for the remainder of the storage period) *etc.* Recent reports of short-term control atmosphere exposure (CAE) techniques indicate a great promise of simulated effect of continuous CA–storage system with particular reference to delay in disease development, delaying the senescence and quality assurance of CA-insensitive climacteric fruits. Other developments include use of atmospheric modification during transport and distribution, improved technologies of establishing, monitoring, and maintaining CA, using edible coatings or polymeric films with appropriate gas permeability to create a desired atmospheric composition around the commodity.

4.2 Modified Atmosphere Packaging

MAP may be defined as the enclosure of food products in gas-barrier materials, in which the gaseous environment has been changed in order to inhibit spoilage agents and therefore either maintain a higher quality within a perishable food during its natural life or actually extend the shelf life. Although the terms Controlled Atmosphere (CA) and Modified Atmosphere (MA) are often used interchangeably, a precise difference exists between these two terms.

4.2.1 Relative Tolerance to O_2 and CO_2

The extent of benefits from CA and MA use depend upon the commodity, cultivar, physiological age (maturity stage), initial quality, concentration of O_2 and CO_2, temperature, and duration of exposure to such conditions. Subjecting a cultivar of a given commodity to O_2 levels below and/or CO_2 levels above its tolerance limits at a specific temperature time combinations will result in stress to the living plant tissue, which is manifested as various symptoms. Tables 14 and 15 include classifications of fruits and vegetables according to their relative tolerance to low O_2 or elevated CO_2 concentrations when kept at or near their optimum storage temperature and relative humidity.

Successful MAP or CA must maintain near optimum O_2 and CO_2 levels to attain the beneficial effects of modified atmosphere without exceeding the limits of tolerance which may increase the risk of physiological disorders and other detrimental effects.

Table 14. Classification of Fruits and Vegetables According to their Tolerance to Low O_2 Concentrations

Minimum O_2 Concentration Tolerated (per cent)	*Commodities*
1.0	Some cultivars of apple and pear, broccoli, garlic, mushroom, onion
2.0	Most cultivars of apple, apricot, Brussels sprout, cabbage, cantaloupe, cauliflower, celery, cherry, green bean, kiwifruit, lettuce, nectarine, olive, papaya, peach, pear, pineapple, plum, strawberry
3.0	Artichoke, avocado, cucumber, pepper, persimmon, tomato
5.0	Asparagus, citrus fruits, green pea, potato, sweet potato

Source: Kader *et al.*, 1989.

Table 15. Classification of Fresh Fruits and Vegetables According to their Tolerance to Elevated CO_2 Concentrations

Maximum CO_2 Concentration Tolerated (per cent)	*Commodities*
2	Apple (Golden Delicious), Pear (Asian and European), apricot, grape, olive, tomato, sweet pepper, lettuce, Chinese cabbage, artichoke, sweet potato
5	Most cultivars of Apple, avocado, banana, brussels sprouts, cabbage, carrot, cauliflower, chilli, egg plant, kiwifruit, mango, nectarine, orange, papaya, pea, peach, plum, radish
10	Asparagus, broccoli, cucumber,, garlic, grape fruit, leek, lemon, lime, okra, onion (dry), onion (green), persimmon, pineapple, potato
15	Blackberry, cantaloupe, fig, kale, mushroom, raspberry, spinach, strawberry.

Source: Kader *et al.*, 1989.

4.3 Creating Modified Atmosphere Conditions

Modified atmosphere can be created either passively by the respiration of commodity also called commodity generated modified atmosphere or active modified atmosphere generation.

4.3.1 Passive Modified Atmosphere

If commodity characteristics are properly matched to film permeability characteristics, an appropriate atmosphere can passively evolve within a sealed package as a result of the consumption of O_2 and the production of CO_2 through respiration. This atmosphere must be established rapidly and without danger of the creation of anoxic conditions on injuriously high level of CO_2.

4.3.2 Active Modified Atmosphere

Because of the limited ability to regulate a passively established atmosphere, an active modified atmosphere can be created by vacuum packaging or replacing the package atmosphere with the desired gas mixture.

a) Vacuum Packaging

Involves placing a product in a film of low O_2 permeability, the removal of air from the package and the application of a hermetic seal. Product compression is unavoidable and it makes vacuum packaging unsuitable for many products.

b) Gas Packaging

Gas packaging overcomes the limitations of vacuum packaging. Oxygen, nitrogen and carbon dioxide are gases generally used which have specific role *viz.* O_2 is essential for respiration of fresh fruits and vegetables, CO_2 is a bacterial and fungal growth inhibitor, nitrogen is inert, tasteless and less prone to pass either into the product or out through the packaging material. It forms an important filler gas.

For a fruit to be compatible with the use of CA it should preferably be characterized by : (i) a long postharvest life (ii) resistance to chilling injury (iii) a large range of non-injurious atmospheres (iv) resistance to fungal and bacterial attack (v) adaptation to a humid atmosphere (vi) a climacteric fruit that can be ripened during or after storage (vii) absence of negative CA residual effect.

4.4 Benefits and Harmful Effects of MA and CA

4.4.1 Beneficial Effects

When used properly, MA and CA can supplement proper temperature management and can result in one or more of following benefits:

i) Retardation of senescence (ripening)

ii) Slowing down of respiration and ethylene production rates

iii) Retards softening and compositional changes

iv) Reduction of fruits sensitivity to ethylene action.

v) Alleviation of certain physiological disorders such as chilling injury of various commodities, russet spotting in lettuce, and some storage disorders of apple.

vi) Have direct or indirect effect on postharvest pathogens and consequently decay incidence and severity.

vii) Modified atmosphere can be a useful tool for insect control in some commodities.

viii) Can replace postharvest chemicals used for control of some physiological disorders. Fungicides and fumigants may be reduced or eliminated.

ix) In specific instances, facilitate picking and marketing more mature (better flavoured) fruits by slowing down their postharvest deterioration.

4.4.2 Harmful Effects

i) Initiation and/or aggravation of certain physiological disorders such as blackheart in potato, brown stain of lettuce, and brown heart in apple and pear.

ii) Irregular ripening of fruits such as banana, pear and tomato at O_2 levels below 2 per cent and CO_2 levels above 5 per cent.

iii) Development of off-flavors and off-odors.

iv) Increased susceptibility to decay when the commodity is physiologically injured by too-low O_2 or too-high CO_2 concentrations.

v) Stimulation of sprouting and retardation of periderm development in some root and tuber vegetables.

vi) MA and CA can be injurious and even deadly to humans entering a room container without proper safety equipment.

vii) MA and CA can cause structural damage to rooms and containers that lack proper pressure relief systems.

4.5 Modern CA Storage System

Modern CA storage system is much safer than the traditional ones due to the use of the following systems for N_2 generations.

4.5.1 Use of Pressure Swing Adsorption (PSA) for Nitrogen Generation

PSA nitrogen gas generator has two sets of columns filled with carbon molecular sieves (CMS). Under Pressure these columns retain all active molecules compounds present in air like (Oxygen, Carbon Dioxide and Water) with the exception of the inert gases like Nitrogen and Argon. The N_2 gas can be produced onsite for direct use or can be stored in nitrogen storage tanks for intermittent use (Nitrogen and Argon). This process is referred as PSA (Pressure Swing Adsorption).

4.5.2 Use of Membranes for Nitrogen Adsorption

In this system compressed air is passed through the specialized membranes and series of filters. The membrane separates the N_2 form the air through exclusion of O_2. The compressed N_2 comes out form the other end which could be stored in tanks for future use.

4.5.3 Safety Considerations

The atmosphere in CA rooms will not support human life. People may die of asphyxia while working inside CA rooms without breathing apparatus. A danger sign should be posted on the door. The access hatch should be large enough to accommodate a person with breathing equipment strapped to his or her back. At least two persons with breathing equipment should work together at all times with one inside and one outside the room watching the first person.

5.0 Conclusion

In appropriate management of horticultural crop produce, first and foremost important priority is the estimation of losses. Mechanisms need to be developed for estimation of near accurate postharvest losses. Postharvest losses could be managed through interventions like choice of varieties, sorting, grading, use of varieties resistant to biotic and abiotic stress, good horticultural practices, market intelligence and storage.

Postharvest management practices that slow down the postharvest deterioration during handling and storage need to be practiced. This would be possible through improved capacity building of stakeholders including extension agencies, up scaling of technologies, developing linkages among people, private and public. Several policy issues for managing postharvest losses also need to be addressed. Amendment of contract farming rules, APMC acts, development of postharvest infrastructure, energy efficient storage system, crop insurance *etc.* are coherently linked with each other for achieving the desired goals for salvaging huge postharvest losses of horticultural produce in India

Among the fruits top priority needs to be given to postharvest management of apple, banana, citrus, guava, mango, papaya, pineapple, pomegranate, strawberry and sapota. The priority crops in vegetables are brinjal, cabbage, cauliflower, okra, onion, tomato, potato, sweet paper, chilli, garlic. Among them rose, carnation, gladiolus, tuberose, lilies, orchids, chrysanthemum should get priority attention. Ginger and Turmeric are the most important spice crops in India requiring priority.

Development of non-destructive methods to determine maturity of fruits and vegetables is need of the hour. Similarly, postharvest research should also focus on understanding fundamentals of fruit ripening including regulation of ethylene biosynthesis in tropical and subtropical fruits of Indian origin for developing commercially practicable technologies. Research on development of sound and economical handling technologies for specific fruit, vegetable, cut flowers, spices and other horticulture crops are some of the research priorities in India. Use of sensors for detecting infected/injured or damaged food including use of multiple chips, nano-biosensors and efficient grading and sorting of healthy produce are also essential. In case of root and tuber crops appropriate curing practices need to be standardized for improving the postharvest quality and reducing the losses during handling. Development of smart/intelligent packaging, punnet, CFBs, plastic trays, stretch films, shrink wraps recyclable plastic crates *etc.* for different types of produce are also emerging area of postharvest research.

Integrated diseases and pest management practices involving minimum chemical use imparting natural disease resistance against plant pathogens and use of bio-agents are some of the novel areas of research. Research on integrated insect pest and disease management in both pre and postharvest phases *i.e.* use of hormone traps for fruit flies, hot water dip, VHT, insecticidal controlled atmosphere, gamma irradiation *etc.* need to be attempted in order to meet quarantine requirements for overseas markets.

In the area of storage of horticulture produce, development of genotype-specific information on use of cold storage and CA/MA storage (conc. of O_2, CO_2 and temperature) are lacking. Similarly, multi-commodity storage needs to be developed for diversified produce. Low temperature storage through intermittent warming, CA, growth regulators improving the efficiency of surface coating and packaging are some of the priority areas. Development of edible and eco-friendly coating (nano-polymers, chitosan) for extending postharvest life also deserves attention. Combined use of computer and engineering tools to estimate precise cost benefit relationship among structural geometric, thermal and ergonomic aspects of commodity packaging and storage is another area.

Genotypes having resistance to physiological disorders and/or decays causing pathogens/pests need to be evolved to minimize the use of chemicals for better shelf life. Studies on genomics, transcriptomics, proteomics and metabolomics in relation to fruit maturity, physiology of ripening, senescence and susceptibility to diseases and pests need to be carried out. Development of nano-biodegradable membrane for smart packaging, nano formulations/molecules/particles for reducing the ethylene and colour development and nano coating/ waxes for better retention of flavor/colour minimizing microbial spoilage are some of the future researchable issues.

Development of holistic quality management system involving safe production practices, harvesting, postharvest handling, storage, transportation, cold chain, distribution and assurance on safe food production is essential. Interdisciplinary approach is the need of the hour to tackle any postharvest problems involving horticulturists, postharvest biologist, plant breeders, molecular biologist, food technologists, microbiologists, bio-chemists, plant physiologists, and agri./food engineers, computer scientists, statistician, economists, management expert and psychologists.

6.0 Important References

Anonymous, 2018. *Horticultural statistics at a glance.* Government of India Ministry of Agriculture & Farmers' Welfare Department of Agriculture, Cooperation & Farmers' Welfare. 490p

Anonymous, 2021. *Release of Final Estimates of 2019-20 and First Advance Estimates of 2020-21 of Area and Production of Horticultural Crops*, Ministry of Agriculture & Farmers Welfare Posted On: 08 MAR 2021 4: 18PM by PIB Delhi

Abeles, F.B., Morgan, P.W. and Saltvett, M.E. 1992. **In:** *Ethylene in plant biology*, 2nd edition, Academic Press, London, UK, pp. 182-186.

Chadha, K.L. 1995. Status of postharvest technology of fruits. In: *Horticultural Renaissance in India.* A compendium of selected lectures of Dr. K.L. Chadha (Eds.) S.M Paul Khurana and S.K. Singh, IARI, New Delhi, 2013, pp. 905-12.

Chadha, K.L., 2000. Pre and Postharvest technology of fruits and vegetables. In: *Horticultural Renaissance in India.* A compendium of selected lectures of Dr. K.L. Chadha (Eds.) S.M Paul Khurana and S.K. Singh, IARI, New Delhi, 2013, pp. 935-952.

Chesworth, J.M., Stuchbury, T. and Scaife, J.R. 1998. **In:** *An Introduction to Agricultural Biochemistry*, Chapman and Hall, London, UK, pp. 388-392.

HLPE, 2014. Food losses and waste in the context of sustainable food systems. A report by the *High Level Panel of Experts on Food Security and Nutrition* of the Committee on World Food Security, Rome.

Kader, A.A. 1986. Biochemical and physiological basis for effects of controlled and modified atmosphere on fruit and vegetables. *Food Technol.* **40**: 99-104.

Kader, A.A., Zagory, D and Kerbal, E.L. 1989. Modified atmosphere packaging of fruits and vegetables. *Critic. Rev. Food Sci. Nutr.* **28**: 1-30.

Lakshminarayana, S., Shetty, M.S. and Krishnaprasad, C.A. 1975. Accelerated ripening of Alphonso mangoes by the application of ethrel. *Trop. Sci.*, **17:** 95-101.

Lalel, H.J.D., Singh, Z. and Tan, S.C. 2003. Maturity stage at harvest affects fruit ripening, quality and biosynthesis of aroma volatile compounds in 'Kensington Pride' mango. *J. Hortic. Sci. Biotechnol.* **78,** 225-233.

Lesham, Y.Y., Halevy, A.H and Frenkel, C., 1986. Fruit ripening, **In:** *Process and Control of Plant Senescence*. Elsevier, Amsterdam, the Netherlands, pp. 162-201.

Ludford, P.M. and Isenberg, F.M.R. 1987. Brassica crops. **In**: Weichman J (ed) *Postharvest Physiology of Vegetables*, Marcel Dekker Inc., New York, pp 497-522.

Mahajan, B.V.C., Kaur Tajender, Gill, M.I.S., Dhaliwal H. S., Ghuman, B. S. and Chahil B. S. 2010. Studies on optimization of ripening techniques for banana. *J Food Sci Technol.* **47:** 315–19.

Mahajan B.V.C., Singh G. and Dhatt, A.S. 2008. Studies on ripening behaviour and quality of winter guava with ethylene gas ethephon treatments. *J Food Sci. Technol.* **45:** 81–84.

Medlicott, A.P., Sigrist M.M., Reynolds S.B. and Thompson A.K. 1987. Effect of ethylene and acetylene on mango fruit ripening. *Ann Appl Biol.* **111**: 439–444.

Nguyen, H.R., McConchie, Hofman, P., Smith, L., Stubbings, B. and Adkins, M. 2002.Effect of ethylene and ripening temperatures on the skin colour and flesh characteristics of ripe 'Kensington Pride' mango fruit. *Acta Hortic.* **575:** 635-642.

O'Connor, Skarshewski P. and Thrower S.J. 1992. *ASEAN Food J.* **7:** 127-136.

Pal, R.K. 1997. Pre-storage exposure to Controlled Atmosphere by using simple gas blending system for better storage of mango. **In:** *Proc. of symposium on Processing, finance and marketing in food industry* (H.O.Gupta, Ed.), AFST (I), Delhi Chapter, October, 1997, pp.96-104.

Pal, R.K. 1998. Influence of ethrel and calcium carbide on respiration rate, ethylene, evolution, electrolyte leakage and firmness of 'Dashehari' mango (*Mangifera indica*) in India. *J. Agric. Sci.* **68:** 201-03.

Pal, R.K. and Buescher, R.W. 1993. Respiration and ethylene evolution of certain fruits and vegetables in response to carbon dioxide in controlled atmosphere (CA) storage. *J. Food Sci. Technol.,* **30**: 29-32.

PFA, 2003. *Prevention of Food Adulteration Act* 1954 with Prevention of Food Adulteration Rules 1955. 19th edn, International Law Book Co, Delhi, p. 140.

Reyes, M.U. and Paull, R.E. 1995. Effect of storage temperature and ethylene treatment on guava fruit ripening. *Postharvest Biol. Technol.* **6**: 357–365.

Saltveit, Mikal E. 1999. Effect of ethylene on quality of fresh fruits and vegetables. *Postharvest Biol. Technol.* **15**: 279–92.

Salunkhe,D.K. and Desai B.B. 1984. *Postharvest Biotechnology of Vegetables.* Vol. I, CRC Press Inc.

Thompson, A.K. and Seymour, G.B. 1982. Comparative effects of acetylene and ethylene gas on initiation of banana ripening. *Ann. Appl. Biol.* **101**: 407- 410.

Valverde, F., Hermadez, R.L. and Sain, M.V. 1986. Effect of ethephon and calcium chloride on the ripening and quality of mango fruit, cv. Kiett in Alajuela, Costa Rice. *Agronomica costarricense* (Costa Rica), **10:** 153-163.

Yang, S.F. 1980. Regulation of ethylene biosynthesis, *HortSci*, **15**: 238–43.

Chapter 2

Fruit Bagging

Ram Asrey

Principal Scientist, Division of Food Science & Postharvest Technology, ICAR-Indian Agricultural Research Institute, New Delhi – 110 012, India

E-mail: ramu_211@yahoo.com

1.0 Introduction

Preharvest fruit bagging is one of the best available alternatives which address the concerns of the farmers, consumers and industry people about residue free quality produce. It is an effective tool to improve the fruit quality and commercial value of the fresh produce. The bag provides a physical protection to the fruit by preventing insect pests and damage due to external injury. For over 200 years, the Japanese tied little bags over developing fruits. They first used sewed white silk bags until the 1960's when plastic bags became readily available. Later on bagging became prevalent throughout Europe in pome fruit orchards during Nineteen Seventies (1970's) and Eighties (80's). Wrapping of fruits before harvest with newspaper is used in several Asian countries for fruit fly control.

Fruit bagging is commercially employed in apple, peach and pear world over. It is also used on a limited scale on late mango cultivars in south-east Queensland (Australia) and there is growing interest for its use in case of litchi. This approach is used to produce high quality, unblemished litchi fruits in South Africa for litchi. In India, fruit bagging is not commercially employed in any crop so far but has been gaining momentum in pomegranate, apple, litchi and mango. Bagging affects pigment synthesis, days to maturity, cell histology, physiology, shelf-life, storage behaviour, biochemical composition, pesticide residue content and other phytosanitary quality aspects of fruit. The impact of fruit bagging depends on several factors and cofactors like genotype, bagging material, colour, time of bagging and covering duration. While bagging has several benefits, its dependency on manual operation limits the practical feasibility of this technology. Therefore, there is a need to design and develop bagging gadgets/ machinery to mechanize this operation.

Bagging affects produce quality through modification in solar light quality–UV rays and other visible rays; Change in microclimate (temperature and relative humidity) around growing fruits; Absorption and translocation of nutrients; Alteration in sink-source relationship; Alleviation of insect pest incidence. It also helps in reducing pesticide residues and advancing maturity in certain crops. Being an Eco-friendly approach, it has become an integral part of organic fruit production in some parts of the world.

2.0 Bagging Materials and Application Time

Fruits can be easily protected by bagging them in newspaper, cellulose or butter paper bags. These bags are available in a range of colours *e.g.*, white, red, blue, orange *etc.* Plastic bags may also be used but these are not ideal due to, moisture and heat accumulation inside the bag that favours fungal growth. Alternatively, bags made of leaves like of like pandanus, betel nut tree may also be used. Different fruit crop behave differently on the use of particular bag colour. For example, bagging of peach gives desirable results with the orange colour bags while apple does better with red coloured bagging. The growers can select the bag colour on the basis of existing available information.

To bagging operation can be performed during sunny bright days when there is no moisture deposition on the fruit surface. Bagging may be done manually by standing on ground or with the help of a tripod ladder. Healthy looking fruit (individually or in bunches) are placed in the bag and its top end is tied up with string or a wire. The bottom of the bag is pushed upwards to make it 'V'-shaped. This prevents damage by rain and keeps the bag inflated keeping the fruit away from touching the sides of the bag.

Timing of fruit bagging is crucial for obtaining better results. Researchers have worked out the ideal bagging time for some commercially important fruit crops. The ideal bagging stage for selected fruit crops and its impact on postharvest quality attributes are given in Table 1.

In pome fruits, bagging is generally performed at an early stage when the fruits are in active growing stage. Tropical and subtropical fruits respond better when they are bagged 60 to 40 days before harvesting.

The following precautions need to be taken while fruit bagging

i) Insertion of leaves, twigs, flower buds and flowers inside the bags should be avoided.

ii) Only disease free and healthy looking fruits should be bagged.

iii) Both the ends should be properly closed to avoid sheltering of insect inside bags. However, for preventing sun scalding one end can be kept open.

iv) Bagging should be done one day after fungicide spray during bright sunny weather.

3.0 Effect of Bagging on Fruit Quality

3.1 Apple

It has been reported that covering of fruits with one, two or three-layered paper bags from early development

Table 1. Impact of Fruit Bagging on Postharvest Quality Attributes

Crop	*Fruit Bagging Stage*	*Impact on Quality*	*References*
Apple	Early stage (4 to 6 weeks after petal fall)	No impact on maturity Higher phenolics and anthocyanin content Low level of procyanidin and quercetin glycosides Higher anthocyanin retention during storage Lower Ca content leading to bitter pit incidence	Ju, 1998 Arakawa, 1998
Banana	After bunch formation	Lower incidence of fruit speckling, anthracnose, cigar-end rot and scaring beetle Higher shelf-life, dry matter content and early maturity	Stover and Simmonds, 1991 Pasberg-Gauhl, 2000
Grape	70 days before harvest	Uniform fruit colour Lower TSS, higher acidity and late maturity Higher sucrose and lower glucose and fructose level High hue angle and low chroma value	Signes *et al.*, 2007
Guava	6-9 weeks before harvesting	Reduction in fruit fly and flying fox incidence	Sharma *et al.*, 2021
Litchi	30-40 days before harvest	Higher fruit retention and colour intensity No effect on TSS/acid ratio	Tyas *et al.*, 1998
Loquat	Marble stage	Improve color and reduce decay loss	Ko *et al.*, 2010
Mango	100 days before harvest	Lower incidence of anthracnose and stem-end rot Reduced fruit Ca content and higher N in fruit Higher dry matter content and early maturity	Hofman *et al.*, 1997
Mangosteen	14 days after full bloom	Bagged fruits more attractive but no statistical difference in peel anthocyanin content Higher PAL activity (known for anthocyanin synthesis) in bagged fruits Prevents fruit fly infestation	Ratanamarmo *et al.*, 2005
Peach	50 days after full bloom	Higher fruit length and firmness Higher TSS Bright red coloured fruit	Bin *et al.*, 2006
Pear	30 days after full bloom	Reduction in the level of blemishes and friction induced discolouration on fruit	Amarante *et al.*, 2002
Pomegranate	80 days before harvest	Higher juice and anthocyanin content Low incidence of bacterial blight and fruit fly	Asrey *et al.*, 2019

stage until harvest did not affect fruit maturity but significantly inhibited anthocyanin accumulation in fruit peel compared with the un-bagged control of apple cv. Delicious (Figure 1). High levels of procyanidin, quercetin glycosides, and simple phenols were detected in fruits covered by one and two-layered bags. When fruits treated with three-layered bags were exposed to light, they started accumulation of anthocyanins rapidly and there accumulation reached maximum after 3 days of light exposure. The anthocyanin content in fruits remained constant during 5 months of cold storage.

Figure 1. Bagging of Apple Fruits

Poor red colouration is often a serious problem in the production of Fuji apples. To overcome this situation, the practice of fruit bagging is common in Japan. It has been reported that the bagging of apple fruits 4 to 6 weeks after petal fall developed a brilliant red or pink colour after removing the bags 4-6 weeks before harvest. However, the ability of the fruits to develop red colour is likely to decrease more sharply with advance in fruit ripening. Therefore, timing of fruit bagging has a paramount importance on overall fruit quality.

3.2 Banana

Generally bagging in banana is performed after bunch formation by covering with a large translucent perforated polyethylene bags to prevent peel blemishes and to create a greenhouse effect around the fruit by raising the temperature about 0.5 °C (Figure 2). Bagging helps the fruit to fill out and mature more quickly. The greenhouse effect can be intensified by reducing size and number of the perforations. However, if the holes are too small and not numerous enough, the ventilation of the bunch is reduced leading to an increased humidity level within the bag. Banana bags with 5 per cent vent holes of about 4 mm in diameter give best result. If early bagging is practiced, it is necessary to clean the bags to avoid the accumulation of fallen bracts and flower remnants. Fruit speckling, anthracnose, cigar-end rot and scaring beetle can be minimized effectively through bagging. Higher shelf-life, finger dry matter content and early maturity has been reported by several workers in bagged banana bunches.

Figure 2. Bagging of Banana Bunches

3.3 Grape

Bagging of grape bunches provide uniform colour to the whole bunch because the grapes ripen in the protected condition against the direct exposure of sunlight. Bunches when covered with cellulose bags had significantly decreased level of soluble solid content and (soluble solid to titratable acidity) ratio of 'Perla' grape. The concentration of sucrose was found higher in bagged fruit while it contained lower level of glucose and fructose than un-bagged fruits. The organic acids, citric, malic and tartaric were found highest in bagged fruits compared to un-bagged fruits. Grapes covered with bags had higher values of lightness, hue angle and lower values of chroma than un-bagged fruit.

3.4 Litchi

Litchi cv. Tai So when covered with paper bags increased the retention of fruit per panicle as compared to control. It also increases fruit size reduces fruit water loss. Bagging litchi reduced the percentage of rejected fruit compared to no bagging. Bagging of panicles at 42 days before harvest showed the lowest percentage of rejected fruits. Longer bagging time increased the per cent rejects. However, bagging had no effect on TSS, acidity or their ratio in litchi.

3.5 Mango

Bagging of Mango (cv. Keitt) with white paper bags, 100-days before harvest influences fruit quality and mineral composition (Figure 3). Bagging reduces the incidence and severity of *Anthracnose* and stem-end rot through a reduction in contact between disease propagules and fruit. Bagging treatment increased the percentage of skin area with yellow colour at the eating soft stage. However, the percentage of red colour and its intensity decreased with increasing duration of bagging.

Figure 3. Bagging of Mango Fruits

Fruit dry matter and Nitrogen concentration was reported to increase, with faster ripening in bagged fruits as against unbagged ones. Fruit mass, flesh colour, total soluble solids, acidity and eating quality were generally not affected by bagging.

3.6 Mangosteen

Mangosteen (*Garcinia mangostana* L.) is known for attractive purplish-red colour which influences consumer acceptance and sales. Fruit fly infestation, otherwise a serious problem in mangosteen could be successfully prevented by bagging. The major pigment associated with the attractive colour is cyanidin-3-sophoroside. Mangosteen attains harvesting maturity when pericarp is light greenish yellow scattering with pinkish spots. Bagged mangosteen fruits at 14 days after full bloom when harvested at commercial maturity stage (when pericarp background turned pinkish) exhibited attractive visual appearance with no statistical difference in peel anthocyanin content. Bagged fruits exhibited higher PAL activity (enzyme known for anthocyanin synthesis) and low TSS over unbagged ones.

3.7 Peach

Preharvest bagging of peach (*Prunus persica*) cv. Hujingmilu at 50 days after full bloom with orange paper bags of single, double and triple-layers improved fruit quality. Fruits covered with triple-layered bags showed higher fruit length and firmness than un-bagged fruits at full-ripe stage. Both values of L (brightness) and hue angle (h °, red to green) were the lowest in un-bagged fruit, indicating darker red colouration of fruit compared with bagged fruit. TSS concentration in juice was found higher in single-layer bagged fruit than in other bagged and un-bagged fruit at the full-ripe stage. However, no consistent difference was found in titratable acidity. Single and triple-layer bagged fruits had higher ethylene production rates than double-layer bagged and un-bagged fruits. Single-layer bagged fruit showed the highest level of dakua-decalactone, a main characteristic aroma of peach and total lactones at the firm-ripe and full-ripe stages. Studies on 'Hakuho' peach revealed that bagging of fruit with orange parchment paper bags 15 days before harvest resulted in higher visual quality than un-bagged fruit. Fruits covered with orange bags had the greatest radiation interception resulting in the highest hue angle. The surface of fruits covered with orange bags had the highest reflectance in the range of 400 nm–650 nm, indicating that the fruit appeared bright red and had a high L value, resulting in good visual quality.

3.8 Pear

Reduction in bird damage (from 28.4 per cent for un-bagged to nil for bagged fruit) and skin blemish (49.5 per cent for un-bagged to 33.4 per cent for bagged fruit) is one of the major advantages of fruit bagging in pear. Bagging also helps in developing resistance of the fruit skin to damage during handling and storage. Preharvest bagging of pear fruits (*Pyrus communis* L. 'Doyenne du Comice') with micro-perforated polyethylene bags 30 days after full bloom has been reported to increase the percentage of acceptable quality fruits for export from 27.2 (in unbagged) to 63.2 per cent. . Bagging reduced the level of blemishes that are mainly caused by fruit friction against leaves and thin twigs during periods of strong winds. When trees were sprayed with captan and azinphos methyl (non-systemic fungicide) on the 3rd day before harvesting, the bagged fruits had 50 per cent less residues compared to unbagged fruits.

Red Chinese sand pears (*Pyrus pyrifolia* Nakai) were covered by light- impermeable paper bags with different levels of light permeable liners during their early development stage followed by re-exposing them totally for 2 or 3 weeks accumulated largest amount of anthocyanin. Bagging treatments did not affect total soluble sugars, but decreased organic acids content in the fruit. With increasing time after the bag removal, the concentration of chlorophyll, carotenoid, flavonoid and total phenols were reported to get enhanced. After the bag removal, the sucrose contents and PAL activities increased gradually and correlation analysis revealed that they were highly correlated with anthocyanin accumulation in two cultivars. This study suggests that anthocyanin biosynthesis in red Chinese sand pears is a highly light dependent process modified by genotypes.

3.9 Pomegranate

Among the fruits, pomegranate is the most ideal crop for fruit bagging to due to its higher individual fruit weight, response to suture pigmentation (due to colour richness) and proneness to disorders (cracking, aril browning, sunburn). The popularity of pomegranate is increasing among growers and consumers in India due to high yielding potential, better keeping quality and high nutraceutical value of fruits. During last decade, pomegranate crop has witnessed tremendous growth potential even in non-traditional areas like Punjab, Himachal Pradesh, Karnataka and south western part of Haryana. Unlike other crops, pomegranate growers are facing several problems like fruit cracking, improper colour development, internal aril breakdown and bacterial blight. Pomegranate (*Punica granatum* L.) is most suitable crop for practicing pre-harvest fruit bagging because of its bushy nature (bagging can be performed by standing on ground or by using tripod ladder). For getting desirable bagging results pomegranate fruits should be bagged with white cellulose/ butter paper at least 85-90 days before harvesting (Figure 4). Zipped polyethylene bags of 15x15cm size may also be used for bagging. Bagging with zipped polythene bags is cheap, rain proof and convenient. Through bagging fruit quality was improved, pathological and physiological problems were also found to be reduced. Fruit sucking moth is a serious problem in pomegranate and the damage may range up to 100 per cent crop loss. It has been reported that 100 per cent protection can be achieved by bagging of fruits immediately after fruit set using non-woven polypropylene bags. It also improves the fruit gloss and overall acceptability. Since bagging may increase the severity of bacterial blight, it may be used for *Haste* or *Ambe Bahar* crops.

Figure 4. Bagging of Pomegranate Fruits

4.0 Conclusion

Nowadays good agricultural practices are gaining the popularity world over to produce quality fruits with least use of pesticides. Of several such practices, pre-harvest fruit bagging has emerged as one of the best tool. It is a physical barrier creation technique, which improves visual fruit quality by promoting peel colouration and reducing blemishes along with enriching internal fruit quality. Pre-harvest bagging of fruits can also reduce disease, insect-pest incidence, sunburn of the peel, fruit cracking, bird damage and pesticide residue load on fruits. Due to its beneficial effects, bagging has become an essential pre-harvest operation of apple, banana, grape, litchi, mango peach, and pear cultivation in countries like Japan, Australia, China and USA. Selection of right kind of bag colour and fruit bagging stage is crucial to harness the desirable effects of pre-harvest fruit bagging. As of now, fruit bagging is practiced in pomegranate and apple fruit crops in a very limited scale in India. Fruit bagging is a laborer intensive practice and much suits to the dwarf statured fruit crops like pomegranate, banana, apple peach and pear which could be attended by the workers while standing on the ground or with tripod ladders.

Table 2. Effect of Pre-harvest Fruit Bagging on Physiological Disorders of Major Fruit Crops

Fruit Crop	*Disorder (S)*	*Reference(s)*
Apple	Minimize the occurrence of bitter pit, cork pit and brown core	Sharma *et al.*, 2013
Banana	Minimize peel bruising and peel cracking	Luyckx *et al.*, 2016
Date palm	Reduction in tip cracked fruit	Kassem *et al.*, 2011
Grapes	Prevents sun burn and overgrowth of the berries	Karajeh, M.R., 2018
Litchi	Reduce fruit cracking and improve color development.	Shah *et al.*, 2020
Mango	Minimize incidence of soft nose, Jelly seed	Haldankar *et al.*, 2015, Singh *et al.*, 2017
Pomegranate	Significant reduce sun burn,fruit cracking and internal breakdown	Asrey *et al.*, 2019, Farzaneh *et al.*, 2019

5.0 Important References

Amarante, C., Banks, N.H. and Max, S. 2002. Preharvest bagging improves packout and fruit quality of pears (*Pyrus communis*). *New Zealand J. Crop Hortic. Sci.* **30**: 93- 98.

Arakawa, O. 1998. Colouring of Fuji apples by bagging. *IDFTA, Compact Fruit Tree*, **31**: 34-35.

Asrey, Ram, Kumar, Kudeep and Sharma,R.R. 2019. Fruit bagging and bag color affects physio-chemical, nutraceutical quality and consumer acceptability of pomegranate arils. *J. Food Sci.Technol. https: // doi. org/10/1007/s3197-019-04182-x.*

Asrey, Ram and Pal, R.K. 2013. Pomegranate fruit-bagging enhances desirable fruit Characteristics, ICAR News, **19**: 15-16

Asrey Ram, Pal, R.K., Kumar, J. and Sharma, R.R. 2011. Get more by bagging pomegranate on trees, *Indian Horticulture*, (ICAR), **56**: 12-13.

Bin, L.I., Hui-Juan, J. and Xiao-Meng, Z. 2006. Effects of fruit pre-harvest bagging on fruit quality of peach (*Prunus persica* Batsch cv. Hujingmilu). *J. Plant Physiol. Mol. Biol.* **32**: 280-86.

Du, C.T. and Francis, F.J. 1977. Anthocyanins of mangosteen (*Garcinia mangostana* L.), *J. Food Sci.* **42**: 60- 67.

Farzaneh, F. H., Moradinezhad, F. and Khayyat, M. 2019. Pre-harvest bagging influences sunburn, cracking and quality of pomegranate fruits. *J. Hortic. Postharvest Res.* **2**: 131–42.

Ganry, J. 1975. Influence du gainage des régimes du bananier avec une housse de polyéthylène sur la température des fruits dans les conditions de Neufchâteau (Guadeloupe). *Fruits*, **30**: 735-738.

Hofman, P.J., Smith, L.G., Joyce, D.C., Johnson, G.I. and Meiburg, G.F. 1997. Bagging of mango (*Mangifera indica* cv. 'Keitt') fruit influences fruit quality and mineral composition. *Postharvest Biol. Technol.* **12**: 83-91.

Huang, C. and Lee, X. 2009. Effects of fruit bagging on colouring and related physiology, and qualities of red Chinese sand pears during fruit maturation. *Sci. Hortic.* 121: 149-158.

Jaroenkit, T., Maichoo, S., Manochai, P. and Ratanamano, S. 2010. Effect of fruit bagging on chemical compositions and skin pigments of fresh longan (*Dimocarpus longan* Lour.). *Acta Hortic.* **863**: 397-402.

Jia, H., Araki, A. and Okamoto, G. 2005. Influence of fruit bagging on aroma volatiles and skin colouration of 'Hakuho' peach (*Prunus persica* Batsch). *Postharvest Biol. Technol.* **35**: 61-68.

Ju, Z. 1998. Fruit bagging, a useful method for studying anthocyanin synthesis and gene expression in apples. *Sci. Hortic.* **77**: 155-164.

Karajeh, M.R. 2018. Pre-harvest bagging of grape clusters as a non-chemical physical control measure against certain pests and diseases of grapes. *Organic Agriculture*, 259-64.

Kassem, H. A., Omar, A. K. H. and Ahmed, M. A. 2011. Response of 'Zaghloul' date palm productivity, ripening and quality to different polyethylene bagging treatments. *American– Eurasian Journal of Agricultural and Environmental Science* **11**: 616–21.

Kitagawa, H., Manabe, K. and Esguerra, E.B. 1992. Bagging of fruit on the tree to control disease. *Acta Hortic.* **321**: 870-875.

Luyckx, A., Lechaudel, M., Hubert, O., *et al.* 2016. Banana peel physiological postharvest disorders: a review. *MOJ Food Process Technol.* **3:** 226-231.

Pasberg-Gauhl, C. 2000. 'Fruit speckling' on bananas in the Atlantic zone of Costa Rica. BASF, Limburgerhof. Germany. San José, Costa Rica, 27p.

Ratanamarno, S., Uthaibutra, J. and Saengnil, K. 2005. Effects of bagging and storage temperature on anthocyanin content and phenylalanine ammonia-lyase (PAL) activity in mangosteen (*Garcinia mangostana* L.) fruit pericarp during maturation. *J. Sci. Technol.* **27**: 711-17.

Shah, Garima, Chand Satish, Srivastava Ranjan, Kumar Rajesh and Sharma Rajat, 2020. Effect of pre harvest fruit bagging on the physicochemical properties of litchi (*Litchi chinensis* Sonn.) CV. rose scented. *J. Pharmacogn. Phytochem.* **9**: 1812-19.

Sharma, R. R., Pal, R. K., Asrey, R., Sagar, V. R., Dhiman, M. R. and Rana, M. R. 2013. On-the-tree fruit bagging influences fruit color and quality of apple cv. Delicious. *Agric. Sci.* **4**: 443–48.

Signes, A.J., Burló, F., Martinez-Sánchez, F. and Carbonell-Barrachina, Á.A. 2007. Effects of preharvest bagging on quality of black table grapes. *World J. Agric.* **3**: 32-38.

Stover, R.H. and Simmonds, N.W. 1991. *Bananas*, Longman, Harlow, England. 468pp.

Tyas, J.A., Hofman, P.J., Underhill, S. J.R. and Bell, K.L. 1997. Fruit canopy position and panicle bagging affects yield and quality of 'Tai So' lychee. *Sci. Hortic.* **72**: 203- 213.

Chapter 3

Maturity Indices

R.K. Pal

Former Director, ICAR-National Research Centre on Pomegranate, NH-65, Solapur-Pune Highway, Kegaon, Solapur – 413 255, Maharashtra, India

E-mail: rkrishnapal@gmail.com

P. Suresh Kumar

Principal Scientist, ICAR-National Research Centre for Banana, Tiruchirappalli – 620 102, Tamil Nadu, India

E-mail: psureshars@gmail.com

K. Usha Kumari

Assistant professor (Hort), College of Horticulture, Dr.YSRHU, Venkata Ramanna Gudem – 534 101, Andhra Pradesh, India

E-mail: ushahrs@gmail.com

1.0 Introduction

Consumption of Horticultural produces (fruits, vegetables, nuts and spices) is important for human health because these foods are primary sources of essential nutrients and contain phyto-chemicals that may lower risk of various chronic diseases. Plantation crops like Coffee, tea, rubber, cocoa *etc.* has the greater role to play in enhancing the agricultural export from the present 30 to 60 billion US$ by 2025 as envisaged in the Agri-export policy document. The ornamentals provide sensory pleasure and serenity, derived from the colours, shape and aroma of individual species. Garden plants, cut flowers, foliage and flowering plants are increasingly used in exterior and interior decoration. Considerable commercial opportunities arise from their role in social, religious and economic ceremonies and special greeting occasions. The desirable selling characteristics of horticultural crops are dictated largely by the market for which it is intended. Traditionally, the fresh market is the one to which the grower turns up and tries to secure an immediate sale of the produce and he/she tries to giving priority to size and eye appeal. Quality, the degree of superiority, is a combination of attributes, properties, or characteristics that give each commodity value in terms of human food. Various stakeholders in the produce handling system demand different attributes from the same produce. To producers a given commodity must have high yield, good appearance, ease of harvesting and feasibility for long distance transportation. Appearance, firmness and shelf life are important from the point of view of wholesale and retail markets. Consumers are interested in freshness of the produce with good value for money.

Maturity is an indication of development of the crop as a marketable product. Maturity indices are important for deciding when, a given commodity should be harvested to provide some marketing flexibility and to ensure the attainment of acceptable eating quality of the consumer. If the produce is to be shipped to distant markets or stored for a better price, it should be picked at mature but unripe stage. However, unlike ripening stage, differentiation between premature and mature stage is difficult to detect. Selection of right stage of maturity for harvest is an important aspect, which has considerable influence on storage life and quality and final acceptance by the consumer. In some cases, if the produce is shipped to distant markets, or stored, to wait for a better price, it should be picked in the mature but unripe stage.

No prominent changes in firmness or colour are evident as harvest indices become arbitrary and subjective. Understanding the meaning and measurement of maturity is therefore, the key postharvest management practice; while maturity is the stage of fruit development, which ensures attainment of maximum edible quality at the completion of ripening process. Basic structure of inflorescence is stem, including pedicel and peduncles, bracts and flowers. Morphologically and physiologically the fruits, vegetables, plantation crops and flowers are highly variable, and commercially useful part may arise from a root, stem, leaf, bud, immature or fully mature seed/nut, cotyledon, endosperm, and ripe fruits. Flowers are basically variations of inflorescence. Artificial intelligence, spectral images are tried to identify the fruit maturity for different crops. This chapter deals in detail with the importance of maturity in commodity handling system and discusses maturity indices for fruits and vegetables to satisfy different stakeholders of the supply chain.

2.0 Maturity

The principles dictating the stage of maturity of a commodity for harvesting are crucial to its subsequent storage and marketable life and quality. Immature produces are more subject to shriveling and mechanical damage and are of inferior quality when ripe. Overripe/matured commodities are likely to become soft and mealy with insipid flavour soon after harvest. Any commodity picked either too early or too late in its season are more susceptible to physiological disorders and has a shorter storage-life than fruit picked at the proper maturity. In general, fruits are harvested when they are ripened on the tree. However, some fruits like mango and banana are usually picked at mature but unripe stage so that they can withstand the postharvest handling system when shipped long distance. Most currently used maturity indices are based on a compromise between the indices that ensure the best eating quality to the consumer and those that provide the needed flexibility in marketing. For export purpose, mandatory quality standards for most fresh horticultural produces include, objective indices of maturity to ensure acceptability of their flavour and visual quality by the consumers. In case of flower crops, postharvest quality is dependent on various pre-harvest factors. Harvesting and grading of flower crops primarily depend upon on purpose (Cut flower/Loose flower), transport (Local/Export) and end use (Decoration/Value addition). Harvesting at right stage of maturity is one such factor that has enormous effect on vase life of cut flowers. Flowers should be examined for proper stage of development and harvested daily to ensure premium quality. Harvesting too early or too late significantly reduces the vase life of the flowers. If harvested after the optimum stage, the developing flowers use the carbohydrates that will be used for the development of smaller flower buds, thus, slowing down the growth of other flowers.

Examples of these objective measurement of maturity indices are total solids concentration (avocado), soluble solids concentration (cherry, grape, grape fruit, kiwifruit, mandarin and pear, palmyrah, cocoa), titratable acidity (pomegranate, lime, lemon), TSS: acid ratio (citrus fruits and grapes) and flesh firmness (apple, pear, banana). Colour is used as the index for apricot, nectarine, peach, persimmon, plum, raspberry and strawberry fruits; tanin content (tea bud and leaves); aroma (Cashew apple and cocoa); color change and oil content in FFB (Fresh fruit bunches, oil palm); caffeine content (Coffee); change of spathe colour in anthurium; reflexed petals in rose *etc.* Many flower species should be harvested at bud stage for export and distant markets. Disorders such as bent neck, improper development of pigmentation, or abnormal opening of the buds can result when flowers are harvested when they are too immature. For short distances and local markets flowers are harvested when they are fully opened.

Postharvest physiologists distinguish three stages in the life span of horticultural crops. They are **maturation,**

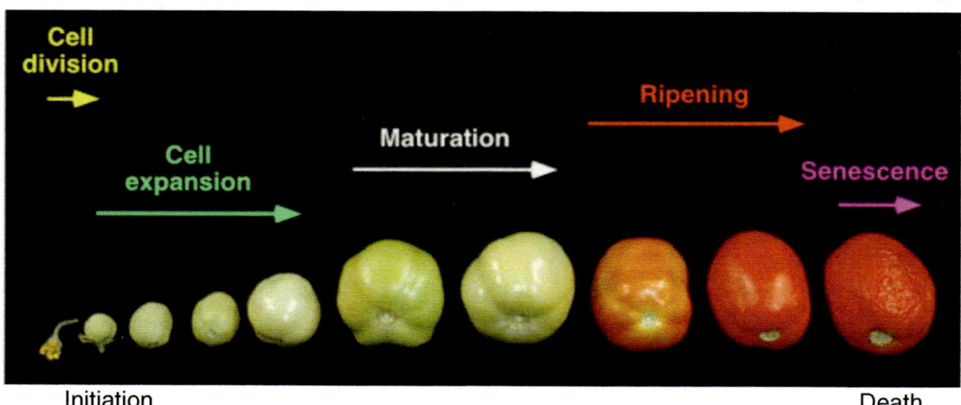

Figure 1. Different phases of fruit growth

Source: http://plantsinaction.science.uq.edu.au/book/export/html/70)

ripening, and senescence. Maturation is indicative of the fruit being ready for harvest (Figure 2). At this point, the edible part of the fruit or vegetable is fully developed in size, although it may not be ready for immediate consumption (FAO, 2001). Ripening follows or overlaps maturation, rendering the produce edible, as indicated by taste. Senescence is the last stage in the ontogeny of the plant organ, characterized by natural degradation of the fruit or vegetable such as in loss of texture, flavour, *etc.*

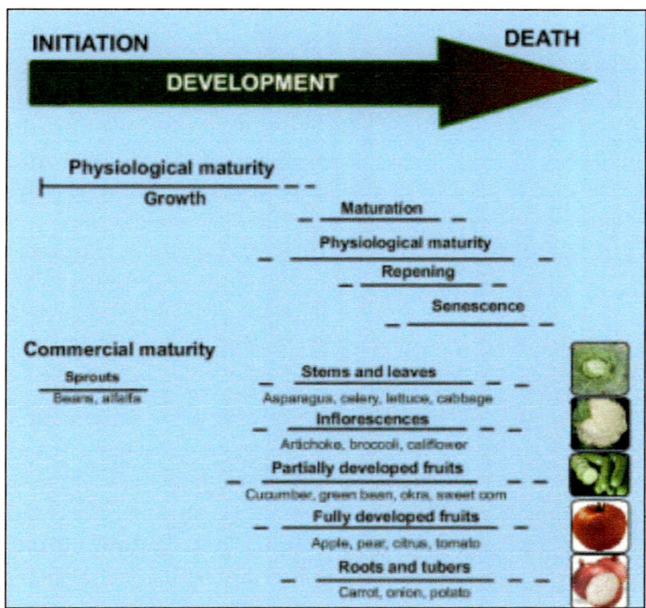

Figure 2. Growth and Development Pattern and Horticultural Maturity of Fruits and Vegetables (Watada *et al.*, 1984).

Similar terminology is applicable for ornamental and flowers, except that ripening stages does not occur. As consequence it is very difficult to delineate the changes from maturation to senescence in some vegetables and ornamentals. Vegetables are harvested over a wide range of physiological ages, that is, from a time well before the commencement of maturation through to the commencement of senescence. Based on this requirement terms like 'physiological maturity' and 'harvestable maturity' are used.

2.1 Types of Maturity

Horticultural crops including fruits, vegetables, flowers and plantation crops are harvested over a wide range of physiological stages, depending upon which part of the plant is used. For example, small or immature vegetables possess better texture and quality than mature or over-mature vegetables. The maturity has been divided into two major categories *i.e.* physiological maturity, horticultural maturity.

2.1.1 Physiological Maturity

It is defined as the stage at which the fruit attains complete maturation and becomes ready for further development of offspring or next generation. In other words it is the final stage of senescence. Although physiological maturity is most important index for maturity of cereals, pulses and oilseed crops, it is also important for certain vegetables in prediction of harvest maturity for their intended use. For example the physiological maturity of commodities *viz.* cucumber is attained when the seeds become hard and have acquired the capability of germination. Similarly, leguminous vegetables and okra are considered physiologically mature when seeds are fully developed and the pod will dehisce with little pressure.

2.1.2. Horticultural Maturity

Horticultural maturity emphasizes the stage of harvesting of fruits and vegetables as per the demand or acceptability by the consumer. Fruits and vegetables are to be harvested when plant or plant part possesses the prerequisites for the utilization by consumer for a particular purpose *i.e.* local or distant or export market (shipping maturity) or exhibition or processing (processing maturity), culinary maturity, dessert maturity. For instance, for local market and for processing, fully coloured tomato fruits are harvested while those which have just started developing colour *i.e.* turning or breaker stage are harvested for long distant marketing. Fruits are harvested at a proper stage of the maturity that exhibit excellent quality. Poor quality and uneven ripening are caused by early harvesting while late harvesting results in extremely poor shelf life. It is imperative that the physical damage to the fruit should be avoided at harvest apart from harvesting at appropriate stage of maturity for quality assurance during marketing. Various Maturity indices such as number of days from fruit set, size, shape, colour, appearances, texture, lenticel number, specific gravity, starch content total soluble solids, sugar-acid ratio and oil content are used to determine maturity of fruits.

A pod vegetable is considered mature when it is tender with maximum size. For example most vegetative tissues, flowers, fruits and other underground organs become horticulturally mature in the mid stage and nuts in the last stage of development, depending on the desired use of the product (Figure 1). In other word, it is a stage of fruit and vegetable at which consumer wants or fruit and vegetable required by market. The horticultural maturity of fruits and vegetables depends upon the purpose for which it is harvested. Horticultural maturity stage of tomato if harvested for long distance transportation would be the "turning stage of skin from green to red", while the optimum stage of harvesting of the same crop for home use or local markets would be "when the fruits have attained full red colour". Again, a qualitative difference in the relationship between maturity and edibility distinguishes many fruits from vegetables. In many fruits such as green banana, the eating quality at maturity will be far less than optimal.

The fruit becomes edible only after ripening. In contrast, in most vegetables, optimum maturity coincides with optimal eating quality.

In many cases of fruit crops *e.g.* Mango, Apples *etc.* physiological maturity coincides with the horticultural maturity.

2.2 Maturity Indices

Generally a single maturity index is not considered to be reliable. In most of the crops more than one or two indices should be made use of determining the exact stage of optimum maturity. Fruits picked at the wrong stage of maturity may develop physiological disorders in storage and may exhibit poor quality. Fruit size is also sacrificed by harvesting too early.

2.2.1. Developing Maturity Index

Some of the common strategies are;

Quantification of the changes in the commodity throughout its development

Identification of physical features (size, colour, shape *etc.*), whose changes correlate well with the stages of the produce development

Conducting storage trials and organoleptic assays to determine the value of the maturity index that defines minimum acceptable maturity.

Examining the maturity index over several years and multiple locations is pre-requisite to standardize the right maturity (*e.g.* heat unit accumulation in days)

2.2.2 Maturity Standards

Many features of commodities have been used to identify the right maturity indices for various commodities. The wide range of methods that have been devised to measure these features are presented in Tables 1 and 2.

Table 1. Maturity Indices for Selected Horticultural Crops

Index	*Example*
a) General characteristics	
Elapsed days (bloom to harvest)	Apple, pear
Heat units	Pea, date, apple and sweet corn, mango, grape, apple and pear
Days from anthesis	Melons, pineapple, coconut, cahsew, areca nut
Days from full bloom	Mango, citrus, apple and pear
Days from fruit set	Banana (90 days for cavendish group and 100-110 days for cooking bananas; mango (110-125 days for Alphonso and Pairi)
Development of abscission layer	Melon, apple
Leaf changes	Potato, bulb crops
Drying of plant parts	Onion, garlic, banana, potato and ginger
Ease of separation	Muskmelon, grape and mango
Ease of snapping	Beans, okra and peas
Tapping	Watermelon, jackfruit and cocoa
b) Surface characteristics	
Cuticle formation	Grape, tomato
Netting	Musk melon
Peel colour	Citrus, papaya, pineapple, tomato, grapes, mango, strawberry and pea
Pulp colour	Tomato, mango, apple, coffee
Skin colour	Rambutan (colour of spintern), coconut, cocoa, cashew, oil palm
Aroma	Jackfruit, cashew apple, cocoa
Specific gravity	Mango (1.0-1.02 for Alphonso; less than 1.0 for dashehari), cherry, water melon, potato, pineapple (0.98-1.02), guava (1.00)
c) Shape	
Angularity	Banana
Fullness of cheeks	Mangoes
Size	Asparagus, cucumber, citrus, apple and pears, cashew apple and nut
Compactness	Broccoli, cauliflower
Flattening of eyes with slight hollowness at centre	Pineapple
Flattening of tubercles	Litchi
Solidity	Cabbage, lettuce, Brussels sprout
d) Textural properties	
Firmness	Apple, pear, stone fruits
Tenderness	Peas, tea
Internal colour	Tomato
e) Compositional factors	
Starch content	Apple, pear and banana
Sugar content	Apple, pear, grape, melon
TSS	Grape (14-16 °Brix for Anab-e-shahi, 18-22 °Brix for Thompson seedless, 12-14 °Brix for Bangalore blue), mandarin 12-14 °Brix, sweet orange-12 °brix; papaya 11-17 °brix, pineapple- 12-14 brix.
Sugar: acid ratio	Citrus, papaya, pomegranate, kiwi, melon
Juice	Citrus (35-50 per cent)
Juiciness (milk exudation)	Sweet corn
Oil	Avocado, oil palm
Astringency	Persimmon, date
Ethylene	Apple pear
Aroma	Guava (β-caryophylene)
Acidity	Citrus (Mandarin-0.4 per cent, sweet orange 0.3 per cent), mango and pineapple (0.5-0.6 per cent)

Adopted from Kader, 2002; Dhatt *et al.*, 2007; Jadhav and Sharma, 2007.

Table 2. Maturity Indices of some Important Horticultural Crops

Crop	*Maturity index*
Banana	Disappearance of angularity in a cross section of the finger
Ber	Colour break stage (light yellow colour appear)
Cherry	TSS: 14-15 per cent, light red colour
Grapes (Table)	Minimum SSC 14-18 per cent; SSC: TA of 20 or higher
Guava	When skin colour changes from dark green to light green
Jack fruit	Tapping gives metallic sound, stopping of milk latex, outer skin turns blunt
Kinnow	TSS: Acid ratio 12-14
Lemon	TSS: Acid ratio; Juice 30 per cent or more juice by volume
Kiwi fruit	TSS 6.5 Brix; firmness- 14 lbs
Litchi	TSS: Acid ratio, skin colour
Mango	Increase fullness of cheeks or bulge of shoulders, flesh colour change from yellow to yellowish orange
Papaya	Fullness, tapping sound, skin colour
Peach	Colour change from green to yellow
Plum	Firmness, change in colour
Pomegranate	The calyx gets curved inside and becomes hard and dry at maturity. Brix acid ratio range from 25-40
Strawberry	2/3rd of berry surface showing pink or red colour
	Vegetables and Spices
Broccoli	Adequate diameter, compact, all florets should be closed
Cabbage	Compactness, Firm head, spreading of wrapper leaves
Cantaloupe	3/4th to full slip under slight pressure, Abscission layer
Carrot	drying of upper leaves, optimum size of the roots, days to maturity
Cauliflower	compactness
Brinjal	full size, firmness, green still tender
Ginger	drying of leaves, days from planting (8-9 months)
Melon	Ground colour changes to white with greenish tint, slightly waxy peel
Mushroom	Caps well rounded, intact
Okra	Full size, optimum maturity can be judged by snapping the tip of fruit, pod 2-4", not fibrous, tips of pods pliable
Pea	Pods well filled but not faded in colour
Lima bean, pigeon pea	Well-filled pods, pods loose their greenness
Pepper	Fruit Size and colour
Onion and Garlic	Tops beginning to dry out and topple down
Watermelon	Dropping of tendril, change of fruit colour, full size, hollow sound when tapped. TSS: 10 per cent
Snake gourd	Full size, thumbnail can penetrate flesh easily
Potato	Haulms dry out and topple down, size of the tuber, starch content, periderm development.
Onion and garlic	Neck fall stage
Lettuce	Head type-Big size, Firmness, Solidity Flavor, Sweetness and bitterness Roamine lettuce- Size and number of leaves
	Flowers
Spike type flowers	Harvest when one-fourth to one-half of the individual florets are open
Daisy type flowers	Harvest when fully open

Source: Kitinoja and Gorny, 1998.

2.3. Methods for Determination of Harvest Maturity

Generally, there are five types of indices to judge the maturity of the commodity.

a. *Visual Means*

Appearance, Skin colour, size, shape, persistence of style portion, drying of outer leaf, drying whole plant body, change in smell or flavour, dropping down of ripe fruits, texture.

b. *Physical Means*

Ease of separation, specific gravity, weight of the fruit, lenticels number, surface morphology, average fruit weight

c. *Chemical Analysis*

TSS, acids, starch, sugar, oil content, sugar-acid ratio, bio-electrical conductance, starch iodine test, tannin content, juice content *etc.*

d. *Computation Methods*

Days for harvesting fruits from fruit set till maturity, calendar date, mean heat units (degree days)

e. *Physiological Method*

Respiration rate, internal ethylene evolution

In recent years, **non-destructive methods** for the measurement of fruit and vegetable characteristics, using principles of NMR (nuclear magnetic resonance), X-rays, sonics and ultra sonic waves, delayed light emission and light reflectance are being tried. Machine language, development of algorithms and artificial intelligence are also being tried to judge the maturity.

2.3.1. Size and Shape

The visual appearance of fruit and vegetable is the most important quality factor, which decides its price in the market. The consumer (wholesaler or retailer) observes the quality of fresh fruits and vegetables with their visual or external appearance. The produce should attain proper shape and size. Size is generally of limited value as a maturity index in fruit, though it is widely used for many vegetables, especially those marketed early in their development. With these produce, size is often specified as a quality standard, with large size generally indicating commercial over-maturity and under-sized produce indicating an immature state. The assumption, however, is not always a reliable guide for all-purpose. Therefore, objective evaluation of size and shape of the produce should be conducted to meet the desired quality characteristics. For banana, the width of individual fingers are used to determine the harvest maturity. Usually a finger is placed midway along the bunch and its maximum width is measured with calipers which is referred as caliper grade. Some cultivars of banana become less angular in cross section as development and maturation progress. Fullness of cheeks adjacent to pedicel may be used as a guide to maturity of mango and some stone fruits. Mangoes also change shape during maturation.

2.3.2. Abscission

Abscission zone develops on the stalk (pedicel) that attaches the fruit to the plant in many fruits during the later stages of maturation and the commencement of ripening. The abscission zone permits the fruit to get detached from the mother plant. Formation of abscission layer is considered as the maturity index for melon, avocado and bael (*Aegle marmelos*) fruit.

2.3.3 Surface Colour

Colour of the skin is commonly applied to fruits, since it changes with the fruit maturity and ripening. Assessment of harvest maturity by skin colour depends on the judgment of the harvester, but colour charts are available for cultivars, such as apples, tomatoes, peaches, chilli peppers, *etc.* There is initially a gradual loss in intensity of colour from deep green to lighter green and with many commodities, a complete loss of green colour with the development of yellow, red or purple pigments. Ground colour as measured by colour charts, is useful index of maturity for apple, pear and stone fruits, but is not entirely reliable as it is influenced by many other factors. For some fruits, development of blush colour, that is additional colour superimposed on the ground colour, can be a good indicator of maturity. Examples are red or red-streaked apple cultivars and red blush on some cultivars of peach. Further, the surface colour of fruits and vegetables can also be measured as an objective evaluation using colour difference meter.

2.3.4 Optical Methods

Light transmission properties can be used to measure the degree of maturity of fruits. These methods are based on the chlorophyll content of the fruit, which is reduced during maturation. The fruit is exposed to a bright light, which is then switched off so that the fruit is in total darkness. Next, a sensor measures the amount of light emitted from the fruit, which is proportional to its chlorophyll content and thus its maturity. Chlorophyll fluorescence is also used in vegetables like broccoli to determine green tissue color. Other methods like NIR spectroscopy for sugar concentration in melons, NMR imaging to identify constituents and internal defects and gene expression for rapid assessment of quality are also in line.

2.3.5 Aroma

Volatile chemicals are synthesized by most fruits as they ripen. Such chemicals give its characteristic aroma and can be used to determine the maturity or ripeness of fruit. These specific aromas may be detectable by human olfactory sense as well as through recent development of e-nose.

2.3.6 Leaf Changes

Quite often marked changes in the leaf quality determines the stage at which the vegetables should be harvested. In most of the root, tuber and bulbous crops, the condition of the leaves can likewise indicate the condition of the crop below ground. For example, if potatoes are to be stored, then the optimum harvest time is soon after the leaves and stems have wilted and dried. If harvested earlier, the skins will be more prone to harvesting and handling damage and become susceptible to storage diseases. Similarly, harvesting of onion is adjudged when 75 per cent of the population exhibit bending of leaf tips that indicate the maturation of bulbs underground.

2.3.7 Firmness

Change in texture is another important index for determining the maturity standard of fruits and vegetables. Excessive loss of moisture may also affect the texture of crops. These textural changes are detected by touch, and the harvester may simply be able to gently squeeze the fruit and judge whether the crop can be harvested. Today sophisticated devices are available to measure texture in fruits and vegetables *e.g.* texture analyzers and pressure testers. A force is applied to the surface of the fruit, allowing the probe of the penetrometer to penetrate the fruit flesh, which then gives a reading on firmness. Commonly used pressure testers to measure the firmness of fruits and vegetables are the Fruit Firmness tester (Figure 2). A more elaborate test, but not necessarily more effective, uses instruments like the Instron Universal Testing Machine (Figure 3), Food Texture Analyzer *etc.* Tenderometer is the device used to test the tenderness of peas based on which the peas are harvested.

2.3.8 Specific Gravity

Specific gravity is obtained by comparing the weights of equal bulks of other bodies with the weight of water. In practice, the fruit or vegetable is weighed in air, then in pure water. This will ensure a reliable measure of fruit maturity. As a fruit matures its specific gravity increases. It is used however to grade crops according to different maturities at postharvest. This is done by placing the fruit in a tank of water, wherein those that float are less mature than those that sink.

Figure 2. Figure 3. Objective firmness measurement (https://www.fao.org/3/y4893e/y4893e04.htm)

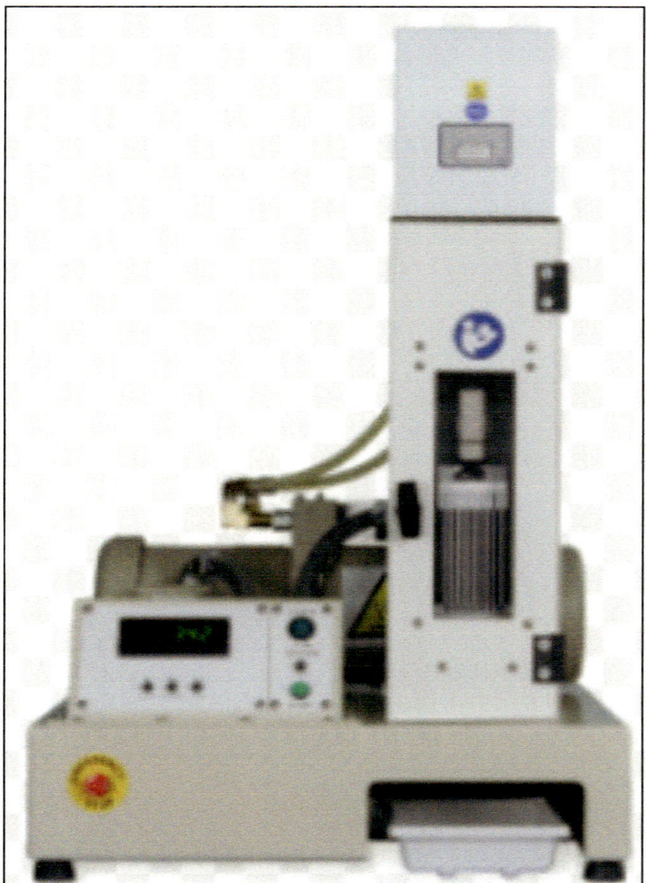

Figure 4. Tendrometer (https://www.textureanalyzers.com/products/tu-tenderometer-pea-tenderometer-solution)

Specific gravity of fruit is proved to be related to their internal characteristics such as dry matter, soluble solids or physical disorders. Commercially it has been used in mangosteen and potato. High specific gravity potatoes (above 1.089-1.080) are better suited for baking, frying, mashing and chipping while low specific gravity (1.079 – below 1.060) potatoes are suitable for boiling and canning. In case of watermelon, density of fruit is related to the degree of hollowness. The density of an immature watermelon is sometimes more than 0.98 g/cm^3. Watermelons with density lower than 0.934 g/cm^3 had cavities or unusually soft flesh of overripe porous tissue. Therefore, watermelons having density of approximately 0.95 g/cm^3 are considered to be optimum mature ones.

2.3.9 Juice Content

In case of citrus fruits, per cent juice content is one of the most important maturity indices. As per international standard no citrus fruit is considered mature if it contains less than 25 per cent juice by weight. Juice content of orange group fruits at optimum maturity ranges from 30-40 per cent whereas in case of limes and lemons the requirement is 25 per cent.

2.3.10 Oil Content

Oil content can be used to determine the maturity of fruits, such as avocados. Avocados at the time of harvest and at any time thereafter, shall not contain less than 8 per cent oil by weight, excluding skin and seed. The oil content is determined by weighing 5-10 g of avocado pulp and then extracting the oil with a solvent (*e.g.*, benzene or petroleum ether) in a distillation column.

2.3.11 Starch Content

Measurement of starch content is a reliable technique used to determine maturity in pear cultivars. The method involves cutting the fruit in two and dipping the cut pieces into a solution containing 4 per cent potassium iodide and 1 per cent iodine. The cut surfaces stain to a blue-black colour in places where starch is present. Starch Pattern Index (SPI) is commonly used to identify the maturity of apple, mango, banana and other starchy fruits and vegetables. Starch converts into sugar as harvest time approaches. Harvest begins when the samples show that 65-70 per cent of the cut surfaces have turned blue-black. Starch content is used as maturity index for pea, potato, apple and other stone fruits.

2.3.12 Sugars

In climacteric fruits, carbohydrates accumulate during maturation in the form of starch. As the fruit ripens, starch is broken down into sugar. In non-climacteric fruits, sugar tends to accumulate during maturation. This factor is used in many parts of the world to specify maturity. The soluble solids content of fruit is determined by shining light on the fruit or vegetable juice and measuring the amount transmitted using refractometer. This technique is used for identifying maturity for citrus, mango, grapes *etc.*

2.3.13 Acidity

In many fruits, the acidity changes during maturation and ripening, and in the case of citrus and other fruits, acidity reduces progressively as the fruit matures on the tree. Normally, acidity is not considered as a measurement of fruit maturity by itself but in relation to soluble solids, it is expressed as brix: acid ratio.

2.3.14 Total Soluble Solids Content

Total soluble Solid (TSS) is the content of total water soluble solids in the juice of the commodity and is one of the major handy measurement techniques for determination of maturity indices in many fruits and vegetables. This is also often termed as Soluble Solid Content (SSC). Fruits and vegetables primarily contain sugars, acids and water soluble polysaccharides as total water soluble solids. Sugars comprise the major ingredient in most of the fruits and vegetables in the TSS reading. Therefore, in general, high TSS value represents the high accumulation of sugars at the time of optimum maturity with the exception in cases where water soluble polysaccharides are present in considerable high amount *e.g.* Ber, Okra, Bael fruit, *etc.* In such cases the high value of TSS does not conform to the values of high sugars. A simple instrument known as Hand or Pocket Refractometer is an important tool used for determination of TSS (Figure 5). Recommended TSS for determination of harvest maturity of different fruits is mentioned below (Table 3).

Table 3. Recommended TSS for Harvesting different Crops

Crop	*TSS (°Brix)*
Apple	10-12.5
Apricot, blue berries, nectarine, melon, peach, water melon	10
Cherry, kiwi fruit	14-16
Grape	14-19
Grape fruit, mango, mandarin, raspberry, orange	8
Papaya	11.5
Pear	13
Persimmon	18
Pineapple, plum	12
Pomegranate	17
Strawberry	7

2.3.15. No. of Days/Heat Units

a) No. of Days

Time of flowering is largely dependent as temperature and the variation in number of days from flowering to harvest can be calculated for some commodities by use of degree-day concept. It has been observed that optimum maturity for harvesting of capsicum is judged at 40 days after flowering. Among the fruit crops banana fruit matures at 90- 110 days after fruit set whereas in case of mango it ranges from 110- 130 days. In case of pomegranate the harvest maturity of cv. Bhagwa reaches between 170-180 days after flowering.

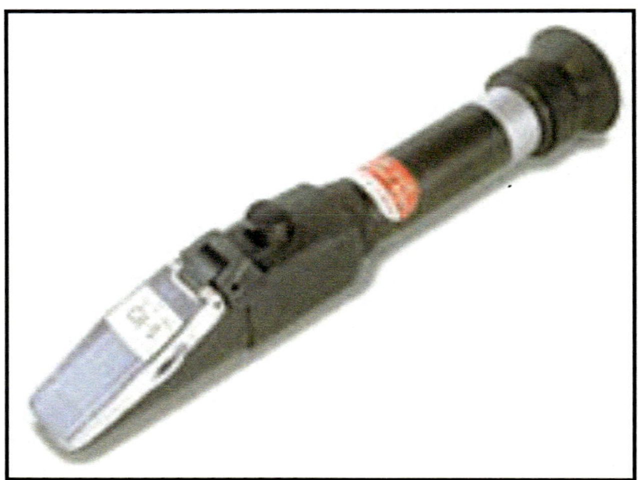

Figure 5. Handheld Refractometer

b) Heat Units

The number of degree days to maturity is determined over a period of several years by obtaining the algebraic sum from the differences, plus or minus, between the daily mean temperatures and a fixed base temperature (commonly the minimum temperature at which growth occurs). The average or characteristic number of degree days is then used to forecast the probable date of maturity for the current year and as maturity approaches, it can be checked by other means. This is an objective measure of maturity expressed as degree days or heat units in a particular environment. It has been found that number of heat unit or degree days is required to mature a crop is almost remains constant. Hence accumulation of particular heat unit will require different number of days with respect to agro-climatic conditions in which the crop is grown. For this reason under warm conditions, maturity gets advanced and under cooler conditions, maturity gets delayed.

2.3.16 Machines assisted Detection of Maturity

With the advent of technologies, researchers and entrepreneurs use various non-destructive methods for determining maturity of fruits and vegetables (Figure 6). Nuclear Magnetic Resonance (NMR), X-rays, ultra sonic waves, delayed light emission and technology, electronic nose, are some of the recent techniques which have vast potential to be used in future. The internal quality of fruit (maturity, firmness, soluble solid content, water content) could be detected with hyper spectral imaging system. Reflectance measurements could also serve as basis for prediction of ripening stages, estimation of harvest window and duration of storage for various fruits. Attempts to apply non-destructive methods based on optical spectroscopy for assessment of fruit quality and their physiological state have been undertaken for several decades. Portable fiber-optic radiometers providing reliable spectral data from very small fruit surface area for monitoring fruit maturity and quality have also gained considerable momentum. Researchers have demonstrated the use of the e-nose for monitoring ripening in a number of fruits including apple, banana, blueberry, grape, mandarin, peach and tomato.

Colour is the human visual perception of light reflected, transmitted, or emitted from an object in the visible portion of the electromagnetic spectrum from 380nm to 780nm.The main factor in the distribution of light energy reflected from the fruit is the presence and concentration of pigments including carotenoids, anthocyanins and other flavonoids, betalains, and chlorophylls in the skin. The colour aspect of visual appearance of the skin can be measured non-destructively using three types of sensors: colorimeters, spectrophotometers, and colour machine vision systems.

3.0 Recommendations for Harvest Maturity

3.1. Fruit Crops

3.1.1 Apple

Apart from attainment of desirable colour and size the TSS, starch content and fruit firmness are widely used as maturity standards for harvesting of apples. A simple iodine test can be used to judge the maturity of apples (Figure 7). A modified scoring pattern of iodine test for Indian apples have been developed for determination of harvest maturity. The starch-iodine test scores have been developed starting from 1 and ending at 4.5 over several weeks during the season. Harvesting of apple depends mainly on its market destination and type of storage is used. Apples destined for long-term controlled atmosphere (CA) storage have to be picked early between 2 and 2.5 with minimum TSS of 10 ° Brix and fruit pressure of approximately 18 lbs. Apples destined for immediate fresh market and processing plant can be picked later and therefore require a higher starch-iodine test score. The selection of the proper score largely depends on each individual operation. No single fixed score can be recommended to guide the whole industry.

The fruits normally mature within 130-150 days after the full bloom stage depending upon the variety. Time/temp (heat units) from anthesis, days from harvest to onset of ethylene production, ground colour, starch disappearance pattern, internal ethylene concentration and changes in firmness are used as maturity index for apple. The fruits at the time of harvest should be uniform, firm and crisp. The colour of the skin at maturity ranges from yellow to red depending on the variety.

3.1.2 Banana

Maturity of banana can be judged by degree of fullness of the fingers, *i.e.* disappearance of angularity in a cross section (Figure 8). Harvest maturity of banana can be decided on the basis of market destination. Normally disappearance of 75 per cent or more angularity of fingers (Full three quarter stage) is considered as the optimum maturity index. When bananas are to be sent to distant urban markets they are best harvested in hard mature but unripe green state, which reduces the risk of deterioration during transport. Bananas are harvested mature-green and ripened upon arrival at destination markets since fruits ripened on the plant often split and have poor texture. The Cavendish bananas are ready for harvest within 11- 14 months after planting, while tall cultivars take about 14-16 months to harvest. A bunch usually takes 90-120 days to mature after flowering, depending on climate and cultural practices. The maturity of banana is indicated by drying of top leaves, change in colour of fruits from dark green to light green and tendency of the floral end of the fruit to fall by slightest touch by hand. The dried parts of flowers at the top of fruit drop off easily. The mature fruit becomes plumpy and all the angles are filled in completely.

30 Days after Flowering

40 Days after Flowering

50 Days after Flowering

Figure 6. Different Stages of Maturity in Capsicum (Source. Behera *et al.*, 2004)

Figure 7. Starch-Iodine Score for Apple Maturity

3.1.3 Blueberry

Colour and soluble solid-acid ratio are maturity indicators for blueberries. Fruits should be blue in colour without any green and soluble solids of 10 to 15 per cent and pH 3.43-3.73. Pickers should handle fruits very carefully, by not holding more than a few berries in their hand at a time and use only shallow containers so that fruits are not bruised due to piling.

3.1.4 Bread Fruit

Breadfruit is usually ready for harvest about 3 months after flowering. Breadfruit has a high starch content and is used as a vegetable when mature but not ripe, and as a dessert when ripe. During ripening, the starch turns to sugar and the fruit develops a sweet custard taste. The surface of breadfruit is patterned with irregular polygon shaped sections that flatten and enlarge when mature. Breadfruit harvested as a starchy vegetable should have sections that are angular and ridged noticeably that stick out on the surface, while the individual segments of dessert-stage fruit should be smoother, flatter, and more round. Harvest maturity can be judged by the occurrence of latex stains on the surface of the fruit and a lack of shine. The amount of brown staining increases as the fruit become fully ripe. The principal internal indices of breadfruit maturity are flesh colour and sugar composition. The flesh of mature but unripe breadfruit is white, starchy, and somewhat fibrous. Fully ripe breadfruit has a pale yellow flesh colour and is somewhat soft and fragrant.

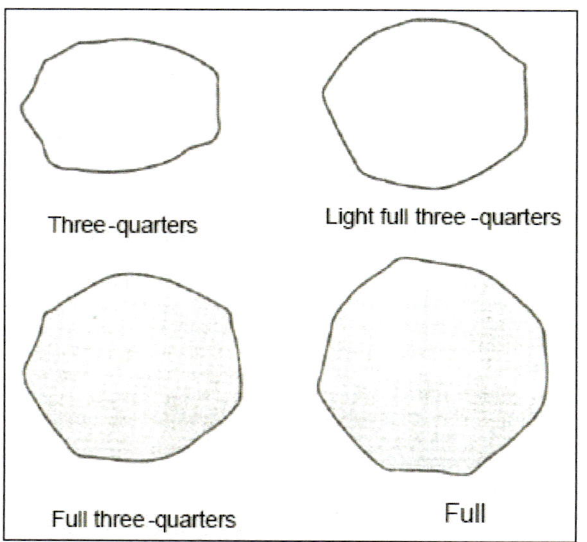

Figure 8. Disappearance of Angularity of Banana with Advancement of Maturity.

3.1.5 Cherry

Skin colour and soluble solids are the most common maturity indicators for sweet cherries. The soluble solids should be at least 16 °Brix and the characteristic skin colour for the variety can range from yellow to black red. Colour is the best indicator for sour cherries - these should be bright red.

3.1.6 Custard Apple

Custard apple fruits require 100 to 120 days for full maturity. The scales on fruits become prominent, plump and well-spaced during optimum maturity. Fruit segments turn creamy white.

3.1.7 Date Palm

Dates ripen in four stages, known throughout the world by their Arabic names: kimri (unripe), khalal (full-size, crunchy),rutab (ripe, soft), tamar (ripe, sun-dried). In India, these stages are called gandora, doka, dang, and pind, respectively. Pind stage rarely occurs in India in India due to commencement of rainy season at that stage. Fruits from most of cultivars should be harvested at doka stage both for raw consumption as well as for its better keeping quality. The days taken from pollination to different stages of maturity and heat summation from spathe opening to maturity are also some important considerations for judging maturity in dates. Apart from it, significant reduction in tannin content coincides with the maturity of date palm.

3.1.8 Fig

Brown scurf on the fruit surface is replaced by yellowish corky brown colour at maturity. The absence of green tissues and latex also indicates maturity. Skin colour changes from light-brown with a tinge of green to light-brown to dark brown. Flesh colour change from green to pinkish brown to reddish brown in some varieties.

3.1.9 Grape

Development of desirable Total Soluble Solids (TSS) concentration of 14 to 17.5 per cent depending on cultivar and production area is the most reliable maturity index of grapes. In some situations, the TSS : acid ratio of 20 or higher is used to determine maturity for early maturing cultivars from early production areas. For red and black coloured cultivars, there is also a minimum colour requirement. The proper stage for harvesting the grapes is adjudged by the development of characteristic colour of a variety. In grapes three major colours, *viz.* white, red and black are found. The criteria used to judge ripening differs according to the uses of grapes. For making raisins, harvesting at a late stage is preferred in order to have more sugar for increasing the weight of the dried product. For all other purposes, ripening is judged on the basis of sugar: acid ratio for getting proper blend required for table purpose or wine making. The correct blend of sugar acid ratio is considered between 25-30. Grapes normally mature within 120 to 135 days after flowering. Easy separation of berries indicates harvesting stage. In addition, mature fruits develop characteristics flavours and aroma. Anab-e-shahi, cultivar matures at 15 to 16 per cent TSS while cv. Thompson Seedless having 18 to 20 per cent T.S.S is considered as optimum stage for harvest.

3.1.10 Guava

Guava fruits are picked at the mature-green stage (color change from dark- to light-green) in some countries where consumers eat them at that stage. In countries where consumers prefer ripe guava, the fruits are picked at the firm-yellow to half-ripe (softer) stage for long-distance transport or at the fully-ripe (yellow and soft) stage for local markets. Guava fruits ripen five months after flowering and they are harvested when the fruits turn greenish yellow. Guava fruits are considered as fully mature when fruits become soft, develops slight depression near the stalk end, ridges on the fruit disappear and it becomes rounded. Mature fruits of guava become lighter in weight with specific gravity less than one. At this stage the force of detachment of the pedicel also becomes low.

3.1.11 Jackfruit

Jackfruit matures in about 180-270 days after flowering. The skin colour of the fruit changes from green to pale green. In matured fruits, the fruit spines become well developed and wider spaced with typical bluntness. A dull, hollow sound is produced when the ripe fruit is tapped by the finger. Though not a common maturity index, in some varieties the last leaf of the peduncle turns yellow with characteristic development of aroma.

3.1.12 Jamun

Fruits required 120-140 days to mature after flowering. In general, the colour of the fruit changes from green to deep purple or black with full size depending upon the variety.

3.1.13 Karonda

Karonda fruit generally require 87 to 90 days to mature after flowering. The colour changes from light green to dark green when the fruits are ready for harvest with optimum maturity. Red blush develops on the fruits of some types. Specific gravity just above 1 (1.012) is also considered as the indicator for maturity along with T.S.S. 13.5 per cent, total sugar 2.35 per cent and acidity 2.62 per cent.

3.1.14 Kokum

Kokum fruits generally require 90-100 days to mature after flowering. The colour of the fruits changes from green to light green and red and purplish red when ripe. Average weight of fruit varies between to 30-35 g depends on the type with the specific gravity of 1.04. Chemical parameters like T.S.S. 13.68 per cent, acidity 3.61 per cent and total sugar 6.35 per cent was considered optimum for judging the maturity.

3.1.15 Litchi

Litchi fruits should be harvested fully-ripe because they do not continue to ripen after harvest. The fruits for local market should be harvested at the full ripe stage as indicated by the attractive skin colour while for distant market the fruit is harvested slightly early, when they have just started turning reddish or pinkish. Red Color of the fruit (due to anthocyanins in the skin) is a good indicator of maturity along with fruit size (minimum of

25 mm in diameter). Similarly the maturity of the fruit for harvesting is judged by flattening of the tubercles on it.

3.1.16 Mandarin and Sweet Orange

Citrus fruits are non-climacteric and they do not ripen further after harvest. To reach their full flavour and sweetness they must be left on the tree to ripen. In the tropics, citrus often remain green when they are fully ripe internally and do not develop an orange-yellow colour on the tree. The development of the orange-yellow skin colour can be artificially induced after harvest (degreening). These facts make it very difficult to assess harvest maturity just from the appearance of the fruit on the tree. Some methods to identify the maturity are;

a) *Skin Colour*

This will be a good guide to maturity where colour develops normally; if normal skin colour does not develop, maturity may be indicated by a change in the shade of green shown by the skin; lemons change from dark green to a silvery-green appearance at maturity.

b) *Size*

Experienced growers may evaluate maturity by considering size as well as with other characteristics, such as slight changes in skin colour.

c) *Internal Condition of Fruit*

A few typical fruit thought to be mature are cut in to two and they can be considered ripe if:

Juice has developed full flavour and is sweet;

Fruit pulp has developed to the normal colour;

Juice drips from the half-fruit when the cut surface is held vertically.

TSS: Acid ratio

In general, citrus fruit mature in about 210 to 240 days. Colour changes from green to pale green or pale yellow or orange. In addition, outer rind looks shiny and one can see oil glands on the fruit. Mandarin has 0.4 per cent acidity and 12 to 14 per cent TSS as maturity whereas orange are matured when 0.3 per cent acidity and 12 per cent T.S.S.

3.1.17 Mango

Mangoes will not ripen normally if they are harvested before reaching the fully mature-green stage on the tree: they will lack sweetness and be poor in flavour. Mature-green fruit left on the tree will ripen and eventually fall. There are no simple reliable tests to indicate when mangoes are ready for harvest. A number of characteristics have been suggested for evaluating maturity, but they are not reliable for all cultivars or for all growing conditions. They must be interpreted in the light of local experience as below:

a) *Fullness in Size*

When fruits attain its full size, which mainly depends upon the variety, by virtue of experience one easily understands the final size and shape and hence its time of harvesting. It can be considered that mangoes are matured when the "cheeks" of the fruit are full. The shoulders of immature mangoes slope away from the fruit stalk; however, on more mature mangoes the shoulders become level with the point of attachment, and with even more maturity the shoulders may be raised above this point (Figure 9).

b) *Change in Peel Colour*

Usually the colour of fruit peel changes from green to light yellow, apricot yellow. In some varieties, attractive red colour appears on shoulders (Alphonso, Kesar, Pusa Arunima, Arka Puneet, Ambika). Sometimes, a variety may develop different colours in different localities. For instance, *Dashehari* in Malihabad region develops attractive yellow colour whereas, it remains green in Saharanpur region. Thus, harvesting has to be done as per size or colour, it attains in a particular locality.

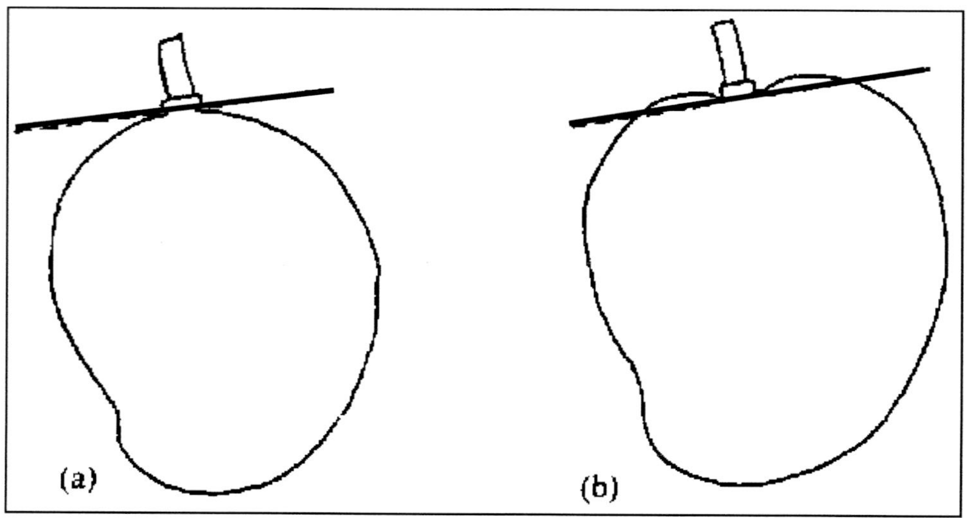

Figure 9. a) Immature and b) Mature Mango Fruits

c) Natural Dropping of Ripe Fruits

Natural falling of ripening fruits from the tree is called as *tapka*. When some fruits start falling down of a tree, it gives clear cut indication that mangoes are mature and must be harvested.

d) Specific Gravity

This is scientific maturity index for harvesting mangoes. If the specific gravity of most of the fruits reaches between 1.01 and 1.02, it indicates that fruits are ready for harvesting. For this, some fruits are randomly harvested from the tree and put in a bucket full of water. Floaters are considered as immature and sinkers as mature (specific gravity > 1.02). However, in case of cv. Baneshan specific gravity method of determination of maturity didn't work due to presence of air cavity in the stone.

e) Number of Days

It depends on the variety and climatic conditions. Usually, mangoes mature in about 90-135 days. Early varieties (Bombay Green) mature in about 90-100 days and late varieties mature in about 130-140 days (Amrapali, Pusa Arunima). In mature fruit the sap flowing from the cut stem at harvest is thick and does not flow freely. Alphanso and Pairi took 110 to 125 days after fruit set for surface colour to change from dark green to olive green and flesh colour from white to pale yellow.

3.1.18 Mangosteen

The mangosteen fruit has an edible aril that is white, juicy, sweet and slightly acidic with a pleasant flavour, located inside a dark purple pericarp that is rich in bioactive secondary metabolites including anthocyanins, oligomeric proanthocyanins and xanthones. Fruit colour is a major criterion used to judge maturity and also for grading of mangosteen fruit. During fruit development the fruit has at least six colour developmental stages. They are pale yellow green, blotchy pink, pinkish red, maroon red, dark maroon violet and violet black. Fruits are usually harvested at different stages according to colour, from light greenish yellow with scattered pink spots to dark purple. Mangoosteen is harvested during pinkish red stage for distant market while it is harvested at maroon red stage for nearby markets. After harvest, the purple colour continues to develop very quickly. For high fruit quality, the minimum harvest colour stage is that of distinct irregular, pink–red spots over the whole fruit. If fruits are harvested at light greenish yellow stage with scattered pink spots, the fruits do not ripen at all.

3.1.19 Papaya

The latex of fruits becomes watery when the fruits are fully matured. Fruits require 125 to 140 days from flowering to maturity. The change of colour from green to yellow and the consistency of the latex from milky to watery indicate that the fruit is ready for harvest. The mature fruit easily gets detached from stem either by turning it upwards or by twisting. Change of skin color from dark-green to light-green with some yellow at the blossom end (color break). Portion of fruit exposed to sunlight becomes dark yellow in colour. Papayas are usually harvested at color break to yellow for export or at ½ to yellow for local markets. A minimum soluble solids of 11.5 per cent is required by the Hawaiian grade standards. Under Indian condition, TSS at harvest should be minimum of six per cent.

3.1.20 Pears

Pears are the only temperate fruit that must be harvested mature and then ripened off the tree. Soluble solids, days from full bloom, iodine starch test, flesh firmness, ground color, and ease of separation from the branch are harvest indicators. Soluble solids are not usually used as maturity index but it should be at least 13 °Brix for marketability. Days from full bloom will give you an idea of when the pears will be ready for harvest and then other methods can be used to pinpoint the best harvest date. The iodine starch test is where the flesh of halved pears is dipped in an iodine solution and approximately 60 per cent of area turning blue-black is accepted for mature fruit. Pear firmness, determined with a pressure tester, should be 23 lb, but can be less if the soluble solids are less than 13 °brix. Ground colour changes are from a green to yellowish green. With ease of separation, the fruit should come off the spur easily without tearing or breaking the fruiting spur.

3.1.21 Pineapple

Pineapple fruit is harvested on the basis of size, colour, or both, and twisted away from the stalk when mature. Shell colour changes from green to yellow at the base of the fruit. Pineapple is a non-climacteric fruit and should be harvested when ready to eat. A minimum soluble solids content of 12 per cent and a maximum acidity of 1 per cent assures minimum flavour acceptability by most consumers. It requires 105 to 130 days for maturity. Eyes on the fruit become smooth or flat and they slightly bulge out. Besides, eye bracts become loose and turn brown in colour with the right maturity. Mature pineapple should have 12 per cent T.S.S. 0.5 to 0.6 per cent acidity, with the specific gravity of 0.98 to 1.02.

3.2.22 Plums

Colour is the major harvest indicator for plums. Each variety has its own characteristic colour change, familiarity with standards for planted varieties is important. Soluble solids should be at least 17 per cent and pressure testing may be useful.

3.1.23 Pomegranate

Pomegranate fruits ripen in about 5-6 months from the time of flowering. Mature fruits should be immediately removed from the plants as delay in harvesting leads to fruit cracking. Fruits harvested at a premature stage have poor keeping quality and are prone to damage during handling and transport. The fruits are harvested when it produces a metallic sound. Complete drying of floral parts could also be considered as an index for maturity. The round fruits become flattened from all the sides and basal beak shaped portion shrinks at the time of maturity. In general, acidity of juice should be below 1.85 per cent. Depending upon the variety, fruits are harvested when the TSS/acid ratio reaches 32-37. Bhagwa and Ganesh the most popular varieties are harvested at 180 and 150 days after anthesis with TSS/acid ratio of 32 and 37, respectively.

3.1.24 Sapota

The erratic flowering habit of sapota and the presence of fruit at all stages of development on the tree make it difficult to determine optimum harvest time. Fruit harvested later than optimum time usually soften very rapidly and become very difficult to handle. Fruit harvested earlier than physiological maturity may not soften, are usually low in sweetness and high in astringency when ripe, with a rather unappealing alcoholic after taste, and form pockets of coagulated latex that lower quality. Unripe fruit are highly astringent and contain large amounts of leucoanthocyanidins. The sucrose content and pulp-to-peel ratio increase during maturation. The fruit shed off brown scaly external material and become smooth when reaching physiological maturity. Fruit ready for harvest will not show a green tissue or latex when scratched with a fingernail. Fully mature fruit will have a brown skin, and fruit will separate easily from the stem without leaking latex. Extent of scruffiness is also a good indicator of maturity. A fruit with a smooth surface, shining potato color and rounded styler end is considered mature. Fruits generally mature in about 240 to 270 days after flowering. Brown scaly material disappears from the fruit surface as the fruits approach full maturity. The dried spine like style at the top of the fruit falls or drops off easily when touched.

3.1.25 Strawberries

Strawberries owing to their non-climacteric nature, can be harvested when two-thirds of the fruit is red in colour. Fruits should be handled very carefully, by not holding more than a few berries in hand at a time of harvest. The TSS acid ratio varies between 5.35 and 9.28 with 25per cent and 100 per cent colour, respectively. Change in colour is the most common method used for determining harvest maturity of strawberries.

3.1.26 Tamarind

Tamarind fruit take about 8 months from fruit set to harvest and growth is a typical sigmoid type. As pods mature, skin develops into a brown, brittle shell, the pulp turns brown or reddish-brown and seeds get covered with dry and sticky pulp. When fully ripe, the shells become brittle and easily broken. Mature fruit can be left on the tree for more than 6 months after ripening without significant spoilage, however birds and insects become pests. Fruit should be harvested when the moisture content is < 20 per cent to facilitate separation of the shell from the pulp.

3.2 Vegetables

3.2.1 Beet

Size is the harvest indicator for beet roots. Most beets are harvested when they are 1 to 3 inches in diameter. Baby size beets can be harvested at a smaller size. Beets prepared with tops are also harvested at a smaller size when the tops are still tender. Standard size beets can be sold with the tops intact, topped with stems or short-trimmed with stems no more than 4 inches long.

3.2.2 Bell Pepper

Generally bell peppers are harvested in the mature green stage. They should be of good shape, waxy, firm and shiny. Peppers for canning are allowed to attain full size and develop a red colour before harvesting.

3.2.3 Brinjal

The fruits should be harvested when they are still tendered but after they have attained a good size and when the surface is bright and glossy. At harvesting the calyx end the stem end are kept attached to the fruit. The fruits become ready for first picking in about 120-130 days of seed sowing depending on the variety. Fruits are harvested when they become greenish yellow or bronze and their flesh turns dry and tough. Pressing the thumb against the side of the fruit can indicate the maturity of the fruit. If the pressed portion springs back to its original shape, the fruit is too immature.

3.2.4 Broccoli

Heads should be harvested when the florets are tight and the color is a rich green with a purplish cast. Terminal head diameter should be at least 3 inches. Smaller heads can be bunched to achieve the specified count for the carton. Stems and heads should be cut at about 8 inches in length. Side shoots and large leaves should be trimmed leaving about a 6-inch bare stem. Any head that is yellowish in color and with open florets should be discarded.

3.2.5 Brussels Sprouts

Harvesting after a moderate freeze will improve the quality of brussels sprouts immensely. They become sweeter and more palatable. Harvest from the bottom of the stem up and only harvest firm sprouts that are 1 to 2 inches in diameter. Outer leaves should be removed.

Figure 10. Ready to harvest-Brussels sprouts

Figure 11. Different maturity stages of Broccoli (https://www.textureanalyzers.com/products/tu-tenderometer-pea-tenderometer-solution)

Figure 12. Purple heading and sprouting Broccoli (https://nicholsgardennursery.com/)

3.2.6 Cabbage

Maturity is indicated by compactness of the head which can be only measured by slightly compressing the head with moderate pressure. Delaying harvest can result in split heads and increased incidence of field disease. Harvesting immature heads, however, reduces yield. They have tendency to burst or loosen the leaves beyond the marketable stage. At low elevations cabbage varieties mature in about 65 to 110 days from field setting and about 80 to 125 days at high elevations (Kader, 2002). Solidity and firmness of head are the usual maturity characteristics used to judge the maturity. Head colour turns a lighter shade of green when full development is attained.

Figure 13. Right Stage of Harvesting Cabbage.

3.2.7 Cauliflower

Mature curds are at least 15 cm in diameter. Loose floral parts are signs of over maturity. The best stage of maturity is determined by curd size and condition. Local growers usually harvest the head upon the desired size and before the curds become discolored, loose or otherwise blemished. Cauliflower is ready for harvest at 90-120 days after planting. Depending upon the variety the curds should be harvested promptly when they are of full size but still compact, white and smooth. Over mature head, becomes too long, flower stalks elongates resulting in loose, leafy condition.

3.2.8 Chilli

The stage of maturity at which chillies are picked depends on the type and purpose for which they are grown. For instance, these are picked either green for pickle making or fully ripe for drying.

Figure 14. Harvesting indices of Sweet pepper and Chilli (https://postharvest.ucdavis.edu/Commodity_Resources/Fact_Sheets/Datastores/Vegetables_English)

3.2.9 Cucumber

Cucumber as vegetable is ready for harvest within 45 to 55 days after sowing. Precisely, in cucumber the proper stage of maturity is judged by size and not by the age of the fruit. Cucumber for slicing should be picked when they are 15 to 35 cm long, whereas for pickling it should be 6 to 15 cm long. In case of slices at marketable stage, spines on fruit becomes soft and fall down. In general cucumber may be picked at any stage of fruit growth, provided yellowish colour has not started.

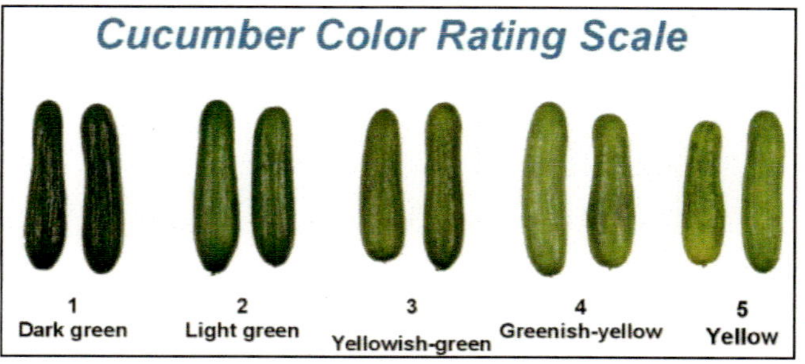

Figure 15. Change in color of cucumber fruits on maturity (https://postharvest.ucdavis.edu)

3.2.10 Drumstick

Drumstick (*Moringa oleifera*) plantation raised from limb cutting come to bearing within one year. Normally size is the harvest indicator for moringa which takes 30- 45 days for maturity from flowering. The maturity index for moringa varies with the region and purpose. In eastern India, very young and soft seeded pods are preferred while in south, people prefer somewhat hard and thick pods.

3.2.11 French Bean

French bean is harvested in different developmental stages depending upon the purpose. The vegetable beans are harvested at 90 days after sowing whereas the seed beans are harvested between 120-130 days after sowing. Vegetable beans must be picked up at the proper stage of maturity, because they start losing their marketability after reaching the edible stage. At maturity, colour of the pods changes from dark green to light green and they are well filled. Quality depends upon sugar content and tenderness. With increasing maturity and size, sugar content declines rapidly with a corresponding increase in starch and protein. High sugar content is indication of high quality. Ease of snapping and days from flowering are the most commonly used indices to determine the maturity.

3.2.12 Gourds

Proper stage of maturity in bottle gourd, sponge gourd and ridge gourd is judged by its fruit size. Bottle gourds are considered mature if gentle press of finger nails easily penetrates the epidermis. In most cases fruits require 12 to 15 days after pollination to attain marketable size.

3.2.13 Knol-Khol

Harvesting is done when it attains its marketable size *i.e.* 5 to 7 cm in diameters and bright colour depending upon variety. At this stage the edible portion is tender and non-fibrous.

3.2.14 Muskmelon

Fruits will be ready for picking in about 110 days depending upon variety. Fruits begin to separate from stem when it is matured. There are two groups of cultivars which behave distinctly. In one group, the fruits when mature slips out easily from the vine with little pressure or jerk or if not it will remain separated from the vine next day. This is called full slip stage. Internal colour, total soluble solids and firmness are some other maturity indices used for musk melon. In some Indian cultivars, green stripes on the skin begin to turn yellow during maturity. Formation of abscission zone, slips and netting are most common maturity indices for musk melon.

3.2.15 Okra

Okra pods are young green fruits. They are ready for harvest in about 45-60 days after seed sowing depending upon variety and season. Okra should be harvested when the fruit is bright green, the pod is fleshy and seeds are small. Therefore, pods should be selected which are 4 to 9 cm in length, green, tender and readily snap at the tip. They are harvested at an interval of 2 to 3 days. The best time of harvest being 6 to 7 days after opening of flower. Mature pods are fibrous, tough, and unfit for human consumption.

3.2.16 Parsnips

Parsnips should be harvested after several moderate freezes to improve their flavor. They should not be less than 1" in diameter but not more than 2 inches at the top and 8" long. Their tops should be trimmed to leave about 1/2" stem.

Figure 17. Parsnips ready for marketing

Tender Knobs **Over Mature**

Figure 16. Varying harvest maturity levels for Knolkhol

3.2.17 Pea

Peas must be picked up at the proper stage of maturity, because they start loosing their sweetness rapidly after reaching the edible stage. The early varieties sown in October are ready to be hand picked in 50 days after planting while the mid season; varieties sown in November are ready in 80-85 days. At maturity, the colour of the pods changes from dark green to light green and they are well filled up. Quality depends upon sugar content and tenderness. With increasing maturity and size, sugar content declines rapidly, with a corresponding increase in starch and protein. High sugar content is indication of high quality.

Green pea maturity and tenderness are usually measured by a tenderometer, an instrument which measures the shearing force required to press a sample of peas through a standard grid. The tenderometer reading (TR) increases with increasing green pea maturity and decreasing tenderness. Yield of green peas per unit area tends to increase with increasing TR. Because of this, processing companies base their price to growers on a sliding scale, peas of a higher TR receive a lower price. Peas with Tenderometer reading between 95-110 and AIS (alcohol insoluble solids)<15 per cent are considered suitable for processing.

3.2.18 Pumpkin

Fruits require 110 to 120 days for maturity from sowing. Fruits should be harvested at full stage of maturity when the rind has hardened. Colour changes from green to yellow or orange. Some varieties remain green only even after full maturity.

3.2.19 Tomato

At any given time, six different maturity stages are generally found during fruiting of tomato. The stage of maturity for harvesting depends on the specific need. For long distance transportation, the fruits are harvested when they are firm, mature and greenish yellow. Fully ripe or nearly red fruits are ready for harvesting in 75-90 days after transplanting.

a) Mature Green Stage

Fruits to be shipped are harvested when dark green colour changes to a reddish pink shade. A simple way to determine maturity is to slice the tomato with a sharp knife. If seeds are cut, the fruit is too immature for harvest and will not ripen properly.

b) Turning Stage

Tannish-yellow, pink or red color shows on over 10 per cent but not more than 30 per cent of the tomato surface.

c) Breaker Stage

Dim pink colour observed on part of the fruit. Fruits are harvested at this stage to ensure the best quality. Such fruit are less prone to damage during shipment often fetch a higher price than less mature tomatoes.

d) Pink Stage

Fruits are stiff and nearly whole fruit turns reddish pink. Fruits for local sale are harvested at this stage.

e) Light Red Stage

Pinkish-red or red color shows on over 60 per cent but red color covers not more than 90 per cent of the tomato surface

f) Red Stage

Fruits are fully ripened and soft having dark red colour. Such fruits are used for processing.

3.2.20 Water Melon

Watermelon are harvested at full maturity as they typically do not develop in internal colour or increase in sugars after being removed from the vine. The ground spot (the portion of the melon resting on the soil) changes from pale white to a creamy yellow at proper harvest maturity. However, size of the fruits and colour of the rind are not good indicators. Fruits are ready for harvest in 90 to 120 days from sowing. Other common indicators for identifying maturity of water melon are;

Figure 19. Maturity Stages of Tomato
(Source. http://www.tomato.org/)

Withering of tendril: The tendril accompanying the fruit withers as fruit ripeness.

Thumping: Ripe fruits when thumped with finger give out heavy dull sound, whereas the immature fruits give metallic sound.

Ripe fruits produce a crisp, cracking noise on being pressed with the flat of the hand.

Figure 18. Growth and development stages of water melon (https://www.researchgate.net/figure/Fruits-of-watermelon-cultivar-Dumara-at-four-different-stages-of-ripening-Ripening_fig1_258500983)

Destructive sampling is used to judge maturity of a population of watermelons. For seeded cultivars, maturity is reached when the gelatinous covering (aril) around the seed is no longer apparent and the seed coat is hard. Cultivars vary widely in soluble solids at maturity. In general, a soluble solids content of at least 10 per cent in the flesh near the center of the melon is an indicator of proper maturity if the flesh is also firm, crisp and of good color.

3.3. Tuber, Root and Bulbous Crops

3.3.1 Arrow Root

The crop attains maturity in 10-11 months after planting. Maturity is indicated by yellowing, wilting and drying up of the leaves. At this stage, the plants are dug out and rhizomes are separated from the stem.

3.3.2 Carrot

Typically carrots are harvested at an immature state when the roots have achieved sufficient size to fill in the tip and develop a uniform taper. Depending upon variety carrot becomes ready for harvesting within 100 to 120 days from sowing. At the time of harvest the roots should not be pithy or solid. Depending in the variety, the desired size is the primary consideration in harvesting the roots. At marketable stage carrots should have at least 2.5 to 4 cm diameter at the upper end.

3.3.3 Elephant Foot Yam

Harvesting of large tuberous corm of elephant foot yam is done in 8-10months at senescence stage. Corms are harvested by digging when green part above soil is completely withered.

3.3.4 Garlic

Garlic is harvested when the bulbs are well mature. It is ready for harvesting when the tops turn yellowish or brownish, show signs of drying up and bend over. Normally it is harvested when 50 per cent of the plants have shown neck fall. Crop gets ready for harvest in about 100 to 140 days after planting.

3.3.5 Ginger

Yellowing of foliage and slowness in growth are signs of maturity. Internal flesh colour should be pale yellow. Delaying harvesting after maturity is reached will lead to very fibrous and pungent rhizomes which in turn reduce the rhizome quality, decrease the storage life and increase the incidence of sprouting during storage.

3.3.6 Onion

Harvesting depends upon the purpose for which the crop is planted and maturity will depend upon the variety being used. "Field-dry" maturity is indicated when bulb neck is completely dry to the touch and tops are fallen over. Best time to harvest *rabi* onion is one week after 50 per cent tops have fallen. However, in *kharif* season, since tops do not fall, bulbs are harvested soon after the colour of leaves change to light yellow and bulbs develop red pigmentation. Onion crop becomes ready for harvesting in five months. However, for marketing as green onion, the crop becomes ready in three months after sowing. Green onions are harvested when they are of pencil stem with a small bulb is formed. Ripe bulb crop is ready for harvest in 3 to 4 months after transplanting. Neck fall is the indication of maturity. The best time to harvest onion is when 60 to 70 per cent of tops show leaf tip fall. Development of red pigment and the pungency pregnancy of the variety are also important harvest indices.

Figure 20. Maturity indices and harvesting of Onions

3.3.7 Potato

Main-crop potatoes which may need to be stored should not be harvested until at least two weeks after the plant tops have died off, by which time the skin of the tubers is fully developed and they are mature. They are then less susceptible to damage than immature potatoes. Potatoes are harvested when they attain sufficient size. Early varieties mature between 80 to 100 days, whereas late varieties are ready for harvest in 120- 160 days. Skin slipping from the tuber, starch content and leaf senescence or top drying are the main harvest indices used for potato.

Figure 21. Readiness for Harvest Based on the Percentage of Tubers with the Desired Size.

3.3.8 Radish

The edible roots become ready for harvesting in about 25-60 days depending on the variety. The temperate types reach harvest maturity 25-30 days after seed sowing while tropical varieties require longer period. At the time of harvest, the roots should not be pithy or solid. European type is ready for harvesting within 30 days from sowing and Asiatic type within 45 to 60 days from sowing. At this stage roots are mild, tender and crisp and usually of the proper marketable size.

3.3.9 Sweet Potato

Sweet potatoes are harvested 3.5 to 4 months after planting. Early variety like Kalmegh or Samrat can be harvested in 90 days. Harvesting is done when the leaves turn pale and start yellowing. Moreover, after cutting the tuber if white gum remains white, then they are ready for harvest, if it turns blackish on green then tubers are said to be immature. In general, delay in harvest after 120 days increases sweet potato weevil infestation.

3.3.10 Tapioca

The tubers are ready for harvest in 7-11 months depending on the variety. The plants would mature in about 7 – 10 months time depending upon the variety. Sree Prakash, H-119, CI-590, Sree Jaya and Sree Vijaya are short duration varieties, mature in 6-7 months. H-97 and H-165 takes 7-8 months; H-226, 8-9 months and Sree Sahya, Sree Visakam, Sree Rekha, Sree Prabha are long duration crops which takes 9-10 months for maturity. When the soil shows cracks a trial digging is done. The plants are cut back before digging of roots.

3.3.11 Taro

Normally it is harvested after 200-210 days. Harvesting at 165 days after planting has been found ideal for States of Andhra Pradesh, West Bengal and Uttar Pradesh when it is grown as a summer crop. Yellowing of leaves is an indication of crop maturity when cultivated in dry condition.

3.3.12 Coleus

Coleus is a short duration crop and yields reasonably good in 4-6 months depending on the cultivar. The tuber can be harvested when all leaves and stem begin to wither.

3.3.13 Winged Bean

Plants take 60 - 80 days from sowing to the first harvest of seedpods. Mature tubers can be harvested 120 - 240 days from sowing.

3.3.14 Yam Bean

Yam bean is harvested after 150 days of sowing. If harvesting is delayed, chances of cracking of tubers are more. This in turn results in the deterioration of tuber quality and thereby affects market value

3.4. Leafy Vegetables

3.4.1 Amaranthus

Uprooting is done 3-4 weeks after sowing. If clippings are to be taken then first clipping is done at 25 to 30 days after sowing followed by other clippings at 20 to 25 days interval. Totally 5-6 clippings are carried out. The fully grown side leaves are removed. The tops may also be cut leaving the lower leaves to produce new shoots.

3.4.2 Asparagus

Harvesting is done from the third week onwards. It is harvested every day during the spring spears may be cut 3-5cm below the soil surface. The green spears being higher in nutritive value are preferred for fresh market. Spears should be at least 1-2 inch in diameter and 8 to 10 inches long with the tips tight and stem bracts adhering tightly to the spear. Spears are bunched with the tips even and cut to a uniform length.

3.4.3 Celery

At the time of harvest the plants are cut just above the soil, trimmed and prepared for the market.

3.4.4 Fenugreek

When common methi is taken, the younger shoots are nipped off in about 3 weeks after sowing. Later on the whole plant is often pulled out, bunched and marketed. In kasuri methi cutting is done on 25-30 days after sowing. Subsequent cuttings were taken at 12 to 15 days interval. After 2-3 cuttings plants are left for seeding. Older leaves become bitter and yellow.

3.4.5 Lettuce

The leafy varieties are harvested when the leaves are immature and tender.

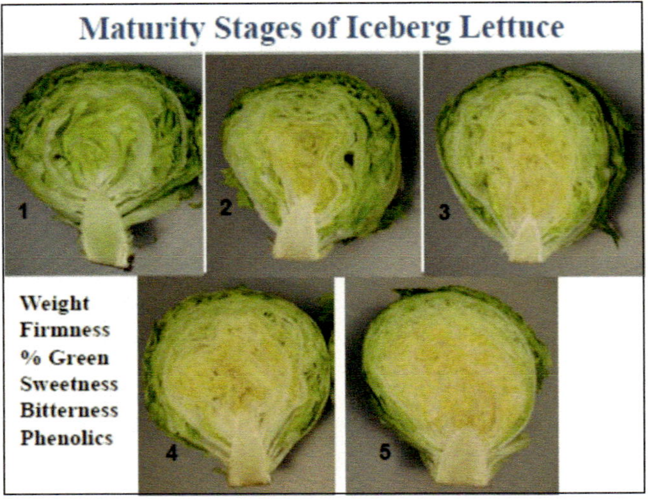

Figure 22. Maturity stages of Head Lettuce (https://www.google.com/postharvest.ucdavis.edu)

3.4.6 Spinach

First cutting is taken 3 to 4 weeks after sowing. Leaves are cut manually from the base. It gives 3 to 4 pickings at 15 days interval.

3.4.7 Globe Artichoke

The buds are harvested before they become loose and fibrous by cutting the stem 2.5-3cm below the base of the bud.

3.5. Commercial Cut and Loose Flowers

3.5.1. Cut Flowers

Maintaining quality in export cut flowers depends on an understanding the factors that lead to deterioration. Understanding these factors allow the grower and shipper to develop and implement optimum postharvest handling technologies. Flower maturity is one of the important factors that influence the postharvest shelf life of the cut flowers. Minimum harvest maturity for a cut-flower crop is the stage at which harvested buds can be opened fully and have satisfactory vase life after distribution. Many flowers are best cut in the bud stage and opened after storage, transport or distribution. This technique has many advantages, including reduced growing time for single-harvest crops, increased product packing density, simplified temperature management, reduced susceptibility to mechanical damage and reduced desiccation. Many flowers are presently harvested at tight bud stage (rose, gladiolus) for long distance transport, although others are harvested at fully open stage(chrysanthemum, carnation). Flowers for local markets are generally harvested when petals are fully open.(Reid 2009). Flowers for any purpose are harvested early in the morning or late in the evening to avoid transpiration losses. Immediately after harvest the cut flowers are kept in the water. Slant cut is recommended to increase the water uptake.

3.5.1.1 Rose

The rose undoubtedly remains the queen of the cut flowers. The historical association of this flower with romance and beauty ensures that roses will continue to be a highly desired cut flower in the future (Reid 2004). Maturity index for harvesting rose differs depending on market and cultivar. For long-distance transport or storage roses are harvested when the sepals reflex begins. Flowers will fail to open if harvested early, however, delayed harvesting may lead to bent neck disorder.

Fast-opening roses, like some yellows and are harvested just before the sepals start to separate from the bud.

For cut flower roses should be harvested when 1-2 petals begin to unfold at tight bud stage with full color envelopment. Because at this stage the petal tissues are still at developing stage that there will be no ethylene production and no transfer of photosynthates to reproductive organs.

In case of pink varieties, roses should be harvested when 1-2 petals unfurl from top.

White colored varieties are cut a little late than pink and red varieties where as yellow slightly earlier.

As loose flower - Cut when fully opened.

3.5.1.2 Carnation

In carnation flowering starts 110-120 days after planting. Flowers are harvested at different stages as given below.

Figure 23. Cut Rose for Export Market

Figure 24. Cut Rose for Local Market

Stages of Harvest

Standard types-Paint brush stage

Spray/Miniature types- When two flowers are open and the remaining flower buds show colour

Table 4. Harvesting Stages of Carnation Based on Market Demand

Sl.No.	Harvest Stage	Target Market
1.	Tight bud stage	For long distance transport. Some flowers may not open after harvest hence not preferred
2.	Paint brush stage	Long distance transport or for use after two days
3.	Semi-open stage	Short distance market
4.	Open stage	Local markets. Ready to use

3.5.1.3 Chrysanthemum

The harvesting maturity of chrysanthemums depends on type of cultivars. Yield starts from 3-4 months after planting. Total duration of crops could be 10 months with 6 months as main crop and 4 months as ratoon crops.

Table 5. Maturity indices of different types of chrysanthemum

Standard	When outer florets fully expanded/ray florets ceases to further devlop.
Spray	Flowers open but before sheddings of pollens/ when two third to three fourth open stage.
Pompons and decorative	Centre of the oldest flower fully open.
Anemones	Open but before central disc florets begin to elongate.
Pot-mums	Harvest with half to fully opened flowers. *Spray mums*-2/3rd -3/4th to full open stage of development *Standard mums*-3/4th to full open stage of development

3.5.1.4 Gerbera

Flowers are harvested when ray florets are completely elongated, 2-3 outer rows of the disc florets are fully expanded and perpendicular to the stalk, 2 -3 whorls of stamens are developed and show shedding of pollens (fully mature).

3.5.1.5 Anthurium

Anthuriums are harvested when the spathe is completely unfurled and the spadix is almost fully developed. Development of true flowers on the spadix is also used as a criterion for harvesting the blooms. The spadix is the true inflorescence of the anthurium and the rough mature flowers are easily distinguished form the smooth, immature flowers. Anthuriams are harvested when one third of the flowers on the spadix matures with visible change of colour from base to tip of spadix.

The proportion of open flowers on the spadix determines the maturity of the anthurium flower. In immature anthuriums, the spadix is smooth. Flower opening starts at the base of the spadix and proceeds upwards: spadices with open flowers are noticeably rough. Flowers that show any signs of chilling injury symptoms viz. purpling of the spathe, browning or wilting of the spadix etc. should be avoided.

3.5.1.6 Gladiolus

Gladiolus takes 60-120 days to produce spikes. The stage at which the spike is to be harvested should depend upon the transportation distance, consumer requirement and prevailing temperature conditions. For export purpose spikes are cut when 1-5 florets from the base of the spike show colour (tight bud stage).For local market spikes are cut when 3-4 florets fully open and remaining florets at the base are start showing colour

The spikes of gladiolus generally exhibit vase life of about 7-15 days. While harvesting, at least four basal leaves should be retained on the plant to ensure proper development of corms and cormels.

Classes of Gladiolus

On the basis of floret size, gladiolus has been put into 5 classes, the description of which is as follows

Class	Designation	Floret Size (cm)
100	Miniature	< 6.4
200	Small or miniature	≥ 6.4 to < 8.9
300	Decorative	≥ 8.9 to < 11.4
400	Standard or Large	≥ 11.4 to 14.0
500	Giant	> 14.0

3.5.1.7 Tuberose

Tuberose flowers are ready for harvesting after 100-120 days of planting. Harvesting is done in cool hours. For local market, cut flowers are harvested when lower one pair of flowers is fully open. For distant market spikes are harvested at the bud burst stage of the lower pair of flower buds. Fully or half opened flowers are used as loose flowers.

3.5.1.8 Orchids

Harvesting is a very important operation and the growers should have thorough knowledge about the flowering behavior of the orchids used for cut-flower production. Proper time, stage and method of harvest determine the quality of the produce. In general, orchid flowers do not mature until 3 to 4 days after they open. Flowers cut prior to their maturity may wilt before reaching the wholesaler. Harvesting should preferably be done in the evening.

Maturity Indicators of different Orchid Species

Dendrobium: The spikes are harvested when few buds at the base open and the buds at the top remain unopened

Cymbidium: Top two buds start opening remaining florets fully open.

Phalaenopsis: When all flowers are fully open (the 40-60 cm long spikes containing at least 8-10 flowers are harvested).

Cattleya: 3-4 days after full opening of florets. (Spikes of 25-40 cm length are harvested).

Oncidium: The spikes are harvested when more than 80 per cent flowers are fully open (a spike length of 60 cm with many florets).

Paphiopedilum: The spikes are harvested after complete opening of flowers.

Vandas: The spikes are harvested when all the flowers are already open (about 10-16 flowers on a stem length of 50-60 cm).

Arachnis: spikes with 50 per cent florets fully opened are harvested.

3.5.2 Loose Flowers

3.5.2.1 Jasmine

Harvested at matured un-opened bud stage. The flowers may be graded according to the corolla tube length, bud size, shape and freshness

3.5.2.2 Marigold

Marigold flower should be plucked when they attain the full size depending upon the variety. Plucking of flowers should be done in cool hours of the day i.e, either in the morning or in evening. Field should be irrigated before harvesting so that, flowers keep well for longer period after plucking. Plucked flowers should be covered with moist gunny bags if kept overnight before taking to market. Productivity of plants increases considerably by regular plucking of flowers.

Flowers are ready for harvest in about 3 to 3 1/2 months of planting. August-September is the peak period of flowering. Depending on the purpose, harvesting is done by cutting the fully-opened spikes from the base or single flowers are harvested as they open by day; the picking of individual flowers should be completed by 8.00 a.m. The flowers have a shelf-life of 3 days.

3.5.2.3 China Aster

Fully opened with short or no stem. For cut flower purpose fully opened with long stem. Flowers are ready for harvesting in 10-12 weeks after transplanting. Harvesting plays an important role in determining the flower quality. China aster is harvested in two different ways. Individual flowers are harvested for decoration and worship purpose whereas flowers along with stalk or the whole plant just above the ground are cut for cut flower purpose.

3.6. Plantation Crops

3.6.1 Coconut

Coconuts are harvested at different stages of maturity based on the purpose. The palm start yielding at the age of 3years (Hybrids) to 7 years (Tall varieties). The longevity ranges from 40-80 years depending on the variety. The most common maturity index is time from flowering, which correlates well with hard white flesh (endosperm) development (Paul and Ketsa 2015).

Table 7. The Stages of Harvesting Coconut

Purpose	*Maturity Stage*	*Maturity Index*
Copra, coconut milk for culinary purpose	10-13 months of flowering	Skin color changes to yellowish or brownish Hollow sound on tapping The skin surface around the calyx (cap) on the top of coconuts is green
Tender coconut water	6-9 months after flowering	Full size the skin is still green Solid sound on tapping Area around cap is creamy white or whitish yellow

3.6.2 Cashew Nut

Cashew starts bearing in 3.5 years to 4 years age. Raw cashew nut is characterized with leathery outer skin (Epicarp) and thin hard inner skin (Endocarp). Between these two walls of the shell is a honeycomb structure, which contains the phenolic material, commercially, known as cashew nut shell liquid (CNSL). The kernel is inside the shell wrapped in a thin brown skin known as testa.

The quality of the raw cashew nuts is tested by visual test, water flotation technique (Sinkers are matured and well filled and floaters are improperly filled) and cutting test (cut open the raw nut and observe the kernel *i.e.* white, shriveled, black spotted or rejects, the percentage of good kernel is calculated) (Balasubramanian 2015).

Table 8. The Stages of Harvesting Coconut

Purpose	*Maturity Stage*	*Maturity Index*
Cashew apple	Two months after flowering	Non climacteric fruit hence fully ripened. Yellow/Red coloured fruits Sweet aroma and taste.
Dry cashew nuts	Two to three months after flowering	Visually size (3.5-5 mm shell thickness 2.5-4 cm long) Kidney shaped and colour (Ash Grey)

3.6.3 Arecanut

The arecanut fruit is a mono-locular one-seeded berry. The colour of the nut is orange red to scarlet when ripe and consists of a thick fibrous outer layer, the husk which encloses a single seed. The mature fruit consists of three zones exocarp, mesocarp and endocarp which are more or less distinct in structure. Harvesting of nuts at right maturity is important for obtaining good quality produce.

Fruits are ready for harvesting in about 180 to 200 days from flowering.

First colour changes take place from green to yellowish or orange.

Two types of final product are seen in arecanut. Vellai choor/Kalipak, prepared out of immature nuts harvested at 6 months of maturity. Chali prepared from fully ripened fruits with 8.6 per cent more weight fetches 72 per cent more price.

3.6.4 Cocoa

Cocoa produces flowers from the 3rd year of planting onwards and economic yield starts from 5th year. Some of the maturity indices are:

Change of colour of pods - Pods that are green when immature turn yellow when mature and reddish pods turn yellow or orange.

The change in colour starts from the grooves on the pods and then spreads to the entire surface.

Rattling sound of beans (25-45 per pod) on ripening.

Hollow sound on tapping-Farmers practice

The harvesting is to be done at regular intervals of 10 -15 days. Over ripening of pods should be avoided. Immature harvest would result in hard cacao beans not suitable for fermentation, while overripe cacao pods lead to fungal-infected, defective, and poor-quality yields.

The maturity indicator may differ from clone to clone. The maturity indicators like the depth of the grooves, apex width, seed diameter, fruit weight, chroma, TSS content, increased with maturity in clones TSH565 where as pH, and moisture content decreased at the maturity. For clone CCN51, the distance between grooves, firmness and weight of the seed are recommended as maturity indicator where as for clone ICS95, only the color parameters and chroma, which increased with maturation, can be recommended as maturity indicators (Karen *et al.*, 2020). The tapping is usually used by knowledgeable farmers or experts to validate their initial measure. If the cacao pod produces a somewhat hollow vibrating sound, the cacao beans inside are losing its intactness, signifying that the pod is slowly becoming ripe.

3.6.5 Tea

Plucking of tea involves cutting newly grown vegetative shoots, comprising an apical bud and 2 or 3 leaves, immediately below it. Best plucking is two leaves and a bud when tea plant attains the age of 3 to 4 years, the plucking is usually started. However, under ideal management conditions it can be started at 18 to 20 months. Young leaves with more of tannins and polyphenols produces better quality tea than old leaves with less tannin content. Tea leaf age, bud and first two leaves, and shade levels influence on the relative concentrations of six major compounds of tea leaf, namely l-theanine, caffeine, and the major tea catechins. The concentration of l-theanine and caffeine decreased as leaf age increased moving from bud to first and then second leaf, while the concentration of the catechins increased from the bud to first and second (Song *et al.*, 2012).

Table 9. Tannin content in Tea leaves at different places

Type of Shoot/Leaves/Bud	*Tannin (per cent)*
Bud	25
I leaf	28
II leaf	21
III leaf	14
Stalk between II leaf and bud	12
Stalk between II and IV leaf	6

Kinds of Plucking Depending on the Extent of Maturity of Shoots

Young leaves with more tannin and polyphenol content produces superior quality tea.

1. **Two leaves and a bud:** Plucking of terminal bud and two most recently opened/ expanded leaves is done. Two leaves and a bud is considered as one plucking unit. Two leaves and a bud is also called fish leaf or Janam plucking.
2. **Fine tea:** Anything less than two leaves and a bud is called as fine tea. It is also known as fish or janam plucking. Young leaves will produce superior quality tea.
3. **Coarse tea:** Anything more than two and a bud. Older leaves will produce coarse quality tea.
4. **Light plucking or light pruning:** It is carried out during initial years to encourage side branches and to make the bush full with tender branches.
5. **Too tender shoot/One leaf and a bud:** Produces very good quality tea but yield goes down drastically.

3.6.6 Coffee

Coffee fruit known as cherries have red skin (called exocarp), pulp (called mesocarp) and a parchment-like covering the bean (called endocarp). Inside these layers are usually two oval shaped beans, with their flat sides facing each other. Harvesting time for coffee cherries will vary by region and altitude. Typically, there is only one harvest per year, which will lasts for 2 to 3 months as cherries ripen.

The ripe coffee cherries have colour ranges from bright red to deep red and yellow, depending on the plant variety. The coffee fruit also contains volatile compounds that are responsible for the aroma and flavour properties of the coffee. These compounds are present at a very low amount at the early stage of the coffee cherry, but later on, it increases as the coffee transformed to the maturity stage. In India there is only one cropping season. Arabica coffee harvested from November to January and Robusta coffee from December to February.

Harvesting is done in stages as follows:

1. Fly picking: Small scale picking of ripe berries from October - November to February
2. Main picking: Well formed ripe berries are harvested in December.
3. Selective picking involves making numerous passes over coffee trees, selecting only the ripe cherries, then returning to the tree several times over a few weeks to pick remaining cherries as they ripen. Selective picking is more expensive due to the labor involved and is only used for Arabica coffee.
4. Stripping: In **S**trip picking is trees are harvested entirely at one time "stripping" all the beans off the branches, ripe as well as unripe cherries. Typically, only Robusta coffee is strip picked. Modernization has provided machine harvest for Robusta coffee which simply shakes the trees knocking of all the cherries at one time.
5. Gleanings: Collection of fruits that have dropped down during harvesting.

3.6.7 Oilpalm

Oil palm comes to bearing in 3-4 years and continues up to 25 years. Over ripe or under ripe fruits should not be harvested as the oil recovery declines due to conversion of fat into fatty acids and glycerol. The following are the maturity indicators for oil palm bunch harvesting.

Time taken from pollination to maturity- It takes about 6 months time from pollination to fruit maturity.

Colour change- Fruit colour changes from black to orange or red or yellowish orange.

Easy separation of fruits from bunch- If ten or more fruits are easily detached from the bunch. This method mostly used by farmers.

Change in fruit texture- Smooth and soft texture of the fruits when ripe.

The ripe oil palm fresh fruit branches (FFB) have high oil extraction rate (OER) while the under ripe and over ripe FFB produce less oil. Blue-to-Red Fluorescence Ratio (BRR_FRF) obtained from blue-green (447 nm) and far-red (685 nm) emission signal by using ultraviolet (UV) light emitting diode as excitation light source is a novel method to classify oil palm FFB into ripe, under-ripe and over-ripe categories (Hazir *et al.*, 2012)

Table 10. Grades of Oil palm

FFB Grade	*Features*
Ripe bunch	Ripe bunches with yellowish and reddish outer layer and its mesocarp are yellow coloured. This bunch has at least ten (10) loose sockets and more than fifty per cent (50 per cent) fruits were still attached bunches and loose fruits should be sent to the mill within 24 hours after harvest.
Un ripe bunch	Unripe Bunch is black or purple-black and fruit mesocarp outer layer of yellowish. These bunches have no fresh fruit socket. Fruit socket on unripe bunch is not caused by the usual process of maturity.
Empty bunches	90 per cent of fruits are fallen before reaching the refinery
Long stalk	Long Stalk Bunch are bunches with stalks more than 5cm measured from lowest level of the bunch stalk.
Dirty Bunches	Dirty bunch are bunches that more than half of its surface is muddy, sandy, and mixed with stones or other impurities.
Poorly pollinated bunches	Poorly Pollinated Bunches is a bunch of fruits that is not normal in size or density or cluster and has more than fifty per cent (50 per cent) of 'parthenocarpic'.
Damaged bunches	Damaged Fruit Bunches is more than thirty per cent (30 per cent) share of the fruit was damaged by animal pests, especially rats
Small bunches	Small Bunch is a bunch of small fruits and weighs less than 2.3 kgs (5 lb)
Under ripe bunches	Under Ripe Bunch is orange in colour and reddish or reddish-purple fruit mesocarp yellowish orange coloured outer layer. These bunches have less than ten (10) loose sockets. Should be sent to the mill within 24 hours after harvest
Over ripe bunches	Over Ripe Bunch are bunches of fresh fruit with dark red and more than fifty (50 per cent) fruits detached, but at least ten per cent (10 per cent) still attached to the fresh fruit bunches. Should be sent to the mill within 24 hours after harvest.

3.6.8 Palmyrah Palm

Palmyrah palm produces clear juice called Neera on tapping which varies with age and sex. Both male and female palms produce Neera (unfermented product) toddy (fermented product). Female palms are considered as more productive because they not only yield 30 -50 per cent more sap than males but also produce edible tubers, tender fruits (Endosperm -Nungu). Sugar, jaggery, vinegar and candy are the value added products from Palmyrah palm. The palms start yielding when they become 10-12 years and continues up to 40-50 years. Neera yield and quality varies with sex, age of the inflorescence, method of tapping.

4.0 Future Researchable Issues

With the changing climatic scenario, and production regime, identifying right maturity for each crop is highly important. Other than visual appearance, nutrient compositional factors become important to provide nutritional security to the growing population. More and more instrumental investigations have been realized to improve evaluation of fruit and vegetable quality by introducing other quality characteristics in addition to sensory ones. The application of NIR spectroscopy to the assessment of the quality of fruits and vegetables, with reference to the determination of postharvest changes in proteins, lipids, carbohydrates, organic acids, ethanol, and chlorophyll in fruits and vegetables, has been studied. A potential application of the results of flavor chemistry, new developments in descriptive analysis, consumer acceptance testing, and the interrelations of chemistry/sensory analysis, may be the trends in quality determination. Following are some of the important researchable areas in maturity studies;

Identification of the important components of quality and the interrelationships among these quality factors for the various fruits destined for the fresh market or for processing.

Developing objective and non-destructive methods for determination of quality attributes, especially those which are related to flavour and nutritional quality of fresh produces.

Conducting consumer acceptance research aimed at relating maturity indices at harvest to the final organoleptic acceptability by the consumer.

Development of new genotypes with better flavour and nutritional quality in all the major fruits. Developing sensor based identification of maturity for different markets and enabling IoT (Internet of Things) and AI (Artificial Intelligence) based platforms for better last mile delivery and reducing post-harvest losses

5.0 Conclusion

Quality determination of fruits and vegetables is at present characterized by the overwhelming use of sensory evaluation. Quality indices include primarily external properties, such as appearance, size, shape, color, texture, and flavor. Conformity with relevant product standards is also an important requirement that is inspected for the trade of fruits and vegetables. Maturity measures to be made by the producers, handlers and quality control personnel must be simple, readily performed in the orchard and require relatively low operational cost. The index should be a objective rather that subjective (an evaluation). The developed index should be nondestructive to avoid loss of produces. The index must be consistently related to the quality and postharvest life of the commodity for all growers and for all through the year. With the increasing competitive and concentrated markets, more important becomes the introduction of systems and technologies to manufacture produce differentiated by their presentation for positioning in these markets. The production of high quality fruits and vegetables and the maintenance and enhancement of this quality in postharvest and distribution operations is associated with careful incorporation of technologies applied throughout the production, harvesting and post harvesting stages. These technologies are crucial to ensure quality and safety. The principle is not to use sophisticated technologies, but to handle efficiently the produce throughout the chain. Only in this way will quality and safety be maintained and postharvest losses can be reduced. A new approach to describe fruit and vegetable quality by an overall evaluation method provides numerical values that could be used for their classification in the future.

6.0 References

Behera, T.K., Pal, R.K., Sen Nita and Singh Manoj, 2004. Effect of maturity at harvest on physico-chemical attributes of sweet pepper (*Capsicum annum* var *grossum*) varieties. *Ind. J. Agril Sci.* **74:** 251- 3.

Deodhar, S.Y. and Dave, H. 2001. Managing Quality of Agri-commodities and Food Products under WTO Regime. In. S. Datta. and S. Deodhar (Eds.). *Implications of WTO Agreements for Indian Agriculture*, Oxford-IBH: Delhi.

Dhatt, A.S., Mahajan, B.V.C., Sandhu, K.S., Garg, A. and Sharma, S.R., 2007. Handbook on Postharvest Handling of Fruits and Vegetables. 3rd edn, PHPTC, PAU, Ludhiana

FAO 2001. Food Quality and International Trade. http: //www.fao.org/ag/ags/Agsi/gtz_html.htm

Goel, A.K., Kumar, R. and Mann, S.S. 2007. Postharvest management and value addition. Daya publishing house. New Delhi-35.

Kader, A.A. 2002. Postharvest Technology of Horticultural Crops. 3rd Edn. University of California, Agricultural and Natural Resources. Publ. 3311. P. 296

Kader, A.A. and M.L. Arpaia, M.L. 1992. Postharvest handling systems: subtropical fruits. In: A.A. Kader (Eds). *Postharvest Technology of Horticultural Crops.* Univ. of Calif., Davis CA, pp. 233-240.

Kitinoja, L. and Gorny, J. 1998. Postharvest technology for fruits and vegetables Produce marketers: Economic opportunities. Quality and Food Safety by, Department of Pomology, University of California, Davis. A joint publication of UC Postharvest Outreach Program and Punjab Horticultural Postharvest technology Centre, USAID/ACE

Kumar, N., Srivastava, G.C. and Dixit, K. 2008. Hormonal regulation of flower senescence in roses (*Rosa hybrida* L.). *Plant Growth Regul.* **55:** 65–71.

Kute, L.S. and Shete, M.B. 1995. Sapota (Sapodilla). In: D.K. Salunkhe and S.S. Kadam (Eds.) *Handbook of Fruit Science and Technology*, Marcel Dekker, Inc., NY, pp. 475-84.

Lakshminarayana, S. 1980. Sapodilla and prickly pear. In: S. Nagy and P.E. Shaw (Eds). *Tropical and Subtropical Fruits. Composition, Properties and Uses.* AVI Pub, Westport CT., pp. 415-41.

Massantini, R., Carlini, P. and Anelli, G. 1997. NIR Spectroscopy to Evaluate Fruit and Vegetable quality During the Postharvest. *Industrie Alimentari* **36:** 321–26.

Meftah Salem M. Alfatni, Abdul Rashid Mohamed Shariff, Helmi Zulhaidi Mohd Shafri, Osama M. Ben Saaed and Omar M. Eshanta, 2008. Oil Palm Fruit Bunch Grading System Using Red, Green and Blue Digital Number. *J. Appl. Sci.* 8: 1444-52.

Pal, R.K. 2008. Final Report of Consultancy Services on Controlled Atmosphere Storage of Fruits and Vegetables submitted to M/S Ministry of Railways, Govt of India, July 2008, 40 p.

Pal, R.K., Babu, K.D., Singh, N.V., Maity, A., Gaikwad, N. 2014. Pomegranate Research in India–Status and future challenges. *Prog. Hortic.* **46:** 184-201.

Petracek, P.D., Dou, H. and Pao, S. 1998. The influence of applied waxes on postharvest physiological behavior and pitting of grapefruit. *Postharv. Biol. Technol.* **14:** 99-106.

Ryall, A.L. and Pentzer, W.T. 1974, Handling, Transportation and Storage of Fruit and Vegetables: Volume 2, Fruits and Tree Nuts, The AVI Publishing Company, Inc. Westport, CN

Sargent, S. 1998. Handling Florida vegetables - Eggplant. http: //hammock.ifas.ufl.edu.

Song, R. and Kelman, D, Johns, K. and Wright, A.D. (2012). Correlation between leaf age, shade levels, and characteristic beneficial natural constituents of tea (*Camellia sinensis*) grown in Hawaii. *Food Chemistry*. 133. 707–14.

Thompson, A.K. 1996. Postharvest Technology of Fruits and Vegetables. 1st Edn. Blackwell Science, Inc. USA.

Wills, R., McGlasson, B., Graham, D. and Joyce, D. 1998. Postharvest: An Introduction to the Physiology and Handling of Fruit, Vegetables and Ornamentals. 4th Edn. CAB International, UK.

Websites

http: //www.postharvest.ucdavis.edu

http: //www.herbs.ph/attachments/article/1610/

http: //www.extension.udel.edu

http: //www.cuke.hort.ncsu.edu

http: //www.extension.missouri.edu

http: //www.www.sakatavegetables.com

https: //www.gapphotos.com/GardensBlog/?attachment_id=3957

http: //postharvest.ucdavis.edu/Commodity_Resources/Fact_Sheets/Datastores/Vegetables_English/?uid=14 and ds=799

https: //www.slideshare.net/

https: //ag.umass.edu/greenhouse-floriculture/fact-sheets/harvesting-handling-cut-flowers

https: //agriinfo.in› maturity-signs-and-harvesting-of-ar.

http: //www.krishisewa.com› id=61: arecanut

http: //www.celkau.in› crops › Arecanut

http: //postharvest.ucdavis.edu

https: //www.krishisewa.com/articles/pht/728-postharvest-management-of-cut-flowers.html

https: //www.slideshare.net/KarthekaThirumugam1/harvesting-indices-and-harvesting-techniques-of-cut-flowers

https: //www.coffeemasters.com/coffee-101/harvesting-of-coffee

https: //www.mjmpom.com/ffb-grading-guideline/

www.Agritech-tnau.in

Chapter 4

Mechanization in Harvesting

S.C. Mandhar

Emeritus Scientist, ICAR-Indian Institute of Horticultural Research, Hessaraghatta, Bengaluru – 560 089, Karnataka, India
E-mail: suresh.mandhar@gmail.com

R.T. Patil

Former Director, ICAR-Central Institute of Postharvest Engineering & Technology, P.O. PAU, Ludhiana – 141 004, Punjab, India
E-mail: ramabhau@gmail.com

1.0 Introduction

Harvesting of fruit from the trees is a cumbersome and time consuming operation. The fruits are generally required to be plucked individually by holding the fruit at their natural break point firmly but gently. Sometimes due to shortage of labour fruits are harvested by shaking the tree manually as is done for mango by many farmers. The fruits fall on the ground and then picked up. This injures the fruit internally and the spoilage increases during ripening. The mango fruits are also harvested manually by using conventional harvesting aids like 1) bamboo stick folded to form elliptical shape frame and tied on one end of a bamboo or any light wooden pole. The hand woven cotton net is tied below the frame to collect the fruits, 2) some conventional harvesters are fitted with a cutting notch to harvest the fruits with pedicel and a canvas collection sack to collect the fruits, 3) bamboo pole fixed with a semicircular or sickle blade and net at the distant end and a net to collect the harvested fruits. The conventional mango harvesters have been modified at Indian Institute of Horticultural Research, Bangalore, Central Institute for Subtropical Horticulture, Lucknow and Dr. Balasaheb Sawant Konkan Krishi Vidyapeeth, Dapoli, to increase the harvesting capacity and increase the shelf life of fruits by harvesting these along with 2 to 3 cm. long pedicle.

Timely harvesting of fruits is important for maintaining the transportability, quality and shelf life. However, the harvesting capacity of these harvesters is very low as the individual fruits are to be harvested and collected. Hence, there has been a need to develop high capacity machines for this purpose. For adopting the mechanical harvesters the tree canopy has to be properly trained. Some of the machines that have been developed and are being used in different countries are presented in this chapter.

2.0 Harvesting Aids

Harvesting aids have been developed to increase the harvesting capacity of the person and to reduce the damage to the fruits. Generally, one person is required to use these aids to harvest the fruits. The harvesting capacity depends upon the yield of fruit tree and type of fruit. The Mango harvester developed at Indian Institute of Horticultural Research, Bangalore (Figure 1) is an improvement on the conventional harvester and harvests the fruit with pedicel. The fruit is lowered in net in the opening in between frame and aluminium pole. While pulling the harvester, the fruit is held between the cushioned frame and the pole and the blade cuts the pedicel of 1-2cm in length to harvest the fruit. There is no sap bleeding, fruits remain shining without blemishes and the shelf life of the fruit is increased by 3-4 days. The harvesting capacity of the harvester is 50 to 150 kg/hr depending upon the variety, yield and height of tree.

Figure 1. Mango Harvester

The sapota, guava and pomegranate harvester (Figure 2) have been developed at IIHR, Bangalore. It consists of harvesting frame, conical in shape and a pole. The harvesting frame is fitted with comb consisting of cushioned pegs to hold the fruit and harvest these while pulling the harvester. The fruits of required size only get harvested while the immature fruits are left on the tree. The capacity of sapota and guava harvesters is 50 – 100 kg/hr. In pomegranate harvester, the distance of pegs on the frame is more to suit the size of fruit to be harvested.

Figure 2. Sapota Harvester

Lime harvester (Figure 3) has been developed at IIHR, Bangalore. A small box of around 10cm dia and 20cm length is fitted with the hook. The harvester is pushed easily in the canopy, the fruit is held in the hook, harvested and collected in the box while pulling the harvester. The box is emptied after 8-10 fruits are collected in it. The lime harvester has the harvesting capacity of 15-20 kg/hr.

Figure 3. Lime Harvester

The cut and hold harvesters have secateurs head fitted at distal end. It is able to hold the fruit after harvesting (Figure 4). These are generally telescopic and the length can be extended to 3 meter. These are suitable for harvesting most fruits including mango and litchi. The pedicel of the fruit is cut and the fruit is lowered to the ground and put in the required container.

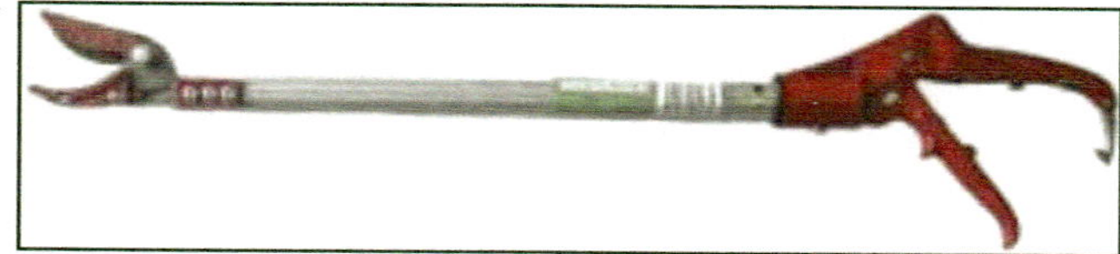

Figure 4. Cut and Hold Harvester/Pruner

2.1 Man Positioning System

Generally the manual labour climbs the tree to harvest the fruits in India. However, this has to be avoided as there is a risk for the person to fall down. Portable aluminium light-weight tripod ladder (Figure 5) can be used as a man positioning system to avoid the climbing of labours on the trees to harvest the fruits. Ladders are also very useful in pruning the tall trees. The double pole telescoping ladder and single pole ladder are suitable aids for harvesting coconut, areca nut, black pepper and betel wine.

A self propelled hydraulically operated man positioning system has recently been developed at CIAE Bhopal (Figure 6). A hydraulically operated three wheel self-propelled elevator system was designed and fabricated at CIAE Bhopal for harvesting of mango, oranges, oil palm *etc.* The unit can be rotated 360 ° and has hand-controls on platform and steering mechanism. It is operated with a 11 hp petrol engine as prime mover. The platform could be elevated up to 6 meter from

Figure 5. Tripod Ladder

Figure 6. Self Propelled Hydraulically Operated Man Positioning

ground level, which makes it suitable for a variety of operations in medium height orchards including oil palm. Self propelled elevating platform can also be used for pruning, spraying and harvesting tall trees. One operator stands in the cage/platform and controls the height and movement of the platform. The fruits are plucked manually and put in the bag provided on the machine. The bag can be emptied after lowering the platform to the ground level. The hydraulic operated chain saw and pruners are also provided for pruning on these platforms.

In tractor operated mango harvester developed at Indian Institute of Horticultural Reserch, Bangalore labourers stand on the platform on the trailer (Figure 7) with harvesting aids for harvesting the individual fruits. The harvested fruits fall on the apron, filled in crates/ bulk bins and kept in the trailer.

Figure 7. Tractor Trailor Mounted Platform

2.2 Ultra Lightweight Harvesters

The light pole aluminum harvesters with pole length of up to 12 meter are available for harvesting coconut and oil palm (Figure 8). The aluminum pole is fitted with a sickle and the fruits are harvested by holding the bunch with the sickle and pulling the pole. The telescopic petrol engine (length 3.7 metre) operated vibrating sickle harvesters are available for harvesting the oil palm (Figure 9). The weight of petrol engine operated harvester is about 7 kg and the fruits can be harvested up to a height of 5 m. About 150 bunches can be harvested per day using this motorized harvester.

2.3 Shaker Harvesters

The telescopic pole harvester with battery operated revolving reel is available and is suitable for harvesting olives (Figure 10a) The beater harvester for olives is operated by 26 cc/900 W petrol engine and produce 1150beats per minute (Figure 10b). The beater harvester has a harvesting capacity of 50-80 kg/hr. The shaker harvester for limb/branch is provided with 48.7 cc/2.2 kW petrol engine, weighs 13.9 kg and the pole length is 2.83 metre (Figures 10 c and d).

2.3.1 Apple

The shaker harvester for apple has been developed in USA (Figure 11). The canopy of the apple trees is trained such that shaker harvester can harvest it. The tree is generally trained into single or two branches (Y-shaped). The individual branch of the tree is shaken by pneumatically operated cylinder by giving one jerk. The fruits get detached and fall on the collecting apron and are collected and conveyed to the truck using belt conveyors.

2.3.2 Mango

The low cost vibrating harvester for mango developed at Asian Institute of Technology, Bangkok

Figure 8. Light Pole Harvester for Oilpalm and Coconut

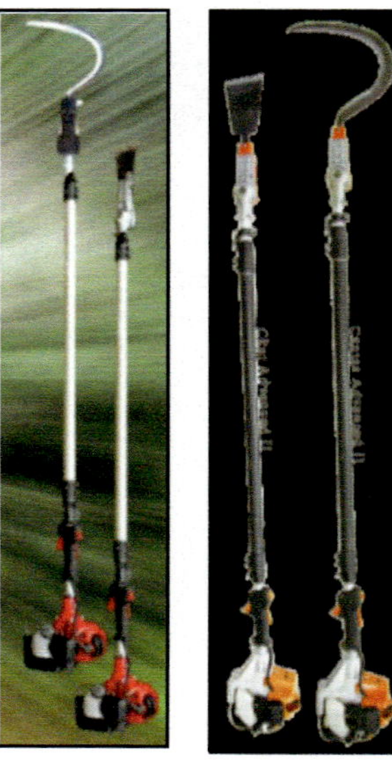

Figure 9. Engine Operated Harvesters for Oilpalm

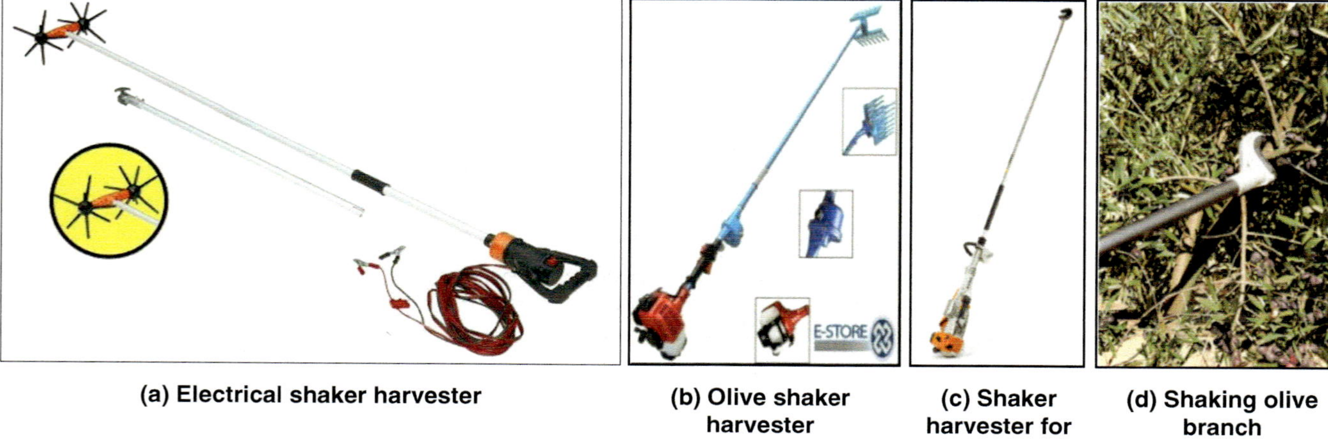

(a) Electrical shaker harvester **(b) Olive shaker harvester** **(c) Shaker harvester for branch** **(d) Shaking olive branch**

Figure 10. Shaker Harvesters

operates with 8.3 hp diesel engine. The individual branches/limbs of mango are held in padded clamps and vibrated with amplitude of 76-102 mm at 11 – 13 cycles per second. About 80-85 per cent fruits are removed. The fruits are collected on catching net supported on portable stands. Two to four trees can be harvested in an hour depending upon the tree size.

Figure 11. Shaker Harvester for Apple

2.3.3 Nuts

The tractor operated harvester shakes the branches of the nut trees to harvest them (Figure 12). The stroke and frequency of the vibration can be adjusted. The fruits are collected by the catching frames.

2.3.4 Orange

Self propelled shaker harvesters (Figure 13) and tractor operated harvesters are available for shaking the trunk of the tree or branches of the tree. The stroke and frequency of the vibration can be adjusted. The fruits fall on the ground and then collected by fruit sweeps. The fruit pick up machines pick up oranges at 15-20 tonnes/hr. at 97 per cent efficiency and caused 3 to 7 per cent split peels from handling. The fruits are collected by horizontal or vertical rotors, conveyed to the bins through conveyor belt and leaves and dust is

Figure 12. Self Propelled Shaker Harvester with Inverted Umbrella Collection Apron and Belt Conveyor

blown away by blowers. The fruits are also collected by the catching frames.

a) Trunk Shake Catch Harvester for Orange

The trunk shake catch harvester (developed at Compton Enterprises Inc., Chico, California, U.S.A.) removed 90-94 per cent at harvesting capacity of 10 tonnes/hour. The orange tree is shaken for 5 to 10 seconds at a frequency of 8 to 10Hz. The shake – catch harvester had 2 sets of unbalanced masses totaled 125 kg had an eccentricity of 22 cm and rotated in same direction at 8 to 10 Hz (480 to 600 rpm).

Figure 13. Self Propelled Tree Shaker

2.4 Canopy Shaker Harvester

2.4.1 Blue Berries

The spiked drum shaker harvester has also been developed for harvesting blue berries (Figure 14). The fruits are collected on the catching aprons/conveyor belts below the harvester and conveyed to the storage bin through conveyor belts. The fruits are also graded by the machine and sorted manually by 2-3 labourers. The harvesting capacity is around 400 – 500 kg/hour. The fruits are suitable for fresh consumption (eating)/ table purpose. The harvester machine can harvest the blue berry grown on trellis. Around 80 per cent yield is marketable. The harvester with spiked tooth type drums for blue berry can be easily modified for harvesting Ber (*Ziziphus* sp.) and Aonla (*Phyllanthus emblica*).

Figure 14. Spiked Drum Shaker Harvester

2.4.2 Orange

Spiked drum type harvester (Figure 15) has been developed in USA and Australia to harvest the orange. The planting of the trees is done at specified spacing and the canopy is shaped. The tree is brought in between the spike drums of 6m dia. and the rotating drums are pushed inside the canopy, which shake the branches of the tree and harvest the orange. The fruits fall on the catching frame/net and from there these are conveyed to the truck by using belt conveyors. The harvested fruits are suitable for processing or juice extraction. The canopy harvester (developed by United States Department of Agriculture) has 2 shaker drums with eight sets of horizontal whorls, 3.2 cm diameter nylon spokes mounted on each of the two vertical shaker shafts (Figure 15). Each horizontal whorl set contained 16 spokes, 1.5 meter long beyond the shaft/hub to which they were bolted. The vertical spacing between horizontal whorls is 38 cm and top set is about 3.7 meter above the ground. Oranges are removed by the 2 shaker drums revolving horizontally in opposite direction as the spokes free wheeled through the tree canopy. Up to 95 per cent oranges are removed (in small trees) when the machine operated at 1.2 to 2.4 km/hour on one side of the canopy and the harvesting capacity was observed up to 25 tonnes/hour. The fruits fall on a conveyor belt provided below the whorls and trash is removed before filling these in tractor trailer.

2.5 Fruit Harvesters

2.5.1 Banana

The banana harvesting machine is a type of lifting crane (Figure 16). It lifts the bunch from the plant and harvest the bunch from the plant by rotary circular blade. The bunch of fruit is then hung on the rope conveyor for carrying to the packaging hall. Generally a blue colour

Figure 15. Canopy Shaker Harvester for Orange

polythene sleeve is put around the banana bunch for uniform ripening. The polythene sleeve is removed in the packaging hall and bunches are cut into hands and packaged into CFB boxes.

Figure 16. Banana Harvesting Machine

2.5.2 Coconut

The palm climber (Figure 17) designed and developed separately by M J Joseph alias Appachan, P K Joseph and Joy Verghese consists of two metal loops of 10 mm MS rod having sub-loops, rubber belt, wire ropes, connecting clamp, MS plates *etc.* One loop is intended for the right leg and the other for the left leg. These are called right leg loop and left leg loop respectively, the left loop (main loop) which is slightly bigger than the right loop. The top of the main loop is bent forward to form handle and just below this part, two metal plates are attached with holes with a long rubber belt. On the rubber belt a wire rope having rings on each end is fixed. The bottom most part is a plate and a clamp is provided above it. A long holed plate is fixed on the main loop, which is used as parking brake.

Ropes attached to the climber are used to fasten it to the palm. The ropes passes around the tree and through hooks provided in the climber. A pedal is provided for resting the foot. The right side peddle is raised a little upward so that the grip of the right hand part of the climber becomes loose facilitating upward movements by using the right hand and the right leg. Once the right part is moved up the weight of the body rests on the right hand part of the climber. The whole process is repeated for the left hand side. In this way one can easily climb coconut or arecanut palms. The equipment costs about Rs 7,000. The labourer does not feel unsafe when he climbs above 10 meter.

The coconut climber designed by TNAU Coimbtore (Figure 18) has two rectangular iron frames. Each frame is fitted with two rubber bands, which hold the trunk of the tree when weight is put on this. The worker climbs up the tree trunk by transferring his weight on lower frame (by legs) and upper frame (by hip) and lifting the frames alternately.

Figure 17. Coconut Climber (Verghese)

Figure 18. Coconut Climber (R-Tech)

2.5.3 Grape

Self propelled grape harvesters (Figure 19) have become very popular for harvesting wine grapes in California, USA and New Zealand. These harvesters harvest grapes raised on trellis. The grape harvester moves over the row or trellis of grapes. The single trellis picking heads consist of two banks of nylon fingers. The fingers can be straight plastic or spring loaded metal rods or curved rods. The diameter and material of the fingers maximizes the amount of fruit harvested. The vines are shaken by the fingers in the harvesting head. The grapes fall on the collection conveyor and these are then conveyed to the tractor towed grape bin moving in the next row. The trash and leaves are blown by blower. About one to two hectare of grapes is harvested in one hour.

Figure 19. Grape Harvester

2.5.4 Pineapple

Generally pineapples are harvested manually by a knife and filled into bag or baskets. As the fruit plant has spiny leaves and fruits have crown with spiny leaves, the harvesting labour has to be very skilled. The protective gears like gloves, spectacles and clothing must be worn. The mechanical harvesting machine (Figure 20) is a collecting device for the fruits. Fruits are harvested manually and placed on conveyor belt. The crown and root portion is cut manually before placing the fruit on conveyor belt. The damaged and small fruits are discarded in the field. The conveyor belt is pulled in the field by the transport truck. The conveyor belts carry the fruits to field bins. The pineapple fruits are filled in the field bins by placing the crowns downward to use these as cushion. The harvested pineapples are then taken to the packaging house for grading and packaging. Lights are also provided on the conveyor belt frame for working at night.

Figure 20. Harvesting Pineapples Manually and Putting on Conveyor

3.0 Conclusion

Efforts have been made in India to develop hand tools to assist the labour for efficient harvesting of fruits. The man positioning platform has been developed for harvesting of tall trees at Central Institute of Agricultural Engineering, whereas tractor mounted lift developed at Indian Institute of Horticultural Research, Bangalore will facilitate mechanical harvesting. These devices are also useful for pruning and spraying of fruit trees. The high capacity self-propelled machines developed overseas are required for large orchards. However, we need to import some machines and work on their scaled down versions. When we systematically and scientifically expand the areas under fruit crops even the bigger machines on custom hire basis can become affordable even under Indian conditions.

4.0 Important References

American Society of Agricultural Engineers, 2003. Outstanding Innovatios. *http//www.asae.org/ resources on 5.8.2003*

Bucello Vince, 2014. Nelson Citrus and Olive Harvester, Video. *http: //www.youtube.com/watch. on 2.5.2014.*

Central Agricultural Engineering Research Institute, 2011. *Development of platform type fruits harvesting system for mango, sapota and citrus,* Annual report 2010-11.

Dole Food Company, Inc, 2013. Harvesting Pineapples. *http: //www.dole.com/https: //www. youtube. com/ watch?v=w FYEeFpvik8, on 3.4.2013.*

Etani Sdn. Bhd.Malasia, 2012. Motorized Oil Palm Cutter, *http: //www.etani.com.my/on 2.5.2012*

Harinath, V.N., 2010. Multi-Tree Climbing Devices, A Boon For Toddy-Tappers, The Hindu. *http: //www. thehindu.com/on 11.10.2010*

Horticulture Australia Ltd., 2014. Mechanical Citrus Harvester; improvement modifications for commercialisation purposes. *http: //www. horticulture.com.au/reports/search_final_reports_ result.asp on 2.5.2014*

Korvan Industries Inc., Oxbo International Corporation, 2010. Oxbo 3220 Citrus Harvester. *http: //www. oxbocorp.com/Products/Citrus.aspx http: //www.korvan. com, as on 5.10.2010.*

Littau Harvester, 2010. Over-The-Row Harvester. *http: //www.littauharvester.com, on 2.5.2010.*

Mandhar, S.C. *et al.* 2003. Mechanization of Fruit Harvesting, **In** *Proceedings of the International Seminar on Downsizing Technlogies for Rural Development* Vol. I,7-9 October, Regional Research Laboratory, Bhubaneswar, India, pp.144-49.

R- Tech Agri Equipments, 2012. Multi Tree Climber. *http: //coconuttreeclimber.blogspot.in/p/what-we-offer.html on 21.5.2012*

OESCO, Inc. 2011. Baldwin Wooden Tripod Apple Ladders, *http: //www.oescoinc.com/baldwin-wooden-tripod-apple-ladders.html on 14.2.2011*

OMC - Orchard Machinery Corporation, 2011. Shock Wave Tree Shakers, *http: //www. shakermaker.com on 11.11.2011*

Parameswarakumar M., and Gupta, C.P. 1991. Design Parameters for Vibratory Mango Harvesting System Transactions of the ASABE. 34: 14-20.

STIHL Incorporated, 2010. Olive Shaker, www.stihlusa.com, on 20.2.2010.

Senthil Kumaran G., 2011. Tractor Trailer Mounted Hydraulic Platform for Harvesting Fruits, *Indian Institute of Horticultural Research,* Bangalore. Personal Communication on 2.7.2011

Verghese Joy, 2011. Kerala Coconut Pluckers, Outlook, *http: //www.outlookindia.com/article.aspx?278543, on 17.10.2011*

Zhejiang Zhongli Tools Manufacture Co., Ltd. 2010. Electrical shaker harvester, *http: //www.made-in-china.com/showroom/grace 955/companyinfo/ Zhejiang-Zhongli-Tools-Manufacture-Co-Ltd-.html on 26.10.2010*

Chapter 5

Postharvest Quality Management

B.V.C. Mahajan

Professor, Punjab Horticultural Postharvest Technology Centre, Punjab Agricultural University, Ludhiana – 141 004, Punjab, India

E-mail: bvc_mahajan@rediffmail.com

Alemwati Pongener

Scientist, ICAR-Indian Agricultural Research Institute, Gogamukh – 787 034, Dhemaji, Assam, India

E-mail: alemwati@gmail.com

1.0 Introduction

Plant parts or organs such as fruits, vegetables and flowers derive water, carbohydrates, and nutrients required for growth, development and metabolism, from their parent plant. As long as they are attached to the plant they continue to receive these metabolites. A critical time in the life of these organs is the time of harvest. Once the produce is harvested,it loses its source of nutrients and metabolites, and there is no possibility for further improvement in such components. But the produce, continues to respire and maintain metabolic activities for which it has to rely entirely on its internal reserves of metabolites. The rate at which the harvested produce exhausts its internal reserves, transpiration losses, respiration rate, and various internal and external factors contribute to decide the quality and shelf life of produce.

Quality at the time of harvest is dictated by a multitude of preharvest factors. These factors span from genetic make-up of the crop through cultural and environmental influences to various preharvest treatments. In addition, after harvesting, the produce can be subjected to several postharvest treatments and practices that interplay with the physiological and biochemical processes within the produce to beneficially prolong storage life and maintain quality in fresh form. Extending the postharvest life of horticultural produce requires the knowledge of all pre and postharvest factors that can lead to loss of quality or generation of unmarketable material.

2.0 Preharvest Factors Affecting Postharvest Life and Quality

Several preharvest factors *e.g.*, the rate at which produce deteriorates progress through its normal developmental pattern of maturation, ripening and senescence, affect postharvest life and quality. It is, therefore, very important to understand these preharvest factors. Environmental factors such as temperature, relative humidity, water availability, light, cultural practices and pest management techniques determine the inherent quality of the produce. However, the ultimate quality is the final manifestation of inter-relation between the commodity and its environment.

2.1 Cultivars

Growers have the choice of selecting preferred cultivars based on the eating needs and desires of the consumers. Within each grouping there is a range of genotypic variation in composition, quality, and potential postharvest life. Cultivars vary among themselves for shape, size, productivity, colour, dry matter, taste, ripening time, nutritional quality, shelf life, susceptibility or resistance to pests and pathogens or postharvest handling damages *etc.* Plant breeders have been successful in selecting carrot, sweet potato, and tomato cultivars with comparably high carotenoids levels and vitamin A content, onion and tomato cultivars with longer shelf-life, sweet corn cultivars that maintain their sweetness longer after harvest; cantaloupe and watermelon cultivars with higher sugar content and firmer flesh, and pineapple cultivars with higher contents of ascorbic acid, carotenoids, and sugars. An important limitation of most markets is that only certain cultivars or varieties of a commodity are traded and others are unacceptable. These cultivars receive considerably more attention during postharvest stage while the unacceptable cultivars do not, even though they might constitute a large part of the total production. Growers can, thus, make use of the knowledge of genetic make-up of cultivars with proven superior postharvest qualities for shelf life and quality, and increase market share by replacing the existing genotypes with those of the desirable varieties some such varieties are listed in Table 1.

2.2 Temperature and Light

Temperature and light play important role in improving the growth, yield and postharvest quality of fruits, vegetable and flowers. The temperatures at which most physiological processes go on normally in plants range from approximately 0 °C to 40 °C. For most horticultural crops, higher the temperature during the growing period earlier is the time of harvest. Based on this principle, the heat units or degree-days concept has been formulated. Warm days and cool nights are prerequisites for full development of colour at ripe stage in fruits. Tomato grown at high temperature has low TSS content than those grown in lower temperature due to higher respiration rate at higher temperature. Fruit colouration is also influenced by temperature; cool nights with hot dry days favour best colour development. Synthesis of lycopene and colour formation in tomatoes is inhibited at temperatures beyond 30 °C, while, in grapes, there is proanthocyanin accumulation and expression of flavonoid pathway genes in response to diurnal variation in temperature. Curd formation in cauliflower and head formation in cabbage does not take place at temperatures below optimum, while in potato, tuberization does not happen if difference in night and day temperature is not wide.

The intensity and quality of light are both required for optimal plant productivity, quality and harvest index. Availability of light during growth period influences the postharvest eating quality. From nutritional point of view, vitamin-C concentration has direct relation with light intensity in many fruits and vegetables, particularly leafy greens. Tomatoes grown in shade conditions have been found to record higher sugars and dry matter as compared to those grown in open conditions. Another effect of light on fruit quality deals with the red pigmentation of skin, through the influence of light on anthocyanin biosynthesis. Light is a key regulatory factor in reddening of skin in apple and mango.

2.3 Water

Water is indispensable for the growth and development of plants. The availability of water, be it excessive, optimum, or scarce, directly reflects on the yield and quality of plant produce. Severe water stress results in increased sunburn of fruits, irregular ripening of pears, tough and leathery texture in peaches, and incomplete kernel development in nuts. Moderate water stress reduces fruit size and increases soluble solids content, acidity, and ascorbic acid content. On the other hand, excess water supply to plants results in cracking of fruits (*e.g.* cherries, plums, and tomatoes), excessive turgidity leading to increased susceptibility to physical damage (such as oil spotting on citrus fruits), reduced firmness, delayed maturity, and reduced soluble solids content; brittle foliage in leafy vegetables and softer fruits in strawberry with increased tendency to decay. Fruit cracking in litchi and pomegranate is a direct result of dry conditions followed by rain or irrigation. Deficit irrigation or water stress has been reported to reduce yield but increase the soluble solids and postharvest quality in horticultural crops such as tomato, strawberry, ber, kiwifruit and mango. A basic understanding of

Table 1. Different Cultivars of Horticultural Crops with Specific Postharvest Traits

Crop	*Scientific Name*	*Cultivar*	*Attribute (s)*
Apple	*Malus* × *domestica*	Ambri	Long storage life
		Granny Smith	Firmer than most cultivars
Chilli	*Capsicum annuum*	Punjab Lal	Rich in capsaisin, resistant to TMV, CMV and leaf-curl virus
Fig	*Ficus carica*	Conardia, Excel, and Deanna	Suitable for drying and canning
Grape	*Vitis vinifera*	Pusa Navrang	Tenturier variety suitable for wine and juice making
		Thompson Seedless	Suitable for raisin making
Litchi	*Litchi chinensis*	Swarna Roopa	Resistant to fruit cracking
		Rose Scented	Rose flavour
Mango	*Mangifera indica*	Pusa Surya	Red peel, good shelf life
		Ratna, Arka Puneet	Free from spongy tissue
Strawberry	*Fragaria* × *ananassa*	Olympus, Hood, Shuksan	High flavour and bright red colour, suitable for ice-cream making

tree-fruit water balance is important in considering the relationship of plant moisture to fruit quality and shelf life.

2.4 Mineral Nutrition

Nutritional status of fruits, vegetables and flowers is one of the important factors that determine quality of produce at harvest. Optimum plant performance depends on a balanced and timely availability of mineral nutrients that may be limiting in many soils. There can be deleterious effects on storage life and quality of horticultural produce due to deficiencies, excesses, or imbalance of various nutrients during the growing period of the crop. Physiological disorders, as a result of excessive or deficient plant nutrient availability, reduce the quality and marketability of fresh horticultural produce.

High nitrogen content is often associated with reduced postharvest-life due to increased susceptibility to mechanical damage, physiological disorders, and decay in many horticultural crops. Increasing the nitrogen and/or phosphorus supply to citrus trees results in somewhat reduced acidity and ascorbic acid content, while increased potassium fertilization results in increased acidity and ascorbic acid content. High nitrogen also increases the susceptibility of apple flesh to core browning during storage, reduces the shelf life of onion, increases the incidence of soft rot in kiwi fruit and adversely affect the quality of mango fruit by anthracnose disease. High nitrogen application or deficiency of potassium results in inferior fruit quality in grapes, while nitrogen deficiency leads to small fruits with poor flavour. In vegetables, high nitrogen levels delay maturity and increase the incidence of several disorders such as internal browning in tomato, hollow stem of broccoli, increased weight loss during storage of sweet potatoes and soft rots in tomatoes. In general, preharvest treatments with higher dosages of nitrogen are reported to reduce the quality and storage life of the horticultural produce.

Phosphorus is involved in energy transfer, maintenance of cell membranes, and brings about a direct effect on yield and tree health. It affects fruit quality by determining fruit size, firmness, colour and storage potential. The role of phosphorus in storage potential of fruits is attributed to its importance in fruit lipid composition. Excessive phosphorus application can lead to lower soluble solids as well as delayed colour development in citrus. Phosphorus deficiency can lead to delayed maturity, reduction in fruit quality, and decreased disease resistance. Phosphorus application has been found to minimize weight loss, sprouting and rotting of onions.

Among the many plant mineral nutrients, K stands out as a cation having the strongest influence on quality attributes that determine fruit marketability, consumer preference, and the concentration of critically important human health-associated phytonutrients. Fairly high K levels in fruit are often reported to increase red colour and to raise fruit acid level, which can improve fruit taste. However, both high and low levels of potassium have been associated with abnormal metabolism. High levels of potassium have been associated with development of bitter pit in apple. On the other hand, low potassium inhibits lycopene biosynthesis in tomato resulting in delayed development of a full red colour. Application of potassium to citrus trees can affect the shape of the fruits and increase their acidity content.

Calcium is one of the cellular regulators of fruit ripening and deterioration; when calcium is low, these processes proceed more rapidly and fruits are likely to ripen earlier, have greater preharvest drop, more water-core at harvest and develop more scald, brown-core, internal breakdown, internal browning, bitter pit, and rots during and after storage. Of these, only internal breakdown and bitter pit may be the direct symptoms of low calcium. Calcium is often related to fruit softening low calcium levels producing softer fruit, whereas, higher calcium increasing firmness.

Several maladies manifest as a result of excess or deficient macro and micro nutrients. Some important physiological disorders in horticultural crops related to inappropriate nutrition are listed in Table 2.

Table 2. Physiological Disorders in Horticultural Crops Associated with Imbalanced Mineral Nutrition

Element	*Crop*	*Disorder (s)*
Calcium	Apple	Bitter pit, lenticel blotch, cracking, internal breakdown, water core
	Avocado	End spot
	Beans	Hypocotyl necrosis
	Brussels sprout	Internal browning
	Chinese cabbage	Internal tip burn
	Carrot	Cavity spot, cracking
	Celery	Blackheart
	Cherry	Cracking
	Lettuce	Tip burn
	Lemon	Red blotch
	Mango	Soft nose
	Parsnip	Cavity spot
	Pear	Cork spot
	Pepper	Blossom-end rot
	Potato	Sprout failure, tip burn
	Tomato	Blossom-end rot, black seed, cracking
	Water melon	Blossom-end rot
Boron	Apple	Internal Cork
	Apricot	Cracking
	Pear	Internal cork, water core

Source: Snowdon, 2010.

While a number of mineral nutrients are essential for plant growth, not all of them have been related to postharvest fruit quality. Market demands high quality fruit. Mineral levels in the fruit as it grows and after it is harvested are highly significant factors in producing and maintaining fruit quality. Many physiological disorders in horticultural crops are attributed to imbalanced mineral nutrition. Both excessive and deficit application of nutrients may have adverse effects on postharvest life and quality of horticultural produce. Maximum benefits can be achieved through integrated nutrient management (INM) as envisaged in the Good Agricultural Practices (GAPs).

2.5 Plant Growth Regulators

The 'classical' phytohormones, identified during the first half of the twentieth century, are auxin, abscisic acid, cytokinin, gibberellin and ethylene. More recently, several additional compounds have been recognized as hormones, including brassinosteroids, jasmonic acid, salicylic acid and nitric oxide.

Plant growth regulators (PGRs) are organic substances, other than nutrients, which in small quantities promote, inhibit, or otherwise modify various plant physiological processes. Every physiological process in the plant can be said to be the result of a delicate balance between such PGRs. Presently, several commercial applications and recommendations are being made to achieve desired effects in plant science, more specifically in crop sciences to achieve better yields. Commercial formulations of PGRs are used in horticultural crops; and PGRs have become so indispensable that any commercial venture in horticulture would not be possible without the use of plant growth substances. There are numerous examples of growth regulators being used commercially to influence some quality attributes of a product. Some compounds such as Naphthalene Acetic Acid (NAA) are used for thinning blossoms in apples. Others such as gibbrellic acid are used to increase fruit size in grape and to maintain rind integrity in citrus. Ethephon is used as a preharvest spray to induce ripening in some crops such as tomato. In order to achieve desired results, it is very important that the correct PGR should be applied in the required concentration at the right time. The commercially followed pre and postharvest applications of PGRs to improve shelf life and quality of horticultural produce are summarized in Table 3.

Table 3. Use of Plant Growth Regulators (PGR) for Enhanced Shelf life and Postharvest Quality

PGR	*Commodity*	*Concentration*	*Desired Effect*
2,4-D	Citrus, Mango	5-25 ppm	Control of fruit drop
	Limes, Lemon	50-100 ppm	Extending shelf life
	Mandarin	12 ppm	Reduces granulation
Cytokinin	Lime and Lemon	10 and 25 ppm	Extend shelf life
Ethephon	Bael	1000 ppm	Ripening can be advanced by 45 days
	Ber	750 ppm	Induces early maturity
	Date palm	1000 ppm	Advances maturity
	Grapes	100 ppm	Reduce uneven ripening
	Peach cv. July Elberta	300 ppm	Fruit thinning
	Persimon cv. Hachiya	500 ppm	Removal of astringency
	Pineapple	25ppm	Induction of uniform flowering
	Walnut	2000 ppm	Promotes dehiscence of hulls
Ethrel (CEPA)	Apple	1000 ppm	Colour enhancement and enhancement of ripening
Ethrel	Mandarin	50 ppm	Degreening
Ethylene	Avocado	10 ppm	Accelerate ripening
	Banana, Mango,	100 ppm	Uniform ripening Papaya
GA	Grapes	10 ppm	Cluster elongation
	Grapes	40 ppm	Berry thinning
	Grapes	60 ppm	Increasing berry size
	Jamun	60 ppm	Control of flower and fruit drop
	Litchi	40ppm	Reduce fruit cracking
Maleic Hydrazide (MH)	Garlic, Onion	3000 ppm	Sprouting inhibition
NAA	Apple	10-20 ppm	Fruit thinning
	Apple	10 ppm	Control of preharvest fruit drop
	Apricot	25-50 ppm	Fruit thinning
	Grapes	50 ppm	Minimize postharvest berry drop

PGR	Commodity	Concentration	Desired Effect
	Pineapple	10-20 ppm	Induction of uniform flowering
	Pineapple	200-300 ppm	Increases fruit size by 15-20 per cent
Paclobutrazol	Loquat	500 ppm	Increasing fruit size
Promalin	Apple	30-60 ppm	Improving oblong shape of Delicious apples
Putrescine	Peach	5-20 mM	Reduced ethylene production, delayed softening, and retained titratable acidity
Salicylic acid	Sweet cherry	2 mM	Induced disease resistance and reduced disease incidence

Source: Chadha, K.L., *Handbook of Horticulture*, ICAR, New Delhi, 2006.

Plants are used to alter the growth behavior of plants. While vital usefulness of foliar applied growth regulators to fruit trees is generally accepted, however, their use growth regulators does not directly influence fruit composition but may indirectly affect it due to delayed or accelerated fruit maturity.

2.6 Cultural Practices

Soil type, rootstock, mulching, irrigation and fertilization influence the water and nutrient supply to the plant, which can in turn affect the nutritional quality and shelf life of the harvested plant part. Good Horticultural Practices (GHP) ensure good yield, and ultimately the postharvest life and quality of produce. Weeds compete with the main crop for soil moisture, minerals and nutrients, and also act as alternate hosts for diseases and pests. Efficient weed management is an important cultural practice that ensures production of quality horticultural produce. In addition, proper crop hygiene should be followed to reduce disease outbreak and carryover of postharvest pathogen inoculum. High crop loads can reduce the dry matter in some fruits such as kiwi fruit and peach and may affect the taste and flavour in many horticultural crops. Fruit thinning is done to increase fruit size, which, however, decreases the overall yield. Traditional horticultural technologies, using size-controlling rootstocks and adaptive pruning and thinning techniques, are good examples of enabling technologies to optimize productivity (crop load) and quality of fruit. Berry crops are known to have better postharvest fruit quality when allowed to grow under open canopy with maximum air circulation. Efficient training and pruning practices are recommended to permit sunlight penetration into the center and lower canopy areas. Girdling is a popular practice followed in grape to strike a balance between vegetative and fruit growth, and to improve bunch shape and berry size. Girdling is also done to improve fruit size in peaches and nectarines. Soil-borne fungi, bacteria and nematodes can build up to damaging levels with repeated cropping of a single vegetable crop. To overcome this, crop rotation is an effective management strategy to prevent inoculum build-up in and minimize postharvest losses by reducing decay inoculum right from the field. Effective preharvest disease control greatly influences disease incidence and severity during postharvest handling of fruits and vegetables. Cultural practices have direct effect on the final quality of the produce. Cultural practices such as pruning and thinning determine the crop load and fruit size, which can in turn influence the nutritional composition of fruit.

3.0 Postharvest Treatments for Enhanced Storage Life and Quality

3.1 Thermal Treatments

3.1.1 Hot Water Treatment

Thermal treatment has been used for more than a century to free plant materials from pathogens. Postharvest heat treatments are used for insect disinfestations, disease and decay control of fruits and vegetables. These modify commodity responses to other stresses and maintain fruit quality during storage. In mango, hot water treatment is recommended at 50-52 °C for 5 minutes to reduce the fungal infection during ripening or storage. This treatment helps in attainment of uniform ripening within 5-7 days. Hot water dips have generally been utilized for fungal pathogen control, since fungal spores and latent infections are either on the surface or in the first few cell layers under the peel of the fruit or vegetables. Many fruits and vegetables tolerate hot water temperatures of 50 to 60 °C for up to 10 min, but even shorter exposures at these temperatures can control many postharvest plant pathogens.

An innovation in hot water treatment has been the development of a hot water spray machine. This machine is designed to be a part of a packaging line, whereby the commodity is moved by means of brush rollers through a pressurized spray of hot water. By varying the speed of the brushes and the number of nozzles spraying the water, the commodity can be exposed to high temperatures for 10 to 50 sec. Heat treatment also induces cold tolerance in a number of produce including citrus, pepper and tomato. Broccoli senesces vary rapidly and florets turn yellow after harvest as a result of chlorophyll degradation. This yellowing can be delayed by treating harvested broccoli with hot water (45 °C) for 10 min. However, if the temperature and time is not monitored properly at the time of hot water treatment, it may lead to internal injury to fruit such as flesh darkening in avocado, citrus, litchi and nectarine, and poor colour development, abnormal softening, lack of starch breakdown and development of internal cavities in mango and papaya.

3.1.2 Vapour Heat Treatment (VHT)

Vapor heat, as a method of heating fruit with warm air saturated with water vapour at temperatures between 40 and 50 °C to kill insect eggs and larvae, is used as a quarantine treatment before fresh market shipment. Vapour heat treatment (VHT) proves very effective in controlling infection of fruit flies in fruits after harvest. The boxes are stacked in a room, is heated and humidified by injection of steam. The temperature and exposure time are adjusted to kill all stages of insects (egg, larva, pupa and adult), but fruit should not be damaged. VHT, as a quarantine measure, is mandatory for export of mangoes. VHT can also reduce the incidence of anthracnose during postharvest storage of mango. To achieve complete control of anthracnose, it is necessary to apply a combination of VHT (95 per cent RH and above) such that fruit core temperature of 46.5 °C is maintained for 10 minutes.

3.2 Curing

Curing is performed immediately after harvest on root, tuber and bulb crops, in which the produce is exposed to relatively high temperature and relative humidity. In root and tuber crops, curing causes wound healing with the development and suberization of new epidermal tissues, which act as an effective barrier against infection and water loss. In bulb crops, curing is the process of drying of neck tissues and of the outer leaves to form dry scales. If tubers are placed immediately after harvest into cold storage, wound healing is delayed. Onions and garlic can be cured in the field. These crops can be undercut in the field and left there to dry for 5-10 days. The crops can be cured either in field or after packing into large fiber or net sacks. The produce must be checked daily until outer skin and neck tissues have properly dried. The optimum conditions for curing root, tuber and bulb crops are given in Table 4.

Table 4. Optimum Conditions for Curing Root, Tuber and Bulb Crops

Crop	*Temperature (°C)*	*Relative Humidity (per cent)*	*Duration (days)*
Cassava	30-40	90-95	2-5
Onion and Garlic	27-35	60-75	1-2
Potato	15-20	85-90	5-10
Sweet potato	30-32	85-90	4-7
Yam	32-40	90-95	1-4

3.3 Irradiation

Ionizing radiation can be applied to fresh fruits and vegetables to control micro-organisms, insects and parasites, and inhibit/or prevent cell reproduction and some chemical changes. Food irradiation is a physical means of food processing that involves exposing the prepackaged or bulk foodstuffs to gamma rays, X-rays or electrons. Foodstuffs are generally irradiated with gamma radiation from a radioisotope source, or with electrons or X-rays generated using an electron accelerator. Radiation doses are measured in Grays (Gy). This technology is regarded as a mature and safe procedure, although its approval for use is country and product specific. Internationally, food irradiation has been considered a safe and effective technology by the World Health Organization (WHO), the Food and Agriculture Organization (FAO) and the International Atomic Energy Agency in Vienna. The Codex Alimentarius standard provides that 'the maximum absorbed dose delivered to a food shall not exceed 10 kGy except when necessary to achieve a legitimate technological purpose'. A combination of hot water treatment (55 °C for 5 minutes) followed by 30 Gy irradiation has been found to be the best treatment for shelf-life extension and quality maintenance of mango. After this treatment the fruit exhibits a storage life of 38 days (at 15 °C) with no irradiation injury. Irradiation can also be used to control postharvest diseases of other fresh fruits and vegetables.

In Hawaii irradiation at a minimum of 250 Gy was recommended for the commercial control of fruit fly in papaya, preferably combined with hot water treatment. A dose of 5-10 krad is recommended for inhibition of sprouting in onion and potato. The role of irradiation to reduce postharvest losses, control insect pests and microorganisms, prolong shelf life and improve quality have been tested and found beneficial in a number of horticultural produce, *viz.*, apple, cabbage, carrot, grapefruit, lettuce, mango, papaya, potato, strawberries, tomato, *etc.*

In recent years the use of electron beam (E-beam) has increasingly been tested for its efficacy in maintaining fresh quality and extending shelf life of fruit and vegetables. Besides being effective, the technology is safe and efficient with no residue. This technology entails the use of high or low-energy electron beam radiation generated by an electron accelerator, that has strong penetrability and high efficacy of disinfection or reduction in population of common microbes of interest such as *Salmonella*, *E. coli*, and *Listeria* etc. Treatment dosage is vital, and a level between 1-3 kGy has been found suitable for extending shelf life of fruits and vegetables. A higher dose of up to 7 kGy may be required to kill pests and parasites deep in fruits or in seeds. However, the dosage is tissue and purpose driven with reports that dose above 1kGy reducing vitamin-C in grapefruit and also limiting vase life of cut roses. The US FDA has approved three radiation sources for use on foods: gamma rays from radioisotopes such as Cobalt-60 or Cesium-137, x-rays, and electron beam technology.

3.4 Biological Control

Increasing public and scientific concerns about presence of synthetic chemicals in food supply and in the environment have provoked the need to look for safer, effective, and more eco-friendly alternatives for postharvest disease control in horticultural crops. Biological control agents or microbial antagonists have been pronounced in control of various postharvest rot pathogens. The yeast *Candida guillermondi* was shown to be antagonistic towards *Penicillium* spp. and could be successfully incorporated into commercial waxes. *Bacillus* spp. on their own or combined with a fungicide could be used to control the postharvest disease of mango, avocado and litchi. *Bacillus subtilis* and *B. licheniform* isolated from mango leaves and applied to fruit as a warm water dip controlled anthracnose levels. Microbial antagonists exert their biocontrol activity through antibiosis, direct parasitism, competition for nutrients and space, or through induction of natural defense mechanisms, such as phytoalexins. Some established reports on postharvest biological control as given in Table 5.

Table 5. Biocontrol Agents Used for Postharvest Decay Control in Horticultural Crops

Commodity	*Disease*	*Biological Control*
Apple	Blue mould	*Pseudomonas syringae, P. cepacia, Cryptococcus* sp., *Pichia guilliermondii*
	Gray mould	*Pichia guilliermondii, Cryptococcus laurentii, C. flavus*
Apricot	Brown rot	*Bacillus subtilis, Pantoea agglomerans*
Citrus	Green mould	*Pichia guilliermondii, Bacillus subtilis,*
	Blue mould	*Pichia guilliermondii, C. laurentii*
	Sour rot	*Bacillus subtilis, Pichia guilliermondii, Trichoderma* sp.
Grape	Gray mould	*Trichoderma harzianum, Pichia guilliermondii*
	Rhizopus rot	*Candida guilliermondii*
Character spacing/ Nectarine	Brown rot	*Bacillus subtilis*
Pear	Blue mould	*Pseudomonas cepacia*
	Gray mould	*Pseudomonas cepacia, P. gladioli*
	Mucor rot	Cryptococcus laurentii, C. flavus
Plum	Brown rot	*Bacillus subtilis, Rhodotorula* sp.
Potato	Soft rot	*Pseudomonas putida*
Strawberry	Gray mould	*Trichoderma* sp.

3.5 Waxing

Preservation of fruits and vegetables is a major concern for the fresh produce industry. Extending shelf life translates to the extension of transport distances and, therefore, markets. Edible coatings are a simple technology by which fruits and vegetables can be physically protected and have their respiration and, in some cases, ripening regulated as with passive modified atmospheric packaging (MAP). This is because, like plastic packaging, coatings create a modified atmosphere within the fruit or vegetable. Use of coatings has been practiced for decades in the United State on apples and oranges as well as on other crops.

Waxing is, therefore, a common postharvest practice in those fruits and vegetables which shrivel rapidly and lose consumer appeal during storage and marketing. Food grade waxes are used to replace some of the natural waxes removed during harvesting and sorting operations and can help reduce water loss during handling and marketing. It also helps in sealing tiny injuries and scratches on surface of fruits and vegetables. It improves cosmetic appearance by providing shine and prolongs the storage life of fruits and vegetables.

Materials used for edible coatings include lipids, polysaccharides, proteins, resins or combinations of these materials or composite coatings. Most fruits and vegetables have a waxy cuticle that resists moisture loss, but often this is partially removed or disturbed during washing. This necessitates a coating or packaging treatment to prevent water loss and subsequent shriveling and shrinking of fresh produce. Any 'Generally Recognized As Safe' (GRAS) material that is approved for use in coatings without restriction are considered by the U.S. Food and Drug Administration (FDA) to be "edible".

Edible coatings have been found beneficial on a number of fruit and vegetable crops such as apple, capsicum, citrus, cucumber, plum, tomato *etc.* to reduce shriveling and improve textural qualities. Shelf life of kinnow and pear fruits can be extended with citrashine coating (Figure 1). Edible coatings have enabled fruits like citrus and apple to be packed and displayed in supermarkets without packaging.

3.6 Chemical Treatments

3.6.1 Chlorine Solution

Clean water containing an appropriate concentration of sanitizers is required to minimize the potential transmission of pathogens from water to produce, from infected to healthy produce within a single lot, and from one lot of produce to another, over time. Chlorine treatment (100-150 ppm available chlorine) can be used in wash water to help control/reduce microbial inoculum build-up during packing operations. The pH of wash solution should be maintained between 6.5 and 7.5 for best results. Now-a-days, Chlorine dioxide (ClO_2) has been increasingly adopted as an alternative to chlorine for sanitization in the fresh and fresh-cut produce industry. ClO_2 used as a disinfectant has several advantages over chlorine, including larger oxidant capacity, lower reactivity with organic matter, and high

Wax Coating **Without Coating**

Figure 1. Effect of Wax Coating on Appearance of Kinnow Fruits

effectiveness at low concentrations. ClO_2 is effective as an antimicrobial agent against such bacteria as *Salmonella* and *Escherichia coli* over a broad pH range (3 to 8).

3.6.2 Calcium Application

It has been established that the disorders of storage organs of fruits and vegetables appear closely related to low calcium content in tissues. Postharvest application of calcium plays an important role in enhancing the storage and marketable life of fruits by maintaining their firmness and quality. Calcium application delays aging or ripening, reduces postharvest decay, controls the development of many physiological disorders and increases the calcium content, thus improving their nutritional value. Vacuum infiltration of mangoes with 2-8 per cent $CaCl_2$ solution results in delayed softening of fruit during storage at 20 °C by 8-12 days. Calcium develops resistance to fungal attack by stabilizing or strengthening cells walls, thereby, making them more resistant to harmful enzymes produced by fungi. Calcium dips have been used as firming agents to extend postharvest shelf life of a wide range of fruits and vegetables *viz.* apples, strawberries, sliced cantaloupe, sliced pears, sliced strawberries, blueberries, peaches, tomatoes, oranges, and pineapples.

3.6.3 1-Methylcyclopropene

1-Methylcyclopropene (1-MCP) is a powerful inhibitor of ethylene action that has been added to the list of options for extending the shelf life and quality of fruits. 1-MCP is a cyclic olefin, non-toxic, odourless and stable at room temperature. This compound has been manufactured in powder form releasing 1-MCP when diluted with water. It has been approved by Environmental Protection Agency (EPA), and the commercial application of 1-MCP was undertaken by Agro fresh, Inc., a subsidiary of Rohm and Hass. Presently, this chemical is being traded under the name of Smart fresh™ and several countries are reaping the benefits of this compound. The magic gas, l-MCP, has been used as a tool to enhance the shelf life of perishables, as it irreversibly binds to the ethylene receptors, thereby bringing about a conformational change in the receptors, and prevents ethylene effects in plant tissues for extended period. It has been patented in 1996, and, since then, numerous scientific reports have been documented establishing the role 1-MCP plays in delaying the process of ripening and senescence in fresh horticultural produce (Table 6).

3.6.4 Diphenylamine (DPA)

Diphenylamine is a colourless, aromatic amine possessing antioxidant properties. Postharvest drenching of fruits with DPA provides good control against scald, a serious physiological disorder of apple and pear during storage manifested as brown discolouration that makes the fruit unmarketable and limits its storage potential. In addition, postharvest treatment of pear with DPA prevents normal rates of respiration and ethylene production during ripening and the fruits remain hard and green during storage. DPA reduces the chilling-induced pitting of green bell pepper. DPA is commercially used in USA and Europe on apple and pear fruit for improving their storage life and maintaining quality.

3.6.5 Fumigation

The fumigation with sulfur dioxide (SO_2) is successfully used for controlling postharvest diseases of grape. This is achieved by placing the boxes of fruit in a gas tight room and introducing the gas to an appropriate concentration from a cylinder. Special sodium metabisulphite pads are available which can be packed into individual boxes of a fruit to give a slow release of SO_2. The primary function of treatment is to control *Botrytis cinerea*. The storage life of grape packed along with sulfur dioxide generating pads can be extended up to 50 days at 0-2 °C and 90-95 per cent RH.

Table 6. Effects of 1-Methylcyclopropene on Postharvest Storage Behaviour and Quality of Horticultural Produce

Crop	*1-MCP*	*Altered Postharvest Physiology and Biochemistry*
a) Fruits Crops		
Apple	1 μL L^{-1}	Reduced softening, delayed ripening, inhibited superficial scald and decay
	2 μL L^{-1}	Reduced softening, and reduced superficial scald
Apricot	0.5 μL L^{-1}	Decreased respiration rate and tissue softening
Banana	0.01-1.0 μL L^{-1}	Delayed peel colour change and fruit softening, extended shelf life
Ber	1 μL L^{-1}	Delayed senescence, maintained quality, reduced decay
Cherry	1 μL L^{-1}	Maintained firmness, enhanced shelf life
Grape	0.5-1.0 μL L^{-1}	Increased shelf life, and improved quality
Guava	0.6 μL L^{-1}	Improved shelf life, ameliorated chilling injury, and reduced decay
Kiwifruit	0.5 μL L^{-1}	Suppressed ethylene production and fruit softening
Lime	1 μL L^{-1}	Maintenance of green colour of fruit
Mandarin	5 μL L^{-1}	Inhibited chlorophyll degradation and senescence
Mango	2 mg kg^{-1}	Delayed ripening, increased antioxidant enzymes
Peach	10 μL L^{-1}	Inhibited degradation of starch, prolonged shelf life
Pear	0.3-0.6 μL L^{-1}	Lower ethylene production and delayed softening
Pineapple	0.1 μL L^{-1}	Effectively controlled internal browning or black heart
Plum	0.5 μL L^{-1}	Slowed fruit ripening, reduced fruit softening
b) Vegetable Crops		
Bell Pepper	1 μL L^{-1}	Delayed senescence through decreased chlorophyll breakdown
Brinjal	1 μL L^{-1}	Delays senescence, maintains quality and reduces browning
Broccoli	0.6 μL L^{-1}	Reduced loss of green colour and chlorophyll pigments
Carrot	1 μL L^{-1}	Decreased respiration rate and loss of sucrose
Coriander	0.5 g m^{-3}	Increased shelf life of leaves, retarded chlorophyll degradation
Muskmelon	1 μL L^{-1}	Delayed mesocarp softening and reduced water soaked spots
Okra	5 μL L^{-1}	Reduced development of chilling injury and extended storage life
Onion	1 μL L^{-1}	Sprout growth was reduced during storage
Parsley	10 μL L^{-1}	Retard senescence of leaves
Spinach	1 μL L^{-1}	Extended postharvest life, maintained higher chlorophyll
Tomato	5-20 μL L^{-1}	Delayed ripening and increased shelf life
Watermelon	5 μL L^{-1}	Protected against ethylene induced water-soaking
c) Flower Crops		
Carnation	1-20 nL L^{-1}	Inhibited wilting, ethylene effects, and extended vase life four times
Freesia	4 nL L^{-1}	Extended vase life up to 9 days
Lilium	500 nL L^{-1}	Superior quality and inhibition of ethylene response
Phalaenopsis	500 nL L^{-1}	Reduced ethylene induced floral bud drop
Tulip bulbs	200 nL L^{-1}	Inhibited ethylene induced splitting

Source: Watkins, 2008; Schotsmans *et al.*, 2009.

Sulphur dioxide treatment in litchi is commercially practiced by exporters to control pericarp browning. Sulphur dioxide is a well-known inhibitor of polyphenol oxidase and it combines with anthocyanins to form a stable complex. Fumigation with 1.2 per cent SO_2 for 10 minutes was shown to be effective in reducing skin discolouration in fresh litchis, especially if it is combined with a 2 minute dip in 1N HCl directly afterwards. Paper pads or wraps impregnated with biphenyl fungicides are commonly applied to citrus fruits. The chemical vaporizes slowly protecting the fruit from fungal infection.

Despite the good efficacy of SO_2 treatment against decay pathogens, there has been growing concerns against the use of SO_2 in food due to the health issues related to sulfite residues. Concerted efforts are therefore needed to find safer alternatives to SO_2 fumigation for overcoming decay during postharvest storage of horticultural produce.

3.6.6 Sprout Suppressants

In potatoes, a range of chemicals have been shown to suppress sprouting. One such chemical, CIPC (3-chlor-iso-propylphenyl carbamate) is commercially used as a

CIPC Treated Potato

Untreated Potatoes

Figure 2. Effect of Postharvest Application of CIPC on Sprouting of Potato

sprout suppressant, which is vaporized and applied as a thermal fog into the stored potato. A periodic fogging of CIPC @ 40 ml/tonne at an interval of 60 days was found optimum for storage of potato at 10±1 °C for 6 months (Figure 2). The potatoes stored at high temperature with periodic CIPC fogging recorded low reducing sugars and negligible sprouting. Sprout suppression in onion bulbs is achieved by preharvest application of maleic hydrazide (MH). This is applied to leaves of the crop some two weeks before harvesting.

3.6.7 Gibberellic Acid

Gibberellins (GAs) possess the capacity to stimulate growth in many plants. The benefit of the activity of GAs has brought about major advances in the field of physiology and applied plant science, and the use of GAs constitutes one of the most important tools in modern horticulture industry. The role of GAs in delaying the process of senescence in plants and plant tissues is well established. Postharvest application of gibberellic acid (GA_3) has been popularly used as a tool to delay the process of ripening and senescence in various fruits, vegetables and flowers. The anti-senescence effect of GAs is manifested through their effect on delay of colour changes, alleviation of chilling injury, reduction in microbial decay, and increase in storage life of fresh produce. The beneficial use of GA_3 as a postharvest application in various horticultural produce is presented in Table 7.

Table 7. Postharvest Application of GA_3 for Enhanced Shelf Life and Quality

Crop	*Concentration*	*Altered Postharvest Physiology and Biochemistry*
a) Fruit Crops		
Banana	1.5 g kg^{-1}	Delay of fruit ripening
Cherry	10 mg L-1	Higher fruit firmness and lower fungal infection
Grapefruit	100 ppm	Maintained peel quality, reduced decay, and increased storage life
Guava	100 ppm	Delayed onset of senescence and maintained fruit quality
Mango	200 mg L-1	Retarded ripening of fruits
Papaya	100 ppm	Delayed ripening and improved shelf life
Peach	100 mg L-1	Maintained fruit firmness and delayed ripening
Pear	2000 µg mL-1	Improved the efficacy of biocontrol against blue mold rot
	50 ppm	Improved palatability and sensory quality of fruits
Persimmon	200 µg mL-1	Induced resistance to black spot caused by Alternaria alternate
Sapota	150 ppm	Maintained fruit quality and improved shelf life
b) Vegetable Crops		
Celery	20 µg mL-1	Increased resistance to pathogens; inhibited decay
Tomato	0.1 per cent	Extended shelf life up to 18 days
c) Ornamental Crops		
Alstroemeria	100 ppm	Inhibited chlorophyll breakdown and improved vase life
Gerbera	50 mg L-1	Enhanced postharvest quality and vase life of cut flowers
Rose	1 mM	Suppressed the development of Botrytis blight

3.6.8 Jasmonates

'Jasmonates' are a class of endogenous plant growth regulators that have unique and potentially useful properties that affect plant growth and development in response to environmental stresses. Jasmonate is a broad term covering numerous linolenic acid-derived compounds. Postharvest application of MeJA as a gas can activate defences and can positively modify postharvest physiology in a number of crops. Some research on of MeJA on horticultural crops has focused on preharvest and postharvest treatments to protect against microbial development on harvested tissue. Postharvest exogenous application of MeJA has also been reported to suppress fruit decay, improve quality, and enhance storage life of a number of horticultural crops such as grapefruit, peach, raspberry, and strawberry. Application of MeJA positively reduces the rate of chilling injury in avocado, grapefruit, guava, pepper, pomegranates *etc.* through enhanced antioxidant potential and reduced ion leakage. Applying 200-600 μM MeJA to cut roses (*Rosa hybrida* L.), through the pulsing solution, or as a spray, reduced the incidence and severity of *Botrytis* and enhanced flower colour persistence.

3.6.9 Ethylene Gas

Ethylene, a gas at physiological temperature, is a natural hormone involved in fruit ripening. The autocatalytic production of endogenous ethylene is considered to immediately trigger off the ripening process in fruits like banana, mango, papaya *etc.* These fruits are commercially ripened by traders with calcium carbide popularly known as *Masala*. This practice is banned under U/S 44-AA of Prevention of Food Adulteration (PFA) Act, 1954 with PFA rules, 1955, due to health reasons. The ripening of fruits with ethylene gas is a safe and internationally acceptable technique. the fruits are exposed to low level of ethylene gas (100 ppm) in an air-tight ripening chamber for 24 hours to induce the fruits to ripen. The most important thing in this technique is temperature and relative humidity control inside the ripening chamber, which should range between 16–24 °C and 90-95 per cent RH, depending upon the fruit type. This technique of fruit ripening is approved by Food Safety and Standard Authority of India (FSSAI), Ministry of Health and Family Welfare. It guarantees uniform ripening with development of good flavour and colour.

3.7. Ozone

Being a natural-occurring oxidizing gas, ozone has a long history of safe use in water treatment in commercial, industrial and public health applications. Studies into application of ozone for postharvest shelf life extension in edible horticultural commodities have gathered momentum over the years. Residual issues and health concerns over use of chlorinated disinfectants, and regulatory acceptance of ozone to qualify for Generally Recognized as Safe (GRAS) in food processing have contributed towards this interest in studying applications of ozone in postharvest horticulture. The US FDA permits the use of ozone as in antimicrobial agent for treatment, storage, and processing of foods in both gas and aqueous forms in direct contact with foods, including fresh and minimally processed fruits and vegetables. Typical ozone use rates for disinfection of postharvest water are 2-3 ppm and although solubility of ozone in water is rather low, chilling the water and lowering the pH increases the solubility. Ozone rapidly attacks bacterial cell walls and is more effective than chlorine with contact time for antimicrobial action at least 4-5 times less than those for chlorine. Besides, use of ozone does not result in production of deleterious by-products. Commercial use of ozone for postharvest disease control has already occurred in horticultural commodities such as apples, cherries, carrots, garlic, kiwifruit, onions, plums, peaches, potatoes, and table grapes. Significant control of *Listeria monocytogenes* has been reported in cold rooms treated with gaseous ozone. Home appliances for surface disinfection of fruits and vegetables using ozone are already commercially available with wide acceptance among masses.

3.8. Ultraviolet (UV) Light

The use of Ultra-violet rays for sterilization and disinfection of bio-safety cabinets and microbiological/ pathological workstations has been commercialized worldwide. Also, UV filters are being widely used in commercial and domestic water purification systems. Easy acceptability of UV lies in it being a non-thermal method of processing that is cost effective, chemical free, eco-friendly and sustainable. UV radiation has a wavelength range varying from 100-400 nm for food processing, out of which the UV-C (200-280 nm) is called the germicidal range for being effective against inactivation of bacteria and viruses. UV rays in continuous and pulsed modes offer tremendous promise for microbiological safety and enhanced functionality of fresh horticultural commodities. Commercially available UV sources include mercury lamps, pulsed lamps, and light emitting diodes (LED). Effective or delivered germicidal UV dose for microbial inactivation is dependent on the total radiant power incident from all directions (fluence or pulse rate) multiplied by exposure time.

UV finds applications in postharvest horticulture through air treatment of storage rooms, cold rooms, and warehouses *etc.* where air disinfection helps in creation of space free of spoilage and disease-causing bacteria, yeasts, and fungi. Fruit, vegetables and the floriculture industry require large volume of water, and use of UV water disinfection not only represents use clean water for sanitation and hygiene but also promises recycling or re-use of wash water. Disinfection of non-food surfaces that come into contact with horticultural commodities also represents an area that benefits from UV technology. This includes disinfection of packaging films and containers, ceilings, walls, floors, and equipment parts such as tanks, cooling coils, knives, conveyors belts, *etc.* UV treatment not only provides benefits such as induction of resistance to postharvest rots, quality maintenance and shelf-life extension of fresh horticultural produce but it is also increasingly being tapped for UV pasteurization of processed foods such as fruit juices, and fresh-cut produce. Reduction in microbial population and improved quality during storage has been reported in zucchini, bell peppers, cantaloupes, watermelons, grapefruit, sweet cherry, grapes, persimmons, apples, pear, mango, pineapple, pomegranate, peach, strawberry, broccoli, tomato, mushrooms, and several other fruits and vegetables.

3.9. Ultrasound

Ultrasound is another non-thermal technology being widely tested for use in microbial safety and increasing shelf-life of fruits and vegetable produce. Besides, sound waves are generally perceived as safe, non-toxic, and environmentally friendly. Ultrasound refers to pressure waves with frequency of 20 kHz or more, and ultrasound equipment uses frequency ranging from 20kHz to 10 MHz. Higher-power ultrasound at lower frequencies (20 - 100 kHz), referred to as 'power ultrasound' is used in food processing to inactivate microorganisms. However, most published data indicates that antimicrobial efficacy of ultrasound is rather low and the best results have been achieved in combination with other safe techniques as is followed in hurdle technology. The principle behind ultrasonication lies in the ability of ultrasound for intracellular acoustic cavitation that can increase thinning and permeability of membranes, and production of free radicals. Removal and decontamination of dirt and microbial inoculum from fruit and vegetables is normally achieved through washing with chlorinated water or added chemicals. Promising results have already been achieved in microbial reduction in the wash water decontamination process in fruits and vegetables namely plums, strawberry, apples, lettuce, bell pepper, shredded carrots, cherry tomatoes, spinach, truffles *etc.* Recently, tomato juice exposed to ultrasonic field with frequency of 20 kHz, with intensity of 40 W cm^{-2} for 10 minutes was shown to be microbiologically pure and free from spoilage microorganisms, thereby, indicating the possible commercialization of ultrasound technology for juice processing. Similar results have also been reported in strawberry juice. Similar Ultrasound also finds application in non-destructive estimation of fruit and vegetable quality.

4.0 Storage

The goals of postharvest research and extension are to maintain quality and safety and minimize losses of horticultural crops and their products between production and consumption. Providing optimal ranges of temperature and relative humidity (RH) is the most important tool for maintaining quality and safety of intact and fresh-cut fruits and vegetables, herbs, and ornamental crops. Thus, increased precision in temperature and RH management is indispensable for quality maintenance from harvest to final consumption. There is no substitute to maintaining the cool chain throughout the postharvest handling system for maintaining quality and safety of horticultural perishables. All other postharvest technology procedures are supplements to proper temperature management. The optimum storage conditions for horticultural produce are summarized in Table 8.

Table 8. Recommended Temperature, RH and Approximate Storage Life of some Important Fruits, Vegetables and Flowers

Commodity	*Temperature (°C)*	*Relative Humidity (per cent)*	*Approx. Storage Life*
a) Fruit Crops			
Apple	–1.0-4.0	90-95	1-12 months
Apricot	–0.5-0	90-95	1-3 weeks
Banana	13.0-14.0	85-90	1-4 weeks
Ber	7.5	90-95	2 weeks
Cherries	–1.0 to -0.5	90-95	2-3 weeks
Coconut	0-1.5	80-85	1-2 months
Fig	–0.5-0	85-90	7-10 days
Grapefruit	10-15	85-90	6-8 weeks
Grape	–1.0 to -0.5	90-95	1-6 months
Guava	5-10	90	2-3 weeks
Kiwifruit	–0.5-0	90-95	3-5 months
Lemon	10-13	85-90	1-6 months
Lime	9-10	85-90	6-8 weeks
Litchi	1.5	90-95	3-5 weeks
Mandarin, Kinnow	4-5	90-95	2 months
Mango	13	85-90	2-3 weeks
Papaya	7	85-90	1-3 weeks
Peach	–0.5-0	90-95	2-4 weeks
Pears	–1.5 to -0.5	90-95	2-7 months
Pineapple	7-13	85-90	2-4 weeks
Plum	–0.5-0	90-95	2-5 weeks
Pomegranate	5	90-95	2-3 months
b) Vegetable Crops			
Asparagus	0-2	95-100	2-3 weeks
Beans, green	4-7	95	7-10 days
Beet	0	98-100	10-14 days
Broccoli	0	95-100	10-14 days
Brussels sprout	0	95-100	3-5 weeks
Cabbage	0	98-100	3-6 weeks
Carrot	0	98-100	4-6 weeks
Cassava	0-5	85-90	1-2 months
Cauliflower	0	95-98	3-4 weeks
Celery	0	98-100	2-3 months
Cucumber	10-13	95	10-14 days

Commodity	Temperature (°C)	Relative Humidity (per cent)	Approx. Storage Life
Eggplant	10-11	90-95	2-3 week
Garlic	0	65-70	6-7 months
Ginger	13	65	6 months
leafy Greens	0	95-100	10-14 days
Lettuce	0	98-100	2-3 weeks
Cantaloupe	0-2	95	5-14 days
Watermelon	10-15	90	2-3 weeks
Mushroom	0	95	3-4 days
Okra	7-10	90-95	7-10 days
Onion (green)	0	95-100	3-4 weeks
Onion (dry)	0	65-70	1-8 months
Peas (green)	0	95-98	1-2 weeks
Chilli (green)	0	95-98	40 days
Potato		90-95	5-10 months
Pumpkin	10-13	50-70	2-3 months
Radish	0	95-100	2-4 months
Spinach	0	95-100	10-14 days
Sweet potato	13-16	85-90	4-7 months
Tomato	8-10	90-95	4-7 days
Turnip	0	95	4-5 months
Yams	16	70-80	6-7 months
c) Flowers			
Alstroemeria	0-1	Sensitive	1 week
Anthurium	12.5-20.0	Insensitive	1-2 weeks
Gypsophila	0-1	Sensitive	1- 3 weeks
Bird of Paradise	6-7	Insensitive	4 weeks
Carnation	0-1	Sensitive	2-4 weeks
Chrysanthemum	0-1	Insensitive	1-3 weeks
Freesia	0-1	Insensitive	–
Gerbera	0-1	Insensitive	1 week
Gladiolus	0-1	Insensitive	1 week
Iris	0	Insensitive	1 week
Lily (Asiatic)	0-1	Sensitive	Up to 4 weeks
Orchids	0-12.5	Sensitive	–
Rose	0-1	Sensitive	1-2 weeks
Snapdragon	0-1	Sensitive	7-10 days
Tuberose	0	Insensitive	1 week
Tulip	0-1	Insensitive	Up to 7 weeks

Sources for compiling this table include Kader *et al.* USDA Handbook 66, 1992.

* Ethylene Sensitivity.

Every fresh commodity has an optimum temperature at which it must be stored to achieve maximum shelf life. The quicker the temperature is reduced to the optimum storage temperature, the longer its storage life. The process of removal of field heat from the produce is called precooling. Depending on the perishability, nature of the produce, and cost factor, cooling can be achieved through room cooling, forced-air cooling, hydro cooling, vacuum cooling, package icing *etc.* (Figure 3).

Figure 3. Effect of Shade on Quality of Capsicum

5.0 Conclusion

Despite modern interventions and the global introduction of good agricultural practices, postharvest losses remain unacceptably high in both developed and developing countries. New innovative technologies are being developed but effective implementation often remains a hurdle. Much is said about postharvest technology but little has been effectively integrated and implemented, and food security remains a challenge as before. Addressing this requires a paradigm shift in how the problem is viewed and handled.

Assuring optimum postharvest life and quality of horticultural produce is dependent on several pre and postharvest factors. Any produce that is of excellent quality at the time of harvest will not store well if not provided with proper postharvest management practices. Similarly, the best postharvest management technologies would be futile if the produce is of poor quality at the time of harvest. Thus, in order to achieve maximum shelf life and maintain quality of horticultural produce, an integrated approach involving both pre and postharvest factors should be taken into consideration. Best results are obtained when crops are grown under sustainable management techniques, harvested at the right maturity, and provided appropriate combination of postharvest management techniques during subsequent storage and distribution. The storage and marketing of high quality produce requires attention to all aspects of the system, from site and cultivar selection, through all cultural practices during production, to harvest maturity, selection and field handling, grading, packing, temperature management as well as distribution. Produce can be maintained and marketed at high price only if it is of high quality at harvest, and if the postharvest facilities are managed to a high standard.

6.0 Important References

Arvanitoyannis, I.S., Stratakos, A.C., and Tsarouhas, P. 2009. Irradiation applications in vegetables and fruits: A review. *Cri. Rev. Food Sci. Nutri.* **49**: 427-462.

Baldwin, E.A. 2003. Coatings and other supplemental treatments to maintain vegetable quality, p. 413-435. In: J.A. Bartz and J.K. Brecht (eds.). *Postharvest Physiology of Vegetables*, 2nd Edition. Marcel Dekker, Inc., New York, NY.

Cohen, S.D., Tarara, J.M., Gambetta, G.A., Matthews, M.A., and Kennedy, J.A. 2012. Impact of diurnal temperature variation on grape berry development, proanthocyanin accumulation, and the expression of flavonoid pathway genes. *J. Exp. Bot.* **63**: 2655-2665.

Cui, N., Du, T., Kang, S., Li, F., Zhang, J., Wang, M., Li, Z. 2008. Regulated deficit irrigation improved fruit quality and water-use efficiency of pear-jujube trees. *Agric. Water Management*, **95**: 489-97.

Daane, K.M., Johnson, R.S., Michailides, T.J., Crisosto, C.H., Dlott, J.W., Ramirez, H.T., Yokota, G.T. and Morgan, D.P. 1995. Excess nitrogen raises nectarine susceptibility to disease and insects. *California Agric.* **49**: 13-17.

El-Samahy, S.K., Youssef, B.M., Aaskar, A.A. and Swailam, H.M.M. 2000. Microbiological and chemical properties of irradiated mango. *J. Food Safety*, **20**: 139–56.

Fallik, E., Grinberg, S., Alkalai, S., Yehkutieli, O., Weisblum, A., Regev, R., Beres, H and Bar Lev, E. 1999. A unique rapid hot water treatment to improve storage quality of sweet pepper. *Postharvest Biol. Technol.* **15:** 25-32.

Gomez-Lopez, V. M., Rajkovic, A., Ragaert, P., Smigic, N. and Devlieghere, F. 2009. Chlorine dioxide for minimally processed produce preservation: a review. *Trends Food Sci. Technol.* **20**: 17–26.

Jin, P., Zheng, Y.H., Cheng, C.M., Gao, H.Y., Chen, W.X., and Chen, H. J. 2006. Effect of methyl jasmonate treatment on fruit decay and quality in peaches during storage at ambient temperature. *Acta Hortic.* **712**: 711-16.

Kader, A.A. 1992. *Postharvest Technology of Horticultural Crops*. 2nd Edition. University of California, Division of Agriculture and Natural Resources. Publication, pp. 3311.

Kays, S.J. 1999. Preharvest factors affecting quality. *Postharvest Biol. Technol.* **15**: 233-47.

Kester, J.J. and Fennema, O. 1986. Edible films on coatings: A review. *Food Technol.* **40**: 47-59.

Léchaudel, M. and Joas, J. 2007. An overview of preharvest factors influencing mango fruit, quality and postharvest behaviour. *Braz. J. Plant. Physiol.* **19**: 287-98.

Lurie, S. 1998. Postharvest heat treatments. *Postharvest Biol. Technol.* **14**: 257-69.

Mahajan, B.V.C., Dhatt, A.S. and Dhillon, W.S. 2004. Effect of prestorage treatments on the quality and storage life of Asian pear. *Indian J. Hortic.* **61**: 324-26.

Mahajan, B.V.C., Dhatt, A.S., Sandhu, K.S. and Garg, A. 2008. Effect of CIPC on the storage and processing quality of potato. *J. Food Agri. Environ.* **6**: 34-38.

Pearce, B.D., Grange, R.I., and Hardwick, K. 1993. The growth of young tomato fruit. I. Effects of temperature and irradiance on fruit grown in controlled environments. *J. Hortic. Sci.* **68**: 1-11.

Purvis, A.C. 2002. Diphenylamine reduces chilling injury of green bell pepper fruit *Postharvest Biol. Technol.* **25**: 41-48.

Rohwer, C.L., and Erwin, J.E. 2008. Horticultural applications of jasmonates: A review. *J. Hortic. Sci. Biotechnol.* **83**: 283-04.

Sayyari, M., Babalar, M., Kalantari, S., Martinez-Romero, D., Guillen, F., Serrano, M., and Valero, D. 2011. Vapour treatments with methyl salicylate or methyl jasmonate alleviated chilling injury and enhanced antioxidant potential during postharvest storage of pomegranates. *Food Chem.* **124**: 964-70.

Schotsmans, W.C., Prange, R.K., and Binder, B.M. 1-Methylcyclopropene: mode of action and relevance in postharvest horticulture research. *Hortic. Rev.* **35**: 263-13.

Sisler, E.C., and Blankenship, S.M. 1996. Method of counteracting an ethylene response in plants. U.S. Patent 5518988.

Sisler, E.C., and Serek, M. 1997. Inhibitor of ethylene responses in plants at the receptor level: Recent developments. *Physiol. Planta.* **100**: 577-82.

Snowdon, A.L. 2010. *Postharvest diseases and disorders of fruits and vegetables*. Manson Publishing Ltd. London, UK.

USDA, 2014. *The Commercial Storage of Fruits, Vegetables, and Florist and Nursery Stocks*. Agriculture Handbook Number 66 (HB-66). Agricultural Research Service, Beltsville Area/BARC. http: //www.ba.ars.usda.gov/hb66//

Watkins, C.B. 2006. The use of 1-methylcyclopropene (1-MCP) on fruits and vegetables. *Biotechnol. Adv.*, **24**: 389-09.

Chapter 6

Postharvest Diseases of Horticultural Crops

R.D. Rawal

Former Head, Division of Plant Pathology, ICAR-Indian Institute of Horticultural Research, Hessaraghatta, Bengaluru – 560 089, Karnataka, India

E-mail: rawal.rd@gmail.com

Jyotsana Sharma

Principal Scientist (Plant Pathology), ICAR-National Research Centre on Pomegranate, NH-65, Solapur-Pune Highway, Kegaon, Solapur – 413 255, Maharashtra, India

E-mail: jyotisharma128@yahoo.com

Manjunatha N

Senior Scientist (Plant Pathology), ICAR-National Research Centre on Pomegranate, NH-65, Solapur-Pune Highway, Kegaon, Solapur – 413 255, Maharashtra, India

E-mail: manju.ars@rediffmail.com

1.0 Introduction

Horticultural crop production in India has revolutionized agriculture sector during the last two decades. Postharvest handling and associated losses due to diseases, however, are major challenges in storage, transport and export. Diseases may reduce quantity as well as quality of the harvested produce. The extent of postharvest losses in horticultural crops depend not only on postharvest handling and storage conditions but also on pre harvest conditions of the crop in the field. Horticultural practices adopted including cultivation practices, plant protection measures, plant nutrition and environmental conditions prior to harvest contribute significantly to postharvest losses due to diseases. Virulence of the causal organism, and susceptibility of the commodity (the host), the storage atmosphere (temperature, relative humidity and composition of the storage air/atmosphere), stage of produce maturity, postharvest hygiene adopted, extent of damage (cuts/bruises/punctures) caused to produce during handling and treatment methods, all contribute to the extent of losses due to postharvest diseases. The most important factors in maintaining quality and extending the shelf-life of fruit and vegetables after harvest are temperature and relative humidity. Most of the physical, biochemical, microbiological and physiological reactions contributing to deterioration of produce quality are largely dependent on temperature and humidity. Due to their high moisture content and nutrient composition, fruits and vegetables are very susceptible to attack by different pathogens. Inoculum loads (population) of the pathogen on the host and in the storage, atmosphere are other important factors responsible for postharvest losses. Among different postharvest pathogens fungi and bacteria are major etiological agents. Fungi species of *Penicillium, Aspergillus, Geotrichum, Botrytis, Rhizopus, Mucor, Fusarium, Alternaria, Colletotrichum, Lasiodiplodia, Phytophthora, Pythium* and *Phomopsis* and bacterial species of *Erwinia, Pseudomonas, Bacillus, Lactobacillus* and *Xanthomonas* are potential postharvest pathogens. A single species of fungus like *Rhizopus stolonifer* is able to infect more than three hundred species of fruits, vegetables and other plants. Viral agents are not the major cause of postharvest diseases; however, some pre harvest viral pathogens may develop faster in the storage conditions.

Although it is very difficult to determine the full extent of postharvest losses due to decay of fruits and vegetables, which varies widely with commodity, production area, season, growth conditions, storage conditions and length of storage, it is well known that these losses are significant, particularly in under-developed and tropical countries. In India, postharvest losses have been estimated to be more than thirty per cent (30 per cent). This not only results in negative impact on our economy but also hampers the efforts in providing food and nutritional security to the masses.

A number of fungal genera such as species of *Penicillium, Alternaria* and *Fusarium*, are known to produce mycotoxins under specific conditions. These toxins have greatest risk when affected produce is used in the production of processed food. Today, the consumers are health conscious and fruit or fruit products are in great demand. Consumers require products that are not only of optimal nutritional quality but are also safe for human consumption, hence, postharvest handling specially treatments to control postharvest diseases assume special significance. It is therefore, of utmost importance to work out effective and economically viable management strategies to reduce postharvest losses.

Postharvest diseases may be categorized as (i) latent/quiescent infections and (ii) those that arise due to infections during or after harvest. In quiescent infections, disease initiates before harvest in the field, but remains dormant for some time in storage. With biochemical changes in the stored produce, the pathogen again becomes active, leading to infection. *Colletotrichum* sp. and *Botrytis cineraria* are two such postharvest pathogens of several commodities. The second type of infection is caused by pathogens like *Penicillium, Rhizopus and Erwinia* spp. We summarize here major postharvest diseases of commercial fruits and vegetables and their management.

2.0 Postharvest Diseases of Fruits

A wide range of microorganism's attack fruits during postharvest period (Table 1). Among different pathogenic microorganisms, fungi are the major organisms deteriorating fruits after harvest. Bacterial pathogens are not common in fruits due to their acidic nature. Postharvest diseases of some important crops are discussed below.

Table 1. Postharvest Diseases of Fruit Crops and their Pathogens

Crop	*Disease*	*Causal organism*
Apple	Soft rot	*Penicillium expansum*
	Bitter rot	*Glomerella cingulata*
	Whisker's rot	*Rhizopus arrhizus*
	Core rot	*Alternaria alternata*
	Pink mould rot	*Trichothecium roseum*
Banana	Anthracnose	*Collectotricum musa*
	Cigar end rot	*Verticillium theobromae*
	Fusarium fruit rot	*Fusarium moniliformae*
	Botryodiplodia fruit rot	*Botryadiplodia theobromeae*
	Pink mould rot	*Trichothecium roseum*
Citrus	Blue mould rot	*Penicillium italicum*
	Green mould rot	*Penicillium digitatum*
	Stem end rot	*Diplodia natalensis*
	Anthracnose	*Colletotrichum gloeosporioides*
	Brown rot	*Phytophthora nicotianae*
	Alternaria fruit rot	*Alternaria citri*
	Aspergillus rot	*Aspergillus niger*
Grape	Grey mould rot	*Botrytis cinerea*
	Blue mould rot	*Penicillium digitatum*
	Cladosporium rot	*Cladosporium herbarum*
	Bitter rot	*Greenaria uvicola*
Guava	*Colletotrichum* fruit rot	*Colletotrichum gloeosporioides*
	Fruit canker	*Pestalotiopsis psidii*
	Phytophthora fruit rot	*Phytophthora nicotianae* var. *nicotianae*
	Macrophoma fruit rot	*Macrophoma allahabadensis*
	Botryodiplodia fruit rot	*Botryodiploidia theobromae*
Mango	Stem end rot	*Botryodiplodia theobromae*
	Anthracnose	*Colletotrichum gloeosporioides*
	Black mould rot	*Aspergillus niger*
	Soft rot	*Rhizopus arrhizus*
Papaya	Anthracnose	*Colletotrichum gloeosporioides*
	Rhizopus rot	*Rhizopus stolonifer*
	Ascochyta rot	*Ascochyta caricae*
	Phomopsis fruit rot	*Phomopsis caricae* pv. *papayae*
	Marcrophomina fruit rot	*Marcrophomina phaseolina*
	Fusarium fruit rot	*Fusarium solani*
Pineapple	Stem end rot	*Ceratocystis paradoxa*
Pome-granate	*Phomopsis* fruit rot	*Phomopsis punicae*
	Soft rot	*Rhizopus arrhizus, R. stolonifier*
	Black Mould	*Aspergillus niger*
	Black Blotch	*Botryodiplodia theobromae*

Crop	Disease	Causal organism
	Dry rot	*Colletotrichum gleosporioides, Botryodiplodia theobromae, Coniella granati.*
	Heart Rot	*Alternaria alternata*
	Fruit rots	*Cladosporium oxysporum, Aspergillus flavus, A. niveus, A. versicolor, Penicillium chrysogenum*
Straw-berry	Grey Mould	*Botrytis cinerea*
	Anthracnose	*Colletotrichum fragerae*

2.1 Apple

More than 20 fungal pathogens have been found associated with postharvest losses in apple from Himachal Pradesh only. Losses between 10 to 18 per cent have been reported. The calculated yearly losses due to fungal pathogens ranged from 3.06 to 20.6 thousand tonnes in Himachal Pradesh during 1969-1977. The cumulative loss for the decade was thus estimated at 89.150 thousand tones. The major postharvest pathogen of apple (Figure 1) is *Penicillium expansum* contributing to major losses followed by other pathogens discussed below.

2.1.1 Soft Rot

This disease is incited by *Penicillium expansum* and is responsible for major postharvest losses in India. In the early stages, the rotten area appear as light in colour and soft and watery in texture. The rot advances rapidly. In humid atmosphere, grey blue cushions of fruiting bodies of the pathogen appear on the diseased skin. Moreover, the attack of *P. expansum* may lead to the accumulation of patulin - a mycotoxin ingestion of which may cause serious health problems.

2.1.2 Bitter Rot

This disease is incited by *Glomerella cingulata* (Ston) Spaud. and Schrenk. The disease is particularly severe on Golden Delicious, however, Ruspippin and Yellow Newton varieties are also infected occasionally. The disease appears as a faint, light brown discolouration beneath the skin. The discolouration increases in surface diameter and extends in a cone shaped manner. The rough, circular lesion becomes depressed. Tiny, raised, black dots appear beneath the cuticle. They give rise to acervuli which break the cuticle. In moist weather, sticky, pink masses of spores are seen which may be arranged in a clearly defined rings. Infected fruits of Richard Red and Golden Delicious show decrease in quantities of amino acid, organic acid and carbohydrates. Trisodium phosphate 2 per cent solution can check the disease in storage.

2.1.3 Whisker's Rot

This disease is caused by *Rhizopus arrhizus* Fischer. The disease pathogen incites a soft rot. The flesh becomes sour in taste without any undesirable odour.

2.1.4 Core Rot of Apple

The core rot incited mainly by *Alternaria alternata*, disease is commonly found in apples having open pore at the calyx end. Typical rot symptoms of the disease are nearly round; brown to black lesions, often centered around a skin break or weakened tissue. The spots are firm, dry and shallow. The surface of spots becomes dark brown to black in the advanced stages, the rotten tissue becomes spongy and the affected flesh turns black.

2.1.5 Pink Mould Rot

This rot incited by *Trichothecium roseum* Link. Ex Fr. is prevalent and destructive. The internal infection becomes evident by a pinkish fluffy growth of the fungus. The rot is usually found associated with scab. In the beginning, rot appears as sunken; brown bands about 1/8 inch wide encircling the scab spots. On the rotten area, fungal threads appear and under moist warm conditions pink spore masses are produced. The second or final stage of rot is characterized by chocolate brown, sunken area with an irregular outline. The rotten area remains firm and dry. The affected tissue has a bitter taste. In the absence of apple scab, the rotting begins from any point of injury on the fruit surface. The rotten flesh becomes musty. The seed cavity contains a pink spore mass.

(a) Bitter Rot (http://www.ipm.iastate.edu)

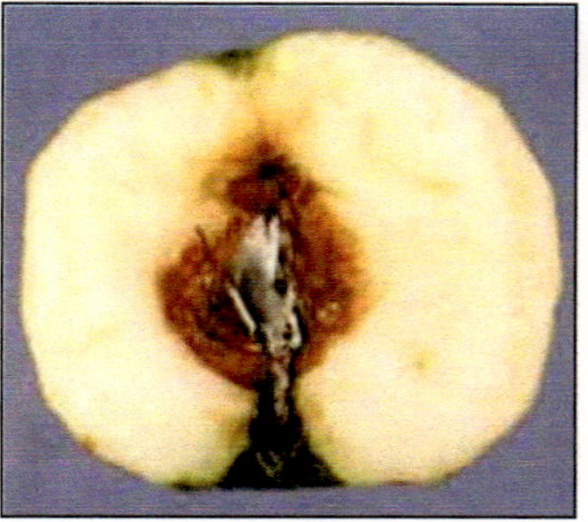

(b) Core Rot (http://postharvest.tfrec.wsu.edu)

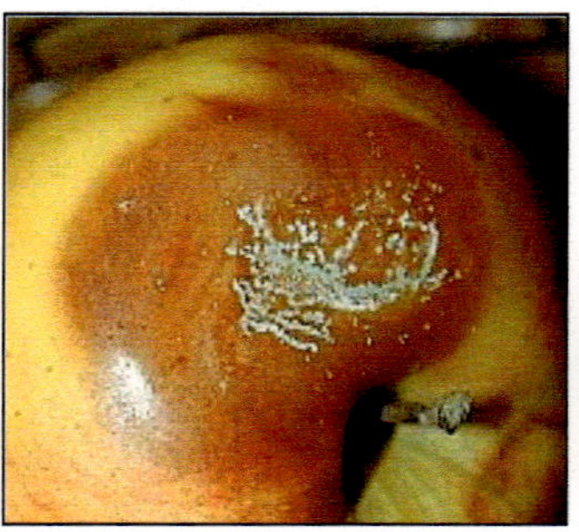

(c) Soft Rot (http://botit.botany.wisc.edu)

(d) Whisker Rot (http://postharvest.tfrec.wsu.edu)

Figure 1. Postharvest Diseases of Apple

2.1.6 Management of Postharvest Diseases of Apple

Postharvest diseases in apple are due to either contamination or incipient infections that occur during the growing season or due to infection through injuries attributable to harvesting, processing, packing and transporting of the produce. The skin injury should be avoided by observance of stringent precautions in picking, grading and packing. The diseased and injured fruits should be sorted out and discarded. Over ripe fruits should not be stored. Fruits should be placed in cold storage after packing. To control the postharvest diseases the fruit temperature should be kept below 10 °C. Growth and spread of the rots are very slow at low temperature and may be almost entirely inhibited at 0 to 1.6 °C.

Several postharvest treatments are applied to optimize the effect of low temperature *e.g.* sulphur dioxide fumigation by use of sodium metabisulphite. Apples coated with paraffin oil and pungent mustard oil withstand rotting. Fruits wrapped with SOPP impregnated wrappers showed some protection as well. Benomyl at 1000 ppm was the best single treatment for control of rot at about 25 °C. Dipping fruits in 1000 ppm diphenyl amine for 5 minutes or in 500 ppm aureofungin for 20 minutes resulted in the best control of the rots. Pathogenic isolates most resistant to thiabendazole are sensitive to imazalil. Use of 2-aminothiozole (5 per cent) plus thiourea (5 per cent) has been reported for best control of *Rhizopus* rot. Maximum protection is achieved by using fumigation with Sodium metabisulphite and wrapping with dichloro nitroaniline wraps. Trisodium phosphate 2 per cent solution can check the bitter rot disease in storage. Treatment of fruits with thioacitamid suppressed development of symptoms of core rot up to 35 days. *Pseudomonas syringae* is recorded to be most effective biocontrol against blue mould pathogen *P. expansum*.

2.2 Banana

The postharvest pathogens in banana generally gain entry into the skin of unripe fruit through bruises, wounds or injuries, but remain quiescent until the fruits start ripening. The pathogens then multiply rapidly resulting in blackening and softening of tissues. The major postharvest diseases of banana (Figure 2) are caused by fungal genera *Collectotricum, Gloeosporium, Fusarium, Botryodiplodia* and some other minor pathogens.

2.2.1 Anthracnose

It is one of the most important and widely distributed disease of maturing bananas caused by *Collectotricum musa*. Symptoms include anthracnose, black spot, stem and crown rot. The infected spots on the surface of the fruits are initially light, moist, but later change in colour to black. Relative humidity of above 85 per cent favours both conidial germination and development of anthracnose on ripe bananas. A high nitrogen dose and total sugars and a low content of ascorbic acid in the peels increases susceptibility to the pathogen.

2.2.2 Cigar End Rot

In India the disease is incited by *Verticillium theobromae*. Fungal pathogens such as *Gloeosporium musarum* and *Verticillium* have also been reported to cause cigar end rot. Infection generally begins from the tip of the finger and slowly spreads backwards causing greyness of the skin. The affected area later gets covered with the white mycelia over which black coloured fructification appear. It becomes corrugated and appears like a burnt cigar tip. Fruits get infected when they are still immature. In advanced stage the pistil end of the fruit shows cracks and the pulp inside the infected zone becomes dry and powdery. Some varieties belonging to Cavendish group of Banana are more susceptible to this pathogen.

2.2.3 *Fusarium* Fruit Rot

Rotting of banana fruits by *Fusarium moniliformae* was first reported from Allahabad and is now prevalent throughout the country. Olive brown spots appear on infected fruit. The spots extend longitudinally and coalesce. The fruit surface turns clove brown or mummy brown. The fruits become pulpy, ooze out a liquid and

Anthracnose
(http://hawaiiplantdisease.net)

Cigar-end Rot
(http://agritech.tnau.ac.in)

Fusarium Fruit Rot
(http://inhabitat.com)

Figure 2. Postharvest Diseases of Banana

emit undesirable odour. Later the rotten tissues get covered with white cottony growth of the fungus.

2.2.4 *Botryodiplodia* Fruit Rot

It is one of the most severe storage pathogens of banana causing heavy losses in Uttar Pradesh, Maharashtra and Andhra Pradesh. The disease is also called as stem end rot and is caused by *Botryadiplodia theobromeae.* The fungus is a wound parasite. Infection originates in or immediately below the decayed perianth or style and spreads uniformly along the fruit causing a progressive brownish black discoloration. The rotting of pulp is comparatively faster than that of the skin.

2.2.5 Pink Mould Rot

The pink mould rot is caused by *Trichothecium roseum* and has been reported from Uttar Pradesh, Delhi, Andhra Pradesh and Bihar. The rot is characterized by pink mould growth over the surface of the fruit. The rot remains limited to peel. The fungus is a week parasite and has a wide host range.

2.2.6 Management of Postharvest Diseases of Banana

The young bunches should be opened up for more light and air so as not to create conducive environment for the pathogens. Cigar-end rot can be checked to some extent by removing of pistil and perianth by hand.

Most of the fungal pathogens require temperature between 25-30°C. Storage at low temperatures reduces postharvest decay. As bananas are sensitive to temperatures below 13°C, it is advisable to store bananas between 13-15°C to slow down decay. Moreover, bananas have short storage life, hence, need to be transported and marketed at the earliest.

Pre harvest sprays of banana with carbendazim (0.1 per cent) and benomyl (0.1 per cent) reduce latent fungal infection in the field. Prophylactic postharvest dip before storage with benomyl (0.1 per cent) and aureofungin (100ppm) give complete control of postharvest rotting due to fungi. Two pre harvest sprays of Burgundy mixture (0.3 and 0.5 per cent) or Bordeaux mixture (0.5 and 1.0 per cent) on 60-75 days old fruits gives best control. Pre harvest spray with azoxystrobin 23 per cent SC (0.06-0.1 per cent) or difenoconazole 25 per cent EC (0.015 per cent) 15 days before harvest followed by dipping in hot water at 42°C has been reported to effectively check fungal diseases like anthracnose. Bio-agent *Burkholderia cepacia* against anthracnose disease and *Trichoderma viride* against pink mould rot have been reported to check postharvest decay.

2.3 Citrus

Citrus fruits have a delicate rind which is liable to injuries and wounds in field as well as after harvest, hence requires proper handling at all stages. Most of the pathogens attacking citrus are moulds like *Penicillium, Aspergillus, Botrytis etc.* which are weak parasites and require injury/wound for entry. In addition some other pathogens also lead to serious losses if postharvest

(a) *Alternaria* Fruit Rot (http://www.plantprotection.hu)

(b) *Aspergillus* Rot (http://www.shouragroup.com)

(c) Brown Rot (http://point09acres.blogspot.in and http://www.shouragroup.com)

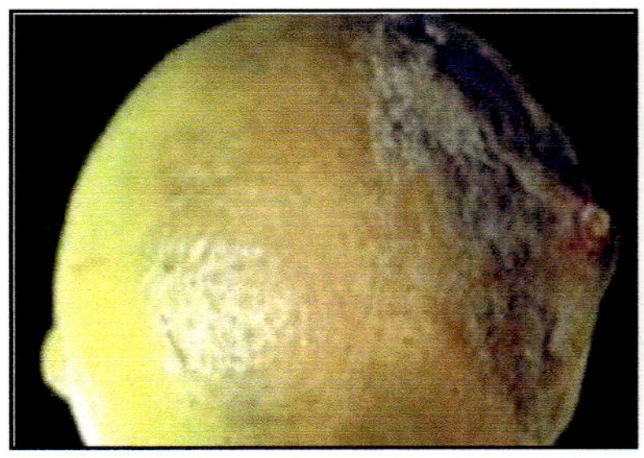

(d) *Colletotrichum* Stem End Rot (http://www.daff.qld.gov.au)

(e) *Penicillium* Fruit Rot (http://www.pestnet.org and http://plantdiseasehandbook.tamu.edu)

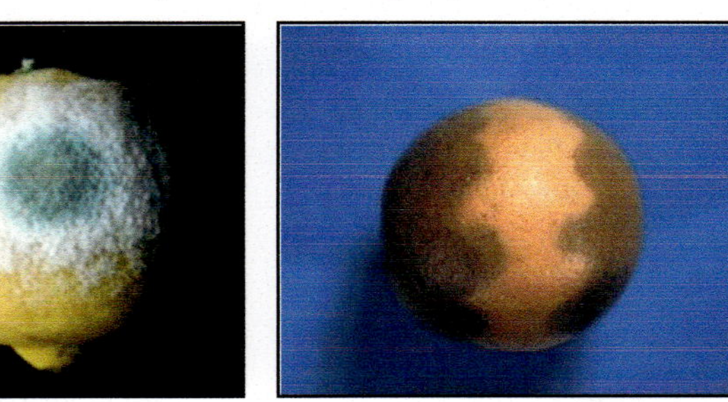

(f) Stem End Rot (http://edis.ifas.ufl.edu)

Figure 3. Postharvest Diseases of Citrus

handling and management are not proper. Important postharvest diseases of citrus (Figure 3) are described below.

2.3.1 *Penicillium* Fruit Rots

The *Penicillium* moulds occur in all the citrus growing countries. In India, their occurrence was first reported by Cheema *et al.* (1937). The fruits are affected both in transit and storage. The rind of affected fruit becomes watery on pressing and breaks easily. The entire area gets covered with the green or blue mouldy growth. Finally the entire fruit rots and emits a foul smell. The blue mould rot is incited by *Penicillium italicum* Wchmer and the green mould rot by *Penicillium digitatum* (Pers.:Fr.) Sacc. Fruits weakened by long cold storage decay more readily than fresh fruits.

2.3.2 Stem End Rot

Diplodia and Phomopsis stem end rot are the two most important fungal diseases of fruits encountered in transit. The earliest symptom is a softening of the rind around the button. The affected area increases rapidly and becomes brown. The discolouration is usually darker and a tendency to show decay at both ends is a characteristic of *Diploidia* stem end rot. Peak decay occurs 2-3 weeks after harvest.

2.3.3 Anthracnose and Stem End Rot

Colletotrichum gloeosporioides infection is known for causing anthracnose in oranges and grape fruits and stem end rot in citrus fruits. Infections usually start from stem end as brown necrotic spots, which gradually spread towards blossom end and cover upper half or the entire fruit in severe cases. This type of rot is observed more in case of Coorg mandarin during main season and to a lesser extent during monsoon crop.

2.3.4 Brown Rot

This disease is common in mandarin oranges grown in heavy rainfall areas of South India *viz.* Coorg, Wayanad and parts of Palni hills. The disease is caused by *Phytophthora nicotianae.* Apparently healthy fruits carry latent infection and develop brown soft fruit rot during storage.

2.3.5 *Alternaria* Fruit Rot

Black rot incited by *Alternaria citri* Ell. and Pierce has been noticed in Punjab, West Bengal, Rajasthan and Madhya Pradesh. The rot develops between 15 to 30 °C. On naturally infected fruits, small, discoloured spots appear which turn black. External injury to rind in var. Malta leads to infection and disease development, which increases with the extent of injury. The core of fruit becomes black. The fungus reduces soluble solids and reducing sugars and increases per cent acidity in the affected fruits.

2.3.6 *Aspergillus* Rot

In India, 10 to 15 per cent of lime fruits are affected by *Aspergillus niger.* The disease is usually noticed in July – August. The rot begins as small, circular water soaked spots which enlarge and turn brown. The soft rot destroys the whole fruit within 8-10 days. The rotten fruits emit a fermented odour. Contact with the diseased fruits can infect injured as well as healthy fruits. Percentage of rotting is greater at 30-35 °C, particularly when humidity is high.

2.3.7 Management of Postharvest Diseases of Citrus

Careful handling and avoiding injury to the soft rind help in prolonging storage life. The rot can be minimised by avoidance of injury through sorting and discarding infected fruits and by storing fruits at low temperature in baskets or open containers. Good results can be obtained by pruning and burning the diseased twigs in winter and by spraying with 1 per cent Bordeaux mixture in February, March and September. Out of various chemicals tried, prochloraz and benomyl give the best results in controlling storage diseases. Copper fungicides (1 per cent) and karathane (1 per cent) as pre harvest sprays can also check infection and disease development.

A 5 minute dip in 2 per cent bleaching powder and 10 minute dip in 0.2 per cent captan followed by storage at 2-4 °C have been reported effective against both the moulds. Thiabendazole was found to be markedly ineffective in reducing the decay as its protective quality remained for a short period. Thiabendazole at 0.1 per cent incorporated in waxol-0-12 has been found useful under refrigeration. Boric acid (4 per cent) is effective against green mould. postharvest treatment with 6-8 per cent borax solution or 300-400 ppm mycostatin is recommended for stem end rot. Imazilil WP and Imazilil EC formulations were highly effective in reducing decay and showed reduction index of 74.16 to 96.66 during a storage period of 36 days. Postharvest application of lipid based hormone jasmonates reduced decay caused by *P. digitatum* in "Marsh Seedless" grapefruit.

Control of decay of citrus fruit caused by *Penicillium spp.* was also reported with bacterial antagonists - *Bacillus subtilis* (Ehrenberg) Cohn, *Burkholderia* (*Pseudomonas*) *cepacia* Palleroni and Holmes, and *Pesudomonas syringe* Van Hall and yeast, *Kloeckera apiculata*(Rees) Janke and *Pichia anomala* (Hansen) Kurtzman. Bacterial suspensions of *Erwinia chrysanthemi* RK-67 and *Bacillus subtilis* RK-6 are reported to reduce rot severity due to *Aspergillus* sp. on lemon cultivars.

2.4 Grape

Grapes being succulent are susceptible to attack by several rot causing fungi like *Botrytis, Penicillium, Alternaria, Cladosporium, Aspergillus, Greenaria, etc.* The important postharvest diseases of grape (Figure 4) are described here.

Grey Mould Rot (http://www.agf.gov.bc.ca) **Blue Mould Rot** (http://agritech.tnau.ac.in) **Cladosporium rot** (www. archive.agric.wa.gov.au) **Bitter rot** (https://www.extension.org)

Figure 4. Postharvest Diseases of Grapes

2.4.1 Grey Mould Rot

Botrytis cinerea Pers.: Fr. (teleomorph: *Botryotinia fuckeliana* (de Bary) Whetzel) is a plant pathogenic fungus distributed world-wide and causes grey mould disease on both pre harvest and postharvest plant tissues. The disease on grapes is incited by *Botrytis cinerea*. In compact bunches, berries on the inside of the attached bunch may split during growth and may be invaded by fungal pathogens resulting in bunch rot. Even a single infected berry can cause 'nest rot' during transit and storage because the pathogen can spread by contact. Mature fruits are more susceptible to *B. cinerea*. Losses due to this mould on grape are dependent on harvest date environmental conditions during or prior to harvest. Bunches picked after rains are more prone to decay.

2.4.2 Blue Mould Rot

As the name suggests the disease is incited by blue mould fungus- *Penicillium digitatum* and is characterised by scanty growth of the pathogen, which is at first white but turns bluish green after sporulation. The affected tissues become soft and watery and develop mouldy flavour.

2.4.3 *Cladosporium* Rot

This disease is incited by *Cladosporium herbarum*. The pathogen causes a black, firm, shallow decay, which is usually localised on one side or blossom end. The fruit becomes flat or wrinkled on the affected side. A sparse surface growth of grey-green fungus may be seen on the fruits kept at room temperatures.

2.4.4 Bitter Rot

The disease produces brown, hard, depressed patches on the berries, which subsequently bear numerous black, dot like fructifications. Harvested grapes with small scars or lesions deteriorate in transit and storage showing typical fruit rot symptoms. The disease is incited by *Greenaria uvicola*.

2.4.5 Management of Postharvest Diseases of Grapes

The rapid growth of pathogens resulting in rotting of grapes can be reduced by preventing field infection through suitable pre harvest sprays, by careful handling to reduce bruises and injuries, pre cooling and refrigeration. Postharvest fumigation is recommended to reduce inoculums load of the pathogen, followed by well planned marketing of stored lots according to their decay potential. Fruits for transit should be pre-cooled to at least 4 °C and then stored at 0.55 – 0 °C. This will delay decaying of berries and reduce postharvest losses.

Sodium bisulphite which releases SO_2 when in contact with air can be used with packing material to control storage diseases. *Vinifera* grapes fumigated with SO_2 in storage, reduce losses due to various *rots*.

Pre harvest infection can be controlled by three applications of captan 50WP (0.25 per cent) at monthly interval before rainfall. Benomyl is equally effective as a pre harvest spray. Postharvest dip with captan (0.2 per cent) and difolatan (0.2 per cent) reduced berry rot above 80 per cent due to *Cladosporium*, showing respectively 12.84 and 10.11 per cent rotten berries as against 78.80 per cent in untreated control.

2.5 Guava

Various organisms associated with storage losses in guava are *Phomopsis psidii, Phoma psidii, Macrophoma allahabadensis, Phytophthora nicotianae var. parasitica, Pestalotiopsis psidii, Colletotrichum gloeosporioides, Cylindrocarpon tonkinense, Rhizopus stolonifier, Curvularia tuberculata, Geotrichum candidum, Botryodiplodia theobromae, Aspergillus niger, Trichothecium roseum, etc.* Some of the important pathogens leading to postharvest rot in guava (Figure 5) are discussed here.

2.5.1 *Colletotrichum* Fruit Rot

This is a serious postharvest disease in north east and other rainy parts of India and results in pin-head sized spots on the unripe fruits which enlarge to form sunken and circular, tan to dark brown to black spots. Infected tissue in unripe fruit becomes hard and corky

Colletotrichum Fruit Rot
(http://agritech.tnau.ac.in)

Fruit Canker
(http://www.insectimages.org)

Phoma Fruit Rot
(http://agritech.tnau.ac.in)

Figure 5. Postharvest Diseases of Guava

2.5.2 Fruit Canker

The fruit canker is widely prevalent in India and incited by *Pestalotiopsis psidii* and *Colletotrichum gloeosporioides*. Infection generally occurs on green fruits. The canker does not penetrate deep into the flesh, but the market value of severely infected fruit is greatly reduced. Small brown, circular, necrotic brown areas with raised margin appear on fruit which later break open epidermis. Safeda variety is highly resistant to canker.

2.5.3 *Phoma* Fruit Rot

Phoma rot incited by *Phoma psidii* is characterised by brown coloured spots produced over the fruit surface. They ooze out cream coloured pycnidiospores. Sucrose in the inoculated fruits is completely hydrolyzed in six days.

2.5.4 *Phytophthora* Fruit Rot

Fruit rot is incited by *Phytophthora nicotianae* var. *nicotianae* is known to occur in Karnataka, Tamil Nadu, Andhra Pradesh, Maharashtra and Rajasthan. Moderate temperature, high humidity and rainy weather favour the disease incidence. Infected area becomes greyish brown and appear water soaked with greyish black center.

2.5.5 *Macrophoma* Fruit Rot

Different species of *Macrophoma* are known to cause diseases in mango, banana, *and other fruits.* A severe fruit rot caused by *Macrophoma allahabadensis* was first described from Allahabad markets.The disease occurs in the field as brown blotches on the fruits which leads fruit to rot. Sucrose is completely hydrolyzed after infection and D-glucose and D-fructose are either consumed or reduced in concentration. Marked changes take place in free and bound amino acids, amides and organic acids

2.5.6 *Botryodiplodia* Fruit Rot

This disease is incited by *Botryodiploidia theobromae,* which causes considerable damage during storage and transit. The affected fruits show brownish discolouration mostly at the stem end which generally proceeds downwards in an irregular wavy manner. In advanced stages, numerous small pycnidia are produced over the entire surface of the fruit. The rot is soft and watery.

2.5.7 Management of Postharvest Diseases of Guava

As in other fruit crops pre harvest operations and environmental conditions, postharvest handling and cold storage are all the more important in guava for deciding extent of postharvest damage. Cold storage and avoidance of fruit injury are the best means of reducing postharvest decay. Careful handling and storage at 15°C or below and quick disposal minimise the losses.

Control of most rots can be achieved by 6 sprays at monthly intervals with dithane Z-78 (0.2 per cent) or aureofungin (10 ppm) during fruiting. thiram, copper-based fungicides, ziram 27 per cent SL and NC 1584 are good protectants in pre inoculation sprays. Field sprays of chlorothalonil or mancozeb at 15 day intervals give good control of the fungal diseases both in the field and in the storage. Sprays of thiophanate methyl and tricyclazole in the field have been found to control this disease effectively. The rot can be checked by treating the fruits with Thiobendazole (TBZ). Some varieties like 'Safeda'are highly resistant to canker, which may be preferred in areas where canker is a problem.

2.6 Mango

In India fungi belonging to around 20 genera are known to attack mango fruits. In addition, some bacteria are also known to decay mango fruits in stores. The important pathogens are *Botryodiplodia theobromae, Colletotrichum gloeosporioides, Rhizopus arrhizus* and *Aspergillus niger*. Some other pathogens of minor importance in mango include - *Alternaria tenuissima (Alternaria rot), Sclerotium rolfsii (soft rot),Boothiella tetraspora (dry* rot), *Actinodochium jenkinsii* (Black spot), *Pestalotiopsis mangiferae and P. glandicola* (Brown rot), *Phomopsis mangiferae* (Fruit rot), *Guignardia mangiferae* (Black rot) *and Pseudomonas mangiferae* (Bacterial rot). The major postharvest rots (Figure 6) are discussed here:

2.6.1 Stem End Rot

Botryodiplodia theobromae earlier name Diplodia natalensis is an important postharvest disease of ripe mango adversely affecting the export of mango. In the initial stages, the epicarp darkens around the base of the pedicel. In the next few hours, the affected area enlarges to form a circular black patch, which under humid conditions extends rapidly and within 2–3 days turns the whole fruit completely black. Profuse black coloured pycnidial sporulation is predominant on the infected part of the fruit. The pulp of the diseased fruit becomes brown and soft. The pathogen incites the soft rot with the aid of pectinolytic and cellulolytic enzymes. The infected fruits lose ascorbic acid rapidly and have lower non-reducing sugar contents in the pulp.

Stem End Rot
(http://www.infonet-biovision.org)

Anthracnose
(http://agritech.tnau.ac.in)

Figure 6. Postharvest Diseases of Mango.

2.6.2 Anthracnose

Anthracnose of mango is caused by *Colletotrichum gloeosporioides* and results in loss of produce before and after harvest. The exporters and farmers are equally concerned about this disease. It appears as black or dark brown spots on the fruits. In advanced stages of the disease, the fungus produces a sexual fruiting bodies called acervuli with abundant orange to salmon pink masses of conidia on the lesions. The pathogen can grow at temperatures above 15°C, the optimum being 25°C.

2.6.3 Black Mould Rot

The rot is caused by *Aspergillus niger*. The affected fruits show brown to pale yellow colour on the damaged part of the fruit with the development of irregular, hazy, greyish spots which coalesce into dark brown or black lesions. The mesocarp of the rotten area gets depressed and softens over which smoky black coloured mouldy growth is conspicuous. Variety 'Totapuri' is highly susceptible to *Aspergillus* rot.

2.6.4 Soft Rot

The pathogen *Rhizopus arrhizus* incites a typical soft watery rot. The disease develops rapidly between 20 and 40 °C. Development of disease is slow and of low intensity at low temperatures (7-8 °C).

2.6.5 Management of Postharvest Diseases of Mango

Dead twigs and bark of mango trees harbour several types of pathogens. With the onset of rains, the atmosphere of the mango orchard gets contaminated with spores of the pathogenic fungi. These are carried to the store as latent infection on the fruits. Hence, orchard sanitation is of utmost importance. Dead discarded twigs and plant debris should be collected and burnt, instead of dumping in or around the orchard.

Pre harvest sprays with chemicals like carbendazim (0.1 per cent), prochloraz (0.1 per cent) are effective in reducing rot in storage. Since carbendazim is under review and prochloraz is not easily available in India, the pre harvest sprays of difenoconazole and tricyclazole are nowadays recommended for controlling the disease. In addition, formulations of Milcop, Fycol, copper oxychloride also have been recommended as sprays. Carbendazim (0.1 per cent), benomyl (0.1 per cent), thiophanate methyl (0.1 per cent) and bitertanol (0.1 per cent) spray at 15-day interval, 4 times before harvest, significantly reduces postharvest rotting in mango. However, pre harvest interval should be taken into consideration to minimize pesticide residue in the fruits.

Short exposure to temperature of 40°C kills several fungal pathogens, hence, hot water treatment of harvested mangoes at 50 to 55°C for 15 minutes is recommended before storage. The treated fruits can be stored for 25 to 47 days at 3.5-10 °C. Dipping mangoes in 6 per cent Borax solution at 43 °C for 3 minutes can also reduce rot infection. Good control of the disease is achieved by dipping of fruits in Benomyl (1500 ppm) and thiobendazole (900 ppm). Among the effect of various postharvest treatments after pre harvest sprays, prochloraz (0.1 per cent) and benomyl (0.1 per cent) dip and hot water treatment at 52°C for 10 minutes give effective control of the disease.

Studies on efficacy of antagonists indicate that four isolates of bacteria, five yeasts and two filamentous fungi are effective *in vitro* against anthracnose (*Colletotrichum gloeosporioides*) in mango. *Brevundimonas diminuta* isolate B-62-13, *Stenotrophomonas maltophilia* L-16-12, a member of Enterobacteriaceae L-19-13, *Candida membranifaciens* F-58-22, and the yeast isolate B-65-23 are effective on naturally infected fruit. The isolates significantly reduce severity of anthracnose on artificially inoculated mango fruit. Biocontrol of black mould rot by *Trichoderma harzianum* has been reported.

2.7 Papaya

Papaya is affected by 8 different types of rots, *Alternaria* fruit rot incited by *Alternaria alternata* and *Botryodiplodia* fruit rot incited by *Botryodiplodia theobromae* are other postharvest diseases of papaya which need more investigation. Over the years these pathogens have become serious and need concerted efforts to work out their management strategies. The important postharvest diseases of papaya (Figure 7) are described below.

2.7.1 Anthracnose

Colletotrichum gloeosporioides is very common in all the papaya growing countries. The disease occurs as lesions on the fruit initially as brown, superficial discolouration of the skin. The spots soon develop into circular, slightly sunken areas, which may be water-soaked. The lesions may coalesce and white mycelial growth appears on the margin of spots.

2.7.2 *Rhizopus* Rot

Fruit rot in papaya caused by *Rhizopus stolonifer* occurs in different parts of India. The rot is characterized by the production of irregular water-soaked lesions which gradually enlarge and get covered by white and dark brown fungal growth and sporangiophores. Finally, the fruit collapses by becoming watery and emits a foul odour. The rot appears when cold stored fruits get exposed to environmental heat during retailing

2.7.3. *Ascochyta* Rot

Ascochyta caricae Pat. causes both fruit and trunk rot. The disease generally occurs on half grown or mature fruits that show small, circular water soaked spots. The lesion may become sunken and brownish-black. The rot usually spreads outwards irregularly. Rotting becomes most severe in high temperature and high humidity.

2.7.4 *Phomopsis* Fruit Rot

Fruit rot caused by *Phomopsis caricae* pv. *papayae* in papaya develops a water spot, which increase in size. The rotten area depresses and cracks at a later stage. The affected portion turns dark brown to black. Water soaking is seen in the area surrounding the depressed portion. In some fruits, the rotten portion is surrounded by raised tissue, white in colour.

2.7.5 *Marcrophomina* Fruit Rot

A fruit rot of papaya incited by *Marcrophomina phaseolina* is responsible for considerable postharvest loss. The disease is characterised by small, water soaked spots, which appear as circular specks on the fruit surface. The spots rapidly enlarge and become sunken. The pathogen advances deep into the fruit and causes rotting and disintegration. The affected fruits turn brown to black. The attacked tissue hardens and become dotted with sclerotia of the pathogen.

2.7.6 *Fusarium* Fruit Rot

Fusarium solani causing fruit rot is reported from Bengal and other parts of northern plains. Affected fruits show small, stem end or surface lesions. Profuse, white hyphae are seen associated with the rot.

2.7.7 Management of Postharvest Diseases of Papaya

Combination of pre harvest sprays, postharvest dips and hot water treatments give best control of papaya

Anthracnose
(http://agritech.tnau.ac.in)

Rhizopus Rot
(http://agritech.tnau.ac.in)

Phomopsis fruit rot
(http://agritech.tnau.ac.in)

Figure 7. Postharvest Diseases of Papaya.

rots. Hot water double dip treatment (42 °C for 30 minute followed by 20 minute at 49 °C) in combination with pre harvest sprays of prochloraz or tricyclazole and storing the fruits at 10 °C or below, have been found to be extremely effective in managing the storage diseases. Spraying the fruits with Bordeaux mixture (2:2:50) at intervals of 21 to 30 day from fruit setting stage, controls several pathogens. Two field sprays either with bavistin (0.1 per cent) at 45 day interval or with daconil (0.2 per cent) at 15 day interval were sufficient to minimize anthracnose after harvest. Pre harvest sprays of thiophanate methyl or difenoconazole provide a perfect control of postharvest rots. The combined effect of prochloraz at 0.1 per cent as pre harvest spray and prochloraz 0.1 per cent as postharvest dip are also highly recommended in controlling the anthracnose disease of papaya. The incidence of *Ascochyta* rot can also be reduced by fruit dip in 2-aminobutane carbonated solution (1 per cent a.i.) or by fumigation with 2-aminobutane. Borax, ziram, captan, and dithane M-45 are effective in preventing symptom development when applied to the detached fruits. Benomyl and difolatan though reduce rot are advised to be avoided for use in the field before harvest and postharvest use. *Trichoderma viride* and *Pseudomonas fluorescence* have been found to suppress *Marcrophomina* fruit rot.

2.8 Pineapple

Pineapple is not affected by many postharvest pathogens, probably because of its tough hard outer rind, which protect from injuries and bruises, thus restricting avenues for entry of rot fungi which are generally weak pathogens. It has only one serious pathogen *Ceratocystis paradoxa* (Figure 8) causing stem end rot.

Figure 8. Pineapple Stem End Rot
(http://www.celkau.in)

2.8.1 Stem End Rot

The stem end rot, core rot or black rot (Figure 8) incited by *Ceratocystis paradoxa* (*Thielaviopsis paradoxa*) is a major fungal pathogen which attacks pineapples after harvest and is the chief cause of deterioration of the stored fruits. The infection generally starts at the cut end of the stem in the form of a small circular, water soaked spots, which are very soft. With the advancement of disease the spots enlarge and coalesce forming a large black patch extending throughout the fruit. The inside tissue is also invaded by the fungus, which becomes very soft, black, watery and emits foul smell.

2.8.2 Management of Postharvest Diseases of Pineapple

Chemical control trials conducted have shown that dipping the fruits for 5 minutes after harvest in thiopendazole (1000 ppm) or benomyl (2000 ppm) would ensure good protection against this disease. Dusting of the stem end with a small amount (0.1 g per cent) of benzoic acid with an inert substance such a kaolin, is highly effective in controlling the stem end rot and is also commercially feasible on account of the low cost of the treatment.

2.9 Pomegranate

Important postharvest disease of pomegranate are rots caused by *Phomopsis punicae, Rhizopus arrhizus, R. stolonifier Aspergillus nidulans, Aspergillus clavalus, Spicaria sp., Aspergillus niger, Syncephalastrum racemosum Botryodiplodia theobromae, Coniella granati, Colletotrichum gloeosporioides* and *Penicillium expansum*. Most of the rot pathogens in pomegranate may gain entry through calyx end and sometimes from stem end or injuries on rind. Important postharvest diseases of pomegranate (Figure 9) are described here.

2.9.1 *Phomopsis* Fruit Rot

Pathogen *Phomopsis punicae* causes fruit rot disease, symptoms of which start from calyx end as dark brown zonate rings and gradually spread over the entire fruit. Pycnidia of the pathogen appear on the affected area. In severe cases the aril also gets rotten and makes the fruit unmarketable. When such fruits with minor infection are packed in a box and transported, adjacent fruits also develop typical rotting within week.

2.9.2 Soft Rot

Soft rot is the major impediment in pomegranate storage and is incited by *Rhizopus arrhizus* and *R. stolonifier*. It starts as small spots that increase in size and coalesce. The disease appears on the rind but the entire fruit also decays internally into a pulpy mass. Under dry conditions, cracking originates from the point of infection.

2.9.3 Black Mould

It is caused by *Aspergillus niger* and is a pathogen of concern which occurs frequently in all seasons and under all storage conditions. The pathogen enters through injuries or calyx end. The pathogen results in brownish soft rots, with a depressed centre in later stages where blackish sporulation of the fungus can be seen. These generally may grow inside without any external symptoms for some time. The rind and underlying pulp become soft, but fruits retain their shape.

Black Mould

Colletotrichum Dry Rot

Alternaria Heart Rot

Figure 9. Postharvest Diseases of Pomegranate.

2.9.4 Black Blotch

It is caused by *Botryodiplodia theobromae.* It causes severe fruit rot and spoils the entire fruit including arils. The disease is characterized as black blotch on any part of the fruit which spreads very fast and under favourable conditions covers the entire fruit. The dark coloured pycnidial fruiting bodies are the typical characteristic of the disease and the pathogen. The disease is most serious in the post monsoon season.

2.9.5 Dry Rot

It is caused by *Colletotrichum gloeosporioides, Botryodiplodia theobromae and Coniella granati.* These are mostly pre harvest infections which may develop in storage as hard dry brown rot. *Colletotrichum* rots develop as tan brown dry rot usually from calyx end. Arils also are affected but rot is dry. This rot is very common in pomegranate growing belts in India, while other species are more important in other countries.

2.9.6 Heart Rot

This disease is basically a pre harvest problem, but remains unnoticed till fruits are cut open. The pathogen enters the host at the time of flowering and develops heart rot symptoms later during fruit maturity or in postharvest period. The infection by the pathogen - *Alternaria alternata* results in black rot of arils, without any exterior symptom. In advance stages the rind may become slightly off colour or paler red.

2.9.7 Other Rots

Some of the fruit rot organisms such as *Cladosporium oxysporum, Aspergillus flavus, A. niveus* and *A. versicolor, Penicillium chrysogenum,* produce orange-red, dull brown, black slimy, soft watery patches on pomegranate fruit, which often emit fermented smell/foul odour

2.9.8 Management of Postharvest Diseases of Pomegranate

Timely disease control and sanitation (removal of old fruit and dead branches), reduce the incidence of the disease. Infected, fruit drop to the ground by gently shaking the tree at the time of harvest. Care should be taken during harvesting and packing to avoid injury to the fruit skin. Treatment of packing straw with SO_2 producing chemicals can protect the fruit from soft rot. Satisfactory control of soft rot has been achieved by treatment of fruits with linseed, mustard and castor oil. Protecting fruits from physical or insect injury save the fruit from rotting to a great extent. Harvested fruits should be stored at 5-6 °C with minimum 80 per cent RH after pre-cooling.

The pre harvest sprays during crop season with carbendazim (0.1 per cent), propiconazole 0.1 per cent thiophanate methyl (0.15 per cent), mancozeb (0.2 per cent), difenoconazole 25 EC (0.05 per cent), captan (0.2 per cent), benomyl (0.2 per cent), ziram (0.25 per cent), Bordeaux mixture (0.5-1 per cent), copper hydroxide (0.15-0.2 per cent) and copper oxychloride (0.25 per cent) take care of most of the postharvest diseases. Combination treatments of carbendazim (0.1 per cent) + mancozeb (0.2 per cent) and benomyl (0.1 per cent)+ mancozeb (0.2) control all postharvest pathogens and also improve fruit quality. The formulation containing tricyclazole 18 per cent +mancozeb 62 per cent WP@2.5g/l has been found effective in reducing *Colletotrichum* rot. Fruit dip with dithane Z-78 (0.1 per cent), captan (0.2 per cent), benomyl (0.2 per cent) also reduces postharvest rots. The damage of *Phomopsis* rot can be greatly reduced by 3 sprays at 10-day interval with 0.5 per cent copper hydrate plus albolium oil.

2.10 Strawberry

Botrytis and *Colletotrichum* are the major pathogens of harvested strawberry (Figure 10). Other fungi capable of infecting strawberries include species of *Rhizopus, Penicillium, Fusarium* and *Alternaria.*

2.10.1 Grey Mould

Botrytis (grey mould) is the most important fungus that has been detected in strawberry fruit samples limiting its storage life. Studies show that *B. cinerea* is

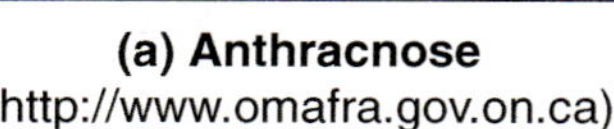

(a) Anthracnose
(http://www.omafra.gov.on.ca)

(b) Grey Mould
(http://www.apsnet.org)

Figure 10. Postharvest Diseases of Strawberry

a necrotrophic pathogen. The fruits disintegrate and grey mould growth is seen on infected fruits. It infects plant tissues through multiple mechanisms, including secretion of cell wall degrading enzymes (*e.g.* cutinase and polygalacturonases) and phytotoxic metabolites (*e.g.* botrydial and oxalic acid).

2.10.2 Anthracnose

Another important postharvest disease of strawberry is caused by *Colletotrichum fragerae.* The disease appears as depressed salmon-coloured spots on mature and ripe fruits, with abundant pink coloured spore mass oozing out. If one such fruit goes in a box overnight more than fifty per cent fruits get infected and they become unmarketable.

2.10.3 Management of Postharvest Diseases of Strawberry

Though many effective chemicals have been reported to control postharvest pathogens in other crops, these cannot be used in strawberry, as the strawberry fruit is consumed immediately and raw. To overcome this problem many microbial agents have been tried and found to be effective in suppression of *B. cinerea* on numerous crops under both pre harvest and postharvest conditions. Commercial products based on *Candida oleophila*, such as Aspire (strain I-182), on *Metschnikowia fructicola*, such as "Shemer",and on *Trichoderma harzianum*, such as Trichodex (strain T39), are now available for protection of postharvest fruits from infection by *B. cinerea.* Exogenous putrescine may have an anti-pathogenic function in strawberry fruits enhancing the shelf life of strawberry. The pre harvest spray with tricyclazole, difenoconazole and thiophanate methyl keeping in mind the PHI of these chemicals to avoid residues, followed by the treatment with *Trichoderma harzianum* is the most effective to check rot.

Cold Plasma (CP) Technology in Postharvest Disease Management

Development of alternative, eco-friendly approaches for control of postharvest pathogens has become a challenge for plant pathologists. Although there are several well established techniques such as heat that are used as postharvest treatments, they often have disadvantages, including alteration of food quality due to physiological responses to the treatment, or environmental pollution. A promising new postharvest treatment is cold plasma (CP), which is a gas-derived mix of atoms, excited molecules and charged particles. CP has no known adverse effects on fresh produce or the environment. CP treatment has been extensively investigated as a non-thermal preservation method for the microbial decontamination. The reactive species in cold plasma, which include ozone, ultraviolet (UV) photons, electrons, positive and negative ions, free radicals, and excited or non-excited molecules and atoms, damage cell membrane components and internal cellular components via different modes of action and ultimately have a lethal effect on different bacterial and fungal species. For horticulture, successful use of CP for bacterial decontamination has been reported for fresh fruit and vegetables. There are also some reports that indicate CP could be a prospective control method for fungal pathogens, particularly those that cause postharvest diseases. Major postharvest fungi include *Penicillium* spp., *Aspergillus* spp., *Botrytis cinerea*, *Monilinia* spp. and *Colletotrichum* spp., all of which have a wide range of hosts and account for significant economic losses globally. For the purpose of treating postharvest fungal pathogens, the CP generation systems will be referred to here as either direct, indirect or hybrid. Whilst CP studies on postharvest fungi are still in the relatively early stages of research, both in vivo and in vitro treatments have demonstrated significant fungicidal ability against a number of postharvest fungi, demonstrating its potential application for the control of postharvest diseases.

Nanotechnology in Postharvest Disease Management

To combat practical problems facing postharvest diseases, nanotechnology is vital and needed. Nanomaterial treatments have found promising to reduce the utilization of synthetic fungicides to control postharvest diseases of fruits and vegetables. It is used to produce nanomaterials for use as antifungal agents in several commodities, including fruits and vegetables. Several nanomaterials have shown potential in postharvest management and have been developed to control diseases in citrus, grapes banana, apple and mango. The nanotechnology approach is also helpful for introducing a new invention for packaging films, preventing impact of gases and unsafe rays, improving packaging appearance and helping for labeling fresh products using the multiple chips (nano-biosensors).

3.0. Vegetables

Most of the vegetables are perishable commodities and hence, require careful handling and storage. Postharvest diseases of vegetables are caused by microscopic fungi and bacteria (Table 2). Vegetables are less acidic than fruit hence, are more vulnerable to bacterial infections than fruits. Postharvest diseases of important vegetable crops are discussed below.

Table 2. Postharvest diseases of vegetables and their causal organisms

Crop	Disease	Causal Organism
Beans	Anthracnose	*Colletotrichum lindemuthianum*
	Grey mould	*Botrytis cinerea*
	Cottony leak	*Phytophthora capsici*
	Pod rot	*Pythium* sp.
Cabbage	Black rot	*Xanthomonas campestris* pv. *campestris*
	Soft rot	*Erwinia carotovora* pv. *carotovora*
	Grey mould	*Botrytis cinerea*
	Bacterial Zonate Spot	*Pseudomonas cichorii*
	Curd rot	*Alternaria brassicae*
Carrot	Black mould/root rot	*Thielaviopsis basicola*
	Black rot	*Stemphyllium* sp.
	Greenish black mould	*Alternaria radicina*
	Sour rot	*Geotrichum candidum*
	Crown rot/Dry rot	*Rhizoctonia* sp.
	Cottony soft rot	*Sclerotinia sclerotiorum*
	Soft rot	*Erwinia carotovora* pv. *carotovora*
	Liquorice rot	*Mycocentrospora acerina*
	Crater rot	*Fibularhizoctonia carotae*
Onion	Neck rot	*Botrytis allii*
	Black mould rot	*Aspergillus niger*
	Bulb rot	*Fusarium solani*
	Soft rot	*Erwinia and Pseudomonas* sp.
	Purple blotch	*Alternaria porri*
	Cephalosporium rot	*Cephalosporium curtipes*
Potato	Brown rot	*R. solanacearum*
	Dry rot	*Fusarium solani var coerulum*
	Late blight	*Phytophtora infestans*
	Charcoal rot	*Macrophomina phaseolina*
	Black sour rot	*Rhizoctonia solani*
	Rhizopus rot	*Rhizopus oryzae* and *R. arrhizus*
	Gilmanella rot	*Gilmanella humicola*
	Cylindrocarpon dry rot	*Cylindrocarpon tonkinensis*
Tomato	Buck eye rot	*Phytophthora nicotianae*
	Phoma rot	*Phoma destructiva*
	Sour rot	*Geotrichum candidum*
	Anthracnose	*Colletotrichum* sp.
	Fusarium rot	*Fusarium roseum*
	Alternaria rot	*Alternaria alternata*
	Sclerotium rot	Sclerotinia sclerotiorum
	Grey Mould	*Botrytis cinerea*
	Rhizopus rot	*Rhizopus nigricans*
	Bacterial soft rot	*Erwinia* spp.
	Bacterial spot	*Xanthomonas vasicatoria*
	Oospora rot	*Oospora lactis*
	Cladosporium rot	*Cladosporium* sp.
	Malustela rot	*Malustella seria*

3.1 Beans

Many rot fungi including *Sclerotinia sclerotiarum*, *Botrytis cinerea*, *Colletotrichum lindemuthianum*, *Phytophthora capsici* and *Pythium* are known to cause postharvest diseases in beans (Figure 11). Watery soft rot and grey mould-rot are the most common maladies affecting beans.

3.1.1 Anthracnose

Anthracnose caused by *Colletotrichum lindemuthianum* is characterised by deep black spots with reddish brown border. One to many, isolated or coalesced spots may be present on the affected pods Pink spores are found on the centre of the spots.

3.1.2 Grey Mould

The disease results in soft rot on bruised tissue with *Botrytis cinerea* growth on affected parts. The mould spreads to neighboring beans and spoils them.

3.1.3 Cottony Leak

It is caused by *Phytophthora capsici* and infects the pods. Most of the time the fungus is not as fluffy as white mould, moreover there is a reddish border between healthy and infected pod tissue,

(a) Anthracnose (http://www.cals.ncsu.edu)

(b) Cottony Leak (https://extension.udel. edu)

(c) Grey Mould-Rot (http://mtvernon.wsu.edu)

(d) Pod Rot (https://extension.udel.edu)

Figure 11. Postharvest Diseases of Beans

3.1.4 Pod Rot

Pythium pod rot is a problem during wet, warm weather. The pathogen is soil borne and pods near ground are attacked/contaminated first which spread the disease to other pods in field as well as in the stores. Occurrence of this disease is extremely hard to predict, unless the field has a history of *Pythium* problems.

3.1.5 Management of Postharvest Diseases of Beans

Refrigeration at 4.4–7.2 °C immediately after harvest prevents spread of most of the diseases in store.

3.2 Cabbage

The major postharvest diseases in cabbage are caused by fungal pathogens *Botrytis cinerea, Sclerotinia and Alternaria, and bacterial pathogens Erwinia carotovora* and *Xanthomonas.*

3.2.1 Black Rot

It is a serious problem of cabbage and is caused by *Xanthomonas campestris* pv. *campestris.* It is a vascular disease, which blackens the veins. It shows vascular discoloration at core of stored cabbage. It is primarily a field disease characterized by darkening of vascular elements forming black network of leaves and also leads to losses in store.

3.2.2 Soft Rot

The disease caused by *Erwinia carotovora* is another serious problem in cabbage and cauliflower. Infection enters in the field through stumps and may remain unnoticed for long time. The infected area inside heads rots under favourable conditions. Rotted parts look cream coloured and later turn brown with foul smell. Injury also helps weak parasites like *Rhizopus stolonifer* to infect cabbage causing soft rot.

3.2.3 Grey Mould

The disease is caused by *Botrytis cinerea* and is a serious disease in prolonged storage under moist conditions. The rot occurs as dark lesion on the outer leaves, and progresses inwards until the whole head is infected. The pathogen appears as a grey fluffy fungal mass, on the outside of the cabbage and as a brown soft rot penetrating deep into the head. Untreated cabbage can lead to losses above 40 per cent and worsens with long-term refrigerated and Controlled Atmosphere stores.

3.2.4 Bacterial Zonate Spot

The disease is caused by *Pseudomonas cichorii* (Swingle) Stapp. First symptoms of the disease are light brown, 2-13 mm diameter irregular lesions on the cabbage head, later turning darker brown. The spots give target board effect. Infected areas are firm and rubber

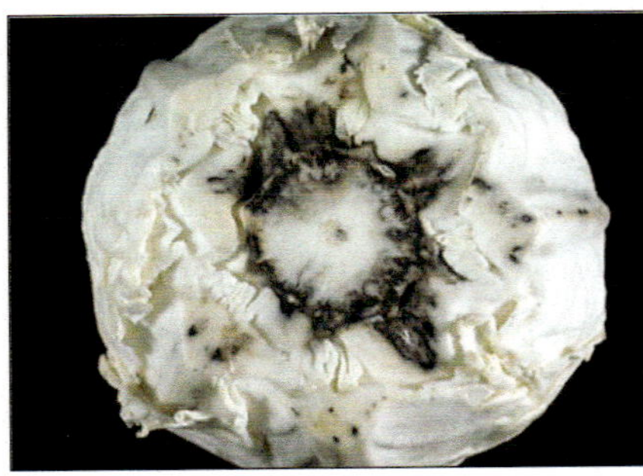

(a) Black Rot (http://vegetablemdonline.ppath.cornell.edu)

(b) Grey Mould (http://www.farminguk.com)

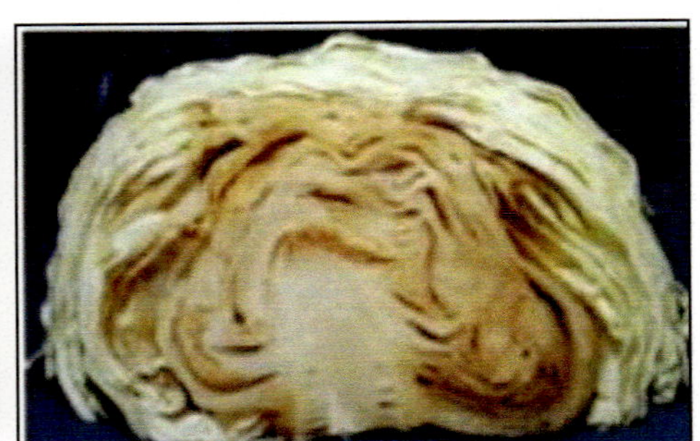

(c) Soft Rot (http://www.agroatlas.ru)

Figure 12. Postharvest Diseases of Cabbage

like. It is an important market disease and is sometimes confused with *Alternaria* rot due to zonate effect. Curd rot due to infection of *Alternaria brassicae* is very common in cabbage, broccoli and cauliflower.

3.2.5 Management of Postharvest Diseases of Cabbage

For control, affected heads or leaves should be discarded during packaging. Use maximum care to reduce bruising of heads. Close trimming of infected leaves before shipment helps reduce removing latent infection.

The best control is storage under high relative humidity and low temperature with good circulation of the storage air, which keeps leaf tissues green and viable, hence less susceptible to attack. Transit temperatures maintained around 4.4 °C and storage at 0 to 0.6 °C but not above 2 °C are required for reducing postharvest losses due to this disease and maintaining good quality.

Field sanitation and pre harvest sprays of copper oxychloride have been found to be the most effective method to control decay. Storage diseases also can be controlled by pre/postharvest application of fungicide such as iprodione, 50 per cent WP at a dose of 67 g/100 liters of water. The heads should be packed in perforated polyethylene bags and stored immediately. Treated cabbages must not be removed from store for at least eight weeks after treatment.

3.3 Carrot

The most common storage diseases in carrot are black mould, black rot and sour rot. The diseases produce discolouration and the affected tissues become soft or mealy. With sour rot fermentative odour is also produced. The Black mould/root rot is caused by *Thielaviopsis basicola*, black rot by *Stemphyllium* sp., greenish black mould by *Alternaria radicina*, sour rot by *Geotrichum candidum*, crown rot by *Rhizoctonia solani* and dry rot by *Rhizoctonia* sp. In cold storage rots due to *Rhizopus* sp. and cottony soft rot due to *Sclerotinia sclerotiorum* are problems and usually appear after some months of storage. At higher temperature, *Fusarium* sp. and *Rhizopus* sp. attack the roots. *Sclerotium rolfsii* is another pathogen which causes heavy yield loss due to the rotting of the roots. Bacterial soft rot is caused by the bacterium *Pectobacterium carotovorum* (*Erwinia carotovora*). Affected carrots develop soft lesions and eventually become a mushy and slimy mass of tissue. Two new postharvest pathogens, *Mycocentrospora acerina* and *Fibularhizoctonia carotae*, which cause liquorice rot and crater rot respectively, during prolonged low temperature storage of carrots have been recorded in Norway. Some postharvest rots of carrot are shown in Figure 13.

Handling of carrots to avoid mechanical injury is the best control measure. Postharvest washing is in chlorinated water and storage temperature of 0 °C is recommended.

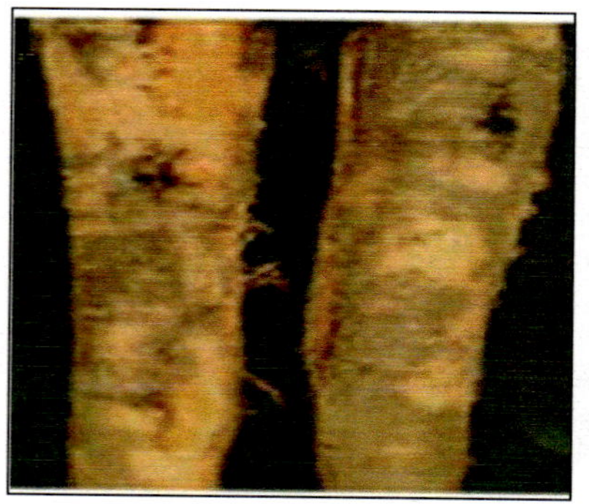

(a) *Alternaria* Rot
(http://agritech.tnau.ac.in)

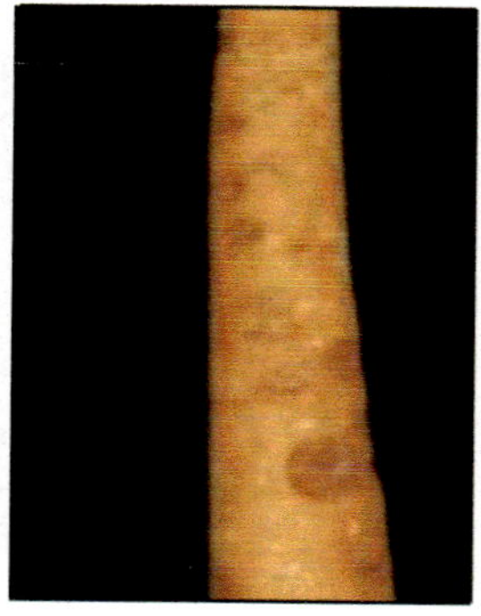

(b) Bacterial Soft Rot
(https://www.agric.wa.gov.au)

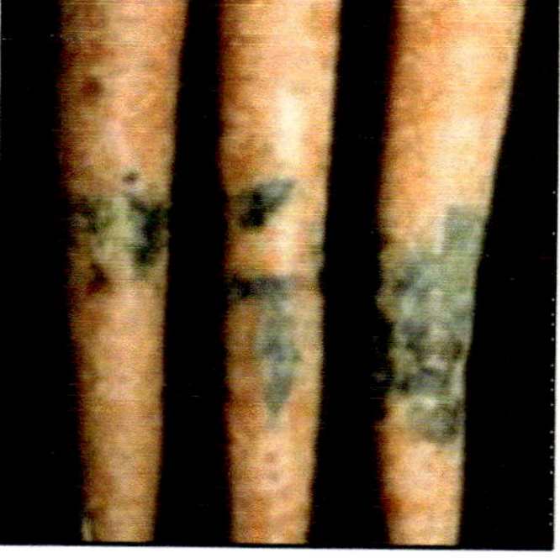

(c) Black Root Rot
(https://extension.umass.edu)

(d) Cottony Soft Rot
(http://agritech.tnau.ac.in)

(e) Crown Rot
(http://agritech.tnau.ac.in)

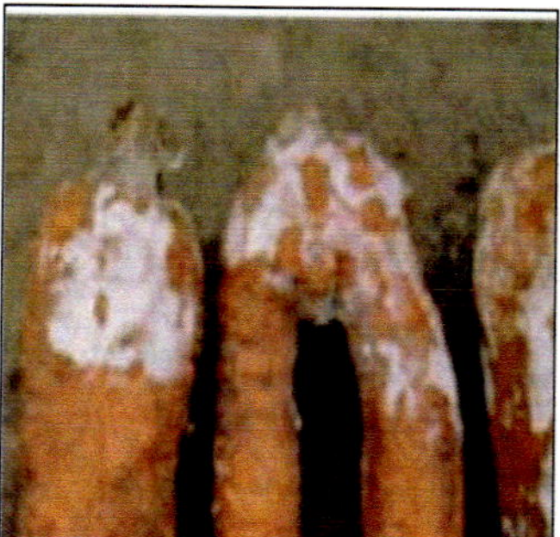

(f) Dry Rot
(http://agritech.tnau.ac.in)

Figure 13. Postharvest Diseases of Carrot

3.4 Onion

Onion is one of the few vegetable crops that can withstand rough handling, distant transportation and storage to an extensive degree. Apart from onion diseases discussed below, other fungi associated with postharvest rotting are *Fusarium oxysporum, Penicillium ducaleeni* and *Rhizopus stolonifier*. Estimated storage losses of up to 60 per cent in varieties like Nasik-Red, Bellary-Red and Pune Selection could be caused due to various fungi like *Sclerotium rolfsii, Aspergillus niger, Alternaria Solani, A. cepulicola* and *Fusarium* sp.(Figure 14)

3.4.1 Neck Rot

The disease is caused by *Botrytis allii* Munn. The fungus causes softening of scales, which become soaked in appearance, later greyish, sporulating mycelial mat develops on the surface of the scales. The disease is severe when moist conditions prevail just before and during harvest and while onions are cured in the field.

3.4.2 Black Mould Rot

In this disease, black powdery spores of the fungus adhere to the exterior scale or in between the outer two layers. The disease is most common in dry conditions and could be very destructive in storage. The main organism involved in this disease is *Aspergillus niger* causing postharvest losses up to 20 per cent.

3.4.3 Bulb Rot

This disease is most common in warm regions and is mainly due to *Fusarium solani.* The base of the bulb is usually the first site of infection. Infected onions become soft and pale-white in colour. In 7-8 days rotting completely consumes the bulb. The rate of decay is very slow below 15 °C.

3.4.4 Soft Rot

Whole onion appears to be water soaked and colour of individual fleshy scale varies from yellow to light brown in this disease. The diseased scales are soft. The micro-organisms involved are species of *Erwinia* and *Pseudomonas.*

3.4.5 Purple Blotch

Alternaria porri (E11) Cif. Infection with this pathogen causes severe storage loss with development of purple blotch. But one thing is established that in the field the disease is initiated with the infestation of onion thrips, *Thrips tabaci.* In storage the infection becomes more serious.

(a) Black Mould Rot
(http://mtvernon.wsu.edu/path_team)

(b) Bulb Rot
(http://vegetablemdonline.ppath.cornell.edu/)

(c) Neck Rot
(https://www.rhs.org.uk and http://omafra.gov.on.ca)

(d) Purple Blotch
(http://gardener.wikia.com/)

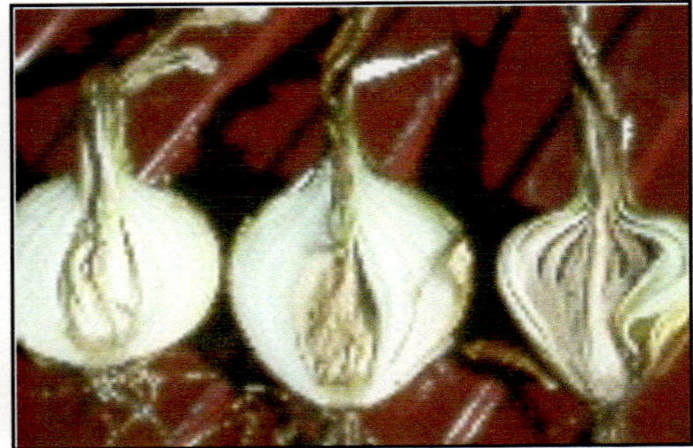

(e) Soft Rot
(http://mtvernon.wsu.edu/path_team)

Figure 14. Postharvest Diseases of Onion

3.4.6 *Cephalosporium* Rot

Cephalosporium curtipes when inoculated in bulbs, caused 26.6 per cent rotting but natural incidence in India is not of economic importance.

3.4.7 Management of Postharvest Diseases of Onion

Postharvest losses in onion can be reduced by proper curing and storing at low temperatures 0 to 2 °C and moderate relative humidity 65-75 per cent under appropriate conditions. Ventilating the onions in special drying chambers with air at 30 °C to reduce the moisture content and then treating them for 8-12 hours at 45 °C before transferring to the store, takes care of several onion pathogens in store.

Pre harvest sprays of tricyclazole or zineb in the field control purple blotch in the storage. National Horticultural Research and Development Foundation, Nashik, has recommended treating of bulbs with acetic acid which controls black mould very efficiently. Pre-inoculation treatments with 2,4-D, potassium metabisulphate, diphenylanine and acetic acid gave 58 per cent reduction in rotting due to *Fusarium* bulb rot, but none of the post inoculation treatments was effective in controlling the disease.

A formulated product of *Coniothyrium minitans* Campbell, biocontrol agent for soil borne pathogens is available in other countries to control onion basal and neck rots

3.5 Potato

In potato an average loss of 22.8 per cent is due to fungal diseases during storage and marketing. Apart from important postharvest diseases (Figure 15) discussed in subsequent paragraphs, some other important storage diseases that have been reported in potato are: dry rot of tubers due to *Aspergillus niger*, charcoal rot due to *Macrophomina phaseoli*, tuber rot due to *Penicillium herquei* and *Pythium aphanidermatum* and black leg due to bacteria *Erwinia caratovora*.

3.5.1 Brown Rot

The brown rot causing pathogen (*R. solanacearum*) survives on tuber surface for months. The tubers harvested from bacterial wilt affected fields carry infection. The infected tubers show ring of water-soaked area in vascular region on cutting, which later turns dark brown to black. Creamy bacterial ooze can be seen oozing from vascular regions in severe infections. Dry cracks can be seen on tuber surface. Isolations carried out from rotten tubers collected from different places yielded species of *Erwinia, Bacillus* and *Ralstonia* as well. The incidence of bacterial rot increases in late blight infected and late stored tubers.

3.5.2 Dry Rot

The disease develops in stores after few months of harvest and is of worldwide occurrence. The pathogens *Fusarium solani* var *coerulum,* enter through proliferated lenticels, injuries or scab pustules. The infected skin

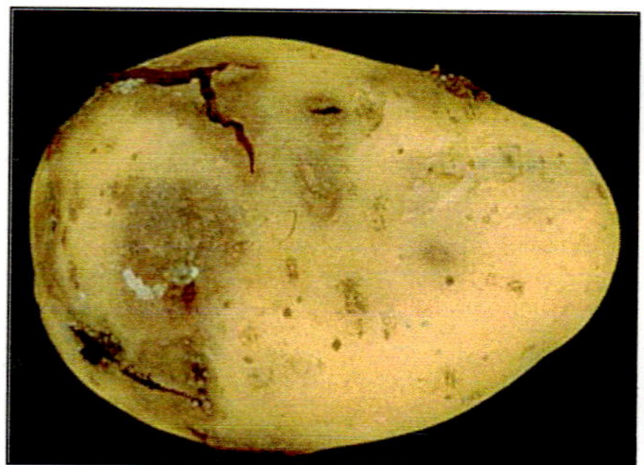
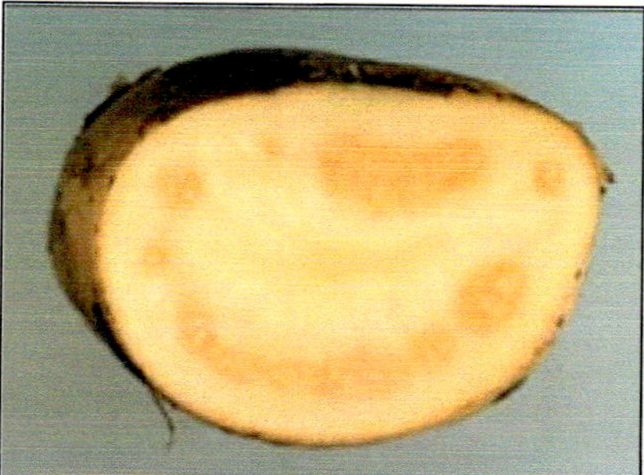
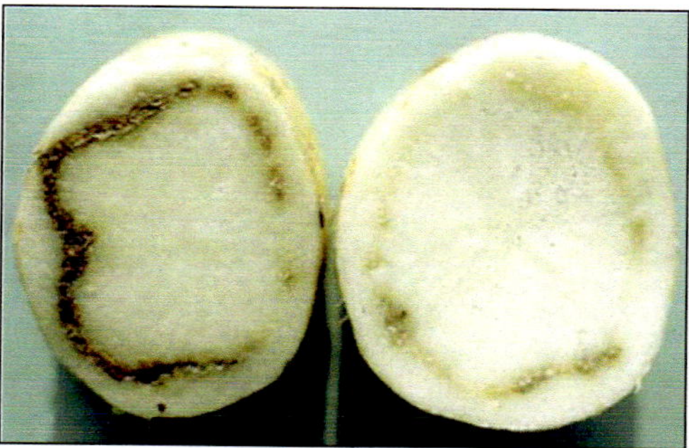

Brown Rot
(http://www.cals.ncsu.edu and http://www.agf.gov.bc.ca/cropprot/brr.htm)

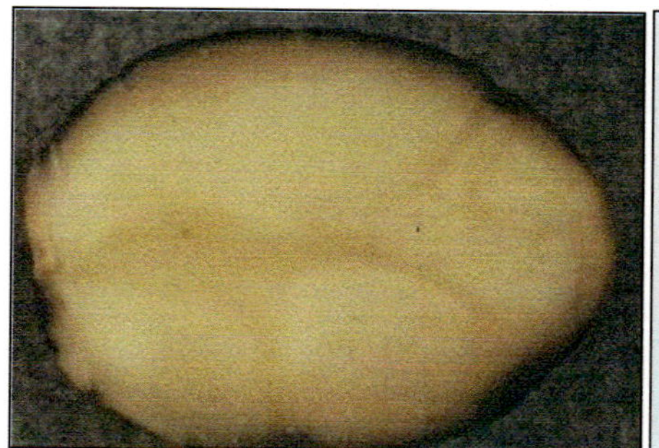

Charcoal rot
(http://www. aardappelpagina.nl)

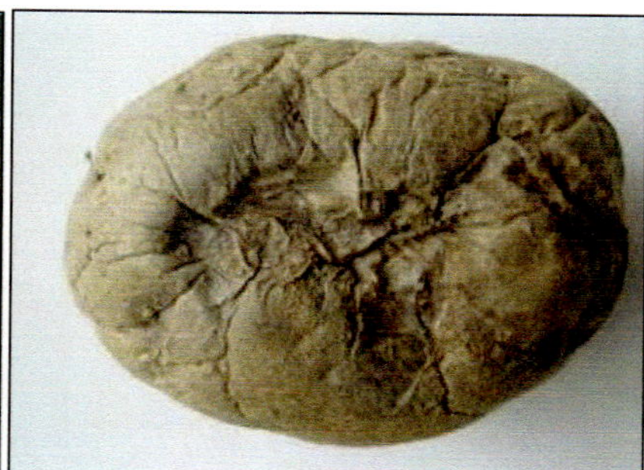

Dry Rot
(https://www.agric.wa.gov.aupotatoes)

Late Blight
(http://archive.bio.ed.ac.uk)

Figure 15. Postharvest Diseases of Potato

becomes dark in colour and as it spreads, the lesions become sunken. The sunken skin becomes wrinkled, which are prominent and arranged in concentric circles around point of infection. The tubers become dry, hard, shriveled and light weight. Inside cavities with fungal mycelium and blue stained wall are observed. The susceptibility of tubers to dry rot increased with the storage period at room temperature and the maximum infection was observed after 4 months of storage in country stores. Among the commercial varieties, the incidence of dry rot was maximum in Kufri Bahar. Some other *Fusarium* species are also reported to cause dry rot.

3.5.3 Late Blight

This disease originates in the field. The tubers start rotting during storage and marketing. Saprophytic bacteria and other fungi are often associated with late blight lesions, leading to further deterioration. The fungus *Phytophtora infestans* infects the tuber still lodged in soil causing dry rot and brown discolouration of the tissues. In wet soil the tubers may rot quickly due to opportunistic infection of soft rot bacteria. Potato variety Kufri Jyoti is resistant to this disease.

3.5.4 Charcoal Rot

The fungus *Macrophomina phaseolina* is the cause of this rot in warm dry periods, leading to heavy losses in different periods and places. The lenticels appear black and the flesh underlying skin also becomes uniformly black up to 2-5 mm depth. Secondary rot causing organisms lead to soft rot, otherwise dry rot with cavities in invaded tissues can be observed in later stages.

3.5.5 Black Sour Rot

Black sour or *Rhizoctonia* rot of potato is caused by its perfect stage *Corticium solani*. Black sclerotial bodies are formed on the diseased tubers. The fungus infects the stem and tubers start rotting in the field which continue in storage also.

3.5.6 *Rhizopus* Rot

It is a damaging postharvest disease resulting in soft rot of tubers. *Rhizopus oryzae* and *R. arrhizus* infections are responsible for the decay of stored potato tubers. In case of high infection, losses can reach up to 100 per cent during transit storage.

3.5.7 *Gilmanella* Rot

Disease caused by *Gilmanella humicola* Bavun has been reported on late blight differential varieties of potatoes in storage. Minute brownish necrotic spots around lenticels and eyes were observed during storage at room temperature (15 °C to 25 °C). Almost all the eyes of affected tubers become shriveled. Affected cells become brown, thick walled and crushed. Such tubers are unfit for seed purpose.

3.5.8 *Cylindrocarpon* Dry Rot

Cylindrocarpon tonkinensis causes dry rot of tuber in storage under normal conditions. It appears as small brown patches on the surface of tubers having extensive white mycelial growth. Internally the flesh becomes dirty grey. The tubers undergo wet rot if the humidity exceeds 60 per cent resulting in pulpy decay with thick brown exudate full of mycelium and spores. In storage the rot may progress until the affected potatoes become quite hard and mummified and are completely destroyed.

3.5.9 Management of Postharvest Diseases of Potato

Potato tubers are harvested from soil, which is an abode for several pathogenic as well as beneficial microflora. Hence, postharvest rots can be greatly reduced by manipulating the soil ecosystem to be more suppressive. General field sanitation, crop rotation and green manuring greatly influence soil and tuber borne diseases by increasing beneficial microflora and improving plant nutrition.

Killing or drying of haulms for a month before harvest and pre storage curing for 15 days is important to reduce storage rot. During harvesting, care should be taken to avoid injuries and damaged tubers should be sorted out before storage.

Potato tubers can be stored at 5-10 °C to reduce rots, however, for table purpose the tubers turn sweet at low temperature. Hence storage in non-refrigerated stores (16-30 °C with 75-95 per cent RH along with chemical treatment for sprout suppressants (CIPC, MH *etc.*) for 3-4 months, till rains start is recommended.

Boric acid (4 per cent), reduced the rot incidence to 7.4 per cent, as compared to 16.6 per cent in control (water) when potatoes were stored for 138 days at average minimum and a maximum fortnightly temperature of 28 °C and 32.3 °C at humidity of 67.5 and 78.5 per cent respectively. Disease severity of *Rhizopus* rot was significantly reduced by calcium chloride and fluorescent pseudomonads used alone or in combination, and it was nearly suppressed by azoxystrobin.

3.6 Tomato

Tomato fruit is the most widely cultivated vegetable, available in all seasons and cherished throughout the world. A large number of fungi and bacteria infect tomato that result in postharvest diseases (Figure 16), which reduce quantity as well as quality. The fruit is highly perishable and most of the infections of tomato occur during unripe stage in field, because almost all postharvest pathogens are also present on foliage parts in field. The infections later develop during postharvest handling and storage. Management throughout the season is therefore, critical for a quality produce and profit.

3.6.1 Buck Eye Rot

This disease is caused by *Phytophthora nicotianae* which is soil and seed borne and favoured by high temperature and humidity. It is more severe in poorly drained soils. On fruits, the symptoms are greyish green or brown water-soaked spots on the fruits near soil level. Lesions enlarge on the surface of infected fruit with characteristic pattern of alternating light and dark brown concentric rings which give this disease its name of buckeye. Buckeye rot lesions have a smooth surface and lack a sharply defined margin. The rot remains firm unless invaded by secondary rot causing organism.

3.6.2 *Phoma* Rot

Rot due to *Phoma destructiva* Plowr causes heavy losses in transit and storage. Lesions appear on wound or insect puncture sites or stem end sçar cracks. The spots on fruit are sunken black of increasing size. The spots develop black pycnidia in later stages.

3.6.3 Sour Rot

It is caused by *Geotrichum candidum* and is the most serious disease of tomato grown in the monsoon season. This is characterized as water soaked lesions followed by the cracks in the fruit skin which ooze out watery excretion. On the cracks white fluffy tuft of the fungus is prominent. If not controlled, the entire lot gets infected overnight emitting characteristic foul smell.

3.6.4 Anthracnose

Black dot disease in tomato is very common and is caused by different species of *Colletotrichum e.g. Colletotrichum phomoides, C. coccodes, C. dematium* and *C. gloeosporioides*. Lesions are small circular, restricted and depressed. They may enlarge to 12 mm or more diameters. The lesions are zonate and surface may appear salmon-colored

(a) *Alternaria* Rot

(b) Anthracnose

(c) Bacterial Soft Rot (http://agritech.tnau.ac.in)

(d) Bacterial Spot (http://ipm.uconn.edu)

(e) Buck Eye Rot (http://pubs.ext.vt.edu)

(f) *Fusarium* Rot (http://agritech.tnau.ac.in)

(g) Grey Mould Rot (http://pnwhandbooks.org and www.interempresas.net)

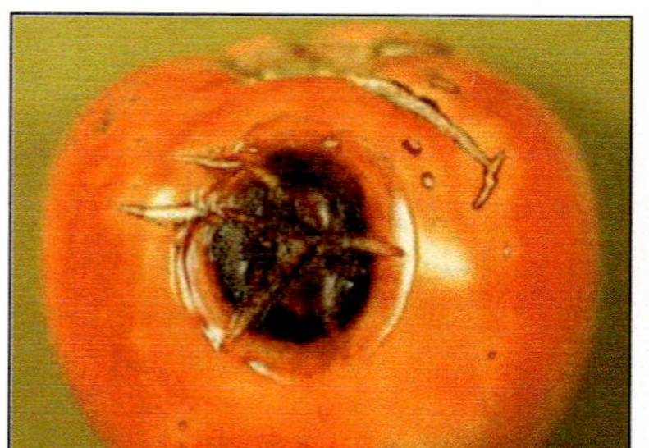

(h) *Phoma* Rot (http://erec.ifas.ufl.edu)

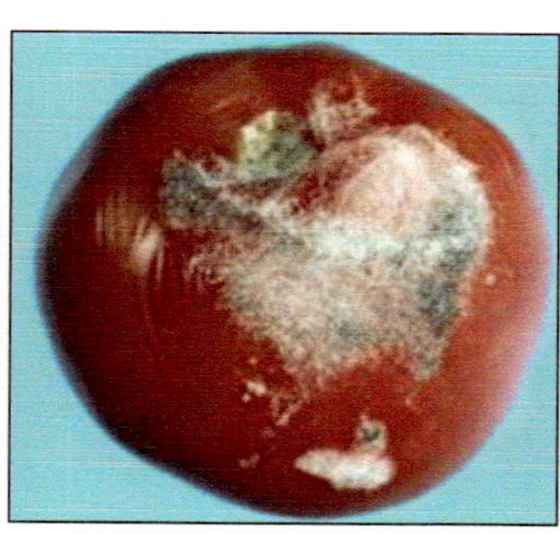

(i) *Rhizopus* Rot (http://agritech.tnau.ac.in)

(j) *Sclerotium* Rot (http://agritech.tnau.ac.in)

(k) Sour Rot (http://allabout growingtomatoes. blogspot.in)

Figure 16. Postharvest Fungal Diseases of Tomato

3.6.5 *Fusarium* Rot

The disease is caused by *Fusarium roseum* Link. and is characterised by the appearance of water-soaked lesions on ripe fruits. The fungus sporulates over the diseased area. Later the decay progresses rapidly causing soft disintegration of the pulp.

3.6.6 *Alternaria* Rot

It is a field as well as storage disease. The fungus - *Alternaria alternata* enters through stem end or injuries and results in black depressed spots with concentric rings around the point of infection. Flesh is also affected.

3.6.7 *Sclerotium* Rot

This is caused by *Sclerotinia sclerotiorum* (Lib.) de Bery. The spots are slightly sunken, yellow on diseased fruit, which decay rapidly and infected tissues collapse. The lesion is covered by a white fungal mass dotted with numerous black sclerotia. One infected fruit can destroy the entire lot.

3.6.8 Grey Mould

Grey mould caused by the fungus *Botrytis cinerea*, an amorphic form of *Botryotinia fuckeliana*, represents one of the main causes of decay in harvested tomatoes. The most common symptom is dirty white growth of the fungus on the lesion, which are soft water soaked. In varieties where infection of fungus is unable to develop further, 3 to 8 mm in diameter faint halo appear on fruits. These halos are white on immature fruits and yellow on ripe fruits. Spots do not develop further, though a small necrotic speck may appear in the spot.

3.6.9 *Rhizopus* Rot

The disease caused by *Rhizopus nigricans* Ebr. causes severe loss of fruits in transit and storage. The fungus attacks the ripe fruits. The affected fruits become soft and watery exuding clear watery fluids. The fungus penetrates deep into the tissue of the pulp and causes a soft rot covered with thick cottony mycelial growth having black sporulation. Favourable temperature for rot development is 18-24 °C and RH above 50 per cent.

3.6.10 Bacterial Soft Rot

Bacterial soft rot of tomato is caused by *Erwinia carotovora* Holland, *E. atroseptica* Jenninson, E. *aroideae* Holland. It causes extensive loss in storage. The bacteria infect the fruits and produce water soaked lesions which enlarge rapidly by making the tissue soft and discoloured. Ultimately the affected tissue disintegrates into a mass of disorganised cells. In controlled atmosphere with 3 per cent O_2 and 5 per cent CO_2, the disease incidence is low in inoculated tomatoes.

3.6.11 Bacterial Spot

The bacterial spot of tomato is caused by *Xanthomonas vasicatoria*. When the relative humidity exceeds 85 per cent and the average temperature is 20°C, the disease spreads rapidly. The bacterial spots generally affect green fruits but sometimes these are also observed on mature fruits. The size of the spot may be 6 mm in diameter. Spots on fruit begin as dark raised spots surrounded by a water-soaked border. As the fruit expands, the spots may appear to be sunken in pits, as the healthy tissue will grow faster than the spotted tissue. Older spots may reach a diameter of 1/3" and are black, slightly raised and have lobed borders. They do not penetrate the seed cavity of the fruit, but are superficial. As they age, the centers of the spots disintegrate and sink, and eventually form a crater. The skin of the fruit is often cracked at the edge of older spots. The water soaked border is absent in older spots.

3.6.12 *Oospora* Rot

This rot is also known as watery rot and is caused by *Oospora lactis*. The fungus attacks the stem end or the growth crack and invades the cellulose walls of the fruit. The decayed tissues are covered with white velvety or granular fungal growth. The affected tissues emit unpleasant odour.

3.6.13 *Cladosporium* Rot

Dry rot of the fruit pericarp region is caused by a species of *Cladosporium* and the spread of the disease is favoured during rainy weather. The rot is confined to pericarp region and does not penetrate the pulp.

3.6.14 *Malustela* Rot

The disease is caused by *Malustella seria* Batista. Irregular lesions produced on the ripe fruits are surrounded by the black spore masses of the fungus. The rot produced by the fungus is firm and not accompanied by any off-smell. The external symptom is similar to *Alternaria* rot.

3.6.15 Management of Postharvest Diseases of Tomato

Tomato being very succulent and juicy is very delicate to handle. Cultural practices like staking the plants and removing the foliage and fruits up to 30 cm from the ground level have been effective in reducing various rots. This enables sunshine and air to circulate and improve microclimate. Sanitation and cultural practices are very important for the control of soft rots. Debris should be removed from the warehouses each season before disinfecting for new storage. Before storage, fruits with scars, cracks, injuries or soft fruits should be sorted out. In controlled atmosphere with 3 per cent O_2 and 5 per cent CO_2, the disease incidence is low. The tomato fruits can be stored for prolonged periods at 7 to 12 °C along with fungicidal treatments. Red fruits are better protected than green at low temperatures.

When the fruits are transported to long distance, they should be dipped in 0.2 per cent mancozeb. Effective check of rots is also achieved with chlorothalonil treatments. Treatment with Thiobendazole 1000 ppm has been found to be effective in controlling the storage disease of tomato caused by *Colletotrichum dematium*. Spraying of streptomycin (Agrimycin 100) @1000 ppm or Bavistin (0.2 per cent) at 10 days interval provides good control of bacterial diseases. A number of bacterial and fungal antagonists have been found to effectively control postharvest rots of tomatoes. They have various modes of action that include antibiosis and/or competition for nutrients and space. This method of control for postharvest rots of vegetables appears to be feasible and may present an alternative to synthetic pesticides. Application of *Trichoderma harzianum* in the field and also as a dip before storage has been recorded to be quite effective.

4.0. Plantation Crops

4.1 Coconut

Many rot fungi are known to cause postharvest diseases (Table 3) in coconut that most commonly lead to rotting of coconut water making it unfit for consumption leading to significant losses.

4.1.1 Stem-End Rot

It is caused most commonly by *Lasiodiplodia theobromae* and its symptoms can be found on the surface as well as inside the fruit. Dark brown necrotic lesions are observed on the still-green surface of the fruits under the bracts (towards the apex) and internally the lesions are formed close to the base (Figure 17).

4.1.2 Mould

Coconut fruits may be affected by mould growing on the surface that does not usually penetrate into the fruit. *Penicillium* is the most commonly associated fungi with surface mould. It is usually caused by condensation of moisture on the surface or when the fruit is stored at a relative humidity more than 90 per cent. Modified atmosphere packaging and storage at low temperatures reduce the incidence of surface mould.

Table 3. Postharvest Diseases of Important Plantation Crops, Spices and Flower Crops their Causal Organisms

Crop	*Disease*	*Causal Organism*
Plantation crops		
Coconut	Stem-end rot/Nut rot	*Lasiodiplodia theobromae*
	Mould	*Penicillium*
	Internal black rot	*Ceratocystis paradoxa*
	Stem-end rot of immature nut	*Pestalotiopsis adusta*
Cocoa	Black pod disease	*Phytophthora palmivora*
	Frosty pod rot	*Moniliophthora roreri*
Rubber	Discoloration of rubber seeds	*Aspergillus candidus, A. tamaril, Penicillium* sp. and *Rhizopus stolonifer.*
Cashewnut	Kernel rot diseases	*Aspergillus, Penicillium, Aspergillus niger* and *Cladosporium, Fusarium, Penicillium* and *Rhizopus* sp.
Spices		
Capsicum	Anthracnose	*Colletotrichum capsici*
	Black rot	*Alternaria alternata*
Seed spices	Moulds	*Aspergillus, Fusarium* and *Penicillium*
Turmeric	Rot	*Fusarium* and *Pythium* sp.
Ginger	Soft rot	*Pythium* spp.
	Dry rot	*Fusarium* spp.
	Blue Mould	*Penicillium*
	Watery rot	*Rhizopus*
	Bacterial soft rot	*Erwinia carotovora*
	Armillaria rot	*Armillaria mellea*
	Sclerotium rot	*Sclerotium rolfsii*
Flowers		
Roses, gladiolus and geraniums	Botrytis blight or rot	*Botrytis cinerea*

Stem End Rot
(http://www.semanticscholar.org)

Nut Rot
(http://www.cpsskerala.in)

Figure 17. Postharvest Diseases of Coconut.

4.1.3 Internal Black Rot

It is caused by *Ceratocystis paradoxa* that infects the inside of the fruit. The symptoms include internal tissue black rot; darkening of the stalk base, exocarps, mesocarps and the endocarps of fruit from pedicel end; bract detachment; disruption of endosperm with extravasation of albumen.

4.1.4 Stem-end Rot of Immature Coconut

it is caused by *Pestalotiopsis adusta.* The symptoms include initially straw-colored lesions in the shell near the peduncle that later progress to entire fruit causing the coconut endosperm to rot rendering it unsuitable for consumption.

4.1.5 Nut Rot

It is also caused by *Lasiodiplodia theobromae* and leads to blackening and cracking of basal parts of the fruits (Figure 17).

4.2 Cocoa

4.2.1 Black Pod Disease

It is one of the most common disease affecting cocoa and is caused by *Phytophthora palmivora*, symptoms are black and mummified pods. Its management includes a combination of canopy management (thinning), tree height and fungicide (copper fungicides and metalaxyl) sprays. *Trichoderma asperellum* is the most effective biocontrol agent against *Phytophthora.*

4.2.2 Frosty Pod Rot

it is caused by *Moniliophthora roreri*. Its symptoms include swelling of the pod followed by formation of a creamy discoloration on the exterior of the pods resulting in white fungal mat on the surface of the pod. This disease can be managed in the field by maintaining canopy height and structure by doing annual pruning (usually once in the dry season). Removal of infected pods to prevent sporulation and reduce inoculum pressure. For pod protection, up to six applications of protectant fungicides (Cu-fungicides) during peak of flowering is recommended.

4.2.3 Charcoal Pod Rot

it is caused by *Lasiodiplodia theobromae* and the symptoms include brown necrotic tissues with dark brown powdery spore masses and greyish black mycelia.

4.3 Rubber

4.3.1 Discoloration of Rubber Seeds

Caused due to infections caused by *Aspergillus candidus*, *A. tamaril*, *Penicillium* sp. and *Rhizopus stolonifer*.

4.4 Cashewnut

4.4.1 Kernel Rot Diseases

These are majorly associated with species of *Aspergillus* and *Penicillium*. In India, presence of *Aspergillus niger* and species of *Cladosporium*, *Fusarium*, *Penicillium* and *Rhizopus* is associated with rotting of both immature and mature kernels. The fungus causing kernel rot have been found to be endophytic to cashew plants or can be introduced into the young kernels by insets or invade ovaries through flowers. These fungi may also produce some toxic metabolites in the kernels.

5.0. Spices

5.1 Capsicum

Postharvest fungal rots are caused by fungus belonging to different genera (Table 3) that include *Pythium butleri*, *Botryodiplodia theobromae*, *Phomopsis equiseti*, *Rhizopus stolonifer*, *Fusarium semitectum* [*F. pallidoroseum*] and *Choanephora cucurbitarum*. Gray mould (*Botrytis cinerea*) is also frequently reported causing brownish spots.

Black Pod Disease

Frosty Pod Rot

(https://www.google.com/images)

Charcoal Pod Rot

(https://www.plantwise.org)

Figure 18. Postharvest Diseases of Cocoa.

5.1.1 Anthracnose

It is caused by *Colletotrichum capsici* and its symptoms include small, slightly sunken dark yellow spots on the surface of the fruit which darken and enlarge and therefore merge during wet conditions. Fungal spores may be visible as pink masses on the spot arranged in rings, that later turn black. Management practices include using seeds that have been treated at 52 °C for exactly 30 minutes to make them disease free. If the previous crop was contaminated then leave a gap of 3 years and avoid plants from capsicum family during crop rotation, plough in or remove the crop residues from infected fields. Applying protective sprays beginning at flowering stage are also recommended, mancozeb or a copper containing fungicide is also effective for chemical control.

Anthracnose Black Rot
(https://ag.umass.edu/vegetable/fact-sheets)

Figure 19. Postharvest Diseases of Capsicum.

5.1.2 Black Rot

It is caused by *Alternaria alternata* and the symptoms include rot lesions or necrotic area visible on the surface. It can also be caused by other fungi such as *Curvularia lunatus [Cochliobolus lunata] and C. pallescens [Cochliobolus pallescens]*. This disease can be managed by spraying Bordeaux mixture 1.0 per cent, Copperoxychloride 0.3 per cent, Difolatan 0.3 per cent and Mancozeb 0.2 per cent. Treatment with hot air (38°C) for 48-72 hours or hot water (50 °C) for 2-3 minutes is also recommended.

5.2 Seed spices

Several spoilage moulds, especially xerotolerant species of *Aspergillus*, *Fusarium* and *Penicillium* grow and contribute to mycotoxin accumulation.

5.3 Turmeric

5.3.1 Rot

Its caused usually by species of *Fusarium* and *Pythium* which multiply rapidly on the rhizomes causing browning and discoloration of tissues that lead to development of small sized lesions. Rotting occurs upon infection by minor pathogens such as *Penicillium* and *Aspergillus* spp.

5.4 Ginger

Ginger rhizome is affected by several fungi causing rot. The major pathogen is *Fusarium oxysporum* while the minor pathogens include *Penicillium brevicompactum* and *Pythium ultimum*. Hot water treatment in combination with carbendazim and biocontrol agent (*Trichoderma harzianum*) can be effective in managing the storage rot and higher recovery of rhizomes.

5.4.1 Soft Rot

Its caused by *Pythium* spp. is one of the major diseases. Its symptoms include small brownish spots on the skin that enlarge into lesions which break down into soft and watery mass. It can be controlled by 30 minutes dip treatments with chemicals such as mancozeb, Ziram, Guazatine, Propineb and Copper Oxychloride. Ridomil MZ (1.25 g/L) also increases survival of rhizome.

5.4.2 Dry Rot

It is caused by *Fusarium* spp. and its symptoms include off-colored dry sunken lesions having a brown border/margin on the rhizome surface. Internally, a pale brown discoloration of vascular tissues occurs. It can be controlled by application of systemic and contact fungicides such as Bavistin, Ridomil gold, Captan, Dithane M-45, Copper oxychloride and Bordeaux mixture.

5.4.3 Blue Mould

Penicillium is the major cause of mould which develops on cut ends and injured tissues resulting in internal decay of tissue.

5.4.4 Watery Rot

It is caused by *Rhizopus* rot, the soft watery rot progresses rapidly and can rotten the entire rhizome quickly. The infected part is mottled brown and soft and under humid conditions white mould covers it quickly which eventually turns black. It can be managed by the use of postharvest fungicide 2,6-dichloro-4-nitroaniline (botran) just prior to packing.

5.4.5 Bacterial Soft Rot

It is caused by *Erwinia carotovora*, and the symptoms include soft wet rot having a strong foul smell. Management practices include harvesting at full maturity, proper handling and efficient curing. Also, the rhizomes should be stored under cool, dry and well-ventilated environment and never in used bags.

5.4.6 *Armillaria* Rot

Caused by *Armillaria mellea*, symptoms include development of tough, dark, string like growth adhering to rhizomes.

5.4.7 *Sclerotium* Rot

It is caused by *Sclerotium rolfsii* whose small spherical resting bodies develop into moulds. They are initially white that turn brown later. Chemical control includes Benomyl (750 ppm) or Gibberellic acid (150 ppm) or Ridomil MZ (0.3 per cent). Also steeping of rhizomes in carbendazim (0.1 per cent) for 60 minutes is effective.

Soft Rot
(apps.lucidcentral.org)

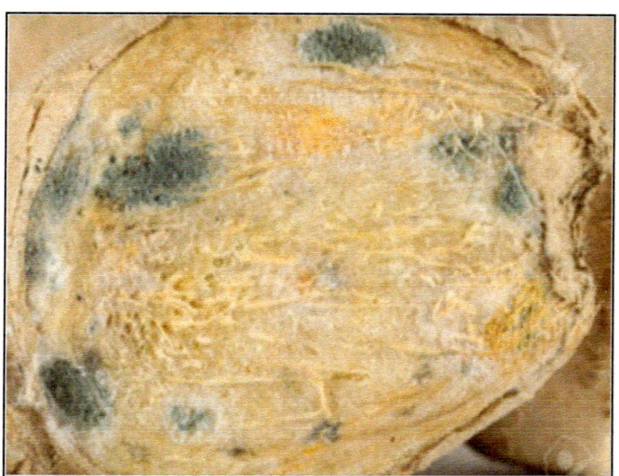

Blue Mould
(https://www.123rf.com/photo ginger-root)

Figure 20. Postharvest Diseases of Ginger.

Botrytis Infection in Roses
(https://www.ludwigsroses.co.za/rose)

Botrytis Infection in Geranium
(https://www.google.com/imgres/geranium)

Figure 21. Postharvest Diseases of Flower Crops

6.0. Flowers

6.1 Botrytis Blight or Rot

It is caused by *Botrytis cinerea* and considered the most serious problem in flowers especially roses, gladiolus and geraniums. In absence of proper storage and transport facilities, cut flowers and flowering stems are attacked by the fungus. During storage and shipping, *Botrytis* causes blight or blast on leaf and flower buds. Healthy and symptomless flowers may also decay during transit or subsequent storage. Phyto-sanitation practices such as dipping the cut flowers in hot water (50 °C) for 20-40 seconds or use of ionizing irradiation (gamma, UV-A, UV-C, electron beam) can manage the disease up to certain limit. Spraying disinfectant through a fog tunnel system can also help.

7.0 Conclusion

A wide variety of fungal and bacterial pathogens cause postharvest disease in horticulture crops. Some of these infect produce before harvest and then remain quiescent until conditions are more favourable for disease development after harvest known as postharvest diseases. These diseases have both nutritional as well as economical significance and the problem is more complex when dealing with management of horticulture crops, which are more perishable than other crops. The knowledge and application methods adopted to reduce postharvest losses are more advanced in developed countries. The developing nations have vast knowledge but are slow to understand and adopt technologies for handling postharvest produce to get high prices.

Integrated management practices should be designed taking into consideration the storage life of the commodity and the extent and type of storage losses the commodity can face. The management methods should include means to avoid the entry of pathogens, through various field operations before harvest which build up resistances in plant and keep pathogen load below economic threshold. Handling methods should be safe so as not to damage the produce, restricting entry of most of the storage pathogens which are weak parasites. Physical methods should be followed to avoid the pathogen damage by storing produce under proper storage conditions suitable for the commodity and using suitable packing, wrapping materials *etc.* Methods to treat the produce using preharvest chemical or biological sprays or postharvest dips, growth regulators, hot water treatment, radiation therapy, chemical/biological wraps should also be followed. The management strategy should be such as to give maximum returns to the investment made.

Safety of consumers should be our first priority. Use of chemicals should be done taking into consideration the pre harvest interval of the chemical and its registration for intended use. However, trends towards reduced chemical usage in horticulture are forcing the development of new strategies such as cold plasma and nanotechnology. This provides an exciting challenge for the 2lst century. Biocontrol organisms also need to be used with precaution and complete knowledge as some of the bioagents like *Burkholderia cepacia* has now been established to be an opportunistic human pathogen. *Pantoea agglomerans* once widely used as bioagent is now banned because it is known to cause skin irritation. In fact, some pathogens produce mycotoxins which can harm human beings and animals. Hence, the methodology used should be best suited to the crop with maximum benefits and human safety.

8.0 Important References

Adebola, M. O., and Amadi, J. E. 2010. Screening three Aspergillus species for antagonistic activities against the cocoa black pod organism (*Phytophthora palmivora*). *Agric. Biol. J. North Am.* 1: 362-65.

Agarwala, R. K. and Sharma, V. C. 1968. Storage rot diseases of apple. *Indian Phytopath.*, **21**: 294-98.

Ali, A., Bordoh, P. K., Singh, A., Siddiqui, Y., and Droby, S. 2016. Postharvest development of anthracnose in pepper (*Capsicum* spp): Etiology and management strategies. *Crop Prot.* **90:** 132-41.

Anonymous, Diseases of fruit crops in Australia 2009. Editors, Tony Cooke, Denis Persley, Susan House. CSIRO Publishing

Aulakh, K. S. and Grover, R. K. 1968. Ripe fruit rots of tomato and their control by oils. *Pl. Dis. Rep.* **52**: 355–59.

Baraldi, E., Mari, E., Chierici, E., Pondrelli, M., Bertolini, P. and Pratella, G. C. 2003. Studies on thiabendazole resistance of *Penicillium expansum* of pears: Pathogenic fitness and genetic characterization. *Plant Pathol.* **52**: 362–70.

Batta, Y. A. 2007. Control of postharvest diseases of fruit with an invert emulsion formulation of *Trichoderma harzianum* Rifai. *Postharvest Biol. Technol.* **43:** 143–50.

Bhargava, S. N. and Singh, A. P. 1975. *Aspergillus* rot of mango and its control. *Indian J. Hortic.* **32:** 190.

Chakravarti, D. K. and Srivastava, D. N. 1964. Stem end rot of mango and orange fruits incited by *Diplodia natalensis* Pole Evens. *Curr. Sci.* **33**: 285.

Chona, B. L. 1933. Preliminary investigations on the diseases of banana occurring in the Punjab and their method of control. *Indian J. Agric. Sci.* **3**: 673-87.

Costa, D. M. de. and Erabadupitiya, H. R. U. T. 2005. An integrated method to control postharvest diseases of banana using a member of the *Burkholderia cepacia* complex. *Postharvest Biol. Technol.* **36:** 31–39.

Dang, J. K. and Gupta, P. C. 1985. Chemical control of storage rot of onion bulbs caused by *Fusarium solani*. *Haryana Agri. Univ. J. Res.* 1984, 403-04.

Dheepa, R., Goplakrishnan, C., Kamalakannan, A., Nakkeeran, S., Mahalingam, C. A., and Suresh, J. 2018. Coconut Nut Rot Disease in India: Prevalence, Characterization of Pathogen and Standardization of Inoculation Techniques. *Int. J. Curr. Microbiol. App. Sci.* 7: 2046-57.

Dhingra, O. D. and Khare, M. N. 1971. A new fruit rot of papaya. *Curr. Sci.* **40**: 612-13.

Dhingra, O. D., Khare, M. N. and Chand, J. N. 1970. Ciger-end disease of banana in India. *Curr. Sci.* **39**: 498-99.

Dikbas, Neslihan and Bostan, Hidayet. 2009. Biological Control of Postharvest Disease Caused by *Aspergillus flavus* on Stored Lemon Fruits**.** *African J. Biotechnol.* **8:** 209-14.

Droby, S., Porat, R., Cohen, L., Weiss, B., Shapira, B., Philosoph-Hadas, S. and Meir, S. 1999. Suppressing green mould decay in grape fruit with postharvest jasmonates application. *J. Am. Soc. Hortic. Sci.* **124:** 184–88.

Droby, Samir. 1991. Biological control of postharvest diseases of fruits and vegetables: alternatives to synthetic fungicides Review Article. *Crop Prot.* **10:** 172-77.

Fravel, D. R. 2005. Commercialization and implementation of biocontrol. *Ann. Rev. Phytopathol.* **43**: 337–59.

Freire, F. C. O., Cardoso, J. E., Dos Santos, A. A., and Viana, F. M. P. 2002. Diseases of cashew nut plants (*Anacardium occidentale* L.) in Brazil. *Crop Prot.*, 21: 489-94.

Govender, V., Korsten, L. and Sivakumar, D. 2005. Semi-commercial evaluation of *Bacillus licheniformis* to control mango postharvest diseases in South Africa. *Postharvest Biol. and Technol.* **38**: 57–65

Grover, R. K. 1965. *Phoma* rot of tomato *Indian Phytopath.*, **18**: 388-89.

Gupta, R. P. and Srivastav, P. K. 1984. Study on storage diseases of onion. *Indian Phytopathol.* **37**: 404.

Hermansen, A., Wanner, L. Nærstad, R and Klemsdal, S. S. 2012. Detection and Prediction of Postharvest Carrot Diseases. *Eu. J Pl. Path.* **133**: 211-28.

Igeleke, C. L., and Omorusi, V. I. 2007. Review of postharvest deterioration of rubber seeds. *J. Agric. Soc. Sci. (JASR),* **7:** 11-19.

Kapoor, I. J. and Tandon, R. N. 1970. A new species of *Macrophoma* causing fruit rot of guava (*Psidium guajava*). *Indian Phytopath.* **23**: 122-25.

Kapur, S. P. and Chohan, J. S. 1974. Factors affecting infection and development of fruit rot in papaya by *Macrophomina phaseoli. Indian Phytopath.* **27**: 252-53.

Kaushik, C. D., Thakur, D. P. and Chand, J. N. 1972. Parasitism and control of *Pestalotia psidii* causing cankerous disease of ripe guava fruits. *Indian Phytopath.*, **25**: 61-64.

Kefialew, Y. and Ayalew, A. 2008. Postharvest biological control of anthracnose (*Colletotrichum gloeosporioides*) on mango (*Mangifera indica*). *Postharvest Biol. Technol.*, **50**: 8–11.

Khanna, K. K. and Chandra, S. 1966. Control of storage diseases of fruits. *Curr. Sci.* **45**: 70.

Khanna, K. K. and Chandra. S. 1974. A new fruit rot of banana in India. *Curr. Sci.*, **43**: 491-92.

Khanna, R. N. and Sharma, Jyotsana. 1993. Soil and tuber borne diseases. In: *Advances in Horticulture. Vol. 7- Potato.* K. L. Chadha and J. S. Grewal. (Eds.), Malhotra Publishing House, New Delhi, pp. 463-90.

Kortekamp, A. 2006. Effectiveness of calcium salts, hydrogen peroxide, azoxystrobin, and antagonistic bacteria to control postharvest rot on tobacco caused by *Rhizopus oryzae. Int. J. Pest Manage.* **52**: 109-15.

Kothari, K. L. 1968. Controlling fruit rot of guava. *Indian Hortic.* **12**: 9-10.

Lahlali, R., Serrhini, M.N. and Jijakli, M.H. 2005. Development of a biological control method against postharvest diseases of citrus fruit. *Comm. Agri. Appl. Biol. Sci.* **70**: 47–58.

Lingaraj, D. S. 1969. Mango diseases and pests. J. Mysore *Hortic.* Soc., 14: 15-18.

Long, C. A., Deng, B.X. and Deng, X. X. 2007. Commercial testing of *Kloeckera apiculata*, isolate 34–9, for biological control of postharvest diseases of citrus fruit. *Ann. Microbiol.*, 57: 203–07.

Mehta, P. R. 1940. Stem end rot and soft rot of pineapple in the United Province. *Curr. Sci.*, **9**: 330.

Mussi-Dias, V., and Freire, M. D. G. M. 2016. Evaluation of Postharvest Stem-End Rot on Coconut Fruits. *J. Agric. Environ. Sci.*, **5:** 25-35.

Nambiar, K.K.N., 1978. Controlling cashew diseases. *Indian Farming,* 28: 17–18.

Nayak, M. L. 1964. A new organism causing dry rot of potatoes. *Sci. Cult.*, **23**: 605-06.

Patel, M. K., Kamat, M. N. and Hingorani, G. M. 1950. *Pestalotia psidii* on guava. *Indian Phytopath.,* **24**: 115-18.

Pinho, D. B., Dutra, D. C., and Pereira, O. L. 2013. Notes on *Ceratocystis paradoxa* causing internal postharvest rot disease on immature coconut in Brazil. *Tropical Plant Pathology,* **38:** 152-57.

Prasad, B. G. R. 1978. An analysis of problems of marketing of few fresh vegetables in Bangalore city. *M.Sc.(Agri.). Thesis,* Univ.Agri. Sci. Bangalore.

Prasad, S. S. and Verma, R. A. B. 1970. Fungal diseases of papaya fruits in Bihar. *Indian Phytopath.*, **23**: 722-24.

Rai, J. N. 1956. Styler end rot of guava fruit (*Psidium guajava* L.). *Proc. Indian Acad. Sci.*, **43**: 55-61.

Rawal, R. D., Muniyappa, N. C. and Ullasa, B. A. 1983. Effect of pre-harvest field sprays of the fungicides on the control of storage rot of papaya. *Indian J. Agric. Sci.*, **53**: 614-15.

Reddy, S. N. and Reddy, C.K.R. 1983. *Greenaria* fruit rot: an endemic diseases of grapes in India. *Indian Phytopath.*, **36**: 110-14.

Roberto, S.R., Youssef K., Hashim, A. F. and Ippolito A. 2019. Nanomaterials as Alternative Control Means Against Postharvest Diseases in Fruit Crops. *Nanomaterials*, **9**: 1752; doi: 10.3390/nano9121752

Rosado, A. W. C., Machado, A. R., and Pereira, O. L. 2015. Postharvest stemend rot on immature coconut caused by *Pestalotiopsis adusta* in Brazil. *Plant Disease*, **99:** 1036.

Sharma, Jyotsana, Sharma, K. K. and Jadhav, V. T. 2012. Diseases of Pomegranate. In: *Diseases of Fruit Crops* A. K. Misra, P. Chowdappa, Pratibha Sharma and, R. K. Khetrapal (Eds.), Indian Phytopathological Society, New Delhi: pp 181-224.

Sharma, R. R., Singh, D. and Singh, R. 2009. Biological control of postharvest diseases of fruits and vegetables by microbial antagonists: A review. *Biol. Control* **50**: 205–21.

Siddique, S. S., St. J. Hardy. G. E and. Bayliss, K. L. 2018. Cold plasma: a potential new method to manage postharvest diseases caused by fungal plant pathogens. *Plant Pathology,* **67**: 1011–21.

Singh, D. 2002. Bioefficacy of *Debaryomyces hansenii* on the incidence and growth of *Penicillium italicum* on Kinnow fruit in combination with oil and wax emulsions. *Annals of Pl. Prot. Sci.* **10**: 272–76.

Spadaro, D., Garibaldi, A. and Gullino, M.L. 2004. Control of *Penicillium expansum* and *Botrytis cinerea* on apple combining a biocontrol agent with hot water dipping and acibenzolar-S-methyl, baking soda, or ethanol application. *Postharvest Biol. Technol.* **33**: 141–51.

Srivastava, M. P. and Tandon, R. N. 1969. Postharvest diseases of guava in India. *Plant Dis. Rept.,* **53**: 206-08.

Srivastava, M. P. and Tandon, R. N. 1971. Postharvest diseases of papaya. *Int. J.Pest Manage.,* **17**: 51-64.

Srivastava, M. P. and Tandon, R. N. 1971. Postharvest diseases of banana in India. *Indian Phytopathol.* **24**: 115-18.

Tandon, R. N. and Mishra, A. N. 1969. Fruit rot diseases of *Carica papaya* and *Musa paradisia* caused by *Rhizopus Stolonifier*. *Indian Phytopath.,* **22**: 334-41.

Valarmathi, P., and Ladhalakshmi, D. Postharvest Diseases of Cocoa. Visit us at: agrobiosonline.com.

Vir, D. and Gaur, A. 1971. Postharvest chemical treatment of apple fruit against rot and deterioration. *Curr. Sci.* **40**: 387.

Xi, L. and Tian, S.P. 2005. Control of postharvest diseases of tomato fruit by combining antagonistic yeast with sodium bicarbonate. *Sci. Agri. Sinica*. **38**: 950–955.

Yu, T., Zhang, H., Li, X., and Zheng, X. 2008. Biocontrol of *Botrytis cinerea* in apple fruit by *Cryptococcus laurentii* and indole-3-acetic acid. *Bio. Cont.* **46**: 171–77.

Zhou, T., Chu, C.L., Liu, W.T. and Schneider, K.E. 2001. Postharvest control of blue mould and grey mould on apples using isolates of *Pseudomonas syringae*. *Can. J. Pl. Pathol*. **23**: 246–52.

Zhou, T., Northover, J., Schneider, K.E. and Lu, X.W. 2002. Interactions between *Pseudomonas syringae* MA-4 and cyprodinil in the control of blue mould and grey mould of apples. *Can. J. Pl. Pathol.* **24**: 154–61.

Chapter 7

Grading and Packaging

M.S. Ladaniya

Former Director, ICAR-Central Citrus Research Institute
Amaravati Road, Shankar Nagar Post Office, Kachimet, Nagpur – 440 010, Maharashtra, India

E-mail: msladaniya@gmail.com

1.0 Introduction

Grading and packaging of fresh fruits is one of the most important steps in the long and complicated journey from grower to consumer and has direct bearing on price realized in the market. These are two important links in entire supply chain from grower to consumer. Plastic crates, baskets, cartons, wooden boxes, bulk bins, and palletized containers are the common containers for handling and transporting fresh fruits.

While grading helps in enhancing value of the fruit, proper packaging can reduce bruising, crushing and moisture loss, prevent re-contamination of the product with spoilage organisms, reduce pilferage and maintain a sanitary environment during marketing and thus improves sale of produce.

2.0 Grading

2.1 Importance

After receiving and preparing fruit in packhouse, grading is next important operation. Fruit grade indicates and assure internal and external quality to the buyer and fetches good price for the producer. Before grading, sorting of fruit is necessary. It helps in removing culled fruits of all types including too large or too small and over ripe and under-ripe fruits. Sorting and grading in large operations are mechanized.

Sorting and presizing operations are done at the same time in advanced systems where very large and very small fruits are also removed as culls along with blemished fruits. Commercial scale sorting conveyers made of aluminium rollers are used where trained people remove culled fruits manually. Machinery of various capacity (from 2 - 5 tonne per hour capacity with 60-150 cm width of the conveyer) is available depending on requirement. The rollers rotate on their axis in order to rotate the fruit to view entire surface while fruit moves forward. The culled fruit needs to be removed quickly from the area and hence such sorting conveyers need extra space around for fruit handling. In case of manual sorting the fruit sorting personnel have to be sufficiently trained to identify culls on the basis of colour, size, shape, blemishes, *etc.* The sorting conveyer should have sufficient light/illumination focused on conveyer and fruit. Supervisors take care for proper sorting to avoid losses (good fruit removed along with the culls) or any "under sorting" (culls being retained in good fruit). In advanced automatic electronic sorting systems, fruits can be sorted on the basis of colour, size, shape and blemishes by cameras (Machine–vision system) which can detect culls and drops and remove them on cross belts running below. These operations are possible with the help of image analysis by the computer which is programmed for the purpose. Fruits can be sorted in different colour categories as desired. For example - lemons can be sorted in 4 colour classes – dark green, light green, silver and yellow by electronic sorters.

2.2 Methods

Grading of fruits is done visually based on surface blemishes and colour or any other deformity as the fruit rolls on the conveyer. Conventionally it is done manually by experienced graders. Graders have to be trained for recognizing percentage of blemished part on the fruit surface since fruits having 5-10 per cent surface blemishes are sometimes allowed for various grades. Most experienced graders should be at last point of the conveyer so that no off-grade fruit is included. Fruits of second grade are removed and packed later and off-grade fruit is sent for processing/juice extraction.

Grading can also be done mechanically using high-speed electronic graders which recognize blemishes and colour using photo–optical grading system. Electro-mechanical grader used in Sunkist Growers Inc. facility in California, USA, uses x-ray scanner (50 Roentgen/min) to scan the fruits for frost damage, granulation and *Alternaria* core rot. Then the fruit is graded for colour blemishes and scars using electronic machine–vision graders. This information is combined (internal and external) and computer operated software system decides the grade and marketability of fruit and it is dropped on proper conveyer for further process. The system is commonly used for lemons and oranges. The efficiency is as high as 480 fruits per min in each line (Johnson, 1991) and there could be several such lines in a packing house.

Sorting and grading are the two most labour intensive activities. These operations engage 40-50 per cent workforce in the packinghouse and even then results are not very satisfactory. There is thus scientific as well as economic interest in developing automatic sorting and grading machines. The 'Videograding TM' is a commercial automatic grading system used in Italy. The reliability and costs are the major factors in adoption of these technologies. New sensors are being developed to evaluate internal quality like sugars and acidity along with colour vision and multiple cameras are being used to locate external defects so that quality in totality can be given to consumers (Blandini *et al.*, 1992).

2.3 Grade/Quality Standards

There are various agencies that have established grade and quality standards including packaging regulations for fruits and vegetables. The most important at international level is Codex Alimentarius Commission of the WHO/FAO. Every country also has its own standards for perishable produce on the basis of which domestic trade is organized. This Commission is established by the Food and Agriculture Organization of the United Nations (FAO) and the World Health Organization (WHO) with the purpose of protecting the health of consumers and ensuring fair practices in the food trade.

(The *Codex Alimentarius* actually means "Food Code" in Latin. The commission's work is to collect internationally adopted food standards, guidelines, codes of practice and other recommendations and evolve a common standards or Code.)

The Commission has published Codex standards for fresh fruits and vegetables to allow their wide use and understanding by governments, regulatory authorities, food industries, retailers, and consumers in the world (Codex Alimentarius Commission, 2007). These standards are also revised/amended time to time particularly after World Trade Organization (WTO) came into existence and need to be followed by Indian fruit producers/packers during export keeping in view guidelines of importing countries. These guidelines include provisions concerning fruit quality, classification/grade of fruit, fruit sizes, packaging and presentation and hygienic/sanitary conditions to be followed.

2.4 Indian Grade/Quality Standards (Agmark Standards and Regulations)

The Indian "Agmark" standards and the standards of the Codex Alimentarius Commission are similar to a large extent. The Agmark standards are established by the Directorate of Marketing and Inspection, Department of Agriculture and Co-operation, Ministry of Agriculture, Government of India. These standards are established under Agricultural Produce (Grading and Marking) Act, 1937 as amended in 1986. It confirms the quality control and the best hygienic condition of the food. The certification also marks the food standards keeping in mind the requirements of WTO and benefits all including the producer and the consumer. Under this certification, the third party guarantee to quality is certified. Under the above mentioned act, rules and Amendments in rules take place from time to time. In recent past these rules were amended in 2004, 2007 and 2012 and these come into force from the date of their Official Gazette notification.

These rules apply to commercial varieties of Fruits and Vegetables. Agricultural Marketing Adviser to the Government of India or his representative is the certificate issuing authority. Producer/packer or trader who wants to follow these rule have apply for this certification and if granted, have to follow all rules laid down therein. The "Authorized packer" means a person or a body of persons who have been granted/issued a certificate of authorization to grade and mark Fruits and Vegetables in accordance with the grade standards and procedure prescribed under these rules. The "Certificate of Authorization" is issued under the provisions of the General Grading and Marking Rules, 1988 authorizing a person or a body of persons to grade and mark Fruits and Vegetables with the grade designation mark. The grade designation mark shall consist of "AGMARK insignia" consisting of a design incorporating the certificate of authorization number, the word "AGMARK", name of commodity and grade designation.

For domestic trade, fruits and vegetables shall comply with the residue levels of heavy metals, pesticides, aflatoxin and other food safety parameters as specified in Prevention of Food Adulteration Rules, 1955 while for export, fruits and vegetables may be graded and marked as per buyer's requirements in that country provided the minimum quality/grade requirements specified are met.

Quality Specifications for some Important Vegetables are Mentioned Below

Bitter gourd	Green, lustrous, more prickly 20-25cms long with short neck
Bottle gourd	Green, tender, cylindrical, 20-25cms with soft hairs
Brinjal	Round, thorn less, dark green calyx, purple with good lustre
Chillies	Green, lustrous, 7-8cms, smooth without wrinkles
Cluster bean	Green, tender, 7-10cm in length
French bean	Tender green, 15 cms in length
Garlic	White round, 3.5-4cms having 15-20 big sized cloves
Okra	Green,tender, 6-8cms, 4-5 ridges, slender & roundish with pointed tips
Onion (Big)	Light – dark red, globose, 4-6cms in size and strongly pungent(Bangladesh, SEA, Gulf) yellow or brown for Europe and Japan; TSS: 8-10%
Onion (Multipliers)	Bigger cloved bulb of bright red colour
Onion (Small)	2-3cm dark red and round shape
Potato	White skin, long & extra long tubers with yellow-light yellow flesh(Iraq, Iran); white & red types oval shapes for Bangladesh
Tomato	Round, medium sized, red colour for Middle East and cherry tomatoes for Europe,TSS acid ratio – 10:1

2.5 Grade Designations

The Agmark grade standards for quality of fruit describe classification of fruit in three grades or classes depending on quality of fruit.

In general for all fruits, the Grade designation/ classification is suggested as follows:

i) Extra class
ii) Class – I
iii) Class - II

Besides Codex Alimentarius Commission and Agmark, several other agencies have also published standards for quality of fruits, particularly for exports to certain countries, *viz.* Agricultural and Processed Food Products Export Development Authority, New Delhi (APEDA), National Horticulture Board (NHB) and Himachal Pradesh Horticultural Produce Marketing and Processing Corporation Ltd. (HPMC).

2.6 Advantages of Grading

Grading is beneficial to the growers, traders as well as to the consumers, as it provides common standards to all and enables farmers to get better price for their produce. It assists the producers and other intermediaries in preparing fresh produce for market with appropriate labeling and helps the consumers to get standard quality produce at fair price. It facilitates the consumer to compare the prices of different qualities of a produce in the market and assures the quality of the produce and also reduces the cost of the marketing and transportation. Produce of similar grade can be stored in bulk and market values are better understood. Commodities can be bought and sold without inspection, through e-trading. Grading also provides an authentic and scientific basis in promoting and managing the marketing system. It serves as a realistic and common basis for market intelligence and reporting. It facilitates the settlement of quality disputes between buyers and sellers.

3.0 Packaging

Packaging is in dispensable for the commodity to be marketed. Either at wholesale or retail or both stages fruit has to be packed for its delivery. Different type of fruit need different types of packages depending on the nature of their physical, anatomical and physiological (mainly transpiration, respiration and ethylene production rate) characteristics and susceptibility to microbial decay. For example strawberries would need different type of packing than mango and papaya. Jackfruit would need totally different packaging. Temperature, relative humidity and ventilation also play a very important role in determining the postharvest life of packaged fruit. Thus, the packaging for fresh fruit need to be designed based on considerations as follows:

i) Providing ventilation for respiration and exchange of gases
ii) Protection against moisture/weight loss
iii) Slowing down of respiration rate, delay in ripening and decay
iv) Control of ethylene concentrations in the package
v) Protection against bruising and physical injury
vi) Protection against microbial contamination and deterioration

All aspects of packaging must be taken into account when considering the introduction of new packaging. These aspects involve, among others, cost of packaging material, labour, acceptance by trader and consumers and changes in product condition. The ultimate goal of grading and packaging is a better quality to easier handling of the produce, and better marketable product.

3.1 Agmark Standards and Regulations for Packaging of Fruits

Agmark rules also prescribe standards for packaging of fruits. As per these rules fruits shall be packed in each container in compliance with the instructions issued by the agricultural marketing adviser, Government of India, from time to time for domestic market and the recommended International Code of Practice for Packaging and Transport of Fresh Fruit and Vegetables (CAC/RCP 44-1995, Amd. 1-2004) for export.

The general common regulations for all fruits are:

i) Fruits shall be packed in such a way as to protect the produce properly.

ii) The materials used inside the package must be new, clean and of such a quality as to avoid causing any external or internal damage to the produce.

ii) The use of materials particularly of paper or stamps bearing trade specifications is permitted provided the printing or labeling has been done with non-toxic ink or glue.

iii) The containers shall meet the quality, hygiene, ventilation and resistance characteristics to ensure suitable handling, shipping and preserving of the fruits. Packages must be free of harmful foreign matter and obnoxious smell.

iv) Contents of each package or lot must be uniform and contain only fruits of same origin, variety and grade designation.

v) The visible part of the contents of the package (if present) must be representative of the entire content.

vi) The grade designation mark shall be securely affixed to or printed on each package in a manner approved by the Agricultural Marketing Adviser or an officer authorized by him.

vii) The following particulars shall be clearly and indelibly marked on each package, namely: (a) Name of the commodity, (b) Variety, (c) Grade designation, (d) Size code, (e) Lot/batch/code number, (f) Country of origin, (g) Net weight/ No. of fruit, (h) Name and address of the packer/ exporter, (i) Best before date (where applicable), (j) Storage condition, (k) Date of packing, Such other particulars as may be specified by the Agricultural Marketing Adviser.

The ink used for marking on packages shall be of such quality which may not contaminate the product. The authorized packer may, after obtaining the prior approval of the Agricultural Marketing Adviser, mark his private trade mark or trade brand on the graded packages provided that the same do not indicate quality other than that indicated by the grade designation mark affixed to the graded packages in accordance with these rules.

3.2 Functions of a Package

Package performs several functions and some of them are described below:

i) An efficient handling unit, a package is easy to be handled and serves as a marketable unit *e.g.* units with the same content and weight.

ii) Package protects fruit from rough handling during loading, unloading and transport. It protects fruit from pressure during stacking and from moisture or water loss with consequent weight and appearance loss. It also facilitates air flow through crate or box via ventilation holes and fumigation is also possible through ventilation holes.

iii) It facilitates identification (a label with country of origin, volume, type or variety of product, *etc.* printed on package serves important purpose in marketing).

iv) It provides recognizable trade name and trademark serves purpose of advertisement and help both packer and consumer. It also guarantees quality.

Proper packaging improves appearance of produce. Standard weight and count of a produce in package increases speed and efficiency of marketing. Packaging reduces cost of transport and handling. Stacking and combining of packages into layered units like pallets make mechanical handling possible. It ensures efficient use of space and reduced losses and also lowers the marketing costs. Labels on packages facilitate inspection, pricing and dispute redressal.

3.3 Types of Packaging

Very diverse type of material is used for packaging and it includes paper, plastic, wood, jute and even iron boxes. Trunks made of iron sheets are used for bulk transportation of high value Alphonso mangoes from farms near Ratnagiri (Maharashtra) to wholesale market in Mumbai. Packaging has several advantages and the most important is hygiene.

Fruit needs to be packed for bulk handling, transportation and wholesaling. It reduces losses considerably as compared to transportation of loose/ unpacked fruit in the truck. For this purpose, package unit is bigger. For retailing, fruit is packed in smaller unit packages which are also called as consumer packages and the process is called prepackaging.

The packages for fresh fruits are classified as:

i) Transport or Bulk packs

ii) Consumer or Retail packs

3.3.1 Bulk Packaging

Transport packages are designed for long distance transportation in capacities ranging from 4 – 5 kgs to 20 – 25 kgs. These packs must withstand impacts, compression and vibration during transport.

a) Sacks and Net Bags

The materials used for sacks and net bags may be woven natural fibres (jute, cotton), woven synthetic fibres (polypropylene, polyethylene), knitted natural fabric (cotton), knitted synthetic fabric (polyethylene) or non-woven synthetic film (propylene).The advantages of using sacks and net bags are merely financial. The sacks and net bags are cheap, have a low weight/volume ratio and, if made of a synthetic material, do not rot. These sacks usually contain 15-20 kg produce.

b) Wooden Boxes

Wooden Boxes provide better protection to the fresh produce against transportation hazards. They have high puncture resistance, good tensile strength and compression strength; but they occupy more space and add on to the tare weight. In addition, the nails cause injuries to the produce during long transportation. A wooden box consists of rigid corners with planks nailed or stretched against those corners. Plank thickness varies normally between 3 and 5 mm.

i) Nailed boxes as generally used for apple or pear are rigid and strong. These are formed just before packing by assembling sides and ends.

ii) Wire bound wooden boxes/crates are stitched with a wire under the stitches which gives extra strength to the container. The wire also serves as a hinge and as a lock for the lid as in case of orange or grapefruit crate.

However, the use of wooden boxes is discouraged due to campaign against deforestation.

c) Fibreboard Boxes

Fibre-board boxes are widely used because of their low weight, their range of sizes and shapes, availability, and versatility (foldable). The other advantages are: a) Low cost to strength and weight ratio, b) Good cushioning properties, c) Smooth and non-abrasive surface, d) Good printability on the outer surface of the board, e) Easy to set up and collapsible for storage, f)Reusability and recyclable, g) Can be manufactured in high volumes and h) Can be provided with ventilation by punching holes.These are of two types *e.g.* Solid fiberboard and Corrugated fiberboard.

Solid fiberboard boxes (cartons) have a thickness between 0.85 and 3 mm. If treated with wax or resin coatings, these boxes are reasonably moisture resistant.

Corrugated fiberboard boxes have a thickness varying from 1.2 up to 8 mm. The strength of corrugated fiberboard is determined by the type of fluting material, the type of facing material and its thickness and a single or double wall. Fluting and facings are kept in place by water resistant glue. These boxes are manufactured in various designs. Most common are one piece Regular Slotted Container (RSC) and two piece full telescopic container (Figure 1).

Corrugated fiberboard boxes can be arranged on the wooden pallet for handling by forklift trucks for mechanized operations. Boxes can be arranged in a cross stacking or column stacking manner.

Figure 1. Telescopic Corrugated Fibre Board Container

d) Plastic Crates

Plastic crates are more expensive than wooden crates or carton boxes, but as a result of their longer life span, the running costs are relatively low. These are usually made of high density polyethylene (HDPE) or polypropylene (PP) by injection moulding. Polyethylene has higher impact strength and a low degradation by ultra-violet radiation while polypropylene has a better scratch resistance. The performance of both materials can be improved by adding antioxidants and UV protectants (for sunlight protection).

The vented plastic crates used for various fruits during handling are of 20-25 kg capacity and hence can be handled manually. Mango, banana, citrus, pomegranate and guavas are handled in field and packinghouses in these crates. These crates are even used for long distance transportation. These are either of stackable, stack-nest or collapsible design. Collapsible plastic crates are the most expensive crates followed by stack-nest and then the stacking crates. The collapsible crates reduce the storage space requirement and transport cost of empty containers. The normal capacity varies between 20 – 40 kgs.

The advantages of plastic crates are: a) Being strong and rigid these can be used for many trips, making the cost per trip relatively low, b) Different sizes and shapes are available, c) They can be easily cleaned and disinfected and d)They are strong and weather resistant and thus they can be used in humid areas

e) Bins

Large size bins or boxes (400-600 kg capacity) are used for more efficient use of available storage spaces compared to smaller crates. Increased speed of mechanical harvesting and handling is another advantage. These bins are handled with forklift trucks for lifting and stacking.

f) Pallet of Corrugated Boxes

These can be mechanized or manual and are unitized for easy mechanical handling with fork-lift trucks. For shipping fresh fruits, the unit size of 120 x 100cm pallet is recommended by organization of economic cooperation and development (OECD). Wood is most preferred material for pallets. Usually pallets designed for 4-way access are preferred. Reusable pallets can be used only if they are made to requirements of EEC standards for use in European Pallet pool system. Non-metal straps for securing and stabilizing the pallet load are used. Corner boards are also used to immobilize the boxes in the pallet (Figure 2). The load is generally 1 metric tonne or little more per pallet.

Figure 2. Wooden Pallet Board at the Bottom with Stacked Corrugated Boxes in Pallet form for Unit Handling

3.3.2 Pre-Packaging or Consumer Packaging

This involves preparation of the fresh produce and packing the same in small marketable units of packages. Fruits like oranges, lemon, banana and grapes can be prepackaged to obtain a one and a half to two times extension in shelf-life in polyethylene bags under normal conditions, without any refrigeration requirements. In the prepackaged condition the product remains fresh without any significant change in nutritional value and taste. The prepackaged produce presents better consumer appeal and has longer shelf-life and considerable handling advantages in transport and marketing.

Prepackaging can be carried out at the farm level or in orchard itself to obtain good results. The major facility required for the prepare at the is a packing shed where these operations can be carried out. Prepackaging in combination with evaporative cooling of the storage space can really benefit the individual farmers or farm cooperatives by extending the marketing period of the fresh produce.

Consumer packages are small in size and designed to hold ½ dozen – 1 dozen fruits weighing about 2-4 kg. Many types of packages in terms of forms and materials are used as consumer packs. The selection criterion for the type of consumer pack depends on fruit characteristic and marketing strategy for the product. The most commonly used packages are listed below:

a) Active Packages
b) Corrugated Boxes/Small cartons
c) Foam Sleeve
d) Modified Atmosphere Packaging (MAP)
e) Paper Bags
f) Plastic Bags made of Flexible Plastic Films
g) Plastic films (Wrapping)
h) Plastic Net/Mesh bags
i) Plastic punnets
j) Smart packages

a) *Active Packages*

Packaging is termed as "Active" when it performs some desired role other than protecting the contents and providing an inert barrier to the external environment. The goal of developing such packaging is to create a more ideal match of the properties of the package to the requirements of the food. Active Packaging can be created by using oxygen scavengers, carbon dioxide absorbents/emitters, ethanol emitters and ethylene absorbents. The appropriate material is placed alongside the fresh produce. It modifies the atmosphere in the package and thereby contributes to the extension of shelf-life of the fresh fruit. Pads are used to provide chemical treatment to reduce decay as with sulphur dioxide pads for grapes and absorb ethylene as with potassium permanganate pads in boxes of bananas.

b) *Corrugated Boxes/Small Cartons*

Corrugated fibre board cartons with or without handle and containing 2-4 kg produce are light weight and are used for packaging of apples, grapes, mandarins and mangoes. These are printed in multi-colours with all desired information about produce and packer and are very handy for retail marketing. These can also be opened by the consumers for inspection while buying fruit. Such cartons are quite popular in case of some fruits as gift packages.

c) *Foam Sleeve*

It is a plastic tubular film made of polyethylene foam available in different colours, diameters and lengths. It can be easily slipped over the individual fruits in a snug fit form. It provides a cushioning effect and protects the fresh produce against abrasion and scratches during transit. It is hygienic, nontoxic and odourless.

d) *Modified Atmosphere Packaging (MAP)*

It is a process which involves modifying the composition of the internal atmosphere in air-tight sealed package by use of permeable film to improve the shelf life of produce. The modification process often tries to lower the amount of oxygen (O_2) from 20 per cent to 8-10 per cent as the respiring fruit consumes O_2. The gaseous atmosphere changes continuously throughout the storage due to factors like respiration and other biochemical changes of the produce and the permeation of gases through the package. This also slows down the growth of aerobic organisms and the speed of oxidation reactions. The CO_2 level inside the package increases by the process of respiration of fruit. CO_2 can also lower the pH and inhibit the growth of bacteria. The film used is permeable to oxygen and carbon dioxide while it is sparsely permeable to water vapour. Therefore, due to higher relative humidity inside the package, produce remains fresh.

MAP reduces respiration, delays ripening, and decreases ethylene sensitivity, decreases production of ethylene, retards textural softening and causes extended shelf life of the product. The crucial factor responsible for modified atmospheric storage of fruit is the permeability of the packaging film and respiration rate of the fruit. The permeability of the packaging film is mainly dependent on the thickness of the film and the type of polymer. As the respiration rate of individual fruit varies from each other, the type of polymer film used for MAP also varies. The packaging material slows down the normal respiration of the product by creating a lowered level of O_2 and a higher level of CO_2 inside the packaging material. Improper packaging material may deteriorate the fruit condition and hence reduce its shelf life. For higher shelf life of the product packaging material should be such that optimum atmosphere would be created in the package within the shortest possible time. Various types of flexible and permeable films differ in their CO_2 and O_2 permeability.

The MAP system is a dynamic process where respiration and permeation occur simultaneously. Factors affecting both of these must be considered when designing a package. The fruit respiration in a package is influenced by: 1) Quantity of the produce, 2) Stage of maturity, 3) Temperature, and 4) Concentration of gases like oxygen, carbon dioxide and ethylene. While the permeation of the film is influenced by: 1) Type and nature of material, 2) Thickness and surface area of the material/package, 3) Temperature and relative humidity and 4) Partial pressure gradients of O_2 and CO_2

All these above factors interact to create equilibrium levels of O_2 and CO_2 inside the sealed package. The selection of a film or a combination of more than one film (laminate) depends on the expected transpiration and respiration rates of the produce. For most of the fresh produce the film selected should be more permeable to CO_2 than to O_2.

e) Paper Bags

Consumer packs for 1-2 kg fruits can also be made of good quality kraft paper. These can be suitably printed and provided with jute fibre strings for handle.

f) Plastic Bags made of Flexible Plastic Films

Plastic bags made of different types of flexible plastic films like LDPE (Polyethylene), PVC (Poly Vinyl Chloride), PP (Polypropylene) and cellulose acetate are the predominant material for fruit consumer packaging. These bags are mostly used as pouches with holes punched at regular intervals to allow respiration. They are available in a wide range of thicknesses and grades and can be engineered to control the environmental gases inside the pouch. LDPE is the most widely used material.

Besides the very low material costs, automated bagging machines further reduce packing costs. Film bags are clear, allowing for easy inspection of the contents, and readily accept high quality graphics for printing. Plastic films are available in a wide range of thicknesses and grades and may be engineered to control the environmental gases inside the bag.

g) Plastic Films (Wrapping)

One of the newest trends in produce packaging is the shrink wrapping of individual produce items or wrapping with stretchable cling film. The greatest advantage of individual shrink wrapping is its ability to control moisture loss. By reducing the transpiration rate and maintaining the fruit firmness the film forms a barrier which increases the resistance to water vapour. The transpiration rate can be reduced 5 to 20 times using selective permeable plastic films. The individual fruit is loosely sealed in a flexible film. The film is then shrunk tightly around the produce by passing these packs through a heat shrunk tunnel where they are exposed to hot blown air for a very short period (few seconds). The fresh produce is then cooled by rapid ventilation. The films most commonly used are LDPE or LLDPE.

Shrink wrapping has been used successfully to package citrus, apple, pomegranate, mango and a variety of tropical fruits which are large in size.

The overall advantages of individual fruit shrink wrapping are:

i) Reduction in weight loss
ii) Maintenance of firmness
iii) Reduction in deformation
iv) Alleviation of chilling injury
v) Reduction of decay from secondary infection
vi) Delay in colour development and senescence
vii) Elimination of the need to maintain high humidity during storage/transit
viii) Extending the market season
ix) Increased consumer appeal
x) Hygiene

Trays with overwrap of film shrunk with heat using heat shrinkable film are also used. The stretch cling film can also be overwrapped on the fruit. The trays used are usually made of moulded pulp tray or plastic material like EPS, PVC and PP. The produce is placed in individual cavities so that abrasion and bruising are avoided during transportation. The trays also provide cushioning effect to the produce. The overwrap cling film is a transparent see through food-grade, odourless plastic film with the property of clinging to the product packed when stretch wrapped. This film can be applied without the application of heat. It is usually made of LDPE, LLDPE or PVC. The films are semi-permeable and allow exchange of gases for respiration of the product.

h) Plastic Net/Mesh Bags

The plastic net bags (Extruded and Woven) have the feature to stretch and accommodate all sizes and shapes of fruit. These are available in roll form or in precut lengths with stretch width of 200 mm – 400 mm. By allowing air to circulate in and around the produce, these net bags prolong the freshness and shelf-life of the fresh produce. There is no water condensation thereby preventing spoilage and wastage. The bags can be loosely tied at about 5 cm below the opening. These bags make a colourful display by allowing clear visibility of the fruit, enhancing the appeal through natural colours of fresh fruit. These are generally made of High Density Polyethylene (HDPE) or Polyamide (PA).The sturdy mesh bags have much wider use. Citrus and ber fruits are commonly sold in mesh bags. In addition to low cost, mesh bags have advantage of uninhibited air flow and can be kept in cool chamber.

i) Plastic Punnets

These small containers made of transparent PET, PVC or PP (polypropylene) material and holding 250-500 g fruit are used for strawberries, sweet cherries and other small fruits. These are strong, versatile, clear, bright containers, which offer product visibility and are provided with holes for ventilation, which keep the produce fresh. These punnets can be bar coded with price labels and all other desired information. These containers are food-grade, odourless, light weight, stackable and recyclable and give good presentation.

j) Smart Packages

These retail containers are coated with certain patented chemical material that is sensitive to volatiles emanated by spoilage causing bacteria and fungi in fruits. As the fruit starts decaying in package due to growth of spoilage organisms, the container colour changes due to coating which indicates that packed produce is no longer good for consumption.

3.3.3 New Material in Packaging

While plastics can be used because they are easy to manufacture, cheap and do not rot,the same qualities make these a huge pollution hazard. The problem comes from indiscriminate use and subsequent disposal, particularly the throwaway plastic material used in wrapping and packaging. Efforts are being made to develop eco-friendly sustainable material for packaging of produce including fruits. Verma (2009) reported jute fibre reinforced laminated paper composite and reinforcement-fibre free paper laminate to combat the war against this pollutant to certain extent. The paper laminate, without reinforcement fibre, exhibited a few fold superiority in tensile properties as compared to single paper strip. Further studies showed that an appreciable improvement in tensile properties can be achieved by introducing continuous jute fibre in paper laminates.

Jute fibres and wood are combined with thermoplastic materials like low density polyethylene (LDPE) and high density polyethylene (HDPE) for making semi-rigid as well as rigid boards. The film layer is placed on both sides of non-woven jute. The composite layers are pressed in a hydraulic press at the required temperature for a certain period of time depending upon the thickness of the end product and the type of film used. These boards have good tensile strength, puncture resistance and moisture barrier property. JRP are now being increasingly used for transport packaging of horticultural products.

4.0 Commodity Specific Grading and Packing

The Codex Alimentarius Commission was established by the FAO and WHO in 1963 for harmonization of International Food Standards, developing guidelines and codes of practice for protect the health of consumers and ensure fair practices in the food trade. The commission also promotes coordination of all food standards work undertaken by international governmental and non-governmental organizations. Increasing number of countries are now aligning their food standards with those of the Codex Alimentarius.

In India, APEDA took due cognizance of the above international developments and embarked on a programme to financially extend assistance to implement GAP, which is an internationally recognized quality assurance system for growing fruits and vegetables. Grading and packing standards as mentioned under

Fruits and Vegetables Grading and Marking Rules of "Agmark " and amended in 2004, 2007 and 2012 are given below for some important fruits grown in India.

4.1 Aonla (*Emblica officinalis*)

4.1.1 Minimum Requirements

Aonla shall be sufficiently developed and display satisfactory ripeness. It shall have the size, shape and colour characteristics of the variety. Aonla shall be fresh in appearance, whole, with characteristic shape of the variety, sound, clean, free of any visible foreign matter, free of damage or any defect or smell or condensed moisture caused by low temperature or high temperature. The stalk shall be removed close to the fruit. Fruit shall comply with the residue levels of heavy metals, pesticides and other food safety parameters as laid down by the Codex Alimentarius Commission.

4.1.2 Grade Designation

Aonla are graded in three grades:

i) Extra class

ii) Class – I

iii) Class – II

Fruit of "Extra class" must be of superior quality without any type of surface defects. The tolerance limit for this class is 5 per cent by number or weight of fruit not satisfying the requirements of the grade, but meeting those of "Class I" or, exceptionally, coming within the tolerances of that grade.

For "Class I", fruit must be of good quality with surface defects covering 10 per cent area but not affecting pulp. Tolerance is 10 per cent by number or weight of fruit not satisfying the requirements of this grade, but meeting those of next class.

For "Class –II", fruit must meet minimum requirement of quality and surface defects covering 15 per cent area are allowed but it should not be affecting pulp. Tolerance is 10 per cent by number or weight of fruit not satisfying the requirements of this grade but meeting the minimum requirements can be included.

4.1.3 Provision Concerning Sizing

Size is determined by the maximum equatorial diameter of Aonla fruit (Table 1).

Table 1. Aonla Size Code with Fruit Diameter

Size Code	*Maximum Equatorial Diameter in mm.*
A	35.1 and above
B	30.1 – 35.0
C	25.0 – 30.0

4.1.4 Size Tolerance

For all grades, 10 per cent by number or weight of aonla fruit corresponding to the size immediately above and/or below than indicated in the package shall be allowed."

4.1.5 Packaging

i) While packaging the development and condition of Aonla fruit shall be such as to enable them to withstand transport and handling and to arrive in satisfactory condition at the place of destination.

ii) Package should be marked as-fruit for table purpose or for medicinal purpose;

iii) The package of Extra class must contain fruit of uniform of colour and ripeness.

iv) Packages must be free of foreign matter.

v) Aonla may be presented – as arranged in layers, *i.e.*, one or two layers separated from each other or loose in the package (including in bulk bins).

vi) Fruit shall be transported in bulk in the means of transport in bulk packaging, except for the 'Extra Class'.

vii) Aonla in 'Extra Class' shall be particularly well packed and presented.

4.2 Apple (*Malus domestica*)

4.2.1 Minimum Requirements

(a) Apples shall be:

i) Whole, the stalk should be intact up to the shoulder height, the cut should be clean and should not damage the adjacent skin;

ii) Firm;

iii) Sound, not affected by rotting or deterioration which makes it unfit for consumption;

iv) Clean, free from any visible foreign matter;

v) Free from scabs;

vi) Free from pests and damage caused by them or diseases affecting the general appearance of the produce;

vii) Free from abnormal external moisture such as condensation following removal from cold storage;

viii) Free from any foreign smell or taste, damage caused by low or high temperature, skin puncture, signs of dehydration and internal blackening or rotting.

(b) These have minimum total soluble solids of 11 degree brix and apples shall comply with the residue levels of heavy metals, pesticides and other food safety requirements as laid down by the Codex Alimentarius Commission or requirement of importing country.

4.2.2 Grade Designation

Apples are graded in three classes as per their grade (Table 2).

In each grade/class, certain tolerance limits have been fixed to accommodate fruit with slight superficial blemishes (Table 3).

4.2.3 Provisions Concerning Sizing

Size code is determined by the diameter in millimeter of the equatorial section or weight of the fruit in grams in accordance with the Table 4.

4.2.4 Size Tolerance

For all classes, 10 per cent by number or weight of apples not satisfying requirements of the size code.

Table 2. Grade Designation for Apples with Grades

Grade Designation	*Grade Requirements*	*Grade Tolerance*
Extra class	Apples in this class must be of superior quality. The flesh must be sound. They shall be characteristic of the variety and they must be free of defects, with the exception of very slight superficial defects, provided these do not affect the general appearance of the produce, the quality, the keeping quality and presentation in the package. (five per cent of the fruits may have the defects as given in table for tolerance of defects)	Five per cent by number or weight of apples not satisfying the requirements of the grade, but meeting those of Class-I or, exceptionally, coming within the tolerances of that grade. Included therein shall be allowed not more than 1% for apples affected by internal breakdown.
Class – I	Apples in this class must be of good quality. The flesh must be sound and they must be characteristic of the variety. Provided that the following do not amount to defect if they do not affect the general appearance of the produce, the quality, the keeping quality and presentation in the package which have: A slight defect in shape and development; A slight defect in colouring; Slight skin or other defects. (ten per cent of the fruits may have the defects as given in table for tolerance of defects)	Ten per cent by number or weight of apples not satisfying the requirements of the grade, but meeting the requirements of grade II. Included therein shall be allowed not more than 1% for apples affected by decay or internal breakdown
Class – II	This class includes apples, which do not qualify for inclusion in the higher classes, but satisfy the minimum requirements. Provided that the following do not amount to defect if the apples retain their essential characteristics as regards the quality, the keeping quality and presentation which have: Defects in shape and development; Defects in colouring; Skin or other defects. (ten per cent of the fruits may have the defects as given in table for tolerance of defects).	Ten per cent by number or weight of apples not satisfying the requirements of the grade, but meeting the minimum requirements, with the exception of apples affected by decay or internal breakdown that should not be more than two percent. Included therein shall be allowed, maximum of 2% by number or weight of fruits which may show the following defects: Cork like blemishes(bitter pit) Slight damage or unhealed broken skin/cracks.

Note: In each class apples must have been carefully picked and have colour that is characteristic of the variety.

Table 3. Tolerance for Defects

Defects Allowed	*Extra Class*	*Class I*	*Class II*
Russetting outside calyx/ stem cavity, net like	1 per cent of surface area	3 per cent of surface area	5 per cent of surface area
Smooth solid russeting	1 per cent of surface area	1 per cent of surface area	3 per cent of surface area
Accumulation for both types of russetting should not exceed	1 per cent of surface area	3 per cent of surface area	5 per cent of surface area
Accumulated Blemishes and Bruising including- Within this limit scar caused by scabs* (*Venturia inaequalis*) shall not exceed.	Nil	0.25 sq.cm area	0.5 sq.cm area
Stem or Calyx cracks (healed or well cured)	Nil	0.5 cm	1 cm
Maximum length of elongated shaped defects	Nil	1 cm	2 cm

Note:

"blemish" means superficial bruising which is not covered under other defects.

"flesh defects" means visible water core and internal or external bitter pit.

"other skin defects" means defects due to hail (healed), limb damage, dark brown or black marks produced by branches, insect pest and disease spot not resulting in skin punctures.

"russetting" means particular type of skin, abnormal roughening or streak and scarring of the fruit surface, which characteristically is smooth.

Table 4. Apple Size Code, Fruit Diameter and Fruit Weight

Size Code	*Fruit Diameter (mm)*	*Fruit Weight (g)*
A	More than 80	More than 241
B	76-80	211-240
C	71-75	181-210
D	66-70	151-180
E	61-65	121- 150
F	56-60	91-120
G	50-55	60-90

Coloured apple varieties shall fulfill colour criteria to be included in to the given Class (Table 5). Following shall be the colour criteria for different grades of apples of different varieties except for green and yellow apple varieties.

Table 5. Colour Criteria for Apples

Grade	Percentage Characteristic Colour of the Variety
Extra Class	85 or more
Class – I	70 or more
Class – II	Less than 70

4.2.5 Packaging

i) The contents of each package must be uniform and contain only apples of the same origin, variety, quality, size and same degree of ripeness.

ii) The Extra Class fruits must be packed in layers only.

iii) The apples in Class-I and Class-II may be presented in one of the following forms, namely:

In regular layers according to size;

In open or closed containers;

Loose in consumer packages;

Shipping packages (master containers or cartons) or

In bulk or pallet bins.

iv) Consumer packages (of net weight not exceeding 5 Kg) may contain mixture of varieties, provided they are of uniform grade designation and each variety is of the same origin. The visible part of the content of the package must be representative of the entire content.

v) Other than bulk or pellet packing, the maximum net weight of the fruits shall not exceed 20Kg.

vi) Stickers shall be affixed to each produce so that, when removed, they neither leave visible traces of glue, nor lead to skin defects.

Himachal Pradesh Horticultural Produce Marketing and Processing Corporation Ltd. popularly known as" hpmc" recommended two Quality Grades of apple *i.e.* Extra Fancy and Fancy Class -1 are to be offered for export. Himachal Pradesh produces premium quality apples of varieties- namely Royal Delicious, Red Delicious, Rich-A-Red Delicious and Golden Delicious. The quality of Himachal Apple meets the entire export standards.

The broad parameters of the two Quality grades for export are that the fruit must be sound, fully developed, clean, without any surface blemishes, coloured, juicy and full of flavour and freshness.

Tray Pack Cartons have specially been developed for export of Himachal Apples (Figure 3). It consists of an inner fibre board carton, a series of moulded trays and an outer telescopic lid which fits right down over the sides and end of the inner carton. Apples in Tray Pack cartons are wrapped as per requirement of the buyer. The net weight of apples in the carton is not less than 17 kg but in practice it is usually over 18 kg depending on the variety. Up to 80 apples of extra large size and 175 apples of extra small size can be packed in these cartons (Table 6). The inner dimensions of the carton are 500x300x278 (ht) mm and inside dimensions of the outer cover are 504x303x281.mm.Apples are also packed in a single layer in open top carton with each fruit placed in polystyrene foam sleeve to avoid rubbing (Figure 4).

Figure 3. Apples Placed on Moulded Trays and Packed in Carton

Figure 4. Apples Packed in Polystyrene Foam Sleeve and Packed in Open Top Carton

Table 6. Apple Size, Fruit per Tray, Number of Trays per Box and Fruit per Carton

Size of Apple	No. of Apple Pocket in a Tray	No. of Trays in a Carton	No. of Apples in a Carton
Extra Large (80 to 85 mm)	20	4	80
Large (75 to 80mm)	20	5	100
Medium (70 to 75mm)	25	5	125
Small (65 to 70mm)	30	5	150
Extra Small (60 to 65mm)	35	5	175

4.3 Banana (*Musa paradisica*)

4.3.1 Minimum Requirements

In all classes, subject to the special provisions for each class and the tolerances allowed, the bananas must be:

i) Whole (taking the finger as the reference);

ii) Sound, produce affected by rotting or deterioration such as to make it unfit for consumption is excluded;

iii) Clean, practically free of any visible foreign matter;

iv) Practically free of pests affecting the general appearance of the produce;

v) Practically free of damage caused by pests;

vi) Free of abnormal external moisture, excluding condensation following removal from cold storage, and bananas packed under modified atmosphere conditions;

vii) Free of any foreign smell and/or taste;

viii) Firm;

ix) Free of damage caused by low temperatures, bruises, malformation or abnormal curvature of the fingers, with pistils removed, with the stalk intact, without bending, fungal damage or desiccation.

In addition, hands and clusters must include:

i) A sufficient portion of the crown of normal colouring, sound and free of fungal contamination;

ii) A cleanly cut crown, not beveled or torn, with no stalk fragments.

Bananas shall comply with the residue levels of heavy metals, Pesticides and other food safety parameters as laid down by the Codex Alimentarius Commission for exports.

4.3.2 Grade Designation

Bananas are graded in three classes as per their quality (Table 7).

4.3.3 Provisions Concerning Sizing

For the purposes of sizing bananas, the length of the fingers is determined along the outside curve from the blossom end to the base of the pedicel where the edible pulp ends and the diameter is defined as the thickness of a transverse section between the lateral faces. The reference fruit for measurement of the length and grade is:

i) For hands, the median finger on the outer row of the hand,

ii) For clusters, the finger next to the cut section of the hand, on the outer row of the cluster.

Table 7. Criteria for Grade Designation of Banana

Grade Designation	*Grade Requirements*	*Grade Tolerance*
Extra class	Bananas shall be of superior quality. They must be characteristics of the variety and/or commercial type. The fingers must be free of defects, with the exception of very slight superficial defects, provided these do not affect the general appearance of the produce, quality, the keeping quality and presentation in the package.	5 per cent by number or weight of bananas not satisfying the requirements of the grade, but meeting those of for Class I grade or, exceptionally, coming within the tolerances for that class.
Class – I	Bananas shall be of good quality. They must be characteristics of the variety and/or commercial type. The following slight defects of the fingers, however, may be allowed, provided these do not affect the general appearance of the produce, quality, the keeping quality and presentation in the package. Slight defects in shape and colour; Slight defects due to rubbing and other superficial defects not exceeding 2 sq.cm. of the total surface area. The defects must not affect the flesh of the fruit.	10 per cent number or weight of bananas not satisfying the requirements of the grade class, but meeting those of Class II or, exceptionally, coming within the tolerances of that grade.
Class – II	This includes bananas which do not qualify for inclusion in the higher classes, but satisfy the minimum requirements. The following defects may be there, provided the bananas retain their essential characteristics as regards the quality, the keeping quality and presentation. Defects in shape and colour provided the product remains the normal characteristics of bananas; Skin defects due to scrapping, scabs, rubbing, blemishes or other causes not exceeding 4 sq. cm. of the total surface area; The defects must not affect the flesh of the fruit.	10 per cent by number or weight of bananas not satisfying the requirements of the grade, but meeting the minimum requirements.

The minimum length of finger should not be less than 14.0 cm and the minimum grade not less than 2.7 cm.

4.3.4 Size Tolerance

For all classes, 10 per cent by number or weight of bananas not satisfying the sizing characteristics, up to a limit of 1 cm for the minimum length of 14 cm.

For export, bananas are to be packed in 13 kg carton (NHB, 2011) and fruit should be green (Table 8).

Table 8. Specification for Packaging of Banana for Export in Middle East

Variety	Grand Naine, Cavendish
Colour	Green
Weight of hand	2.5 kg
Packing	13 kg
Storage	13-14 °C
Transport	By Sea

4.3.5 Packaging

i) While packaging development and condition of the bananas must be such as to enable them to reach the appropriate stage of physiological maturity corresponding to the particular variety to withstand transport and handling, and arrive in satisfactory condition at the place of destination in order to ripen satisfactorily.

ii) Bananas must be packed in hands and clusters (parts of hands) of at least four fingers. Bananas may also be presented as single finger.

iii) Clusters with no more than two missing fingers are allowed, provided the stalk is not torn but clearly cut, without damage to the neighboring fingers.

iv) Not more than one cluster of three fingers with the same characteristics as the other fruit in the package may be present per row.

4.4 Cherry (*Prunus avium*)

4.4.1 Minimum Requirements

Cherries should have stem attached.

Missing stems are allowed in the case of :

i) "**Sweet varieties**" which naturally lose the stem at harvest, on condition that the skin is not damaged.

ii) "**Sour varieties**" on condition that there is no severe leakage of juice.

Cherries should comply with the residue levels of heavy metals, pesticides and other food safety parameters as laid down by the Codex Alimentarius Commission for exports.

4.4.2 Grade Designation

Cherries are graded in three grades:

i) Extra class

ii) Class – I

iii) Class – II

Fruit of Extra class must be of superior quality.

4.4.3 Provisions Concerning Sizing

Size is determined by the maximum diameter of the equatorial section of the fruit (in millimeters)

Cherries must have following minimum sizes: 20 mm for Extra Class and 17 mm for Class I and II.

4.4.4 Size Tolerance

For all grades, 10 per cent by number or weight of cherries not conforming to the minimum size specifies, provided, that the diameter is not less than

i) 18 mm. in the "Extra class"

ii) 15 mm. in class I and Class II.

4.4.5 Packaging

i) At the time of packaging, cherries must have been carefully picked and have reached an appropriate degree of development and displayed satisfactory ripeness in accordance with criteria proper to the variety and to the area in which they are grown.

ii) While packing, development and condition of the cherries must be such as to enable them to withstand transport and handling, and to arrive in satisfactory condition at the place of destination.

iii) Cherries marked as 'Extra' class on the package must be of uniform colouring and ripeness.

iv) The packages must be free from foreign matter.

v) For sour cherries, the package may be marked 'sour cherries'.

4.5 Custard Apple (*Annona squamosa*)

4.5.1 Minimum Requirements

At the time of maturity, custard apple colour between the areoles should be creamy white.

Fruit shall comply with the residue levels of heavy metals, pesticides and other food safety parameters as laid down by the Codex Alimentarius Commission for exports.

4.5.2 Grade Designation

Custard apple are graded in three grades:

i) Extra class

ii) Class – I

iii) Class – II

Fruit of "Extra class" must be of superior quality without surface defects. The tolerance limit for this class is 5 per cent by number or weight of fruit not satisfying the requirements of the grade, but meeting those of "Class-I" or, exceptionally, coming within the tolerances of that grade.

For "Class I", fruit must be of good quality with surface defects covering 10 per cent area but not affecting pulp. Grade tolerance is 10 per cent by number or weight of fruit not satisfying the requirements of the grade, but meeting those of next class. For "Class –II", fruit must meet minimum requirement of quality and surface defects covering 15 per cent area are allowed but not affecting pulp. Grade tolerance is 10 per cent by number or weight of fruit not satisfying the requirements of the grade but meeting the minimum requirements can be included.

4.5.3 Provision Concerning Sizing

Size is determined by the weight of the fruit (Table 9).

Table 9. Custard Apple Size Code and Fruit Weight

Size Code	*Weight (g)*
A	150-200
B	201-300
C	301 and above

Minimum weight of fruits for "Extra class" shall not be less than 200 gms.

4.5.4 Size Tolerance

For all classes, 10 per cent by number or weight of Custard apples not conforming to the minimum size specifies, for the grade, but meeting the size requirement for that grade immediately below.

4.5.5 Packaging

i) Fruit should have been carefully picked and have reached an appropriate degree of development in accordance with criteria proper to the variety and to the area in which they are grown.

ii) The development and condition of the custard apples must be such as to enable to withstand transport and handling, and to arrive in satisfactory condition at the place of destination.

iii) In case of "Extra class" and "Class I" packages, extra care is necessary to avoid any disintegration into segments.

4.6 Grape (*Vitis vinifera*)

4.6.1 Minimum Requirements

i) Bunches and berries of Table grapes shall be: (a) clean, sound, free of any visible foreign matter; (b) free of pests, affecting the general appearance of the produce; (c) free of damage caused by pests or diseases; (d) free of abnormal external moisture; (e) free of any foreign smell and/or taste; (f) free of all visible traces of moulds;

ii) Berries shall be intact, well formed and normally developed;

iii) Table grapes shall comply with the residue levels of heavy metals, pesticides and other food safety parameters as laid down by the Codex Alimentarius Commission for Exports.

iv) Table grapes shall have minimum soluble solids of 16 degrees Brix.

v) Table grapes shall have minimum sugar/acid ratio of 20: 1.

vi) Pigmentation due to sun is not a defect.

4.6.2 Grade Designation

Grapes are classified in three grades depending on their quality (Table 10)

4.6.3 Provisions Concerning Sizing

Size is determined by the weight of bunches (Table 11). The following minimum (in gms.) requirements per bunch are laid down for large and small berries grapes.

4.6.4 Size Tolerance

Extra class, Class I, Class II : 10 per cent by weight of bunches not satisfying the size requirements for the grade, but meeting the size requirements for the grade immediately below (Table 11).

After harvesting, preliminary sorting is necessary to remove culled damaged bunches. Bunches of specified weight are packed in CFB cartons of 1, 4.5 or 9 kg capacity (NHB, 2011) depending on the country to which fruit is exported (Table 12).

4.6.5 Packaging

While packaging, grapes must have been carefully picked and have reached an appropriate degree of development and ripeness in accordance with criteria proper to the variety and/or commercial type and to the area in which they are grown. The Development and condition of the Grapes must be such as to enable them to withstand transport and handling, and to arrive in satisfactory condition at the place of destination.

Different types of containers are used for packing grapes in India. These include: i) bucket shaped baskets (36cm. diameter at the top, 20 cm. at base and 25cm. height) of 5 kg capacity and made of thin bamboo strips are commonly used in Andhra Pradesh and Karnataka. ii) Wooden boxes rectangular in shape and capacity to accommodate 5 to 8kg. of fruits are also used. iii) Ventilated corrugated fibre board (CFB) cartons.

Grapes are packed with paper clippings as cushioning which absorb moisture released by the grapes and provide insulation against minor fluctuations

Table 10. Criteria for Grade Designation of Grape

Grade Designation	Grade Requirements	Grade Tolerance
Extra class	Grapes must be of superior quality. The bunches must be typical of variety in shape, development and colouring and have no defects. Berries must be firm, firmly attached to the stalk, evenly spaced along the stalk and have their bloom virtually intact.	5 per cent by weight of bunches not satisfying the requirements for the grade, but meeting those of Class I grade or exceptionally coming within the tolerances of that grade.
Class – I	Grapes must be of good quality. The bunches must be typical of variety in shape, development and colouring. Berries must be firm, firmly attached to the stalk and, as far as possible, have their gloom intact. The may, however, be less evenly spaced along the stalk than in the extra class. Following slight defects may be there, provided these do not affect the general appearance of the produce and keeping quality of the package. A slight defect in shape, A slight defect in colouring	10 per cent by weight of bunches not satisfying the requirements for the grade, but meeting those of Class II grade or exceptionally coming within the tolerances of that grade.
Class – II	The bunches may show defects in shape, development and colouring provided these do not impair the essential characteristics of the variety. The berries must be sufficiently firm and sufficiently attached. They may be less evenly spaced along the stalk than Class I grade. Following defects may be there, provided these do not affect the general appearance of the produce and keeping quality of the package. Defects in shape, Defects in colouring, Slight sun scorch affecting the skin only, Slight bruising, Slight skin defects.	10 per cent by weight of bunches and satisfying the requirements of the grade, but meeting the minimum requirements.

Table 11. Grape-Grades and Minimum Bunch Weight

Grade	Bunch Weight (g)	
	Varieties with Large Berries	Varieties with Small Berries
Extra class	200	150
Class I	150	100
Class II	100	75

Table 12. Specifications for Grape Export to Europe and Middle-East

Variety	Colour	Berry Size (mm)			Packing (kg)		
		Holland/ Germany	U.K.	Middle East	Holland/ Germany	U.K.	Middle East
Thompson Seedless	White/Amber	16	18	15	4.5 or 9	4.5 or 9	1
Sharad Seedless	Black	16	18	15	4.5 or 9	4.5 or 9	1
Flame Seedless	Pink	16	18		4.5 or 9	4.5 or 9	1

in temperature during cold storage. Size of the box varies to accommodate 2 to 4 kg. of fruit. The inner dimensions of the CFB carton accommodating 4 kg. grapes are 37 cm. in length, 25.5 cm in width and 11.5 cm. in height, and those of the carton accommodating 2kg. grapes are 25.5x18.5x11.5 cm. Irrespective of the size of the carton, each one contains two holes of 2cm. diameter on either side.

Table grapes for overseas markets are packed in five-ply corrugated fibre board boxes, printed and labeled attractively. The standard dimensions of the carton are 50 cm. x 30 cm. x 12 cm. to accommodate 5 kg. grapes.

Cartons made from recyclable materials, without wax and printed with water based inks in custom design and graphics are now available. Cartons with die cut ends require less energy and save on paper waste also.

Table grapes meant for local market are picked and packed directly in containers in the field. Table grapes meant for long distance markets and those for overseas markets are packed differently in the packing shed.

Grapes are also packed in plastic punnets with a net weight of 500 or 400 g. Trimmed and sorted fruit is punnet packed thereafter punnets are placed into 5 kg cartons. Pallet units contains 115 or 125 cartons.

4.7 Guava (*Psidium guajava*)

4.7.1 Minimum Requirements

The guavas must have been carefully picked and have reached an appropriate degree of development and ripeness in accordance with criteria proper to the variety and to the area in which they are grown. The development and condition of the guavas must be such as to enable them to withstand transport and handling, and to arrive in satisfactory condition at the place of destination.

4.7.2 Grade Designation

Guavas are graded in three grades:

i) Extra class

ii) Class – I

iii) Class – II

"Extra class" fruit must be of superior quality without surface defects. The tolerance limit for this class is 5 per cent by number or weight of fruit not satisfying the requirements of the class/grade, but meeting those of "Class I" or, exceptionally, coming within the tolerances of that grade.

For "Class I", fruit must be of good quality with surface defects covering 10 per cent area but not affecting pulp. Tolerance is 10 per cent by number or weight of fruit not satisfying the requirements of the class/grade, but meeting those of next class.

"Class –II" fruit must meet minimum requirement of quality and surface defects covering 15 per cent of the fruit area are allowed but it should not affect pulp. Ten per cent by number or weight of guavas satisfying neither the requirements of the class nor the minimum requirements, with the exception of produce affected by rotting or any other deterioration rendering it unfit for consumption.

4.7.3 Provisions Concerning Sizing

Size is determined by the weight or maximum diameter of the equatorial section of the fruit, in accordance with the Table 13.

4.7.4 Size Tolerance

For all grades, 10 per cent by number or weight of guavas corresponding to the size immediately above and/or below that indicated on the package.

Table 13. Guava Fruit Size Code with Corresponding Weight and Diameter

Size Code	*Weight (g)*	*Diameter (mm)*
A	> 350	> 95
B	251-350	86-95
C	201-250	76-85
D	151-200	66-75
E	101-150	54-65
F	61-100	43-53

4.7.5 Packaging

For export from India, guavas are being packed in 5 kg capacity open top carton. There are about 170 cartons in one pallet and gross weight of each pallet is about 850 kg. Twenty such pallets are accommodated in 40 feet long refrigerated container (Reefer) with weight of 17 tonnes.

4.8 Grapefruit (*Citrus paradisi*)

4.8.1 Minimum Requirements

Grapefruit shall be fresh in appearance, whole, with characteristic shape of the variety, sound, clean, free of any visible foreign matter, free of damage or defect or smell or condensed moisture caused by low temperature or high temperature. The stalk shall be removed close to the fruit keeping the button intact. Fruit shall comply with the residue levels of heavy metals, Pesticides and other food safety parameters as laid down by the Codex Alimentarius Commission for exports. Grapefruits shall have minimum diameter of 70 mm.

4.8.2 Grade Designation

Grapefruit are graded in three grades:

i) Extra class

ii) Class – I

iii) Class – II

Fruit of "Extra class" must be of superior quality without surface defects. The tolerance limit for this class is 5 per cent by number or weight of fruit not satisfying the requirements of the grade, but meeting those of "Class I" or, exceptionally, coming within the tolerances of that grade. For "Class I", grapefruit must be of good quality with surface defects covering 10 per cent area but not affecting pulp. Tolerance is 10 per cent by number or weight of fruit not satisfying the requirements of the grade, but meeting those of next class. For "Class –II", fruit must meet minimum requirement of quality and surface defects covering 15 per cent area are allowed but not affecting pulp. Tolerance is 10 per cent by number or weight of fruit not satisfying the requirements of the grade but meeting the minimum requirements can be included.

4.8.3 Provision Concerning Sizing

Size is determined by the maximum diameter of the equatorial section of the fruit in accordance with the Table 14.

Table 14. Grapefruit Size and Diameter

Size Code	Diameter (mm)
A	71-80
B	81-90
C	91-100
D	101-110
E	111-120
F	121-130
G	131 and above

4.8.4 Size Tolerance

i) For all classes and types of presentation: 10 per cent by number or weight of Grapefruits corresponding to the size immediately below and/or above that mentioned on the packages provided the maximum difference between the small and largest fruit does not exceed the following maxima:

Maximum difference between fruit in A,B,C and D size code should not be 7 mm and for Size code E,F and G, it should not be more than 10mm.

ii) In case of bulk consignments, 10 per cent only applies to Grapefruits with a diameter of not less than 70 mm.

4.8.5 Packaging

i) At the time of packaging grapefruits shall have been carefully picked and have reached an appropriate degree of development and display satisfactory ripeness in accordance with criteria proper to the variety and to the area in which they are grown.

ii) The development and condition of the Grape fruits shall be such as to enable them to withstand transport and handling, and to arrive in satisfactory condition at the place of destination.

iii) The content of each package must be uniform and contain only grapefruits of the same origin, variety, quality, size code, same degree of ripeness and development.

iv) During packaging, use of any substance tending to modify the natural characteristics of the grapefruits, especially its taste or smell is prohibited.

v) Fruits for "Extra Class" shall be arranged in regular layers in packages. This type of fruit arrangement is optional for Class I and II.

vi) If required, fruits of Classes I and II can be jumble packed.

4.9 Lime (*Citrus aurantifolia*)

4.9.1 Minimum Requirements

Limes shall be fresh in appearance, whole, with characteristic shape of the variety, sound, clean, free of any visible foreign matter, free of damage or defect or smell or condensed moisture caused by low temperature or high temperature. The stalk shall be removed close to the fruit keeping the button intact. Fruit shall comply with the residue levels of heavy metals, Pesticides and other food safety parameters as laid down by the Codex Alimentarius Commission for exports. Limes shall have a minimum diameter of 33 mm.

4.9.2 Grade Designation

Limes are graded in three grades:

i) Extra class

ii) Class – I

iii) Class – II

Fruit of "Extra class" must be of superior quality without surface defects. The tolerance limit for this class is 5 per cent by number or weight of fruit not satisfying the requirements of the grade, but meeting those of "Class I" or, exceptionally, coming within the tolerances of that grade.

For "Class I", limes must be of good quality with surface defects covering 10 per cent area but not affecting pulp. Tolerance is 10 per cent by number or weight of fruit not satisfying the requirements of the grade, but meeting those of next class. For "Class –II", fruit must meet minimum requirement of quality and surface defects covering 15 per cent area are allowed but not affecting pulp. Tolerance is 10 per cent by number or weight of fruit not satisfying the requirements of the grade but meeting the minimum requirements can be included.

4.9.3 Provision Concerning Sizing

Size is determined by the maximum diameter of the equatorial section of the fruit in accordance with the Table 15.

Table 15. Acid Lime Fruit Size and Diameter

Size Code	Diameter (mm)
A	33-37
B	38-42
C	43-47
D	48-52
E	53 and above

4.9.4 Size Tolerance

For all classes, 10 per cent by number or weight of limes corresponding to the size immediately below and/or above that mentioned on the packages. In any case, the tolerance of 10 per cent applies only to fruit not smaller than the minimum of 33 mm.

4.9.5 Packaging

i) While packaging, limes shall have been carefully picked and have reached at an appropriate degree of development and display satisfactory ripeness in accordance with criteria proper to the variety and to the area in which they are grown.

ii) The development and condition of the Limes must be such as to enable them to withstand transport and handling, and to arrive in satisfactory condition at the place of destination.

iii) The content of each package shall be uniform and contain only limes of the same origin, variety, quality, size code, same degree of ripeness and development.

iv) The use of any substance tending to modify the natural characteristics of the limes, especially its taste or smell is prohibited.

v) The limes shall be arranged in regular layers in packages; it is mandatory for "Extra Class" and optional for Class I and II.

vi) For Class I and II, it is allowed to pack the fruit without regular layers in packages or in bulk bins. This is called as jumble pack.

4.10 Lemon (*Citrus limon*)

4.10.1 Minimum Requirements

Lemons shall be fresh in appearance, whole, with characteristic shape of the variety, sound, clean, free of any visible foreign matter, free of damage or defect or smell or condensed moisture caused by low temperature or high temperature. The stalk shall be removed close to the fruit keeping the button intact. Fruit shall comply with the residue levels of heavy metals, Pesticides and other food safety parameters as laid down by the Codex Alimentarius Commission for exports. Lemons shall have minimum diameter of 45 mm.

4.10.2 Grade Designation

Lemons are graded in three grades:

i) Extra class

ii) Class – I

iii) Class – II

Fruit of "Extra class" must be of superior quality without surface defects. The tolerance limit for this class is 5 per cent by number or weight of fruit not satisfying the requirements of the grade, but meeting those of "Class I" or, exceptionally, coming within the tolerances of that grade. For "Class I", lemons must be of good quality with surface defects covering 10 per cent area but not affecting pulp. Tolerance is 10 per cent by number or weight of fruit not satisfying the requirements of the grade, but meeting those of next class. For "Class –II", fruit must meet minimum requirement of quality and surface defects covering 15 per cent area are allowed but not affecting pulp. Tolerance is 10 per cent by number or weight of fruit not satisfying the requirements of the grade but meeting the minimum requirements can be included.

4.10.3 Provision Concerning Sizing

Size is determined by the maximum diameter of the equatorial section of the fruit in accordance with the Table 16.

Table 16. Lemon Fruit Size Code and Diameter

Size Code	*Diameter (mm)*
A	45-50
B	51-56
C	57-62
D	63-68
E	75 and above

4.10.4 Size Tolerance

For all classes, 10 per cent by number or weight of lemons corresponding to the size immediately below and/or above that mentioned on the packages. In any case, the tolerance of 10 per cent applies only to fruit not smaller than the minimum of 45 mm.

4.10.5 Packaging

i) At the of packing, lemons shall have been carefully picked and have reached an appropriate degree of development and display satisfactory ripeness in accordance with criteria proper to the variety and to the area in which they are grown.

ii) Fruit development and condition must be such as to enable them to withstand transport and handling, and to arrive in satisfactory condition at the place of destination.

iii) The content of each package shall be uniform and contain only lemons of the same origin, variety, quality, size code, same degree of ripeness and development.

iv) The use of any substance tending to modify the natural characteristics of the lemons, especially its taste or smell is prohibited.

v) The lemons shall be arranged in regular layers in packages; it is mandatory for "Extra Class" and optional for Class I and II.

vi) For Class I and II, it is allowed to pack the fruit without regular layers in packages or in bulk bins. This is called as jumble pack.

4.11 Litchi (*Litchi chinensis*)

4.11.1 Minimum Requirements

Litchis shall be carefully picked and must be sufficiently developed and matured. The development and condition of the Litchis must be such as to enable them to withstand transport and handling and to arrive in satisfactory condition at the place of destination.

4.11.2 Grade Designation

Litchis are graded in three grades:

i) Extra class

ii) Class – I

iii) Class – II

Fruit of Extra class must be of superior quality without surface defects. The tolerance limit for this class is 5 per cent by number or weight of fruit not satisfying the requirements of the grade, but meeting those of Class I or, exceptionally, coming within the tolerances of that grade.

For Class I, fruit must be of good quality with surface defects covering 10 per cent area but not affecting pulp. Grade tolerance is 10 per cent by number or weight of fruit not satisfying the requirements of the grade, but meeting those of next class. For "Class –II", fruit must meet minimum requirement of quality and surface defects covering 15 per cent area are allowed but not affecting pulp. Tolerance is 10 per cent by number or weight of fruit not satisfying the requirements of the grade but meeting the minimum requirements can be included.

4.11.3 Provisions Concerning Sizing

Fruits are graded depending on minimum diameter (Table 17).

Table 17. Litchi Fruit Size and Diameter

Grade	*Minimum Equatorial Diameter (mm)*
Extra class	33
Class – I	28
Class – II	23

4.11.4 Size Tolerance

i) A maximum size range of 10 mm. between fruits in each package is permitted.

ii) For all grades, 10 per cent by number or weight of litchis not satisfying the requirements as regards the minimum size of grade are allowed, provided, that the diameter is not less than 23 mm.

4.11.5 Packaging

Packed litchis shall be presented under one of the following forms:

i) Individually - In this case the pedicel must be cut at the first knot and the maximum length of the stalk must not extend more than 2 mm. beyond the top of the fruit. Litchis in special grade must be presented individually.

ii) In bunches — In this case the bunch must include more than three attached and well formed Litchis. The branch must not exceed 15 cm. in length.

Fruits are packed in printed corrugated printed fibre - board boxes of required strength. Boxes should be able to withstand high humidity in cold storage during stacking.

4.12 Mandarin (*Citrus reticulata*)

4.12.1 Minimum Requirements

Mandarins shall be fresh in appearance, whole, with characteristic shape of the variety, sound, clean, free of any visible foreign matter, free of damage or defect or smell or condensed moisture caused by low temperature or high temperature. The stalk shall be removed close to the fruit keeping the button intact. Fruit shall comply with the residue levels of heavy metals, Pesticides and other food safety parameters as laid down by the Codex Alimentarius Commission for exports. Mandarins shall have minimum diameter of 54 mm.

4.12.2 Grade Designation

Mandarins are graded in three grades:

i) Extra class

ii) Class – I

iii) Class – II

Fruit of "Extra class" must be of superior quality without any type of surface defects. The tolerance limit for this class is 5 per cent by number or weight of fruit not satisfying the requirements of the grade, but meeting those of "Class I" or, exceptionally, coming within the tolerances of that grade.

For "Class I", mandarins must be of good quality with surface defects covering 10 per cent area but not affecting pulp. Tolerance is 10 per cent by number or weight of fruit not satisfying the requirements of the grade, but meeting those of next class. For "Class –II", fruit must meet minimum requirement of quality and surface defects covering 15 per cent area are allowed but not affecting pulp. Tolerance is 10 per cent by number or weight of fruit not satisfying the requirements of the grade but meeting the minimum requirements can be included.

4.12.3 Provision Concerning Sizing

Size is determined by the maximum diameter of the equatorial section of the fruit in accordance with the Table 18.

4.12.4 Size Tolerance

For all classes and types of presentation: 10 per cent by number or weight of Mandarins corresponding to the size immediately below and/or above that mentioned on the packages, provided the maximum difference between the smallest and largest fruits does not exceed the following maxima.

Table 18. Mandarin Fruit Size and Diameter

Size Code	*Diameter (mm)*
A	54-61
B	62-69
C	70-77
D	78-85
E	Above 85

Maximum difference allowed between fruit in the same package is 7 mm for size Code of A,B,C and D and difference of 10 mm is allowed for size code E. In any case, the tolerance of 10 per cent applies only to fruit whose diameter is not less than 54 mm.

4.12.5 Packaging

The content of each package shall be uniform and contain only Mandarins of the same origin, variety, quality, size, same degree of ripeness and development.

The use of any substance tending to modify the natural characteristics of the Mandarins, especially its taste or smell is prohibited.

The Mandarins shall be presented:

i) Arranged in regular layers in packages. It is mandatory for "Extra Class" and optional for Class I and II.

ii) Not arranged in regular layers in packages. This presentation is only allowed for Class I and II.

4.12.6 Bulk Packing Containers

In India, mandarin fruits are packed in wooden boxes for long distance shipments/transportation. It is estimated that 8-9 million wooden boxes are required annually to pack 'Nagpur' mandarin fruit produced in central India alone. The common lining material is newspaper in the box with paddy straw as a cushioning. With unavailability of wood over the period of time wooden boards and batons have become thinner (thickness reduced from 5-6 mm to 3-4 mm) resulting in damage to minimum 5-10 per cent boxes during transport. Anandswamy and Venkatsubbiah (1976) tried to improve wooden cases (42x32x29 cm internal dimensions) to hold 110-120 mandarins and also developed ventilated corrugated fibre board boxes (46.5 x 27.5 x 27.5 cm internal dimensions) with 18 kg gross weight for export of 'Coorg' mandarins. For transportation of Kinnows, gunny bags and bamboo baskets were reported to be cheaper than CFB and wooden boxes but losses were higher in bags and baskets (Jain and Chauhan, 1994).

To a limited extent cotton plant stalks (CPS) are utilised as domestic fuel by the rural masses in central India, while the bulk of the stalks are disposed off by burning in the field itself. The material is rich in cellulose and is much similar to hard wood in fibre dimensions. It has been found to be suitable raw material for the preparation of pulp and paper. The stacking strength of the box can be achieved with kraft paper of desired grammage. As a substitute for wooden boxes, corrugated fibre board boxes from commercially available kraft paper and the kraft paper from cotton plant stalks (CPS) have been designed and developed (Ladaniya *et al.*, 1999; Shaikh *et al.*, 2003).

Since the water absorption is a major problem in CFB boxes under high humidity conditions of refrigerated storage, water vapour barrier options such as 1). lamination with PP film, 2).coating with wax, 3).use of bitumen paper and 4). resin coating can be utilized.

The cost of the CFB box has been a major constraint in its wide acceptance by the mandarin industry in India and hence boxes with cost comparable to conventional wooden box have been developed. The one piece regular slotted type cartons cost comparatively less. The cost of CPS kraft paper boxes is also on par with wooden boxes. Corrugated fibre board (CFB) cartons of following specifications have been suitable for packing, transport and storage of 'Nagpur' mandarin fruits. The specifications are as follows: Two piece, 5 ply, full telescopic, laminated with polypropylene film from outside, B flute vertical, 48.5 x 28.5 x 28.5 cm inner dimensions, 50 x 30 x 30 cm outer dimensions, bursting

Table 19. Fruit Size and Count of 'Nagpur' Mandarin in Conventional Wooden Boxes and CFB Cartons

Wooden Box with Paddy Straw Cushion (45.5 x 32.5 x 32.5cm)			*Telescopic CFB Box with Paper Board Dividers (50x30x30cm)*		
Fruit Size	*Fruit Diameter Average (cm)*	*Count*	*Fruit Size*	*Fruit Diameter (cm)*	*Count*
Extra large	7.55	80-120	Extra large	7.31-8.00 (Avg. 7.65)	96-104
Large	7.00	120-140	Large	6.71-7.30 (Avg. 7.00)	112-120
Medium	6.40	150-180	Medium	6.11-6.70 (Avg. 6.40)	140-175
Small	5.80	170-200	Small	5.50-6.10 (Avg. 5.80)	200
Very small	5.00	225-270	Very small	4.70-5.49 (Avg. 5.00)	225-250

Source: Ladaniya *et al.*, 1999.

strength 10 kg/cm^2 (5 ply), compression strength 460kg weight of box (at 25-30 °C; 50-60 per cent RH) about 1.10 kg with cobb value less than 50 or nil. These boxes are most suitable with reference to fruit count and utilization of space in transport vehicle. Number of fruits packed in the box varies with the size of the fruit. The count of fruit in the carton is similar to that of conventional wooden box as given in Table 19. Boxes with 19.25 kg/cm^2 bursting strength of the board are desirable for export.

Aeration vents at 4-5 per cent of side areas punched as 4 long slits (9.5 cm x 1.75 cm) and 1.65 per cent of area punched at ends with one slit (8 x 1.7 cm) as a handling slot are sufficient. Ventilation up to 6 per cent of the side areas of the box provided better aeration during precooling in shipping containers.

Nearly 40 to 50 per cent of 'Nagpur' mandarin fruit produced in Central India is packed in conventional wooden boxes (48x 35 x 35 cm) for dispatch to distant markets either from orchard or assembly markets. Earlier gunny bags or bamboo baskets with paddy straw were used for packing of Kinnows produced in Punjab and Rajasthan and transported by truck (Chauhan *et al.*, 1987; Chopra and Dhar, 1990). Since 1995-96 CFB cartons are increasingly being used for packing of Kinnow fruit as the fruit is waxed in modern packing houses. Presently more than 80 per cent Kinnow mandarin fruit from Rajasthan and Punjab states (India), is packed in 10 kg capacity boxes (48-84 per carton count; carton size 45 x 23 x 18 cm) for export (Table 20). Ten kg capacity carton was also used by some exporters for 'Nagpur' mandarin.

Table 20. Size and Number of Kinnows in 10 kg Box

'Kinnow' Size (mm)	*Number of Fruit in 10 kg Box*
60 – 64	84
65 – 69	72
70 – 72	60
72-74	54
75-79	51
80-85	48

For export of Nagpur mandarin to countries in Middle East, fruit of specified grade and weight need to be packed in CFB cartons (NHB, 2011). The packaging and other details are given in Table 21:

Table 21. Fruit Specifications for Export

Colour	Light orange
Fruit weight	150-170 g for 65-70 mm fruit,90-100g for 45-50 mm fruit
Size (diameter)	65-70 mm to be packed in 7 kg box
	45-50 mm to be packed in 2.5 kg box
Temperature for shipping	5-7 °C
Shipping	By sea (Reefer)

The CFB cartons used for mandarins in California are two piece full telescopic type half slotted with side vents. The dimensions are 41.6cm (L) x 27.2cm (W) x 19.0-21.6cm (H) and 41.6cm (L) x 27.2cm (W) x 26cm (H) with volume nearly 28 lit. The fruit count is : Jumbo – 100 count (6.35 – 6.98cm diameter), Large – 135 count (5.72 – 6.35cm diameter), Medium – 180 count (5.08 – 5.72cm diameter. Small – 230 count (4.44 – 5.08cm).

4.13 Mango (*Mangifera indica*)

4.13.1 Minimum Requirements

In all classes, subject to the special provisions for each class and the tolerances allowed, the mangoes must be:

i) Whole;

ii) Sound, produce affected by rotting or deterioration such as to make it unfit for consumption is excluded;

iii) Clean, practically free of any visible foreign matter;

iv) Practically free of damage caused by pests;

v) Free of abnormal external moisture, excluding condensation following removal from cold storage;

vi) Free of any foreign smell and/or taste;

vii) Firm;

viii) Fresh in appearance;

ix) Free of damage caused by low temperatures;

x) Free of black necrotic stains or trails;

xi) Free of marked bruising; and

xii) Sufficiently developed and display satisfactory ripeness.

When a peduncle is present, it shall not be no longer than 1.0 cm.

Mangoes shall comply with the residue levels of heavy metals, pesticides and other food safety parameters as laid down by the Codex Alimentarius Commission for exports.

4.13.2 Grade Designation

Mangoes are classified in to three grades/classes depending on fruit quality (Table 22).

4.13.3 Provisions Concerning Sizing

Size is determined by the weight of the fruit, in accordance with the Table 23.

4.13.4 Size Tolerance

For all grades, 10 per cent by number or weight of mangoes in each package are permitted to be outside (above or below) the group size range by 50 per cent of the maximum permissible difference for the group. In the smallest size range, mangoes must not be less than 90

Table 22. Criteria for Grade Designation of Mangoes

Grade Designation	*Grade Requirements*	*Grade Tolerance*
Extra class	Mangoes must be of superior quality. They must be characteristic of the variety. They must be free of defects, with the exception of very slight superficial defects, provided these do not affect the general appearances of the produce, the quality, the keeping quality and presentation in the package	10 per cent by number or weight of mangoes not satisfying the requirements for the grade, but meeting those of Class II grade or, exceptionally coming within the tolerances of that grade.
Class – I	Mangoes must be of good quality. They must be characteristic of the variety. Mangoes may have following slight defects, provided these do not affect the general appearance of the produce, the quality, the keeping quality and presentation in the package. Slight defects in shape; Slight skin defects due to rubbing or sunburn, suberized stains due to resin exudation (elongated trails included) and healed bruises not exceeding 2,3,4,5 sq. cm. for size groups A, B, C, D, respectively	10 per cent by weight of bunches not satisfying the requirements for the grade, but meeting those of Class II grade or exceptionally coming within the tolerances of that grade.
Class – II	This grade includes mangoes which do not qualify for inclusion in the higher grades, but satisfy the minimum requirements. Mangoes may have following defects, provided they retain their essential characteristics as regards the quality, keeping quality and presentation. Defects in shape, Slight skin defects due to rubbing or sunburn, suberized stains due to resin exudation (elongated trails included) and healed bruises not exceeding 4,5,6,7 sq. cm. for size groups A, B, C, D respectively.	10 per cent by number or weight of mangoes not satisfying the requirements of the grade, but meeting the minimum requirements.

In Class I and Class II, scattered suberized rusty lenticels, as well as yellowing of green varieties due to exposure to direct sunlight, not exceeding 40 per cent of the surface and not showing any signs of necrosis are allowed.

gms. and for those in the largest size range a maximum of 925 grams applies, as follows (Table 24).

Table 23. Mango Fruit Size and Code

Size Code	*Weight (g)*	*Maximum Permissible Size difference between the Fruit within the Package (g)*
A	100-200	50
B	201-350	75
C	351-550	100
D	551-800	125

Table 24. Size Range, Permissible Size difference

Size Code	*Normal Size Range*	*Permissible Size Range (10 per cent of fruit/package exceeding the normal size range)*	*Maximum Permissible difference between Fruit in each Package*
A	100-200	90-220	50.00
B	201-350	18-425	112.50
C	351-550	251-650	150.00
D	551-800	429-925	187.50

4.13.5 Packaging

i) At the time of packing, development and condition of the mangoes must be such as to enable them to ensure a continuation of the maturation until they reach appropriate degree of maturity corresponding to the varietal characteristics.

ii) Fruit shall be firm enough to withstand transport and handling, and to arrive in satisfactory condition at the place of destination.

For export purpose, packaging of mangoes is done in corrugated fibre board boxes having outer ply of duplex board as detailed below in the Table 25. Only packing material of food grade should be used for packing mangoes at the pack-house facility and the package boxes having dimensions of 370 X 275 X 90 mm should be used for packing export mangoes, as approved by US Department of Agriculture- Animal and Plant Health Inspection Service (USDA-APHIS), Each package must be either preprinted or affixed with a specified label, which is duly approved by the US Department of Agriculture- Animal and Plant Health Inspection Service (USDA-APHIS). The label should be appropriately marked/stamped on left-half side indicating Production Unit Code Number (PUC), Packinghouse Code Number (PHC), Date of Packing, and Lot Number.

While packing suitable holders (polystyrene foam individual fruit holders and bubble sheets etc need to be used to avoid any injury to the fruit. Recommendation of APEDA for packaging of mango for export purpose are as follows: 1)Each individual fruit of mango shall be enclosed in a clean, white, soft, expandable and netted type polystyrene sleeve to prevent bruising before

Table 25. Specifications for Corrugated Fibre Board (CFB) Boxes for Packing Mango for Export

	Characteristics	*Requirements*			
		Ring and Flap (4 kg)	*Full Telescopic (4 kg)*	*Ring Flap (8 kg)*	*Full Telescopic (8 kg)*
i)	Material used	3 Ply Corrugated fibre board	3 Ply Corrugated fibre board	5 Ply Corrugated fibre board	5 Ply Corrugated fibre board
ii)	Grammage (g/m sq), Min (outer to inner)	*230/140/140	*230/140/140	*230/140/140	*230/140/140
iii)	Bursting strength kg/cm sq, Min	6.50	6.50	10.50	6.50
iv)	Puncture resistance, ozs inches/tear inch Min	110.00	110.00	280.00	110.00
v)	Compression strength, kgf, Min	275.00	275.00	250.00	250.00
vi)	Cobb (30 minutes) g/m sq, Max	130.00	130.00	130.00	130.00

* Outer ply of duplex board ; Source: Postharvest Manual for Export of Mangoes, APEDA, New Delhi.

packing in a box. 2) The mangoes must be packed in insect-proof boxes. If ventilated boxes are used, all the ventilator openings of the box should be covered with insect-proof screen of minimum 30 mesh and all the sides of box should be sealed with adhesive tape to prevent any entry of pests. 3)The materials used inside the package must be of food grade, new, clean, and of a quality such as to avoid causing any external or internal damage to the produce. 4)The use of materials, particularly of paper or stamps bearing trade specifications is allowed, provided the printing or labeling has been done with non-toxic ink or glue(APEDA, 2007).

For domestic market, usually mangoes are packed in wooden boxes, details of which are given below (Table 26). However, Alphonso and Kesar are packed in CFB boxes.

Table 26. Specifications of Wooden Boxes for Domestic Market

Type of Carton	*Inner Dimension (cm)*	*Capacity (kgs)*
Wooden crates	45 × 30 × 30 (Ratnagiri)	16-18
	21.6 × 21.6 × 42 (Malihabad)	10-11

For export of mangoes to countries in Europe and Middle –East, fruit of specified weight need to be packed in CFB cartons (NHB, 2011). Variety and other details are given in Table 27.

4.13.6 MAP/Shrink Wrapping

Green Alphonso mangoes wrapped with heat-shrinkable film were stored up to 5 weeks at 8 °C in acceptable condition (Figure 5). Fruits ripened normally at room temperature during one week ripening period and further had shelf life of one week (Sudhakar Rao, Personal communication, 2013).This indicated that shelf life of Alphonso mangoes can be up to 7 weeks by shrink wrapping.

4.14 Papaya (*Carica papaya*)

4.14.1 Minimum Requirements

Papaya shall be whole, fresh in appearance, firm, sound, clean, free of any type of damage or smell.

The peduncle, if present, should not exceed a length of 1 cm.

Table 27. Specifications for Export of Mango to Europe and Middle-East

Variety	*Weight (g)*			*Packing (kg)*			*Shipment*			*Storage Temperature*
	Netherlands/ Germany	*U.K.*	*Middle East*	*Netherlands/ Germany*	*U.K.*	*Middle East*	*Netherlands/ Germany*	*U.K.*	*Middle East*	
Alphonso	250-300	250-300	250-250	1 Doz/2.5	1 Doz/2.5	1 Doz/2.5	By Air	By Air	By Sea	13 °C
Kesar	225-250	225-250	200-250	1 Doz/2.5	1 Doz/2.5	1 Doz/2.5	By Air	By Air	By Sea	13 °C

Figure 5. Shrink Wrapped Green Alphonso Mangoes after Ripening

Papaya shall comply with the residue levels of heavy metals, pesticides and other food safety parameters as laid down by the Codex Alimentarius Commission for exports.

4.14.2 Grade Designation

Papayas are graded in three grades:

i) Extra class

ii) Class – I

iii) Class – II

Fruit of Extra class must be of superior quality without surface defects. The tolerance limit for this class is 5 per cent by number or weight of fruit not satisfying the requirements of the grade, but meeting those of Class I or, exceptionally, coming within the tolerances of that grade.

For Class I, papayas in this class must be of good quality with surface defects covering 10 per cent area but not affecting pulp. Tolerance is 10 per cent by number or weight of fruit not satisfying the requirements of the grade, but meeting those of next class. For Class –II, papayas must meet minimum requirement of quality and surface defects covering 15 per cent area are allowed but not affecting pulp. Tolerance is 10 per cent by number or weight of fruit not satisfying the requirements of the grade but meeting the minimum requirements can be included.

4.14.3 Provisions Concerning Sizing

Size is determined by the weight of the fruit with a minimum weight of 200 g. in accordance with the Table 28.

Table 28. Papaya Size Code and Corresponding Weight

Size Code	*Weight (g)*
A	200-300
B	301-400
C	401-500
D	501-600
E	601-700
F	701-800
G	801-1100
H	1101-1500
I	1501-2000
J	>2001

4.14.4 Size Tolerance

For all grades, 10 per cent by number or weight of papayas corresponding to the size immediately above and/or below that indicated on the package, with a minimum of 190 g. for those papayas packed in the smallest size range.

4.14.5 Packaging

i) At the time of packing, papayas must have been carefully picked and have reached an appropriate degree of development and ripeness, account being taken of the characteristics of the variety and/or commercial type and to the area in which they are grown.

ii) The development and condition of the papayas must be such as to enable them to withstand transport and handling, and to arrive in satisfactory condition at the place of destination.

a) Shrink Wrapping

Different types of polymeric films *viz.*, LDPE, PP, Pebax® and storage temperatures *viz.* 7, 13 °C and 27-32 °C have been tried for individual fruit packing of green "Solo" papaya. MAP of individual papaya fruits with Pebax® or LDPE film could extend the storage life and maintain the quality up to two weeks over one week in non-packed fruits under ambient conditions (27-32 °C). At low temperature (13 °C), the storage life of papaya could be extended to one month when packed in LDPE or Pebax® film without any chilling injury (CI) symptoms and the fruits ripened normally in one week under controlled conditions (20 ±1 °C and 75-80 per cent RH).The fruits stored at 7 °C failed to ripen normally and showed severe CI symptoms and disease incidence (Sukhvinder Pal Singh, 2003).

For export of papaya fruit to Europe and Middle-East, packaging and fruit specifications have been recommended by National Horticulture Board (NHB, 2011) as given in Table 29.

4.15 Pear (*Pyrus pyrifolia*)

4.15.1 Minimum Requirements

Fruit should be wholesome, sound, clean and free from defects

4.15.2 Grade Designation

Pears are graded in three grades.

i) Extra class

Table 29. Specifications for Export of Papaya

Variety	*Weight*		*Packing (kg)*		*Shipment*	*Storage Temperature*
	Europe	*Middle East*	*Europe*	*Middle East*		
Solo	400-500 g/fruit		5	8	By sea	10-13 °C
Taiwan 785	400-500 g/fruit	1-1.25 kg/Fruit	5	8	By sea	10-13 °C
Taiwan 786	400-500 g/fruit	1-1.25 kg/Fruit	5	8	By sea	10-13 °C

ii) Class – I

iii) Class – II

Fruit of Extra class must be of superior quality without surface defects. The tolerance limit for this class is 5 per cent by number or weight of fruit not satisfying the requirements of the grade, but meeting those of Class I or, exceptionally, coming within the tolerances of that grade.

For Class I, pears must be of good quality with surface defects covering 10 per cent area but not affecting pulp. Tolerance is 10 per cent by number or weight of fruit not satisfying the requirements of the class/grade, but meeting those of next class. For Class –II, fruit must meet minimum requirement of quality and surface defects covering 15 per cent area are allowed but not affecting pulp. Tolerance is 10 per cent by number or weight of fruit not satisfying the requirements of the grade but meeting the minimum requirements.

4.15.3 Provisions Concerning Sizing

i) Size is determined by maximum diameter of the equatorial section.

The minimum size by diameter is required for all grades as follows:

Extra Class: 60 mm; Class-I: 55mm ; Class-II: 50 mm

The difference in diameter between fruit in the same package shall be limited to:

i) 5 mm for Extra Class, Class-I and Class-II fruit packed in rows and layers.

ii) 10 mm for Class-I, Class-II fruit packed in bulk in the package.

4.15.4 Size Tolerance

For all classes, 10 per cent by number or weight of fruit not corresponding to the size with a maximum variation of 5mm below the minimum.

4.15.5 Packaging

i) At the time of packing, development and condition of the pears must be such as to enable them to continue their maturing process and to reach the appropriate stage of physiological maturity corresponding to the particular characteristics of the variety.

ii) Fruit shall be able to withstand transport and handling and be able to arrive in satisfactory condition at the place of destination.

iii) The contents of each package must be uniform and contain only of the same origin, variety, quality, size and same degree of ripeness.

iv) In the case of, fruit colour shall be uniform.

Pears are packed in corrugated fibre-board boxes of required strength for bulk packaging and transport. Paper pulp trays are used for fruit of "Extra Class". Consumer or retail packaging is done in small trays and wrapped with heat shrinkable film.

4.16 Pineapple (*Ananas comosus*)

4.16.1 Minimum Requirements

Pineapples must have been carefully picked and have reached an appropriate degree of development and ripeness in accordance with criteria proper to the variety and/or commercial type and to the area in which they are grown.

When peduncle is present, it shall be no longer than 2.0 cm. and the cut must be transversal, straight and clean. The fruit must be physiologically ripe *i.e.* without evidence of unripeness (opaque, flavourless, exceedingly porus flesh) or over ripeness (exceedingly translucent or fermented flesh). Trimming consists in tearing some leaves off the top of the crown.

4.16.2 Grade Designation

Pineapple are graded in three grades:

i) Extra class

ii) Class – I

iii) Class – II

Total soluble solids in Brix degrees (Minimum) shall be 12 in all the classes.

Fruit of "Extra class" must be of superior quality without surface defects. The tolerance limit for this class is 5 per cent by number or weight of fruit not satisfying the requirements of the grade, but meeting those of Class I or, exceptionally, coming within the tolerances of that grade. The crown, if present, shall be simple and straight with no sprouts and shall be between 50 and 150 per cent of the length of the fruit for pineapples with trimmed or untrimmed crowns.

For "Class I", pineapple in this class must be of good quality with surface defects covering 10 per cent surface area but not affecting pulp. Grade tolerance is 10 per cent by number or weight of fruit not satisfying the requirements of the grade, but meeting those of next class. The crown, if present shall be simple and straight and slightly curved with no sprouts, and shall be between 50 and 150 per cent of the length of the fruit for pineapples with trimmed or untrimmed crowns.

For "Class –II", pineapple must meet minimum requirement of quality and surface defects covering 15 per cent area are allowed but not affecting pulp. Grade tolerance is 10 per cent by number or weight of fruit not satisfying the requirements of the grade but meeting the minimum requirements can be included. The crown, if present, shall be simple or double and straight or slightly curved, with no sprouts.

4.16.3 Provisions Concerning Sizing

Size is determined by the average weight of the

fruit with a minimum weight of 700 g. except for small size varieties (such as victoria and queen) which can have a minimum weight of 250 g. in accordance with the Table 30.

Table 30. Pineapple Fruit Size, Code and Weight

Size Code	*Average Weight (g) (+ or – 12 per cent)*	
	With Crown	*Without Crown*
A	2750	2280
B	2300	1910
C	1900	1580
D	1600	1330
E	1400	1160
F	1200	1000
G	1000	830
H	700	560
I	< 700	< 560

4.16.4 Size Tolerance

For all classes, 10 per cent by number or weight of pineapples corresponding to the size immediately above and/or below that indicated on the package.

4.16.5 Packaging

At the time of packing, development and condition of the pineapples must be such as to enable them to withstand transport and handling, and enable fruit to arrive in satisfactory condition at the place of destination.

Fruits are packed in printed corrugated fibre–board boxes of required strength. Ventilation is provided to boxes. Fruit shall not be over-ripe at the time of packing.

4.17 Plum (*Prunus domestica*)

4.17.1 Minimum Requirements

Plums must have been carefully picked and have reached an appropriate degree of development and ripeness in accordance with criteria proper to the variety and/or commercial type and to the area in which they are grown.

4.17.2 Grade Designation

Plums are classified in three classes depending on fruit quality (Table 31).

4.17.3 Provisions Concerning Sizing

Sizing is determined by the maximum diameter of the equatorial section (Table 32).

Table 31. Criteria for Grade Designation of Plums

Grade Designation	*Grade Requirements*	*Grade Tolerance*
Extra class	Plums shall be of superior quality. They must have the shape, development and colouring typical of the variety. They must be Free from defects; Practically covered by bloom, according to the variety; Of firm flesh.	5 per cent by number or weight of Plums not satisfying the requirements of the grade, but meeting those of Class I grade or, exceptionally, coming within the tolerances for that class.
Class – I	Plums must be of good quality. They must be characteristics of the variety and/or commercial type. The following slight defects of the fingers, however, may be allowed, provided these do not affect the general appearance of the produce, quality, the keeping quality and presentation in the package. A slight defect in shape, A slight defect in colouring; A slight defect in development; Skin defects of elongated shape that must not exceed in length one-third of the maximum diameter of the fruit; Other skin defect, of which total area affected must not exceed 5 per cent of the whole surface; The stem may be damaged or missing, provided that there is no risk of the fruit rotting in consequence.	10 per cent by number or weight of Plums not satisfying the requirements of the grade, but meeting those of Class II grade or, exceptionally, coming within the tolerances of that grade.
Class – II	Plums which do not qualify for inclusion in the higher classes, but satisfy the minimum requirements. Defects in shape, development and colouring are allowed provided that the plums retain their characteristics. Skin defects not liable to impair the external appearance of the fruits or its keeping quality are allowed provided that they do not exceed 10 per cent of the whole surface.	10 per cent by number.or weight of mangoes not satisfying the requirements of the grade, but meeting the minimum requirements.

Table 32. Plum Fruit Grade and Corresponding Fruit Diameter

Grade	Large Fruit Varieties (Min.) (mm)	Other Varieties (mm)	Mirabelles and Damsons (mm)
Extra class	40	28	20
Class I	35	28	20
Class II	30	25	17

4.17.4 Size Tolerance

For all classes, 10 per cent by number or weight deviating from the minimum size or the size stated on the packages the deviation not exceeding 3 mm. above or below.

4.17.5 Packaging

i) At the time of packing, plums must be sufficiently developed and display satisfactory ripeness. The development and condition of the plums must be such as to enable them to withstand transport and handling and to arrive in satisfactory condition at the place of destination.

ii) Plums are prone to damage during transport and hence boxes and cushioning shall be such as to provide necessary protection during long distance transport.

4.18 Pomegranate (*Punica granatum*)

4.18.1 Minimum Requirements

Pomegranates must be carefully picked and have reached an appropriate degree of development and ripeness in accordance with criteria proper to the variety.

4.18.2 Grade Designation

Pomegranate are graded in three grades:

i) Extra class

ii) Class – I

iii) Class – II

Fruit of Extra class must be of superior quality without surface defects. The tolerance limit for this class is 5 per cent by number or weight of fruit not satisfying the requirements of the grade, but meeting those of Class I or, exceptionally, coming within the tolerances of that grade.

For Class I, pomegranate must be of good quality with surface defects covering 10 per cent area but not affecting pulp. Tolerance is 10 per cent by number or weight of fruit not satisfying the requirements of the grade, but meeting those of next class. For Class –II, fruit must meet minimum requirement of quality and surface defects covering 15 per cent area are allowed but not affecting pulp. Tolerance is 10 per cent by number or weight of fruit not satisfying the requirements of the grade but meeting the minimum requirements can be included.

4.18.3 Provisions Concerning Sizing

Size is determined by the weight or maximum diameter of the equatorial section of the fruit, in accordance with the Table 33.

Table 33. Pomegranate Fruit Size Code, Weight and Diameter

Size Code	Weight (in grams) (Minimum)	Diameter in mm (Minimum)
A	400	90
B	350	80
C	300	70
D	250	60
E	200	50

4.18.4 Size Tolerance

i) For all grades, 10 per cent by number or weight of pomegranate corresponding to the size immediately above and/or below than indicated on the package.

ii) The maximum size range of 8 mm between fruit in each package is permitted

4.18.5 Packaging

At the time of packing, development and condition of the fruit must be such as to enable them to withstand transport and handling and to arrive unsatisfactory condition at the place of destination.

The main problems associated with prolonged storage of pomegranate fruit are weight loss and shrinkage, decay development, appearance of skin blemishes, especially scald, and impaired quality and taste. Artes *et al.* (2000) stored Spanish 'Mollar de Elche' sweet pomegranates at 5 °C for 12 wks in un-perforated polypropylene (UPP) film of 25 µm thickness in modified atmosphere packaging (MAP). MAP strongly reduced water loss and chilling injuries without incidence of decay. Porat *et al.* (2008) reported simple and low-cost MAP technology to alleviate storage problems and maintain fruit quality for up to 3-4 months after harvest. For moderate storage up to 2 months, packing of fruit in Xtend® Easy-Tear® bags in small 4-5 kg retail cartons has been recommended. This type of bag includes a notch, which allows the final user to easily tear and remove the top part of the bag for retail display. The main advantage of packing the fruit in Xtend® Easy-Tear® bags in export cartons is maintenance of fruit quality and reduction in shrinkage and scald development both during the storage period and subsequent marketing. For prolonged storage up to 3-4 months after harvest, packing of fruit in 20 kg bulk bags in plastic crates or in 80 kg bulk bags suitable for storing fruit in harvest bins

Table 34. Fruit and Packaging Specifications for Export of Pomegranate

Variety	Colour	Weight (g)			Packing (kg)			Shipment			Storage Tempe-rature
		Nether-lands/ Germany	U.K.	Middle East	Nether-lands/ Germany	U.K.	Middle East	Nether-lands/ Germany	U.K.	Middle East	
Ganesh	Red	250-300	250-300	300-450	3	3	5	By Sea	By Sea	By Sea	5 °C
Arakta, Mrudula	Deep Red	200-250	200-250	200-250	3	3	5	By Sea	By Sea	By Sea	5 °C

has been suggested. For retail storage, repacking of fruit in to Xtend® Easy-Tear® bags is necessary.

Shelf life of 'Bhagawa' pomegranate was extended up to 3 months with heat–shrinkable film wrapping and storage at 8 °C. Under ambient storage condition, shrink wrapping extended storage life of fruit up to one month (Figures 6a and b) (Sudhakar Rao, 2013, personal communication).

For export of pomegranate to some European and Gulf countries, packaging and fruit specifications have been suggested by NHB, (2011) as per preference of countries for the fruit (Table 34). Fruit shall have pleasant flavour and aroma and higher TSS (16-17 Degree Brix) is preferred. Fruit shall be free from any type of blemish on the surface and uniformity of size and shape in the package is of paramount importance.

4.19 Sapota (*Achras sapota*)

4.19.1 Minimum Requirements

Sapota shall have been carefully picked and have reached an appropriate degree of development and displayed satisfactory ripeness in accordance with criteria proper to the variety and to the area in which they are grown. The development and condition of fruit must be such as to enable them to withstand transport and handling, and to arrive in satisfactory condition at the place of destination. Sapota shall not be very soft, shriveled or water soaked.

Sapota should comply with the residue levels of heavy metals, pesticides and other food safety parameters as laid down by the Codex Alimentations Commission for exports.

4.19.2 Grade Designation

Sapota are graded in three grades:

i) Extra class

ii) Class – I

iii) Class – II

Fruit of "Extra class" must be of superior quality without surface defects. The tolerance limit for this class is 5 per cent by number or weight of fruit not satisfying the requirements of the grade, but meeting those of Class-I or, exceptionally, coming within the tolerances of that grade.

For Class I, fruit must be of good quality with surface defects covering 10 per cent area but not affecting pulp. Grade tolerance is 10 per cent by number or weight of fruit not satisfying the requirements of the grade, but meeting those of next class. For "Class –II", fruit must meet minimum requirement of quality and surface defects covering 15 per cent area are allowed but not affecting pulp. Grade tolerance is 10 per cent by number or weight of fruit not satisfying the requirements of the grade but meeting the minimum requirements can be included.

4.19.3 Provisions Concerning Sizing

Size is determined by the weight of the fruit (Table 35).

Table 35. Sapota Fruit Grade and Corresponding Weight

Class	Weight (g)
Extra class	101 and above
Class – I	100
Class – II	80

The difference in weight between the largest and the smallest fruit in any one package must not exceed:

i) 15 gms. for fruits weighing less than 100 gms.

ii) 20 gms. for fruits weighing between 100 gms. and above.

4.19.4 Size Tolerance

For all grades, 10 per cent by number or weight of sapota not conforming to the minimum weight and or the size range specified.

As a practice of sapota marketing in Gujarat, fruit grading is mainly based on size and shape of the fruits. Sizes are only three *viz.* large, medium and small. The study by Varshney *et al.* (2005) revealed that small farmers grade sapota manually into 2 to 3 sizes before marketing. The farmers' cooperative marketing societies do the grading and packaging operations. The prices of fruits are paid to the growers on the basis of fruit count per 10 kg. Traders also grade the fruits manually and supply them to meet the requirement of the cities at short distances.

Figure 6. Shrink Wrapped Pomegranate (a) Ambient Condition (Above) (b) Refrigerated Condition (Below)

4.19.5 Packaging

While the packaging of fruits is done in ventilated corrugated fiberboard boxes by cooperative societies and traders, gunny bags are used by the farmers in Gujarat for packaging of sapota fruits (Varshney *et al.*, 2005). Fruits are packed in cardboard boxes of 10 kg. capacity with rice straw as padding material and ethylene absorbents. Bamboo baskets with banana leaves as lining material and covered with net are used for carrying the produce from farm to local market. In some cases, produce is packed in wooden containers and transported to the urban areas for sale. Wooden boxes prevent damage of ripened fruits and fetch better price in urban areas. Sapota are also packed in corrugated fibre board boxes with paper clippings as cushioning material (Figure 7) (Sudhakar Rao, personal communication, 2013).

Figure 7. Sapota Fruit Packed in Corrugated Fibre Board Boxes

4.19.6 MAP

Based on the respiration rate and permeability of the packaging material, the suitable packaging films reported for the modified atmospheric packaging of sapota fruit were low density Polyethylene, Polyvinyl chloride, Polypropylene and Polystyrene film (Dash, *et al.*, 2012). Saran and polyester films were found to be unsuitable for MAP storage of the sapota fruit. These researchers have developed a model that provides an effective way for selection of the proper packaging material for sapota fruit MAP and this model can also be employed for designing packaging films for other fruits also as per the respiration rate of the fruit.

For export by sea rout to countries in Middle-East, packaging of Cricket Ball and Kalipatti sapota fruit in 3 kg carton and temperature of 15-17 °C has been suggested by National Horticulture Board (NHB, 2011).

4.20 Strawberry (*Fragaria* spp.)

4.20.1 Minimum Requirements

Strawberries should be sound, clean and free from any defect when harvested.

4.20.2 Grade Designation

Strawberries are graded in three grades:

i) Extra class

ii) Class – I

iii) Class – II

Fruit of Extra class must be of superior quality.

4.20.3 Provisions Concerning Sizing

Size is determined by the maximum diameter of the equatorial section of the fruit.

Strawberries must have following minimum sizes.

Extra Class : 28 mm and Class I and II: 18 mm.

4.20.4 Size Tolerance

For all grades, 10 per cent by number or weight of strawberries not conforming to the minimum size provided, however, that the diameter is not less than 16 mm.

4.20.5 Packaging

Strawberries have very short shelf life and need to be carefully handled and packed. Generally for retailing fruit is packed in plastic punnets holding 9-12 fruits (Figure 8).

Figure 8. Strawberry Packed in Plastic Punnet

For bulk transport small punnets can be packed in master containers. Following care shall be taken:

i) For packing, strawberries must have been carefully picked and have reached an appropriate degree of development and display satisfactory ripeness in accordance with criteria proper to the variety and to the area in which they are grown.

ii) The development and condition of the strawberries must be such as to enable them to withstand transport and handling, and enable fruit to arrive in satisfactory condition at the place of destination.

iii) Strawberries in the 'Extra Class' must be well presented.

iv) Packages must be free from all foreign matter.

4.21 Sweet Orange (*Citrus sinensis*)

4.21.1 Minimum Requirements

Oranges shall be fresh in appearance, wholesome, with characteristic shape of the variety, sound, clean, free of any visible foreign matter, free of damage or defect or smell or condensed moisture caused by low temperature or high temperature. The stalk shall be removed close to the fruit keeping the button intact. Oranges shall comply with the residue levels of heavy metals, Pesticides and other food safety parameters as laid down by the Codex Alimentarius Commission for exports. Oranges shall have minimum diameter of 53 mm.

4.21.2 Grade Designation

Sweet oranges are graded/classified in to three classes depending on quality (Table 36).

4.21.3 Provision Concerning Sizing

Size is determined by the maximum diameter of the equatorial section of the fruit in accordance with the Table 37.

4.21.4 Size Tolerance

For all classes, 10 per cent by number of oranges corresponding to the size immediately below and/or above that mentioned on the packages, provided the maximum difference between the smallest and largest fruits does not exceed the following maxima (Table 38).

4.21.5 Mechanized Sizing and Grading

Sizing of sweet oranges both on the basis of weight and dimensions is better as compared to only weight-sizing or only dimension-sizing. Disorders like granulation and frost damage may create problem for correct determination of sizing as the fruit lose juice and weight in these disorders but the fruit size (dimensions) remains same. The microprocessor based (computer aided) modern sizers can weigh the individual fruit in the specially designed cups. These cups weigh the fruit and accordingly drop it on cross belts below the cups. If fruit is of higher weight then it moves further to next cup. Fruit with highest weight falls last. Weight sizers may be useful in case of oranges and grapefruit.

Dimension sizer has to be designed in order to measure and segregate fruit on the basis of 2 or 4 contact points. Measurement on the basis of 4 contact points is most accurate since diameter is usually greater or smaller in citrus fruit as compared to height or length. When fruit is in contact at four points it is called as volumetric sizer. These are best suitable for lemons which are oblong to ovate in shape with greater length and relatively smaller diameter. Sizing of fruits should be fairly accurate for it may affect volume and lead to problems in uniformity of packaging. The packers will face problem maintaining uniformity of fruit in package and that could slow their speed in searching right sized fruit.

The 2 point contact sizers can be of two types. In roller sizers, singular fruit is dropped on the divergent rollers which are positioned with increasing spacing. As the fruit moves from the beginning to the end of the roller, small fruit falls on the pull-out belt below the roller while larger fruit continues along the rollers until

Table 36. Criteria for Grade Designation of Oranges

Grade Designation	*Grade Requirements*	*Grade Tolerance*
Extra class	Extra class oranges shall be of superior quality. They shall be characteristic of the variety and/or commercial type in shape, external appearance, development and colouring. They must be free of defects, with the exception of very slight superficial defects provided these do not affect the general appearance of the produce, the quality, the keeping quality and presentation in the package; Shall be uniform in colour; Shall have minimum diameter of 70 mm.	5 per cent by number or weight of Oranges not satisfying the requirements of the grade, but meeting those of Class I grade or, exceptionally, coming within the tolerances of that grade.
Class – I	Grapes Class I oranges shall be of good quality. They must be characteristic of the variety or commercial type (in shape, external appearance, development and colouring). Following slight defects may be there, provided these do not affect the general appearance of the produce, the quality, the keeping quality and presentation in the package; Slight defects in colouring, Slight defects in shape, Slight skin defects occurring during the formation of the fruit, such as silver scurf, russetting etc., Slight healed defects due to mechanical cause such as hail damage, rubbing, damage from handling.	The defects shall not, in any case, affect the pulp of the fruit. 10 per cent by number or weight of Oranges not satisfying the requirements of the grade, but meeting those of Class II or, exceptionally, coming within the tolerances of that grade.
Class – II	This grade includes Oranges which do not qualify for inclusion in the higher grades, but satisfy the minimum requirements. Following defects may be there, provided the Oranges retain their essential characteristics as regards the quality, the keeping quality and presentation in the package. Defects in shape; Defects in colouring; Skin defects occurring during the formation of the fruit, such as silver scruffs, russetting etc.; Skin defects, healed defects due to mechanical cause such as hail damage, rubbing, damage from handling; Superficial healed skin alteration; Rough skin.	The defects shall not, in any case, affect the pulp of the fruit. 10 per cent by number or weight of Oranges not satisfying the requirements of the grade, but meeting the minimum requirements. Within this tolerance, a maximum of 5 per cent may be fruits showing slight superficial unhealed damage, dry cuts or soft and shrivelled fruit.

Table 37. Size Code and Fruit Diameter for Pranges

Size Code	*Diameter (mm)*
A	53-60
B	61-68
C	69-76
D	77-84
E	85-92
F	93-100
G	Above 100

Table 38. Sweet Orange Size Code and Permitted difference in Diameter

Size Code	*Maximum difference between Fruit in the Same Package (mm)*
A,B,C and D	7
E, F and G	10

appropriate spacing between the roller allow the fruit to drop. The rollers can be adjusted for different sizes of the citrus. In another method fruit are conveyed to different size openings, the small opening first and increasing gradually to separate small, medium, big and large sizes. These sizers have been fairly appropriate for oranges and mandarin with near spherical shape. In case of flat shaped fruit such as mandarin/tangerine with greater diameter and smaller height (length between stem and stylar-end) the sizing could be erroneous and would solely depend on orientation of fruit at contact points.

The divergent roller type sizers or rotating disc type dimension sizers (with 2 contact points) have given fairly good results in case of Kinnow and 'Nagpur' mandarin, respectively provided that contact takes place at equatorial diameter. The rotation speed of the disc has to be adjusted according to the feeding of fruit or else it may lead to congestion and mixing of grades. At the edge of rotating disc, the opening for different sizes of fruit is given which is adjustable. Small fruit is segregated first followed by medium sized and large fruit. Reduction of speed of rotary disc of the sizer from 10 to 8.5 rpm resulted in satisfactory sizing at the feeding rate of 1 tonne/hr. in case of 'Nagpur' mandarin. This type of sizer could segregate Mosambi orange fruit on dimension basis at two points with more than 90 per cent accuracy in sizing. There was no injury to fruit in mechanical sizing. Divergent roller grader with capacity to grade 400 kg acid lime (*C. aurantifolia* Swingle) fruit per hour can be operated manually also for small scale operations. A similar grader with a capacity of 200-250 kg/h is reported to segregate Kinnow mandarin fruits on the basis of diameter.

More accurate electronic sizers using optical technology are now available which segregate on the basis of the dimensions and shape. These sizers use light beams and on the basis of blockage of light determine the shape and size. The singulation and orientation of fruit when light is focused on it is very important for accuracy of the sizing. In Japan, fully automated opto-electric-sizing units are used for more than 25 years now. These units use the optical properties of the fruit *viz.* reflection, transmittance or delayed emittance. With these sizers dropping impact is reduced. About 20-25 tonne fruit are sorted per hour in large and efficient packing houses. In opto-electrical sizers, light source and detector combination is used. The horizontal sizing is done by relating fruit size to a time length required for a fruit to continue intersecting a light beam which traverses conveyer belt (Iwamoto and Chuma, 1981).

It was reported electro–optical sorter for colour and size sorting of citrus fruits. This system includes 5 microprocessors to control feeding, size detection, colour detection, discharge and master control. Size grading is accomplished by number of LED light beams blocked by projected area of fruit, while colour grading is conducted by using dual wavelength IVRF quality index at 880 and 650 nm. Accuracy of the system is reported to be 91-92 per cent at 24-48 fruit/m speed.

The most important requirement of good sizer is its ability to adjust the sizes and flow of fruit even when in operation if required. After sizing, fruit of only required grade can be washed, waxed, dried and packed. The other grade can be waxed and packed later or small sizes can be sent for juice processing.

4.21.6 Packaging

The content of each package must be uniform and contain only oranges of the same origin, variety, quality, size, same degree of ripeness and development.

The use of any substance tending to modify the natural characteristics of the oranges, especially its taste or smell is prohibited.

The oranges shall be presented as follows:

i) Arranged in regular layers in packages. It is mandatory for "Extra Class" and optional for Class I and II.

ii) Not arranged in regular layers in packages or in bulk bins. This presentation is only allowed for Classes I and II.

10 per cent tolerance only applies to fruit whose diameter is not less than 53 mm.

Packing is the last operation on packing line and place packing (manually placing fruits in the containers and arranging in layers) is most common for oranges and mandarins. For hand packing fruits are collected in tubs, chutes, bins or in long sliding tables. Fruits are arranged in the box to sit with desired orientation. The padding or paper board separators or paper pulp tray can be used between fruit layers. Like sorting, packing is another operation where trained personnel are required since this is the operation which makes direct impact on buyers when box/container is opened in the market. The convenience of packer is of utmost importance in designing of packing table and sitting/standing facility for easy reach of fruit and container/tray. The position of the box should be as near as possible to the body of the packer. This allows packer to look at the bottom and place the fruit properly and conveniently. The filled boxes are lifted and put on the conveyers/rails traveling sideways. The empty boxes and other materials like trays and separators should be available to packer at a stretch of the arm or at the most with one or two step movement. Automatic place-packing units are also operational in many packing houses in the US and elsewhere. The carefully sized and sorted fruit of desired number are lifted and placed automatically in the box in specific orientation. Box is then inspected and closed. In orange packing machine used in some of the facilities of Sunkist growers, California, fruits are metered in to pattern formers where they are picked up by spring loaded suction cups. The fruit is then moved over the carton and placed automatically. Machines with capacity as high as 315 cartons of 18 kg capacity per hour (Johnson, 1991) are operational. The actual speed is up to 200 cartons per hour The machine reduces labour requirement by 75 per cent.

Limes and Lemons are jumble-packed/volume-filled sometimes. Jumble pack can be automatic or semi-automatic. Fruits for local markets are usually jumble-packed where prices are to be kept low. The automatic volume–fill or jumble–pack machine fills the box with pre sized fruit to the desired weight and shakes it to immobilize the fruit. The vibration of box helps in settling the fruit below the desired top level of the box so that any fruit protruding above the top edge is not pressed while closing. The box is weighed on the scale manually and fruit is added/removed so that box is filled to desired weight. Then the box is closed and sealed automatically. The fully automatic large machines have facilities as high as 400-600 boxes per hour.

Whatever may be the method of packing, it is to be ensured that fruit are immobilized and fitting in to the container space without pressing.

4.21.7 Marking and Labeling

Labeling with individual lable/sticker on individual

orange fruit is done just before packing. The singulated fruit in cups are synchronized to pass beneath the cassette of pregummed labels. Each fruit gets a label with gentle pressure. The inks made of food–quality dyes are also developed to stamp on the fruit, however, its acceptance depends on rules and regulations of the country where fruit is to be marketed. Recently new laser technology is developed to tattoo each fruit with which product look-up (PLU) code, 'country of origin' labeling (COOL), grower lot number and any other requested security information can be marked on the fruit directly. This helps in tracking and tracing of fruit source and settling any problem. It is designed by Durand-Wayland Inc. USA (Anonymus,2005).Most grocers require PLU labeling as the number can be punched in to get the price.

Pre gummed labels with all required information are also stuck to the package box. The marking stamps of the date of packing or any special treatment to fruit, and packers name and address are also used.

In the state of Florida, all citrus fruit sold or shipped for consumption in fresh form need to be graded in a registered packinghouse in the state according to standards established by the Department of Citrus, and the grade of such fruit shall be indicated/stamped on the carton. Brands or trademarks need to be properly registered with the department to represent state or U.S. grades. In accordance with prescribed rules of Department of Citrus, licensed citrus fruit dealers in the state shall be entitled to register labels, brands, or trademarks for grade identification purposes. The department maintains a record of all labels, brands, or trademarks registered for grade identification purposes.

6.0 Conclusion

Grading is a value addition and results in better returns to growers as compared to sale of non-graded fruit. Grading also provides an authentic and scientific basis in promoting and managing the marketing system. It serves as a realistic and common basis for market intelligence and reporting. It facilitates e-trading and also settlement of quality disputes between buyers and sellers. Packaging reduces bruising, crushing and moisture loss, prevents re-contamination of the product with spoilage organisms, reduces pilferage and maintains a sanitary environment during marketing and thus improves sale of produce. Grading is done manually as well as mechanically. Electronic machine vision technology, high-speed x-ray scanners and Near–Infrared (NIR) technology is used for automatic grading on the basis of external and internal fruit quality. This ensures perfect fruit quality to the buyers. The Indian "Agmark" standards and the standards of the Codex Alimentarius Commission are similar to a large extent. For export, producers and exporters have to follow Codex standards apart from standards set by the country where fruit is being exported. There is a considerable development in the field of packaging. In consumer /retail packages, the concept of 'Smart packages' involves coating with certain patented chemical material that is sensitive to volatiles emanated by spoilage causing microbes. As the fruit starts decaying in package due to growth of spoilage organisms, the container colour changes due to coating which indicates that packed produce is no longer good for consumption. The intensity and hue of colour also indicates how long produce can be stored and safe for consumption. 'Active Packages' can be created by using oxygen scavengers, carbon dioxide absorbents /emitters, ethanol emitters and ethylene absorbents. The appropriate material is placed alongside the fresh fruit that modifies the atmosphere in the package and thereby contributes to the extension of shelf-life of the fresh fruit. New materials are being tried to develop eco-friendly sustainable material for packaging of produce. Continuous jute fibre as reinforcement in paper laminates to increase tensile properties of paper bags could be the substitute for plastic bags.

7.0 Important References

Anandswamy, B. and Venkatsubbaiah, G. 1976. Wooden and corrugated shipping containers for the export of Coorg oranges. *Indian Food Packer*, **30**: 44-49.

Anonymous. 1995. Central Institute on Postharvest Engineering and Technology (CIPHET), Ludhiana. *Annual Report for 1993-94 and 1994-95*, 31pp.

Anonymous. 2005. Tatooed fruit is on the way. *Processed Food industry*, New Delhi. **8**: 40-41.

APEDA, 2007. Guidelines for export of Indian mangoes to USA. Agricultural and processed Food Products Export Development Authority, New Delhi. p.45.

Artés, F., Villaescusa, R. and Tudela, J.A. 2000.Modified atmosphere packaging of pomegranate. *J. Fd Sci.* **65**: 1112-16.

Blandini, G., Levi, P., and Pappalardo, R. 1992. Problem analysis of an automatic citrus grading machine. *Proc. Intn. Soc. Citric.* Italy pp. 1033-35.

Chauhan, K.S., Sandooja, J.K., Sharma, R.K. and Singhrot, R.S. 1987. A note on assessment of certain prevailing practices for marketing of commercial fruits. *Haryana J. Hortic. Sci.* **16:** 229-32.

Chen, S.M., Fon, D.S., Hong, S.T, Wu, C.J., Len, K.C, Tien, B.T. and Chang, W.H. 1992. Electro-optical citrus sorter. *Am. Soc. Agric. Engg. No.* 92-3520.19 pp.

Chopra, S.K. and Dhar, R.P. 1990. Packing and storage of citrus with special reference to Kinnow. *Proc. Seminar on Problems and Prospects of Kinnow cultivation*, PAU, Ludhiana, pp 205-209.

Codex Alimentarius Commission, 2007. Codex standards for fresh fruits and vegetables, (First Edition), WHO and FAO of the United Nations, Rome. Italy, p. 185.

Dash, K.K., Goswami,T.K. and Deka, S.C. 2012. Modeling for the modified atmosphere packaging of sapota fruit. *Agric. Engg. Int.* **14**: 246-55.

Agmark, 2012. Fruits and Vegetables Grading and Marking Rules. 2004, 2007. and 2012. Gazette of India, GSR, 220, and GSR 171.

Iwamoto, M. and Chuma, Y. 1981. Recent studies on development in automated citrus packinghouse facility in Japan. *Proc. Int. Soc. Citric,* **2**: 831-34.

Jain, P.K. and Chauhan, K.S. 1994. Influence of packaging and modes of transportation on marketability of Kinnow. *Indian J. Hortic.* **51:** 251-54.

Johnson, T.M. 1991. Citrus postharvest technology to control losses. *Proc Intn Citrus Symp, China*, pp. 704-08.

Ladaniya, M.S. 2001. Response of 'Mosambi' sweet orange (*Citrus sinensis*) to degreening, mechanical waxing, packaging and ambient storage conditions. *Indian J. Agric. Sci.* **71:** 234-39.

Ladaniya, M.S., Naqvi, S.A.M.H. and Dass, H.C. 1994. Packing line operations and storage of Nagpur mandarin. *Indian J. Hortic.* **51**: 215-21.

Ladaniya, M.S. and Shyam Singh. 2000. Influence of ventilation and stacking pattern of corrugated fibreboard containers on forced-air precooling of 'Nagpur' mandarin.*J. Fd Sci. Technol.,* **37**: 233-37.

Ladaniya, M.S., Sonkar R.K., Shaikh, A.J. and Varadarajan, P.V. 1999. Corrugated fibre board containers for packaging, transport and storage of 'Nagpur' mandarin in domestic market. *Indian Food Packer,* **53**: 5-15.

NHB. 2011. Indian Horticulture database 2011. National Horticulture Board, Gurgaon, India, p. 281.

Pandiarajan, T, Pugaledhi, S, Devadas, C.T. and Gothandapani, L. 1993. Development and testing of lime fruit grader. *National Seminar on optimization of production and productivity of acid lime.* Horticulture College and Research Institute, Periyakulam, Tamil Nadu.

Porat, R., Weiss, B., Fuchs, Y., Sandman, A., Ward, G. and Kosto, I. 2008. Keeping quality of pomegranate fruit during prolonged storage and transport by map: new developments and commercial applications. *Acta Hort.* **804**: 115-20.

Shaikh, A.J, Ladaniya, M.S, Varadarajan, P.V. and Shyam Singh. 2003. Paper and corrugated boxes from cotton plant stalks for effective packaging of oranges. *J. Sci. Ind. Res.* **62***:* 311-18.

Sukhvinder Pal Singh, 2003. Modified atmosphere packaging of papaya (*Carica papaya* L.) fruits for extension of storage life. M.Sc. Thesis. University of Agricultural Sciences, Bangalore.

Varshney, A. C., Mangaraj, S. and Birla, S. L. 2005. Grading, packaging and marketing of sapota in south Gujarat. *Agric. Eng. Today,* **29**: 40-47.

Verma, B.B. 2009. Continuous jute fibre reinforced laminated paper composite and reinforcement-fibre free paper laminate. *Bull. Mater. Sci.*, Indian Academy of Sciences, **32**: 589–95.

Chapter 8

Insect Disinfection and Quarantine

Abraham Verghese

Former Director, ICAR-National Bureau of Agricultural Insect Resources, Bengaluru – 560 024, Karnataka, India and Former Head, Entomology, ICAR-Indian Institute of Horticultural Research, Bengaluru – 560 089, Karnataka, India

E-mail: abraham.avergis@gmail.com

M.A. Rashmi

Senior Technical Officer (Entomology), Regional Plant Quarantine Station, Directorate of Plant Protection Quarantine and Storage (DPPQS), Bengaluru – 560 024, Karnataka, India

E-mail: rashmigowda.ento@gmail.com

B.R. Jayanthi Mala

Scientist, Division of Entomology, ICAR-Indian Institute of Horticultural Research, Bengaluru – 560 089, Karnataka, India

E-mail: entjaya@gmail.com

1.0 Introduction

Fruit flies are a large and important group of insect pests that globally attack a wide range of fruits and vegetables, with significant economic implications for horticultural production and market access. Fruit flies do not recognise commercial/domestic or state, regional boundaries and jurisdictions. They are highly mobile, capable of flying short distances but can be transported to long distances through fruits and vegetables. They also have the potential to infest a wide range of crops from mango to apple and tomato to cucurbits.

Recently the Asian Pacific regions have witnessed rapid growth in horticultural development. Changes in dietary habits owing to increasing incomes continue to accelerate demand for fruits and vegetables in the region. Among the important pests of horticultural crops, fruit flies (Tephritidae: Diptera) rank among the world's most serious pests. Although many insect pests attack fruits and vegetables, none have garnered greater notoriety than Tephritid fruit flies which are recognized worldwide as the most important threat to the horticultural industry (Allwood and Drew, 1997; Barnes, 2004). Besides causing direct losses in yield and marketability, they are barriers to the international trade of fresh fruits and vegetables.

Fruit flies may gain entry to a region where they are not yet present by several pathways. The risk of introduction through imports is generally considered low because imports of horticultural produce are regulated by strict quarantine requirements. The unauthorized movement of fruits and vegetables coming from untreated areas and carried by sea and airline passengers present the highest "risk pathway" for fruit fly introductions. This paper aims to review the quarantine risks posed by major exotic fruit flies, to create awareness of the potential threats posed by these pests, and to highlight the management and phytosanitary issues.

1.1 Importance of Quarantine

The volume of global travel and imports are growing, so does the threat of invasive species. Each year, invasive pests destroy up to 40 per cent of food crops around the world and cause billions of dollars in production and trade losses. Climate change is also an added phenomenon that eases the spread of invasives. The world is facing an unprecedented climate crisis. It's directly impacting the plants because weather change is the most important trigger for insect pests. Higher temperature, humidity, light have a great impact on the life cycle of the pests and their habitat where they can spread. (Rashmi *et al.*, 2020)

Inspection of plants and plant material meant for export as per the requirements under the International Plant Protection Convention (IPPC) 1951 of FAO to facilitate pest free trade becomes very crucial in the above context. These include detection of exotic pests and diseases for their containment by adopting domestic quarantine regulations; Quarantine inspection and laboratory testing of plants and plant material to ensure freedom from exotic pests; Phytosanitary certification (for issuance of Phytosanitary Certificates (PSCs); to undertake fumigation/disinfestations/disinfection of commodities to control Infestation/infection.

2.0 Mango Exports

Mango (*Mangifera indica* Linn) is the most important fruit in India and is known as the "King of fruits".) In India fruit is cultivated in the largest area *i.e.* 2,312 thousand ha (4946 thousand ha in world) and the production is around 15.03 million tons (>37 million tons in the world), contributing 40.48 % of the total world production of mango.) China and Thailand stood at second and third position among mango producing countries in the world respectively. The main Mango producing states in India are Uttar Pradesh (23.86%), Andhra Pradesh (22.14%), Karnataka (11.71%), Maharashtra (8.79%), Gujarat (6.00%) and Tamil Nadu (5.09%).)The total export of mangoes from India is >60 thousand tons, valuing Rs. >165 crores. India exports mango to over 40 countries worldwide (APEDA, 2021).

India is the largest producer of mangoes in the world and has a large opportunity to tap global trade. Besides this, a relatively long period of availability (March to August), diversity of varieties, economic liberalization and priority to the export of fresh products have opened up the possibility of boosting mango exports from India. Both fresh mango and processed mango products are exported to different countries namely the United Arab Emirates, Bangladesh, United Kingdom, Australia, Saudi Arabia, Kuwait, Qatar, Bahrain, Nepal, United States of America, Oman, Singapore *etc.* and the foreign exchange earned from such exports amounts to about Rs. 94795391. Mango production worldwide, but only 2 per cent of its production is exported. Other major mango producing countries, like Mexico, Brazil, and Pakistan, export much higher quantities of their production. The reason may be that majority of India's mango plantations are owned by small farmers and 72 per cent of the farms are less than three hectares. The other reason may also be that nearly 15 per cent of India's mango production is wasted due to the lack of adequate infrastructure facilities.

In addition, India's mango exports also suffer from other factors such as heterogeneous quality and phytosanitary issues due to the lack of adequate postharvest operations. Indian farmers are not properly informed about internationally practiced postharvest operations for mangoes, which result in not only heavy postharvest wastage but also the low level of price realization in international markets.

The scenario of export of mangoes may be altered by efficient orchard management with GAP, improvements in infrastructure, use of advanced technologies, and meeting the food quality standards such as ISO9001, HACCP, with India becoming the leading exporter.

2.1 Importance of Quarantine

Mango exports have a huge market to capitalize on its potential. Exporters will have to comply with the quarantine requirements by adopting quality standards throughout the value chain *i.e.*, preharvest and postharvest requirements like plant quarantine, pre shipment treatment, phytosanitary certification, surveillance and monitoring, irradiation treatment, verification, and certification of vapor heat treatment, hot water dip treatment, and registration of packing house facilities. For the convenience of exporters, Government of India has instituted an elaborative mechanism, which includes all these operations. The early harvested crop of mangoes (April- May) in India is well suited to international demand, as there is less competition from other countries during that period. Before the shipment the consignment is inspected by officers of the Directorate of Plant Protection, Quarantine and Storage to ensure compliance with Phytosanitary requirements of importing country. The National Plant Protection Organization of India (NPPO) is headquartered in Faridabad, while there are seven regional quarantine stations located at Amritsar (North), Delhi (North), Kandla (North), Chennai (South), Bengaluru (South), Kolkata (East) and Mumbai (West). The Ministry of Agriculture undertakes regular pest survey and monitoring of mangoes, meant for export under the National Fruit Fly Surveillance Programme. Total mango exports to different destinations in the past 3 years from 2019-2022 and 15 major countries importing mangoes from India are listed in Tables 1 and 2.

Though India is the chief mango producer, Indian mangoes are grossly inadequate in the international market as the export of fresh fruits is less than 5 per cent. Although the cultivation of mango is very high, its consumption in local and foreign markets is

Table 1. Total Mango Exports to different Destinations in the Past 3 Years

Year	*Major Destination Countries*	*Quantity (Wt)*	*Value (Rs.)*
21-22	Australia, Canada, Chile, France, Germany, Indonesia, Ireland, Malaysia, Kuwait, Netherlands/ Holland, Poland, Qatar, Switzerland, United Arab Emirates, United Kingdom	7,859.91 Mts	94795391
20-21	Australia, Canada, Netherlands/Holland, Poland, Qatar Ethiopia, Thailand, China, Oman, Hong Kong, Saudi Arabia, Singapore, United Arab Emirates, United Kingdom, Switzerland	21,033.56 Mts	83190377
19-20	Italy, USA, Portugal, Bahrain, Belgium, Australia, Canada, Netherlands/Holland, Poland, Qatar Ethiopia, Thailand, China, Oman, Hong Kong, Saudi Arabia, Singapore, United Arab Emirates, United Kingdom, Switzerland	49,658.68 Mts	187903712

APEDA, 2021.

Table 2. Country-wise Export of Mango (Quantity in kg) and value in Rs. Lakhs

Sl.No.	*Year*					
	2020-2021			*2019-2020*		
	Country	*Qty*	*Value*	*Country*	*Qty*	*Value*
1	UAE	12,756.40	15,104.24	UAE	16,567.22	14,286.80
2	U K	2,471.69	5,128.25	U K	4,356.26	6,761.57
3	Qatar	1,762.11	1,916.17	U S A	1,095.42	3,056.49
4	Oman	1,478.27	1,684.36	Oman	3,636.81	2,756.97
5	Kuwait	515.96	962.3	Qatar	2,744.95	2,740.20
6	Singapore	373.41	593.38	Nepal	11,975.04	2,451.14
7	Baharain	382.63	421.26	Kuwait	1,170.85	1,605.08
8	Germany	158.06	207.68	Saudi Arab	1,517.39	1,299.53
9	Canada	141.28	204.99	Bangladesh	3,038.11	903.51
10	Saudi Arab	166.41	183.4	Singapore	687.55	735.86
11	Nepal	384.89	99.43	Baharain	700.33	668.5
12	Switzerland	62.64	98.39	Canada	492.95	580.91
13	Maldives	56.08	76.44	Malaysia	240.36	237.18
14	Hong Kong	39.38	74.87	France	196.75	233.88
15	Italy	42.32	59.96	Germany	185.67	232.98

APEDA, 2021.

comparatively less due to infestation by insects and poor shelf life in mangoes.

2.2 Major Insect Pests of Mango

a) The Oriental Fruit Fly, *Bactrocera dorsalis* (Hendel) (Diptera: Tephritidae)

It is a major pest of commercially important varieties of mangoes. It is distributed throughout India and undergoes hibernation during winter in north India; it is serious and active throughout the year in southern parts of India especially when the minimum temperature surges in summer. It is widespread in different parts of the world and is reported on more than 150 hosts. It was reported that the yield loss due to *B. dorsalis* (Figure 1) in India ranged from 1 to 31 per cent with a mean of 16 per cent.

Figure 1. Fruit Fly, *Bactrocera dorsalis* Damage in Mango

European and Mediterranean Plant Protection Organization (EPPO) lists *Bactrocera dorsalis* as an A1 quarantine pest (OEPP/EPPO, 1983) within Plant Health Committee of the broad category "non-European Tephritidae"; it is also of quarantine significance to Asia and Pacific Plant Protection Commission (APPPC). Southern Cone (COSAVE), Caribbean Plant Protection Commission (CPPC), Inter-African Phytosanitary

Council (IAPSC), Cartagena Agreement Commission (JUNAC) and International Regional Organization for Agricultural Health (OIRSA). *Bactrocera dorsalis* is indigenous to Asia, but like other *Bactrocera* spp. is known by experience to have the potential to establish adventive populations in various other tropical areas.

Besides *B. dorsalis*, *B. zonata* is quite important especially in the western and northern parts of India. (Appendix I mentions some important fruitfly of global significance).

b) Mango Stone Weevil, *Sternochetus mangiferae* (Fabricius)

Mango stone weevil is another important pest of mango with quarantine importance (Figures 2). It is a monophagous pest of with wide geographic distribution. It has univoltine life cycle, most of which, in India, is spent as resting adults from May/June to February/ March. When mango fruits are developed to pea size (0.5- 1cm diameter) the adults become active and begin feeding and mating. Oviposition on mango fruits begins when the fruits are lime size (2-4 cm diameter). The adult female makes a hollow, boat shaped depression, and spreads over the eggs.

Figure 2: Stone Weevil, *Sternochetus mangiferae* Damage by Mango

The larva hatches after an incubation period of 5-7 days and burrows through the flesh to reach the seed coat, leaving a visible track as a streak of water area leading to seed. Infected fruits are difficult to detect since no damage is visible externally. Incisions made by the ovipositing females are small and heal as the fruit develops. Adults emerge by burrowing through the seed, pulp and rind, congregate in small numbers in cracks and crevices of mango stems, branches, leaf litter, clods, *etc.*, and remain inactive until the next flowering season.

Direct loss due to mango stone weevil varies from 5 per cent to 80 per cent in different mango varieties. Infestation may cause indirect losses by stimulating the premature fall of developing fruits. If the weevil survives up to harvest, it affects the processing industry and export of fruit. Longer-range dispersal occurs largely through transport of fruit and seeds containing the larvae, pupae or adults. The pest has been intercepted in mango fruits and seeds in international trade. *Sternochetus mangiferae* is a quarantine pest for the CPPC, IAPSC, NAPPO and OIRSA. It is not currently listed as a quarantine pest by the EPPO. However, as mango production increases in the south of the EPPO region, there is no doubt that *S. mangiferae* could establish and be of economic importance. Fruits of mango containing larvae or pupae pose major risk of introducing *S. mangiferae* through the market chain. Although mango seeds with diapausing adults could also be a danger, such materials from countries where the pest occurs should be prohibited by mango-growing countries (EPPO) (Verghese *et al.*, 2005).

2.3 Major Sucking Pest Hindering Exports

Thrips are one of the major sucking pests difficult in post harvest management. Thrips, *Thrips palmi* (Thysanoptera: Thripidae) are important group of sucking insect pests causing substantial yield losses directly or indirectly. In recent past thrips on vegetables mainly on bitter gourd and okra are hampering their exports. Though they are not the major pests of these vegetable crops their presence is a bottleneck for exports.

Figure 3. Thrips Palmi Adult (*Source.* Rachana, R.R ICAR-NBAIR)

Adults are pale yellow or whitish in color, but with numerous dark setae on the body. The slender fringed wings are pale. The hairs or fringe on the anterior edge of the wing are considerably shorter that those on the posterior edge. They measure 0.8 to 1.0 mm in body length, with females averaging slightly larger than males. Adult longevity is 10 to 30 days for females and seven to 20 days for males. Development time varies with temperature, with mean values of about 20, 17, and 12 days at 15, 26, and 32°C. Females produce up to about 200 eggs, but averaging about 50 per female. Heavily infested plants are characterized by a silvered or bronzed appearance of the leaves, stunted leaves and terminal shoots. At high densities, feeding by *T. palmi* may cause damage to fruits as well, such as scarring, discoloration and deformation in developed fruits or fruit abortion in an early stage. Nymphs and adults scrapes or nibbles on the fruits of the bitter gourd and okra causing silvery patch appearances (Kawai, 1986).

2.3.1 Pre-harvest Management of Thrips in bitter Gourd and Okra

Removal of alternate weed host plants (Solanaceae and compositae families) for minimizing the thrips incidence

Thrips pupates in soil can be controlled by soil application of neem cake @250 kg/ha.

Monitoring and mass trapping through use of blue sticky traps to attract thrips (Sridhar and Naik, 2015)

Spraying of botanicals like Azadirachtin 1000ppm 0r Azadirachtin 3000ppm

Spraying of Spinosad @ 0.3 ml/L

At severe incidence of thrips may be sprayed with Spraying of Fipronil 25 SC @ 1 ml/L and followed 10 days waiting period.

Fruit bagging: Few studies have been performed on the use of fruit bagging in reducing the incidence of thrips pests: Affandi *et al.* (2008) found a reduction in scarring of mango fruits (caused by an unspecified Commodity risk assessment of *Momordica charantia* fruits from Sri Lanka www.efsa.europa.eu/efsajournal 11 EFSA Journal 2021;19(2):6397 species of thrips) of 32–42% in Indonesia using double-layered bags of plastic and paper. According to de Lopez *et al.* (2020) bagging alone of bananas reduced losses by 90–100% by the red rust thrips (*Chaetanaphothrips signipennis*) compared to bunches with no bags.

2.3.2 Post Harvest Management of Thrips

Bitter gourd export to the European union was banned in 2014-15 because of many Non-compliances which arose due to the presence of thrips in most of the consignments export. Therefore post harvest treatments to control the thrips is crucial for exports. Bitter gourd fruits are harvested at 12 days maturity (12 days after pollination). The storage life of fruits at room temperature (25-30°C) is 3 days. The storage life of bitter gourd fruits is extended to 6 days at room temperature (25-32°C) and 10 days at 12oC by modified atmosphere packaging with ethylene absorbent. The storage life is extended to 14 days by post harvest treatment with 1-MCP (250 ppb) and packaging in polyethylene bag (100 gauge) and storing at 12°C (Anonymous, ICAR publication 2017).

Field infested thrips will be carried over to the bitter gourds and okra which are being interfered with exports. As of now, there are no standard protocols identified for post harvest management of thrips for vegetables. Similar to the post harvest treatment of fruit flies in mangoes can be adopted or standardized for thrips management also. For example, hot water treatment at 45°C at 60 min or irradiation of vegetables need to be standardized for bitter gourd and okra. The bagging of fruits can be standardized to reduce the thrips on fruits.

2.4 Requirements for Exporting Mangoes

Import conditions define the biosecurity requirements which must be met for any import. Every country has its unique requirement for the import of mango from India, list of important countries along with the Treatment requirements is given in Table 1.

2.4.1 Requirements for Exporting Mangoes to United States

The importation of mangoes from India is regulated under the Fruits and Vegetable Quarantine, 7 CFR 305 and 319 [Docket No.APHIS-2006-0121] RIN 0579-AC19, USDA-APHIS)The mangoes must be treated in India with irradiation by receiving a minimum absorbed dosage of 400 Grays). Importers must secure the USDA-Import Permit at least 30 days in advance of the arrival of irradiated commodities at the scheduled port of entry to facilitate the transmission of the permits to the inspector at the U.S. Port of entry in advance of the arrival of shipments)The mangoes must be given post-harvest hot water fungicidal dipping (Sodium hypochlorite at 200 ppm) at 52°C for 3-4 minutes. Each consignment of mangoes must be inspected jointly by USDA-APHIS and NPPO of India inspectors as part of the required pre-clearance inspection activities as determined by mutual agreement between USDA-APHIS and NPPO of India. Each consignment of mangoes must be accompanied by a phytosanitary certificate issued by NPPO of India certifying that the fruits received the required irradiation treatment. The phytosanitary certificate must also bear the following two additional declarations confirming that:)the mangoes were subjected to post-harvest mitigation options described as above and)the mangoes were inspected during pre-clearance activities and found free of *Cytosphaera mangiferae, Macrophoma mangiferae, and Xanthomonas campestris pv. Mangiferae indicae*

2.4.2 Requirements for Exporting Mangoes to Australia

Export of Mangoes to Australia requires Vapour heat treatment (VHT) or Irradiation based on the place of the production site. Vapour heat treatment is only permitted for mangoes grown from designated places of production or production sites within Uttar Pradesh. VHT is recommended only for mangoes from areas certified as pest free for mango stone weevil. The requirements for mango exports are mentioned below (BIOCON,2021):

a. Prior to export, the plants or plant products must be inspected or tested by the National Plant Protection Organization (NPPO) according to appropriate procedures and be considered free from biosecurity pests.
b. To demonstrate compliance with this requirement a phytosanitary certificate with the following declaration "This is to certify that the plants, plant products or other regulated articles described herein have been inspected and/or tested according to appropriate official procedures and are considered to be free from the quarantine pests specified by the importing contracting party and to conform with the current phytosanitary requirements of the importing contracting party, including those for regulated non-quarantine pests" is mandatory.
c. If the phytosanitary certificate is issued after dispatch the date of inspection must be identified as an additional declaration.
d. The consignments of mangoes must be grown in Uttar Pradesh, India and found to be free from quarantine pests. The registration, treatment and consignment details must be provided.
e. The Phytosanitary certificate must endorse:
    1. For goods entering Western Australia, the additional declaration "The mangoes in this consignment have been sourced from a designated place of production or production site in Uttar Pradesh which is free of *Deanolis sublimbalis, Sternochetus mangiferae* and *S. frigidus*.
    2. "For goods entering all states other than Western Australia, "The mangoes in this consignment have been sourced from a designated place of production or production site in Uttar Pradesh which is free of *Deanolis sublimbalis* and *S. frigidus*."
f. The certification must be stamped by an authorised officer of the Department of Agriculture and Cooperation (DAC) and should have all details mentioned below.
    1. Treatment details, including date of treatment, temperature and duration of treatment.
    2. Packing house/treatment facility registration numbers.
    3. Number of cartons per consignment.
    4. Container and seal number (sea freight only).
g. Orchards, treatment facilities and packing sheds supplying and handling mangoes for Australian territory must be registered with India's National Plant Protection Organization (NPPO).
h. Mangoes must be sourced from treatment facilities/packinghouses that are registered with India's Department of Agriculture and Cooperation (DAC) and approved by the Department of Agriculture, Water and the Environment to treat and pack fresh mangoes for Australian territory.
i. All Indian mangoes must undergo pre-export vapour heat treatment (VHT) to protect against fruit flies.
j. DAC must supervise all VHT treatments performed on mangoes destined for Australian territory.
k. The treatment schedule for VHT treatment is as follows:
    1. The Pulp core temperature of 46.5°C maintained for a minimum of 30 minutes, or
    2. The pulp core temperature of 47.5°C maintained for a minimum of 20 minutes.
    3. Treatment commences when the fruit pulp temperature of all probe-monitored fruit reaches or is above, the required temperature.

The total treatment time will be for a minimum of two hours, including the warming and cooling periods to bring the fruit pulp to the target temperature. (BIOCON,2021)

2.5 Sanitary and Phyto-sanitary Issues

Quarantine problems cannot be looked at in isolation. Regional and International collaborations are vital if progress is to be achieved in reducing risks of fruit fly introduction and spread. The introduction and spread of plants and pests of plants from one geographical area to another is an issue of worldwide concern, and is addressed at the international level by several agreements. The principal agreement aimed at preventing the spread and introduction of pests of plants and plant products is the International Plant Protection Convention (IPPC).

A quarantine pest is defined by the IPPC as "a pest of potential economic importance to the area endangered thereby and not yet present there, or present but not widely distributed and being officially controlled" (International Plant Protection Convention, 1997). In other words, it is any organism that is injurious or potentially injurious, directly or indirectly, to plants or plant products or by-products of plants and includes bacteria, fungi, insects, mites, molluscs, nematodes, other plants and viruses, not present in a specified area at risk or, if present, being controlled by an NPPO (National Plant Protection Organization). The invasion potential of fruit fly is determined by its likelihood to be

Table 3. Mango Treatment and Pest Free Requirements from the Importing Countries from India

Country	*Treatment and Pest Free Declaration*
Australia	The fruit in this consignment has been produced in India in accordance with the conditions governing the entry of fresh mango fruit to Australia and in accordance with the Irradiation Operational Work Plan between India and Australia. 2. Name of treatment Facility: 3. Irradiation treatment Certificate No.
Japan	This is to further certify that the mango fruits covered by this certificate are free from *Bactrocera dorsalis* species complex and melon fly (*Zeugodacus cucurbitae*).
South Korea	This consignment complies with import requirements agreed between QIA and NPPO of India, and is free from *Cytosphaera mangiferae, Xanthomonas campestris pv. mangiferaeindidae, Sternochetus frigidus and s. mangiferae*". registration code of the orchards ___ AND Pack House Code ____
European Union	Consignment complies with Annex VII, 61 d of Commission Implementing Regulation (EU) 2019/2072 : (61d) have been subjected to an effective post-harvest treatment to ensure freedom from Tephritidae (non-European), details of the treatment method are indicated on the phytosanitary certificate referred to in Article 71 of Regulation (EU) No 2016/2031, provided that the treatment method have been communicated in advance in writing to the Commission by the national plant protection organization of the third country concerned.
USA	The mangoes were subjected to one of the pre based or post harvest mitigation options described in 7cfr319.56-46 (Irradiation)2. The mangoes were inspected during pre clearance activities and found free of *Cytophaera mangiferae, Macrophoma mangiferae and Xanthomonas campestris pv. mangiferae indicae.*
New Zealand	The mangoes in this 1st commercial treatment, consignment have (i) been visually inspected in accordance with appropriate official procedures and found free from regulated organisms and (ii) been produced in accordance with terms of, and treated as per section 4.1 of the official assurance programme between MAF and the Directorate of Plant Protection, quarantine and storage (PPQS), Ministry of Agriculture of India.
Mauritius	The consignment of Mango treatment was supervised by Regional Plant Quarantine Station, Mumbai and found free from Tephritidae, Plant Debris and Soil. Pack House Registration Code: APEDA_________. Import Permit No.___________ Date-_______
IRAN	The consignment of mango has been subjected to Hot Water Immersion Treatment (HWT) at 46.1 degree Celsius temperature for 70 minutes at the facility at _________. Treatment was supervised by Regional Plant Quarantine Station, _________ and Imports permit Number:_________ and found free from 1. *Aleurodicus disperses*, 2. *Aulacaspis tubercularis*, 3. *Bactrocera spp.*, 4. *Ceroplastes rubens*, 5. *Cryptoblabes gnidiella*, 6. *Cytosphaera mangiferae*, 7. *Dysmicoccus brevipes*, 8. *Elsinoe mangiferae*, 9. *Hendersonia creberrima*, 10. *Macrophoma mangiferae*, 11. *Nipaecoccus viridis*, 12. *Phomopsis mangiferae*, 13. *Parasaissetia nigra*, 14. *Planococcus lilacinus*, 15. *Pseudaulacaspis cockerlli*, 16. *Pseudaonidia trilobitiformis*, 17. *Rastrococcus invadens*, 18. *Sternochetus frigidus*, 19. *Sternochetus mangiferae*, 20. *Thrips palmi* and 21. *Xanthomonas campestris pv. mangiferae indicae.*

PQIS, 2021.

transported and carried from one place to the other, and by its ability to get established in the new environment. Once a fruit fly reaches the new environment it requires to have the plasticity to adapt to the new physical conditions, and be able to reach its hosts to reproduce.

Phytosanitary measures to enhance mango exports in terms of pest free areas (PFA) have been undertaken at IIHR, Bangalore for *Bactrocera caryeae* (Kapoor) and commodity based PFA for *Bactrocera spp.* including *B. dorsalis.* As part of the trade efforts there is a need to bring simple identification keys for quarantine pests. At IIHR, biosystematics of common fruit flies has been taken up and a key has already been published (Madhura and Verghese, 2004). Further acceptable postharvest treatments are essential to export and import fresh mango fruits.

2.5.1 Strategy for Export

The strategies for boosting up of mango exports suggested by the Expert Group of the Planning Commission and APEDA are as under:-

i) Trans-shipment of mangoes by sea to make them competitive in price sensitive markets of Europe and Middle East.

ii) Promotion of non-traditional varieties like Chausa, Safeda and Badami.

iii) Augmentation of supply during seasonal gaps. (February/March in Middle East and March/April and July/August in Europe).

iv) Development of postharvest management facilities/practices (DMI, 2013).

3.0 Globally Accepted PHM Practices

The presence of arthropod pests in or on horticultural commodities has caused major disruptions in the shelf-life, processing, shipment of fruits and vegetables and exports. Postharvest disinfestation treatments can be used to control the presence of insects, so that the risk of rejection at importing countries as a result of insect contamination can thereby be reduced. These treatments should control the pest species without damaging the crop. However, there is normally only limited opportunity between combating the insects and damaging the fruit. The challenge of research on postharvest quarantine treatments is that the treatment must not only be effective against virtually all of the pest individuals present on the fruit, but also minimize damage to the commodity being treated. Additionally, the treatment has to be commercially viable *i.e.* it must be practical, economical and amenable to regulatory approvals at the importing destinations. The main techniques available are fumigation, usually with methyl bromide, heat treatment, either with hot vapour or hot water, cold treatments, insecticidal dipping and irradiation. Current technology has led to the development of numerous means to control (chemical, mechanical or procedural) these pests, at preharvest. For the most part, postharvest pest control is focused on trade and exports. The accidental introduction or fear of the introduction of a pest into a place where it is not known to be present has resulted in the establishment of quarantine restrictions and export treatment requirements. Postharvest heat treatments are being used for disinfestations and disinfection of an increasing variety of flowers, fruits, vegetables and flower crops. Postharvest management determines food quality and safety, competitiveness in the market, and the profits earned by producers. Postharvest treatments can include conventional chemicals, physical, irradiation and organic treatments.

3.1 Physical Methods

Physical methods of quarantine and phytosanitation treatments include mechanical, ultrasound, vacuum and temperature extremes. Fumigants are the most prevalent, but are losing support because of health, environmental and safety concerns. Quarantine systems using fumigants such as ethylene di-bromide (EDB) and methyl bromide (MB) are not acceptable anymore by almost all importing countries. Vacuum treatments, applied primarily to durable commodities, are gaining interest due to the development of flexible, low- cost treatment containers. Vacuum treatments appear to cause mortality in insects by creating a low-oxygen atmosphere and in low-moisture environments and may cause rapid dehydration. For entry into USA, the mangoes need to be treated in India with irradiation by receiving a minimum absorbed dosage of 400 Grays. Each consignment of mangoes must be accompanied by a phytosanitary certificate issued by the National Plant Protection Organization (NPPO) of India.

3.2 Ionising Radiation

The development of non-chemical techniques to maintain post harvest quality of fruits and vegetables is increasingly emphasized in many researches. Perhaps the most successful and widely accepted nonchemical quarantine treatment is the use of ionising radiation. Although irradiation is considered a physical treatment, it is not commonly considered an 'organically compliant' measure, even though it does not render the commodity or affected arthropods radioactive nor does it leave any detectable residues deemed to be harmful to human or animal health.

3.3 Chemical Treatment

Chemical treatments such as fumigants, pesticide dips and detergents are considered conventional and undesirable since they do not fit into the organic health and environmental standards of many countries. Among the many chemical fumigants identified for controlling postharvest arthropods are phosphine and sulphuryl fluoride (SF). But these are not advocated for fruit flies in mango.

4.0 Postharvest Treatments

4.1 Cold Water Treatment

Cold temperature treatments particularly cold storage have practical application for many commercial operations such as holding commodities before marketing, as well as pest control. Thermal treatments must be precise because of the narrow margin between efficacy and commodity tolerance to the temperature, particularly in fresh fruit and vegetables.

The fruits are subjected to cold storage quarantine treatment when shipped to areas outside of the known range and where the fly could survive. Rapid cooling in cold water increases the mortality of Caribbean fruit fly *Anastrepha suspensa* (Loew), larvae in carambolas.

In India, cold treatment experiments at 8 °C exposure for a week, showed zero per cent infestation in Banganapalli and Totapuri mango, while Alphonso showed 10 per cent infestation. This has treatment the potential to act as an ovicide. Varieties levels were 25, 40 and 40 per cent in the Banganapalli, Alphonso and Totapuri respectively.

4.2 Hot Water Treatment

Hot water immersion quarantine treatments are relatively easy to engineer. Its use as a quarantine system was intensified after the banning of ethylene di bromide. In Mexico, this quarantine system started in 1988 for the control of fruit fly (*Anastrepha ludens* and *A. obliqua*), mango fruits exported to USA and Japan, and consists of immersion of fruit in water at 46.1 °C (115 °F) for 65, 75 or 90 minutes, depending on the weight of the fruit. This system is now widely used in different mango growing regions in Mexico, Central and South America, and the West Indies. This system did not reduce the mango exports as anticipated and rather increased it. Other countries outside of America have also started experimenting and using hot water quarantine systems. It is effective in the disinfestation of several insects, other than the genus *Anastrepha*. Hot water treatments can damage the fruit. In the first few seasons the treatment caused major fruit injury in Mexico. However, now the treatment is used much more properly and therefore the damage has been controlled. After quarantine treatment, fruit must be introduced to an isolated, insect-proof area to prevent re-infestation.

From here on (packaging, transport, *etc.*) and until the fruit is received at the importing end, secure measures are implemented to prevent reinfestation. After heat treatment, fruit is usually cooled with water at ambient temperature and/or forced air cooling. In Mexico, fruit is first cooled in ambient water (at temperatures of < 21 °C), packed, and then cooled with forced-air cooling.

Figure 4. Hot Water Treatment (*Source.* DPPQS, 2021)

For exports of mango, the treatments for the Oriental fruitfly, *Bactrocera dorsalis* should ensure 99.99683 per cent control (By 'probit 9' standards). Heat treatment of infested fruits is known to ensure this standard (Verghese *et al.*, 2011). It has been reported that the eggs showed total mortality at temperature- time regimes of 45 °C -60 min and 48 °C – 45 min and 60 min. The first instar maggot was completely affected at 48 °C -60 min while the second instar maggot succumbed to 100 per cent mortality at 45 °C and 48 °C for 45 °C min and 60 min. Further the temperature regime 47 °C ± 1 °C for 75-90 min was validated in different varieties namely Alphonso, Banganapally, Neelam and Totapuri. This is the most cost effective method to disinfest exportable fresh, provided it is done within 24hrs after harvested and de-sapping (Verghese *et al.*, 2011).

Table 4. Registered Hot Water Immersion Treatment Facilities in Karnataka, India

Sl.No.	*Registered Hot Water Immersion Treatment Facilities as per NSPM – 15*
1	M/s. Innova Agri Bio Park Ltd. Plot No. B-12, 3rd Phase, KIADB Industrial Area, Malur, Kolar Dist. - 653 130, Karnataka (India)
2	M/s Karnataka State Mango Development and Marketing Corporation Ltd. Madikere, Chintamani- 563125, Chickaballapur District, Karnataka State (India)

PQIS, 2021.

Hot water plants for fruit fly are commercialized in India and are about 43 units of which there are 2 registered facilities in Karnataka (Table 4). The approximate cost of a one tonne capacity thermal tank will range vary depending on type of material used (zinc sheet, steel, fibre glass *etc.*,) and source of energy (electricity/solar). To treat one kg of fruit the approximate cost, including labour has been reported to be less than Rs.2.0.

4.3 Vapour Heat Treatment

Vapour heat, the oldest of the quarantine heat treatment, consists of heating the host fruit by moving hot air saturated with water vapour over the fruit surface. When the temperature of the mango fruit is at or below the dew point, condensation of atmospheric moisture occurs on the surface of the fruit. In this way, fruits are heated by conductive energy transfer. The heat from the fruit surface is transferred towards the fruit centre. Commercial facilities operate in Philippines, Thailand, the United States and Australia. The VHT disinfestations protocol accepted for mango access to the high-value markets in Japan include

i) 46 ° C fruit core temperature held for 10 min for Philippine 'Carabao' mango

ii) 46.5 ° C fruit core temperature held for 30 min for Taiwanese 'Irwin' and 'Haden' mangoes

iii) 46.5 ° C fruit core temperature held for 10 min for Tailand 'Nang Klang Wun' mangoes

iv) 47 ° C fruit core temperature held for 10 min for 'Nam Doc Mai', 'Pimsen Dang' and 'Rad' mangoes.

Vapour heat treatment requires that mango be treated with vapor heat until the surface temperature is 46 °C and seed temperature reaches 46.5 °C and held that way for 10 minutes. A vapour heat treatment has also been approved for the control of the Queensland fruit fly in Kensington mango exported from Australia to Japan. The treatment consists in raising the 46.5 °C core temperature held at 10 minutes. The USA also approved a quarantine vapour heat treatment for the control of Mexican fruit fly (*Anastrepha ludens*) and other *Anastrepha* species in Manila mango, and for mango from Taiwan infested with the oriental fruit fly. Vapour heat treatment has also been approved for cv Alphonso, for export to Japan from India. Through efforts of ICAR and APEDA Vapour Heat Treatment Plants have been set up at important mango belts to ensure export to Japan. The current cost of the equipment is Rs.2-4 crores, and the equipment is supplied by Japan.

4.4 Forced Hot-Air Heating Treatment (FHAT)

Forced hot-air, also known as high-temperature forced air, is a modification of the vapour heat treatment developed by Armstrong *et al.* (1989) in to kill Mediterranean fruit fly, melon fly and oriental fruit fly eggs and larvae in papaya. It is essentially the same as vapour heat except that the fruit surfaces are dry during forced hot-air treatment. Improvements in temperature and moisture monitoring and air delivery have advanced forced hot-air treatments, leading forced hot-air treatments to be developed for commodities previously treated with vapour heat and also being developed for new commodities. Forced hot-air treatment appears to be as effective in controlling internal pests as vapour heat, and provides better fruit quality, becoming the treatment of choice for many fruit previously treated with vapour heat. The fruit skin temperature remains cooler during forced hot air treatments than during vapor heat treatment while the tissue just below the skin heats to lethal temperatures because of the occurrence of evaporative cooling on the fruit surface during forced hot air treatment at lower relative humidity. The hot air is usually forced over the surface of the fruit which slowly heat the pulp. When air at 50 °C is forced over mango surface all stages of mango fruit fly, West Indian fruit fly, and *Anastrepha serpentina* are killed when the seed surface temperature reaches 48 °C. Mango fruit weighing over 700 g cannot be treated with this system. However, commercial usage of FHAT (forced hot air treatment), currently occurs only for Papaya grown in Hawaii and exported to mainland United States, and in the Cook Islands where papaya is treated before being exported to New Zealand. This is not a treatment available in India, hence economics of this are not available.

4.5 Irradiation

Irradiation has been amply shown to have potential as a quarantine treatment. The reason it is not currently being used is frequently attributed to public distrust. Although the application of ionizing irradiation as a quarantine treatment against tephritid fruit flies has been modest, the research record is extensive. Koidsumi in 1930 seems to have been the first to study ionizing radiation as a quarantine treatment over 70 yr ago. The measure of the efficacy of an ionizing radiation disinfestations treatment against tephritids is normally prevention of adult emergence from eggs and larvae that are irradiated in fruits. The efficacy of quarantine treatments against fruit flies usually must be confirmed by treating large numbers of insects of the most tolerant stage present in the commodity with no survivors. In the United States, this requirement has historically been interpreted to mean that at least 93,600 fruit flies of the most radio tolerant stage should be treated with no survivors to achieve the level of control of LD 99.9968 (referred to as probit 9) at the 95 per cent confidence interval.

Irradiation of fruit flies for the development of quarantine treatments has been done in three ways: (i) *In vitro* with or without a semi-artificial rearing medium, (ii) rearing on one medium followed by insertion into fruit for irradiation, and (iii) rearing and irradiation in the fruit. Fruit flies irradiated using the first two techniques have often been reported to be more susceptible to irradiation than those reared and irradiated inside fruits; the latter technique, being the closest to the natural situation, is more appropriate for irradiation quarantine treatment research. Research using gamma radiation for the disinfestation of oriental fruit fly, *B. dorsalis*, in mangoes and its effect on the overall quality and acceptability of the treated fruit undertaken in the Philippines showed that mature larvae of the fruit fly were the most tolerant to irradiation, with the young eggs being the most sensitive. A minimum dose of 100 GY was required to prevent the emergence of adult fruit flies and to maintain quarantine security against the possibility of introducing this pest into the importing country. In contrast to vapour heat treatment, no internal breakdown was observed, even if irradiated at 350 GY. Therefore, use of irradiation at a minimum dose of 100 GY as a quarantine treatment for the oriental fruit fly in mature green mango fruits has been recommended. To reduce fruit damage and speed treatment quarantine treatment doses should be as low as is possible, provided that sufficient substantiating data support efficacy.

In Hawaii, packaged mangoes were treated with cobalt-60 gamma irradiation and forced that minimum doses of 20.6 and 32.9 krad killed insects of all stages within the fruits which the few surviving adults were sterile and short-lived. In South Africa, irradiation of ripe, marketable fruit protected it from damage and prevented adult emergence. The most effective dosages ranged from 0.5 to 0.85 kGy. There has been some variation in the precise definition of the efficacy of irradiation as a quarantine treatment against fruit flies. Efficacy with irradiation has been defined as prevention of emergence of adult flies capable of flight and prevention of flies capable of reproducing (APHIS, 1989). These seemingly minor discrepancies may result in different doses required. Of the above three criteria, prevention of reproduction may be achieved with lower doses than prevention of adult emergence with prevention of adults capable of flight requiring a dose between those two. Currently APHIS bases efficacy on the prevention of adults capable of flight to prevent any flies surviving a quarantine treatment from reaching detection traps and needlessly triggering eradication programs. Data on cost benefit ratio are not available, though if trade to US and Australia, the benefit accrued will be high.

Few Irradiation facilities are also available in India for commercial purpose one at Innova Agri Bio Park Ltd., Malur, Kolar District (Karnataka) India.

4.6 Microwave or Radio Frequency Treatment

The microwave region of the electromagnetic spectrum is from 1 to 100 GHz, between infrared and FM radio, and is close to the radio frequency range. Radio frequency (RF) waves are at the lower frequency range of the electromagnetic spectrum, with longer wavelengths. Accepted frequencies for industrial purposes are 13.56, 27.12, and 40.68 MHz. RF energy generates internal heat by agitating molecules in the fruit with a very rapid change in charge within the electrical field. The advantages of RF heating are that it is very fast, can penetrate deep, and can sometimes heat insects more than fruit.

The time required to increase fruit centre temperature to 47 °C in vapour heat treatments can be 45 minutes or longer, resulting in fruit susceptibility to heat damage. Rapid heating by microwave or radio frequency energy can reduce the potential for fruit damage. The concept of high-temperature-short-time treatment is possible to shorten the treatment time while retaining fruit fly control at the Probit 9 (quarantine security) level. This is not a treatment available in India; hence economics of this are not available.

5.0 Integrated Pest Management

5.1 Fruit Fly

The Indian Institute of Horticultural Research (IIHR), Bangalore, has worked out cost-effective and residue-free control strategies for the integrated pest management (IPM) of fruit flies and the technical know-how is as follows:

i) Fruit fly infestation begins 45- 60 days prior to harvest. Initial breeding of the fruit flies takes place in fallen fruits. So, collect and destroy [by deep burying (Figure 3) (at least 2 feet) or burning] all fallen fruits at weekly interval, from two months prior to harvest.

ii) Place fruit fly traps, developed with the know-how of IIHR @ 15-20 per hectare. Before placing the traps in the field add 5 drops of Deltamethrin or Chlorpyrifos on the plywood pieces impregnated with pheromone. Traps should be fastened well on lower branches, between 2 meter height, at least 30 days prior to harvest or earlier.

iii) Traps should be kept in shade, to enhance their life.

iv) After 12- 15 days of trap placement, apply again 5 drops of Spinosad or Malathion on the plywood using ink filler to reactivate the trap (Figure 6).

v) Replace the used plywood with fresh plywood lure every 3-4 weeks, if harvest is prolonged, or if there are late varieties like Neelum still in fruiting. (If trap is full, count and empty the dead flies).

vi) If trap monitoring shows more than 5 flies/day, there is a need to give three bait splashes on trunk (Figure 5), starting 3 weeks prior to harvest

vii) Bag with cloth lined covers 3 weeks prior to harvest to prevent infestation

viii) Postharvest disinfestation (within 24hr of harvest/after desapping) in water at 47 °C±1 for 75 – 90 min.

Figure 5. Bait Splash on Trunk.

Figure 6. Fruit Fly Trap.

Figure 7. Field Sanitation.

Figure 8. Burying of Infested Fruits.

5.2 Stone Weevil

i) The adults spend nearly 6-8 months in hiding injunctions of trunk and primary branches between July and February where mud and leaf litter gather and decompose. A good cultural method to control the weevils is to clean these junctions with a hard broom any time after July/ August. This may be repeated in December/ January.

ii) When any chemical sprays like imidacloprid is given for hopper (*Idioscopus* or *Amritodus)* control, farmers should ensure good coverage on trunk, branches and canopy. Not only this would control the hoppers (even those hidings in the barks) but also kill resting adult stone weevils (Note: Adult weevils rest on mango trunk junctions approximately from July to Feb

iii) When 50 per cent of the fruits have attained lime size (2-4cm diameter) give a spray of Chlorpyriphos 50 EC. (Note: Wherever fruits are intended for export, should be very careful on choosing of chemicals as most of the importing countries require Maximum residue Limit (MRL) report) (Verghese and Shivananda, 2010).

6.0 Pest Risk Analysis

Pest Risk Analysis (PRA) is the application of sound scientific principles to identify, evaluate and manage pest risks associated with trade and travel and is intimately associated with quarantine regulations that may hinder free trade and movement. Pest Risk Analysis (PRA) is done to protect the country's agriculture from damages that can be caused by harmful (quarantine) pests that can be brought in along with imported commodities. A process consisting of three stages (i) initiation of the PRA through identification of a pest or pathway, or review or revision of an existing phytosanitary policy, (ii) pest risk assessment, and (iii) pest risk management. Risk communication is an integral component that occurs throughout each step.

PRA examines whether the criteria for the quarantine pest status of an animal or disease are justified. PRA examines the whole sequence of events that pests must survive if they are to become established in a new area and makes explicit and open the scientific judgements used. Such procedures are increasingly important as measure to liberalize world trade increase in importance. Agreement reached by signatories to the World Trade Organization requires that "phytosanitary measure(s) are based on scientific principles and (are) not maintained without sufficient scientific evidence" (WTO, 1999).

A sound, rational and internationally justifiable and defendable quarantine policy must be based on pest risk analysis. PRA relies on a full understanding of the biology, ecology, behaviour and control of the potential pest species under consideration and places considerable technical demands in the gathering and interpretation of the data needed to support accurate assessments. As any assessment may be limited by the amount of information available about each species PRA should be used cautiously and critically, its very openness to examination and criticism being crucial to ensuring that assessments are soundly based.

The FAO define a quarantine pest as "a pest of potential economic importance to (an) area and not yet present there, or present but not widely distributed and being officially controlled". Clearly a major component of justifying quarantine regulations against a pest is the demonstration of clear economic importance. For this to occur a pest must become established and spread in an area and it is the evaluation of the risks associated with this potential establishment and spread that form the core of PRA.

The PRA results of risk of fruit flies in the Indian Ocean Region shows that exotic fruit species predominate and are more widely distributed than native species. These introduced species account for most of the damage caused by fruit flies to fruit crops and vegetables. Exotic fruit fly species have not only gained entry to the Region, but have been more successful in colonising and have led to the ecological suppression of native species such as *Ceratitis catoirii* and *Dacus demmerezi*. Of fruit flies occurring outside the Indian context the fruit flies *Bactrocera*, *B. invadens ?* (Drew, Tsura and White), *B. papaya* (Drew and Hancock), *B. tryoni* (Froggatt), *B. neohumeralis* (Hardy), *B. philippinensis* (Drew and Hancock), *Ceratitis capitata*, *Dacus* spp, *Anastrepha* sp, *Rhagletus* spp *etc.* are of most quarantine concern. Among these the genus with the greatest quarantine significance is *Bactrocera*. For example *B. dorsalis* is a global quarantine pest with complex of 52 sibling species, of which 40 are considered new species and at least 8 are of economic importance. This is a matter of great concern to fruit producing countries where these species occur, requiring them to re-evaluate and revise their control strategies. In countries where these species have not been recorded, quarantine measures have been considerably strengthened to minimize the risk of their entry, establishment and spread. In Australia, a country where quarantine occupies top priority, 80 species of fruit flies, both endemic and exotic, have been identified as significant economic threats to Australian horticultural crops. The majority of these species are exotic to Australia, however many are found in close proximity throughout South-East Asia and the South Pacific.

In addition, Australia has a number of endemic and established species. Two, in particular, the Mediterranean fruit fly (*Ceratitis capitata*), which permanently inhabits the southern growing regions of Western Australia, and the Queensland fruit fly (*B. tryoni*), found in parts of the Northern Territory, Queensland, New South

Wales and the eastern corner of Victoria. These species have a significant economic impact increasing costs of fruit production and market access. In the 1970s the Asian species *Bactrocera carambolae* (Drew and Hancock) was found for the first time in the Western Hemisphere, in Surinam probably introduced through passenger luggage from Indonesia, and is the subject of an international control and eradication campaign. This in turn demonstrates that economically important American fruit fly species such as *Anastrepha ludens*, *A. fraterculus* and *A. obliqua* must also require consideration by the Quarantine Services of the Indian Region.

6.1 Bottlenecks for Export

i) Complex supply chain

ii) Weather fluctuations, causing the interrupted supply of fruits to the industry.

iii) Lack of supply of mangoes.

iv) Lack of forward and backward linkage between growers/suppliers and buyers/exporters

v) Inconsistent and inadequate supply of quality of mangoes.

vi) Disaggregate and multi-layered marketing channels, resulting in inefficiencies in procuring of raw material.

vii) Due to highly perishability and to pests and diseases in High postharvest losses

viii) Inadequate transport facilities high freight cost.

7.0 Conclusion

Complex supply chain is one of the main barriers to development of the Indian Mango industry. In markets several small producers, commission agents are operating. Therefore, for a bulk purchase of lot, these numbers of functionaries are to be dealt with, causing inconvenience and more postharvest losses. Mango fruit flies and Stone weevil are the major quarantine pests from India. Of these two, fruit flies are serious pests of agriculture throughout the world and represent a threat to the horticulture and ecology of India. The presence of arthropod pests in or on horticultural commodities has caused major disruptions in the shelf-life, processing, shipment of fruits and vegetables and exports. The fruit is cultivated in the largest area *i.e.* 2,312 thousand ha and the production is around 15.03 million tons, contributing 40.48 per cent of the total world production of mango. More than thousand varieties of Mangoes are grown in India. However, only about 30 varieties are grown on commercial scale in different states.

During 2020-21, export of Mangoes was 21,033.56 metric tonnes. Australia, Canada, Netherlands/Holland, Poland, Qatar Ethiopia, Thailand, China, Oman, Hong Kong, Saudi Arabia, Singapore, United Arab Emirates, United Kingdom, Switzerland were the important countries where Mangoes were exported. Postharvest disinfestation treatments can be used to control the presence of insects, so that the risk of rejection at importing countries as a result of insect contamination can thereby be reduced. As India imports apple from Chile, US, Australia, South America, New Zealand *etc.*, inadvertent introduction of fruit flies and other pests should be expected. The major potential invasives could be *Ceratitis capitata*, *B. aquilonis*, *B. jarvisi*, *B. neohumeralis*, *B. papyae*, *B. psidii*, *B. tryoni* being threat from Australia and New Zealand, *Anastrepha* sp. from South America, *Rhagoletis*, from US.

It's important that India maintain high vigilance of the imported fruits and they constantly ensure strict compliance of quarantine and comprehensive post harvest control programs by the exporting countries.

8.0 Important References

Affandi A, Emilda D and Jawal MAS, 2016. Application of fruit bagging, sanitation, and yellow sticky trap to control thrips on mangosteen. *Indonesian J. Agric. Sci.*, **9**: 19–23

Allwood, A. J. and Drew, R. A. I. 1997. *Fruit fly management in the Pacific*. ACIAR Proceedings No 76. 267pp.

Amstrong, J.W. 1994. Heat and cold treatments. **In**: Paull, R.E. and Amstrong, J.W (eds) *Insect pest and horticultural products*: treatments and responses CAB International, Wallingford. UK pp. 103-119.

Animal and Plant Health Inspection Service. 1989. Section III, 9 and section VI-T106. Plant protection and quarantine manual. US Dept. Agric., Wash. DC.

Animal and Plant Health Inspection Service. 1996. Schedules for propagative plant material T200d. Dipping. PQ Treatment Manual. US Dept. Agric., Wash., DC.

Anon, 2010. Second Annual Report of DBT Scheme, *"Prevention of Postharvest Spoilage Due to Fruit Flies in Mango"*.

Anonymous, 2006, Food irradiation update. *Food Technol.* **60**: 73–75.

APEDA, 2021. Agricultural and Processed Food Products Export Development Authority (https: // apeda.gov.in/apedawebsite/)

Armstrong, J.W., Hu, B.K.S., Brown, S.A., 1996. Single-temperature forced hot-air quarantine treatment to control fruit flies (Diptera: Tephritidae) in papaya. *J. Econ. Entomol.* **88**: 678–82.

Armstrong, J.W., J.D. Hansen, B.K.S. Hu, and Brown, S.A. 1989. High-temperature forced-air quarantine treatment for papayas infested with tephritid fruit flies (Dipterna: Tephritidae). *J. Econ. Entomol.* **82**: 1667-74.

Balock, J.W and Kozuma, T.T. 1964. Notes on the biology and economic importance of the mango weevil, *Sternochetus mangiferae* (Fabricius), in Hawaii

(Coleoptera: Curculionidae). *Proceedings of the Hawaiian Entomological Society*, **18**: 353-64.

Barnes, B. N., 2004. **In:** Fruit Flies of Economic Importance. *Proc. 6th Intl. Symp.* Isteg Scientific Publications, Irene South Africa.

Bicon, 2021 Australian Biosecurity Import Conditions,(https: //bicon.agriculture.gov.au/ Bicon)

CFFP. 1999. Carambola Fruit Fly Programme. (http: // www.carambolafly.com)

De Lopez MA, Corozo-Ayovi RE, Delgado R, Osorio B, Moy on D, Rengifo D and Rojas JC, 2020. Red rust thrips in smallholder organic export banana in Latin America and the Caribbean: pathways for control, compatible with organic certification. *Acta Hortic.* **1272**: 153–61

Drew, R. A. I, and Hancock, D. L., 1994. The *Bactrocera dorsalis* complex of fruit flies (Diptera: Tephritidae: Dacinae) in Asia. *Bull. Entomol. Res. Supple.* 2.

DPPQS, 2021, Directorate of Plant Protection Quarantine and storage, (http: //ppqs.gov.in/)

Ekesi, S. and Billah, M. K., 2007. A field guide to the management of economically important tephritid fruit flies in Africa. ICIPE, Nairobi, Kenya.

FAO, 1999. *Guidelines for Pest Risk Analysis*. Food and Agriculture Organisation. (http: //www.fao.org/ docrep/x5585e/x5585e0g.htm

FAO, 2002. International Standards for Phytosanitary Measures. Glossary of Phytosanitary Terms, Publication No. 5. UN-FAO-IPPC, Italy.

FAO, 2007. FAOSTAT. Food and Agriculture Organization of the United Nations.

FAO Statistics Division, Rome, Italy. http: //faostat.fao. org/(accessed 18 May 2007).

Fields, P. G and White, N. D. G. 2002. Alternatives to methyl bromide treatments for stored-product and quarantine insects. *Ann. Rev. of Entomol.* **47**: 331–59.

Follett, P. A. 2009, Generic radiation quarantine treatments: The next steps. *J. Econ. Entomol.* **102**: 1399–06.

Hansen, J.D. and Johnson, J.A. 2007. Introduction. **In**: Tang, J., E. Mitcham, S. Wang and S. Lurie, Eds., *Heat Treatments for Postharvest Pest Control: Theory and Practice*, CAB International, Wallingford, Cambridge.

Jordan, R. A. 1993. The disinfestation heat treatment process. Plant quarantine in Asia and the Pacific. A Report of an Asian Productivity Organization Study Meeting, Taipei, Taiwan, 17-26 March 1992. *Asian Productivity Organization*, Tokyo, pp. 53-68.

Kawai A, 1986. Studies on population ecology of Thrips palmi Karny. XI. Analysis of damage to cucumber. *Jpn. J. Appl. Entomol. Zool.*, **30**: 12–16. https: //doi. org/10.1303/jjaez.30.12

Koidsumi, K. 1930. Quantitative studies on the lethal action of x-rays upon certain insects. *J. Soc. Trop. Agric. (Jpn.)*. **2**: 243-63.

Krista, C., Robert, S. L. and Mangan. 2000. Postharvest disinfestation heat treatments: response of fruit and fruit fly larvae to different heating media, *Postharvest Biol. Technol.* **21**: 51–60.

Madura, H.S. and Verghese, A. 2004. A guide to identification of some common fruit flies (*Bactrocera* spp.) (Diptera: Tephritidae: Dacinae). *Pest Management in Horticultural Systems*, **10**: 1-10.

Manoto, E. C., Resilva, S. S., Rosario, S. E. D., Casubha, L. C., Lizada, C. C., Esguerra, E. B., Brena, S. R. and Fuentes. R. A. 1992. Effects of gamma radiation on the insect mortality and fruit quality of Philippine ÔCarabaoÕ mangoes, pp. 91- 116. **In**: *Use of irradiation as a quarantine treatment of food and agricultural commodities*. International Atomic Energy Agency, Vienna, Austria.

Navarro, S. and Calderon, M. 1979, Mode of action of low atmospheric pressures on *Ephestia cautella* (Wlk.) pupae. *Experientia*. **35**: 620- 621.

Navarro, S., Donahaye, J. E., Dias, R., Azrieli, A., Rindnere, M., Phillips, T., Noyes, R., Villers, P., Detrain, T., Truby, R. and Rodrquez, R. 2001. Application of vacuum in a transportable system for insect control **In:** Proceedings of international conference of controlled atmosphere and fumigation in stored products, Donahaye, E. J., Navarro, S and Leesch, J.G (eds) Fresno, California, 29 october-3 nov 2000. Executive printing services, Clovis, California, pp. 308-315.

OEPP/EPPO, 1983. Data sheets on quarantine organisms No. 41, Trypetidae (non-European). *Bulletin OEPP/ EPPO Bulletin* **13**.

PQIS, 2021, Plant Quarantine Information System (https: //plantquarantineindia.nic.in/)

Rashmi, M.A.,. Abraham Verghese, P.D. Kamala Jayanthi, A.K. Chakravarthy, Arathi Kirubha and Subhash B. Kandakoor, 2020. Effect of climate change on biology of oriental fruit fly, *Bactrocera dorsalis* hendel (Diptera: Tephritidae). *J. Entomol. Zool. Stud.*, **8**: 935-40.

Schneider, S. M, Rosskopf, E. N, Leesch, J. G, Chellemi, D. O, Bull, C. T. and Mazzola, M. 2003. United States Department of Agriculture – Agricultural research service research on alternatives to methyl bromide: preplant and postharvest. *Pest Manag. Sci.* **59**: 814–26.

Shellie, K. C. and Mangan, R. L. 2000. Postharvest disinfestation heat treatments: response of fruit and fruit fly larvae to different heating media, *Postharvest Biol. Technol.* **21**: 51–60.

Sridhar V and Onkara Naik, 2015. Efficacy of colour sticky traps for monitoring chilli thrips, *Scirtothrips dorsalis* Hood (Thysanoptera: Thripidae). *Pest Manage. Hortic. Ecsyst.* **21**: 101-03.

Stonehouse, J. M., Verghese, A., Mumford, J. D., Thomas, J., Jiji, T., Faleiro, R., Patel, Z. P., Jhala, R. C., Patel, Shukla, R. K. R. P., Satpathy, S., Singh, H.S., Singh, A. and Sardana. H.R. 2005. Research conclusions and recommendations for the on– farm IPM of Tephritid fruit flies in India, *Pest Manage. Hortic. Ecsyst.* **11**: 172-80.

Tang, J., J.N. Ikediala, S. Wang, J.D. Hansen, and R.P. Cavalieri. 2000. High temperature-short-time thermal quarantine methods. *Postharvest Biol. Technol.* **21**: 129-45.

U.S. Department of Agriculture [USDA]. 2005. Animal Plant Health Inspection Service (APHIS), Plant Protection and Quarantine (PPQ) Treatment Manual: regulating the importation of fresh fruits and vegetables, Riverdale, Maryland, 306pp.

Uma Maheshewari and Purusotham, U.T., 1999. A simple method to diagnose nut weevil incidence in mango. *Insect Environ.* **5**: 101.

Varith, J., W. Sirikajornjaru, and R. Kiatsiriroat. 2006. Quarantine treatment on mango using microwave-vapor combined process. *Proceedings of the American Society of Agricultural and Biological Engineers Annual International Meeting*, Portland, Oregon, 9–12, July, 2006.

Verghese, A. and Devi, S. 1998. Relation between traps catch of *Bactrocera dorsalis* Hendel and abiotic factors. Pp. 15-18. **In**: *Proc. First Nat. Sym.* Pest Management in horticultural crops: Environmental Implications and Thrusts (Eds., Reddy, P.P., Krishna kumar, N.K. and Verghese, A). Association for advancement of pest management in horticultural ecosystem, Bangalore.

Verghese, A., 1998. Fallen fruit sampling helps timing of first spray for control of mango stone weevil, *Sternochetus mangiferae* Fab. (Coleoptera: Curculionidae). *Insect Environ.* **3**: 91-92.

Verghese, A., 2000. Status and management of the mango stone weevil *Sternochetus mangiferae* Fab. (Coleoptera: Curculionidae) in India. *Pest Manage. Hortic Ecosys.* **6**: 15-21.

Verghese, A., Madhura, H. S., Jayanthi, P. D. K and Stonehouse, J. M. 2002. Fruit flies of economic significance in India, with special reference to *Bactrocera dorsalis* (Hendel). *Proceedings of 6th International Fruit fly Symposium* held during 6 – 10 May 2002, Stellenbosch, South Africa. pp. 317–324.

Verghese, A., Madhura, H.S., Jayanthi, P. D. K, Stonehouse, J. M., 2003. Fruit flies of economic significance in India with special reference to *Bactrocera dorsalis* Hendel. In: Barnes, B., Addison, M. (Eds.), Proceedings of the Sixth International Symposium on Fruit Flies of Economic Importance, 6–10 May 2002, Stellenbosch, South Africa, in press.

Verghese, A., Sreedevi, K., Sudhadevi, D. K., Nagaraju, B. R. and Jayanthi, M., 2006, Pest Free Areas (PFA) of *Bactrocera caryeae* (Kapoor) in Major Mango Belts of South India *Pest Manage. Hortic. Ecsyst.* **12**: 75-84.

Verghese, A., Shivananda, T.N. and Hegde, M.R., 2010, Integrated pest management of fruit fly in mango. Hand bill.

Verghese, A. and Shivananda, T.N., 2010, Integrated pest management of mango stone weevil in mango. Hand bill.

Verghese, A., Nagaraju. D. K. and Sreedevi, K, 2011. Hot water as an effective postharvest disinfection for the oriental fruit fly, *Bactrocera dorsalis* (Hendel) on mango. *Pest Manage. Hortic. Ecsyst.* **17**: 63-68.

Vijaysegaran, S., Lam, P. F., Yon, R. M., Karim, M. N., Ramli, A. H. and Yusof, N. 1992.Gamma irradiation as a quarantine treatment for carambola, papaya and mango, pp. 53-76. **In**: Use of irradiation as a quarantine treatment of food and agricultural commodities. International Atomic Energy Agency, Vienna, Austria. Wall, M. 2008. Quality of postharvest horticultural crops after irradiation treatment. *Stewart Postharvest Review.* **4**: 1.

White, I.M., and Elson-Harris, M. M. 1992. *Dacus* (Didacus) *ciliatus* Loew. **In**: Fruit Flies of Economic Significance: Their identification and Bionomics, pp. 329-331.CAB International, Wallingford, U.K.

World Trade Organisation. 1999. *Agreement on the Application of Sanitary and Phytosanitary Measures.* (http: / /www.wto.org/wto/legal/finalact.htm)

Yokoyama, V.Y. 1994, Fumigation. **In**: Quartantine treatments for pests of food plants. Sharp, J. L and Hallman, G. J (eds) Westview press, Boulder, Colourado, pp. 67-87.

Appendix I

Some examples of selected global fruit flies of quarantine importance

Bactrocera dorsalis (Hendel) Oriental fruit fly: India, Sri Lanka, Nepal, Bhutan, southern China (Sichuan, Guizhou, Hunan, Fujian, Yunnan, Guangxi, Guangdong, Hong Kong, Hainan), Burma, Thailand, Laos, Vietnam, Cambodia, Taiwan.

Bactrocera papaya (Drew and Hancock): Papaya fruit fly. Southern Thailand, Malaysia, Singapore, Brunei, Indonesia (Kalimantan, Sulawesi, Java, Bali, Lombok, Sumbawa, Flores, Timor, Maluku, Irian Jaya).

Bactrocera carambolae (Drew and Hancock) Carambola fruit fly: Southern Thailand, India (Andaman and Nicobar Is.), Singapore, Malaysia (Sabah), Brunei, Indonesia (Java, Lombok, Sumbawa).

Bactrocera cucurbitae (Coquillett) Melon fly: Probably originates from Asia; present in Afghanistan, Pakistan, India, Sri Lanka, Nepal, Bangladesh, Burma, southern China, Thailand, Laos, Cambodia, Vietnam, Malaysia, Singapore, Taiwan, Philippines, Brunei, Indonesia (Sumatra, Java, Kalimantan, Sulawesi, Sumba, Timor, Irian Jaya).

Bactrocera latifrons (Hendel) Solanum fruit fly: Pakistan, India, Sri Lanka, Burma, China (Fujian, Yunnan, Hong Kong, Hainan), Thailand, Laos, Vietnam, western Malaysia, Singapore, Taiwan, Brunei.

Bactrocera minax (Enderlein) Chinese citrus fruit fly: East India (Sikkim, W. Bengal), Bhutan, China (Jiangsu, Sichuan, Hubei, Hunan, Guangxi, Guizhou, Yunnan).

Bactrocera occipitalis (Bezzi): Philippines, Brunei, Malaysia (Sabah).

Bactrocera philippinensis (Drew and Hancock): Philippines (Luzon, Panay, Negros, Cebu, Mindanao).

Bactrocera tau (Walker): India (Andaman and Nicobar Is., northeast India), Bhutan, Burma, China (north to Xizang, Sichuan, Hubei and Zhejiang), Taiwan, Thailand, Laos, Cambodia, Vietnam, Malaysia (western Malaysia and Sarawak), Singapore, Brunei, Indonesia *Bactrocera umbrosa* (Fabricius): Southern Thailand, Malaysia (western Malaysia and Sabah), Singapore, Philippines, Brunei, Indonesia (Sumatra, Java, Kalimantan, Sulawesi).

Bactrocera zonata (Saunders): Peach fruit fly. Pakistan, India, Sri Lanka, Bangladesh, Burma, Thailand, Laos, Vietnam.

Other economically important genera of Dacinae include:

Dacus – species particularly in Africa and in some parts of Asia.

Ceratitis – mainly in Europe and Africa, though the most damaging species (the Mediterranean fruit fly or Medfly, *Ceratitis capitata* Wiedemann), has spread to some other areas including America and Australia.

Anastrepha – a genus with a number of economically important species attacking various fruits in North and South America. Some other genera also have economic importance in temperate to cold areas of Europe, North America and North Asia.

Chapter 9

Pesticide Residues in Horticulture Crops

Ahammed Shabeer T.P.

Senior Scientist

Kaushik Banerjee

National Fellow and Principal Scientist

ICAR-National Research Centre for Grapes, P.O. No.3, Manjari Farm, Pune – 412 307, Maharashtra, India

E-mail: shabsnrcg@gmail.com; E-mail: kbgrape@yahoo.com

1.0 Introduction

India is a major exporter of fresh fruits and vegetables. It ranks second in fruit and vegetable production in the world, after China. Amongst the fruits, the country ranks first in production of banana, papaya, mango, lemon and lime. India is also the largest producer of ginger and okra and ranks second in production of potato, onion, cauliflower, brinjal, cabbage, *etc.* The vast production base offers India tremendous opportunities for export. During 2020-21, India exported fruits and vegetables worth Rs. 9,940.95 crores/1,342.14 USD Millions which comprised of fruits worth Rs. 4,971.22 crores/674.53 USD Millions, and vegetables worth Rs. 4,969.73 crores/667.61 USD Millions (APEDA, 2021). Among the horticultural commodities, the major exported fruits are table grape, mango and pomegranate and vegetables like okra, green chilli, bitter gourd. The export of table grapes alone is about 313 million US dollars. Despite all these facts, India's share in the global food market is still < 5%, which is mainly because of the challenges in complying with the strict quality control standards in the importing countries like European Union, USA, GCC *etc.* Compliance with international food standards is a prerequisite to gain a higher share of world trade. At the same time, there is a growing consciousness among Indian consumers for safe food due to recent Government initiatives through the Food Safety Standards Authority of India (FSSAI) on setting up of quality and safety standards for different food products at the national level.

Since many decades, a wide range of chemical pesticides are applied worldwide in agriculture. India being a tropical country, infestations of pests and diseases are frequent and regular use of plant protection chemicals is therefore inevitable to safeguard the crops. However, injudicious and indiscriminate use of pesticides is likely to have negative impact on human health through exposure to residues in food and drinking water besides via occupational, bystander, and residential exposure, *etc.*, as well as through environmental contamination via wind drift, leaching, runoff, *etc.* The general public is exposed to pesticides mainly through residues in agricultural food crops. Management of residues in fruits and vegetables is a difficult task and processing could be a way to minimize pesticide residue loads initially present in fresh fruits and vegetables. This chapter presents a general overview of pesticide residues, their analysis, dissipation, monitoring and management in fruits and vegetables, with special emphasis to India.

2.0 Pesticides and their Residues

2.1 Pesticides

The term pesticide has been defined by the World Health Organization (WHO) as any substance or mixture of substances intended for preventing or controlling any unwanted species of plants and animals and includes any substances or mixture of substances intended for use as a plant-growth regulator, defoliant or desiccant. It also includes any substance used for the control of pests during production, storage, transport, marketing or processing of food for man or animals or which may be administered to animals for the control of insects or arachnids in or on their bodies. It however, does not apply to antibiotics or other chemicals administered to animals for other purposes, such as to stimulate their growth or to modify their reproductive behavior; nor does it apply to fertilizers. IUPAC also adopted the same definition for pesticide.

2.2 Pesticide Residue: A Trade Barrier

Pesticide residue is any substance or mixture of substances in food for man or animals resulting from the use of a pesticide and includes any specified derivatives, such as degradation and conversion products, metabolites, reaction products and impurities which are considered to be of toxicological significance. The term also includes residues from unknown sources (i.e., background residues) as well as those from known uses of any chemical in question.

Article 20 of the General Agreement on Tariffs and Trade (GATT) allows governments to act on trade in order to protect human, animal or plant life or health, provided they do not discriminate or use this as disguised protectionism. Sanitary and Phytosanitary Measures Agreement (SPS) allows countries to set their own standards. However, regulations must be based on science, but "precautionary principle" may be applied. The Technical Barriers to Trade Agreement (TBT) tries to ensure that regulations, standards, testing and certification procedures do not create unnecessary obstacles. It encourages countries to recognize each other's procedures for assessing whether a product conforms to desired standards. Although TBT encourages removal of unnecessary obstacles in trade flow by avoiding over-stringent regulations, it simultaneously recognizes individual countries' rights to adopt their own standards which they might consider appropriate with respect to the life and health of human being, animal population, plants and environment.

Residue limits (MRLs) are the trade standards that are set in a way that there are no concerns for public health, especially with regard to vulnerable subpopulation groups (as children and the unborn). Good Agricultural Practices (GAP) take into account the application of minimum quantities of pesticides necessary to achieve adequate pest control in such a manner that the amount of residues remaining in food are smallest possible. At the international level, the Codex Alimentarius Commission of the FAO/WHO decides the MRL. The Codex Committee on pesticide residues (CCPR) was formed by United Nations with primary mandate to establish MRL for pesticides in food. In India, the MRL of different pesticide-commodity combinations are prescribed under the Food Safety and Standard Authority of India Act, 2006 (FSSAI Act, 2006).

Although the Codex MRLs are applicable to all the signatory nations, individual countries are free to have their own MRL regulations. For agro-export, it is thus essential that the commodities should comply with the latest MRL regulations of the importing country. For example, the MRLs set by the European Union (EU) have to be complied for export of any food commodity (*e.g.* fruits and vegetables) from India to any of the EU member countries. Earlier, before September 2008, individual EU countries used to have their own MRLs. Since 1st September 2008, the MRLs have been harmonized across the member countries of the European Union, which has expanded the market access across the Europe and minimized trade barriers out of conflicting MRLs within the same territory. In addition to such Official MRLs, the Indian exporters often have to comply with specific private standards imposed by supermarkets, where MRLs are typically at 50-70 per cent of the official value rendering product compliance further challenging. Such private standards often raise apprehensions of combined toxicity in case of multiple pesticide detections in the same sample and even when the individual residue levels are well below the MRL set.

3. Pesticide Residue Analysis

Pesticide residue analysis involves several steps such as sampling, extraction, clean-up, analysis and data processing. The major steps are described below.

3.1 Sampling

Sampling is one of the crucial steps in pesticide residue analysis. It should be done in such a way that it is representative of the whole population and is without any bias. The sampling size and techniques, *i.e.* the size of sample, transport, storage, *etc.* might vary with commodities. The collected sample should be as homogeneous as possible before further processing. Random method of sampling most widely used and a detailed sampling procedure is given in the Codex guideline CAC/GL 33-1999 entitled "Recommended methods of sampling for the determination of pesticide residues for compliance with MRLs". An example of sample size required for plant products as per codex guidelines is given in Table 1.

The samples should be transported to the laboratory in clean containers with robust packaging. Ventilated polythene bags are acceptable for most samples but low permeability nylon film bags must be used for samples to be analysed for residues of fumigants. Samples of commodities prepacked for retail sale should not be removed from their packaging before transport. Very fragile or perishable products (*e.g.* ripe raspberries) may have to be frozen to avoid spoilage and then transported in "dry ice" *etc.*, to avoid thawing in transit. Samples that

Table 1. Plant Products. Description of Primary Samples and their Minimum Size

Commodity	Examples	Nature of Primary Samples	Minimum Size of Paboratory Sample
Small sized fresh fruits and vegetables	Berries, Peas	Whole units or packages or units taken with sampling device	1 Kg
Medium sized fresh fruits and vegetables	Apples, Brinjal, Mangos, Oranges	Whole units	1 Kg(At least 10 units)
Large sized fresh fruits or vegetables	Cabbage, Cucumber, Grape Bunches	Whole units	2 Kg(At least 5 units)
Herbs	Leafy vegetables	Whole units	0.5 Kg

are frozen at the time of collection must be transported without thawing. Samples that may be damaged by chilling (*e.g.* bananas) must be protected from both high and low temperatures. Sample preparation, sample processing and sub-sampling to obtain analytical portions should take place before visible deterioration occurs. This is particularly important when the analytical result is considered to assess fitness for consumer intake. Canned, dried or similarly processed samples should be analysed within the stated shelf life. Where there is evidence that comminution (cutting and homogenisation) at ambient temperature has a significant influence on the degradation of certain pesticide residues, it is recommended that the samples are homogenised at low temperature (*e.g.* frozen and/ or in the presence of "dry ice"). Where comminution is known to degrade residues (*e.g.* dithiocarbamates or fumigants) and practical alternative procedures are not available, the test portion should consist of whole units of the commodity, or segments removed from large units. For all other analyses, the whole laboratory sample (in most cases 1-2 kg) is thoroughly comminuted. As a thumb rule, residue analysis is carried out within the shortest time practicable after sampling to minimise any spoilage induced degradation of residues. In case, analysis is not immediately possible, the samples should be appropriately stored (usually at 0 °C) and analyzed within the shortest possible time.

3.2 Sample Extraction

Analysis of pesticides in food matrices (fruits and vegetables) is a difficult task, because of the complexity of the matrix and the low concentrations at which pesticide residues are usually present. Thus, despite advances in the development of highly efficient analytical instrumentations for their final determination, sample pretreatment remains an important part of obtaining accurate quantitative results. Many choices have been proposed for pretreatment and/or extraction of pesticide residues in foods. The extraction procedure usually involves sample homogenization with an organic solvent, alone or mixed with water or pH-adjusted water, using a homogenizer, blender, or sonicator. Some of the residue analysis techniques include solvent extraction (QuEChERS, Ethyl acetate method), supercritical fluid extraction (SFE), pressurized-liquid extraction (PLE), microwave-assisted extraction (MAE), matrix solid-phase dispersion (MSPD), solid-phase extraction (SPE), solid-phase micro extraction (SPME), and stir-bar-sorptive extraction (SBSE). Most of these techniques have resulted in new possibilities in sample treatment and advantages such as a substantial reduction in extraction time and incorporation into on-line flow-analysis systems. Each technique has its advantages and disadvantages and the choice should depend on the analytical problem.

Solvent extraction (SE) which may be followed by solid phase extraction (SPE) cleanup is still one of the most widely used techniques, mainly because of its ease of use, possibilities of analyte enrichment and wide ranging applicability. Such methods typically require a sample size of 50-100 g, which is extracted with solvents in the ratio of 1: 2, followed by cleanup by liquid-liquid partitioning and/or column chromatography. These methods have several disadvantages as they are laborious, time-consuming, expensive, and subject to problems arising from evaporation of large volumes of solvents and the disposal of toxic or inflammable solvents. To overcome the drawbacks mentioned above, new trends in sample treatment and miniaturization of time-tested SE have resulted in improved SE techniques that use much smaller amounts of organic solvent, enable more efficient extraction and faster cleanup ensuring easier laboratory automation and higher sample throughput. Good example for such SE methods are QuEChERS and ethyl acetate methods which because of increasing interest by many laboratories in recent years, are described in detail in the next section.

3.2.1 QuEChERS Method of Sample extraction

QuEChERS means "quick, easy, cheap, effective, rugged and safe" method for the multiclass, multiresidue analysis of pesticides in fruits and vegetables. The QuEChERS approach takes advantage of the wide analytical scope and high degree of selectivity and sensitivity provided by gas and liquid chromatography (GC and LC) coupled with tandem mass spectrometry (MS) for detection and quantification. A limited number of GC amenable pesticides were evaluated in the original QuEChERS study. Although this version has been demonstrated to yield excellent results for hundreds of pesticides across a wide variety of commodities, subsequent experiments showed stability and recovery problems for some pesticides depending on pH of the matrix necessitating modifications of the original QuEChERS protocol by use of relatively strong acetate buffering conditions or weaker citrate buffering conditions in terms of ionic strength.

QuEChERS method went through extensive inter-laboratory trials for numerous pesticides (both fortified and incurred samples) in different matrices and using different types of GC–MS and LC–MS conditions and instruments. These efforts successfully met statistical criteria for acceptability from independent scientific standards organizations, with the acetate buffering version becoming AOAC official method 2007.01 and the citrate-buffering version being named European Committee for Standardization (CEN) Standard Method EN 15662.

QuEChERS is a quick and convenient replacement for liquid-liquid extraction (LLE) which furnishes high-quality results in a minimum number of steps and with low consumption of solvent and glassware. The original version of QuEChERS consists of extraction of homogenized samples (10 g) by hand-shaking or vortex mixing with 10 mL of acetonitrile (MeCN) to furnish a final extract sufficiently concentrated to avoid the need for solvent evaporation. Gram quantities of salts (4 g anhydrous magnesium sulfate and 2 g sodium chloride, a combination which affords well defined phase separation without dilution with hazardous non-polar organic solvents) are then added to the sample, with mixing, to drive partitioning of the analytes between the aqueous and the solvent phases. After simple vortex mixing and centrifugation, which results in perfect physical separation of the phases, cleanup and removal of residual water is performed simultaneously by use of a rapid procedure, called dispersive solid-phase extraction (DSPE), in which a primary secondary amine (PSA) adsorbent and anhydrous $MgSO_4$ are mixed with the sample extract. In dispersive SPE, the adsorbents are added directly to the extract without conditioning and the clean-up is easily performed by shaking and centrifugation. The latest procedure requires less time than traditional SPE and simultaneously removes residual water and many polar matrix components, for example organic acids, some polar pigments, and sugars.

The QuEChERS method has the advantages of high recovery, accurate results, high sample throughput, low solvent (no chlorinated solvents) and glassware usage, less labour and bench space, lower reagent costs, and high degree of ruggedness. Organic acids and other potential contaminants are removed during clean-up. The main disadvantage of QuEChERS is that for 1 g sample per mL of final extract the concentration of the extract is lower than the extracts obtained through traditional procedures, which uses much higher quantity (5-10 times or even more) of sample for extraction. Thus, the final extract must be concentrated to a greater extent to furnish the necessary sensitivity and to achieve the desired limits of quantification (LOQ). The success of this method also relies on application of highly sensitive and selective GC-MS and LC-MS techniques especially tandem mass spectrometry (MS/MS).

3.2.2 Ethyl Acetate Method of Extraction

In a similar work flow as QuEChERS, ethyl acetate also provides high recoveries for most of the pesticides in fruits and vegetables. Ethyl acetate extract is more suitable for injection into GC-MS and for LC-MS analysis, a solvent exchange step is carried out to methanol-water. It is especially suitable for the extraction of high-sugar commodities like grape, mango, *etc.* since sugar has limited solubility in ethyl acetate. Furthermore, substitution of acetonitrile by ethyl acetate can substantially reduce the input cost of analysis. The sample preparation method follows extraction of homogenized samples (10 g) with 10mL ethyl acetate (+10 g anhydrous sodium sulphate), cleanup by dispersive SPE with PSA (25 mg) or/and other dSPE sorbents and final analysis by LC– MS/MS after solvent-exchange to methanol–water (1: 1). Selection of ethyl acetate offered precise advantages over acetonitrile in minimizing the matrix components in the final extract and reducing the cost of analysis. This method provided an overall savings to the tune of around 30% per sample when compared to the QuEChERS technique. This method was extended and validated in other matrices and now this method is used by food testing laboratories in India for residue analysis of fruits and vegetables for export.

3.3 Estimation of Pesticide Residues

Once the sample preparation step is over, the sample has to be analysed by appropriate instrumentation techniques for the detection and quantification of the target analyte/analytes. Nowadays, by using GC combined with MS, simultaneous determination and confirmation of pesticide residues can be obtained with one instrument in a single analytical run. In most cases, the sensitivity obtained with GC–MS is similar to that of classical GC detectors with the support of large volume injection and selective techniques like selected ion monitoring (SIM) and MS/MS. The commercial availability of atmospheric pressure ionization has caused a spectacular change in LC-MS applicability. Compared to traditional detectors, electrospray (ESI) and atmospheric pressure chemical ionization (APCI) in combination with tandem-MS instruments have increased the sensitivity of LC detection by several orders of magnitude. Moreover, with such progress in instrumentation techniques, the requirement of cumbersome procedures like HPLC column switching techniques and extensive sample cleanup requirements was unnecessary especially if tandem mass spectrometers are used and operated in the selected reaction monitoring (SRM) mode. Due to the high level of selectivity by LC–MS/MS in the SRM, the signal-to-noise ratio increases distinctly and the full sensitivity range of LC–MS instruments can be utilized. Several researchers have reviewed the application of

GC-MS/MS and LC-MS/MS technology for analysis of pesticide residues in fruits and vegetables.

4.0 Dissipation of Pesticides

After pesticides are applied to the crops, they may interact with the plant surfaces, get exposed to the environmental factors such as wind and sun and may be washed off during rainfall. The pesticide may be absorbed by the plant surface (waxy cuticle and root surfaces) and enter the plant transport system (systemic) or stay on the surface of the plant (contact). On the surface of a crop, a pesticide can undergo volatilization, photolysis, chemical and microbial degradation. Dissipation can be defined as the loss of pesticide residues in one compartment (*e.g.* plant tissue matrix) due to degradation and mass transfer to other compartments. Volatilization, photolysis and microbial degradation are the main processes by which pesticides get dissipated on fruits and vegetables. These processes are illustrated in Figure 1. All these processes can reduce the original pesticide's concentration load, but can also introduce some metabolites in the crops.

4.1 Dissipation Kinetics

A large number of kinetic models are used to describe the change in pesticide concentrations with time. This list of core models comprises the sample first-order model, and also a number of models that are able to describe bi-phasic degradation kinetics and lag phase models. In case of pesticide dissipation or degradation in environmental systems, the above mentioned models represent simple and sensible approaches to mathematically describe the experimental data, and do not represent actual chemical reactions. Also, it may be noted that the more complex the dissipation/ degradation pathway and the type of kinetics used, the more parameters the model will require, and the more data points are needed for adequate parameter estimation. Sample first-order kinetics (SFO) is expressed through exponential equation with only two parameters (total amount of chemical present at time t=0 and rate constant k). It assumes that the number of pesticide molecules is small relative to the number of factors microorganisms and their enzymes or number of water molecules in case of hydrolysis. As a result, the rate of the change in pesticide concentration (dM/dt) is at any time directly proportional to the actual concentration remaining in the system. For SFO kinetics, the time for a decrease in the concentration by a certain percentage is constant throughout the experiment and independent of the initial concentration of the pesticide. First-order kinetics has also frequently been used to describe degradation in pesticide fate models. A number of publications have shown that simple first order kinetics cannot adequately explain the degradation behaviour of pesticides in natural systems, where cannot the degradation pattern may follow a nonlinear path. A non-linear two compartment first + first order model can adequately fit to the degradation pattern of many pesticides, and could also predict the halflife (DT50) and preharvest interval (PHI) in a more realistic manner than a linear first-order model.

First-order model: $[A]t = [A]1 \exp(^{-}k1t)$

First + first-order model: $[A]t = [A]1 \exp(^{-}k1t) + [A]2 \exp(^{-}k2t)$

where [A]t is the concentration (mg kg^{-1}) of A at time t (days), [A]1 and [A]2 are the initial concentrations of A at time 0 degraded through first-order processes 1 and 2 and k1 and k2 are the degradation rate constants for 1 and 2. The units of k depend on the model used.

4.2 Dissipation Studies: Importance in Residue Management

Information on dissipation kinetics of pesticides in fruits and vegetables is a key aspect in residue management risk assessment practices. This is because human exposure to pesticides is predominantly caused by residues in agricultural crops grown for human and animal consumption. Dissipation kinetic models are used to determine the DT50 and PHI of pesticides in fruits and vegetables. The DT50, which is the time at which the concentration of initial deposits reaches the 50 per cent level in the applied fruits or vegetables is

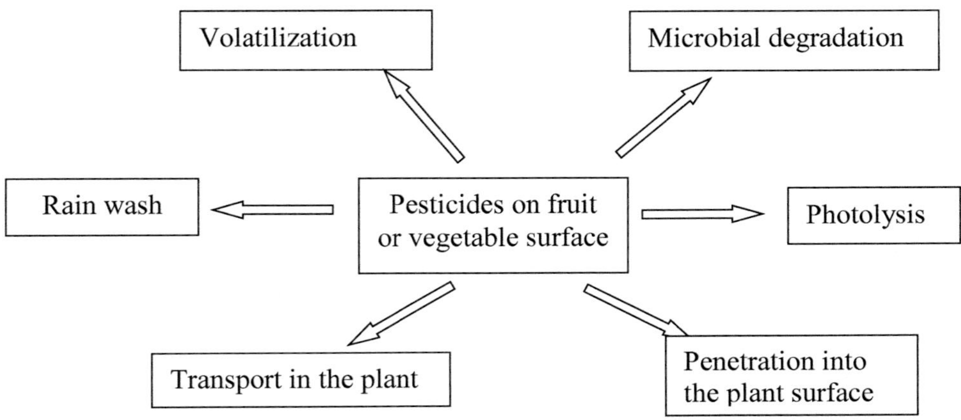

Figure 1. Dissipation Routs of Pesticides on Fruit or Vegetables Surfaces.

calculated first order kinetics by the equation ln(2) k^{-1}. DT_{50} is an important parameter that signifies the speed of degradation. PHI is the time period (in days) required for dissipation of the initial residue deposits to below the MRL. DT_{50} and PHI data are very important in managing pesticide residues in any food commodity.

In India, several researchers have studied the fate of pesticides in/on fruits and vegetables and established crop-specific DT50 and PHIs. Several researchers investigated the dissipation of pesticides in a variety of fruits and vegetables in India and such data are successfully utilised for residue management in respective crops. Some recently reported dissipation studies on a variety of fruits and vegetables are summarised in Table 2.

Table 2. Dissipation of Half-life of Selected Pesticides in Fruits and Vegetables

Commodity	*Pesticide*	*Half-life (Days)*	*References*
a. Fruits			
Apple	Kresoxim-methyl	4.58–4.77	Malhat *et al.*, 2013
	Trifloxystrobin tebuconazole	19.38–25.99	Oatyal *et al.*, 2013
	Propargite	1.66–2.61	Kumar *et al.*, 2005
	Emamectin benzoate	2.75–3.09	Wang *et al.*, 2012
	Mancozeb Carbendazim	2.82-3.15 1.88-2.38	Sapna *et al.*, 2017
	Tebuconazole	1.30-2.60	Dubey *et al.*, 2020
Grape	Tetraconazole Difenoconazole	4.5–9.5 4.5–5.5	Banerjee *et al.*, 2008
	Carbendazim Mancozeb	7.3–8.1 4.6–5.7	Banerjee, 2005
	Azoxystrobin	5.4–11.2	Gajbhiye *et al.*, 2011
	Forchlorfenuron Benzyl aminopurine Gibberellic acid Ethephon	5 3–3.5 2-3 12–15	Ugare *et al.*, 2013
	Chlorantraniliprole	8-10	Saini *et al.*, 2018
	Hexythiazox Bbifenazate	6-10 5-6	Shabeer *et al.*, 2020
Mango	Imidachloprid Carbosulfan	38.0 7.0	Bhattacherjee A. K., 2013
	Imidachloprid Heaxaconazole	1.86 1.97	Reddy *et al.*, 2013
	Spirotetramat Imidacloprid	3.3 5.2	Mohapatra *et al.*, 2012
	Fluopyram Tebuconazole	4.3–5.4	Mohapatra *et al.*, 2018
	Azoxystrobin Buprofezin Dinocap Hexaconazole	10.4–12.1 5.8–8.5 5.4–6.2 4.4–6.1	Mohapatra *et al.*, 2020
b. Vegetables			
Brinjal	Thiacloprid	0.47 and 0.50	Sahoo *et al.*, 2013
	Profenofos Cypermethrin Chlorpyrifos Cypermethrin	2.15–2.31 0.91–1.86 3.27–3.10 2.19 to 3.27	Mukherjee *et al.*, 2012
	Quinalphos	2-3	Pathan *et al.*, 2012
	lubendiamide	0.62 and 0.54	Takkar *et al.*, 2012
	Monocrotohos	2.75-3.01	Patil *et al.*, 2020
	Spiromesifen	1.32 - 2.18	Vinothkumar *et al.*, 2018
	Emamectin benzoate	1.53 - 1.57	Vinothkumar *et al.*, 2018
Cabbage	Emamectin benzoate	1.34–1.72	Wang *et al.*, 2012
	Fipronil	3.21 -3.43	Bhardwaj *et al.*, 2012
	Trichlorfon	1.80	Li *et al.*, 2011
	Quinalphos Fubendiamide	4.8 - 5.3 3.4 - 3.6	Soudhamini and Deepa, 2013
	Flubendiamide	3.4 to 3.6	Paramashivam and Banerjee, 2013
	Thiacloprid	12.3–13.1	Dutta *et al.*, 2012
	Spinosad	1.4 and 1.5	Singh and Battu, 2012
	Metaflumizone	1.7–2.1	Chatterjee and Gupta, 2013
	Emamectin Benzoate	<5	Singh *et al.*, 2013
	Novaluron Lambda-Cyhalothrin	1.77–3.38 1.36–3.82	Sharma *et al.*, 2021
	Carbendazim Tebuconazole	2.17 - 2.99 2.74 - 2.81	Pallavi *et al.*, 2021
	Chlorpyriphose Profenophos	1.91 2.01	Beevi *et al.*, 2018
Cauliflower	Chlorantraniliprole	1.36	Kar *et al.*, 2013
	Cypermethrin Deltamethrin Profenofos Triazophos	1.5–2.1 2.9–3.3 2.6–3.0 2.2–2.6	Gupta *et al.*, 2013
Chilli	Chlorpyriphos	4.43 and 2.01	Jyot *et al.*, 2012
	Cypermethrin	2.51 and 2.64	

Commodity	Pesticide	Half-life (Days)	References
	Trifloxystrobin Tebuconazole	1.81 and 1.58 1.37 and 1.41	Sahoo *et al.*, 2012
	Deltamethrin	0.36 to 1.99 0.38 to 2.06	Pandher *et al.*, 2012
	Chlorpyriphos Cypermethrin	4.43 and 2.01 2.51 and 2.64	Jyot *et al.*, 2013
	Acetamiprid	2.24–4.84	Sanyal *et al.*, 2008
	Chlorfenapyr	2.93 to 2.96	Ditya *et al.*, 2010
	Tebuconazole	1.30-2.60	Dubey *et al.*, 2020
	Kresoxim-methyl	5.30-6.30	Majumder *et al.*, 2020
	Dinotefuran	1.34-3.45	Visal *et al.*, 2019
Okra	Deltamethrin Alphamethrin Triazophos Ethion Cypermethrin Profenophos	2.09 2.09 1.68 1.27 2.59 1.88	Parmar *et al.*, 2012
	Flubendiamide	4.7 to 5.1	Das *et al.*, 2012
	Spiromesifen	1.68	Raj *et al.*, 2012
	Imidacloprid Beta-Cyfluthrin	1.07 - 2.41 1.98 - 3.30	Banerjee at al, 2012
	Chlorpyriphos	0.6	Samriti *et al.*, 2012
	Cypermethrin Spiromesifen	1.90 2.0	Kavitha *et al.*, 2021
	Acetamiprid	3.06-3.09	Ratnamma *et al.*, 2021
Tomato	Flubendiamide Thiacloprid	0.72-1.32 0.83-1.79	Sharme and Parihar, 2013
	Flubendiamide	1.64 - 1.98	Paramashivam and Banerjee, 2013
	Bifenthrin	1.83-2.32	Chauhan *et al.*, 2012
	Chlorpyrifos	4.38- 4.43	Rani *et al.*, 2013
	Mancozeb Metalaxyl	3.76 - 4.14 0.41 - 1.29	Rani *et al.*, 2013
	Fenamidone Mancozeb Iprovalicarb Propineb	2-3 2-2.5 1.5-2 2-3	Kumar *et al.*, 2020
	Iprovalicarb Propineb	1.08 - 4.67 3.36 - 11.41	Tripathy *et al.*, 2021
	Chlorantraniliprole	1.26 - 1.77	Paramasivam, 2021
	Tebuconazole	1.30-2.60	Dubey *et al.*, 2020

Human exposure to pesticides is predominantly caused by residues in agricultural crops grown for human and animal consumption. Available information on pesticide dissipation in plants from experimental studies only covers a small fraction of possible combinations of substances authorized for use in agriculture. Additionally, aspects and processes influencing dissipation kinetics are still not fully understood. Therefore, Fantke and Juraske (2013) systematically reviewed 811 scientific literature sources providing 4513 dissipation half-lives of 346 pesticides measured in 183 plant species. They focused on the variability across substances, plant species and harvested plant components and finally discussed different substance, plant and environmental aspects influencing pesticide dissipation. Measured half-lives in harvested plant materials ranged from around 1 hour for pyrethrins in leaves of tomato and pepper fruit to 918 days for pyriproxyfen in pepper fruits under cold storage conditions. Ninety-five per cent of all half-lives fall within the range between 0.6 and 29 days. These results emphasize that future experiments are required to analyze pesticide–plant species combinations that have so far not been covered and are highly relevant for assessment of human exposure.

5.0 Monitoring of Pesticide Residues

The purpose of pesticide monitoring programs is to bring out the food safety status and ensure that the residue loads in fruits and vegetables do not exceed MRLs prescribed by the regulatory bodies. It also deters misuse of pesticides that could result in unexpected residues in food and facilitates implementation of good agricultural practices (GAP). The outcome of the residue monitoring programme is used by regulatory bodies for future and risk assessment exercises for public health. In most countries, the monitoring programs are organised by a single agency designated as the competent authority. The agency designs a monitoring plan based on the previous data available from dietary consumption and risk assessment exercises or pesticide usage in the available fruits and vegetables. Unfortunately, in India an ideal institutional frame work for this purpose is still lacking. Pesticide legislation in developing countries is generally lacking or not implemented and this also affects pesticide monitoring since it relies on legislation to be effective. In the European Union (EU) countries there is a coordinated programme under European Food Safety Authority (EFSA) for all the member countries as per the European Commission guidelines and the individual member states have their national residue monitoring programs. It is important to note here that the monitoring program will give an insight on the problematic pesticides, *i.e.*, it will only identify the problem. Taking the outcome of the monitoring program, strategies need to be adopted for the management of residues in fruits and vegetables.

In a comprehensive study at national level in 2010, out of 34027 samples analyzed under the project "Monitoring of Pesticide Residues at National Level", 543 samples (1.6 per cent) were found above MRL. Commodity-wise break-up shows 3.1 per cent samples of vegetables, 0.8 per cent of fruits, were found to be above respective MRLs. Several research publications or reports are available on pesticide residue monitoring in fruits and vegetables in India. In view of the detection of pesticide residues, several rapid alerts were being issued by the importing countries and the situation is precarious enough where import of certain fruits and vegetables from India could be banned. More than the presence of pesticide residues above MRLs, a serious fact revealed was the detection of non-recommended pesticides in vegetables. Moreover, the pesticides which have a label claim are mostly old generic molecules exhibiting a higher mammalian toxicity. These molecules do not have any label claim and beckon an immediate replacement/substitution with chemicals having a low mammalian toxicity and green chemistry. Thus, extensive data generation regarding the bioefficacy, residues dissipation, dose standardization, PHI and application technology of these molecules needs to be conducted at research institutions for developing a safe package of practice nucleating upon rigorous stringent and methodical research activities.

6.0 Risk Assessment

Risk assessment of chemicals is described as a process intended to calculate or estimate the risk to a given target organism, system or (sub) population, including identification of attendant uncertainties, following exposure to a particular agent taking into account the inherent characteristics of the agent of concern as well as the characteristics of the specific target system (OECD 2003). Risk assessment process includes four steps: hazard identification, hazard characterization (dose-response assessment), exposure assessment and risk characterization.

6.1 Hazard Identification

It is the first step in risk assessment and it involves the identification of the type and nature of adverse effects that an agent has as inherent capacity to cause in an organism, system or (sub) population (OECD2003). It is generally performed before a pesticide is approved for usage in agriculture or other areas and the information on hazards posed is readily available from the pesticide registering bodies and on their websites for public usage. Most of such information is also available from international organizations such as JMPR, OECD and EC. For hazard identification, studies are carried out in test organisms (microbial, cells or animals) and the exposure level is increased until an adverse effect is produced. The highest dose of a pesticide that does not cause detectable toxic effects on the test organisms is called the no-observed adverse-effect-level (NOAEL) and is expressed in terms of mg per kg of body weight per day (WHO 1997). It is used in calculation of the acceptable daily intake (ADI).

6.2 Hazard Characterization

Hazard characterization is the qualitative and, wherever possible, quantitative description of the inherent properties of an agent or situation having the potential to cause adverse effects. This should, wherever possible, include a dose response assessment and its attendant uncertainties (OECD 2003). Hazard characterisation involves comparing the pesticide exposure concentration with the ADI. The ADI is the estimate of the amount of a substance in food (mg/kg body weight/day) that can be ingested daily over a lifetime without appreciable health risk to the consumer (WHO 1997). ADI is calculated by dividing the NOAEL for animal studies with an uncertainty factor of 100 to convert to a safe level for humans.

6.3 Exposure Assessment

It is the evaluation of the concentration or the amount of a particular agent that reaches a target organism, system or (sub) population in a specific frequency for defined duration (OECD 2003). The potential intake or consumption of pesticide residues is divided by the body weight and compared to ADI.

Exposure = (Concentration of pesticide residue x Food consumed)/body weight

Dietary exposure assessment is very important since the food consumption pattern varies from country to country. So, as much as possible, the food consumption data used for dietary exposure should be the average daily consumption of the particular commodity in the targeted population. If such information is not available, the WHO data on average daily consumption of food can be used although it could be too much generalized.

6.4 Risk Characterization

The qualitative and, wherever possible, quantitative determination, including attendant uncertainties, of the probability of occurrence of known and potential adverse effects of an agent in a given organism, system or (sub) population, under defined exposure conditions (OECD 2003). The international estimated daily intake (IEDI) has been used to characterize the risk of pesticides and it is expressed as:

$$IEDI = \Sigma\ STMR \times E \times P \times F$$

where, STMR = supervised trial median residue level, E = Edible portion, P = processing factor, F = consumption of the food commodity

When the IEDI is more than the ADI, the food involved is considered a risk to the concerned consumers. In India, an institutional framework is being established under the aegis of FSSAI for the risk assessment of pesticide residues in food commodities.

7.0 Pesticide Residues: Issues to be Addressed in Horticulture Crops

Some of the key issues identified below need to be addressed on priority for the effective and scientific management of pesticide residues in horticulture crops.

7.1 Stringent Ever-changing Country Specific MRLs

The Technical Barriers to Trade Agreement (TBT) encourages removal of unnecessary obstacles to trade flow by avoiding over-stringent regulations, it simultaneously recognizes individual countries' rights to adopt their own standards which they might consider appropriate with respect to the life and health of human being, animal population, plants and environment. Although the Codex MRLs are applicable to all the nations, individual countries have their own MRL regulations. The MRL established for a pesticide varies with crop and the country. For agro-export, it is thus essential that the commodities should comply with the latest MRL regulations of the importing country. As the PHI or waiting period established for various pesticides varies with MRLs, it is almost impossible to adopt a universal PHI for a pesticide for market specific residue complied production.

7.2 Lack of Label Claim Chemicals for Use

As per the regulation of CIB-RC, a pesticide can be officially recommended or allowed to use against a particular pest or disease in a specific crop only if the pesticide have a label claim with CIB and RC for the use against that particular pest in the specific crop. The pesticides which have a label claim for use in specific horticulture are very few and they are not sufficient to manage all pests and diseases in that particular crop. In such cases the growers are left with no option other than to rely on the available chemicals in market to protect their crop and their nonscientific use lead to more residues in the harvested products.

7.3 Spurious Pesticides in Market

As per Pesticide Management Bill, 2008, a pesticide shall be deemed to be spurious if it is not registered or licensed in the manner required, if it shows active ingredient higher or lower the limits prescribed, if it is an imitation of another pesticide, if the container bears the name of the manufacturer either fictitious or does not exist, if the chemical composition as approved by the Registration Committee is not adhered, if it has outlived its shelf-life or if its import, manufacture, use or sale is prohibited. The presence of spurious pesticides in the market and the lack of awareness among the farmers about them became a major reason for pesticide residues. It is being estimated that India is facing crop losses worth over Rs 25,000 crore a year because of the use of spurious pesticides and insecticides, pegging the annual sale of such fake products at over Rs 1,500 crore. Strict actions to be taken against such manufacturers to safeguard our farmers and proper legal provision need to be established for this.

7.4 Uncontrolled Use of "Bio-pesticides"

Use of bio-pesticides in organic horticulture are very common in commercial horticulture. However, the registration and use of these bio-pesticide are not officially regulated. One of the important issue associated with this bio-pesticides are their purity with respect to any chemical or spurious pesticides . They are hardly tested or monitored at state level. The use of such bio-pesticides at later stages of crop or near to harvest results in pesticide residues in harvested produce.

8.0 Pesticide Residue Management

Considering the national and international food safety issues, it is very important to adopt residue management strategies to tackle the pesticide residue problems in fruits and vegetables and thereby increase the export of these commodities and also to ensure safety to domestic consumers. Some strategies for pesticide residues management in fruits and vegetables are described below:

8.1 Processing

Fruits and vegetables like other foods often pass through culinary and food processing treatments before they are consumed. The effects of these culinary and food processing techniques have been investigated by various researchers and they have been found to reduce the pesticide residue levels except in cases where there is concentration of the product like in juicing, drying and oil production. Some toxic metabolites may be produced during processing treatments, especially thermal processing. However, the consumers can still be encouraged to employ those processing methods that reduce pesticide residues. Food processing studies often results in transfer factors or food processing factors (PF) of the pesticide residue in the transition from raw agriculture commodity to the processed product. These processing factors are expressed as the concentration of pesticide after processing divided by the concentration before processing. Some processing factors are available in public literature while others are only available from the pesticide registering bodies. Processing studies have become a part of pesticide registration requirements. Some processing techniques and average processing factor are summarized in Table 3. Kaushik *et al.* (2009) reviewed the effect of processing on dissipation of pesticide residues in fruits and vegetables.

It is important to note here that some processing techniques especially which involves dehydration of the fresh commodities may increase the pesticide residues in the final processed products (Shabeer *et al*, 2015). Such cases the processing factor will be more than one and termed as concentration factor. It is mainly applicable in case of commodities like spices, pulses, dry fruits *etc.*

Table 3. Processing Techniques and Average Processing factor

Processing Technique	*Processing Factor*
Baking	1.38
Blanching	0.21
Boiling	0.82
Canning	0.71
Frying	0.1
Juicing	0.59
Peeling	0.41
Washing	0.68

Adopted from: Keikotlhaile, B. M and Spanoghe P., 2011.

8.2 Preharvest Interval (PHI) a Tool for Residue Mmanagement

When a pesticide is applied in the field, the initial residue (on the day of application present on crop) will degrade by various factors such as microbial degradation, chemical degradation (hydrolysis, oxidation, reduction *etc.*) and by physical means (photolysis *etc.*) and ultimately the residue will come down to zero. PHI is most important in field level management of pesticide residues. On the basis of the MRL, the PHIs of pesticides are calculated. PHI is also termed as the safe waiting period, which is the minimum time in days that must be provided between last application of a pesticide and harvesting of the produce so that its residue level at harvest reaches below the MRL. Estimating PHI of a pesticide ideally involves multi-location field trails wherein the pesticides are applied following the guidelines of GAP. Although first order rate kinetics is largely used for estimating PHI, nonlinear kinetics, *e.g.* first + first order often governs the dissipation of pesticides, where the residues dissipate with time following a non-linear relationship. In most of the dissipation studies described in the above section, the PHI was also established for respective pesticide in studied crops. When the PHIs are strictly followed, the pesticide residue loads are expected to dissipate to below MRL levels in the harvested produce. So for effective residue management at field level, it is very important to follow the waiting period prescribed for the particular pesticide with respect to target MRL. It is also important to note here that once the MRL of a pesticide changes, the PHI prescribed for that pesticide also will change. If the MRL of a pesticide is reduced, then the PHI/waiting period will increase and vice versa. The schematic graph presented below clearly depict the relationship of MRL and the PHI. The PHI is considered as the most effective tool for the management of pesticide residues below the target MRLs in the produce at harvest.

8.3 Adoption of Precision Spraying Techniques

Appropriate spraying technology improves field coverage with lesser quantity/volume and therefore minimizes the actual amount of pesticides. Most of the pesticides which are sprayed in the field don't reach the target. So, the use of sophisticated spraying technologies like electrostatic sprayers as a precision farming technology can reduce the amount of pesticide usage to some extent.

8.4 Integrated Pest Management with minimal Pesticide

To some extent, IPM can reduce the usage of chemical pesticides in the field. As much as possible, use environmental friendly strategies and involve the less persistent synthetic pesticide in the IPM package.

8.5 Field Level Management

The major thing strategies to take care at field level management are:

- Use only the recommended pesticides
- Do not spray banned as well as non-recommended chemicals
- Strictly adhere to the recommended dose of application including the recommended volume of water to be sprayed per hectare or per acre.
- Strictly maintain recommended Pre-Harvest Interval to minimize consumer risk regarding detection of residues

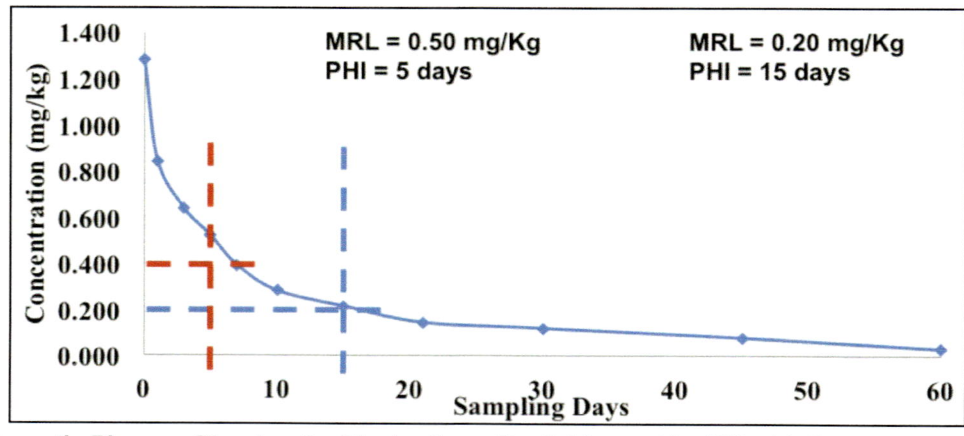

Figure 2. Schematic Diagram Showing the Dissipation a Pesticides and its PHI with Respect to Change in MRL.

Integrated Pest Management involving minimal usage of highly toxic chemicals

Sequencing the pesticides in relation to their persistence particularly in the last 2 months period before harvest

Adoption of Bio-control measures during last month before harvest

"Combination Products" which contains two or more active ingredients as given in the above list can be used. However, the dose of the individual active ingredient should not exceed the recommended dose.

8.6 Promote the Use of Non-regulated Alternate Chemicals

One of the strategy for pesticide residue management could be to use alternate chemicals who's MRLs are not prescribed by any of the regulatory body. For examples; Sulphur, Chitosan, Silver complex of H_2O_2, Chlorine di oxide, Ozonized water, phosphonic acid *etc.* may be used whenever possible. As MRL are not prescribed for these chemicals, their residue will not be tested for the regulatory compliances. It is worth to note here that use of salts of phosphonic acid in some horticulture crops may be restricted even though this chemical is not directly regulated. This is because of the fact that the residue definition of Fosetyl-Al also include the phosphonic acid residues. Hence, the use and the residues of phosphonic acid salt will accounted as Fosetyl-Al in residue test report. In order avoid this, PHI based on phosphonic acid residue with respect to Fosetyl-Al MRL should be followed for phosphonic acid salts.

8.7 Regulatory Control

Knowingly or unknowingly farmers keep applying banned or restricted use chemicals for pest and disease management. It is extremely important to replace the usage of highly toxic chemicals with those having safer mammalian and environmental toxicity profiles. In many cases, the detection of the pesticide residues in fruits and vegetables include restricted use chemicals orthose which are not having label claim. Strict regulatory control need to be established to minimize the use of such restricted or unauthorized (without official label claim) chemicals. The regulatory authority should ensure the availability of label claim for optimal control of various pests and diseases in targeted crops which will avoid violation of GAP recommendations. This can be achieved through extension programmes involving all the stakeholders *e.g.* farmers, state government agricultural/horticultural department officials and exporters under the overall supervision of Food Safety and Standard Authority of India and Central Insecticide Board.

9.0 Conclusion

Consumer awareness is growing with respect to the right to have safe food. No one in the society is ready to tolerate pesticide residues in the food intended for his/her nurture. On the other hand, being in tropical belt we do not have the luxury of avoiding the use of agrochemicals in pest management. It is the dose and time of application, which differentiate the safe and unsafe use of pesticides. If the recommendations of the GAP and PHI are sincerely followed, it is definitely possible to minimize the residue load of agrochemicals in fruits and vegetables at field level, which in turn, will ensure safety to the health of the consumers and ecosystem as a whole. The monitoring data reveals that the market samples of fruits and vegetables have residues above the prescribed MRLs. Processing can be used as the best tool for postharvest management of pesticide residues. Risk assessments associated with pesticide residues need to be scientifically executed in a structured manner. A well-structured residue monitoring system with an institutional frame work can possibly ensure field-level traceability and management of agrochemical residues in fruits and vegetables through implementation of good agricultural practices.

12.0 Important References

Anastassiades, M., Lehotay, S.J., Stajnbaher, D. and Schenck, F.J., 2003. Fast and easy multi-residue method employing acetonitrile extraction/partitioning and "dispersive solid-phase extraction" for the determination of pesticide residues in produce. *J. AOAC Int.* **86**: 412-31.

Anastassiades, M., Scherbaum, E., Tas¸ B., delen, Stajnbaher, D., Ohkawa, H., Miyagawa, H. and Lee, P.W., 2007 (Eds.). *Crop Protection: Public Health, Environmental Safety*. Wiley-VCH, Weinheim, Germany, p. 439.

Banerjee, K., Oulkar, D.P., Dasgupta, S., Patil, S.B., Patil, S.H. and Savant, R., 2007. Validation and uncertainty analysis of a multi-residue method for pesticides in grapes using ethyl acetate extraction and liquid chromatography–tandem mass spectrometry. *J. Chromatogr. A* **1173**: 98–09.

Dubey, J.K., Patyal, S.K., Katna, S. *et al.*, 2020. Persistence and dissipation kinetics of tebuconazole in apple, tomato, chilli and onion crops of Himachal Pradesh, India. *Environ. Sci. Pollut. Res.* **27**: 11290–302.

Hernandez, F., Sancho,J.V. and Pozo, O.J., 2005. Critical review of the application of liquid chromatography/mass spectrometry to the determination of pesticide residues in biological samples. *Anal. Bioanal. Chem.* **382**: 934–46.

Kaushik, G., Satya, S. and Naik, S.N., 2009. Food processing a tool to pesticide residue dissipation: A review. *Food Res. Int.* **42**: 26–40.

Kavitha, K., Anitha, V. and Aruna. M, 2021. Dissipation and persistence studies of certain pesticides in/on Okra. *Int. J. Chem. Stud.* **9**: 716-19.

Keikotlhaile, B.M. and Spanoghe, P., 2011. Pesticide Residues in Fruits and Vegetables; Pesticides - Formulations, Effects, Fate, Edited by Prof. Margarita Stoytcheva, p. 243-252.

Kumar, B. and Mukherjee, D.P., 2012. Organochlorine residues in vegetables. *Int. J. Veg. Sci.* **18**: 121-36.

Kumar, S.M., Abrar, F.J. and Kalbey, R.S., 2018. A Simple LC-MS/MS Method for Estimation of Chlorantraniliprole Residues and its Dissipation Kinetics in/on Grape Berries and Cropped Soil. *Pestic. Res. J.* **30**: 45-50. 10.5958/2249-524X.2018.00008.0.

Kumar, Y.B., Shabeer, T.P.A., Jadhav, M. *et al.*, 2020., Analytical method validation, dissipation and safety evaluation of combination fungicides fenamidone+mancozeb and iprovalicarb+propineb in/on tomato. *J. Food Sci. Technol.* **57**: 2061–69.

Kumari, B.R., Ranga, Rao, G.V.R., Sahrawat, K.L. and Rajasekhar, P., 2012. Occurrence of insecticide residues in selected crops and natural resources. *Bull. Environ. Contam. Toxicol.* **89**: 187-92.

Lehotay, S.J., 2002. Application of gas chromatography in food analysis. Trends in ***Anal. Chem.***, **21**: 686-97.

Lehotay, S.J., Hiemstra, M., van Bodegraven, P. and de Kok, A., 2005. Validation of a fast and easy method for the determination of residues from 229 pesticides in fruits and vegetables using gas and liquid chromatography and mass spectrometric detection. *J. AOAC Int.*, **88**: 595-614.

Majumder, S., Verma, C.K., Rani, V., Rani, A. T., Pandey, K. K. and Singh, J. 2020.Residue dynamics and food safety evaluation of fungicide kresoxim-methyl in green chilli (*Capsicum annum* L.), *Int. J. Environ. Anal. Chem.* https: //doi.org/10.1080/03067319.2020.1830986

Mohapatra, S., Siddamallaiah, L., Buddidathi R. and Matadha, N.Y., 2018., Dissipation kinetics and risk assessment of fluopyram and tebuconazole in mango (*Mangifera indica), Int. J. Environ. Anal. Chem.* **98**: 229-46.

Mohapatra, S., Siddamallaiah, L., Matadha, N.Y. *et al.* 2020. Persistence and dissipation study of azoxystrobin, buprofezin, dinocap and hexaconazole on mango (*Mangifera indica* L.). *Environ. Sci. Pollut. Res.* **27**: 32820–828.

Naseema, B.,Ambily, P.,Thomas, G., Pratheeshkumar N, George, X. and Shaija, V.R., 2018.Dissipation of Chlorpyriphos and Profenophos in Cabbage (*Brassica oleracea* var. *capitata* L.). *Pestic. Res. J.* **30**: 121-25.

OECD, 2003. Descriptions of selected key generic terms used in chemical hazard/risk assessment. OECD Series on Testing and Assessment, No. 44.

Pallavi, M.S., Harischandra N. K., Ratnamma, Nidoni, U., Bheemanna, M., Pramesh, D., 2021.

Paramasivam, M. 2021. Dissipation kinetics, dietary and ecological risk assessment of chlorantraniliprole residue in/on tomato and soil using GC–MS. *J. Food Sci. Technol.* **58:** 604–11.

Patil, C.S., SA Landge, BV Deore and YS Saindane. 2020., Residues and dissipation of monocrotohos 36% SL in/on brinjal and soil. *J. pharmacogn. phytochem.* **9**: 2439-41.

RASFF Annual Report, 2012.TheRapidAlert Systemfor FoodandFeed (ec.europa.eu/food/food/rapidalert/./rasff_annual_report_2012_en.pd).

Ratnamma, Harischandra, N. R., Prabhuraj, A., Nagaraj, M. Naik. and Bheemanna, M. 2021. Determination and dissipation of acetamiprid using LC-MS/MS in okra. J. *Entomol. Zool. Studies,* **9**: 110-16.

Sapna, K., Kumar, P.S., Kumar, D.J., Tanuja, C.B., Avinash, C., Nisha, D. and Ajay, S., 2017., Dissipation and risk assessment of combi product of mancozeb and carbendazim on apple at different locations in north India. *Pestic. Res. J.* **29** : 48-54.

Shabeer A.T.P., Girame, R., Hingmire, S. *et al.* 2020., Residue dissipation, evaluation of processing factor and safety assessment of hexythiazox and bifenazate residues during drying of grape to raisin. *Environ. Sci. Pollut. Res.* **27**: 41816–823.

Sharma, K.K., Tripathy, V., Mohapatra, S., Matadha, N.Y., Pathan, *et al.*, 2021. Dissipation kinetics and consumer risk assessment of novaluron + lambda-cyhalothrin co-formulation in cabbage, *Ecotoxicol. Environ. Saf.* **208**: 111494. .

Tripathy, V., Sharma, K.K., George, T. *et al*, 2021. Dissipation kinetics and risk assessment of iprovalicarb + propineb fungicide in tomato under different agro-climates. *Environ. Sci. Pollut. Res.* **28**: 31909–919.

Vinothkumar, B., R Kaviya and PV Shibani., 2019. Dissipation pattern of Emamectin benzoate in/on brinjal fruits. *Int. j. chem. Stud.*, 6: 2630-34.

Vinothkumar, B., Shibani, PV., Kaviya R. and Kowshika J. 2018, Dissipation pattern of spiromesifen in/on brinjal fruit. *Int. J. Chem. Stud.* 6: 2485-88

Visal S.K., Subhashchandran K.P., George, T., Ambily, George, X., Vijayasree V and Suryamol, S., 2019. Dinotefuran Residues and their Dissipation in Chilli Pepper (*Capsicum annuum* L.) and Soil and their Risk Assessment. *Pestic. Res. J.* **31**: 211-19.

Fruit Crops

Chapter 10

Almond

Desh Beer Singh

Former-Director,

ICAR-Central Institute of Temperate Horticulture, Old Air Field, Rangreth, Srinagar – 5, Jammu & Kashmir, India

E-mail: deshbsingh@yahoo.co.in

Nazeer Ahmed

Former-Vice Chancellor

Sher-e-Kashmir University of Agricultural Sciences and Technology, Srinagar, J&K, India

E-mail: dnak59@rediffmail.com

1.0 Introduction

Almond (*Amygdalus communis*) is native to Mediterranean region and is considered as one of the oldest tree nut in the world. It is closely related to peach which probably evolved from the same ancestral species in South-Central Asia:- from there it spread along the shores of the Mediterranean in Northern Africa and Southern Europe by Egyptians, Greeks and Romans. It's an elixir for good health and part of India's cultural heritage, used extensively during festive seasons and weddings. Almond is nutritionally dense being a rich source of vitamin E, containing 26 mg per 100 g. Almonds are also rich in dietary fiber, B vitamins, essential minerals such as magnesium, copper, manganese, calcium, potassium as well as monounsaturated and polyunsaturated fats, fats which potentially may lower LDL cholesterol.

2.0 Status of Cultivation

Almonds were bought to California in the 17th century by Spanish where the industry progressed and production increased several folds due to cultivation of superior varieties accompanied by prudent scientific production and protection technologies making California the world leader in almond production.

Major producers of almond include the U.S, Spain, Iran, Italy, Morocco, Syrian Arab Republic and Turkey. In Spain, numerous commercial cultivates of sweet almond are produced, most notably the Jordan almond and the Valencia almond. As far as the production is concerned, the EV almond accounts for about 32 per cent of the world's total production. Spain is the main EV almond producer with 62 per cent of EV almond production, while Italy is the second main European producer, followed by Greece and Portugal. In 2019, world production of almonds was 3.5 million tonnes, led by the United States providing 55 per cent of the world total (table). Other leading producers were Spain, Iran, and Turkey. Data on the production of Almond in 2020 are presented in Table 1.

Table 1. Top Twenty Almond Producing Countries (FAOSAT-2020)

Country	*Production (metric tonnes) 2020*
United States	1,802,785.00
Spain	319,448.00
Iran	136,893.00
Morocco	120,075.00
Turkey	100,637.00
Italy	77,061.00
Australia	75,140.00
Tunisia	67,840.00
Algeria	60,463.00
China	52,211.00
Greece	38,878.00
Chile	36,678.00

Country	Production (metric tonnes) 2020
Afghanistan	32,362.00
Lebanon	30,774.00
Libya	30,243.00
Syria	29,004.00
Pakistan	20,402.00
Portugal	16,317.00
Uzbekistan	16,094.00
Israel	8,732.00

In India almond was introduced in Kashmir during 16th century by Persian settlers but in spite of its great potential in the region, the crop could not develop on commercial scale. Almond cultivation in India is at present restricted to selected hilly areas of Jammu and Kashmir, Himachal Pradesh and Uttar Pradesh. In J&K, it is cultivated in an area of 18,870 hectares with total production of 11,000 metric tonnes. In Himachal Pradesh, it is gaining popularity and is cultivated in an area of 2,400 hectares with production of around 1,700 tonnes. Recently, the J&K government has formulated almond development program for bringing an additional 12,000 hectares of area under its cultivation in a phased manner. Green almond is widely grown in the forest of West Orissa. In India, several varieties and hybrids are grown. These include early varieties like Ne plus ultra, Pareiel, Fascionello, Peerles, mid season varieties like Cristomarto, Vesta, Marcona and late varieties like Genco and Texas. The soft-shelled cultivars are commercially more important than the hard-shelled ones.

Spain has one of the most diverse commercial cultivars of almonds. The most cultivating types of almonds in Greece are Ferragnes and Texas (mission) which are known for their sweet taste and premium quality.

Table 2. Indian Production of ALMOND-2017-18

Sl.No.	State	Production (000 tonnes)
1.	J &K	12.90
2.	Himachal Pradesh	0.76
	Total	13.69

Source: National Horticulture Board (NHB).

3.0 Major Varieties

There are number of varieties available in almond. But there are four major varieties of export quality *viz.*, Non Pareil, California Paper Shell, IXL and Merced. In addition to these exotic cultivars, recently released indigenous cultivars like Shalimar, Makhdoom and Waris have also shown very promising results under Kashmir conditions. The salient features of important cultivars are:

3.1 Major Types

There are three types of almond:

3.1.1 Sweet Almonds

These are commercially produced, with an edible seed. There are innumerable varieties, some with paper shells and some with hard shells. Jordan almonds are a hard-shelled type with a thinner integument on the seed, a finer flavour and are used for confections.

3.1.2 Bitter Almonds

They contain a bitter glycoside called amygdalin which readily breaks down into prussic acid hence it is not used as food. These are grown as a source of oil.

3.1.3 Green Almond

It is a small tree and a native of west Asia and has been cultivated in the Mediterranean countries for nearly 40,000 years. It is chiefly grown in Iran, Afghanistan, the Southern U.S. and California. It is also cultivated in the Northern parts of India. The fruit is drupe and the seed contains two large green cotyledons with a reddish covering. These find great use in mixed nuts and as a flavouring material in ice-creams, candy and sweetmeats. It is also eaten after being salted and roasted. It is considered to be a digestive, a sedative and a tonic.

3.2 Varieties

3.2.1 Varieties Grown in California (Figures 1a-j)

a) BUTTE (BT): Shell: Semi-hard shell, light colour, smooth surface, low suture opening. Nut: Small, short plump shape, wrinkled surface.

b) CARMEL (CR): Shell: Soft shell, good shell integrity, fair suture opening. Nut: Medium, narrow shape, slightly wrinkled surface.

c) FRITZ (FR): Shell: Semi-hard shell, good shell integrity, low suture opening. Nut: Small, medium plump shape, fairly wrinkled surface.

d) MISSION (MI): Shell: Hard shell, good shell integrity, no suture opening. Nut: Small, short wide shape, dark brown, deeply wrinkled surface.

e) MONTEREY (MT): Shell: Hard shell, smooth surface, low suture opening. Nut: Large, long narrow shape, deeply wrinkled surface.

f) NONPAREIL (NP): Shell: Soft shell, light colour, high suture opening. Nut: Medium, flat shape, smooth surface.

g) PADRE (PD): Shell: Hard shell, good shell integrity, no suture opening. Nut: Small, short wide shape, wrinkled surface.

h) PEERLESS (PL): Shell: Hard shell, good shell integrity, smooth surface, no suture opening. Nut: Medium, wide shape, fairly wrinkled surface.

i) PRICE (PR): Shell: Soft shell, dark brown colour, rough surface, high suture opening. Nut: Small, short narrow shape, fairly wrinkled surface.

j) SONORA (SN): Shell: Soft shell, dark brown colour, rough surface, high suture opening. Nut: Large, long narrow shape, light colour, smooth surface.

3.2.2 Major Varieties in India

a) California Paper Shell

Regular bearer, bloom during 2nd week of March and ready to harvest after 152 days after full bloom. Tree is upright suitable for high density plantations, bears flowers and nuts on both spurs as well as on long shoots with good ability to renew fruiting wood. Nut and kernel are longer in size with extra light colour, papery shell that gives shelling percentage of 50 per cent. It is also suitable for export market. The average productivity is 2.0 t/ha.

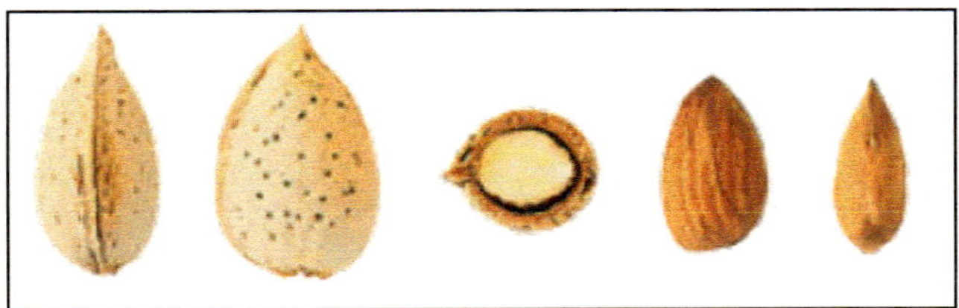

(a) BUTTE (BT)

(b) CARMEL (CR)

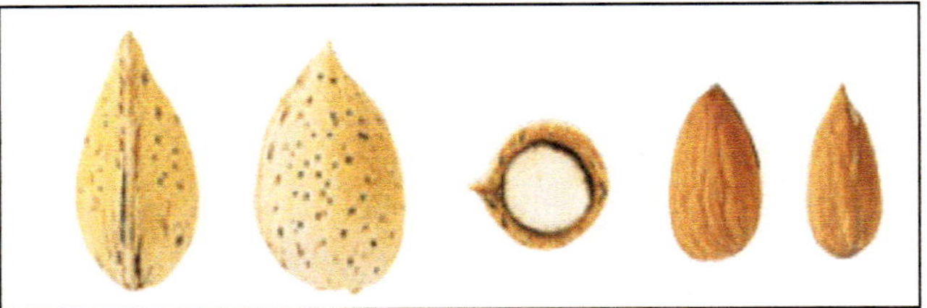

(c) FRITZ (FR)

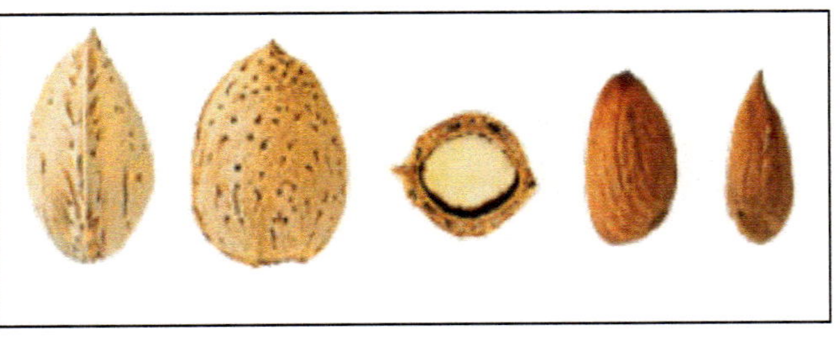

(d) MISSION (MI)

(e) MONTEREY (MT)

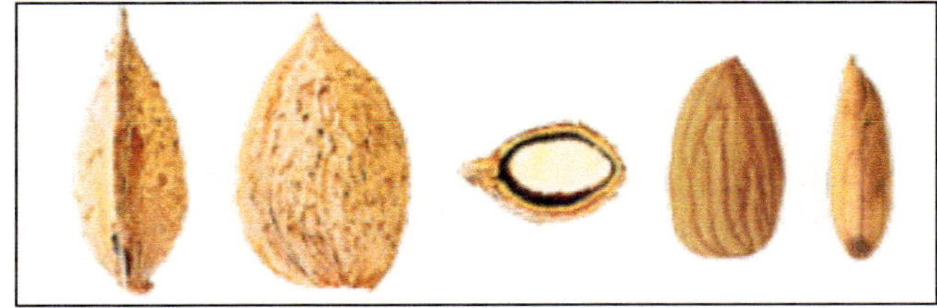

(f) NONPAREIL (NP)

(g) PADRE (PD)

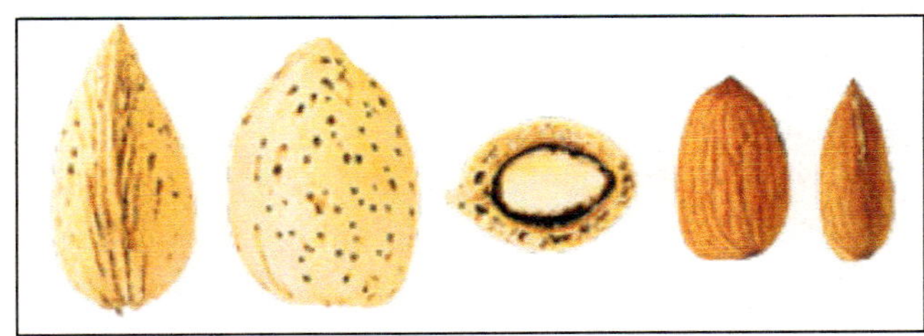

(h) PEERLESS (PL)

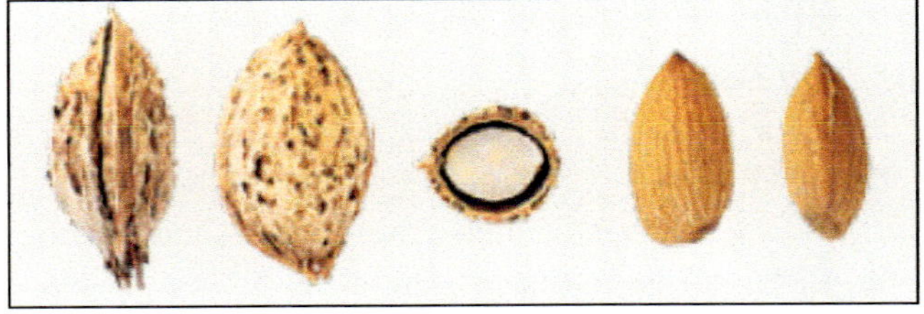

(i) PRICE (PR)

(j) SONORA (SN)

Figures 1a-j. Varieties Grown in California

b) Drake

Trees low in vigour, spreading, mid blooming. Nuts small to medium in size, bold and roundish with pointed apex and light creamy whitish brown in colour. Semi soft shelled and mid season maturity.

c) IXL

Regular bearer, bloom during 3rd week of March and ready to harvest after 151 days of full bloom. Tree is spreading type and of intermediate vigour which bears flowers and nuts on both spurs as well as on long shoots with good ability to renew fruiting wood. Nuts and kernels are medium, shell colour intensity intermediate. Shells are soft with high shelling percentage (55 per cent). Average productivity is 2.5 t/ha.

d) Makhdoom

Regular bearer, bloom during 1st week of March and ready to harvest after 141 days of full bloom. Tree is spreading/ drooping type, bears flowers and nuts on long shoots and spurs with good ability to renew fruiting wood. Shell colour is medium, soft type plump with shelling percentage of 42 per cent. Average productivity is more than 2.0 t/ha.

e) Merced

Regular bearer, blooms during 3rd week of March and ready to harvest after 152 days of full bloom. Tree is upright suitable for high density plantations, bears flowers and nuts on both spurs as well as on long shoots with good ability to renew fruiting wood. Shell colour intermediate with papery shell, shelling percentage 56 per cent. The variety is suitable for export of kernels. Average productivity is 2.0 t/ha.

f) Ne-Plus Ultra

Trees vigorous and spreading, mid blooming. Nuts medium to large flattened, bold and light brown in colour, paper shelled and mid season maturity.

g) Non Pareil

Regular bearer, mid bloomer (3rd week of March) and early maturing (141 days after full bloom). Tree is upright to spreading. Which bears flowers and nuts on both spurs as well as on long shoots having good ability to renew fruiting wood, relatively resistant to frost. Nut light in colour with papery shell that gives a high shelling percentage (60 per cent). It has a medium size kernel. The variety is suitable for export both as nut and kernels. The average productivity is 2.5 t/ha.

h) Pranyaj

Regular bearer, blooms during 3rd week of March and ready to harvest after 144 days of full bloom. Tree is upright, suitable to grow under high density orcharding, bears flowers and nuts on long shoots and spurs. Shell colour light, very soft shelled, nuts are medium, with plump kernels and shelling percentage of 44 per cent. The average productivity is more than 2.0 t/ha.

i) Primorskij

Trees spreading and moderately vigorous, mid to late blooming. Nuts medium to large, bold, slightly flattened and brown in colour, kernel medium to large. Soft paper shelled and late season maturity.

j) Shalimar

Regular bearer, bloom during 2nd week of March and ready to harvest after 143 days of full bloom. Tree is spreading/ drooping type, bears flowers and nuts on both long shoots and spurs with good ability to renew fruiting wood. Shell colour light, papery type with a high shelling percentage (50 per cent). Variety suitable for export. Average productivity is 2.0 t/ha.

k) Waris

Regular bearer, blooms during 3rd week of March and ready to harvest after 145 days of full bloom. Tree upright, suitable to grow in high density orchard, bears flowers and nuts on long shoots and spurs. Shell colour medium, soft shelled, plump kernels and gave shelling percentage of 48 per cent. Average productivity is more than 2.0t/ha.

4.0 Nature and Causes of Postharvest Losses

4.1 Damage Caused by Insects

Early harvest prevents the damage caused by Navel Orange Worm (NOW) as it reduces the number of eggs laid on the shaken almonds. NOW moth prefers to lay its eggs on almonds that are still within the tree canopy, thus almonds that are shaken before the emergence of the third generation larvae will suffer less damage than almonds shaken after this period. With Ants, it is just the opposite. The longer the nuts are on the ground to dry, the more ant damage can be expected. Many growers therefore place baits (Pyriproxyfen, Abamectin, and/or Methoprene) 6-8 weeks prior to harvest to selectively reduce the ant populations that fed on the almonds. If the populations of the almond feeding ants are still high at harvest, it is possible to control ants by making an application of broad spectrum insecticide (Chlorpyrifos) on the soil surface.

4.2 Damage Caused by Diseases

Almonds, like other tree nuts, are susceptible to aflatoxin producing moulds. Aflatoxins are potent carcinogenic chemicals produced by moulds such as *Aspergillus flavus* and *Aspergillus parasiticus*. The mould contamination may occur from soil, previously infested almonds, and almond pests such as navel orange worm. High levels of mould growth typically appear as grey to black filament like growth. It is unsafe to eat mould infected tree nuts. Some countries have strict limits on allowable limits for aflatoxin contamination on almonds, and require adequate testing before the nuts can be marketed to their citizens. The almond industry not only tests and processes almonds to ensure infected almonds do not reach the market, the industry also takes steps to prevent sources that cause contamination.

5.0 Postharvest Management

5.1 Maturity Indices

Almond maturation can be monitored externally by evaluating the extent of hull dehiscence. In the absence of pressure from insects like the naval orange worm (*Amyelois transitella*), harvest can be delayed until the two halves of the hull are fully bent back to expose the shell. At this point, hulls readily separate. Yield is maximized because the kernel's dry weight is no longer increasing and nut removal is close to 100 per cent. Nut maturation on a given tree is not uniform; development tends to be most rapid on the south and south-western faces of the tree higher in the canopy.

Early harvested nuts contain more water than is acceptable and must be dried on the orchard floor for 1 to 2 weeks before they are picked up and hulled.

5.2 Harvesting

As nuts split birds, squirrels, and worm pests find them especially attractive and become more troublesome the longer harvest is delayed. The unripe fruits are shed in large quantity by strong winds blowing during May-June. The fruits ripen towards the end of July and are harvested in the first or second week of August. A sheet of cloth is spread under the tree and the nuts are shed by giving the branches a vigorous shake. The more persistent nuts are knocked off by striking the branches with bamboo poles. The nuts are then dried in the sun for a short period. Hard-shelled varieties are bleached with sulphur fumes which make the shells bright and improve appearance. The dry nuts are usually graded into two categories *viz.*, thin-shelled and hard shelled.

5.3 Sorting and Grading

The nuts are roughly sorted out for empty, highly blighted, shrivelled, muddy and darkened nuts that are removed. During the sorting process almonds are separated based on diameter and weight (Figure 2). For example "18/20" denotes between 18 and 20 almonds kernels per ounce to qualify for this size.

In-shell and shelled grades are defined by USDA standards. For the in-shell product, emphasized characteristics include the integrity, shape and hardness of the shell as well as the brightness and uniformity of its colour. Freedom from foreign material and signs of insect damage or decay are also important. The shelled almonds should be free of shell debris and foreign material, and insect or fungal damage. The kernel skin should be intact and should show no shriveling or discolouration. Double, split or broken kernels are negative factors. There should be no indication of rancid flavour.

5.3.1 USDA Grade Descriptions

a) U. S. Fancy

The highest grade; typically appropriate for products where the visual appeal of the almond is critical to the application. The grade is not widely used.

b) U. S. Extra No. 1

Similar to U.S. Fancy; ideal for food applications where the appearance of the almond is very important.

c) U. S. No. 1 Supreme

Typically used for whole almond applications or for further processing, such as blanching and roasting.

d) U. S. Select Sheller Run

Mid-quality grade; after minimal sorting/processing can be incorporated with other ingredients; for example, inside a confectionery product, higher level of chipped and scratched kernels are accepted. Also, appropriate for further processing such as blanching, grinding, roasting, dicing, and slicing.

e) U.S. Standard Sheller Run

Good grade for further processing such as blanching, dicing, or making paste; particularly where a higher level of split and broken kernels is not a concern.

f) USDA Grading Parameters (Figure 3a-h)

(a) **Dissimilar:** Typically used for whole almond applications or for further processing, such as blanching and roasting.

Figure 2. Grading of Almonds
(http://www.ams.usda.gov/standards.nutpdct.htm.)

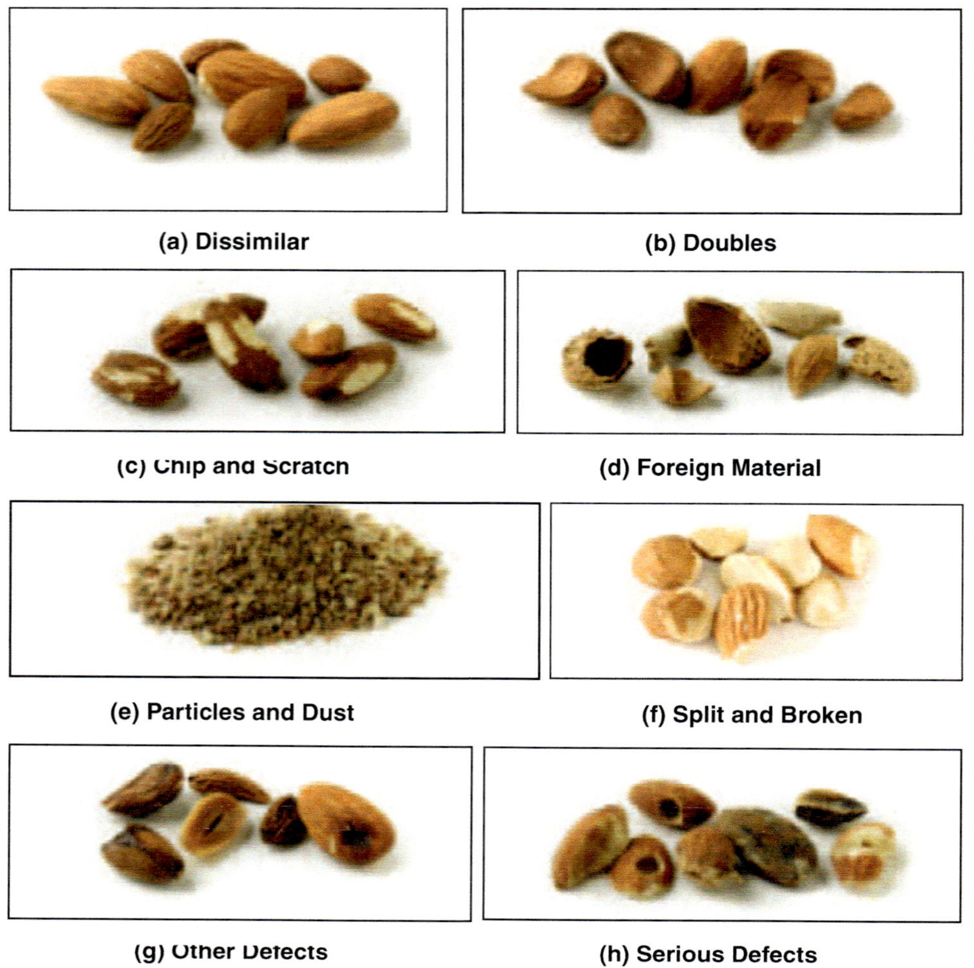

Figure 3. USDA Grading Parameters

(b) **Doubles:** Two kernels developing in one shell. One side of a double kernel is flat or concave.

(c) **Chip and Scratch:** Loss of kernel skin as a result of mechanical processing. Greater than 1/8" (3.2mm) in diameter, it is defined as injury; if affecting, in aggregate, greater than 1/4" (6.4mm) in diameter, it is defined as defect.

(d) **Foreign material:** Pieces of shell, hulls, or other foreign matter that will not pass through a round-mesh screen measuring 8/64" (3.2mm) in diameter.

(e) **Particles and dust:** Fragments of almond kernels or other material that will pass through a round-mesh screen measuring 8/64" (3.2mm) in diameter.

(f) **Split and broken:** Seven-eighths or less of whole kernels that will not pass through a round-mesh screen measuring 8/64" (3.2mm) in diameter.

(g) **Other defects:** Any defect that materially detracts from the appearance of the individual kernel or the edible or shipping quality of the almonds. The defects include gum, shrivel, brown spot, discoloured, chipped and scratched kernels with area of defect greater than 1/4" (6.4mm) in diameter.

(h) **Serious defects:** Any defect that makes a kernel or piece of kernel unsuitable for consumption (includes decay, rancidity, insect injury, and damage by mould).

5.4 Drying

Once early-harvested green nuts have been shaken from the trees they must be spread out to dry to prevent the development of mould. If wet nuts are left in a pile, bag or container, they will develop mould very quickly. Nuts should be spread no more than one or two nuts deep on a firm surface from where they can be easily picked up (patios, garage floor, deck) (Figure 4). They dry equally well in the sun or the shade, when weather is sunny and not humid, within two days. Protection from ants and birds should be provided while they are drying. Drying proceeds quicker and takes less space if the hull is removed first and it's often easier to remove the hull while it's still green. If a commercial huller is used, the nuts must be dry before hulls can be removed. Once hulled and dried, nuts can be stored inshell or

Figure 4. Drying of Almonds

cracked out and stored as nut meats. In storage they must be protected from worms and ants and kept cool and dry to preserve product quality.

Almonds are dry when they rattle while being shaken.

5.5 Packaging

Almonds are packed in 11.33 and 22.6 kg cartons of kernels, and 1 ton inshell fiber bulk sacks. The thin nuts are packed in wooden boxes, while the medium shelled nuts are packed into gunny bags

5.6 Storage

Inshell almonds can be stored for about 7-8 months at room temperature below 70 per cent relative humidity (RH), if the nuts are dried properly. For prolonged storage, inshell almonds should be held at 10 °C or below. Almonds can remain in excellent marketable condition for as long as 2 years if stored at 0 °C with about 75 per cent RH. Although shelling generally increases the perishability of all nuts, shelled almonds are comparatively less perishable than most other nuts. Air storage at 0 °C with 60-75 per cent RH is satisfactory to hold almonds for 15-16 months.

Modified atmosphere (vacuum, CO_2 or N_2) at 0 °C has been reported to be superior to air for maintenance of flavour during prolonged storage. However, at 21.1 °C, vacuum-packed almonds were superior in colour to those stored in CO_2, N_2 or air after 16 months of storage. Good quality was maintained for 8 months in vacuum CO_2 or N_2 at 26.7 °C, whereas almonds stored in air at this temperature turned black in 8 months. Controlled-atmosphere (CA) storage was found to be equal to or better than normal-atmosphere storage for almonds in the shell or as meats. Flavor was maintained for 12 months at 18 and 27.5 °C in insect controlling atmospheres of < 1 per cent O_2 and 9 to 9.5 per cent CO_2. Reduction in the O_2 content of the storage atmosphere will improve oil stability. Stored almonds may be infested with *Paramyelcis transitella* Walker, which initiates its infestation in the orchard. During storage, various storage insects may infest the nuts Insect control is accomplished through safe means *viz.* subjecting the kernels to insecticidal CA condition, irradiation *etc.* Use of chemical fumigants is banned in many countries due to their environmental and health hazards. Almonds are best stored in dry, cool places. If kept under 40 °F and 65 per cent relative humidity, almonds can store for nearly two years and retain their quality. The California Almond Board also suggests avoiding exposure to strong odors such as onion, garlic, and fish; as almonds absorb odors from their surroundings. It is also important to protect them from prolonged, direct sunlight during storage. Few easy tips are recommended for storage:

Store under cool and dry conditions (<10 °Cand <65 per cent relative humidity)

i) Moisture content should be maintained at 6 per cent or less

ii) Avoid exposure to strong odors

iii) Protect from insects and pests

iv) Rotate stock to optimize shelf life.

The low water and high fat content of the kernel make it relatively metabolically stable to tolerate low temperatures. Federal regulations define a safe moisture level for nuts as a water activity of < 0.7 at 25 °C (77 °F) to retard microbial growth. The recommended storage RH is 65 per cent because too low a water contents negatively impact flavour, colour and texture. Cold storage is useful to minimize lipid oxidation (rancidity) in inshell almonds which can be stored for up to 20 months at 0 °C, 16 months at 10 °C, and 8 months at 20 °C. Shelled nuts can be stored for about half as long as nuts in the shell Recommended air exchange rate for storage of almond is at least 10 changes/hour.

6.0 Domestic Marketing and Export

The European Union introduced a mandatory requirement since 2007 that all almond shipments to EU must be tested for aflatoxin. If aflatoxin level does not meet strict safety regulations, the entire consignment may be reprocessed to eliminate the aflatoxin or it must be destroyed.

Most of the domestic demand of Almond in India is fulfilled by import from the US, account-ing for 74 per cent of total imports, but only 2 per cent of the domestic demand is fulfilled through in house production. Afghanistan supplies 5 per cent of total Indian imports. No import tariffs are applied to the Almonds from Afghanistan.

Apparently enough focus should be given also to non tariff barriers in the form of market access of goods from developing countries. The most important legal requirements concern food safety, tracking and tracing, maximum residue levels (MRLs), food additives, food labeling, *etc.*

7.0 Conclusion

India's dependence on imported almonds is rising due to lower production and also due to a steady growth in the domestic demand for use in confectionery, snack and health products. Increasing awareness of the nutritional and health properties of almonds, particularly, among middle class consumers calls for an increase in the domestic production. However, there exist several problems in this sector which should be addressed without any further delay. Only then can the country minimize its imports. The problems of this sector are;

i) Problems of spring and frost cause great damage to this crop.

ii) Problems of the almond tree are prone to attack of disease.

iii) Non-availability of late-blooming varieties for cultivation.

iv) Poor marketing practices.

Generally, all the fruits harvested from the orchards are collected before dehulling, which results in the mixing together of nuts of all grades, shapes, sizes and tastes. Thereby, what comes to the market from the orchard is a mixture of several nuts types which fetches low returns to the farmers

Even though the cultivation of almond fetches economic returns to the growers, still the above problems have forced the farmers to opt for substitute crops. Hence, to revive the interest of the growers there is a need to provide late blooming varieties and varieties requiring lesser chilling. So as to overcome the marketing problems, the state-owned agencies should rescue the small growers by undertaking the responsibility of marketing. R and D should be undertaken to tap certain alternative uses of bitter nuts which may be an incentive to the growers. To meet the domestic demand, cultivation of almond should be extended vigorously. The same types of efforts are needed in the other growing belts so as to minimize the demand and supply gap. Since, these nuts contain a high level of protein; the government should take proper steps to solve the problems and to expand the production of this in our country.

Although strong demand and shorter supply have kept prices high for almond growers in recent years, much attention will be on future changes to supply and demand and the effect of the recent period of high prices which has promoted industry expansion. Additional concerns include the honeybee supply which is being decimated by mites and caused the price of rented hives to double in 2005, pesticide use as it affects water quality, and environmental concerns regarding production of airborne dust during.

8.0 Important References

Amsterdam, Elana. 2009. *The Gluten-Free Almond Flour Cookbook: Breakfasts, Entrees, and More*. Random House of Canada. ISBN 978-1-58761-345-6.

Berryman, C.E., Preston, A.G., Karmally, W., Deckelbaum, R.J. and Kris-Etherton, P.M. 2011. Effects of almond consumption on the reduction of LDL-cholesterol: a discussion of potential mechanisms and future research directions. *Nutr. Rev.* **69**: 171–85.

Bradley, Ryan. 2013. "The Bee Business". *Fortune* (paper). pp. 11-13.

Cavaletto, C.G., Kader, A.A. and Kester, D.E. 1985. Quality in relation to marketability of almond varieties. Report to the *California Almond Board*. Pomology Department, Univ. of Calif., Davis.

Connell, J.H., Sibbett, G.S., Labavitch, J.M. and Freeman, M.W. 1996. Preparing for harvest. **In:** *Almond Production Manual*, W.C. Micke (ed) Univ. Calif., DANR Pub. No. 3364, pp. 254-259.

Davis, P.A. and Iwahashi, C.K. 2001. Whole almonds and almond fractions reduce aberrant crypt foci in a rat model of colon carcinogenesis. *Cancer Lett.* **165**: 27–33.

Deutscher Transport-Versicherungs-Verband e.V.: Die Ware in der Transport versicherung, Hamburg 1990-1994.

FAOSAT, 2020. Food and Agriculture Organization of the United Nations.

Gradziel, T.M. 2011. Origin and dissemination of Almonds. In: *Horticultural Reviews*, Vol. 38, J. Janick (Ed), Wiley Blackwell

Guadagni, D.G., Soderstrom, E.L. and L. Storey. 1978. Effect of controlled atmosphere on flavour stability of almonds. *J. Food Sci.* **43**: 1077-80.

http: //www.modbee.com/2014/01/11/3129738/ water-battles-loom-as-almond-rchards.html

Jenkins, D.J, Kendall, C.W., and Marchie, A., *et al.* 2002. Dose response of almonds on coronary heart disease risk factors: blood lipids, oxidized low-density lipoproteins, lipoprotein(a), homocysteine, and pulmonary nitric oxide: a randomized, controlled, crossovertrial. *Circulation* **106**(11): 1327-32.

John, Labavitch. 2001. *Almond*. United States Dept of Agriculture.

Kader, A.A. 1996. In-plant storage. In: Almond Production Manual, W.C. Micke (ed) Univ. of Calif., Div. Ag. and Natural Res., Pub. No. 3364, pp. 274-77.

Kester, D.E. and Ross, N.W. 1996. History. In: *Almond Production Manual*, W.C. Micke (ed) Univ. of Calif., Div. Ag. and Natural Res., Pub. No. 3364, pp. 1-2.

King, Jr. A.D. and J.E. Schade. 1986. Influence of almond harvest, processing and storage on fungal population and flora. *J. Food Sci.* **51**: 202-05, 215.

King, Jr., A.D., Halbrook,W.U., Fuler, G. and Whitehand, L.C. 1983. Almond nutmeat moisture and water activity and its influence on fungal flora and seed composition. *J. Food Sci.* **48**: 615-17.

Ladunki okretowe. Poradnik encyklopedyczny, Polskie Towarzystwo Towaroznawcze. Oddia Morski, Sopot 1994

Liu, Z., Lin. X., Huang, G., Zhang, W., Rao, P. and Ni, L. 2014. Prebiotic effects of almonds and almond skins on intestinal microbiota in healthy adult humans. *Anaerobe* **26**.

Mahesh, Venkatachalam. and Shridhar Sathe. 2006. Chemical Composition of Selected Edible Nut Seeds. *J. Agric. Food Chem.* **54**: 4705–14.

Mandalari, G., Tomaino, A., Arcoraci, T., Martorana, M., Turco, V. Lo., Cacciola, F., Rich, G.T., Bisignano, C., Saija, A., Dugo, P., Cross, K.L., Parker, M.L., Waldron, K.W. and Wickham, M.S.J. 2010. Characterization of polyphenols, lipids and dietary fibre from almond skins (*Amygdalus communis* L.). *J. Food Compos. Anal.* **23**: 166.

Phillips, D.J., Mackey,B., Ellis, W.R. and Hansen, T.N. 1979. Occurrence and interaction of *Aspergillus flavus* with other fungi on almonds. *Phytopathology* **69**: 829-31.

Reil, W., Labavitch, J.M. and Holmberg, D. 1996. Harvesting. **In**: *Almond Production Manual*, W.C. Micke (ed) Univ. of Calif., Div. Ag. and Natural Res., Pub. No. 3364, pp. 260-264.

Schatzki, T.F. 1996. Distribution of aflatoxin in almonds. *J. Ag. Food Chem.* 44: 3595-97.

Soler, *et al.* 1988. Oil content and fatty acid composition of developing almond seeds. *J. Agric. Food Chem.* **36**: 695–697.

Thompson, J.F., Rumsey, T.R. and Connell, J.H. 1996. Drying, hulling and shelling. **In:** Almond Production Manual, W.C. Micke (ed) Univ. of Calif., Div. Ag. and Natural Res., Pub. No. 3364, pp. 268-273.

USDA. 2010, Tree nuts Annual Report - Turkey.

Chapter 11

Apple

R.R. Sharma

Principal Scientist, Division of Food Science and Postharvest Technology, ICAR-Indian Agricultural Research Institute, New Delhi – 110 012, India

E-mail: rrs_fht@rediffmail.com

R.K. Pal

Director, ICAR-National Research Centre on Pomegranate, NH-65, Solapur-Pune Highway, Kegaon, Solapur – 413 255, Maharashtra, India

E-mail: rkrishnapal@gmail.com

K.K. Paramanick

Principal Scientist and Head ICAR-IARI

RS Amartara Cottage, 4, Cart Rd, Near R.T.O, Shimla – 1710 04, Himachal Pradesh, India

E-mail: kallolpramanick@gmail.com

1.0 Introduction

The cultivated apple is usually referred as *Malus pumila* L. (earlier called *M. communis* or *Pyrus malus*). Now it is referred to as *Malus* x *domestica* Borkh. Most of the cultivated apple varieties are diploid with chromosome number 2n = 34. Some cultivars are triploids (*e.g.* Suntan, Jupiter and Jonagold) and some are tetraploids (*e.g.*, Alpha-68). The primary center of origin of cultivated apple is thought be the region, which includes the Caucasus, Asia Minor, Soviet Central Asia and Himalayan regions of India and Pakistan where maximum diversity in wild apple exists. Ambri, a dessert apple variety is considered to be indigenous to Kashmir (India). All other commercial varieties have been introduced from various places and grown in India. Captain A.A Lee, an Englishman, established the first apple orchard in India in Himachal Pradesh at Bandrole (Kullu valley) in 1870. The famous varieties under Delicious group were introduced by Samuel Nicholas Stokes, a resident of Philadelphia, USA in 1918 at Kotgarh, Shimla (H.P). Later, apple spread to different parts of our country. In Nilgiri hills, apple was introduced as early as 1897, but was destroyed by woolly apple aphid.

Apple is the most important and premier temperate fruit of the world. It has long been the staple fresh fruit in temperate part of the world. It is considered as one of the most delicious fruits. The old proverb 'an apple a day keeps the doctors away' focuses its importance in human diet as it has a high nutraceutical value, primarily because it is a rich source of anthocyanins, flavonoids and polypehnols and contains high fibre content. It is considered as a good source of carbohydrates and vitamin C and contains appreciable amount of thiamine and riboflavin. Although, it is a popular table fruit, but it can also be canned or can be used for juice, wine, chutney, jam, vinegar and cider making. Apple is also highly recommended for balancing blood sugar levels. Apples remain alive and continue metabolic processes after harvest. Without appropriate handling at all stages along the supply chain, from producer to consumer, fruit will continue to ripen and deteriorate. The main causes of post-harvest losses and poor-quality fruit are bruising and other mechanical injury, water loss, under- or over maturity at harvest, and chilling injury.

Apples should be harvested when they are physiologically mature but before their peak of ripeness. Apples for eating fresh or for short-term storage (2–3 weeks) should be left on the tree until they are fully ripe. Store only those sound apple fruit that is free from all types of insect or disease damage.

The apple is a temperate fruit crop. However, in India the apple growing areas do not fall in temperate zone but the prevailing temperate climate of the region is due to the Himalayan ranges and high altitudes. The average summer temperature should be around 21-24 °C during active growth period. Apple succeeds best in regions where the trees experience uninterrupted rest in winter and abundant sunshine for good colour development. It can be grown at an altitude of 1500-2700 m above the sea level. Well-distributed rainfall of 1000-1250 mm throughout the growing season is most favourable for optimum growth and fruitfulness of apple trees. Himachal Pradesh, which lies in the western Himalayas, is considered as model state for the development of mountain/hill states. Agriculture contributes around 9 percent of state gross domestic product (SGDP) and about 40 percent of the agricultural GDP is contributed by the horticulture alone. Horticulture is thus, considered as the backbone of state economy. The area under fruits, which was 792 hectares in 1950-51 with total production of 1200 tones increased to 1,12,634 hectares by 2017-18. The total fruit production in 2017-18 was 4.46 lakh tones. Apple is the most important fruit crop of Himachal Pradesh, which constitutes about 49 percent of the total area under fruit crops and about 79 per cent of the total fruit production. An estimated 3500 crore rupees apple economy has strong backward and forward linkages generating employment to lakhs of people.

2.0 Status of Cultivation

The Europe is considered to be the largest producer of apple. In recent years, China has emerged as a leading apple producing country. China contributed nearly 47 per cent (35,987 thousand metric tons) of total world production during 2013-14. This is followed by USA (4,273 thousand metric tons) and India (2,891 thousand metric tons). Among the 95 apple producing countries in the world, USA is the largest producer with 49.97 million MT during 2019 followed by China (42.42 million MT), Turkey (36.18 million MT), Poland (30.08 million MT) and India (23.27 million MT). In India, apple is predominantly grown in Jammu and Kashmir, Himachal Pradesh and hills of Uttarakhand, accounting for more than 90 per cent of the total production. Of total apple production, J&K alone contributes 64.1 per cent followed by Himachal Pradesh (30.9 per cent). Leading apple producing states in India is presented in Figure 1. Its cultivation has also been extended to Arunachal Pradesh, Sikkim, Nagaland and Meghalaya in north-eastern region and Nilgiri hills in Tamil Nadu. The agro-climatic conditions in these states are not as conducive as in north-western Himalayan region. Early and continuous rains from April onwards do not favour the production of quality fruits besides resulting in high incidence of diseases. As per latest data, the highest productivity of apples had been observed in J&K (11.4 t/ha) followed by H.P. (3.9 t/ha) and Uttarakhand (2.31 t/ha) (Table 1).

Table 1. State-wise Area, Production and Productivity of Apple in India

State	*Area (000'Ha)*	*Production (000'MT)*	*Productivity (MT/Ha)*
Arunachal Pradesh	4.66	7.35	1.6
Himachal Pradesh	112.63	446.57	3.9
Jammu and Kashmir	158.15	1808.3	11.4
Uttarakhand	25.3	58.66	2.31
Others	0.26	0.6	
Total	**301**	**2327**	**7.7**

Source: Indian Horticulture Database, NHB 2017-18.

The major apple producing belts in India (Figure 1) are as under:

i) ***Arunachal Pradesh***: West Kameng, Tawang, Lower Subansiri, East Kameng, Upper Siang, Upper Dibang Valley

ii) ***Jammu and Kashmir***: Srinagar, Ganderbal, Budgam, Anantnag, Kulgam, Pulwama, Shopian, Baramulla, Bandipora, Kupwara, Kathua, Udhampur, Reasi, Doda, Kishtwar, Ramban, Rajouri, Poonch, Leh, Kargil

iii) ***Himachal Pradesh***: Shimla (Kotkhai, Jubbal, Rohroo, Rampur, Theog, Chopal, Kumarsain, Kotgarh, Dodra, Kewar), Kullu (Kullu, Anni,

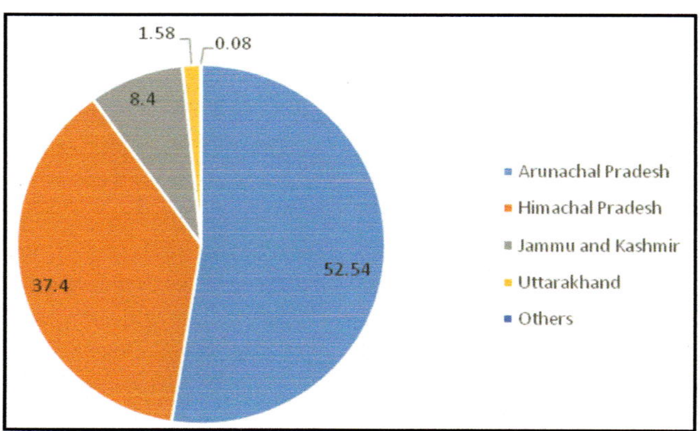

***Source*. Indian Horticulture Database, NHB 2017-18.**

Figure 1. Leading Apple Producing States in India

Manali, Nirmand, Sainj), Mandi (Karsog, Janjhali, Banjar), Chamba (Pangi, Bharmour, Saluni), Kinnaur (Rekong, Peo, Ropa Valley, Sangla Valley, Chango, Pooh), Lahul & Spiti (Kaza, Lahul), Kangra (Multhan, Bara Bhangal, ChotaBhangal), Sirmour (Rajgarh)

iv) ***Mizoram***: Champhai

v) ***Nagaland***: Zunheboto, Phek, Kiphire

vi) ***Sikkim***: North Sikkim (Lachung), West Sikkim (Ribdi, Okhrey)

vii) ***Tamil Nadu***: Krishnagiri, Nilgiris, Dindigul

viii) ***Uttarakhand***: Nainital, Dehradun, Chamoli, Tehri, Uttarakashi

3.0 Major Commercial Varieties

Important commercial varieties of apple grown in the world are Red Delicious, Golden Delicious, McIntosh, Rome, Beauty, Jonathan, York Imperial, StaymanWinesap, Winesap, Yellow Newton, Cortland, Northern Spy, Gravenstein, Baldwin, Grimes Golden, Black Ben Davis, Wealthy, Cox's Orange Pippin and Rhode Island Greening. On the basis of time of harvesting, apple varieties can be categorized as early, mid season and late in India (Table 2).

A number of bud sports of the popular varieties have been selected which have similar characters as of the parent except that they develop attractive fruit colour on maturity and a few matures little early. Some of these bud sports having high market values are Starking Delicious, Richared, Hardeman, Royal Red, Vance Delicious, Top Red and Jonagold. Some mutants of the standard varieties have also been selected which are becoming quite popular. These mutants have compact tree, shorter internodal length and high spur development potential. Furthermore, the trees of spur type mutants remain 2/3rd of the size of standard varieties. Some of such mutants bear better-coloured fruits and have early maturity. Important spur types of Delicious group are Red Chief, Oregon Spur, Red Spur, Starkrimson and of Golden Delicious are Starkspur Golden Delicious and Golden Spur.

A farmer from Himachal Pradesh has developed an innovative self-pollinating apple variety that does not require long chilling hours for flowering and fruit setting. This has spread to plain, tropical and subtropical areas in various parts of India, where the temperature is as high as 40-45°C during summer. Commercial cultivation of this apple variety has been initiated in Manipur, Jammu, low lying areas of Himachal Pradesh, Karnataka Chhattisgarh, and Telangana, and fruit setting has been expanded to 23 states and UTs so far. Shri Hariman Sharma, a progressive farmer, hailing from Paniala village of Bilaspur district, Himachal Pradesh, who developed this innovative apple variety – HRMN 99. Amartara Pride and Pusa gold variety of apple are developed by ICAR-IARI, RS, Shimla, H.P. As a result of extensive work on breeding, some apple hybrids such as Lal Ambri (Red Delicious x Ambri), Sunehari (Ambri x Golden Delicious), Shalimar-1 (Sunehari x Prima), Ambred (Red Delicious x Ambri), Ambroyal (Starking Delicious x Ambri), Chaubattia Anupam (Early Shanburry x Red Delicious) Swarnima (Benoni x Red Delicious) *etc.*, have been released in our country.

The chief characteristics of important cultivars, which are grown in India are briefly described hereunder (Figure 2).

a) Ambri

Indigenous variety, grown commercially in Kashmir valley, red streaks over a greenish yellow background, very high keeping quality.

b) Gold Spur Delicious

Bud mutant of Golden Delicious, dwarf trees, polliniser for all spur-types.

c) Golden Delicious

Attractive yellow colour, high juice content, pleasant flavour, high keeping quality, suitable for processing.

d) Hardeyman

Fruits are similar to Starking Delicious in shape and colour, but matures 10-15 days in advance.

Table 2. Recommended Varieties of Apple in different States of India

Season	*Himachal Pradesh*	*Jammu and Kashmir*	*Uttarakhand*
Early season	Tydeman's Early (P), Michael Molies, Delicious Schlomit, Starkrimson	Irish Peach Benoni	Early Shanburry (P), Fenny Benoni, Chaubattia Princess,
Mid-season	Staking Delicious, Red Delicious, Richared Vance Delicious, Top Red, Lord Lambourne (P), Red Chief, Oregon Spur, Red Spur, Red Gold (P)	American Mother, Razakwar Jonathan (P), Cox's Orange Pippin, Red Gold (P), Queen's Apple, Rome Beauty, Scarlet Siberian	Red Delicious, Starking Delicious, McIntosh (P), Cortland, Golden Delicious (P)
Late season	Golden Delicious (P), Yellow Newton (P), Winter Banana Granny, Smith (P)	King Pippin, American, Apirouge, Kerry Pippin, Lal Ambri, Sunhari, Golden Delicious (P), Red Delicious, Ambri Baldwin, Yellow Newton (P), White Dotted Red	Rymer Buckingham (P)

P: Pollinizer.

Ambri

Gold Spur Delicious

Golden Delicious

Hardeyman

Oregon Spur

Red Chief

Red Delicious

Royal Delicious

Top Red

Figure 2. Commercial Varieties of Apple

Figure 3. Bitter Pit Affected Apple

Figure 4. Colour Development in Bagged vs. Non Bagged Fruits

e) Oregon Spur

Bud support of Delicious, fruits similar to Delicious, but mature 10-14 days earlier, crunchy sweet and a bit hard.

f) Red Chief

Bud sport of Delicious, fruits similar to Delicious with prominent red streaks, matures 15-20 days early, excellent flavour.

g) Red Delicious

Most popular, peel red with yellow strips, flesh firm, sweet and juicy.

h) Royal Delicious

Peel red with yellow covered stripes all over the surface flesh firm, sweet juicy and aromatic.

i) Top Red

Bud sport of Shot Well Delicious, fruits with red streakes, mature 15-20 days before Starking Delicious.

4.0 Nature and Causes of Postharvest Losses

4.1 Preharvest Factors

Several preharvest factors contribute directly or indirectly to postharvest losses in apple. For example, non-adoption of proper method of training, faulty pruning, inadequate and inappropriate supplementation of nutrients and water, excessive fruit drop, incidence of insect- pests especially Sanjose scale, wooly apple aphid, mites, blossom thrips, root borers, tent hairy caterpillar, codling moth and diseases such as scab, premature leaf fall, powdery mildew, root rot and attack by birds are major preharvest factors causing severe losses to apple. Among these factors, faulty training and pruning contribute indirectly to losses by affecting bearing potential and quality. Some factors such as water stress, incidence of insect-pests and diseases affect quality and thereby cause severe losses. Malnutrition may favour the development of several physiological disorders (*e.g.*, bitter pit, brown core, cork pit) which is also responsible for losses. In addition, improper colour development distracts consumers (Figures 3 and 4) .

4.2 Postharvest Factors

Occurrence of several physiological disorders such as bitter pit, brown core and scald can cause severe postharvest losses in apple. Similarly, blue mold (*Penicillium expansum*) and grey mold (*Botrytis cinerea*) are major postharvest diseases in apple, causing spoilage of fruits during storage and marketing. Besides, Alternaria rot (*Alternaria alternata*) bitter rot (*Glomerella cinguata*), pink mold rot (*Trichothecium roseum*) and core rot (*Fusarium* sp.) are also found in apple during storage and can contribute significantly to the spoilage of fruits.

5.0 Postharvest Management

5.1 Preharvest Crop Management

For reducing postharvest losses and getting quality produce, adequate preharvest management practices should be adopted. For example, in the initial years of establishment, trees should be trained scientifically on a specific training system such as central leader system, modified leader system or spindle bush. If not trained properly, there may be several problems in the fruiting years. During bearing period, pruning plays important role on the production of quality apples. Similarly, an orchardist needs to adopt GAP (good agricultural practices)in terms of application of manures and fertilizers, water management, weed control and integrated disease and pest management.

In some apple growing areas, zinc, magnesium, iron, copper, boron, manganese and calcium deficiencies are common. Such deficiencies should be determined by soil and leaf analysis, and for correction of deficiencies, recommended doses of the salts should be used in time (Table 3).

Continuous sprays of calcium chloride (0.5 per cent) or calcium nitrate (0.5 per cent) at biweekly intervals (4-6 sprays) from walnut stage of fruit development till harvesting are recommended for production of high quality and better coloured apples. Moreover, preharvest calcium sprays helps to control bitter pit incidence during storage. An orchardist should take every care not to use higher doses of calcium chloride because higher chloride causes russeting and burning in apple leaves. Deficiency of calcium results in a severe postharvest disorder known as biterpit.

Table 3. Correction of Nutrient Deficiencies in Apple through Foliar Sprays

Element	*Salt*	*Concentration (per cent)*	*Time of Spray*	*No. of Sprays and Interval*
Boron	Boric acid	0.1	June	1-2 at 5 days interval
Calcium	Calcium sulphate	0.5	June-July	2, at 15 days interval
Copper	Copper sulphate	0.3	May-June	1-2 at 10-15 days interval
Iron	Ferrous sulphate	0.5	May-June	1-2 at 15 days interval
Magnesium	Magnesium sulphate	1-2	June-July	1-2 at 10-15 days interval
Manganese	Manganese sulphate	0.4	June	1-2 at weekly interval
Zinc	Zinc sulphate	0.5	May-June	1-2 at 5 days interval

5.1.1 Colour Improvement

In Delicious varieties of apple, colour development is generally poor in marginal warmer areas and hence such apples fetch low price in the market. Application of 1,000 ppm 2-chloroethyl phosphonic acid (Ethrel/ CEAP/Ethephon), an ethylene releasing hormone, about 10 days before harvest improves fruit colour substantially but impairs shelf life. Since this chemical accelerates fruit abscission, NAA (10 ppm) should be added to check fruit drop. In very high elevations due to low temperature, the fruits develop intense red colour even before they are mature, which also do not fetch optimal price in the market. An application of ethephon (500 ppm) + NAA (10 ppm), 3 weeks before harvesting can enhance maturity of apples by 7-10 days.

Preharvest fruit bagging technology has been standardized for apple. Bagging should be done with light yellow coloured recyclable non-woven polypropylene bags about 45-60 before the expected date of harvesting. These bags should be removed 2-3 days before harvesting. Bagged fruits develop attractive red colour than non-bagged apples, and have good postharvest quality attributes. Similarly, bagged fruits are less prone to diseases like fly speck and sooty mold. In addition, such apples tend to develop less bitter pit and other disorders during storage.

5.2 Maturity Indices

Apple fruit undergoes several physical, chemical and physiological changes after which it becomes ready for harvesting. The time required for maturity depends on cultivar and climatic conditions. However, the degree of maturity for harvesting is of prime importance because the concerted efforts put in by the grower for producing quality fruits are wasted if the fruits are not harvested at appropriate stage of maturity. Several maturity indices have been standardized for harvesting apple at right stage of maturity. However, in practice, more than one methods are used as no index used alone is reliable measure of readiness of fruits for harvest.

5.2.1 Days from Full Bloom to Harvest

The number of days from full bloom (DFFB) to harvest is the most reliable maturity index for harvesting apple at right stage of maturity. The number of days after full bloom remains almost constant under a wide range of climatic and cultural conditions. This period varies from 90±4 days (Tydeman's Early Worcesster) to 180 ± 5 days (Granny Smith) in apple depending cultivar, climatic conditions and the elevation (Table 4).

5.2.2 Fruit Firmness

Fruit firmness declines gradually as apples mature. It is considered as one of the most reliable maturity indices for harvesting apples at right stage of maturity. Fruit firmness varies from variety-to-variety. However, fruit firmness of 6.8-8.7 Kg/cm^2 is considered optimum for all the varieties. It is also used as an indicator for determining their ultimate destination (Tables 4 and 5). A simple hand held penetrometer is required for determination of fruit firmness.

Table 4. Recommended Maturity Indices for Apple

Cultivar	*DFFB (days)*	*Firmness (kg)*	*TSS (ºB)*
Golden Delicious	148±6	8.4 + 0.40	12.0-14.5
Granny Smith	180±5	8.7 + 0.30	11.5-13.0
Lord Lambourne	103±3	8.2 + 0.20	12.5-13.5
McIntosh	135±6	6.8 +0.25	11.5-13.5
Red Delicious	134±5	8.4 + 0.40	11.0-14.0
Red Gold	122±3	8.3 + 0.20	12.0-15.0
Rich-a-red	128±3	8.6 + 0.25	12.0-13.0
Royal Delicious	120±5	8.2 + 0.40	13.0-15.0
Starkrimson	103±3	8.2 + 0.20	11.0-13.5
Tydeman's Early Worcester	90±4	7.8 + 0.15	12.0-13.0

Table 5. Recommended Firmness of Apples in Relation to Destination

Destination	*Firmness (lbs)*
Local markets	5.9-6.4
Short distance markets	6.6-7.0
Long distance markets	8.2-9.1
Export markets	9.1-10.0

5.2.3 Calendar Dates

Based on the experience of the orchardist over the years, optimum maturity dates can be specified for harvesting the crop in the following years. Orchardists usually start harvesting their crop on a particular date. It is one of the most commonly used maturity index and is reasonably accurate. For example, orchardists start harvesting 'Royal Delicious' after 15th August in Kullu Valley (H.P.) and by 25-30th August in Shimla hills (H.P.) and by 2nd week of September in high hills (Table 6). Similarly this period of harvesting is different in J&K and Uttrakhand.

Table 6. Calendar Dates for Harvesting Apples in India

Cultivar	*Calendar Dates*	*Place*
Ambri	25th October to 15th November	Anantnag, Shopian, Baramulla (J&K).
American	10th September	Anantnag, Shopian, Baramulla, Pulwama (J&K)
Early Shanburry	4th week of June to 1st week of July	Chaubatia (UK)
Golden Delicious	28th August5th October	Mashobra (HP) Anantnag, Shopian, Baramulla, Pulwama, Srinagar (J&K)
Maharaji (White dotted Red)	15th October- 5th November	Anantnag, Shopian, Baramulla, Pulwama (J&K)
Mollies Delicous	15th July – 15th August	Anantnag, Shopian, Baramulla, Pulwama (J&K)

Cultivar	Calendar Dates	Place
Red Delicious	25-30th August	Mashobra (HP) Anantnag, Shopian, Baramulla, Pulwama, Srinagar (J&K)
Rich-e-Red	5-10th September	Rohru (HP)
Royal Delicious	15th August	Kullu (HP)
	4th week of August	Chaubatia (UK)
	20-25th September	Kinnaur (HP)
Star Crimson	15th August	Patta (Baramulla), Shopian, Pulwama, Srinagar (J&K)

5.2.4 Peel Colour

There is increase in red colour in most of the apple cultivars (excluding Golden Delicious and Granny Smith) as the fruits approach maturity. Although, colour is not considered as an index of enough practical value for several apple cultivars, yet it is recommended that when fruits attain 70-85 per cent of red colour on their peel, they are ready for harvesting. However, in practice, orchardists start harvesting apples when they attain 50 per cent or more red colour.

5.2.5 Starch Content

Now-a-days, starch pattern index (SPI) is used as an important maturity index. For judging maturity, starch-iodine test is performed to correlate the correct picking time of apples. A starch index of 4.5/10 for Delicious, 4.0/10 for Golden Delicious has been recommended (Figure 5).

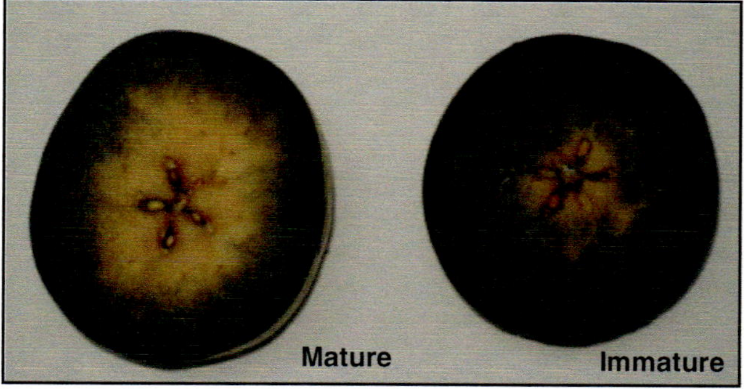

Figure 5. Testing of Apple Maturity by Iodine Test

5.2.6 Total Soluble Solids (TSS)

The total soluble solids is another important index for assessing maturity of apples. TSS depends on several factors including the variety, elevation and climatic conditions. For instance Red, Royal, Richared and Golden Delicious apple cultivars in orchards at 1500-2000 m above mean sea level become ready for harvest at 13.3, 15.01, 14.03 and 12.1 °Brix, respectively, whereas at 2,250-2,750 m above mean sea level these cultivars mature at values 14.1, 14.4, 15.0 and 11.5 °Brix, respectively. Hence, some researchers don't consider TSS as a reliable maturity index. However, TSS of 11-14.5 per cent is considered best for harvesting several varieties at appropriate maturity. A simple hand refractometer (0-32 °B) would serve the purpose of determination of TSS. In order to determine the TSS, the apples are cut into pieces and a drop of juice is poured on the prism for recording the TSS values on the scale.

5.2.7 Titratable Acidity

A few researchers consider acidity as a good maturity index for harvesting apples. The acidity is greatly influenced by several factors such as variety, elevation and locality. For example, the acid content of Delicious group of apples grown at different elevation varies from 0.20 – 0.41.

5.3 Harvesting

The time of harvesting apples largely depend on the variety and location where it is grown. For example, Red June matures in the first week of June whereas Tydemans Ealy Worsbter matures in October-November (Table 4). Apples are usually harvested manually. However, in advanced countries mechanical harvesting is preferred. Picking is done by lifting the fruits from bottom followed by gentle twisting with fore-fingers. Apples during this operation should be handled carefully to prevent bruise injury. The harvested apples should be put in baskets lined with gunny cloth bags or blistered polyethylene sheets or foam and placed into similarly lined crates and carried further to packinghouse. In general, apples are graded into different grades depending upon their soundness, visual appearance, typical varietal characteristics *etc.*, followed by size grading.

5.4 Handling, Sorting and Grading

The harvested apples are brought to the packing sheds manually using *Kilta*. All the cull, diseases or insect-pest infected apples are then removed and the healthy apples are graded (Figure 6). However, the apples should be handled carefully through all stages of harvesting, packing, storage and marketing. These may get bruises and punctures which may mean serious losses in storage. Some practices - washing, trimming and curing - also reduce the severity of infection. Boxes and bins to collect apples in the orchard should be cleaned periodically. Packing houses and surroundings should be kept free from cull and decaying fruits. Chief means of avoiding blue mould rot involves careful handling of fruits and strict sanitation in packing houses. Squashed and rotting fruits should be destroyed before they become a source of *P. expansum*. Washing water should be frequently changed and disinfected. Careful handling and prevention of mechanical injuries help reduce brown rot, Mucor rot, pink mould and Altemaria rot.

Figure 6. Manual Sorting and Grading of Harvested Apples

Medium-sized fruits are stored well for a longer period, while large-sized ones are prone to a number of infections and disorders. Storage of large-sized fruits for a longer period should be avoided. Ripening fruit is a source of ethylene production. If some ripening fruits are present, ethylene produced from these may trigger ripening of other fruits. Similarly, ethylene from wounded fruits may be sufficient to trigger ripening of unwounded fruits. Wounded fruits are attacked by a number of postharvest pathogens and large quantities of ethylene are evolved from these rotting fruits. Besides rotting fruits, many pathogens also produce ethylene. Such wounded and rotting fruits should be removed from healthy ones. The production of ethylene by rotting fruits is perhaps the basis of traditional saying 'one bad apple spoils the lot'.

Proper grading of the produce is one of the most important operations to bring uniformity, avoid wastage, fetch better returns and to establish trade language in the market. The efficiency of grading depends upon careful handling of fruits during harvest and sorting.

5.4.1 Quality Grading

Depending upon the shape, colour, soundness, freedom from defects like injuries, blemishes, disease spots, bruises *etc.*, apples are commonly graded into 3 recognized quality grades, *viz.* 'A, 'B' and 'C'.

i) **Grade A:** Apples of 'A' grade have the best shape and colour of the variety. They are sufficiently mature, free from blemishes, injuries or bruises.

ii) **Grade B:** Apples with slightly abnormal shape, 50 per cent colour characteristics of the variety are graded into 'B' grade.

iii) **Grade C**: Apples, which are not graded into 'A' or 'B' grades and can't fetch good price in the markets are classified as 'C' grade apples. Such fruits are mainly used for processing purposes.

5.4.2 Size Grading

In apple, the following proportion of fruit size of different grades is found on a tree.

i) Large, super and extra large sized apples: 20-24 per cent

ii) Medium sized apples: 35-40 per cent

iii) Small sized apples: 15-20 per cent, and

iv) Extra small apples, pittoo and culls: 20-25 per cent.

Therefore, quality grading should always be followed by the size grading. Each quality grade is separated into different size grades before packing. In India, apples are usually graded Into 7 size grades with the conventional method of measurement with finger and thumb by the orchardists (Table 7).

Table 7. Conventional Method of Size Grading in Apple in India

Grade	*Minimum Fruit Diameter (+2.5mm)*
Super Large	85
Extra Large	80
Large	75
Medium	70
Small	65
Extra Small	60
Pittoo	55

5.4.3 Method of Grading

Manual size grading consists of holding the fruit in left hand at the broadest point and placing the fingers of the right hand in between the uncovered space. The size grade is then determined by referring to Table 7, which depends upon the number of fingers that can be accommodated in the uncovered space.

The method of size grading has been further improved by the introduction of hand-grader. This operation consists of sliding or passing the fruit gently through the hand-grader from the broader side and taking out the fruit, when it reaches the appropriate size grade However, manual methods of grading and packing are both time-consuming as well as labour intensive. Grading and packing alone add about 15 per cent of the total cost of standard wooden box. These operations can however be simplified by using mechanical graders. Now mechanical graders have been established at several locations.

5.5 Precooling

Precooling of Delicious apples is essential, which otherwise show poor shelf life during August and September when temperature is high. Ambient temperatures in apple-producing areas normally range between 15 to 25 °C during August-September, while in Delhi (a major terminal market), it ranges from 40 to 44 °C. Simple and practical methods of precooling for apples are: picking fruit in early morning, evaporative cooling, and forced air cooling. Keeping fruits over night near a tree basin is another practical way to remove their field heat.

5.6 Postharvest Treatments

In order to control postharvest losses caused by storage disorders and diseases, several postharvest treatments have been recommended in apple.

5.6.1 Physiological Disorders

During storage, bitter pit and scald are serious disorders of apple. Bitter pit of apples can be controlled by postharvest dipping of fruits in 0.5 per cent calcium chloride solution. The CA storage and waxing of apples also reduce the extent of bitter pit. A new product, Topcal (a form of $CaCI_2$ chelated with polyphenolic acids extracted from plant tissue), applied as a spray, has also been found effective in reducing bitter pit. Coating of apples with paraffin, vaseline or olive oil retard the development of apple scald in storage. Further, certain chemicals with anti-oxidant properties like diphenylamine (DPA) and ethoxyquin are used in some advanced countries to inhibit the development of apple scald.

5.6.2 Postharvest Diseases

Blue mold and grey mold are major diseases, causing spoilage of apples during storage and marketing. In addition, bitter rot, pink mold rot and core rot can also infect apples during storage. Chemical control of blue mold and bitter rot usually involves postharvest application of fungicides such as benomyl, thiobendazole, or SOPP, in addition to prompt cooling and low temperature storage. Incidence of grey mold rot (*Botrytis* sp.) and Alternaria rot (*Altemaria tenius*) can be controlled by careful fruit handling, prompt storage and maintenance of recommended low temperature. Blue mold rot (Figure 7a) can be effectively controlled by dipping apples in bavistin. Use of bavistin has also been found to completely check Rhizopus rot and black rot (Figure 7b) of apples as well as blue mold rot. Similarly, hot water treatment of apples at 45%C for 10 minutes helps in checking Botrytis and Penicillium rot effectively during storage.

5.6.3 Postharvest Chemical Treatments

Postharvest chemical treatment is effective and one of the important means of controlling apple rots. The fruits are dipped for one to several minutes in a fungicidal suspension before packing. Dip treatment with fungicides is effective to control Alternaria rot. Dip treatment with iprodione, thioacetamid, aureofungin and maleic hydrazide provided effective control to Altemaria rot. Dipping apple fruits in carbendazim and thiabendazole controlled blue mould rot effectively.

The effectiveness of postharvest dipping in fungicides or drenching can be increased if applied in conjunction with calcium treatment. The time of application is critical and its delaying for a few hours can result in huge losses. Use of the same or closely related chemicals, before and after harvesting, is not advised because it can lead to the rapid development of resistant strains of *P. expansum*. It is preferable to use a range of fungicides with different modes of action. There are reports of increased losses due to Altemaria rot following the use of systemic fungicides to control blue mould from abroad.

(a) Blue Mold

(b) Rhizopus Rot

Figure 7. Postharvest Diseases of Apple

5.7 Packaging

Packaging is one of the major factors influencing quality of the produce during transportation, storage and marketing. Proper fruit packaging not only reduces bruising and compression injuries but also checks moisture loss, microbial spoilage and pilferage. It also helps in maintaining the freshness and eye appeal of the fresh apple. Thus, packaging is a coordinated system of preparing goods for transport, distribution, storage, retailing and end use. Different types of packing materials used in apple trade are:

5.7.1 Gunny Bags

Packing of low-grade culled apples in gunny bags is an age-old practice. Such fruits are meant for processing purpose. Before filling apples in gunny bags, a layer of pine needles is placed at the bottom to serve as a cushioning material. After packing, the bags are shaken to settle fruits, padded at top with cushioning material, and then sewed tightly to prevent vibration

Table 8. Arrangement of different Grades of Apples in a Wooden Box

Grade	Size of Box (cm)	Size of Wrapping Material (cm)	No. of Fruits/ Box	No. of Layers
Super Large	45.7 x 30.5 x 27.9	27.9 x 27.9	54-57	3
Extra Large	45.7 x 30.5 x 25.4	26.7 x 26.7	60-63	3
Large	45.7 x 30.5 x 30.5	25.4 x 25.4	96	4
Medium	45.7 x 30.5 x 27.9	24.1 x 24.1	112	4
Small	45.7 x 30.5 x 25.4	22.8 x 22.8	128-132	4
Extra Small	45.7 x 30.5 x 25.4	21.5 x 21.5	160	5
Pittoo	45.7 x 30.5 x 25.4	Not wrapped	Variable	Loose

bruises during transport. However, this method is not recommended for packing quality apples for fresh consumption, since packing in gunny bags give rise to high incidence of bruising and rotting.

5.7.2 Wooden Boxes

Substantial quantity of apple is conventionally packed in wooden boxes. Generally, these boxes are 30, 28 and 25 cm in height for packing large, medium and small sized apples, respectively. The fruits are packed by lining the inside of box with newspaper sheets and keeping margins for overhanging the flaps. The wrapped fruits are initially padded with wood wool/pine needles at the bottom of box and later in between well arranged intervening layers. The top layer of fruits is covered with paper by bringing together the over-hanging flaps followed by nailing. However, packing apples in wooden boxes results in high incidence of bruising and more loss in fruit weight (Table 8). The loss in fruit weight is due to its water absorption by the timber and padding material.

5.7.3 Corrugated Fibre Board Cartons

Corrugated Fibre Board (CFB) boxes are made of fluted paper, which is pasted on flat paper sheets. The corrugating medium nowadays is also prepared from raw materials other than wood like bagasse and paddy husk. Similarly, fruit packing trays can; also be made from non-woody materials (thus leading to substantial saving in timber wood. In India, tray-packed telescopic cartons (CFB) owe their introduction in apple trade during 1980's. These cartons are suited to withstand various transportation hazards besides causing minimum weight loss and bruising damage to fruits. For packing in tray packed telescopic CFB containers, apples should be arranged in the order given in Table 8. Some of the common mistakes in packaging is overfilling and non removal of vent holes in CFB packages that results in poor quality at destination due to bruise injury of the top layer of apples

5.7.4 Plastic Crates

Plastic crates (collapsible and non-collapsible) have also been recommended and Introduced to some extent in apple trade in Himachal Pradesh; as field boxes for collection of fruits, for stacking of fruit in cold stores; carriage to nearby markets; and to supply fruits to processing units. However, impact bruising of fruits packed at the bottom of these crates has been noticed by many orchardists. Bottom padding with suitable material like thin foam sheet can possibly overcome this problem.

Figure 8. Overfilled CFB Boxes

Table 9. Arrangement of different Grades of Apple in CFB Carton

Grade	No. of Layers	No. of Fruits/ Layer	No. of Trays/ Layer	No. of Fruits/Box
Super Large	4	18	5	72
Extra Large	4	20	5	80
Large	5	20	6	100
Medium	5	25	6	125
Small	5	30	6	150
Extra Small	5	35	6	175
Pittoo	Loose			Loose

In India, CFB boxes were introduced in apple trade during 1980's (Figure 8). These cartons are suited to withstand various transportation hazards besides causing minimum weight loss and bruise damage to fruits. For packing in tray packed telescopic CFB containers, apples should be arranged in the order given in Table 9.

5.7.5 Shrink-wrap Packaging

For easy handling, reduction in losses during transportation and easy carriage, shrink-wrap packaging has been standardized in apple. This technique delays physiological deterioration of fruits and also prevents condensation of droplets within the package. Individual shrink-wrapping of the produce provides optimum gas and humidity for maintaining quality of the produce during the transit and storage. As a result, it doubles or sometimes triples storage life of the fruits under proper storage conditions. Such unit packs also provide protection against abrasion and maintain attractive appearance of the product. This technique involves over-wrapping of the produce with heat-shrinkable films of the desirable thickness with the help of a machine. For individual shrink-wrapping, initially individual fruit is loosely packed and then sealed in a desirable heat-shrinkable film (9μ) with the help of an impulse sealer. The produce is then placed inside the machine where the temperature is maintained at 120 °C or so. Then upper cover of the machine is pressed for 10-15 seconds to activate fan, which circulates hot air around the produce, and the film shrinks tightly around the produce. In case of tray-wrap packaging, produce is first sealed in a consumer pack of suitable size and then passed through the machine. Shrink-wrapped (Figure 9) produce is immediately removed from the machine and cooled for 2-3 hours at 5-10 °C. Finally, the produce can be packed in plastic crates for further storage or transportation.

Figure 9. Shrink-Wrapped Apples

Between different fruit layers, separators/liners are kept to avoid the bruising injury. In India, newsprint paper, tissue paper, polyethylene liner or bags, paper moulded trays are commonly used in apple.

5.8 Storage

Different methods of storing apples are given as under.

5.8.1 Low Temperature Storage

For most apple cultivars, optimum storage temperature is -1 to 0 °C with 90-95 per cent relative humidity. Delicious apples exhibit higher storage life at -1 °C than at 0 °C. However, somewhat higher storage temperatures (2–3 °C) are recom-mended for cultivars like Jonathan and McIntosh.

5.8.2 Air-cooled Storage

Air-cooled stores should be used to keep apples on the farms, in areas experiencing low night and high day temperatures. The temperature in such on-farm stores is brought down by opening air inlets/vents near the floor and the exhaust openings near the ceiling across the opposite walls to admit cool air at night.

5.8.3 Zero Energy Cool Chambers

An on-farm, low cost, low energy, cooling chamber has been developed at the IARI, New Delhi, by using locally available materials, which can also be used for short-term storage of apples. The double-walled chamber constructed by bricks, sand and bamboo cover is kept soaked with water. The evaporative cooling effect reduces the inside temperature by as much as 17-18 °C with a relative humidity of more than 90 per cent during peak summer.

5.8.4 Controlled Atmosphere Storage

Controlled atmosphere (CA) storage at 3.3 °C in 2-5 per cent CO_2 and 3 per cent O_2 prolongs the storage life of apples cultivars such as Mcintosh, Newton, and Cortland, which do not keep well at -1.1 °C to 0 °C. However, it can also be used for Royal Delicious, Golden Delicious, Rome Beauty and Stayman Early Worcestor (Figure 10).

5.9 Transportation

After packaging in CFB or wooden boxes, apples are usually transported to plains either for storage or marketing (Figure 11). Bulk transportation is done in trucks (Figure 12).

6.0 Domestic Marketing and Export

In India the harvesting season of apples begins in July and continues up to November. State wise availability of apples from different states is presented in Figure 13.

India is facing a supply gap in its domestic apple market due to increased demand from the growing middle class and the limiting factors such as seasonality, geographical separation, and limited infrastructure. In domestic market, the demand for apples in the apple producing areas is limited as the consumers are spread throughout India. Thus, the producers are forced to depend on the non-producing area markets for selling their produce. The major neighbouring markets include Delhi, Agra, Chandigarh, Ambala, Karnal, Jallandhar, Amritsar, Ludhiana and Pathankot while the major distant markets include Mumbai, Chennai, Kolkata and

Figure 10. Inside View of CA Store

Figure 11. Specially Designed Liners for Apple Packing in CFB Boxes

Figure 12. Apples Loaded on a Truck

STATE/UT'S	JUN	JUL	AUG	SEP	OCT	NOV
ARUNACHAL PRADESH				P	L	
HIMACHAL PRADESH	P	P	L	L	P	P
JAMMU & KASHMIR		P	P	L	L	P
UTTARAKHAND		P	P	L		

P Peak Season   L Lean Season

Source. Director of Horticulture/Agriculture of respective State/UT's.

Figure 13. Harvesting Season of Apple in India

Bangalore (AERC, 1985) (Figure 14). The most important terminal market with high absorptive capacity is Delhi (Azadpur), where >75 per cent of the marketed surplus of the apples are dispatched by the orchardists.

Figure 14. Display of Apple Varieties in a Local Market

Efficient marketing strategies of apple are very much needed to create the consumer utilities (time, place and form) which ensure satisfaction and profitability to both consumer as well as the producer. Efficient marketing includes, preparation of fruits for the market, selection of the markets, packaging, transportation and an efficient marketing channel to reduce rate of moisture loss and subsequent shriveling of fruits resulting in significant economic losses. Marketing channels, being one of the major component of marketing strategy comprises of various marketing functionaries that are involved in the distribution of the produce from producers to the consumers. Some of the channels through which the apples are dispatched include:

Orchardist → Preharvest contractor → Commission agent/Wholesaler → Retailer → Consumer

Orchardist → Commission agent/Wholesaler → Retailer → Consumer

Orchardist → Processing unit → Consumer

India has exported about 23.8 mMT of apple during the year 2012-13 which is of worth Rs. 5151 Lakh. The major countries to which the apples are exported from India are Bangladesh (73.53 per cent) and Nepal (26.39 per cent). The quantity of apples exported from India has significantly decreased by 49.4 per cent from the year 2010 which might be due to increasing demand from the domestic market. The major apple varieties exported from India are Red Delicious and Golden Delicious.

7.0 Conclusion

Apple is the most important fruit of the world. In India, it is the 5th most important fruit as well. Although the availability of apple has increased tremendously in the recent years but its productivity is quite low. The problem of non-availability of apple to non-traditional areas can be increased significantly if the postharvest losses are reduced by adopting recommended preharvest as well as postharvest technologies. Although several preharvest practices have been suggested but adoption of GAP (Good Agricultural Practices) can be very useful in thiis direction. In addition, use of safer, environment friendly chemicals (calcium chloride) or technologies (*e.g.*, fruit bagging), postharvest treatments such as edible coatings or heat treatments cane help in producing consumer-friendly produce and reducing postharvest losses significantly.

8.0 Important References

AERC, 1985. Market Survey Reports for various markets. Agro-economic Research Centre, HP University, Shimla.

Anonymous, 2014. National Horticulture database. National Horticulture Board, Gurgaon, Haryana, India.

Armugam, R. and Srinivasan, P.S. 1983. Maturity standards in apple. *South Ind. Hortic.* **31**: 40-42.

Chadha, K.L. and Awasthi, R.P. 2005. The Apple. Malhotra Publishing House, New Delhi, India.

Chadha, T.R. 1990. Textbook of Temperate Fruits. Indian Council of Agricultural Research, New Delhi, India.

Chattopadhyay, T.K. 2012. A textbook on Pomology, Vol IV: Temperate Fruits. Kalyani Publishers, Ludhiana, India.

Dilley, D.R. and Dewey, D.H. 1973. Hypobaric storage of apples. *Hortic. Sci.* **8**: 273-75.

Fidler, J.C. and Mann, G. 1972. Refrigerated storage of apples and pears – A practical guide, Horticultural Review No.2.Commonwealth Bureau of Horticulture and Plantation crops. East Malling, Kent, England.

Garg, R.C., Ram, H.B., Singh, S.K. and Singh, R.V. 1979. Effect of regulators on storage behaviour of apples cv. Red Delicious (*Malus pumila* Mill.). *Prog. Hortic.* **11**: 63-69.

John, D. and Wurn, S. 1987. Rationalization of apple grading for supply and storage. *Gartenbau.* **34**: 180-81.

Kaul, J.L. and Munjal R.L. 1982. Apple losses in Himachal Pradesh due to postharvest fungal pathogens. *Ind. J. Mycol. Pl. Pathol.* **12**: 209-12.

Lal, B.B and Sharma, S.D. 1980. A note on the maturity standards of apple. *Haryana J. Hortic.* **9**: 132-33.

Mahajan, B.V.C. 1994. Biochemical and enzymatic changes in apple during storage. *J. Food Sci. Technol.* **31**: 142-44.

Mitra, S.K., Bose, T.K. and Rathore, D.S. 1991. Temperate fruits. Horticulture and Allied Publishers, Kolkata, India.

Negi, Y.S.,Prasher, R.S. and Tewari, S.C. 1987. Marketing of Himachal apples: A spatio-temporal analysis. *Agri. Econ. Res. Rev.* **10**: 88-94.

Patil, R.T., Singh, D.B. and Gupta, R.K. 2008. Postharvest Management of Horticultural Produce. Daya Publishing House, Delhi, India.

Sharma, R.R. and Krishna, H. 2014. Fruit Production: Major fruits. Daya Publishing House, Delhi, India.

Sharma, R.R., Pal, R.K., Asrey, R., Sagar, V.R., Dhiman, M.R. and Rana, M.R. 2013. Preharvest fruit bagging influences fruit colour and quality of apple cv. Delicious. *Agric. Sci.* **4**: 443-48.

Sharma, R.R., Pal, R.K., Singh, D., Samuel, D.V.K., Kar, A. and Asrey, R. 2010. Storage life and fruit quality of individually shrink-wrapped apples (*Malus domestica*) in zero energy cool chamber. *Ind. J Agric. Sci.* **80**: 338-41.

Sharma, R.R., Singh, D. and Singh, R. 2009. Studies on transportation losses in apples packed in different containers. *Ind. J. Hortic.* **66**: 245-48.

Sharma, R.R., Singh, D., Singh, R., Singh, D.B. and Saharan, V.K. 2010. Effect of modified atmospheric packing on the quality and shelf-life of apple (*Malus domestica*).*Ind. J Agric. Sci.*, **80**: 222-26.

Singh, R.S. 2000. Diseases of Fruit Crops. Oxford and IBH Publishing Company Pvt. Ltd., New Delhi, India.

Chapter 12

Banana

C.K. Narayana

Principal Scientist, Division of Postharvest Technology and Agricultural Engineering, ICAR-Indian Institute of Horticultural Research, Hessaraghatta, Bengaluru – 560 089, Karnataka, India

E-mail: narayana.ck@icar.gov.in

Vijay Rakesh Reddy, S.

Scientist, ICAR-Indian Institute of Horticultural Research, Hessaraghatta, Bengaluru – 560 089, Karnataka, India

E-mail: rakesh.reddy@icar.gov.in

1.0. Introduction

Banana (*Musa* sp. L.) is one of the cheapest sources of wholesome nutrition to many in the world. Southeast Asia is the home to bananas, which later spread to almost every part of the world. For a western consumer, banana is just a fruit in supermarket, while it is a matter of livelihood for many in poorer nations. After wheat, rice and maize, it is the fourth important staple food, providing the necessary calories through carbohydrates besides important nutrients. Reference to banana is found in some of the oldest histories of human civilization and along with the human settlements it moved across the world. Though the word 'banana' is of Egyptian origin, in ancient Indian literature it was referred to as 'Kadali'. In typical Hindu religious ceremonies or rituals, banana occupies a place of prime importance. Banana rose from the status of a wild species of the genus *Musa* growing in the jungle to the present, where it is one of the most important internationally traded fruits of the world, which can make or break a nation.

Banana and plantains are presently cultivated in the country with a total production of 31.50 million tonnes from an area of 8.78 lakh hectares followed by China, Philippines and Ecuador. The other leading banana producing countries are Brazil, Indonesia, Tanzania, Guatemala, Mexico and Colombia (NHB Database). Though India produces one third of the total world bananas and plantains, almost all of this is consumed domestically. Unlike the corporate banana cultivation in major exporting countries like Ecuador, Costa Rica and Colombia, most of the banana farms in India are small and marginal. Dissimilar to perennial banana plantations of Africa, it is grown as an annual crop in India with a few exceptions of one or two ratoons. We find three systems of banana cultivation in India, *viz.* irrigated or wetland cultivation, homestead or backyard cultivation and rainfed hill cultivation.

1.1 Nutritional Composition

Banana fruit are highly nutritious in terms of their carbohydrates and sugars. Most of the calories are from carbohydrates (93 per cent), followed by proteins (4 per cent) and fats (3 per cent). Banana acts as a good source of vitamins (C, B6), dietary fiber and minerals. The fruits are low in saturated fat, cholesterol and sodium. The glycemic Index (GI) of banana fruit is relatively high *i.e.* 62 compared to other fruits such as apple (39), orange (40) and pear (38) and thus considered as a natural source of instant energy. The fruits are also a rich source of various phytochemicals *viz.* carotenoids, phenolics, flavonoids, and amines (Table 1)

Table 1. Nutritional Composition of Raw Banana Fruit (per 100g)

Nutrient	Quantity	Per cent Daily Value
Proximate composition		
Water (g)	74.91	—
Energy (kcal/kJ)	89/371	4
Protein (g)	1.09	2
Total lipid (fat) (g)	0.33	—
Ash (g)	0.82	—
Carbohydrate, by difference (g)	22.84	8
Fiber, total dietary (g)	2.6	10
Sugars, total (g)	12.23	—
Sucrose (g)	2.39	—
Glucose (dextrose) (g)	4.98	—
Fructose (g)	4.85	—
Starch (g)	5.38	—
Minerals		
Calcium (mg)	5	1
Iron (mg)	0.26	1
Magnesium (mg)	27	7
Phosphorus (mg)	22	2
Potassium (mg)	358	10
Sodium (mg)	1	0
Zinc (mg)	0.15	1
Manganese (mg)	0.27	13
Selenium (μg)	1	1
Vitamins		
Vitamin C, total ascorbic acid (mg)	8.7	15
Thiamin (mg)	0.03	2
Riboflavin (mg)	0.07	4
Niacin (mg)	0.67	3
Pantothenic acid (mg)	0.33	3
Vitamin B6 (pyridoxine) (mg)	0.37	18
Folate, total (μg)	20	5
Choline, total (mg)	9.8	—
Vitamin A, RAE (μg)	3	—
α-Carotene (μg)	26	—
β-Carotene (μg)	25	—
Vitamin A (IU)	64	1
Lutein + eaxanthin (μg)	22	—
Vitamin E (α-tocopherol) (mg)	0.1	1

Source: USDA (2019).

2.0. Status of Cultivation

Banana is presently cultivated commercially in almost 135 nations across the globe, with a global production of close to 116.72 million tonnes from an area of 5.15 lakh hectares. India is the largest producer of banana with a production of 30.46 million tonnes (FAO Database).

Banana occupies 11.1 per cent of the total area under fruit crops in India and contributes 32.6 per cent of total fruit production in the country. The productivity of banana in India is one of the highest in the world. The major banana producing states are Tamil Nadu, Maharashtra, Gujarat, Andhra Pradesh, Karnataka, Bihar, Madhya Pradesh, Uttar Pradesh, West Bengal, Assam, Odisha and Kerala. The statewise area, production and productivity of banana in India is presented in the Table 2

Table 2. State-wise Area, Production and Productivity in Banana, 2017-18

State	Area (000' ha)	Production (000'MT)	Productivity (MT/ha)
Andhra Pradesh	88.96	5003.07	56.24
Arunachal Pradesh	2.21	14.08	6.37
Assam	53.08	913.27	17.21
Bihar	31.07	1396.39	44.94
Chhatisgarh	26.57	745.78	28.07
Gujarat	68.15	4472.32	65.62
Himachal Pradesh	0.08	0.35	4.38
Jharkhand	9.17	32.06	3.50
Karnataka	110.55	2328.9	21.07
Kerala	109.26	1119.16	10.24
Madhya Pradesh	26.38	1834.03	69.52
Maharashtra	82.00	3600.00	43.9
Odisha	27.49	521.31	19.0
Tamil Nadu	111.36	5136.20	46.1
West Bengal	44.70	1077.80	24.1
Others	139.49	1637.47	11.7
Total	776.00	26509.10	34.2

Source: Horticultural Statistics at a Glance, 2018.

Table 3. Trends in Area, Production and Productivity of Banana in India during the Last Three Decades

Year	Area (000'ha)	Production (x000MT)	Productivity (t/ha)
1991-92	383.9	7790.0	20.3
2001-02	466.2	14209.9	30.5
2002-03	475.3	13304.4	28.0
2003-04	498.6	13856.6	27.8
2004-05	589.5	16744.5	28.4
2005-06	569.5	18887.8	33.2
2006-07	604.0	20998.0	34.8
2007-08	658.0	23823.0	36.2
2008-09	709.0	26217.0	37.0
2009-10	770.3	26469.5	34.4
2010-11	830.0	29780.0	35.9
2011-12	796.5	28455.1	35.6
2012-13	776.0	26509.1	34.2
2013-14	802.6	29724.0	37.0
2014-15	821.8	29211.5	35.6
2015-16	841.2	29134.9	34.6

Year	Area (000'ha)	Production (x000MT)	Productivity (t/ha)
2016-17	860.0	30477.3	35.4
2017-18	883.8	30807.5	34.9
2018-19	866.0	30460.0	35.2
2019-20	878.0	31504.0	35.9

Source: Indian Horticulture Database, NHB 2018; www. nhb.gov.in.

There has been a consistent increase in area under cultivation of banana at an average rate of 129.24 per cent during the last 30 years, while the production has been increasing at an average rate of 304.42 per cent and the productivity at 76.85 per cent during the same period (Table 3). This has been possible due to availability of elite planting material, improved production practices, advanced plant protection measures and timely intervention of government in providing subsidies.

3.0 Major Commercial Varieties

Though there are several varieties of banana belonging to different genomic groups, different regions in India have preference for certain varieties. Of more than 50 varieties commercially cultivated all over India, Cavendish bananas called 'Grand Naine' occupy the largest area. Other important cultivars are Poovan, Neypoovan or Elakki or Nzyalipoovan, Rasthali or Morthaman, Karpuravalli or PisangAwak, and Red banana that are dessert varieties grown in plains. Virupakshi and Sirumalai are hil banana varieties, while Monthan, and Batheesa are cooking types. Nendran is a dual purpose plantain mainly grown in Tamil Nadu and Kerala. The varietal distribution of banana in Indian states is summarized in the Table 4.

4.0 Nature and Causes of Postharvest Losses

Though the productivity of banana in India is one of the highest in the world it varies with variety and location from 12 tonnes/ha to 100 tonnes/ha. One of the major problems is non-availability of disease free planting material. Suckers still remain the widely used planting material by majority of farmers since availability of tissue cultured plants are mostly limited to one variety (Grand Naine-Cavendish). Non-availability of quality planting material is thus one of the major bottlenecks. Among the abiotic bottlenecks cold, heat and winds limit its cultivation in certain areas. As banana is a water loving plant declining water level, particularly in summer is responsible for lower yields. Salinity, poor drainage, shallow soil depth and nutrient deficiencies are the other important abiotic problems in cultivation of banana in the country. Among the biotic constraints, viral diseases like banana bunchy top virus (BBTV), banana bract mosaic virus (BBMV), banana streak virus (BSV) and cucumber mosaic virus (CMV) are spreading through planting material (suckers) and vectors, limiting the productivity. Due to entry of unscrupulous tissue culture plant suppliers, often farmers end up with planting infected material as many tissue culture companies do not have virus indexing facilities.

Fusarium wilt poses a major threat to some of the choice varieties grown in India. It has almost wiped out the hill varieties like Virupakshi and some silk group bananas like Nanjangud Rasbale. Sigatoka leaf spot is rapidly spreading in high productivity states like Maharashtra bringing down the banana productivity in high yielding varieties like Grand Naine and Robusta.

Table 4. Varietal Distribution of Banana in India

State	Cultivars
Andhra Pradesh	Grand Naine (AAA), Robusta (AAA), Amritpani (AAB –Silk group),Thella Chakkrakeli (AAA), Karpura Chakkrakeli (Poovan, AAB), Monthan (AAB), Yenagu Bontha (AAB), Dwarf Cavendish (AAA), Rasthali (AAB).
Assam	Jahaji (AAA), Borjahaji (AAA), Honda (AAA), Chinia (AAB), Manohar (ABB), Kachkola (ABB), Chini Champa (AB), Bhimkol (BBB), Attikol (BBB), Jatikol.
Bihar	Grand Naine (AAA), Alpan (AAB), Chinia (AAB), Chini Champa (AB), Malbhog(AAB), Kothia (ABB), Muthia (ABB) and Gauria (ABB).
Gujarat	Grand Naine (AAA), Robusta (AAA), Dwarf Cavendish (AAA), Lacatan (AAA),Harichal (AAA), Gandevi Selection (AAA), Basrai (AAA), Shrimati (AAA).
Karnataka	Grand Naine (AAA), Dwarf Cavendish (AAA), Robusta (AAA), Elakkibale (AB), Poovan (AAB), Rasabale (AAB), Monthan (ABB), Hill banana (AAB).
Kerala	Nendran (AAB), Palayankodan (Poovan, AAB), Poovan (Rasthali, AAB), Monthan (ABB), Nzhalipoovan (Ney Poovan, AB), Red Banana (AAA).
Maharashtra	Basrai (AAA), Robusta (AAA), Bhusavali (AAA), Mahalakshmi (AAA), GrandNaine (AAA), Lal Velchi (AAB), Safed Velchi (Ney Poovan, AB), Ardhapuri (AAA), Hanuman (AAA) and Rajeli (AAB, Nendran), Dwarf Cavendish (AAA), Shreemanti (AAA), Red Banana (AAA).
Orissa	Champa (AAB), Morthaman or Amritpani (AAB, Rasthali), Monthan (ABB), Chinia (AAB), Amrit Sagar (AAB), Lacatan (AAA).
Tamil Nadu	Grand Naine (AAA), Robusta (AAA), Poovan (AAB), Rasthali (AAB), Nendran(AAB), Karpuravalli (ABB), Sakkai (ABB), Peyan (AA), Matti (AA), Virupakshi (AAB), Red banana (AAA), Monthan (ABB)
Uttar Pradesh	Grand Naine (AAA), Dwarf Cavendish (AAA), Alpon (AAB), Chinia (ABB), Chini Champa (AAB) *etc.*
West Bengal	Champa (AAB), Morthaman (AAB, Rasthali), Amrit Sagar (AAB), GiantGovernor (AAA), Lacatan (AAA) and Monthan (ABB).

Source: Indian Horticultural Database, NHB 2013; Horticultural Statistics at a Glance, 2018.

Nematode attack is one of the common problems faced all over the country limiting banana production and productivity. The most damaging and widespread nematodes attacking banana are: the burrowing nematode, *Radopholus similis*, root-lesion nematode, *Pratylenchus coffeae*, spiral nematode, *Helicotylenchus multicinctus*, root-knot nematode, *Meloidogyne incognita* and cyst nematode, *Heterodera oryzicola*. Banana crop losses caused by nematode are estimated to be about 20 per cent. Some of the regional varieties like Poovan and Rasthali face physiological disorders like Kottavazhai (seed formation) and 'hard lump' formation respectively.

Banana being a highly perishable crop, postharvest losses are high and farmers are forced to sell off the produce at whatever price being offered by the commission agents. Fluctuating market prices, spiraling cost of inputs and dwindling labour availability are some of the other reasons discouraging the farmers from taking up banana cultivation.

Postharvest losses are very common if the harvested banana is not transported and stored at optimum temperature conditions. Zhang *et al.* (2005) specified that ~ 20 per cent of total harvested bananas might turn into discards and thus become unmarketable. Kader (2005) testified that estimates of postharvest losses in developing countries are generally much higher than those in the US, and can be up to 50 per cent for some fresh commodities. These losses could be due to sorting of bananas too small for shipping and damaged, injured or spoiled fruits that could induce microbial contamination of the full bunch in the collection stations (Bello-Pérez *et al.*, 2012).

5.0 Preharvest Management

Being a perishable commodity the postharvest losses in banana are quite high in India. As per one of the estimates, it is reported that 4.18 per cent postharvest losses occur during farm operations in banana and 2.42 per cent during storage. Preharvest care plays a major role in postharvest shelf life and quality of banana. One of the most important preharvest care for maintaining the postharvest quality is the bunch management.

5.1 Fruit Bagging or Bunch Covering

Proper bunch management, particularly bagging the bunches with polyethylene covers has been found to improve the yield and quality of banana. Covers protect the bunch from being damaged by birds, and help to increase the weight and improve the quality of the fruit. There are several advantages of bunch covers. These increase the yield, bring about uniformity in size or grade, protect from damages of insects, bats and birds and improve the colour, look and the overall quality. The highest harvest index was observed in bunches having covers with 6 per cent ventilation. Covers made of PVC are advantageous in areas where temperature is a limiting factor. Covering banana bunches with blue polyethylene sleeves a fortnight prior to harvest ensured uniform colour development and avoided the rupture of skin. Though thickness of the covers did not significantly affect the quality of fruits, based on its re-usability, 0.5mm thick perforated polybags are normally recommended. The bagging is done when all hands have emerged. The fruits in covered bunches were found to be more attractive, free from blemishes, early to mature with increased size of fingers.

Variation in temperature within the covers was considered responsible for such changes. The performance of different bunch cover materials in improving the quality of Cavendish banana fruits in Thailand was experimented by Chaiwong *et al.* (2021) and found that water proof non-woven bags has the potential to be used as bunch cover for improving the fruit quality and is found to be eco-friendly as it is reusable in nature.

5.2 Growth Regulator/Chemical Sprays

Preharvest application of different chemicals was found to have pronounced influence on the shelf-life, quality and chilling injury tolerance of the banana fruits during storage. Preharvest application of benzothiadiazole (100 µg/mL) two weeks prior to harvest was found to control anthracnose of harvested banana fruits, retained firmness and delayed ripening process when stored at room temperature (Li *et al.*, 2013). Preharvest application of calcium was found to be ineffective in extending the green life of Cavendish bananas (Tixer *et al.*, 2010). Preharvest foliar application of Glycine-Betaine (100 mM) to banana plants reduced the biochemical and physiological alterations caused by chilling injury. The harvested fruits were first stored at 10°C for 6 h and then transferred to 23°C ± 1°C until they reached commercial ripening, where the Glycine-Betaine treated fruits exhibited tolerance for chilling injury (Rodríguez-Zapata *et al.*, 2015).

5.3 Banana Bunch Management

Banana bunch feeding technology was developed by ICAR-IIHR, Bengaluru for enhancing the size of banana fingers in the bunch prior to harvest. It involves de-navelling and post-shooting feeding of N, K and S through the distal stalk-end of Rachis. De-navelling saves mobilization of nutrients into the unwanted sink of banana plant and earns additional income when the excised male bud is used as a vegetable. The technique involves blending 15 g of (Approximately 7.5g of Urea and 7.5 g of Sulphate of Potash dissolved in 100 ml water) in 500 g of fresh cow dung and applying the slurry to the de-navelled stalk-end of 'Robusta' bunch soon after fruit set. About 10-15cm long rachis should be available after the last hand to tie the plastic bag (used and milk bag is convenient) with a strong string. Though the experimental studies showed 67 per cent increase in bunch weight over the control plants with male bud retained till harvest, field demonstration of this technology exhibited only about 22 per cent increase in bunch weight of 'Robusta' banana. However, the technology is gaining popularity among the banana farmers of Karnataka as it is very cost effective Since the cost involved is about Rs. 2/- per bunch while the improvement is of 4 – 5 kg of 'Robusta' and 2 – 3 kg of 'Ney Poovan' banana bunch. Bunch thinning through dehanding was standardized by Horticulture Research Station, BCKV, West Bengal. The distal hands were removed in three different intensities at various times of bunch emergence and its effect on bunch weight and fruit quality was studied. In general, dehanding decreased bunch weight over control and the intensity of reduction increased with delayed dehanding after opening of last hand. But removal of three hands immediately after opening of last hand increased fruit weight, fruit length, fruit circumference and pulp weight to the maximum tune of 53.9, 34.5, 13.9 and 58.6 per cent, respectively over the control. No significant changes in the biochemical composition was observed except for reducing sugars. Dehanding also shortened the crop cycle by 18 days. Thus, dehanding was found to help in improving market appeal and customer attraction of banana (Sarkar, 2015).

6.0 Postharvest Management

6.1 Maturity Indices

The quality of fruits are influenced by many factors, of which the stage of maturity is of prime importance. Commercially, bananas and plantains must be harvested while mature green, and transported to destination markets where bananas are ripened under controlled conditions. Though the duration to maturity varies based on variety, season, diurnal variations in temperature, soil moisture content and various other biotic and abiotic stresses, there are some broad indicators based on which farmers generally harvest these bunches. These indicators have been scientifically validated by correlating with physiological events in the life of a banana plant.

Bananas to be marketed locally have to be harvested at a more mature stage than those to be transported over longer distances. Even though harvesting before attaining full maturity results in slightly lower yield, early harvest is resorted to delay the postharvest ripening of banana during transshipment. Generally fully mature bunches are harvested for domestic market where fruits have to be ripened and consumed within a week. For regional markets where it may take about 2 weeks, the fruits are to be harvested at 90 per cent maturity wherein the angles are fairly well defined. For long distance transportation and export, the harvesting is done at 75 per cent maturity where the fruits have plainly visible angles. Subjective judgment of maturity is highly variable with individual's experience and often results in admixtures of fruits with different maturity stages.

One of the widely adopted objective method of judging maturity in banana is shooting to harvest period or bunch emergence to harvest period.

Under Indian conditions, Grand Naine banana reach full maturity in about 90-100 days. It is necessary to have the data for different varieties on days from shooting to harvest throughout the year. For this purpose the bunches should be marked at the time of shooting by tagging with coloured stripes or ribbons and subsequently harvested at appropriate time taking into consideration other parameters like finger size and angularity. This practice is followed in Ecuador and several other Central American countries. Bunches are harvested between 98 and 112 days in Grand Naine or Valery cultivars in southern Ecuador. Some studies indicate that plantain reaches proper stage for harvesting in about 90 days after shooting irrespective of time of planting. The best practical and objective method standardizing harvest maturity is a combination of flowering to harvest duration and caliper measurement of finger diameter. Pisang Awak group bananas take longer duration, beyond 130 days to attain full maturity. Similarly when flowering and fruit development stage happens to occur during winter months, variations in maturity are observed in the same variety too. Though lot of research has been carried out on this aspect, the results have been mostly location specific. These indices also differ considerably among different varieties and even among strains of fruits of same group.

Some people confuse maturation with ripening. Often maturation phase is described as the time between growth and senescence. Development and maturation of fruit are completed only when it is attached to the plant, but ripening and senescence may proceed on or off the plant. The details about ripening changes are discussed elsewhere in this chapter.

Angularity is one of the visual methods of assessment of maturity. Visual inspection of the fullness of fingers or disappearance of angularity is one of the most commonly used methods. Fruits are classified as three quarter (with clearly visible angles), full three quarter (less prominent angles) and full mature where angles virtually disappear. However, angularity as a basis of maturity cannot be fully relied as cooking bananas have visible angles even at full maturity. Caliper grade is one of the most widely accepted method of judging maturity in banana exporting countries of Latin America.

In export trade, the maturity requirements depend on the specification provided by the importing country. Caliper grade and length of the fingers are the most commonly used indices. Middle finger in the outer whorl of the second hand is taken and is measured for the caliper grade at the thickest part of the finger. The caliper grade in millimeters is used in Caribbean Islands and Africa, while 40/32 or 48/32 of an inch is practiced in Central and South America. In French Antilles, heat units are correlated with caliper grade and used as an index. A fruit, which attains 34 mm caliper grade is supposed to have accumulated 900degree days of heat units in cv. Williams. Besides, heat units accumulated during the growth and development stage of fruits, several preharvest factors like ratooning, application of fertilizers, growth regulators and spacing have also been observed to influence the maturity.

Other objective methods like TSS, starch content, solids to acid ratio, specific gravity, pulp to peel ratio, firmness, though have been studied and recommended, being destructive in nature and non-user friendly, are not widely adopted.

Non-destructive sonic techniques have also been developed for testing the firmness. Visible and Near-Infrared Spectroscopy in reflectance mode could be used for predicting the maturity and quality of whole banana fruits through determination of fruit chlorophyll and sugar content (Zude, 2003). In India a minimum pulp to peel ratio of 1:1 has been accepted as being the acceptable level of maturity for ripening of Dwarf Cavendish bananas. However, farmers largely look for drying of floral ends on the fingers, which when touched falls instantly, indicating its full maturity.

6.2 Harvesting

Faulty method of harvest results in injury to the fruit rendering it vulnerable to postharvest diseases and rotting. These are ranging from primitive ways to highly mechanized system of harvesting practices, across different locations of the world. While large plantations like in Costa Rica, Ecuador and Colombia are amenable for mechanized harvesting and bunch handling, small farms like in India and other southeast Asian countries, are not suitable for mechanization. However, careful harvest, collection and field packaging, can reduce the losses and maintain the quality. The improved method of harvest followed is similar in pattern in most of the commercial plantations in the world. In this method a 'carrier' stands in position below the bunch to catch it and the 'cutter' cuts a notch into the pseudostem so that it falls slowly towards the carrier which guides the bunch on to a shoulder pad. The cutter severs the peduncle about 30 cm above the proximal end and carries the bunch to a trailer or hangs on to a cable wire depending on the transport method used. In order to reduce the losses due to bruises and mishandling, cushioned pads of either PUF or blistered polyethylene sheets can be used.

In several parts of India the method of harvest of banana involves giving a slant cut to the pseudostem and bending the upper portion of plant bearing bunch towards ground and severing the bunch from pseudostem leaving about 12-13 inches of peduncle above the first hand. The size of peduncle retained varies depending on the size of the bunch. The bunches are then carried to a collection centre in field either on shoulder of a carrier person or in a wide basket held on head. In some cases the harvested bunches are carefully placed on 10-12 cm thick bed of banana leaves on ground, to allow the latex to flow from the cut end. After latex flow ceases, bunches are carried to the packing shed one or two at a time on a prototype of medical stretchers of 150 cm long and 75 cm wide, made of thick smooth cloth. For weighing the bunches, a heavy string is tied to the stalk near the distal end and bunches are weighed on a spring balance in field.

A number of methods are employed to transport bunches from field to the packhouse. In small farms, the bunches are carried on heads or shoulders to the edge of the field, where normally the transport vehicle is parked. There these are weighed and loaded into the truck or trolley. Sometimes, it is common to see the stacking of bunches on the road side to facilitate easy loading into the truck. In either case the bunches are stacked horizontally in the trucks which are lined with banana leaves on the sides and floor. Regardless of method of harvest, the repeated handling of bunches at several stages like collection, loading, unloading, *etc.* damages the fruits resulting in reduced shelf life. In Latin American countries, where bananas are cultivated in thousands of hectares contiguously, the overhead cableways are employed to transport the bunches from farm to the packhouse. More often, the bunches are protected with PUF sheets and also PUF pads between the hands to prevent bruises during harvesting and handling.

Until 1960's Gros Michel, the then export cultivar of banana was exported as whole bunches. But later, when Cavendish group replaced it, the practice of dehanding came into vogue. Dehanding of bananas is cutting of the hands from bunch stem. It is an important operation essentially done in export trade of banana. However, in domestic trade of banana in India, the bunches are directly loaded into the trucks. The disadvantage of this method is huge postharvest losses, besides transportation of unwanted peduncles to the market. At the destination market, this is discounted from the net weight of the produce. Dehanding reduces the wastage and also helps in easy handling of fruits in crates or boxes.

6.3 Sorting, Washing and Grading

Dehanding is also an essential postharvest operation done in order to avoid crown rot and finger drop. During this operation the hands are cut with crescent like pad of tissue, which attaches the hand to main stem. The hands may further be cut into required number of fingers based on the size and shape of the packaging cartons. The cut hands or fingers are to be placed on the banana leaves or rexin sheets on tables or on ground in such a manner that the tips of the fingers and cut end portion of the hand rest directly on the floor or table. The latex from the cut end should not be allowed to drop or stain the fruits. The de-latexing can be done physically by allowing the latex to ooze out or it can be floated in a tank containing detergent or sodium hypochlorite solution where the latex gets washed out. In India for domestic market, dehanding is not commonly practiced. In recent past the Cavendish varieties like Grand Naine are dehanded to facilitate ripening by exposure to ethylene gas in ripening chambers. However, non-Cavendish bananas are still transported as bunches. For export trade, dehanding is essential during which, defective, damaged and underdeveloped hands are sorted and culled out. Selected bigger hands are then washed in tanks containing 10 ppm chlorine and treated with 500 ppm benomyl solution for 2 minutes as preventive measure against crown rot and anthracnose during transport and storage. After washing and fungicidal treatment the hands are dried under the blast of air from pedestal fans or blowers to ensure that no droplets of water are sticking on the surface or in between fruits.

For export purpose 75-80 per cent mature, only green bananas are suitable. Further after harvest the sorting and grading is essentially done to adhere to the minimum quality standards. However, the specific quality requirements including the levels of pesticide residues, *etc.*, vary depending on the importing country. The sorting should be done to remove immature or over mature and undersized hands, fruits with excessive scars, discolouration and any other form of mechanical damage. Deflowering of fingers, if any present, should be done before washing of the fruits.

As per the Codex standards export quality bananas should be minimum 14 cm long with 2.7 cm diameter. The reference fruit for measurement of length and girth is the medium finger on the outer row of the hand. The minimum length and girth of finger depends upon the customer requirement. Generally customers in international market demand bananas of 7 inches length and 3.8 to 4.8 cm diameter. The general postharvest protocol followed for export of bananas is indicated in the flow chart of operations (Figure 2).

6.4 Ripening

Banana being a climacteric fruit, ripens naturally after attaining the full maturity. On attaining full maturity or due to abiotic or biotic stresses the fruit shifts into senescence phase, associated with a rise in rate of respiration and ethylene evolution. Fully mature fruits respond vigorously to the auto-catalytic property of endogenously evolved ethylene, while the response of less mature fruits to ethylene is dose dependent. Therefore, less mature fruits have longer green life compared to full mature. Fruits with 75 per cent maturity can be stored up to 3 weeks in green condition at low temperature (13±1 °C; 90 per cent RH), while the more mature fruits have relatively lower storage life. However, taking the yield and quality into consideration, harvesting banana fruits with less than 75 per cent maturity is not advisable. Studies using Cavendish banana fruits harvested from 65 to 105 days after flowering showed that after the onset of climacteric, little differences in ripening characteristics were found among the fruits harvested at different maturities. When bananas are used for processing, commercially acceptable pulp quality is attained only after a minimum of 80 days after shooting. Studies have revealed that when both early and late harvested Giant Governor (AAA) bananas were ripened and compared for their quality, late harvested fruits showed higher percentage of total soluble solids, reducing sugars and total sugars and lesser starch and acidity. Poovan (Mysore, AAB) and Karpuravalli (PisangAwak, ABB) varieties of banana also showed similar results. However, less mature fruits had more weight loss during storage compared to full mature fruits.

In Robusta banana, by harvesting the fruits at early stage of maturity (at 100 days) the shelf life in ambient conditions could be extended from 16-21 days, while cv. Poovan when harvested at 75 per cent maturity, de-handed and either kept at room temperature (23-28 °C, control) or covered with moist sand (20-26 °C) could be stored up to 30 days. Control fruits ripened in 10-13 days in Poovan and in 13-15 days in Robusta compared to 20-21 and 22-30 days respectively for those kept in sand.

Climacteric fruits in general and bananas in particular are harvested while still green and firm but in un-ripened condition and are then subsequently allowed to ripen. These fruits ripen naturally, but this may happen slowly, at a time which is not predictable, and/or unevenly. The time of onset of ripening in the absence of exogenous ethylene may be determined by a change in sensitivity to endogenous ethylene during maturation. In many fruits heterogeneity in ripeness and quality within commercial consignment is frequently observed as a result of variation in physiological maturities at harvest. Equalization of this is enabled both by acetylene and ethylene treatments in postharvest phase through initiation of simultaneous ripening. Ethylene releasing compounds are applied after harvest to initiate ripening of fruits. In the postharvest phase, endogenous ethylene production accelerates ripening and subsequent senescence of fruit tissues by stimulating physiological processes such as respiration, softening and chlorophyll degradation. Although ethylene is generally considered as the main triggering factor, an interaction between different hormones is necessary for the control and regulation of the fruit ripening process. The harvested bananas pass through three physiological development stages, namely the preclimacteric or 'green life' stage, the climacteric or ripening stage which covers the eat-ripe, and finally the senescence stage when the fruits are over-ripe and dying.

Fruit ripening involves a complex network of metabolic processes that include pigment, carbohydrate and aroma metabolism. The golden yellow colour of the ripe fruit is due to chlorophyll breakdown, which unmasks carotenoid pigments in the plastids. However, when ripening at warm temperatures above 24 °C, bananas fail to develop a fully yellow peel as they retain high levels of chlorophylls in their peel. Thus, the peel of bananas remains a greenish yellow colour. These "green-ripe" fruits are perceived to be of poor quality and consequently fetch a lower price. The response of banana fruit to temperature appears to be abnormal compared to the majority of other fruits. In plantain, chlorophyll breaks down in the peel is reported to be faster at higher temperatures within the range of 24–35 °C. Chlorophyll breakdown in banana peel was exclusively suppressed at a temperature of 30 °C. The inhibition of chlorophyll (Chl) breakdown under tropical temperature conditions in banana fruit has been studied in various aspects. Studies in the plastid ultrastructure showed that ripening at tropical temperature resulted in a retention of thylakoid membrane, which might prevent the release of chlorophyll from membranes for degradation. However, the biological basis for the effects of elevated temperature on thylakoid disassembly in banana is still unclear.

Starch constitutes the major carbohydrate in banana, which forms 20-25 per cent of the fresh weight of the pulp of unripe fruit. During ripening this starch is degraded rapidly and the sugars like sucrose, glucose and fructose accumulate. Alpha amylase, β- amylase, α 1,6-glucosidase and phosphorylase are the principal enzymes involved in starch degradation during ripening of banana.

The banana fruit softens rapidly once ripening is initiated. Fruit softening is generally attributed to cell wall disassembly, particularly due to pectin solubilisation. Involvement of polygalacturonase (PG) or/and pectin methyl esterase (PME) in enzymatic disassembly of cellular wall has been widely reported. In addition, pectatelyase (PL), cellulose, b- galactosidase, transglycosylases or expansin proteins may play a role in cell wall disassembly during fruit softening.

Banana peel tissues are rich in phenolic compounds like 3,4-dihydroxyphenyl- ethylamine and 3,4-dihydroxyphenylalanine, which on oxidation by PPO forms brown pigments. These phenols are also responsible for the formation of flavour volatiles. The banana pulp during ripening shows a marked increase in organic acids such as malic, citric and oxalic.

A wide variety of volatile compounds are formed during the ripening process which includes esters, alcohols, ketones, aldehydes and phenol esters. The result of principal aroma component analysis of banana in cold storage revealed that these compounds are more strongly affected during cold storage. Esters such as 2-pentanol acetate, 3-methyl-1-butanol acetate, 2-methylpropyl butanoate, 3-methylbutyl butanoate, 2-methylpropyl 3-methylbutanoate and butyl butanoate were drastically reduced in the cold stored Nanicão cultivar of banana.

Bananas (*Musa acuminata,* AA Group) show peel spotting when the peel is just about as yellow as green, which coincides with optimum eating quality. Consumers might relate the spotting to overripe fruit. Studies with Sucrier' banana (*Musa acuminata*) have shown that the spotting in the over-ripe bananas is related to the increased Phenylalanine Lyase (PAL) activity in the peel.

Technically, ripening process can be induced by introduction of ethylene gas. Storage conditions such as relative humidity and temperature are very crucial in deciding the final quality of fresh fruit as they influence the physiological and chemical activity in the live biological commodities. It is reported that the relative humidity of above 80 per cent and temperature of about 20 °C are the ideal for ripening of bananas.

In any given consignment, as there is wide heterogeneity in level of maturity of banana they need to be treated with a ripening aid like ethylene gas which is either applied through ethylene gas cylinders or catalytic ethylene generators or ethylene releasing compounds like CEPA (Ethrel, Ride, Ethephon, *etc.*).

Other conventional methods of ripening of bananas like smoking, burning of kerosene stove, use of incense sticks, *etc.* in closed chambers are also found to release ethylene or acetylene or mixture of both which aid in uniform ripening of fruits. However, in these methods other harmful gases like carbon monoxide are also released. The use of ethylene gas has been reported to be more beneficial over acetylene gas for the purpose of ripening of fruits by several research workers. However, both acetylene and ethylene being explosive gases, mishandling or inappropriate usage in large quantities can cause harm or accident.

Comparative studies of different ripening methods revealed that smoking induced rapid ripening resulting in attainment of full ripe stage within 6 days with reduced shelf life while it was 10 days with ethrel treatment (Mebrate *et al.*, 2015). The search for natural agents/chemicals as alternatives for calcium carbide (CaC_2) has resulted in identification of two natural plant sources *i.e.* Bilimbi fruit and Venna leaf which showed good results to act as an alternative ripening agent (Basak and Akter, 2018).

6.4.1 Use of Ethylene Gas or Ethylene Releasing Compounds

Although dipping of fruits in diluted ethrel solution is recommended for enhancing ripening, it is a cumbersome process and may cause some problems if commercially available ethrel contains chemical impurities. Hence, the Food Safety and Standards Authority of India (FSSAI) has banned the application of ethrel solution in the form of dip or spray in 2009. However, use of ethylene gas at a concentration of 100 ppm is allowed. To overcome these problems ethylene gas is commercially used in modern ripening chambers. Catalytic reactors based ethylene generators are also available which produce ethylene gas using ethanol or methanol or ethrel. After 18-24 hours of exposure, the fruits are taken out for completing the ripening process at room temperature or 18-24 °C especially for Cavendish banana. The commercial ripening chambers follow a set procedure for ripening of bananas.

Comparison of aromatic compounds released from naturally ripened Vs. ethylene treated banana revealed that the total aromatic concentration was high in naturally ripened fruits (60,437 µg/kg) compared to artificially ripened fruits (55,243 µg/kg). Majority of the aromatic compounds *viz.* n-butyl acetate, isopentyl isobutanoate, 2-pentyl formate, 2-cyclohexanol acetate, ethyl isobutanoate, and 2-pentenyl butanoate were found to present in higher concentrations in the naturally ripened fruits. Likewise, the level of sucrose was also significantly higher in the naturally ripened fruits (98.40 g/kg) compared to ethylene treated banana fruits (89.40 g/kg) (Sonmezdag *et al.*, 2014). Ripening of banana fruits was carried out by different methodologies like naturally, stressed by fumes, placing in hot condition, using calcium carbide, calcium chloride, potash, ethylene, ethrel, ethylene glycol, methyl jasmonate, etc. to observed the morphological and physiological changes like colour, sweetness, firmness, acceptability and nutritional quality. The order of best treatments based on various beneficial effects is in the order: natural > ethylene > ethrel with sodium hydroxide > methyl jasmonate > ethephon > calcium chloride > ethylene glycol > calcium carbide (Chander and Smith, 2019).

6.4.2 Commercial Method of Ripening

Bananas are received at the ripening plant at colour stage 1 (refer colour chart). The product is precooled to the pulp temperature of 18 °C and after stabilization of temperature ethylene is introduced at a concentration of 100 ppm. Most convenient and safe method of introducing ethylene is through ethylene generators. Cylinders containing 5 per cent mixture of ethylene gas with nitrogen may also be used. After 24 hours, the chambers are vented. During this period the fruit starts producing ethylene by itself at the rate of 2 - 4 µl/kg/hr. Excess ethylene and carbon dioxide must be removed for faster and uniform ripening. For this purpose the chambers are vented after 24 hours of loading. Sometimes the ripening chamber suppliers provide the venting system along, which is capable of removing excess carbon dioxide in order to maintain concentration level below 10,000 ppm. Bananas are ripened to colour stage 3 or 4 depending on mode of retailing for dispatch to market. Depending on cultivar and market requirements, ripening cycle may be spread from 4 to 8 days.

6.4.3 Colour Chart Guide for Ripening

Colour changes in banana during ripening are based mainly on the peel colour rather than the pulp colour and hence colour of banana peel has been used in the assessment of the stages of ripeness of banana. Commercial standard colour charts are available in which there are 7 stages of peel colour which are translated to a numerical scale: Stage 1=all green, 2= green with trace of yellow, 3= more green than yellow, 4= more yellow than green, 5= yellow with trace of green, 6= full yellow, 7= full yellow with brown spots. This colour chart is more applicable to Cavendish bananas than any other (Figure 1).

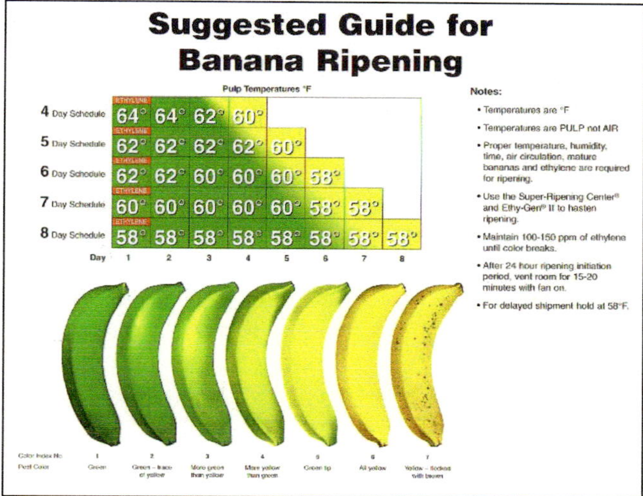

Figure 1. Colour Chart for Banana Ripening.

These stages also indicate the level of ripeness and are a useful guide for commercial ripening service providers. The fruits are generally taken out of the ripening chamber at stage 3-4 so that these can be transported to the retail stores, where these complete the ripening process and become edible ripe in one or two days. At the same time seller gets sufficient time to dispose off the produce. For short distance transportations, which may involve 5-6 hours of journey by road, the fruits can be taken out at stage 2-3.

Recently non-invasive method of prediction of ripeness of banana was also developed based on capacitive property of banana fruit. Relative permittivity was correlated with quality parameters of banana fruit. Green-ripe banana fruits have larger permittivity than the full-ripe ones. A quadratic regression equation had higher prediction power to assess the ripening of banana.

Real time *in-situ* techniques were also identified to determine the ripening stage of banana fruits. Where in a combination of three devices namely DA meter (measures chlorophyll degradation), colorimeter (measuring CIE-Lab) and spectrometer (measuring light reflectance troughs at 494 nm and 679 nm). A novel banana ripening index (BRI) was developed and was proposed by Ringer and Blanke (2021) to identify and distinguish the ripening stages of banana with values starting at 4 at R1 and peaking at 8.1 at ripening stage R7 (Figure 1).

6.5 Irradiation

Irradiation, also known as radurisation, is the process of exposure of fruits to ionizing radiations. Though doubts have been raised on the prospects of using ionizing radiation for extending the shelf life of fruits and vegetables, it is a well accepted method in developed countries like USA. The ripening process in bananas can be effectively delayed by irradiation at lower dose (0.2 kGy with a dose rate of 7.35 kGy). This effect is brought out through retardation of softening and colour change. Irradiation decreases sensitivity of the banana to its own endogenous ethylene without causing any phytotoxicity. The fruit maturity at harvest and post-irradiation storage temperature significantly influence the response to irradiation. The ability of banana fruit to withstand higher doses of gamma irradiation depends on the physiological status of the fruit at time of irradiation. Irrespective of varietal differences, irradiation of preclimacteric bananas to doses above 0.5 kGy resulted in severe skin discoloration and fruit splitting. Fruits at climacteric phase could tolerate up to 2 kGy with no effect on rate of ripening. At the same time, it does not affect ripening using high concentrations of exogenous ethylene. Irradiation seems to decrease the sensitivity of banana fruit to the ripening action of exogenously added ethylene.

Irradiation of bananas at 0.85kGy increased the storage life in the ripening room at 13 °C from 14 to 29 days. Gamma irradiation of green bananas at 0.2-0.3 kGy delayed soluble solids formation, starch disappearance and respiratory activity. Although a number of countries in Asia irradiate fruits, the consumer acceptance of such a treatment is highly questionable. In some European countries like Germany, the irradiated fruits are not allowed.

6.6 Precooling

Rapid cooling after harvest is generally referred to as precooling and particularly benefits rapidly respiring fruits like banana. Precooling is essential as refrigerated ships, land vehicles and containers are not designed to handle the full load of field heat but are designed to merely maintain precooled produce at the selected carriage temperature.

Field heat generated due to the harvest stress can cause overheating of the fruits. This may result in damage of the plant tissues and acceleration of the biochemical activities, causing spoilage. Several methods like hydro cooling, airblast cooling, vacuum cooling and liquid carbon dioxide cooling have been in use, separately or in combination, to take care of this heat load from horticultural crops. Hydrocooling is achieved by dipping/drenching, rinsing/immersing or spraying cold water over the banana bunches for effective field heat removal. In air blast cooling method, banana bunches are subjected to a jet of cooled air for removal of field heat whereas in vacuum cooling the fruits are treated under vacuum for few minutes.

Out of various methods of precooling, aircooling and hydrocooling are widely practiced in fruits. Bananas after packaging in cartons should be placed in precooling chambers till the core temperature of banana attains the storage temperature. Care should be exercised to ensure that precooling is done gradually for a period of 4-6 hours to avoid internal damage of the tissue. Precooling treatment of 5 and 10 minutes combined with storage at 21 °C temperature for the first 24 hours or storage at 13 °C straight away maintained the green colour of the fruit for 20-22 days. Presently banana trade has developed in tandem with the development of specialized ' Reefer ship' or 'Reefer container' which have the capacity to cool fruit loaded at close to field temperature down to shipping temperature (13± 1 °C). These unique features of banana trade has necessitated specific development of banana packaging systems and materials. Precooling, at times gives additional shelf life if held at low temperatures. Alternatively, shelf life advantage could also be obtained if stored under evaporative cool storage conditions following precooling using either forced air cooling or hydro cooling. Precooling rarely has helped in extending the shelf life of banana under ambient condition unless stored or held in shelves having relatively lower temperature and higher relative humidity.

Pre-cooling of banana, by forced air cooling/ hydro-cooling/dipping at temperatures of 5 °C, 8 °C or 10 °C gave additional 5 days of shelf life in cold storage. Alternatively, shelf life advantage of 3 days could also be obtained if pre-cooled using either forced air cooling or hydro cooling at temperatures of 5 °C, 8 °C or 10 °C and stored under evaporative storage conditions of Zero Energy Cool Chamber (Doshi and Sutar, 2015).

6.7 Storage

6.7.1 Pretreatments with Chemicals

Pretreatments with fungicides, wax emulsions have also shown beneficial effect in extending the storage life at ambient as well as low temperature. Dipping of bananas in 200 ppm thiabendazole was an approved and recommended postharvest treatment for export of bananas. A double coating of 12 per cent wax emulsion was found to prolong the storage life of Dwarf Cavendish bananas by 10-12 days at 14.5 °C. When effect of various pretreatments like dipping in thiabendazole at 400 ppm or 5 per cent Quinolate, wrapping in PE bags before packing in cartons were evaluated for banana cv. Magrabi, it was found that all packaged fruits were in good condition after one month of storage and had a shelf life of 5-7 days under ambient conditions. Treatments with GA_3, Blitox, Bavistin, Dithane M-45 or 6 per cent waxol have also shown beneficial effect in extension of storage life of Giant Governor bananas.

Weight loss is one of the common changes observed in fresh produce during storage and ripening due to transpiration and respiratory loss of carbohydrates. Cool chamber storage significantly slowed down the weight loss (PLW) and decrease in firmness of Cv. Dwarf Cavendish fruits compared to rapid changes in fruits stored in ambient conditions. When Taijiao bananas were harvested at 70-80 per cent maturity and sprayed with Jiaoxuan-1 or Jiaoxuan-2 (preservative agents) or with carbendazim (fungicide), wrapped in PVC film and stored at 11-25 °C, they could be stored up to 90 days with least weight loss. Green bananas coated with sucrose ester formulation (0.5 – 2 per cent) in the quick dip method delayed the ripening up to 10 days under ambient condition. A prestorage dip in E 9267 emulsifiable mineral oil (0-4 per cent) reduced decay caused by *Colletotrichum musae* and prolonged the storage life. Silver iodide was also found to retard ripening of banana at 800-ppm concentration and extended storage life up to 28 days at 15 °C. Chemical compounds such as Sodium dichloroisocyanurate (NaDCC) were found to delay ripening and senescence of banana fruit during storage. Application of NaDCC in the concentrations of 200 mg L^{-1} significantly reduced the ethylene production and down regulated the xyloglucan endotransglucosylase/hydrolase and pectinesterase genes. Additionally, it was found to significantly affect the accumulation of ripening related primary metabolites *viz.* sugars and organic acids (Wu *et al.*, 2018).

The fruits meant for domestic markets are not stored. The dehanded fruits are arranged in plastic crates and sent to ripening chambers for ripening. In traditional system of marketing, the bunches are straight away subjected to smoking up to 12 hours (particularly non-Cavendish varieties) before retailing. But in export trade the dehanded fruits of Cavendish bananas are precooled and stored at 13±1 °C with 90-95 per cent relative humidity.

6.7.2 Shelf Life

Shelf life of banana can be divided into two stages, as green life and yellow life. Most of the banana varieties turn yellow on ripening at ambient temperature, while Cavendish bananas require a ripening temperature of 16-21 °C for development of yellow colour. The period for which it remain green is called 'green life' and from turning yellow till unmarketable stage is called 'yellow life'. Cavendish bananas have a longer green life and shorter yellow life, while others have the reverse. The fully mature fruits of Cavendish banana have a shelf life of 6-9 days depending upon the ambient temperature. In local trade, banana fruits are not subjected to precooling or cold storage as it is ripened and consumed immediately. However, in export trade a green life of more than 2 weeks is required to reach the destination market and for its further distribution. Therefore, it undergoes all the postharvest export protocols as detailed elsewhere in this chapter. In order to extend the storage life, several methods have been studied by various researchers and a combination of these are used in practice by the traders. Harvesting of bunches a few days before they reach full maturity, storage at reduced temperatures (temperature management), modified atmosphere storage, packaging in film bags, pretreatments with fungicides, chemicals and coating the skin with waxes/emulsions are some of the methods reported.

6.7.3 Temperature Management

Optimum temperatures for storage and holding of green bananas are 13.3 to 14.4 °C (56 to 58 °F). Some banana cultivars can be handled at 12.8 °C (55 °F) too. Optimum RH for holding and transport of fruit is 90 to 95 per cent. Ripe bananas cannot be held for long as compared to green. Storage temperature influences the postharvest physiology of banana and its interaction with the microbes in the storage environment. As the ripening is hastened at higher temperatures, the complex carbohydrates like starch degrade into sugars to sustain the high rate of respiration. Further the self-defense of fruit weakens, and the microbes in the storage environment attack the fruits resulting in spoilage and rots. Shelf life was found to decrease exponentially with increasing temperature. A reduction in acid-phosphotase activity and failure to ripen were noticed when mature green bananas were stored at a constant 40 °C in contrast with controls stored at 20 °C.

Lowering of temperature slows down the catabolic process in the fruit and also makes the environment noncongenial for growth and proliferation of spoilage microorganisms. Storage of bananas at 13 °C considerably suppressed respiration rate, ethylene production and ripening. Normally bananas are stored at a critical temperature of 13±1 °C with a relative humidity of 85-95 per cent. At temperature lower than this fruits suffer chilling injury. However, variation in varietal response to chilling temperatures and manifestation of chilling injury symptoms has also been reported.

The produce for overseas markets are carried in refrigerated ships or 'reefer ships', which maintain a circulating cool air temperature of 12-13 °C. Under a normal low temperature storage condition (14.5±1 °C; 80-90 per cent RH), after 4 weeks, 64 per cent of Robusta banana were found to be fully ripe and 100 per cent of Dwarf Cavendish overripe. Therefore, ideally these can be kept green only for 3 weeks.

Green unripe fruits at 2 stages of maturity (100 and 115 days after fruit set) when held at room temperature or at 15 and 20 °C for 1-4 weeks followed by ripening at ambient conditions showed that the quality of fruits after 1-4 weeks of holding at 20 °C was the best in quality. As the storage duration at low temperature prolongs, the post storage ripening quality decreases.

6.7.4 Modified Atmosphere Storage (MAS)

Modified atmosphere storage is achieved by using polymeric films and sometimes using selectively permeable films which have a specific rate of exchange for oxygen, carbon dioxide and water vapour. However, in this system there is no attempt to control the atmospheres of oxygen, carbon dioxide and nitrogen at specific concentrations unlike control atmosphere storage.

Storage life of banana can be extended by packaging in sealed polyethylene bags with ethylene absorbent. The objective of using polyethylene bags is to maintain high humidity and create a modified atmosphere around the produce in the package. Though bananas release a remarkable level of ethylene, inclusion of an ethylene-absorbing compound, such as potassium permanganate on an inert carrier, in the package can reduce accumulation of ethylene considerably. A simple polyethylene bagging was adequate for extending the storage life by about a week at warm ambient temperatures, while longer storage could cause ripening of green bananas due to accumulation of ethylene gas. When the whole banana bunches were packed in sealed polyethylene tubes and held at 29, 20 and 13 °C, the average storage life increased 2- to 3-folds (over control). Further increase of 3- to 4-folds was observed when an ethylene absorbent ($KMnO_4$ on vermiculite) was packed with the bunch in PE tubes. Use of fruit coatings (Tal-Prolong – a sucrose ester of fatty acids) at 0.5 per cent concentration were also found to extend the storage life of banana. The storage life of bananas (Dwarf Cavendish and Williams) was extended when fruits were held in a modified atmosphere containing 10 per cent carbon dioxide for 1 to 3 days, followed by storage under regular atmosphere conditions at 12.5 °C.

Studies conducted to determine the colour and suitable thickness of polyethylene films used for extending the shelf life of banana cv. Giant Governor, showed that non-perforated polyethylene packs of 300 or 400 gauge thickness was ideal where the physiological loss in weight (PLW) was least, and fruits remained in marketable condition up to 28 days after harvest. During a 28-day storage period at 13.5 °C, rates of fungal rot development and ripening were considerably slowed down when polybags were kept sealed.

Use of ethylene scrubbers were found to have complimentary effect in extending the shelf life and storage life of banana when used in polybags or storage chamber. Potassium permanganate is an effective ethylene absorbent and several commercial formulations of ethylene scrubbers based on this are available in the market. Banana fruit stored with ethylene absorbent at 30 °C was reported to have a similar storage life to that of conventionally stored fruits in air at 12.8 °C. This is a very useful technique especially for developing countries where refrigeration is not readily available or is expensive. Use of vermiculite block soaked in potassium permanganate solution to remove ethylene from sealed bags containing banana is very economical. Purafil (a silicate carrier impregnated with alkaline $KMnO_4$) is a commercial formulation available for this purpose, which extended the storage life of preclimacteric bananas up to 16 days at 13 °C and 8 days at 24 °C when kept in sealed polyethylene bags. While various materials like vermiculite, coke, alumina, lime stone, pumice stone, sand, brick pieces and sawdust were used as a carrier for $KMnO_4$ based ethylene absorbent, use of other chemicals like charcoal and palladium chloride was also found to prevent the accumulation of ethylene in banana packs.

The banana fruits packed in LDPE pouches under modified atmosphere packaging and stored at 13°C temperature given a shelf-life of 25 days in comparison to only 5 days for unpacked bananas under ambient conditions (Venkateswarlu and Satyanarayana, 2016). Robusta banana when stored under active (8.2 per cent CO_2 and 2.6 per cent O_2) and passive MAP (8.6 per cent CO_2 and 2.8 per cent O_2) at 12 ± 1°C and 85–90 per cent RH could extend their shelf-life by 7 weeks and 5 weeks, respectively against 3 weeks shelf-life of openly kept control fruits (Kudachikar *et al.*, 2011). The treated fruits have shown very good sensory quality after 5 days of ethrel dip treatment.

6.7.5 Controlled Atmosphere Storage (CAS)

Commonly CA is the term used for a gaseous environment with high carbon dioxide, decreased oxygen and high nitrogen levels as compared to normal atmosphere. Controlled atmosphere storage is a system for holding produce in an atmosphere that differs substantially from normal air in respect to CO_2 and O_2 levels. Controlled atmosphere storage requires constant monitoring and adjustment of the CO_2 and O_2 levels within gas tight stores or containers. The gas mixture will constantly change due to metabolic activity of the respiring fruits in the store and leakage of gases through doors and walls. The gases are therefore measured periodically and adjusted to the predetermined level by the introduction of fresh air or nitrogen or passing the air in store atmosphere through a chemical to remove CO_2.The concentration of carbon dioxide, oxygen and nitrogen are manipulated to attain various gas compositions as desired depending on the commodity. CA storage also has adverse effects on the produce, at O_2 levels below 1 per cent. In the absence of O_2, anaerobic conditions can prevail with the consequent formation of alcohol, aldehydes and related physiological changes. Also high CO_2 and low O_2 may cause abnormality in metabolism. Bananas were found to suffer carbon dioxide injury at more than 7 per cent concentration. It was observed that reduced O_2 and increased CO_2 inhibited the action of ethylene so that the ripening processes of ethylene treated bananas were delayed.

For CA storage of Dwarf Cavendish and Williams bananas 2 per cent oxygen and 7 per cent carbon dioxide was ideal to keep the fruits in green condition for six weeks at 12.5 °C. Storage of bananas in CA shipping containers or packing in polyethylene bags and keeping in standard shipping containers was also successful for export of bananas by sea.

6.7.6 Vacuum Packaging

Vacuum packaging is a special type of modified atmosphere packaging, wherein a partial vacuum is created in the head space, leaving an altered initial atmosphere. By vacuum packaging (in polyethylene bags), the storage life of Golden bananas was reported to be extended up to six weeks under refrigerated condition (17 °C) and up to three weeks at room temperature. Though there are few reports on usefulness of vacuum packaging in extending the shelf life of bananas, studies at NRC for Banana, Trichy showed that it is not suitable for long term storage of green bananas, as the fruits failed to ripen after removal from vacuum due to damage of the ripening apparatus of fruit beyond recovery. The fruits also suffered carbon dioxide injury characterized by surface blackening and fermented off flavour. Partial vacuum may have beneficial effect when coupled with low temperature storage. The normally available films are not meant to maintain the absolute vacuum, as they are permeable to air and water vapour. Therefore, over a period of time such packs pull air from atmosphere due to negative pressure inside the pack and lose vacuum. Later due to buildup of carbon dioxide inside the pack, the fruits suffer carbon dioxide injury.

However, vacuum packaging with low temperature storage at 13 °C was found to control crown rot of organically-grown Balangon (*Musa acuminata* AAA Group) banana during simulated domestic and export shipment studies carried out at Postharvest Horticulture Training and Research Center (PHTRC), College of Agriculture, Philippines. The vacuum packed banana fruits remained green for a period of 25 days (holding period) and ripened normally when treated with ethrel and held at 23 °C (Esguerra *et al.*, 2013).

6.7.7 Hypobaric Storage

Hypobaric storage, also known as low pressure storage, is a type of CA storage with emphasis on reducing the pressure exerted on the stored material. This method reduces oxygen concentration in the storage environment and also evacuates the ethylene in the tissues by diffusion. Produce is stored in a partial vacuum and chamber is vented continuously with water-saturated air to maintain oxygen levels and minimize water loss. The reduced partial pressure of oxygen in hypobaric storage retards ethylene action and its production as well as ripening in some fruits. Extended storage life of 120 days has been reported when bananas are stored at a partial pressure of 150 or 80 mm Hg. The rate of ripening was found to be inversely related to the pressure during storage. When Dwarf Cavendish bananas were pretreated with ethylene and stored for 28 days in $1/10^{th}$ atmospheric pressure at 14 °C, it remained green and firm until the end of storage, but started ripening almost immediately after being placed in 21 °C without additional ethylene treatment. This indicates that sub-atmospheric storage (hypobaric storage) retains the produce in fresh condition much longer than conventional cold storage.

6.7.8 Chilling Injury (CI)

Low temperature storage cannot be utilized to its full potential, in extending storage life of tropical fruits like mango and banana, because these are susceptible to chilling injury (CI) when stored at <13 °C. Physiological events related to CI have been extensively described. Modifications in membrane composition, especially the saturation degree of their constitutive lipids, which regulate membrane fluidity and permeability, are the first notable effects of CI. An increased level of oxidative stress due to the overproduction of reactive oxygen species (ROS) has been considered a secondary effect of CI. The most common visual symptoms of CI in fruits are dark colouration of the skin, uneven ripening, development of off flavour and poor fruit quality. The symptoms of CI are strongly dependent on the duration and temperature of chilling treatment. Sometimes, symptoms may only develop when the produce is placed at higher temperatures and may appear immediately or may take several days to appear. Chilling injury in bananas occur due to browning of cortical tissues at temperature below critical (12.8 °C). Several theories have been put forth with regard to the mechanism of development of chilling injury. Increase in the level of oxaloacetic acid in the fruits kept at lower temperature is one such earliest theories, but it is not clear whether this change is the result or cause of injury. Changes in membrane permeability due to exposure of tissues to chilling temperatures and electrolyte leakage have been reported to be responsible for manifestation of chilling injury in Cavendish banana. Variety, maturity, growing conditions and genomic constitution have been found to influence the vulnerability of fruits to chilling injury. The symptoms of chilling injury appear to vary among banana cultivars and are related to the genomic group. The cooking banana cv. Bluggoeare most susceptible to chilling injury compared to the Pisang Awak (ABB). Pome group of bananas are more susceptible than Cavendish or silk group. In general it has been observed that the genome of B group (B- *Musa balbisiana*) cultivars confers greater resistance to low temperatures compared with the AAA group (A-*Musa acuminata*). However, the Prata cultivar (AAB group) seems to be more tolerant to cold storage (around 12 °C) than the Nanicão cultivar (AAA) (similar to Grand Naine), which makes it feasible to transport this variety over long distances without impairing the quality of the fruit.

Less mature fingers are more susceptible to chilling injury than over mature fingers of banana. Higher sugar content was reported to confer resistance against chilling injury. Chilling injury in banana increases as the duration of exposure at 0 °C increase from 5 to 20 days. Postharvest treatment with dimethyl polysiloxane, vegetable oil, mineral oil, polyamines, higher relative humidity, modified atmosphere storage, cold adaptation or cold conditioning, intermittent warming are some of the methods found to ameliorate the effect of chilling injury in banana. Decrease in the activities of beta-amylase and increased polyphenol oxidase activity was found associated with chilling injury in banana. It is reported that cooling the fruit slowly, reducing the level of oxygen, and maintaining high relative humidity can reduce the chilling injury. However, the most practical method is to avoid exposure of banana to chilling temperatures below critical level depending on the sensitivity of the cultivar.

Of late fibroin, a hydrophilic polymer protein from silkworm is being used for delaying chilling injury in cold stored banana fruits. The fibroin treated banana fruits exhibited lower peel browning index and ROS levels during overall storage and reduction in the contents of hexadecanoic acid and octadecanoic acid towards the end of storage compared to control fruits. The antioxidant activities of the treated fruits were found to be enhanced significantly during cold storage (Liu *et al.*, 2019).

6.8 Packaging

Packaging is a system for preserving the safety and quality of produce throughout the distribution chain up to consumer by reducing damage, maximizing shelf life, retaining quality and carrying important information on the label. Besides unitization, packaging discharges many functions including product identification, traceability, advertisement and also the ease of handling and transportation. Packaging can impair or improve the quality of a saleable product. A good package is designed to deliver the contents conveniently and cheaply. Packaging is a vital component of postharvest management to assemble the produce in convenient units and to protect it from deterioration during handling and marketing. Adequate packaging provides protection to the fruits from physiological, pathological and physical deterioration in marketing channels and retains their attractiveness. For different types of products, pack size, strength requirements and qualities dispatched by different types of transport need to be carefully assessed before deciding on the type of packaging. The need to reach distant markets particularly for exports and use of refrigerated transport has posed new challenges for developing appropriate packaging materials.

In the past, bananas for export from Fiji to New Zealand were detached individually from the hands and packed tightly in 33 kg wooden boxes. But this resulted in bruising of upper layer and the fruits in contact with sides. Reduction of fruit quality was found to offset the economic advantages of filling all the shipping space with fruits. Later wooden boxes were abandoned and suppliers resorted to packing fruits in cartons. Therefore, in the modern systems of handling and transport, quality control and good container design have made carton packing not only feasible but also necessary. First the hands are graded for size and quality and then packed in layers in special ventilated cartons with plastic (PUF) padding to minimize bruising.

In Indian banana marketing system the concept of packaging is yet to be practically adopted. A large quantum of bananas is traded uncovered or just wrapped with banana leaves. No systematic grading and packaging is followed. As the consumer preference and buying behaviour is towards a standardized and fresh looking produce, appropriate packaging can lead to high price realization. Research work carried out at CFTRI, Mysore and NRCB, Trichy has resulted in development of suitable CFB boxes for packing of banana both for export and domestic trade. The cartons are a two-piece telescopic type made of virgin, 5 ply corrugated fibreboard with the dimensions of 65 X 30 X 25 cm. Each carton is provided with ~6 per cent ventilation holes equally distributed on all four sides and side grooves for easy lifting of the carton. The banana hands are packed into the cartons in a lengthwise fashion with their cushion resting on the bottom of the carton and fingertips of the fruits pointing towards the upper lid. The cartons are packed with 6 to 7 hands (Cavendish variety), without any bruise. Between each layer of fruits a polyurethane foam sheet is provided as cushion, which avoids rubbing of hands against each other (Table 5).

Modified Atmospheric Packaging (MAP) is another system, wherein the product respiration and gas permeation of film continue to take place simultaneously. In a MAP system, the fruits are enclosed in a plastic bag such that a limited exchange of air is allowed, leading to reduction in oxygen levels and build up of carbon dioxide takes place. It normally takes 3-4 days for the system to stabilize. MAP of bananas slows down the respiration and delays the ripening thereby extending the green life. Studies using mature banana fruits of cultivar Kalibabu, showed extended shelf life of up to seven days when packed in polythene bags. The film thickness, permeability and ventilation plays an important role in maintenance of green colour. Mature green Embul banana fruits showed no change in peel colour even after 15 days when stored at 26±3 °C in sealed polyethylene bags, while the same fruits packed in perforated or open bags showed full yellow colour development by ninth day. The MAP is more effective only when combined with low temperature storage. Under ambient conditions occurrence of 'green soft' disorder and carbon dioxide injury are common in long term MAP storage of banana.

Table 5. Packaging Specifications for Export of Banana Fruits from India

Minimum Specifications	*5 kg Capacity*	*7 kg Capacity*	*13/18 kg Capacity*
Material of Construction	Corrugated Fibre Board	Corrugated Fibre Board	Corrugated Fibre Board
External Dimensions (mm)	333 x 300 x 200	333 x 390 x 250	545 x 330 x 255 (13 kg) 750 x 300 x 200 (18 kg)
Style of Box**	Telescopic- 0300	Telescopic - 0300	TELESCOPIC
No. of Plies	Lid :3 Tray : 3	Lid :3 Tray : 3	Lid :3 Tray : 5
Type of Flutes	Lid: B (Narrow) Tray: B/B (Narrow/ Narrow)	Lid :B (Narrow) Tray: B (Narrow)	Lid :B (Narrow) Tray : B/B (Narrow/ Narrow)
Direction of Flutes	Vertical	Vertical	Vertical
Grammage (g/m²)	Lid: 250/150/150 Tray: 150/150/150	Lid : 250/150/150 Tray : 250/150/150	Lid : 250/150/150 Tray : 250/120/120/120/120
Burst Factor of Paper (Kraft) (Minimum)	20.00	20.00	20.00
Bursting Strength of Board (Kg/ cm²) (Minimum)	Lid :11.0 Tray : 11.0	Lid :11.0 Tray : 11.0	Lid :11.0 Tray : 14.0
No. of Pieces per Box	Not more than two	Not more than two	Not more than two
Manufacturer's Joint	By gluing	By gluing	By gluing
Compression Strength of Box (Kgf) (Minimum)	350	450	500
COBB (30 Minutes)*** (g/m²) (Maximum)	60	60	60
No. of Ventilation Holes	16.0	16.0	16.0
Diameter (mm) of ventilation holes	20.0	20.0	20.0
Position of Holes			
Two Sides Each	3	3	3
Two Ends Each	1	1	1
Top and Bottom Each	4	4	4

Source: Specifications developed by IIP for APEDA.

6.9 Transportation

Bananas are transported by road, rail and ship. In India for domestic trade they are transported naked in trucks. The bunches are transported from field to the truck (parked on nearest road) on head loads. Some times baskets made of locally available material are used to take bunches from field to the collection centre or truck. Based on the volume of harvest and destination of market either a small three wheeler automobile or four wheeler mini-truck or normal 10 tonne capacity trucks are used. The fruits are transported to the nearest or long distance market in one of these vehicles. After unloading in the market for auctioning, the whole bunches are sometimes again packed in banana leaves to prevent damage and are re-loaded into trucks for distant markets by traders. These trucks are not normally refrigerated. The trucks take the produce up to 1500 – 2000 kilometers by road. Another method of transportation is by train wagons. The green banana bunches are brought from different localities to the identified wagon loading centres, where they are loaded into railway wagons. The green banana leaves are used as padding inside the wagons. Based on the dimensions of the wagon, each of the wagons holds about 5 to 20 tonnes of produce. Containers in ships transport the export consignments of banana. These containers are called controlled atmosphere reefer containers and refrigerated reefer containers. The capacity of these containers varies from 27.5 to 28 tonnes.

A green banana starts ripening without any external trigger based on the conditions of harvesting, handling and transportation with no visible indicators for rest of the green life. Fruits with minimum shelf-life are not detected during handling in the transportation network; however they are disposed-off early at the store/ consumer level prior to reach of the expected storage period. Since quality inspection at each step of supply chain is practically impossible, hence we can depend on biological or/and thermodynamic models for continuous and remote supervision of temperature abuse as well as other risks causing hidden transportation losses. Thus, the consignment handler would know the expected quality of foods inside the container well in advance to its arrival and opening at the destination market. Accordingly, the products/ consignments with limiting shelf-life could be sent to adjacent markets or to shops with greater turn-over rates (Jadermann *et al.*, 2015).

7.0 Domestic Marketing and Export

Almost 99 per cent of bananas produced are consumed domestically while only around 56,000 tonnes are exported. In domestic trade there are mainly three marketing channels in operation.

i) Farmer ⟶ Cooperative societies ⟶ Traders (distant market) ⟶ Retailer ⟶ Consumer

ii) Farmer ⟶ Commission Agent ⟶ Trader (distant market) ⟶ Retailer ⟶ Consumer

iii) Farmer ⟶ Commission Agent ⟶ Local Retailer ⟶ Consumer

After harvesting the bunches are brought near the truck, where very small and damaged fruits/ hands are removed, bunches weighed and loaded into the trucks with leaves as cushioning material on the sides, at the bottom and between the bunches. Only good grade (A and B) of bunches are sent to distant market and the left over (C grade) are sold to the local retailers. While in states like Maharashtra, which have farmers' cooperative societies, majority of farmers sell the produce to these cooperative societies and rest to commission agents. The commission agents mediate the sales of bananas to distant traders and retailers. The market traders facilitate the ripening of fruits. In some states like Madhya Pradesh the farmers sell the produce to contractors who in turn sell these to the local retailers or commission agent or distant traders. The retailers procure the bunches or hands which are then subjected to ripening treatment. During the course of retailing, the fruits ripe completely.

In export trade the exporters procure the produce from the farmers directly either with or without a preharvest agreement. The fruit after harvest are subjected to series of operations like dehanding, postharvest treatments like washing, fungicidal treatment, sorting, grading packaging and precooling. The precooled fruits in boxes are loaded into the refrigerated reefer containers and then the containers are transported to shipping port. At the destination the fruits are subject to ripening treatment and then retailed. Protocol for export of banana is mentioned in Figure 2.

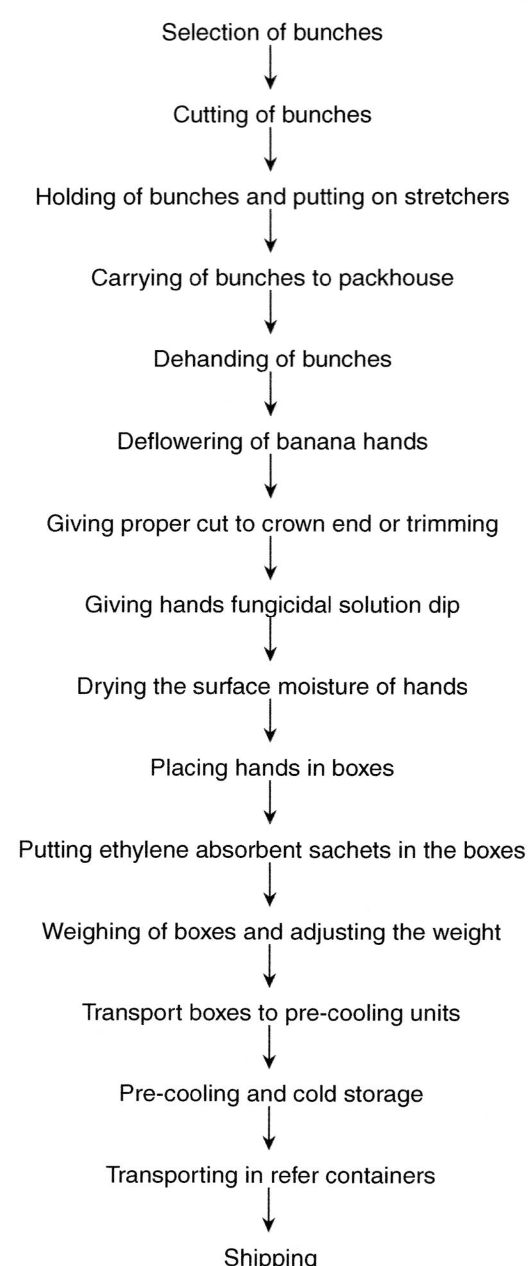

Figure 2. Flow Chart of Postharvest Operations for Export of Banana.

7.1 Codex Standards for Export

The Codex standards for export are general in nature and often the specific requirements in terms of physical quality and the permissible chemical residue varies from one country to another. Based on the importing country's' specifications these requirements may vary. A premium quality banana:

i) Must be whole, firm, sound, clean, practically free from any visible foreign matter and free from decay

ii) Should have an adequate finger length and diameter with no excess curvature

iii) Have the desired uniform bright yellow colour and sensory attributes of flavour and aroma upon ripening

iv) Free from malformation or abnormal curvature of the finger

v) Practically free from bruises

vi) Practically free from pests and damage caused by pests

Table 6. Grade Specification of Banana

Grade Designation	*Grade Requirements*	*Grade Tolerance*
Extra Class	Bananas must be of superior quality. Characteristics of the variety and/or commercial type. The fingers must be free of defects, with the exception of very slight superficial defects, provided these do not affect the general appearance of the produce, quality, the keeping quality and presentation in the package.	Five per cent by number or weight of bananas not satisfying the requirements of the grade, but meeting those for Class I grade or, exceptionally, coming within the tolerances for that class.
Class I	Bananas shall be of good quality. Characteristics of the variety and/or commercial type. The following slight defects of the fingers, however, may be allowed, provided these do not affect the general appearance of the produce, quality, the keeping quality and presentation in the package. slight defects in shape and colour; slight defects due to rubbing and other superficial defects not exceeding 2 sq.cm of the total surface area.	Ten per cent by number or weight of bananas not satisfying the requirements of the grade class, but meeting those of Class II or, exceptionally, coming within the tolerances of that class.
Class II	This includes bananas which do not qualify for inclusion in the higher classes, but satisfy the minimum requirements. The following defects may be there, provided the bananas retain their essential characteristics as regards the quality, the keeping quality and presentation. defects in shape and colour provided the product remains the normal characteristics of bananas; skin defects due to scrapping, scabs, rubbing, blemishes or other causes not exceeding 4 sq.cm. of the total surface area; The defects must not affect the flesh of the fruit.	Ten per cent by number or weight of bananas satisfying neither the requirements of the class nor the minimum requirements, with the exception of the produce affected by rotting, major imperfections or any other deterioration rendering it unfit for consumption.

vii) With the stalk intact, without bending, fungal damage or desiccation

viii) With pistils removed

ix) Free from damage caused by low temperature

x) Free from any strange smell

Grade specifications and tolerance limits within each grade are given in Table 6.

7.2 Domestic Marketing and Export

Bananas are among the most widely consumed and globally traded foods after rice, wheat and maize. Total world exports has raised by 26.67 % since 2008 and accounted for 23.18 million MT during 2017.Two thirds of the exports were generated by only five countries *viz.*, India, China, Philippines, Costa Rica and Gautemala. The top three producing countries do not contribute much to export trade while Costa Rica and Guatemala along with Ecuador and Colombia contributes 66 per cent to total world exports. Among the top ten producers only the Philippines has a consistent position both as a producer and exporter (9 per cent) from Asia (Tables 7 and 8). As far as import is concerned, the United States, the European Union, Russia, German, China, The Netherlands and Japan during 2020 accounted for about 59.28 per cent of the total world's fresh banana imports. The exports of conventional bananas to developed countries is on rise.

Bananas are among the most widely consumed and globally traded foods after rice, wheat and maize. Total world exports has raised by 26.67 per cent since 2008 and amounted for 23.18 million metric tonnes during 2017. Two thirds of the exports were generated by only five countries viz., India, China, Philippines, Costa Rica and Gautemala. The top three producing countries do not contribute much to export trade while Costa Rica and Guatemala along with Ecuador and Colombia contributes 66 per cent to total world exports. Among the top ten producers only the Philippines has a consistent position both as a producer and exporter (9 per cent) from Asia (Tables 7 and 8). As far as import is concerned, the United States, Russia, German, Japan, China and Netherlands during 2020 accounted for about 59.28 per cent of the total world's fresh banana imports. The exports of conventional bananas to developed countries are getting settled, while the export of organic bananas to these countries is on raise. Organic fruits command a significant premium price in the foreign market; on average, during the period 2013–2017, organic bananas at the US retail level commanded a premium of $0.29/ lb over the price paid for conventional bananas (USDA-AMS 2018).

Though India is the largest producer of banana with 20 per cent of the global production, its contribution to global banana exports is only 0.3 per cent. The Indian banana exports are mainly to UAE, Saudi Arabia, Iran, Kuwait, Bahrain, Qatar, Oman, Nepal and Maldives, bringing in approximately Rs.1000 million (~ US$ 20 million). The export from India has been consistently increasing for the last ten years. In year 2000 India

Table 7. Trends in Exports of Banana from Far East during the Last 10 Years

Country	*Year/Quantity Exported ('000 tonnes)*									
	2011	*2012*	*2013*	*2014*	*2015*	*2016*	*2017*	*2018*	*2019*	*2020*
China	10	8	11	4	8	8	16	19	23	20
Malaysia	23	18	19	21	22	25	26	24	24	28
Pakistan	59	30	85	43	53	50	46	65	116	102
Philippines	2047	2648	3293	3680	1795	1734	2726	3126	4403	3808
Thailand	25	22	23	32	34	25	29	39	24	16
Vietnam	8	3	8	22	20	90	90	127	255	255
India	40	55	28	54	80	112	90	116	174	212
Indonesia	2	-	1	2	3	19	2	15	17	8
Far East Total	2214	2784	3468	3858	2015	2063	3025	3531	5036	4449

Source: http://www.fao.org/economic/est/est-commodities/bananas/banana-exports/en/

exported 8,600 tonnes of banana, which increased to 14,400 tonnes in 2005 and 58,100 tonnes in 2010. With the end of "Banana War" in 2012 the world markets have opened up without any preference, hence the scope for increase in export of banana from Asian countries like Philippines and India is great. Besides, our neighbor, China, is emerging as one of the importers of banana.

Table 8. Top Ten Exporters of the Banana/ Plantains in the World (2019)

Country	*Exports Worth (Billion $)*	*Percentage of World Exports*
Ecuador	3.60	29.0
Philippines	1.60	12.80
CostaRica	1.08	8.69
Colombia	0.99	7.93
Guatemala	0.96	7.65
Others	4.27	66.07
Total	12.5	100

Source: http://www.fao.org/economic/est/est-commodities/bananas/banana-exports/en/

Minimization of postharvest losses in banana is one of the crucial factors for achieving sustainability in banana cultivation. These losses could be curtailed to certain extent by following specific pre-harvest as well as postharvest practices and it is imperative to control each step of postharvest handling from farmers' field to consumer. Hence, adequate pre-harvest treatments, as well as the correct stage of maturity, proper harvesting method, transportation, washing, cleaning, grading, packing, cold storage, ripening process, air and relative humidity during storage or in the ripening chambers, and efficient marketing are crucial in curtailing postharvest losses

8.0 Conclusion

Banana is one of the most important fruit that needs appropriate attention right from harvesting, handling, packaging, ripening, and storage for both domestic marketing and export. Spotless golden colour banana is the most desirable character in western countries and it fetches premium price. Here lies the importance of proper harvesting and postharvest handling. Green banana at the time of harvest contains tannins and polyphenol oxidase enzymes in its exudates. These exudates come out on the skin on very little bruising during harvesting and postharvest handling operations. These tannins get oxidized in presence of oxygen and polyphenol oxidase enzyme that makes the black spots on the surface of the skin. As long as the banana skin remains green these spots are not much visible. However, on ripening these spots are seen conspicuously leading to poor quality. Therefore, postharvest handling plays a crucial role in quality assurance of banana.

8.0 Important References

Anonymous. 2007. Annual Report- (2006-07), National Research Centre for Banana (ICAR), Trichy, Tamil Nadu, India.

Anonymous. 2009a. Final Project Report (RPF-III). Studies on handling, storage and processing of banana. National Research Centre for Banana (ICAR), Trichy, Tamil Nadu. India.

Anonymous. 2009b. Final Project Report (RPF-III). Standardization of storage conditions of banana. National Research Centre for Banana (ICAR), Trichy, Tamil Nadu. India.

Basak, P., and Akter, B. 2018. Comprehensive investigation on banana ripening by using natural agents and chemicals. *Ann. Food Sci. Technol.* **19**: 307-15.

Blackbourn, H., Jeger, M. and John, P. 1990. Inhibition of degreening in the peel of bananas ripened at tropical temperatures. III. Changes in plastid ultrastructure and chlorophyll–protein complexes accompanying ripening in bananas and plantains. *Ann. Appl. Biol.*, **117**: 147–61.

Chaiwong, S., Saengrayap, R., Ogunsua, J. M., Kitazawa, H., and Prahsarn, C. 2021. Performance of different bunch cover materials to improve quality of cavendish banana cultivated during winter and summer in Thailand. *Agronomy*, **11**: 610.

Chandar, M. M., and Smith, A. A. 2019. Artificial Ripening of Banana-An Overview. *Res. J. Pharm Technol.* **12**: 5101-06.

Doshi, J. S. and Sutar, R.F. 2010. Studies on Precooling and Storage of Banana for Extension of Shelf Life. *J. Agr. Eng. Res.* **47**: 14-19.

Esguerra, E., Del Carmen, D., Reyes, R. D., and Lualhati, R. A. 2017. Vacuum Packaging Controlled Crown Rot of Organically-Grown Balangon (Musa acuminata AAA Group) Banana. *Horticulturae*, **3**: 14.

Evelin Mary, A. 1997. Preharvest treatments, packaging and storage studies in banana cvs. Rasthali (AAB) and Sanna Chenkadali (AA). *M.Sc Thesis*, Tamil Nadu Agricultural Univesity, Coimbatore, Tamil Nadu.

Facundo, Heliofabia Virginia de Vasconcelos; Deborah dos Santos Garruti; Carlos Tadeu dos Santos Dias; Beatriz Rosana Cordenunsi, and Franco Maria Lajolo. 2012. Influence of different banana cultivars on volatile compounds during ripening in cold storage. *Food Res. Int.* **49**: 626–33.

Floros, J.D. 1990. Controlled and modified atmospheres in food packaging and storage. *Chem. Eng. Prog.*, **26**: 25-32.

Hörtensteiner, S. 2006. Chlorophyll degradation during senescence. *Ann. Rev. Plant Biol.* **57**: 55–77.

Jedermann, R., Praeger, U., Geyer, M., Moehrke, A., and Lang, W. 2015. The intelligent container for banana transport supervision and ripening. *Acta Horticulturae*, **1091**: 213-20.

Kudachikar, V. B., Kulkarni, S. G., and Prakash, M. K. 2011. Effect of modified atmosphere packaging on quality and shelf life of 'Robusta' banana (*Musa* sp.) stored at low temperature. *J. Food Sci. Technol.* **48**: 319-24.

Leibibet,D., Metzidakis, I and D.Gerasopoulos.1995. Effects of storage temperatures on the ripening response of banana fruit grown in the mild winter climate of Crete. *Acta Hortic.*, **379**: 521-26.

Li, X. P., Shi, J.Y., Zhu, X. Y., Wang, J. H., Yuan, Z. X., Luo, J., Liu, T., Wang, R., Rao, S. and Chen, W. 2013. Pre-harvest and postharvest application of benzothiadiazole for controlling anthracnose and extending shelf life of harvested banana. *Int. J. Agric. Biol. Eng.* **6**: 76-83.

Liu, F.W. 1976. Storing ethylene pretreated bananas in controlled atmosphere and hypobaric air. *J. Amer. Soc. Hortic. Sci.* **101**: 198-01.

Liu, J., Li, F., Liang, L., Jiang, Y., and Chen, J. 2019. Fibroin delays chilling injury of postharvest banana fruit via enhanced antioxidant capability during cold storage. *Metabolites*, **9**: 152.

Lyons, J.M. 1973. Chilling injury in plants. *Ann. Rev. Plant Physiol.*, **24**: 445-66.

Marriott, J. 1980. Bananas-physiology and biochemistry of storage and ripening for optimum quality. *CRC Crit. Rev. Food Sci. Nutr.* **13**: 41-88.

Mebratie, M. A., Woldetsadik, K., Ayalew, A., and Haji, J. 2015. Comparative study of different banana ripening methods. *Sci. Technol. Arts Res. J.*, **4**: 32-38.

Mohapatra Debabandya, Sabyasachi Mishra and Namrata Sutar. 2010. Banana Postharvest Practices: Current Status and Future Prospects- A Review. *Agric. Rev.* **31**: 56-62.

Nanda, S.K., Vishwakarma, R.K.,H.V.L.Bathla, H.V.L., Anil Rai and Chandra, P. 2012. Harvest and Postharvest Losses of Major crops and livestock produce in India. All India Coordinated Project on Postharvest Technology, (ICAR), Ludhiana, pp. 69.

Narayana, C.K. and Mustaffa, M.M. 2006. Improved Postharvest Handling Technology in Banana. Tech. Bull No.15. National Research Centre for Banana, Trichy, Tamil Nadu.

Narayana, C.K. and Mustaffa, M.M.2007. Influence of maturity on shelf life and quality changes in banana during storage under ambient conditions. *Indian J. Hortic.* **64**: 12-16.

Narayana, C.K. and Singh, B.P. 1998. Effect of ameliorative treatments on reduction of chilling injury in guava fruits. *Prog. Hortic.* **30**: 148-52.

Narayana, C.K., Mustaffa, M.M. and Sathiamoorthy, S. 2002. Effect of low temperatures on chilling injury and quality changes in Pachanadan (Pome, AA group) banana. Paper presented in 'Global Conference on Banana and Plantain' held at Bangalore on 29-31st October 2002. Abstract No. P.P.09.5., pp. 208.

NHB Database. 2011. Indian Horticulture Database-2011. National Horticulture Board, New Delhi, pp.254.

Palmer, J.K. 1971. The Banana. **In**: *The Biochemistry of Fruits and their Products*, A.C. Hulme(Ed.), Vol.2, Academic Press, London, 65-101.

Pantastico, E.R.B. 1975. Postharvest physiology, handling and utilization of tropical and subtropical fruits and vegetables. The AVI Publishing Company Inc, Connecticut, USA., pp. 314-337.

Paunescu, C. and Bratucu, Gh.2010. Research regarding the storage of fruits and vegetables in controlled atmosphere. *Bulletin of the Transilvania University of Brasov.*, **3**: 177-182.

Peacock, B.C.1980. Banana ripening-Effect of temperature on fruit quality. *Qld. J. Agric. Anim. Sci.*, **37**: 39-45.

Ringer, T., and Blanke, M. 2021. Non-invasive, real time in-situ techniques to determine the ripening stage of banana. *J. Food Meas. Charact.*, **1**: 12.

Robinson, J.C. and Human N.B. 1988. Forecasting of banana harvest ('Williams') in the subtropics using seasonal variation I bunch development rate and bunch mass. *Sci. Hortic.* **34**: 249-63.

Rodríguez-Zapata, L. C., y Gil, F. L. E., Cruz-Martínez, S., Talavera-May, C. R., Contreras-Marin, F., Fuentes, G., Sauri-Duch, E., and Santamaría, J. M. 2015. Preharvest foliar applications of glycine-betaine protects banana fruits from chilling injury during the postharvest stage. *Chem. Biol. Technol. Agric.*, **2**: 1-10.

Sarkar, S. 2015. Dehanding in improving fruit quality of banana. *Int. J. Bio-resour. Stress Manag.* **6**: 198-01.

Satyan, S., Scott, K.Y. and Graham, D. 1992. Storage of banana bunches in sealed polyethylene tubes. *J. Hortic. Sci.* **67**: 283-287.

Scott, K.J. and Soertini Gandanegara. 1974. Effect of temperature on the storage life of bananas held in polyethylene bags with ethylene absorbent. *Trop. Agric.* **51**: 23-26.

Sevillano, L., Sanchez-Ballesta M.T., Romojaro, F. and Flores, F.B. 2009.Physiological, hormonal and molecular mechanisms regulating chilling injury in horticultural species. Postharvest technologies applied to reduce its impact. *J.Sci. Food Agric.* **89**: 555–73.

Seymour, G.B. 1993. Banana. **In**: *Biochemistry of fruit ripening*. Seymour, G.B., Taylor, J.E., and Tucker, G.A. (Eds.), Chapman and Hall, 2-6 Boundary Row, London, Pp. 83-101.

Shantha Krishnamurthy and Krishnamurthy, S. 1989. Storage life and quality of Robusta banana in relation to their stage of maturity and storage temperature. *J. Food Sci. Tech.* **26**: 87-89.

Soltani, M., Alimardani, R. and Omid, M. 2010. Prediction of banana quality during ripening stage using capacitance sensing system. *AJCS*, **4**: 443-47.

Sonmezdag, A.S., Kelebek, H. and Selli, S. 2014. Comparison of the aroma and some physico-chemical properties of grand naine (*Musa acuminata*) banana as influenced by natural and ethylene-treated ripening. *J. Food Process. Preserv.* **38**: 2137–45.

Stover, R.H. and Simmonds, N.W.1987. Bananas. Longman Scientific and Technical. UK, pp. 269-273.

Tapre A.R. and Jain R.K. 2012. Study of advanced maturity stages of banana. *IJAERS*, **1**: 272-74.

Thompson, A.K. and Burden, O.J. 1995. Harvesting and fruit care. **In**: *Bananas and Plantains*. S. Gowen (Ed.), Chapman and Hall, London. pp. 403-432.

Tixier, P., Bugaud, C., Duguet, R., and Salmon, F. 2010. Effect of preharvest and postharvest application of calcium on banana green-life. *Fruits*, **65**: 201-08.

Venkateswarlu, A., and Satyanarayana, C. V. V. 2016. Studies on modified atmospheric packaging of banana. *Int. J. Proc. Post Harvest Technol.* **7**: 67-72.

Wang, Chien Y.I. 1993. Approaches to Reduce Chilling Injury of Fruits and Vegetables. *Hortic. Rev.* **15**: 63-95.

Wang, Y., Lu, W. J., Jiang, Y. M., Luo, Y. B., Jiang, W. B. and Joyce, D. 2006. Expression of ethylene-related expansin genes in cool-stored ripening banana fruit. *Plant Sci.* **170**, 962–67.

Wu, Q., Li, T., Chen, X., Wen, L., Yun, Z., and Jiang, Y. 2018. Sodium dichloroisocyanurate delays ripening and senescence of banana fruit during storage. *Chem. Cen. J.* **12**: 1-11.

Zude, M. 2003. Non-destructive prediction of banana fruit quality using VIS/NIR spectroscopy. *Fruits*, **58**: 135-142.

Chapter 13

Ber

Sunil Pareek

Associate Professor, Department of Agriculture and Environmental Science, NIFTEM, Plot No. 97, Sector 56, HSIIDC Industrial Estate, Kundli, District-Sonepat – 131 028, Haryana, India

E-mail: sunil.pareek@niftem.ac.in

1.0 Introduction

Ber (*Ziziphus mauritiana* Lamk, Rhamnaceae) is a tropical/subtropical fruit native to the northern hemisphere. *Ziziphus jujube* Mill and *Z. mauritiana* Lamk. are the most important species in terms of distribution and economic significance. The former is native to China, where it is known as the Chinese date or Chinese jujube. It is the less tropical of the two species, tolerating temperatures as low as –29 °C, and is deciduous. *Ziziphus mauritiana* is semi-deciduous or evergreen, and is commercially most important in India, where it is known as Indian jujube or ber. It is also known as desert apple or Indian plum. Ber is considered an underutilized fruit crop in semi-arid regions of the world and can be successfully cultivated in the marginal ecosystem of the subtropics and tropics. Ber can be grown on most saline soils and water conditions such as near Dead Sea in Israel. The fruit is a drupe, globose to ovoid in shape, up to 6 4 cm in size; skin smooth or rough, glossy, thin but tough, yellowish to reddish or blackish; flesh white, crisp, juicy, sub acid to sweet, becoming mealy with fully ripe fruit, whereas, Chinese ber is brownish to reddish in colour. It is richer than apple and mango in vitamin C, protein and minerals, and contains higher phosphorus and iron than orange. Ber is generally eaten as dessert fruit, but has good potential for making processed products such as candy, squash, powder, and syrup.

The storage life of ber fruit is very short and its high perishability is a problem. At ambient temperature conditions (15-25 °C) a shelf-life of 2 to 4 days is common. Due to the glut of fruits in the local markets during peak season, a substantial quantity is wasted, resulting in heavy postharvest losses. This chapter provides information on maturity indices, ripening, pre and postharvest treatments to extend the postharvest life, and postharvest diseases, among other important postharvest information.

2.0 Status of Cultivation

Ber is cultivated in many countries of the world. India, Pakistan, Bangladesh, Sri Lanka, Afghanistan, Iran, Iraq, Israel, Zimbabwe, South Africa, many other African countries and some South East Asian countries are growing ber fruit. Whereas, Chinese ber is cultivated on subtropical to temperate climate such as China, USA, Spain, Vietnam *etc.* China produces about 90 per cent of the world's ber for food and pharmaceutical application. Chinese production itself rose from 1.1 million tonnes in 1998 to 2.0 million tonnes in 2008.

In India, ber fruit covers an area of 51,000 hectares with production of 5,42,000 tonnes [Table 1 as per the 3rd revised estimate of 2019-20 (NHB)]. Gujarat (12,160 ha) occupied the highest area under ber cultivation followed by Jammu and Kashmir (7,900 ha), Chattisgarh (4,720 ha), Haryana (4,260 ha), and Madhya Pradesh (4,150 ha in 2013). Gujarat leads in total ber production in the country with 1,28,630 tonnes followed by Chattisgarh (1,16,920 t), Haryana (44,580 t), Madhya Pradesh (44,000 t), and Punjab (26,600 t). Gujarat (29.37 per cent) and Chattisgarh (26.69 per cent) contribute more than half (56.06 per cent) of country's ber production. However, the productivity is exceptionally high in Karnataka (28.78 t ha^{-1}) and Chattisgarh (24.77 t ha^{-1}). The average productivity of ber in the country is 10.81 t ha^{-1}. Punjab,

Madhya Pradesh, Gujarat, Haryana, Rajasthan, joint Andhra Pradesh are other states with the average productivity of 16.48, 10.60, 10.58, 10.46, 10.31, and 10.00 t ha^{-1}, respectively.

Significant increase in the percentage share in production in 2015-16 was observed in Andhra Pradesh (from 0.89 to 5.60 per cent) whereas the increase was found to be marginal in M.P. (from 10.05 to 13.13 per cent). Similarly, a sharp decline in the per cent share in production during the same period was observed in Chattisgarh from 26.69 to 16.67per cent. However, the variation in the per cent share in total production in other states remained marginal (Table 1).

Table 1. Area, Production and Productivity of Ber in India in 2012-13 and Change in Production Shares in 2013-14 and 2015-16

State	*Area (hectares)*	*Production (tonnes)*	*Productivity (t ha^{-1})*	*Share in per cent (2012-13)*	*Share in per cent (2015-16)*
Andhra Pradesh	390	3,900	10.00	0.89	5.60
Bihar	470	2,780	5.91	0.63	-
Chattisgarh	4,720	1,16,920	24.77	26.69	16.67
Gujarat	12,160	1,28,630	10.58	29.37	28.34
Haryana	4,260	44,580	10.46	10.18	9.30
Himachal Pradesh	40	10	0.25	0.002	-
Jammu and Kashmir	7,900	13,200	1.67	3.01	2.54
Jharkhand	1,110	1,210	1.09	0.28	0.67
Karnataka	900	25,900	28.78	5.91	3.00
Madhya Pradesh	4,150	44,000	10.60	10.05	13.13
Nagaland	20	120	6.00	0.03	0.20
Punjab	1,830	30,160	16.48	6.89	6.60
Rajasthan	2,580	26,600	10.31	6.07	1.59
Total	40,530	4,38,010	10.81	100	

Source: Indian Horticulture Database, NHB, 2013, 2015-16

3.0 Major Commercial Varieties

'Gola', 'Seb', 'Umran', 'Mundia', 'Jogia', 'Tikadi', 'Kaithli', 'Goma Kirti', 'Kathaphal', 'Thar Sevika', and 'Thar Bhubaraj' are important cultivars developed at various institutes in India (Figure 1). According to the fruit maturation period cultivars can be classified as early-, mid-, and late- cultivars. The salient features of some of these cultivars are described here:

"Apple ber" fruit is another variety from Thailand and it is also known as the Indian jujube or Chinese date is gaining its popularity for commercial cultivation in India. This variety is becoming popular in India due to its precocity in bearing, fruit size and high yield (even up to 200 kg per tree depending on the age of the tree).

The taste of this Apple ber fruit is Sweet, Crispy and Juicy. Apple ber cultivated areas in India are West Bengal, Telangana, Andhra Pradesh, and Maharashtra.

a) Gola

An early-maturing cultivar, it starts bearing fruit in the first week of January. Fruits are very attractive, roundish in shape and golden yellow in colour with an average weight of 20 g. Flesh white and very juicy, semi-soft with delicious taste. Fruit pulp has a TSS of 17 to 19 per cent and 0.46 to 0.51 per cent acidity.

b) Goma Kirti

An early maturing clonal selection from 'Umran' cultivar. Physicochemical characters and keeping quality of fruits are similar to those in 'Umran'. Fruit maturity 3 weeks earlier.

c) Jogia

Fruits with light purple tinge when unripe, but still edible and have an extended harvest period up to early February. Skin surface with coarse ridges, greenish yellow in colour. Flesh white, soft, juicy and sweet with TSS of 19 per cent. Pulp to stone ratio 14.

d) Kaithli

Mid-season cultivar, ripens during March. Fruit medium in size with average weight of 18 g, oval in shape with tapering apex. Fruit pulp quite soft, with TSS of 18 per cent and 0.5 per cent acidity. Fruit does not withstand transportation and has poor keeping quality.

e) Kathaphal

Late ripening cultivar of ber with small to medium size fruits. At maturity, the fruits are green on one side while the other side develops a reddish yellow tinge. Average fruit weighs 10 g with high TSS (23 per cent) and an acidity of about 0.77 per cent.

f) Mundia

An early cultivar that matures during mid-January. Fruits large, juicy and bell shaped with average weight of 40 g. Skin yellowish green with smooth depressions. Flesh white and soft with a TSS of 20 per cent and pulp to stone ratio of 23.

g) Seb

A popular mid season cultivar grown in throughout North and West India. Fruits mature in mid January to mid February. Fruits round, resembling crab apple, styler end round with depression at the centre, stem end broad, deeply grooved, skin yellowish green with brown specks, white pulp, hard, size around 3.6 x 3.3 cm, weight 24 to 25 g, pulp : stone ratio 14 to 15 and TSS is 16 to 18 per cent.

h) Thar Bhubharaj

Fruits ready for harvesting during last week of

Figure 1. Important Varieties of Ber Grown in India

December-first week of January. Fruits very juicy, sweet with a TSS content of 22-23 per cent.

i) Thar Sevika

'Thar Sevika' released from Central Institute for Arid Horticulture, Bikaner, Rajasthan is developed by hybridization from a cross between Seb x Katha. It is 10–15 days early in maturity. Fruits juicy with high TSS (22 per cent), high ascorbic acid (90 mg 100g^{-1}), total sugar (5.0 mg 100g^{-1}) and protein (16 mg g^{-1}). Fruits do not cause throat soar. Average fruit yield is 30-32 kg tree^{-1}.

j) Tikadi

Late maturing ripens in February and March. Fruits small, weighing on average 10 g, edible, with a creamy soft flesh, when the skin turns to a red colour. Ripe fruit has a high TSS (25 per cent) with a large stone.

k) Umran

Fruits large sized, average between 30 and 80 g, oval in shape with roundish apex. Attractive golden yellow colour which later turns into chocolate brown at full maturity. Ripens during mid-March to mid- April. Fruits are sweet with 19 per cent TSS and 0.12 per cent acidity, pleasant flavour and excellent dessert quality and good keeping quality suitable for long transportation. The main reason for its popularity is long shelf life (10 to 15 days) and excellent organoleptic qualities.

l) Apple Ber

Apple Ber is a variety from Thailand. It looks similar to green apple. The fruits are very sweet and the average fruit weight varies from 100-200 g. This variety can tolerate very high temperature and aridity since it undergoes cessation of growth, dormancy and leaf fall.

A list of important early, mid season and late varieties grown in different states of India is given in Table 2.

Table 2. Commercial Cultivars of Ber Grown in Different States of India

State	*Early Season*	*Mid Season*	*Late Season*
Andhra Pradesh	Gola	Dudhia	Umran
Gujarat	Gola	Mehrun	Ajmeri, Randeri
Haryana	Gola, Seb, Safeda	Kaithli, Sanaur-5, Banarasi Kadaka	Umran
Karnataka	Gola	–	Umran
Maharashtra	Gola, Badami, Shamber	Mehrun, Darakhi, Kharki	Umran
Punjab	Gola, Seb, Safeda	Dandan, Kaithli, Sanaur-2	ZG-2, ZG-3, Umran
Rajasthan	Gola	Mundia, Seb, Jogia, Ilaychi	Tikadi, Umran, Maharwali, Bagwadi
Tamil Nadu	Gola	–	Umran
Uttar Pradesh	Delhi Gola, Narma, Banarasi Peondi	Banarasi Kadaka, Mundia, Sasni	Aliganj, Katha

4.0 Nature and Causes of Postharvest Losses

Ber is a hardy fruit crop of desert and subtropics. It flourish well even in most marginal ecosystems of the world. However, fruit drop and size reduction is a problem with water stress. Therefore, 2-3 irrigations after fruit set are crucial for postharvest quality and size of fruit. Likewise, pruning is most important in India. Pruning intensity up to 5 secondary branches or 50 per cent is desirable for better flowering and fruiting and ultimately for good quality fruits. Recently, in North India, chilling injury on tree due to the frost is also seen and adversely affects the postharvest life, appearance, ripening and overall quality of fruits (Figure 2).

4.1 Insect Pests

Fruit fly and fruit borer are important pests of ber which affect the postharvest quality of fruits. Fruit fly (*Carpomyia vesuviana*) damages the fruits as much as 80 per cent of the crop under severe infestation. It is the most serious pest of ber. Infestation starts with the onset of fruit setting. The adult female lays eggs single by inserting its ovipositor in the young developing fruit. After 2 to 5 days, the larvae hatch, start feeding on the pulp and make galleries in it. Infested fruits become deformed. A large number of such fruits drop off. Cultivars 'Tikadi' and 'Umran' are found to be tolerant to fruit fly. Damage by larvae of the fruit borer moth has also been observed mainly in southern and western India. The reddish larvae bore into the fruit and feed on the pulp (Figure 3).

Figure 2. Preharvest Chilling Injury (Frost Damage) Deteriorate Fruit Quality

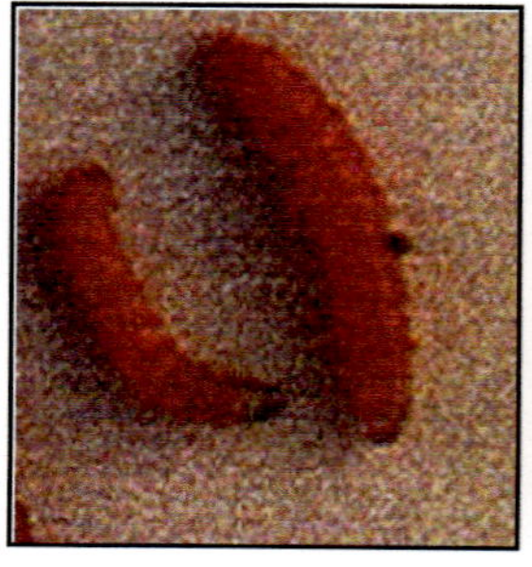

Fruit Borer

Fruit Borer Infestation (Outer)

Fruit Borer Infested Ber (Inner)

Figure 3. Fruit Borer Infestation in Ber

Ber is a climacteric fruit and thereby ethylene production is continued even after harvest from the tree. High respiration as well as ethylene evolution rate observed during postharvest. Due to high temperature during last February to March-April water is occurs from fruits and weight loss as well as shriveling (Figure 4) is noticed in fruits, which deteriorate the quality. It is soft fruit and damage due to faulty packaging, and transportation. Compression bruising is commonly seen in market on ber fruits. Severe bruising is one of the causes of rotting also.

Figure 4. Shriveling in Ber Fruit Due to Water Loss

Some of the postharvest problems of ber fruit include variation in ripening time, failure of immature fruit to ripen after harvest, and poor retention of mature fruit on the tree. Ethephon is reported to enhance ripening and improve ber fruit quality. However, the response to ethephon has been variable depending on concentration, cultivar, time of application, and temperature during the ripening period.

4.2 Diseases and Disorders

Ripe fruits are susceptible to attacks by various pathogenic fungi and bacteria responsible for postharvest diseases (Figure 5). Fruit may be exposed to various decay-causing microflora during packaging, storage, and transport. *Botryodiplodia theobromae Phoma hissarensis, and Ulocladium chartarum* caused decay losses more frequently in packages. When fruits are weakened by senescence and chilling injury, they are attacked by several pathogens that cause fruit rot. The *Alternaria* soft rot starts as light brown lesions at the site of a cut or incision, and later changes to dark brown or black colour with the increase in severity of the symptoms. The outer skin of the spot remains intact but the flesh inside turns dark brown, and appears visibly macerated and water soaked. As the time of incubation advances rotting penetrates inside the tissue up to the seed in the center and quickly encircles it, emitting a foul smell. Bavistin and Difolatan could be recommended as fruit dips at 500 and 1000 ppm, respectively. Integration of *Trichoderma viride* isolate T.v-CIAH240 with several fungicides was very efficient in the control of *A. alternata*. The latent infection by a few isolates of *Trichoderma* spp. in over ripened fruit was suppressed completely when fungicides were added in combination with a bioagent even at low concentration. *A. alternata* can be managed by combined application of *T. viride* (T.v-CIAH240) with an inexpensive fungicide like mancozeb, alcidine, and thiophanate methyl at low doses of 50-100 $\mu g\ g^{-1}$.

Figure 5. Postharvest Rot in Ber Fruits

Blue mold rot, caused by *Penicillium expansum*, is one of the most important diseases in ber fruit. Application of 5μM brassinosteroids effectively reduced incidence of blue mold rot by stimulating defence related enzymes, such as phenylalanine ammonia-lyase, polyphenol oxidase, catalase, and super oxide dismutase.

4.3 Fruit Cracking

Considerable reduction in yield and quality of ber fruit occurs due to preharvest physiological disorders like fruit cracking, fruit skin discolouration, and fruit drop. These disorders pose serious problems during fruit development. Preliminary studies on the mechanism of fruit cracking indicated that susceptibility increases at maturity, which appears to be due to accumulation of soluble sugars, a decrease in water potential, and a decline in peel toughness.

4.4 Chilling Injury

Being a subtropical fruit, ber is very susceptible to low temperature injury. The common physiological disorder in ber fruit is pitting, discolouration, brown lesions on surface, water soaked lesions, and uneven ripening (Figure 6). Chilling injury symptoms are dependent upon the storage temperature, as well as the duration of these temperatures. These symptoms were greater in fruit held at 0 °C than those at 4 °C; in addition they become more serious as the time of the storage at 0-4 °C was extended. Surface pitting appeared after 5, 7, and 15 days at 0, 4 and 10 °C.

5.0 Postharvest Management

5.1 Preharvest Treatments

Preharvest application of 1 per cent $CaNO_3$ at colour turning stage and 1.7 gL^{-1} $CaCl_2$ with 1 per cent Teepol

Figure 7. Maturity and Ripening Stages of Ber

10 days before harvest increased shelf life at room temperature, reduced weight loss and delayed colour development during storage. Preharvest application (at colour break stage) of $CaCl_2$ (2 per cent) can retain storage life of ber fruit cv. Umran for 20 days at 3- 5 °C with 85-90 per cent. Preharvest sprays with 500 ppm Captan was more pronounced in reducing weight loss and that of TBZ in reducing decay during storage.

Figure 6. Chilling Injury Symptoms in Ber Fruits after 28 Days Storage at 6 °C

5.2 Maturity Indices

The time required for ber fruit to develop from fruit set to maturity is dependent on cultivar and location. The development of fruit skin colour is one of the most reliable maturity indices (Figure 7). Fruits are harvested at the mature golden yellow stage in cv. Umran and Kaithli, and at the mature green stage in cv. Mallasey and Bambawi. In general, the harvest period in North India is mid January to March for cv. Gola, mid February to last week of March for cv. Seb, mid March to last week of March for cv. Kaithli, and March to April for cv. Umran. Availability periods of ber fruit in India vary from December to April.

5.3 Harvesting

The most common method of harvesting ber fruits is by manually shaking or beating the tree branches to cause the ripe or mature fruits to fall to the ground. Harvesting can also be carried out by mechanical shaking of the tree. Neither of these methods of harvesting is very satisfactory as they cause considerable damage to the fruit and the harvest includes a mixture of mature and immature fruits. Other more suitable methods of harvest include plucking the fruit retaining

Table 4. Grading Standards for Dry Red Jujube

Grade	*Shape and Size*	*Quality*	*Mechanical Injury and Blemish*	*Moisture Content*
Top	Uniform size (≤300 fruits kg^{-1}) and shape, and cultivar characteristics	Plum flesh, bright red, dryness, impurity ≤0.5 per cent	No mildewed, serous part, immature, diseased and wormy fruits, skin crack and oiled skin spot fruit ≤3 per cent	≤28 per cent
1	Uniform size (≤360 fruits kg^{-1}) and shape, and cultivar characteristics	Plum flesh, bright red, dryness, impurity ≤0.5 per cent	No mildewed, serous part, immature, diseased and wormy fruits, skin crack and oiled skin spot fruit ≤5 per cent	≤28 per cent
2	Uniform size (≤420 fruits kg^{-1}) and shape, and cultivar characteristics	Plum flesh, bright red, dryness, impurity ≤0.5 per cent	No mildewed, serous part, immature. Diseased, wormy, skin crack, oiled skin spot or dried immature fruit ≤10 per cent and diseased and wormy fruit ≤5 per cent	≤28 per cent
3	Fruit with normal shape and cultivar characteristics	Fruit flesh uneven, fruit with uneven bright red ≤10 per cent, dryness, impurity ≤0.5 per cent	No mildewed fruit, serous part, diseased, skin crack, oiled skin spot or dried immature fruit ≤15 per cent and diseased and wormy fruit ≤5 per cent	≤28 per cent

Source: Sheng and Shen, 2011.

pedicel using a clipper (an iron hook attached to a long bamboo pole) and hand picking the individual fruit. Since ber fruits mature at different rates and do not ripen simultaneously, four to five or even seven pickings are needed to complete the harvest, especially for late maturing varieties such as Umran.

5.4 Grading

So far, proper grading has not been standardized in ber. Ber fruit may be graded according to maturity, size, shape and colour into three grades as given in Table 3.

Table 3. Grading Criteria for Ber Fruit

Grade	*Standard*
A	Shining yellow, large (> 35 mm) to medium size (25-35 mm) fruits of uniform size with no blemishes.
B	Uneven yellow or yellow red, large (> 35 mm) to medium (25-35 mm) fruits to uniform shape with some blemishes.
C	Red, large (> 35 mm) to small (< 25 mm) fruits. Uneven yellow, small (< 25 mm) fruits.

Source: Pareek and Yahia, 2013.

Grading can either be carried out manually or by passing through sieves of different mesh sizes. Grading standards for dry red jujube is different and given in Table 4.

5.5 Postharvest Treatments

Postharvest dipping of ber fruit in cold water reduced respiration, ethylene production and enzymatic activities, whereas hot water treatment controlled the development of pathogens, thereby prolonging the shelf-life and maintained the quality.

Hot water (40 °C) dipping of fruit has been reported to reduce water loss of fruit during storage. It also prevents development of surface mould. Postharvest hot water dipping of ber fruit cv. Umran at 50 °C for 5 min can significantly increase the shelf life and retain quality during 8 days storage. Postharvest dipping of fruits of cv. Gola in 1000 ppm $KMnO_4$ at colour turning stage prolonged the keeping quality up to 14 days at room temperature. It has been reported that exposing Chinese ber fruit to 100 ppm ethylene at 20 °C for 4 h or dipping in 2000 ppm aqueous solution ethephon for 2 min induced rapid and uniform ripening. Ethephon at 500 mg L^{-1} advanced the ripening by 6–7 days compared with the control.

5.6 Packaging

Ber can be packaged in small packages of 1–2 kg in perforated polyethylene bags, nylon net bags (Figure 8) and in cardboard cartons. Fruit are also packed in bulk packages of 10–20 kg in gunny bags, wooden or plywood boxes with holes or slits. For transportation, corrugated fibre board cartons holding about 5 to 10 kg are the most suitable packaging material. Shredded paper is the best material for cushioning and protection during transport. In India use of nylon net bags are quite common as consumer packaging (Figure 8). Premium quality ber fruits are marketed in Israel in 250 g punnet or trays as consumer packs.

Figure 8. Road Side Marketing of Ber in Consumer Package of 1 kg in Nylon Bags

5.7 Storage

Ber fruits are commonly stored at ambient/ room temperature (25-35 °C) from harvest until their consumption. Depending on maturity stage, ber fruits

Harvesting Stage

After 2 Weeks at 5 °C

After 2 Weeks at 7.5 °C

After 2 Weeks at 10 °C

Figure 9. Storage of 'Umran' Ber Fruits at Different Temperatures

could be stored for up to 4 to 7 days at room temperature. Under ambient conditions, fruits showed high degree of pathological infection and loss in colour and could be stored only for 9 days. The normal shelf life of ber cv. Umran is up to 8 days at room temperature however, in pusa zero energy cool chamber (developed at IARI, New Delhi) it could be stored up to 15 days. However, fruits of cv. Kaithli and Gola cultivars could be stored up to 14, and 18 days, respectively in this chamber.

Ber fruits are susceptible to chilling injury. The success of low temperature storage ber fruits are dependent of cultivar. Fruits of cv. Gola and Seb can be stored at 10-12 °C for 2 weeks. However, the ber fruits of cv. Umran harvested at colour break stage can be stored at 7.5±1 °C and 90-95 per cent RH for two weeks with acceptable colour and quality (Figure 9). Further, storage at 6 °C results in chilling injury to fruits. Modified atmosphere packaging with 5 per cent O_2 and 5 per cent CO_2 was found suitable to reduce chilling injury up to some extent.

6.0 Domestic Marketing and Export

In India, ber fruits are marketed through a marketing channel that differs according to the quality of produce, distance from the market and available storage and marketing infrastructure. Marketing of ber fruits is carried out through the two channels:

Producer-commission agent-wholesaler-outside markets; and Producer-preharvest contractor-commission agent-wholesaler-outside markets. Ber has remained almost unknown in the export market.

Other producing countries in Asia (Afghanistan, Bangladesh, Myanmar, China) produce fruit largely for the local rural people and local markets. Thailand contrasts with these countries because enhanced production has been geared to potential exports. Chinese ber is intensively grown in China, Vietnam and Azerbaizan. These countries export the fruit to the rest of world.

7.0 Conclusion

Studies on harvesting, grading, and packaging of fruit have resulted in a number of recommendations for growers and packers. However, standard grading related to end use has hardly been implemented. Ber fruits have a climacteric respiration behaviour. Various maturity indices have been characterized including colour, total soluble solids, days from full bloom to maturity. The postharvest quality and shelf-life of ber fruit are influenced by both pre and postharvest factors particularly plant growth regulators, stage of maturity, composition of fruit, storage conditions, storage type, and packaging. Low storage temperatures (6-10 °C) can extend the shelf-life beyond 2–3 weeks, but chilling injury is a serious problem, Therefore, some treatments to reduce chilling injury such as salicylic acid, acetylsalicylic acid, 1-MCP, brassinosteroids, and polyamines need to be tested at low temperatures.

Standardized conditions for commercial storages have yet to be developed. Controlled and modified atmospheres have not been commercially applied. More attention should be focused on characterizing the physiology of ber fruit, including ripening, gene expression, protein synthesis, and regulation of enzymes catalyzing sugar formation and translocation. Reliable maturity and international standards, with respect to quality, should be developed. In order to extend the storage period, delay ripening, and maintain fruit quality low temperature storage in combination with controlled and modified atmospheres should be evaluated along with methods of ethylene control.

8.0 Important References

Abbas, M.F. 1997. Jujube. **In**: *Postharvest physiology and storage of tropical and sub tropical fruits,* S.K. Mitra (Ed.), CAB Intl., Wallinhgford, Oxfordshire, UK. p. 405-15.

Jat, L., Pareek, S. and Shukla, K.B. 2013. Physiological responses of Indian jujube (*Ziziphus mauritiana* Lamk.) fruit to storage temperature under modified atmosphere Packaging. *J. Sci. Food Agric.* **93**: 1940-44.

Jawanda, S.K., Randhawa, J.S., Mahajan, B.V.C. and Gill, M.I.S. 2009. Effect of postharvest treatments on the cellulase activity and quality of ber fruits under cold storage. *J. Food Sci. Technol.* **46**: 87-88.

Kader, A.A., Li, Y. and Chordas, A. 1982. Postharvest respiration, ethylene production, and compositional changes of Chinese jujube fruits. *HortScience,* **17**: 678-79.

Lal, G., Fageria, M.S., Gupta, N.K., Dhaka, R.S. and Khandelwal, S.K. 2002. Shelf-life and quality of ber (*Ziziphus mauritiana* Lamk.) fruits after postharvest water dipping treatments and storage. *J. Hortic. Sci. Biotechnol.* **77**: 576-79.

Nallathambi, P., Umamaheswari, C., Thakore, B.B.L. and More, T.A. 2009. Postharvest management of ber (*Ziziphus mauritiana* Lamk.) fruit rot (*Alternaria alternata* Fr. Keissler) using *Trichoderma* species, fungicides and their combinations. *Crop Prot.* **28**: 525-32.

Pareek, S. 2013. Nutritional composition of jujube fruit. *Emr. J. Food Agric.* **25**: 463-70.

Pareek, S. and Yahia, E.M. 2013. Postharvest biology and technology of ber fruit. *Hortic. Rev.* **41**: 201-40.

Pareek, S., Fageria, M.S. and Dhaka, R.S. 2003. Genetic variability and association analysis for fruit fly (*Carpomyia vesuviana* Costa) infestation in ber. *Ind. J. Plant Prot.* **31**: 89-90.

Pareek, S., Kitinoja, L., Kaushik, R.A. and Paliwal, R. 2009. Postharvest physiology and storage of ber. *Stewart Postharvest Rev.* **5**: 1-10.

Purohit, A.K., Rawat, T.S. and Kumar, A. 2003. Shelf life and quality of ber (*Zizyphus mauritiana* Lamk.) fruit cv. Umran in response to postharvest application of ultraviolet radiation and paclobutrazol. *Plant Foods Human Nutr.* **58**: 1-7.

Sheng, J.P. and Shen, L. 2011. Chinese jujube (*Ziziphus jujuba* Mill.) and Indian jujube (*Ziziphus mauritiana* Lam.). **In**: *Postharvest Biology and Technology of Tropical and Subtropical Fruits*. Yahia, E.M. (Ed.). Woodhead Publishing Limited, UK.

Siddiqui, S. and Gupta, O.P. 1988. Effect of postharvest application of calcium chloride and diphenyl package on the storage behaviour of ber (*Ziziphus mauritiana* Lamk.) fruits. *Haryana Agric. Univ. J. Res.* **18**: 337-40.

Zhu, Z., Zhang, Z., Qin, G. and Tian, S. 2010. Effects of brassinosteroids on postharvest disease and senescence of jujube fruit in storage. *Postharvest Biol. Technol.* **56**: 50-55.

Chapter 14

Citrus Fruits

D.V. Sudhakar Rao

Principal Scientist, ICAR-Indian Institute of Horticultural Research, Hessaraghatta, Bengaluru – 560 089, Karnataka, India

E-mail: sudhadvrao@gmail.com

M.S. Ladaniya

Former Director, ICAR-Central Citrus Research Institute, Amaravati Road, Shankar Nagar Post Office, Kachimet, Nagpur – 440 010, Maharashtra, India

E-mail: msladaniya@gmail.com

1.0 Introduction

Citrus fruits belonging to the genus *Citrus* of the *Rutaceae* family are highly popular for their refreshing, thirst-quenching ability and for providing adequate nutrients and phytochemicals like Vitamin 'C', carotenoids, vitamin-'B' complex, limonoids, flavanones (naringin, hesperidin) *etc.* These are gaining importance worldwide due to their nutritive, therapeutic value and increase in health awareness among all categories of people. Citrus fruits are fast becoming an important part of daily diet of many people and are also extensively used by beverage and food industries that require flavour of citrus. Citrus cultivation is not only remunerative but also generates employment and thus plays an important role in the horticultural wealth and economy of India.

2.0 Status of Cultivation

Citrus fruits rank first in the world with respect to production and consumption. China, Brazil and USA are top ranking citrus producers in that order with China contributing almost 20 per cent with 22.94 million tonnes to world production. India ranks fourth in citrus fruit production in the world and contributes 7.15 per cent in total world citrus production of 115.52 million tones. Nearly 7.5 million tonnes of citrus fruits are produced in India from an area of 8.5 lakh hectares. The per capita annual citrus consumption varies from 40kg in Europe to 4 kg in Asian and African countries. Almost all citrus fruits are consumed fresh with 1-2 per cent being utilized in processing. In India, Citrus fruits are cultivated in four major zones, *viz.*, north-west zone comprising arid western plains of Punjab, Haryana, Rajasthan, Gujarat and western Utter Pradesh; central zone comprising semi-arid tracts of Maharashtra and Madhya Pradesh; southern zone comprising hot and humid areas of Andhra Pradesh, Tamil Nadu and Karnataka and north-eastern zone comprising Assam, Meghalaya, Sikkim, Tripura and North Bengal. Thus it is widely adapted to different soil and climatic conditions.

The commercially important citrus fruits include sweet orange, mandarin, lime, lemon, pummelo and grapefruit. Among sweet oranges, 'Mosambi' is important in Maharashtra while 'Sathgudi' is a major crop in Andhra Pradesh. Acid lime is grown throughout the country. Among mandarins, Nagpur mandarin is important in Vidarbha region of Maharashtra while Coorg mandarin is grown in Coorg, Wynaad and Palani hills region regions of Karnataka, Kerala and Tamil Nadu. Kinnow mandarin is acclimatized very well in semi-arid irrigated zones of Punjab and adjoining states while Khasi mandarin is adapted to Darjeeling area of West Bengal and the entire north-eastern hill region.

that is commonly known as ethephon and commercially available as Ethrel is used as a source of ethylene. The fruits are dipped in this solution at concentrations of 500 ppm and above for a period of 5-10 minutes. Ethrel concentration of 3000 ppm is the best treatment for bright yellow colour development in 'Mosambi' orange. Mature green acid lime fruit (*Citrus aurantifolia* Swingle) can be degreened by ethephon (500-2000 ppm) dip for 1 min at room temperature. In three days about 70 per cent of the fruits turn yellow following ethephon treatment at 750 ppm and above. The major drawback of ethephon dip application is that effective concentration of ethylene gas is not maintained in the atmosphere around fruit and it takes 5-8 days for degreening after the treatment. Further, since the fruits come in contact with the chemical solution its use is governed by food regulations in different countries

b) Ethylene Gas Method

Ethylene gas is commercially available as compressed gas cylinders and the fruits are exposed to this gas at recommended concentrations in an air-tight chamber for required time. Catalytic reactor based ethylene generators are also available which produces ethylene gas using ethanol or some other liquid compounds. Two methods of gas exposure are followed: shot method and trickle method. In shot method, fruits are exposed to higher concentration of ethylene gas (250-1000 ppm) for 1-12 hours in a closed chamber with no environmental control. In the trickle method, citrus fruits are exposed continuously to a very low concentration of ethylene gas (7-10 ppm) till degreening, in a continuously ventilated chamber at a controlled temperature and RH.

In continuous flow method (trickle method) of degreening, "Nagpur" mandarin and "Mosambi" orange fruits with light green colour on 1/2 of the surface can be degreened completely in 48 h at 26-30 °C with ethylene gas. Degreening unit with one tonne fruit holding capacity has been developed at NRCC, Nagpur and this technique can be used for 'Sathgudi' oranges also. At present, continuous process (Trickle) of ethylene gas application is getting replaced with intermittent ethylene application. Low concentration of ethylene or intermittent ethylene application at high RH in ventilated degreening chamber gave good results. Fruits grown in northern part and NEH region of India do not require degreening since the attractive colour develops on the fruit naturally.

As the application of ethylene during degreening treatment is also associated with calyx senescence, new degreening treatments combining different periods of exposure to ethylene with periods without addition of ethylene was evaluated in order to reduce this disorder. The introduction of periods without ethylene exposure during degreening treatments significantly reduced calyx senescence. The optimum degreening treatment for 'Clemenules' mandarins was found to be 72 h ethylene exposure and 48 h without ethylene when fruits were harvested with a colour index of between 9 and 3. However, for fruits harvested with an external colour index higher than 2, a treatment of 24 h with ethylene plus 48 h without ethylene was sufficient. For 'Navelina' orange, which is highly sensitive to calyx senescence, a treatment consisting of 24 h with ethylene plus 48 h without ethylene is recommended, when the fruits are harvested with colour index of over 0.

Degreening (2 ml/LC_2H_4 for 120 h at 21 °C and 95 per cent RH) combined with a cold-quarantine treatment (1 °C for 16 days plus shelf-life of 7 days at 20 °C, 95 per cent RH) in commercial early-season citrus cultivars of 'Navelina' oranges and 'Clemenules', 'Clemenpons', 'Oronules', 'Prenules', 'Basol', 'Clemenrubi' and 'Orogros' clementines did not induce detrimental changes in DPPH and FRAP antioxidant capacities, ascorbic acid, total ascorbic acid or total phenolic contents. There was also no loss in the total content of flavanones and flavones after ethylene degreening and the subsequent quarantine treatment.These findings indicated that, in addition to improving the external quality of the early-season citrus fruit, the ethylene degreening treatment under commercial conditions does not cause loss of nutritional value in the fruit.

5.4.2 Growth Regulator Treatment

The fruit button (receptacle + sepals) constitutes a major barrier to the development of stem-end rots. A green button is also a sign of freshness. The chemical treatments and environmental conditions that maintain its vitality will also delay decay development. Low temperature and high humidity delays button abscission. Field sprays of 2,4-D increased the storage life of citrus by delaying the abscission and senescence of the fruit button. Postharvest treatment of 2,4-D has been extensively reported to retard calyx abscission, drying and browning, that occur as the result of degreening processes. Application of 2,4-D (100 to 500 ppm) in a water-wax formulation to lemons before storage delays button deterioration, rate of colouration, and water loss by fruit during storage. Today virtually all California lemons are treated with a water wax containing 2,4-D (250-500 ppm). It is also recommended in South Africa and Israel to control stem-end rots during long-distance shipment, and prolonged storage of oranges and grapefruit. In Morocco, Clementine fruits are harvested into buckets containing water + 2,4-D (4-8 ppm) to maintain fruit button vitality and prevent its abscission during the degreening process.

However, as the use of 2,4-D is restricted in many countries. Recently, the effect of postharvest application of a new growth regulator, S-ethyl-4-chloro-O-tolilooxytioacetate (HF-Calibra®), to avoid calyx senescence of early season citrus varieties ('Oronules', 'Clemenrubi' and 'Clemenules' mandarins and 'Navelina' oranges) was evaluated on postharvest degreening. The postharvest dipping application of HF-Calibra® (10, 20, 40 or 60 ml/L), prior to the commercial degreening treatment, has been shown to effectively reduce calyx senescence caused by exogenous ethylene. The postharvest application of HF-Calibra® treatments did not affect the quality (firmness, total soluble solids, acidity, flavour) of fruit.

Postharvest salicylic acid application by vacuum infiltration (at 30.55 kPa for 7 min) effectively maintained fruit firmness, reduced weight loss and efficiently delayed the reduction of the chlorophyll a contents of lime cultivars ('Persian lime' and 'Tahiti') during cold storage at 4°C and 85 per cent RH for 60 days. Further, postharvest treatment with methyl jasmonate (0.001 µmol L^{-1}), 1-MCP (1.5 µL L^{-1}) fumigation and salicylic acid (0.002 µmol L^{-1}) extended cold storage life and maintained quality of 'Kinnow' mandarin up to 75 days. Salicylic acid (4 mM) was also found to be effective to control disease incidence and maintain total phenolic contents, antioxidant activity and antioxidant enzymatic activities; whereas Hydroxypropyl Methyl Cellulose (HPMC) wax coating was observed to be the best treatment to minimize weight loss of Kinnow mandarin fruits during 90 days of storage at 5±1°C and 90±5 per cent RH.

5.4.3 Hot Water Treatment (HWT)

Hot water treatment of citrus fruits is done to disinfect and to enhance their tolerance to chilling injury during low temperature storage. Other advantages of HWT are that it cleans the fruit and improves its general appearance, and requires a short exposure time of 20 s to 2 min. In addition, it is simple to apply in the citrus industry since it can be incorporated into the packing-house sorting line and does not require any special handling. Besides, marketing of hot water-treated fruit is not encumbered by fungicide residues, and it can qualify for specialty classifications such as 'organic', which command much higher prices.

Hot water dipping could be applied to Satsuma mandarin as an effective pretreatment to maintain postharvest quality during storage and marketing. HW dip treatments (52 °C for 2 min, 55 °C for 1 min, and 60 °C for 20 seconds) significantly reduced postharvest decay in Satsuma mandarin fruit of an early harvesting cultivar (*Citrus unshiu* Marc., cv. Gungchun) stored at 5 °C for 3 weeks and subsequently at 18 °C for 1 week without impairing any of the other fruit quality parameters. Sensory evaluation showed that HWT at 60 °C for 20 s markedly improved fruit appearance, making them cleaner and glossier.

The effect of hot-water dip at 53 °C for 2-3 min on chilling injury and decay in various citrus fruits has been compared with the effect of curing at 36 °C for 2 h. Prestorage hot water dips significantly reduced the sensitivity of fruits to chilling and also reduced decay. Polyamine content increased as a result of HWT that significantly reduced chilling injury. HW dip has been suggested as a more practical and easier way to reduce chilling injury and decay compared with curing.

Hot water treatment at 50°C for 5 min could be a useful method to delay chlorophyll degradation and to maintain postharvest quality of mature green Thai lime fruit (*Citrus aurantifolia* Swingle cv. Paan) during storage at 25°C and prolonged storage life by 10 days.

5.4.4 MCP Treatment

Mandarins, especially small fruit types, are usually distributed and marketed with attached leaves in Southeast Asia. Attached leaves improve commercial value of the fruit because of a more fresh appearance. However, the attached leaves deteriorate rapidly after the fruits are detached from the tree, exhibiting abscission, degreening/browning and decay, resulting in reduced market value. Currently, abscission and browning of leaves attached to the harvested fruit has not been controlled by common postharvest technologies. As leaf abscission and browning are associated with senescence, delaying senescence could be an important way to extend the shelf-life of mandarin fruit with attached leaves. Exposure of "Shatangju" mandarin fruits with attached leaves to 1-MCP (5 µl L^{-1}) for 16 h inhibited leaf senescence (Li., 2012) without affecting respiration metabolism, chlorophyll degradation, flavour and nutritive quality.

5.4.5 Postharvest Disease Control

Losses from postharvest diseases caused by different pathogens account for nearly 50 per cent of the total wastage in citrus fruits. The important diseases that cause citrus postharvest decay are green mold (*Penicillium digitatum*), blue mold (*Penicillium italicum*), *Alternaria* rot (*Alternaria citri*), stem end rots (*Botryodiplodia theobromae, Phomopsis citri*), brown rot (*Phytophthora citrophthora*), *Asperigillus* rot (*Asperigillus niger*) *Rhizopus* rot (*Rhizopus oryzae*) and sour rot (*Geotrichum candidum*).The general control strategies for postharvest decay control are:

i) Minimizing physical damage during harvesting and handling.

ii) Treatment with postharvest fungicides and/or biological antagonists. Also, heat treatments may be used.

iii) Prompt cooling and subsequent maintenance of optimum temperature and relative humidity throughout marketing operations.

iv) Removal and/or exclusion of ethylene.

v) Effective sanitation procedures throughout postharvest handling.

Steps are to be taken for wastage control at various points in packing house after the fruit is received. Fruits should be treated and packed as quickly as possible except unavoidable holding after its receipt in packing house. Delayed use and improper concentration of fungicide (s) can lead to higher wastage. Timely applications are as important as fungicide selection itself. The use of benomyl and carbendazim on harvested fruit has been banned by many countries although they can be used in India. In many countries benomyl and carbendazim can be applied 15-30 days before harvest. The residue limits for these fungicides have been set and fruit should not have residues of these compounds more than MRL.

After receipt in packing-house, trash is removed from fruit and the fruit is drenched with fungicide before preliminary sorting. Fruit has to be accumulated in packing house to manage timely operations without loss of energy on idling machines and this improves efficiency. It also allows degreening and packing on time.

Various disinfectants popular in citrus pack houses include chlorine (200-1000 ppm) and sodium orthophenyl phenate (SOPP).Commercially, postharvest citrus decay is controlled by applying fungicides to fruits on packing lines. Due to the presence of fungicide resistant strains and concerns about impact of fungicides to the environment and human health, sustainable alternative methods to control postharvest decay are being actively investigated.

The use of fumigation for decay control purposes offers significant advantages over fungicide application in packing line as fumigation can be applied to packaged fruits in mass. Postharvest green mold (*Penicillium digitatum*) and blue mold (*Penicillium italicum*), were effectively controlled by fumigation of lemons and oranges with two applied dosages of 3000 µL L^{-1} of ammonia gas for 6 h at 22 °C, that was injected initially and again 2 h later. Thus, ammonia could replace synthetic fungicides or augment Imazalil (IMZ) performance in citrus postharvest decay management. Its capacity to control IMZ resistant isolates of *P. digitatum*, common in citrus packing-houses, is particularly valuable. However, fumigation of lemons with 6000 µL L^{-1} of ammonia slightly accelerated the natural transition of rind colour from green to yellow.

In recent years several alternatives to chemical control of postharvest spoilage have been investigated but satisfactory levels of control, with a single system, has not been achieved yet. Curing of hybrid mandarin fruits ("Fremont" and "Fairchild) at 36 °C for 36 h with 95 per cent RH followed by fumigation with 50 and 75 µL/L of acetic acid vapours (15 min.) for late and early harvested fruits, respectively resulted in lowest decay percentage during 2 weeks storage at 20 °C and 80 per cent RH. Fumigation alone with higher concentration of acetic acid vapours (100µL/L) resulted in rind damage as pitting.

One of the most effective systemic fungicides registered for citrus green mould control is Imazalil (IMZ), which has been used for over two decades. Double application with dip (45 seconds in IMZ sulphate at 500 µg mL^{-1}) followed by 2000 µg mL^{-1} IMZ in wax coating resulted in higher protective as well as curative control of the sensitive isolate. However, the resistant isolate could not be controlled which highlights the importance of fungicide resistance management strategies and alternative methods of green mould control in citrus fruits.

Sodium carbonate and potassium carbonate (2 per cent) applied as a preharvest sprays (7 days before harvest) or combination of pre and postharvest applications were more effective than postharvest dipping alone in controlling the postharvest decay on 'Comune' clementine and 'Valencia Late' orange fruits. A bio-control agent, *Paenibacillus polymyxa* strain SG-6 isolated as an endophyte from the root tissue of *Sophoraton kinensis* was found to be highly effective in reducing postharvest green mold decay of citrus fruit. The bacterium also significantly reduced the natural development of green mold decay of citrus fruits following storage at 25 °C for 4 weeks or at 6 °C for 4 weeks followed by 25 °C for 2 weeks without affecting the quality of fruits.

Postharvest dip treatment of Sodium ethyl paraben (80 mMat 20 °C for 60 seconds) controlled green mould and blue mould of oranges (cvs. 'Valencia' and 'Lanelate') more effectively than that of mandarins (cvs. 'Clemenules', 'Nadorcott' and 'Ortanique'). The effectiveness increased when Sodium ethylparaben treatment was combined with Imazalil 25 and was recommended to be included as a non polluting alternative in citrus postharvest disease control programs.

Many herb extracts have been screened for antifungal activity against *P. digitatum* and *P. italicum* and clove extract was able to significantly inhibit both *Penicillium* species. Clove extract (dip in 100mg/mL for about 3 min) significantly reduced the acceleration of the ripening process and fungal spoilage, maintained the quality but had little effect on weight loss. It was suggested that clove extract might be a viable alternative to synthetic fungicides to extend the postharvest storage period and maintain fruit quality of navel orange when amended with some edible coating material such as chitosan and sodium alginate, which have good ability to retard moisture.

The use of wax coatings enriched with antifungal compounds has significantly contributed to quality maintenance of citrus fruits. On the other hand, interest in essential oils (EOs) as an alternative to synthetic fungicides has recently gained momentum. Excellent disease control against green and blue rots was achieved when *Cinnamomum zeylanicum* essential oil was incorporated in shellac and/or carnauba wax compared to other essential oil-wax formulations. The preventive and curative effects of *C. zeylanicum* essential oil could make it an excellent bio fungicide, because as a postharvest treatment this oil would prevent the development of both latent preharvest and postharvest infections.

Ozone is a powerful oxidizing agent and has been used during postharvest treatments, in order to reduce growth, sporulation and mycotoxin production of pathogenic microorganisms, and to extend the shelf-life of various produce. Exposing tangerine (*Citrus reticulata* Blanco cv. Sai Nam Pung) fruit to gaseous ozone (200 µL L^{-1} for 4 and 6 h) delayed the green mold incidence and reduced severity without affecting the quality. Ozone treatment did not cause any phytotoxicity and can be recommended as an effective residue-free alternative to traditional postharvest fungicide practices.

Essential oil treatments have the potential to control green and blue mold diseases and among them lemon grass oil was the best for storage of Kinnow mandarin without affecting the colour texture at 5±1°C and 85–90 per cent RH. Similarly, essential oils of *Zataria multiflora* and *Thymus vulgaris* (300-400 µl l^{1}) were effective in preventing fruit decay (Alternaria decay, black rot) development without impairing the quality traits of 'Washington Navel' oranges stored at 5 °C for 60 days. Volatile compounds of *Cinnamon zeylanicum* (CEO) and *Zataria multiflora* (ZEO) could also reduce the rots caused by *Botrytis cinerea* and *Rhizopus stolonifer* fungi in Nectarines.

The disease reducing activity of various resistance inducers against postharvest green mold of citrus fruit was investigated in commercial cultivars of orange, lemon and grapefruit. Chitosan (0.5%) significantly reduced green mold decay (6–20% disease incidence) on 'Valencia' and 'Washington Navel' orange, 'Femminello' lemon and grapefruit cv. Marsh Seedless whereas -aminobutyric acid (1000 mmol) provided the best results (12–16% disease incidence) on 'Tarocco Scirè' and 'Valencia' orange and grapefruit cv. Marsh Seedless. Citrus fruits treated with oligochitosan induced disease resistance to anthracnose disease caused by *C. gloeosporioides* by significantly increasing the contents of lignin and other antioxidative enzymes. Satsuma mandarin fruit treated with 24-epibrassinolide (5 mg L^{-1}) and stored at 12-16 °C showed significantly reduced postharvest rots (by more than 78%) attributed mainly to the accumulation of H_2O_2 and stress-related metabolites and the induction of stress-related genes.

Among the different food additives (dehydroacetic acid, dimethyl dicarbonate, ethylene diamine tetracetic acid, sodium acetate, and sodium benzoate) tested, sodium benzoate effectively controlled the green and blue molds on citrus fruit. Dip treatments in hot (at 50°C) solution of 3 per cent sodium benzoate for 60 seconds resulted in about 90 per cent reduction of green and blue mold incidence on 'Valencia' oranges, 'Lanelate' oranges, 'Fino' lemons and 'Ortanique' mandarins. Further, sodium benzoate combined with low doses (25 or 50 µL L^{-1}) of the fungicide imazalil (IMZ) greatly improved the efficacy of stand-alone treatments. Decay control and quality preservation of lemons stored at room temperature can be achieved for 8 weeks by a sequential treatment with NaOCl and a heated water emulsion of imazalil at 50 mg L^{-1}, followed by wrapping the fruit with more permeable plastic films.

Biological control of postharvest diseases of fruits is gaining importance recently to overcome chemical residue problem. Among the different non-pathogenic bacteria strains, *Bacillus subtilis* TV-17C and *Agrobacterium rubi* RK-33 strains, significantly reduced disease severity on mandarin fruits during storage. However, it is necessary to conduct tests for studying the complete cytotoxicity using human cell, ecotoxicity and toxicity of these bacterial strains on target organisms. Among twenty five isolates of epiphytic yeasts isolated from fresh horticultural produce two Isolates HAB-31 and HAB-53 showed significant control of Penicillium rot on kinnow fruit stored at the room temperature (25–30°C) with minimum quality changes.

5.4.6 Radiation and Light Treatments

The change in the color of the lime peel from green to yellow is the main postharvest problem that is unacceptable to consumers in many countries. Shellac coating (10 per cent) combined with E-beam irradiation (1kGy) was found to be an alternative approach to delay chlorophyll degradation and maintain the quality of lime fruit. Pre-storage UV-C treatment (3.4 to 10.5 kJ m^{-2}) maintained peel green colour and retained calyx attachment of Tahitian limes after 28 days storage at 10 °C and 80% RH without effecting fruit weight loss or TSS or TA contents. Further, UV-C treatment (7.2 kJ m^{2}) when combined with 1-MCP (0.5 µL L^{-1}) treatment prior or after UV-C irradiation resulted in further improvement in quality by delaying the peel degreening, maintaining the attachment of the calyx and improved the acceptability index during storage at 20°C.

Peel yellowing is a major postharvest problem of lime fruit. Irradiation of mature green lime fruit with UV-B doses at 8.8 kJm^{-2} effectively retarded the degradation of chlorophyll during storage at 25 °C. Furthermore, UV-B treatment effectively delayed the accumulation of pheophorbide especially during the development of yellowing.

UV treatments have been reported to maintain postharvest quality of many horticulture crops. UV-B treatment (19.0 kJm^{-2}) was effective in retarding degreening of lime fruitsby suppressing chlorophyll degradation through the control of chlorophyll-degrading enzyme action. Thus UV-B treatment could be a useful method to prolong the postharvest quality of lime fruit during storage.

Blue light treatments were reported to show promise as an alternative to fungicide application for postharvest decay control and blue light at wavelengths from 390 to 500 nm reduced decay caused by *P. digitatum*. Inhibition of postharvest decays by blue light may be due to a combination of inhibition of fungal growth and induction of defensive responses in the host. The effect of blue light exposure on pathogenicity of a variety of fungal postharvest pathogens including green mold, blue mold, stem end rot, and anthracnose that cause significant decay losses in citrus was very recently evaluated by Liao (2013). Direct exposure of mature 'Fallglo' tangerine fruit to blue light with a photon fluence rate 40 μmol m^{-2} s^{-1} reduced symptom development of blue mold (*Penicillium italicum*), green mold (*Penicillium digitatum*), and stem end rot (*Phomopsis citri*) postharvest decays

Packing-house machinery automation has seen great developments in last few decades in developed countries. The Journey is from manual to semi-automatic and now fully automatic packing-houses with huge capacity to pack fruits every day. The developments in packing houses machinery have witnessed a sea change in the last 50 years and fully automatic citrus packing houses handling up to 300 tonne fruits every day are common. In India the growth in this area has been slow since1990-91. Much needs to be done, most of which depends on available finances, growers co-operative activity and government policies.

5.4.7 Washing

Citrus fruits are washed to remove adhering dirt, residue and surface organisms if any using spray washers. Soft bristles are to be used for washers not to cause any abrasions and bruises on the fruit while washing. Washing can be done with detergent or disinfectant like chlorine followed by plain water rinse to remove any residue. Fruits are then run on sponge rollers to remove excess water so that fruits are just damp at the time of waxing.

5.4.8 Wax Coatings

Waxing is done on citrus to reduce water loss, thereby reducing shriveling. The application of edible waxes will also enhance the appearance of the produce. In addition to provide cosmetic shine to fruit surface, wax coatings also act as a carrier to fungicides and plant growth regulators (2,4-D or GA) to reduce decay and to retain greenness of buttons (Figure 1-3).

Waxes coatings are available in various forms such as solvent waxes, aqueous emulsions, resin solutions, *etc.* Wax coatings should be of Food Grade material and edible and bio-degradable coatings are preferred. The common waxes are carnauba, bee wax, paraffin, and oxidized polyethylene. Carnauba is extracted from leaves of *Copernica prunifera* palm which is native to Brazil and found in tropical forests. Oxidized polyethylene is an ethylene polymer with acidic groups for emulsification. Paraffin is hard white crystalline wax, mostly containing saturated hydrocarbons and is obtained in petroleum industry. Edible coating can be developed from protein, polysaccharides, lipids or from blend of these compounds. Sucrose ester based 'edible' coating formulations are also commercially available (*e.g.* Semperfresh™).

Figure 1. Washing of Kinnow Mandarins

Figure 2. Kinnow Coating in Pack-House

Figure 3. Wax-Coated Citrus Fruits

Ability of these coatings to extend shelf life is due to differential permeability to CO_2, O_2 and water vapours which reduce metabolic rate and water loss. Permeability of citrus coatings should be high to O_2 and low to water vapour to reduce transpiration as much as possible and not overly restrict respiration. From the fruit physiology point of view, coatings that excessively restrict respiration are harmful in the way they result in off-flavour. Many wax formulations chalk or blush under humid conditions. This can be avoided if direct contact of water with fruit is reduced and fruit is properly dried before wax coating.

Minimum weight loss (6.46 per cent) was recorded in Stayfresh high shine wax + carbendazim (2000 ppm) treated Nagpur mandarin fruits stored in polyethylene lined cartons up to 21 days under ambient condition as compared to 16.03 per cent in untreated fruit. Less physiological weight loss in wax treated fruits was due to reduction in water loss and respiration rate. The organoleptic characteristics such as taste, aroma and colour were not affected by coating.

In recent times solvent waxes have been replaced by water-based waxes. A lac-based formulation was developed at Indian Institute of Natural Resins and Gums (IINRG), Ranchi for fruit coating applications on citrus fruits like Kinnow mandarin and oranges. The formulation yielded good results in respect of gloss and firmness to Kinnow. To evaluate and assess the performance of fresh coat on Kinnow fruit at commercial level, about 40 litres of the formulation was tried at 7 Kinnow waxing and grading plants at Abohar (Punjab). Approximately 83 tonnes of Kinnow fruits were coated with the above formulation. Application of fresh coat produced better results in respect of gloss, spread area and firmness to the fruits as compared to commercial wax. Other advantages of lac-based formulation over the available commercial ones observed are: it is natural, non-toxic and its composition is well specified. No fungicide addition is required since lac has inherent antifungal properties, Lac-based formulation does not produce any obnoxious smell during waxing operation. It is water-based and thus makes the packing line machinery more versatile. There are more than 17 Kinnow waxing and grading plants at Abohar in Punjab, and they use commercial formulations such as NU Coat Flo, Citrashine, Stay fresh 451 and Stayfresh high shine for coating Kinnow fruits.

Commercial fruit coatings enriched with some essential oils have been reported to control postharvest pathogens, while maintaining fruit quality. Coating of Tomango oranges either with pure (d)-limonene and R-(")-carvone or essential oils (*Mentha spicata* and *Lippia scaberrima*) incorporated in commercial citrus coatings gave excellent disease control. Moisture loss decreased significantly in fruits treated with essential oil enriched coatings. This approach eliminates the need for synthetic fungicides, thereby meeting consumer preferences, organic requirements and reducing environmental pollution.

Edible composite coatings based on hydroxypropyl methylcellulose (HPMC), hydrophobic components (beeswax and shellac), and food preservatives (potassium sorbate, sodium benzoate, sodium propionate and their mixtures) with antifungal properties were evaluated on 'Valencia' oranges during long-term cold storage. The potassium sorbate (PS) + sodium propionate (SP) based coating was the most effective at inhibiting both molds during storage at 20 °C after a cold storage period of 60 days at 5 °C. HPMC–lipid edible composite coatings containing antifungal organic acid salts and mixtures such as PS + SP could be a promising treatment for oranges that are stored in cold storage.

Some salts can be applied in combination with waxes in order to provide an extra barrier to decay as substitutes for fungicides. Potassium sorbate incorporated in commercial wax significantly reduced the incidence of postharvest decay in oranges (cvs. 'Tarocco' and 'Valencia Late') and Elementines (cv. 'Comune'). However, salts like sodium carbonate and bicarbonate, potassium carbonate and bicarbonate and potassium sorbate interfered with the action of the wax in retarding weight loss.

Different edible coating materials like 1 per cent carboxymethyl cellulose, 5 per cent gum Arabic, 1 per cent Persian gum, 0.5 per cent beeswax and 1 per cent carnauba wax were comparable with commercial wax coating in maintaining the quality of 'Kinnow' mandarins during 3 weeks storage at 5°C and 90–95 per cent RH. All coatings could reduce the weight loss as compared to the control and Persian gum was found to be the best coating. In case of Satsuma mandarins, carnauba wax coating significantly reduced weight loss, respiration rate, and firmness loss during storage at 25°C and incorporation of grapefruit seed extract (1 per cent) reduced *P. italicum* incidence. Carnauba wax coating of 'Salustiana' orange could prolong the shelf life up to 40 days at 4±1°C and 90 per cent RH, by reducing weight loss and preserving their chemical and sensorial profile. Coating of Eureka' lemon with carboxymethyl cellulose incorporated with moringa leaf extract followed by CMC or chitosan alone was found to be effective in reducing the incidence of peteca spot in 'Eureka' lemon stored at 3°C for 12 weeks. Grapefruit seed extract (1 per cent) incorporated carnauba wax (18.1 per cent) coating was useful in extending the postharvest shelf life of Satsuma mandarins (Citrus unshiu Marc.) by inhibiting the growth of P. italicum without altering the physicochemical properties of fruits during storage at 25°C. Shellac coating (10 per cent) enriched with 0.5 per cent cinnamon essential oil reduced weight loss and maintained firmness and ascorbic acid content of 'Thomson navel' orange up to 21 days at 5°C and being recommended as an efficient formulation to serve as an effective alternative to synthetic fungicides to control green mold decay in citrus.

Juice sac granulation is a serious problem in pummelo fruits during postharvest storage and chitosan (1.5 per cent) treatment can efficiently reduce this problem by mitigating ROS accumulation by maintaining higher active antioxidant enzymes (superoxide dismutase, catalase, ascorbate peroxidase and glutathione reductase and delayed fruit senescence during storage at room temperature (20 ± 2 °C). Novel edible composite coatings based on pea starch and guar gum (PSGG) was reported to be a beneficial substitute to common commercial waxes for maintaining quality and storability, as well as extending shelf life of Valencia oranges for up to four weeks at 5 °C with an additional 7 days at 20 °C.

5.5 Sorting and Grading

5.5.1 Sorting

Sorting is done to remove immature, undersized, insect damaged, diseased or bruised fruits. Sorting can be done manually on sorting conveyers made of rollers that rotate on their axis in order to rotate the fruit to reveal the entire surface while the fruit moves forward. The sorting conveyers should have sufficient illumination focused on the fruit rolling on the conveyer. The fruit sorting personnel have to be sufficiently trained to identify culls on the basis of size, shape, colour, blemishes, *etc.* and the cull fruits need to be removed quickly from the area.

In automatic electronic sorting systems, fruits can be sorted on the basis of size, shape, colour and blemishes by the cameras (machine-vision system) that can detect culls and drop/remove them on cross belts running below.

5.5.2 Grading

Grading of citrus fruits has to be done to maintain uniform size and quality to fetch better prices in domestic as well as international markets. Conventionally grading is done manually by experienced graders mainly based on visual size. Now grading is being done automatically using high speed electronic weight and size graders to reduce labour costs. Grading of fruit on the basis of size dimensions is better than weight grading in most of the citrus cultivars.

Sizing on the basis of weight is usually not followed in citrus since the non-uniform shape (height or diameter) can result in difficulties in place packing although weight is similar. In place packing arrangement, fruit is packed in layers in the box. If height or diameter is not uniform and the fruits are not of same shape and size, the packed fruits do not appear uniform to the buyer.

For getting premium price and assuring quality to consumers, the fruits are graded for different sizes and variation within one grade range should be minimum, not more than 10 per cent of the fruit or as permitted by the importers. The various grades recommended for Kinnow fruits by APEDA are showing in Table 1.

The recommended grade standards for different citrus fruits in Andhra Pradesh are furnished in Table 2.

Table 1. Grades of Kinnow Fruit

Grade	*Size Range (mm dia.)*	*No. of Fruits in 10 kg Pack*
a	60-64	84
b	65-69	72
c	70-72	60
d	72-74	54
e	75-79	51
f	80-85	45
g*	50-60	96
h*	45-50	120

* Only for domestic market.

Table 2. Grades of different Citrus Fruits in Andhra Pradesh

Fruit		*Grade*	
	I	*II*	*III*
Acid lime	4.5 cm and above	3.5–4.5 cm	3.5cm and less
Lemon	8 cm and above	6.5–8 cm	6.5cm and less
Orange (Sathgudi)	6.5cm and above	5–6.5 cm	5cm and less

Earlier all operations of sorting, sizing and grading were on mechanical conveyers with manual operators. With developments in electronic sorting (machine vision technology), now industry has fully automatic packing houses. Hand sorters have been found to reject fairly large quantity of good fruit. Now, with the help of electronic sorting, consistently accurate sorting is ensured and the rejections are reduced up to 8 per cent or less. Mechanical sizers are likely to bruise the fruit while electronic sizers can reduce this damage. The electronically (computer aided) controlled volumetric sizers can give very good size segregation of even oblong or ellipsoidal lemon fruits. The singulation and orientation of fruit is necessary in these machines. With increasing labour cost, and cost of the packing material, automation is surely to give some relief to pack-houses.

In order to separate sound fruits from internally damaged/hollow fruits in the same size lots, a grader of 300-400 kg/h capacity suitable for oranges and other round shaped fruits has been developed at Pantnagar Centre. It uses the principle of counter balancing of predetermined weight with the weight of the fruit using mechanically operated lever and flap system (go and no-go logic). The machine is operated by 1.0 KW electric motor. Another model of 600-800 kg/h has also been developed that costs Rs. 25,000/- and the machine is now commercially available.

Mechanized sorting, washing, wax coating, and size grading has been standardized for "Nagpur" mandarin, "Mosambi" orange and "Kagzi" acid lime way back in 1990s. The process is now being used by the growers and industry for domestic and export marketing in India. In case of 'Kinnow' mandarin grown in Punjab and other states, mechanization of sorting, coating and grading is being followed on commercial scale and 40 such units are operational in that area. This practice has reduced losses as the handling and contamination is reduced considerably. On similar lines efficient application of these postharvest technologies to maintain quality and freshness of citrus fruit during marketing have to be planned in other parts of the country. In case of 'Nagpur' mandarin in Maharashtra, much needs to be done as the marketing is still in the hands of traders which is done on conventional lines This technology, if followed, coupled with proper orchard management (timely fungicide sprays, sanitation), packaging and transportation would bring down losses from 25 per cent at present to about 5 per cent.

5.6 Packaging

In India, due to lack of government policies, wooden boxes are still being used. Technology is however available for CFB boxes while the corrugated cartons are slightly costlier or on par with price as compared to wooden boxes. The lighter weight of CFB boxes allows more cartons to be transported and hence more fruit can be transported per trip of the truck (Figure 4). Thus it is cheaper to use cartons and little higher cost of the box gets compensated in freight (Figure 5).

5.6.1 For Wholesale Marketing

Corrugated fiber board cartons (50x30x 30 cm size) have been designed and developed long back for "Nagpur" mandarin at NRCC, Nagpur. These corrugated boxes (Universal and Telescopic type) are suitable for distant transport and have sufficient strength to withstand storage conditions at 90-95 per cent RH. The strength and capacity of the box have already been tested. These containers can be manufactured from available kraft paper in the market using required specification of the box. Corrugated boxes (universal RSC type and telescopic type) were made from Kraft paper that is made of cotton plant stalks with collaboration between CIRCOT, Mumbai and NRCC, Nagpur. This kraft paper is cheaper than commercially available kraft paper. Thus the cost of the box is less while its strength is similar to box made from commercially available kraft paper.

CFB boxes are preferred for citrus packaging due to their characteristics like lightweight, ease in handling, attractiveness, good printability and less damage due to bruising. Thus, it provides a safe delivery of perishable contents from farm to consumer. Various materials

Figure 4. Packing of Mandarin Fruit in CFB Boxes

Figure 5. Wrapping with Heat–Shrinkable Film (Individual fruit, 3-7 fruits in a tray)

are used for packaging of Kinnow fruit. These include wooden carton, bamboo, gunny bag, plastic crates and corrugated fibre board (CFB) box. The kind of material and structure adopted depends on the method of transportation, destination distance and the value of fruit and type of market. For the Kinnow fruit gunny bags and plastic crates are used for local market (distance limit to about 25 km) and CFB boxes and wooden cartons are used for distant market during transportation. Generally, CFB box is preferred over wooden carton due to its light weight ease in handling, good printability and less damage due to bruising in comparison to wooden cartons and the extent of damage varies in different containers.

Kinnow and all other citrus fruits are generally graded on the basis of size. However, for all grades only one type of CFB box (43.5 x 23.0 x 20.2 cm) is used. Thus, for some grades, it is tight and for some there is free space causing higher damage to the fruits during transportation. CFB box used for Kinnow transportation should have sufficient mechanical strength (5 to 7 ply). Although CFB box is superior over the other types of packages namely, wooden carton or gunny bag, but due to faulty packaging practices like long pedicle with the fruit or improper arrangement of fruits, the transportation and handling losses are also significantly high. At CFTRI, it was found that three different size boxes namely, 44 x 25 x 21 cm; 41 x 25 x 21 cm and 48 x 25 x 21 cm could accommodate economically all the grades of Kinnow as shown in Table 3. In general, the weight of fresh fruit in these boxes will be about 10 kg. These sizes were found to be the most suitable for the packaging of Kinnow to minimize the postharvest losses.

Table 3. Optimized Dimensions of CFB Boxes for different Grades of Kinnow

No. of Kinnows per Box	*Size of Box (cm)*			*Weight of CFB Box Filled with Fruit (Kg/box)*
	L	*B*	*H*	
84, 72 and 66	44	25	21	9.6-10.4
45 and 36	41	25	21	9.9-10.6
54 and 30	48	25	21	9.7-10.5

In general, during transit fresh fruits lose weight, which is loss of water in the form of vapours. Corrugated fibre boards being hygroscopic in nature, absorb this moisture and thereafter the mechanical strength of the box reduces greatly and box collapses during transportation. To avoid the collapse, the box could be laminated but then the box cannot be recycled and hence at will not be environment friendly practice. Therefore, it is suggested to paint the corrugated fibre sheet with waterproof paints before making the box. This was found to reduce the collapse of box during transportation and even the box could be recycled.

5.6.2 For Retail Marketing

a) Modified Atmosphere Packaging

Modified atmosphere (MA) packaging could be very useful for extending the marketing period of citrus fruits but the package has to be sealed (air tight). Modified atmosphere packaging results in a reduction of O_2 and elevation of CO_2 concentrations and increased RH around the fruit inside the film. Effects of this reduced O_2 and increased CO_2 levels on fruit would vary with gas concentration, exposure time and variety of fruits. It, however, reduces respiration rate and softening of fruits, may retard fungal infection and can inhibit the effects of ethylene in causing deterioration. Temperature for storage of such packaged fruit should not be too high. MAP relies on the permeability properties of the packaging film to generate an atmosphere which contains the desired concentrations of O_2 and CO_2. This is called as passive MAP since it is slowly created due to respiring fruit. It is possible to find an appropriate MAP by testing different films. The films used in MAP include various kinds of plastic polymers. These plastic films must provide protection, strength, clarity, and a printable surface. However, their unique function is to restrict the movement of O_2 and CO_2 through the package and allow the establishment of a modified atmosphere. This modified atmosphere (MA) can be beneficial to the storage of fruit provided the films are properly selected. The MAP retains the natural colour and firmness during storage in comparison to control (unpacked) fruits. Automatic high-speed machines for MAP are available to pack ½ dozen or 1 dozen fruit and the capacities vary greatly.

b) Shrink Wrapping

Individual shrink-wrap packaging is a new technique for postharvest handling of citrus fruits. National Research Centre for Citrus (NRCC),Nagpur has standardized this packaging technique for Nagpur mandarin and Mosambi sweet orange while CIPHET (Central Institute of Postharvest Engineering and Technology), Ludhiana reported work on Kinnow. In general 20-30 per cent reduction in transpiration rate is possible by shrink-wrap packaging under ambient condition. Such unit pack provides protection against abrasion and maintains an attractive appearance of the produce. It avoids condensation of droplets within the package. One of the biggest advantages of individual shrink-wrap packaging is that it prevents secondary infection, which is important for long-term storage. Individual fruit wrapping also provide the optimum gas and humidity condition for maintaining the quality during transit and storage. As a result, it doubles or at times triples the shelf life of fruits without any refrigeration. The extent of benefit from shrink-wrap packaging depends upon the type of produce, its physiological maturity and initial quality. The machine with window size of 350x200 mm and conveyer speed of 1-2 meter per minute can give an output capacity of 80-100 kg fruits per hour. However, the cost of film packaging varies with the commodity to be wrapped or films used for wrapping. In practice, it adds a packaging cost of Rs 2-3 per kg of fruit in a tray and for individual fruit, cost is much less. The shelf life of Kinnow fruit shrink wrapped in 25µ film was extended to 30 days compared to 15 days in control at room temperature. Shelf life could be further extended to 10-12 weeks compared to 6 weeks in control under refrigerated conditions.

The technology of packaging 'Nagpur' mandarin and Mosambi orange in permeable plastic films has been standardized at NRCC, Nagpur. The individual fruit wrapping and tray-wrapping of "Nagpur" mandarin in "Cryovac" heat-shrinkable films (D-955 "Impact" 15 and 25 micron and cling film) and shrink-film wrapping of "Mosambi" sweet orange (individual and tray-wrap) has given very promising results and can be used on commercial scale. The automatic high speed machines can reduce packing cost considerably.

The control of moisture loss is more in wrapping with plastic films as compared to even wax coating. The other advantages of wrapping of fruits are:

i) Spread of decay is effectively reduced in plastic film wrapping of individual fruit.

ii) Allows fruit to be stored at ambient temperatures if individually seal packed.

iii) Seal-packing allows fruit transport at ambient conditions from few weeks (mandarins) to few months (pummelo, grapefruit, lemon, sweet orange).

Net bags and polyethylene bags are used for prepacking or consumer packaging. These are small packages of 1-2 kg or 4 kg each and are used for direct retail marketing.

Labelling inks and gums used in packaging are considered as food additives. The adhesives and inks used on labels of boxes on fruits and packages must be acceptable to importing countries. Proper stamps/labels allow "Brand" identification and it is necessary to promote marketing of quality citrus fruits.

5.7 Precooling

Precooling is the first step in good postharvest management of fresh produce. It is the rapid removal of heat from the produce to reduce the rate of respiration, transpiration, microbial activity and refrigeration load. In case, pack-house operations are delayed by 3-4 days, it is a normal practice to keep fruits at lower temperature. In precooling terminology, half-cooling time is defined as the time needed to reduce the temperature difference between the initial produce temperature and the cold room temperature by half.

Forced air-cooling method has been most suitable for citrus fruits. Fruits should be precooled at 6-8 °C and 90-95 per cent RH in order to avoid postharvest losses in distant transportation. The circulation of forced air should be ensured for adequate cooling of produce by proper alignment of side holes of the boxes. The produce should be kept in precooling chamber for a period until the temperature at the centre of box reaches 6-8 °C. The temperature should not fall below 5 °C, otherwise it will result in chilling injury. The palletization and strapping should be done in cold rooms in order to maintain the cool chain.

5.8 Storage

Storage temperature depends on cultivar and growing conditions/climate. Sathgudi oranges grown in Andhra Pradesh, India, can be stored up to 16 weeks at 5.5-7.2 °C. Navel and Valencia oranges grown in Australia develop surface pitting after a few weeks at temperature below 4.5 °C. Storage at 7 °C for Valencias and early–picked Navels and 4.5 °C for mid or late season Navels has been suggested. Careful harvesting and handling is necessary to minimize oleocellosis rind blemishes in Washington Navel oranges and a carriage temperature of 10 °C for 2 weeks followed by 5 °C is recommended to minimize mould spoilage, weight loss and rind blemishes caused by chilling during shipment to USA.Israel grown Shamouti orange develops chilling injury at 6 °C and 90 per cent RH during storage up to 12 weeks followed by 2 weeks at 17 °C. At 12 °C, fruit did not develop chilling injury and Shamouti is more sensitive to CI than Valencias.

Late harvested Kinnow mandarins grown in Punjab (India) do not store well whereas fruits harvested in mid-January suffer least spoilage and weight loss. Temperature of 4.4-6.1 °C is better than 2.2-3.9 °C for Kinnow mandarin grown at Saharanpur. Completely orange coloured Kinnows can be stored longer and retain higher TSS and Vitamin 'C' content than green and yellow-green fruits.

Lemons grown in different areas have different temperature optima. Australian lemons are stored at 9-12 °C depending on duration of storage. California lemons are stored at 12.5–15 °C while Cuban lemons are stored at 12 °C. Relative humidity in storage should be maintained at 85-90 per cent.

The mandarin and sweet orange fruits should generally be stored in cold store at 5-6 °C and 90-95 per cent RH. It must be ensured that temperature of the cold store should not fall below 5 °C otherwise it can result in chilling injury. Un-bruised and mature fruits can be stored up to 8 weeks only.

Storage life of citrus fruits varies depending on species and growing conditions (Table 4). Although storage life varies with climate under which the fruit is grown, in general, storage life as given below can be achieved for fruits from healthy orchards.

Table 4. Storage Life of Citrus Fruits at Optimum Temperatures

Citrus Fruit Type	*Storage Life (Weeks)*	*Optimum Temperature (°C)*
Grapefruit	12-16	10-12
Kinnow	8-10	5-6
Lemons	9-12	10-12
Limes	6-8	8-9
Mandarin Nagpur	6-7	5–6
Sweet Oranges	8-12	5- 6

Citrus fruit are susceptible to chilling injury (CI) when exposed to temperatures of less than 2–5 °C. Chilling injury is expressed as pitting, staining and necrotic areas in the peel that increase in number and size over time. Several postharvest treatments have been reported to reduce chilling injury symptoms in citrus including intermittent warming low temperature conditioning and high temperature treatments using water. Hot water dip (HWD) treatment at 47.4–50 °C for 2 min reduced the chilling injury incidence (proportion of unacceptable fruit) and severity (level of damage) of Satsuma mandarin fruit (*Citrus unshiu* Marc) stored at 2 °C for 8 weeks. Hot water temperatures higher than 50 °C increased fruit peel damage.

Hot water dipping (HWD) treatment (41 °C for 20 min) reduced chilling injury in Washington Navel and Valencia Late oranges during cold quarantine at 1 °C and 85–90 per cent RH for 20 days, subsequent storage at 10 °C and 85–90 per cent RH for 20 days (as a transit period) and an additional 20 days of simulated marketing period at 20±2 °C and 40–65 per cent RH. Enhancement of peroxidase and catalase activities in both fruit peel and juice by HWD contributed to reduction of CI without affecting other postharvest qualities such as weight loss, juice content, soluble solids content/total acidity, ascorbic acid, or reducing sugars.

The effect of postharvest vacuum infiltration of calcium chloride on chilling injury, firmness and quality of Lisbon lemon at yellow-green stage was studied. Vacuum infiltration (-33 Kpa for 10 min) of $CaCl_2$ at 1.5 per cent was found effective in reducing the severity of CI by 53.2 per cent following 12 weeks storage at 1.5 °C (85-90 per cent RH) and one additional week at 20 °C.

Influence of chilling temperature, intermittent warming (IW) and fungicidal wax coating was evaluated during storage of 'Nagpur' mandarin (*Citrus reticulate* Blanco). 'Nagpur' mandarins stored at 3.5 °C with intermittent warming cycle (2 weeks at 3.5 °C followed by 1 week at 19.5 °C) with wax coating resulted in complete alleviation of chilling injury, orange colour development, and reduction of fungicide residue (Carbendazim) during storage up to 75 days. Constant chilling temperature resulted in lower reducing and total sugars, TSS and carotenoids with higher chilling injury and chlorophyll content in peel while IW at 3.5 °C resulted in higher sugars, juice and TSS contents and carotenoids. Thus, 'Nagpur' mandarin storage life could be extended up to 75 days at 3.5 °C with IW cycle with or without wax coating. This technology resulted in increased storage life of 15-30 days over normal storage life of 45-60 days at 5-6 °C constant temperature (without IW).

Mandarins are very prone to losing flavour quality during storage and, as a result, often have a short shelf life. Influence of storage temperature on sensory attributes, quality parameters and aroma volatile profiles of two mandarin varieties (Murcott' and 'Owari') stored for 0, 3 and 6 weeks at either 0 °C, 4 °C, or 8 °C plus 1 week at 20 °C was studied. Flavour quality was reduced in 'Owari' following 4 weeks of storage as off-flavour increased, while for 'Murcott' the hedonic score decreased after 7 weeks of storage. Large increases in alcohols and esters occurred during storage in both varieties, a number of which were present in concentrations in excess of their odour threshold values and are likely contributing to the loss in flavour quality. 'Murcott' stored at 8 °C had slightly superior flavour than fruit stored at either 0 °C or 4 °C, and the better flavour was associated with higher SSC/TA and lesser tartness.

Mandarins suffer from short 'flavour-life' compared with other citrus species and the recommended minimum safe temperature for mandarin storage is 5-8 °C. Further, mandarin varieties differ in their degree of chilling tolerance or susceptibility affecting the colour and flavour quality. Changes in quality and flavour of chilling-tolerant 'Or' and chilling-sensitive 'Odem' mandarins after 4 weeks of storage at 2, 5, or 8 °C followed by 3 days at 20 °C were evaluated. Lower storage temperatures resulted in loss of orange peel colour in fruit of both varieties, which became paler and yellowish. The flavour of 'Or' mandarins was not affected by different storage temperatures, whereas 'Odem' showed severe flavour loss at low storage temperatures. Massive accumulation of terpenes is most likely the cause for the decrease in flavour acceptability of 'Odem' mandarins after storage at low chilling temperatures. Storage temperature was found to be a fundamental factor affecting colour and flavour of mandarins and, therefore, it is crucial for mandarin packing-house operators and exporters to determine and enforce the optimal minimum safe temperature for each mandarin variety.

The low cost evaporative cool chamber with 1.5 tonne fruit storage capacity and costing about Rs.15000/- constructed from locally available material was developed at NRCC, Nagpur in 1990-91. Shelf life of mandarins, Mosambi and acid lime in this chamber can be extended up to 30-40 days due to lower temperature and higher RH. At CIPHET, Ludhiana, evaporative cooled (EC) rooms (3 x 3 x 3m size) with relatively higher cost and storage capacity of 2 to 5-7 tonne were developed a few years back. The shelf life inside the room was 34 days for early Kinnow and 23 days for late Kinnow compared to 21 and 11 days, respectively for the fruits stored at room temperature.

In India, it is now established practice in case of Kinnow handling to wash, wax, grade and pack the fruit in packing-houses. Punjab Agro-Industries Development Corporation has played an important role in this regard by subsidizing some of the operations in supply chain. Recently, with addition of more such units under private sector, the number of mechanized handling units/packing-houses has gone up to 40 (small and big units with capacity of 2-5 tonnes per hour) in Punjab, Haryana and Rajasthan.

In Vidarbha (Maharashtra), there are three mechanized handling units of mandarin fruit under government sector and four under private sector. Several small and big cold storage units (ranging from 100–5000 tonnes capacity) have been established during the last 10-15 years. Thousands of boxes each holding 20 kg fruit are being stored during every season at Nagpur, Hyderabad and other places near metro cities in the country. There are 3- 4 units for handling and degreening of the sweet oranges in Andhra Pradesh. All these developments have taken place in India during the last 15-20 years. Before 1990s, citrus fruit handling was mostly on traditional lines.

Pre-storage neem leaf extract (30 per cent) application was effective in maintaining various physiochemical properties and increasing the antioxidative responses (SOD, POD and CAT) of Shamber grapefruit with extended shelf life up to 43 days at 10°C. Postharvest methyl jasmonate (0.25 mM) dip (1 min) treatment reduced chilling injury and maintained fruit quality of Midknight Valencia orange during 90 days storage at 4°C or 7°C.

5.9 Transportation

Loading in trucks should be done in such a way that there is proper air movement. The corrugated boxes should be gently and properly stacked to avoid vibration, bouncing and load shifting. The temperature inside the refrigerated van should be constantly and continuously maintained at 6-8 °C during the entire transit period. The fruits should not be transported in ordinary trucks as it will result in loss of weight and quality of fruits.

6.0 Domestic Marketing and Export

In India, marketing of the citrus fruits is organized mainly in three ways. i) The preharvest contractors: traders purchase the orchard at flowering stage or just before harvest and pack the fruit in orchard itself or at their packing houses located in production areas and send it to distant markets. ii) Growers send the fruit directly to distant city markets as packed or loose by truck and train. iii) Growers sell the produce in local wholesale markets through auction process where buyers from distant markets or local traders purchase it and send it to distant markets either loose or packed. This is the important channel of marketing for citrus fruits in the country. The commission agent is an important functionary in all marketing channels.

The commercially important channels of citrus fruit marketing in the country are:

i) Grower ⟶ local trader/preharvest contractor/sub contractor ⟶ wholesaler ⟶ retailer ⟶ consumer

ii) Grower ⟶ wholesaler ⟶ retailer ⟶ consumer

iii) Grower ⟶ commission agent ⟶ wholesaler ⟶ retailer ⟶ consumer

iv) Grower ⟶ buyer (distant market) ⟶ wholesaler ⟶ retailer ⟶ consumer

Growers, crop contractors and wholesalers also sell produce to processing units. The above mentioned three marketing systems exist since last many decades. The unpredictable fluctuations in the prices result in huge losses to growers in bumper crop seasons. Traders hamper the correct assessment of fruit quality and price and hence creation of alternate buying agency and regulatory system is considered necessary.

Considerable improvements are needed in marketing process, auctioning and distribution of citrus fruit in India. Due to faulty handling and long traditional marketing chain with many middlemen losses are high. Losses are relatively less in citrus fruits grown in NEH region due to low temperatures during harvest season.

Most of the citrus fruits produced in the country are consumed in domestic market as a fresh commodity. However, a small quality of sweet oranges, mandarins, limes and lemons are being exported since long. Presently, exports of citrus fruits from India is limited to Bangladesh and Nepal by road. India exported 33,575 tonnes of oranges (mainly to Bangladesh), 22,262 tonnes of lemons and limes (mainly to UAE), 464 tonnes of other mandarins (including tangerines and Satsumas) and 277.5 tonnes of grapefruit during 2012-13 (APEDA, DGCIS Annual Export statistics). Small quality of fruit is also exported to UAE, Saudi Arabia and other gulf countries. These fruits are not treated with any fungicides or coatings to reduce decay neither cooled. These are conventionally packed in wooden boxes with paddy straw as cushioning materiel. Export of properly treated and packed fruit by reefer container to developed countries of Europe is the real challenge and these markets can offer premium price for export to be profitable. However, these markets are very demanding. Several successful attempts have been made to export Kinnow and Nagpur mandarins in reefer containers by APEDA and some containers have been sent successfully to European Union (EU).

There is a greater demand globally for supply of healthy, attractive and high quality citrus fruits. The potential of exports, particularly mandarins and acid limes, has increased with increase in production in India. The harvesting season can also provide some advantage for the export of citrus fruits from India. The harvesting season can also provide some advantage for the export of citrus fruits from India, particularly in the case of monsoon crop. Similarly limes and lemons can be exported during most part of the year. Considering the geographical location of India, most of the distant markets in Europe can be reached within 28 days and markets of Southeast Asia and Gulf countries can be reached within 10 days by sea routes. At present, postharvest packaging and storage technologies are available to keep the harvested produce fresh, longer than these durations.

7.0 Conclusion

There is a tremendous development in postharvest handling, storage and marketing of citrus fruit in India since 1980s. Modern mechanized packinghouses coupled with cold storages and marketing infrastructures are established in most parts of the country and this has led to reduction in postharvest losses. Awareness among all stakeholders have increased and benefits of modern postharvest management practices are visible in terms of longer availability of fruit, better returns and reduced losses. With changing times, alternative eco-friendly but effective and sustainable postharvest treatments need to be developed and efforts are already underway in this direction.

8.0 Important References

Alferez, F., Liao, H. L. and Burns, J. K. 2012. Blue light alters infection by *Penicillium digitatum* in tangerines. *Postharvest Biol. Technol.* **63:** 11–15.

Bahareh, S., John B.G.J., Marques. R., Penta, P., Suwimol, C., Christopher, J.S. and Costas E.S. 2018. Application of biocomposite edible coatings based on pea starch and guar gum on quality, storability and shelf life of 'Valencia' oranges. *Postharvest Biol. Technol.* 137: 9-20 .

Bassala, M. and El-Hamahmy, M. 2011. Hot water dip and preconditioning treatments to reduce chilling injury and maintain postharvest quality of Navel and Valencia oranges during cold quarantine. *Postharvest Biol. Technol.* **60**: 186–91.

Bassan, M. M., Francisco de Assis, A. M. F., Caron, V. C., Couto, H. T. Z. and Jacomino, A. P. 2013. The harvesting system affects the quality and conservation of the 'Tahiti' acid lime. *Scientia Hortic.* **155**: 72–77.

Baswala, A.K., Dhaliwal, H.S., Singh, Z., Mahajan B.V.C and Gill, K.S. 2020. Postharvest application of methyl jasmonate, 1-methylcyclopropene and salicylic acid extends the cold storage life and maintain the quality of 'Kinnow' mandarin (*Citrus nobilis* L. X *C. deliciosa* L.) fruit. *Postharvest Biol. Technol.* 161.

Boonkorn, P.,Gemma, H., Sugaya, S., Setha, S., Uthaibutra, J. and Whangchai, K. 2012. Impact of high dose, short periods of ozone exposure on green mold and antioxidant enzyme activity of tangerine fruit. *Postharvest Biol. Technol.* **67**: 25-28.

Chundawat, B. S., Gupta, A. K. and Singh, A. P. 1978. Storage behaviour of different grades of Kinnow fruits. *Punjab Hortic. J.* **18:** 156-60.

Clara, M.H., Pedro A.M. and Lluís, P. 2016. Evaluation of sodium benzoate and other food additives for the control of citrus postharvest green and blue molds. *Postharvest Biol. Technol.* 105: 72-80.

Clara, M-H., Smilanick, J. L., Tebbets, J. S., Walse, S. and Palou, L. 2011. Control of citrus postharvest decay by ammonia gas fumigation and its influenceon the efficacy of the fungicide imazalil. *Postharvest Biol. Technol.* **59**: 85–93.

Codex Alimentarius website: http: //www. codexalimentarius. org/

Cronjé, P. J. R., Crouch, E. M. and Huysamer, M. 2005. Postharvest calyx retention of citrus fruit. *Acta Hortic.* **682**: 369–74

de Carvalho, D.U., da Cruz, M.A., Colombo, R.C., Tazima, Z.H. and Neves, C.S.V.J. 2020. Fruit quality of 'Salustiana' seedless oranges during cold storage: effect of carnauba-based wax and rootstocks. *Food Measure* **14**: 3397-07 .

Devkota, R. P., Grewal, S. S., and Dhatt, A. S. 1982. Maturity determination in Kinnow mandarin. *Punjab Hortic. J.* **22**: 131-35.

Dhillon, B. S. and Randhawa, J. S. 1983. Harvesting maturity and cold storage of Kinnow oranges. *Proc. Int. Citrus Symp.* 17-22 December, 1977, Bangalore, Horticultural Society of India, pp. 235-38.

Fallik, E. 2004. Prestorage hot water treatments (immersion, rinsing and brushing). *Postharvest Biol. Technol.* **32**: 125–34.

FAO. 2012. Citrus fruit – Fresh and processed. *Annual Statistics*, FAO, Rome.

Gautam, D. R., Dhar, R. P., Bhutani, V. P. and Dhuria, H. S. 1977. Ethephon for postharvest degreening of Kagzi lime. *Indian J. Agric. Sci.* **47**: 282-84.

Ghasemnezhad, M., Marsh, K.,Shilton, R., Babalar, M. and Woolf, A. 2008. Effect of hot water treatments on chilling injury and heat damage in 'Satsuma' mandarins: Antioxidant enzymes and vacuolar ATPase, and pyrophosphatase. *Postharvest Biol. Technol.* **48**: 364–71.

Hofman, P. J., Stubbings, B. A., Adkins, M. F., Corcoran, R. J., White, A. and Woolf, A. B. 2003. Low temperature conditioning before cold disinfestations improves 'Hass' avocado fruit quality. *Postharvest Biol. Technol.* **28**: 123–33.

Hong, S. I, Lee, H. H. and Kim, D. 2007. Effects of hot water treatment on the storage stability of Satsuma mandarin as a postharvest decay control. *Postharvest Biol. Technol.* **43**: 271–79.

Jawanda, J. S, Arora, J. S, and Sharma, J. N. 1973. Fruit quality and maturity studies of Kinnow mandarin at Abohar. *Punjab Hortic. J.* **13:** 3-12.

Kaewsuksaeng, S., Urano, Y., Aiamla-or, S., Shigyo, M. and Yamauchi, N. 2011. Effect of UV-B irradiation on chlorophyll-degrading enzyme activities and postharvest quality in stored lime (*Citrus latifolia* Tan.) fruit. *Postharvest Biol. Technol.* **61**: 124-30.

Kluge, R. A., Luiza, M., Jomori, L., Jacomino, A. P., Carolina, M., Vitti, D. andPadula, M. 2003. Intermittent warming in 'Tahiti' lime treated with an ethylene inhibitor. *Postharvest Biol. Technol.***, 29**: 195–03.

Kouassi, K. H. S., Bajji, M. and Jijakli, H. 2012. The control of postharvest blue and green molds of citrus in relation with essential oil–wax formulations, adherence and viscosity. *Postharvest Biol. Technol.* **73**: 122-28.

Ladaniya, M. S. 1996. Standardization of fruit maturity indices in Spring blossom (*Ambia*) crop of Nagpur mandarin (*Citrus reticulata* Blanco). *J. Maharashtra Agric. Univ.* **21**: 73-75.

Ladaniya, M. S., Naqvi S. A. M. H. and Dass, H. C. 1994. Packing line operation and storage of Nagpur mandarin. *Indian J. Hortic.* **51**: 215-21.

Ladaniya, M. S. 2001. Response of 'Mosambi' sweet orange (*Citrus sinensis*) to degreening, mechanical waxing, packaging and ambient storage conditions. *Indian J. Agric.* Sci. **71**: 234-39.

Ladaniya, M. S. and Shyam Singh. 2000a. Maturity indices for acid lime (*Citrus aurantifolia*) cultivar Kagzi grown in Central India. *Indian J. Agric.* Sci. **70:** 292-95.

Ladaniya, M. S. 2000. Response of 'Nagpur' mandarin to degreening by ethylene trickle method, mechanical wax coating and storage at ambient condition. *J. Food Sci. Technol.* **37:** 103-10.

Ladaniya, M. S. 2007. Quality and carbendazim residues of 'Nagpur' mandarin fruit in modified atmosphere package. *J. Food Sci. Technol.* **44**: 85-89.

Ladaniya, M. S. 2004. Standardization of temperature for long term refrigerated storage of 'Mosambi' sweet orange (*Citrus sinensis* Osbeck). *J. Food Sci. Technol.* **41**: 692-95.

Ladaniya, M. S. and Mahalle, B. 2011. Fruit maturation and associated changes in Mosambi orange (*Citrus sinensis*). *Indian J. Agric. Sci.* **81:** 494 -99.

Ladaniya, M. S. and Shyam Singh 2000b. Influence of ventilation and stacking pattern of corrugated fibre board containers on forced-air precooling of Nagpur mandarin. *J. Food Sci. Technol.* **37:** 233-37.

Ladaniya, M. S. and Sonkar R. K. 1999. Effect of curing, wax application and packaging on collar breakdown and quality of stored Nagpur mandarin (*Citrus reticulata*). *Indian J. Agric. Sci.* **67**: 500-03.

Ladaniya, M. S., Sonkar, R. K., Shaikh, A. J. and Varadarajan, P. V. 1999. Corrugated fibre board containers for packaging, transport and storage of 'Nagpur' mandarin in domestic market. *Indian Food Packer* **53:** 5-15.

Ladaniya, M. S. 2011. Physico-chemical, respiratory and fungicide residue changes in wax coated mandarin fruit stored at chilling temperature with intermittent warming. *J. Food Sci. Tech.* **48:** 150-58.

Lai, K., Chen, S., Hu, M., Hu, Q., Geng, P., Weng, Q. and Jia, J. 2012. Control of postharvest green mold of citrus fruit by application of endophytic *Paenibacillus polymyxa* strain SG-6. *Postharvest Biol. Technol.* **69**: 40-48.

Li, Q., Wu, F., Li, T., Su, X., Jiang, G., Qu, X., Jiang, Y. and Duan, X. 2012. 1-Methyl- cyclopropene extends the shelf-life of 'Shatangju' mandarin (*Citrus reticulate* Blanco) fruit with attached leaves. *Postharvest Biol. Technol.* **67**: 92-95.

Liao, H-L., Alferez, F. and Burns, J. K. 2013. Assessment of blue light treatments on citrus postharvest diseases. *Postharvest Biol. Technol.* **81**: 81-88.

Mayuoni, L., Tietel,Z., Patil, B. S. and Porat, R. 2011. Does ethylene degreening affect internal quality of citrus fruit. *Postharvest Biol. Technol.* **62:** 50-58.

Mann, S. S. 1978. Harvesting and storage of Kinnow fruits. *Punjab Hortic. J.* **18**: 154-55.

Mukherjee, P. K. and Singh, S. N. 1983. Storage behaviour of citrus fruits grown in Northern India- A review. *Proc. Int. Citrus Symp.* Bangalore, 17-22 December, 1977, Horticultural Society of India, pp 223-229.

NHB. 2011. Indian Horticulture database, 2011. *National Horticulture Board*, Gurgaon. India. pp. 278.

Njombolwana, N. S., Erasmus, A. and Fourie, P. H. 2013. Evaluation of curative and protective control of *Penicillium digitatum* following imazalil application in wax coating. *Postharvest Biol. Technol.* **77**: 102-10.

Nong, Z. R., Tang, Z. P., Liang, Y. P. and Lu, L., 2008. Study on the effects of Shatangju (Citrus reticulate Blanco) fruit attached leaves storage under the normal temperature. *J. Anhui Agric. Sci.* **36:** 9714–16 (in Chinese with English abstract).

Obenlanda,D., Collin, S., Mackey, B., Sievert, J. and Arpaia, M. L. 2011. Storage temperature and time influences sensory quality of mandarins by altering soluble solids, acidity and aroma volatile composition. *Postharvest Biol. Technol.* **59**: 187-93.

Ortiz, C., Blasco, J., Balasch, S. and Torregrosa, A. 2011. Shock absorbing surfaces for collecting fruit during the mechanical harvesting of citrus. *Biosyst Eng.* **110:** 2-9.

Pedro A. M-R., Clara M-H. and Palou, L. 2013. Control of citrus postharvest *Penicillium* molds with sodium ethylparaben. *Crop Prot.* **46**: 44-51.

Porat, R., Weiss, B., Cohen, L., Daus, A. and Cohen, E. 2003. Effects of intermittent warming and temperature conditioning on the postharvest quality of 'Oroblanco' citrus fruit following long term cold storage. *Hortic. Technol.* **13**: 70-74.

Rodov, V., Ben-Yehoshua, S., Albagli, R. and Fang, D. Q. 1995. Reducing chilling injury and decay of stored citrus fruits by hot water dips. *Postharvest Biol. Technol.* **5**: 119-27.

Safizadeh, M. R., Rahemi, M., Tafazoli, E. and Emam, Y. 2007. Influence of vacuum infiltration with calcium on chilling injury, firmness and quality of Lisbon lemon fruit. *Am. J. Food Technol.* **2:** 388-96.

Salvador, A., Navarro, P., Monterde, A. and Martínez-Jávega, J. M. 2008. Postharvest application of auxins to control calyx senescence in clementines submitted to degreening treatment. *Proceedings of the International Society of Citriculture* **11:** 1377-82.

Sanchez-Ballesta, M. T., Lluch, Y., Gosalbes, M. J., Zacarias, L., Granell, A. and Lafuente, M. T. 2003. A survey of genes differentially expressed during long-term heat-induced chilling tolerance in citrus fruit. *Planta,* **218**: 65–70.

Sanghvi, K. U. and Patil, A. V. 1983. Effect of Ethrel on nucellar Mosambi fruit (*C. sinensis* Osbeck). *Proc. Int. Citrus Symp.* 17-22, December 1977, Bangalore, Horticultural Society of India, pp 322-325.

Sdiri, S., Navarro, P. and Salvador, A. 2013. Postharvest application of a new growth regulator reduces calyx alterations of citrus fruit induced by degreening treatment. *Postharvest Biol. Technol.* **75**: 68-74.

Sdiri, S., Navarro, P., Monterde, A., Benabda, J. and Salvador, A. 2012a. New degreening treatments to improve the quality of citrus fruit combining different periods with and without ethylene exposure. *Postharvest Biol. Technol.* **63**: 25-32.

Sdiri, S., Navarro, P., Monterde, A., Benabda, J. and Salvador, A. 2012b. Effect of postharvest degreening followed by a cold-quarantine treatment on vitamin C, phenolic compounds and antioxidant activity of early-season citrus fruits. *Postharvest Biol. Technol.* **65**: 13-21.

Silvia A., Valencia-Chamorro, Pérez-Gagoa, M. B., Río, M. A. D and Palou, L. 2009. Effect of antifungal hydroxypropyl methylcellulose (HPMC)–lipid edible composite coatings on postharvest decay development and quality attributes of cold-stored 'Valencia' oranges. *Postharvest Biol. Technol.* **54**: 72–79.

Singh, K. K. 1971. Storage behaviour of sweet orange and mandarins. Technical Bulletin (Agric series), ICAR, No 35, pp. 106.

Singh, H. K. P. P., Singh, S. N. and Dhatt, A. S. 1998. Studies on fruit growth and development in Kinnow. *Indian J. Hortic.* **55:** 177-82.

Srilaong, V., Aiamla-or, S., Soontornwat, A., Shigyo, M. and Yamauchi, N. 2011. UV-B irradiation retards chlorophyll degradation in lime (*Citrus latifolia* Tan.) fruit. *Postharvest Biol. Technol.* **59**: 110–12.

Tietel, Z., Lewinsohn, E., Fallik, E. and Porat, R. 2012. Importance of storage temperatures in maintaining flavour and quality of mandarins. *Postharvest Biol. Technol.* **64**: 175-82.

Tugwell, B. L. and Chvyl, W. L. 1995. The effect of storage temperature on development of rind blemishes on Washington naval orange. *Conference on Science and Technology for the fresh food revolution*, Melborne, Australia, pp. 18-22.

Vendittia,T.,Dorea, A., Molinua,M. G.,Agabbiob,M. and D'hallewina. G. 2009. Combined effect of curing followed by acetic acid vapour treatments improves postharvest control of *Penicillium digitatum* on mandarins. *Postharvest Biol. Technol.* **54**: 111–14.

Wilma, D. P., Regnier, T. and Combrinck, S. 2009. Essential oil amended coatings as alternatives to synthetic fungicides in citrus postharvest management. *Postharvest Biol. Technol.* **53**: 117–22.

Youssef, K., Ligorio, A.,Nigro, F. and Ippolito, A. 2012a. Activity of salts incorporated in wax in controlling postharvest diseases of citrus fruits. *Postharvest Biol. Technol.* **65**: 39-43.

Youssef, K., Ligorio, A., Sanzani, S. M., Nigro, F. and Ippolito, A. 2012b. Control of storage diseases of citrus by pre and postharvest application of salts. *Postharvest Biol. Technol.* **72**: 57-63.

Zeng, R., Zhang, A., Chen, J. and Fu, Y. 2012. Postharvest quality and physiological responses of clove bud extract dip on 'Newhall' navel orange. *Scientia Hortic.* 138: 253-58.

Zhang, J. 2007. The potential of a new fungicide fludioxonil for stem-end rot and green mold control on Florida citrus fruit. *Postharvest Biol. Technol.* **46:** 262–70.

Chapter 15

Custard Apple

Sunil Pareek

Associate Professor, Department of Agriculture and Environmental Science, NIFTEM, Plot No. 97, Sector 56, HSIIDC Industrial Estate, Kundli, District-Sonepat – 131 028, Haryana, India

E-mail: sunil.pareek@niftem.ac.in

1.0 Introduction

Custard apple (*Annona squamosa* Linn.) belongs to the family Annonaceae and is native to South America. All sweet *Annona* species including hybrids which, show creamy-custard like flesh are sometimes included as custard apple. However, *A. squamosa* L. is botanically classified as sugar apple, whereas those of *A. reticulata* L. as custard apple. In India, *A. squamosa* is popularly known as custard apple. Therefore, in this chapter custard apple refers as *A. squamosa*. The important commercial species, synonyms and common names are given in Table 1. Custard apple produces sweet-sour and pleasant tasting fruits containing a variety of nutrients for either fresh consumption or the juice and pulp industries. The postharvest limitation for distant markets include quick ripening and softening that makes fruits difficult to handle and prone to damage and loss.

Custard apple could be consumed fresh as a dessert fruit and added into some sweet recipes. The fruit is good source of carbohydrates, antioxidants especially ascorbic acid (vitamin C), and some minerals such as magnesium and potassium. Ripe fruit of 'Fai' and 'Nang' contain 25 to 26.5 per cent of total soluble solids (TSS), but the later comprises high titratable acidity (TA). The TSS/TA ratio of 'Fai' and 'Nang' fruit is 223.4 and 114.9, respectively. Levels of both vitamin C and glucose in fruit reach a maximum at the climacteric peak, but decrease in over ripe fruit.The amounts of sugar was found to be quite high (58 per cent of dry mass). Glucose and fructose are the major sugars in the fruit. The portion of glucose: fructose: sucrose in ripe pulp is 5.8: 4.5: 0.8 g 100 g^{-1} FW. Fruit pulp is rich in vitamin C (70.8 mg 100 g^{-1} FW). While flesh of 'Green' fruit contains vitamin C at 6.8 mg 100 g^{-1} FW, 'Brown' custard apple fruit accumulates as high as 21.3 mg 100 g^{-1} FW. Custard apple flesh also contains some phenolic compounds such as eugenol, catechin, chlorogenic and gallic acids (0.58 to 0.65 mg gallic acid ml^{-1}). Phenolic compounds are 3-4 times higher in purple type than green type varieties of sugar apples on ripening.

Table 1. Botanical or Specific, Common and Synonyms of Five *Annona* Species

Botanical Name	*Synonyms*	*Common Name*
A. cherimola Mill.	*A. tripetala* Aiton, *A. pubescens* Salisb.	Cherimola, Cherimoya
A.muricata L.	*A. muricata* L., *A. bonplandiana* Kunth; *A. cearensis* Barb. Rodr.; *A.macrocarpa* Werckle; *A. muricata* var. *borinquensis* Morales; *Guanabanus muricatus* Gomez	Soursop
A. reticulata L.	*A. excelsa* Kunth; *A. laevis* Kunth; *A.longifolia* Moc. and Sesse; *A.riparia* Kunth.	Custard apple
A.senegalensis Pers.	*A.arenaria* Thonn.; *A. chrysophylla* Boj.; *A. Chrysophylla var. porpetac* Bail.; *A.senegalensis var. porpetac* Bill. wild	Wild soursop
A.squamosa L.	*A. asiatica* L.; *A. cinerea* Dunal; *Guanabamus Squamosus* Gomez	Sugar apple

Source: Pinto *et al.*, 2005.

2.0 Status of Cultivation

Tropical regions such as Brazil, Mexico, India, Philippines, Taiwan and Thailand are important production areas for custard apple. Thailand can generally produce and supply custard apple fruits for about 6-7 months between June and December, but normal seasonal production is between June and August. Custard apple fruits are seasonally harvested between September and October in Taiwan, while in Florida, fruit matures from late June to October, with a small second crop maturing in December and January. However, since the flowering behaviour is practically forced by pruning techniques and irrigation control, custard apple can be programmed to bear fruits all year round.

In the world market, demand for custard apple has however, failed to increase. In Asia, custard apple is one of Taiwan's 'Top 10' fruits; in Thailand and Southeast Asian countries, the volumes of export have not been increased due principally to its short postharvest life. Some physiological disorders such as fruit cracking, splitting, and blackening as well as difficulty to eating fruit containing a lot of seeds are additional factors for export failure. Furthermore, consumers of custard apple are limited because of inconsistent fruit quality, quick ripening (within a few days after harvest), and lack of proper advertisement or promotion in many countries. According to a commercial survey, consumers are largely unaware of varietal differences but the more common and less expensive varieties are more in demand. In terms of fruit quality, consumers desire fruit with minimum seeds, more consistent flavour and better external shape and appearance, and without disorders such as woodiness or skin blackening.

In India, custard apple fruit covers an area of 46,000 hectares with production of 4,01,000 MT during 2018 as per the Horticulture statistics at a glance published by Dept of Agril Cooperation, Min of Agriculture and FW, Govt of India. However, this is not a comprehensive picture of production. Custard apple is recently included in National Horticulture Board Database. The major producing states are Chhatisgarh, Gujarat, Andhra Pradesh, Karnataka and Rajasthan. The productivity per hectare is however highest in Gujarat (11.0 t ha^{-1}) followed by Rajasthan (8.2 t ha^{-1}), Karnataka (8.0 t ha^{-1}) and Andhra Pradesh (6.0 t ha^{-1}). During 2015-16 Gujarat recorded the highest production of custard apple fruits (61,180 t) with 20.53 per cent share in total production (Table 2).

Table 2. Production of Custard Apple in India (2015-16)

State	*Production (t)*	*Share (per cent)*
Chhatisgarh	39,580	13.28
Gujarat	61,180	20.53
Andhra Pradesh	11,350	3.81
Karnataka	14,450	4.85
Rajasthan	6,120	2.05

Source: APEDA, Agri exchange (2015-16).

3.0 Major Commercial Varieties

In Thailand, custard apples are divided into two main groups: 'Fai' and 'Nang' types. 'Fai', a local Thai type, contains a lot of seeds. The flesh turns very soft soon after the onset of ripening and the fruit carpels quickly split out during ripening. Two cultivars of 'Fai' type are named 'Green Fai' and 'Purple Fai' based on their peel colours. The 'Nang' type, which spread to Thailand from neighboring Vietnam, retains consistency of pulp during ripening and the carpels of the fruit remain stuck together. Three cultivars of the 'Nang' type are 'Green Nang', 'Purple Nang', and 'Golden Nang' (Table 3). In Florida, USA, the fruit of 'Seedless' custard apple, apparently identical to 'Cuban Seedless' and 'Brazilian Seedless' cultivars, splits when it matures on the tree and its quality is slightly lower than normal. The 'Purple' cultivar from Cuba has an attractive appearance with purple skin and pinkish flesh, but inferior internal quality.

Table 3. Important Cultivars in Major Producing Countries

Country/Region	*Name of Cultivar*
Brazil	'IPA Selections'
Cuba	'Cuban Seedless'
Egypt	'Abd El-Razik', 'Beni Mazar'
India	'Balanagar', Arka Sahan', 'Mammoth'
Philippines	'Molate', 'Lobo'
Taiwan	'Ruan-zhi', 'Cu-lin', 'Da-mu', 'Xi-lin', 'Ti-nong No. 1'
Thailand	'Fai Kaew', 'Fai Krung', 'Nang Kaew', 'Nang Sir Krung', 'Nang Thong'
USA	'Thai Lessard', 'Kampong Mauve', 'Red Sugar', 'Cuban Seedless', 'Brazilian Seedless'
Vietnam	'Nabo', 'Na Dai'

Source: Modified from Wanichkul (2009a), Chen and Paull (2008).

In India, 'Arka Sahan', 'Balanagar' and 'Mammoth' are promising cultivars of custard apple. The salient features of the important cultivars are described as under:

a) APK (Ca)-1

It is a clonal selection from high yielding type in State Horticultural Farm, Courtallam, Tirunelveli District, T.N. and developed at Regional Research Station, Aruppukkottai. The cultivar is high yielding in rainfed vertisol (Black soil) and adapted to semiarid plains. Average number of fruits would be 72 $tree^{-1}$. Average yield 14.90 kg $tree^{-1}$, 30.7 per cent more than Balanagar. Fruit weight 207 g, TSS 24.5 °Brix, acidity 0.2 per cent.

b) Arka Sahan

Developed at IIHR, Bangalore, is the progeny of Island gem (*Annona atemoya* Hort.) x Mammoth (*A. squamosa* L.). Fruits are harvested in September-October and take about 6-7 days to ripe. The fruit skin is light green in colour, moderately thick (0.5 cm) with large, flat eyes and waxy bloom on the surface. The pulp is creamy white colour, juicy with mild pleasant aroma, scarce seeds (9 per 100 g) and large segments. Edible pulp is remarkable for its sweetness with 22.8 per cent total sugars and 30 °Brix as against 24 °Brix in 'Mammoth'. Average yield is 12 t ha^{-1} (Figure 1).

Figure 1. Fruits of Arka Sahan

c) Balanagar

It is a promising selection from Mahboob-nagar District (A.P.). Fruit shape is pyramidal or chordate; areoles very rough, pitted, usually remains green when ripe. Pulp is bright white, very sweet, juicy with medium granular mesocarp. TSS is 20.7 °Brix. Bears 50-60 fruits per tree. Each fruit weights 130-140 g. Average seed content is 25 per fruit.

d) Mammoth

Plant is dwarf compared to other varieties. Bears fruits with smooth, prominent and round areoles. Pulp bright white and sweet. Fruit weight about 125 g, TSS 20 °Brix with a better keeping quality.

e) Red Sitaphal

Fruits pinkish dark with erythrite red pulp. Average fruit weight 150-160 g with TSS 22.3 °Brix.

f) NMK-01 (Golden)

This variety has vary large fruits approximately 700g with very less seeds (around 15 seeds/fruit) as against 70 seeds per fruit in cv. Balangar. The average yield is 30 t/ha as against 12-15 t/ha in other commercial varieties. This was developed by Mr. Kaspate from Solapur, Maharashtra.

Figure 2. NMK-01 (Golden)

4.0 Nature and Causes of Postharvest Losses

4.1 Preharvest Factors

Custard apple is hardy tree and thrives well in low rainfall regions with rocky and poor fertility soils. It requires very less care with respect to orchard management. Insects such as mealy bug and seed borer and postharvest fungal disease are causing the problem in custard apple cultivation. Mealy bug (*Pseudococcus* sp.) and *Annona* seed borer (*Bephratelloides cubensis*) are predominant pests and attack custard apple during fruit development in the field, affecting the postharvest quality. Mealy bug attacks fruit at the groove knobs of young fruit after fruit setting. Although the attaching mealy bugs are removed from fruit after harvest, the lesion scrap symptoms remain between the grooves. Custard apple seed borer infests fruit at all stages of fruit development. The larva develops in the seeds of young fruit and transforms to be an adult emerging at the mature stage of the fruit.

Harvested *Annona* fruits are susceptible to fungal attack when stored at temperatures above 25 °C. Postharvest diseases are mostly latent infections or fruits are infected by diseases during development on trees. Anthracnose caused by *Colletotrichum gloeosporioides* is the major concern of custard apple fruit during both pre and postharvest handlings. *C. gloeosporioides* could cause diseases of fruit russeting, fruit rot, and mummified fruit in infected fruit. Anthracnose produces pink spore-masses, causing a dry rot which cause serious losses in the orchards in South Asia. *Botryodiplodia* rot caused by *B. theobromae* Pat. severely affects custard apple fruits grown in Africa and Southeast Asia. The symptoms are at first purple, developing to black pimpled skin when the flesh under infected lesion is rapidly raided and then become either soft or there is brown hardening, probably depending on environmental conditions. Black canker caused by *Phomopsis anonacearum* which spread out in Australia during wet production seasons expresses as purple spots on the fruit near the stem end. The symptoms expand as fruit development proceeds and the surface becomes hard and cracks. Pink mold rot, caused by *Trichothecium roseum* is generally found on fruit in the orchard and market places in India.

4.2 Fruit Setting and Time of Harvest

The productivity of some custard apples is reduced with natural pollination due to failure of fruit setting. Although spraying to blooming flowers with 250 ppm 2,4,5-trichlorophenoxyacetic acid (2,4,5-TP) spectacularly improves initial fruit setting and checks 100 per cent of the fruit drops at premature stage. Artificial hand pollination of custard apple and atemoya dramatically increases fruit setting, fruit shape and fruit development. Thus artificial hand pollination should form part of a 'Good Agricultural Practice' to improve the yield of existing cultivars.

Some *Annona* fruits often crack during postharvest ripening. Harvesting in early stages of maturation not only extends the shelf life, but reduces fruit cracking. 'African Pride' fruit harvested at commercial maturity of 17-18 weeks after anthesis (WAA) took several days for full ripening with skin turning black rapidly. However, the fruits harvested at 16 WAA ripened normally in good quality whereas when harvested before that, the fruit ripened normally with longer shelf life, but with lower eating quality. Harvesting custard apple fruits several weeks before the commercial maturity effectively reduced fruit cracking, but the fruit had higher weight loss and lower levels of TSS and vitamin C.

4.3 Plant Growth Regulator Treatments

Fruit cracking and splitting on and off the tree is one of the main problems in custard apple postharvest management. Preharvest gibberellic acid (GA_3) treatment @ 50 and 100 ppm twice at 7 and 14 days before harvest reduced postharvest cracking of 'African Pride' atemoya.

4.4 Fruit Bagging

Bagging custard apple fruit during early stages of fruit development could prevent pest and disease infestation. In Brazil, bagging fruit with a white plastic bag reduces infestation from fruit borer *Cerconota anonella* without any effect on size, weight, and TSS content, but it showed low fruit firmness. Fruits bagged in yellow bags had the highest brightness. Furthermore, fruits covered with a paper bag after fruit set showed larger fruit and reached maturity faster than non-bagged fruits.

5.0 Postharvest Management

5.1 Harvesting

Harvest time based on anthesis in custard apple is impractical, because flowering can occur during many months. On the other hand some cultural practices, such as pruning or time of hand pollination can alter the time of harvesting. In Taiwan, fruits are harvested from July to September or October to March if pruned in January-February or in June, respectively. Premature harvesting can promote poor fruit quality. Fruits left to ripen on the tree are often eaten by birds and bats, and over-mature fruits have a tendency to undergo cracking and decay.

Harvesting of fruits at 108 days after pollination showed high pulp content with poor shelf life of 3 days. A slight early harvesting *viz.* 104 days after pollination could result in better shelf life.

The time of harvesting is commonly determined by the fruit skin colour, which changes with the proximity of physiological maturity. Changes in skin colour from dark green to light green or greenish-yellow is the primary maturity index for custard apple. Custard apple fruits are considered to be mature and reach their harvesting point when the skin changes colour and the segments spread far apart, exposing a creamy yellow skin. In India, it is harvested at fully mature stage when it develops a creemish-green skin colour with valleys between aereols turning yellowish. For local markets, fruits must be harvested when mature, with 20 to 40 per cent yellowish-green skin, while for export market, 10 to 20 per cent yellowish-green skin is satisfactory. When the fruits are harvested with more than 75 per cent yellowish-green skin, they will ripen in 1 to 3 days, while fruits harvested at less than 5 per cent yellowish-green skin colour do not ripen completely at all.

Production of *Annona* fruits throughout the year can be managed by production practices. Pruning branches followed by drought stress for 5-6 weeks induces flowering in custard apples. Fruits can be harvested within four months after flowering. For good quality, fruits should be picked at a mature stage and ripened after harvest. Local farmers generally harvest sugar apple fruit at the suitable maturity by noticing the areoles swelling and the fruit grooves expanding. The physiological maturity stage is between 16-17 weeks after anthesis (WAA) when the fruit specific gravity at maturation should be less than 0.94.

Custard apple is usually harvested during the summer months, from July to November, and can be extended to March in USA. Atemoya is harvested in autumn in the southern hemisphere and in winter in Australia. In India, custard apple harvesting time differs in south, west and north India. In Andhra Pradesh and Karnataka fruits harvested during August to October, in Maharashtra from September to January, in Gujarat from September to December and in Rajasthan from October to mid November.

Fruit is hand harvested and put into lug boxes or baskets. Sometimes the fruit is harvested using pruning shears, in order to leave a portion of the peduncle attached to limit stem end rots. Fruit should be harvested preferably early in the morning, when temperatures are still cool.

5.2 Sorting and Grading

Custard apple rind is very susceptible to mechanical damage mainly due to protruding areoles of its carpel segments. Gentle handling is required throughout the postharvest chain. Quality requirement of custard apple fruit includes proper fruit maturity and shape of the cultivar without physical damages and diseases or pests on the fruit. Fruit size of custard apple is divided into four groups in Thailand's domestic markets, including extra-large size (average weight 350 g), large size (average 300 g), medium size (average 250 g) and small size (average 160 g). For international standards, the Thai agricultural standard (TAS 21-2013) and the European standard of the United Nations/Economic and Social Council (UN/ECE, 2002) or the East African Standard (CD/K/022: 2010), custard apples are graded according to their sizes as shown in Table 4. The criteria for grade designation in India is given in Tables 5 and 6.

Table 4. Standard Sizes of Custard Apple Fruits

Size	*TAS[1] (g)*	*Size*	*UN/ECE[2] or EAS[3] (g)*
1	>650	1	>825
2	550-650	2	425-825
3	450-550	3	225-425
4	350-450	4	100-225
5	250-350		
6	200-250		
7	200-250		
8	100-150		

1: Thai Agricultural Standard (TAS 21-2013).

2: Standard of the United Nations/Economic and Social Council (UN/ECE, 2002).

3: East African Standard (CD/K/022: 2010).

5.2.1 Grade Designations and Quality of Custard Apples in India

a) Definition of Quality

In India, APEDA define the custard apple quality and grades as below:

i) Custard apple shall be obtained from varieties (cultivars) of *A. squamosa* of Annonaceae family.

b) Minimum requirements

i) Custard apple shall be:

(a) intact, firm, sound and clean;

(b) fresh in appearance;

(c) free from any visible foreign matter;

(d) practically free from bruising;

(e) free of pests affecting the general appearance of the produce;

(f) free of damage caused by pests;

(g) free from damage caused by low and high temperature;

(h) free of abnormal external moisture;

(i) free of any foreign smell and/or taste;

(j) with or without the stem attached.

ii) At the time of maturity, colour between the areoles should be creamy white,

iii) They shall comply with the residue levels of heavy metals, pesticides and other food safety parameters as laid down by the Codex Alimentarius Commission for exports.

Table 5. Criteria for Grade Designation in India

Grade Designation	*Grade Requirements*	*Grade Tolerance*
Extra class	Custard apples must be of superior quality. They must be well developed and have all the characteristic and colouring typical of the variety. Following slight defects may be there, provided they do not affect the general appearance of the produce, the quality, the keeping quality and presentation in the package; Slight skin defects (blemishes, scars, sunspots) not exceeding 2 per cent of the total surface area.	5 per cent by number or weight of custard apples not satisfying the requirements of the grade, but meeting those of Class I grade or, exceptionally, coming within the tolerances of that grade with the exception of over-riped fruit.
Class I	Custard apples must be of good quality. They must be well developed and have all the characteristics and colour typical of the variety. Following slight defects may be there, provided they do not affect the general quality, the keeping quality and presentation in the package; Slight skin defects (blemishes, scars, scrapes, scratches, sun-spots) not exceeding 5 per cent of the total surface area.; Slight defects in colour. The defects should not affect the pulp of the fruit.	10 per cent by number or weight of custard apples not satisfying the requirements of the grade, but meeting those of Class II or, exceptionally, coming within the tolerances of that grade.
Class II	This grade includes custard apples which do not qualify for inclusion in the higher grades, but satisfy the minimum requirements. Custard apple may have the following defects, provided they retain their essential characteristics as regards the quality, the keeping quality and presentation in the package. Defects in shape and colour Small healed surface scars, not likely to impair significantly the appearance or conservation of the fruit, Skin defects (*i.e.* scratches, scars, scrapes bruises and blemishes) not exceeding 10 per cent of total surface area. The defects should not affect the pulp of the fruit.	10 per cent by number or weight of custard apples not meeting the requirements of the grade, but meeting the minimum requirements. Within this tolerance, not more than 2 per cent in total may consist of overripe and/or split fruit.

c) Other Requirements

i) The custard apples must have been carefully picked and reached an appropriate degree of development in accordance with criteria proper to the variety and to the area in which they are grown. The development and condition of the custard apples must be such as to enable them

to withstand transport and handling, and

to arrive in satisfactory condition at the place of destination.

ii) In case of Extra class and Class I – extra care is necessary to avoid any disintegration into segments.

d) Provision Concerning Sizing

Table 6. Size Grades of Custard Apples in India

Size Code	*Weight (g)*
A	150-200
B	201-300
C	301 and above

Minimum weight of fruits for Extra class shall not be less than 200 g.

Size Tolerances

For all classes, 10 per cent by number or weight of custard apples not conforming to the minimum size specified, for the grade, but meeting the size requirement for that grade immediately below.

5.3 Packaging

After harvest, fruits should be immediately transferred to a packing house and then graded according to size and shape. In Thailand, local containers both for retail and bulk are usually made from bamboo sheets that easily damage the fruit. For export, fruits are recommended to be wrapped with foam net and laid with stem end down to prevent the fruits from mechanical damage (Figure 2). Fruits are packed in a single layer 4.5 or 9 kg fibre board boxes with foam sleeves or paper wrapping. In Australia, custard apple is commonly packed in single layer trays of 6-8 kg. In Brazil, custard apples are often packed in wooden boxes with weight varying according to the state or local market; smaller fruits are often sold in packages with 12 to 24 fruits. The fruit arrangement in the box varies according to the fruit size and number of fruits per box; a type 9 carton has fruits with average weights of 500 g, which is the best commercial weight, and the package has 9 fruits (Table 7). However, in India, mostly custard apple fruits are sold in wooden baskets of inferior quality. Cardboard packages are also in use. In modern markets, now fruit is packaged individually in paper wraps or in foams and about 12 fruits packed in single layer tarys. Custard apples are also packed in India in 3-Ply CFB boxes of RSC type or EPS (Expanded Polystyrene) boxes. The capacity of the box is 3 kg. The fruits are placed in one layer inside the box and depending upon the size the number of fruits vary. The box is closed by application of pressure sensitive tape. Individual film sealing of atemoya cv. 'PR3' extended their shelf life. At 15 °C, non-wrapped atemoyas had a shelf life of 13 days against 17 days of those wrapped with PD-955 (copolymer). Low density polyethylene (LDPE) film was found to be inadequate to pack atemoya. Fruits should be unwrapped before ripening at room temperature to avoid off flavour development.

Figure 2. Custard Apple Fruits Wrapped up with Foam Net and Laid with Fruit Stem End Down

Table 7. Carton Types and Classification and Packing of Custard Apple Fruit in Brazil

Fruit Characteristics			*Fruit Arrangement in the Package*		*No. of Fruits/ Carton*
Length (cm)	*Diameter (cm)*	*Weight (g)*	*Columns*	*Rows*	
>10.5	10.5-11.5	600-620	2	4	8
9.5-10.5	10.0-10.5	480-520	3	3	9
8.5-9.5	9.0-10.0	360-390	3	4	12
8.0-8.5	8.5-9.0	280-320	3	5	15
7.5-8.0	7.5-8.5	210-215	3	6	18

5.3.1 Modified and Controlled Atmosphere

Coating 'Nang' custard apples with 0.5 and 1.0 per cent chitosan or individually wrapping the fruits with linear low density polyethylene (LLDPE) did not delay fruit softening when the fruits were stored at 13 °C and 95 per cent RH. Modified atmosphere packaging (MAP) with 6 μM and 15 μM PE bags, though, reduced weight loss and maintained skin and pulp colour. Fruits kept in 6 μM PE bags had a storage life of at least 18 days whereas after day 12 fruit kept in 15 μM PE bags showed skin darkening after removal from the bags due to the high CO_2 inside. Fully mature and freshly harvested custard apple stored in fibre board boxes had a shelf life of four days at ambient (27±2 °C) and eight days in cool storage conditions (15±2 °C). The postharvest combination of coating with 6 per cent waxol + 0.1 per cent carbendazim and forced air-precooling (at 10 °C) extended shelf life up to 14 days under cool storage. There are reports of the synergistic effects of ethylene absorbents and MAP when used in combination to extend storage life. Custard apples placed in PE bags containing $KMnO_4$ had a storage life of nine days while untreated fruit could be stored for five days. Custard apple packed in 0.1 mm thick polyvinylchloride (PVC) film plus $KMnO_4$ absorbent and stored at 16 °C with 90-100 per cent RH delayed fruit ripening.

5.4 Storage

5.4.1 Low Temperature Storage

Custard apple fruit become soft and ready-to-eat within several days at ambient temperature (24-32 °C). Low temperature storage is essential to extend the shelf life. However, since they are tropical fruits, temperatures used for storage of custard apples should be above 13 °C, principally to avoid chilling injury.

The marketing potential of cherimoya is hampered by its high perishability (at 18-20 °C it will ripen in 3-6 days) and susceptibility to chilling injury. The optimum temperature for prolonged cold storage of cherimoya, depending on the cultivar, ranges between 8 and 15 °C. Reports on low temperature storage of *Annona* indicate that the minimum storage temperature varies from 7 to 12 °C depending upon cultivar. The ripening period ranges from 5 to 8 days in cherimoyas, 7 days in custard apple and about 5 day in atemoya at 20 °C. Storage temperature at 20 °C gave the most acceptable fruit for fresh consumption. The safe range of storage temperature for 'Balanagar' custard apple was found to be between 15 and 20 °C, with maximum shelf life at 15 °C. The ripening of fruits was observed on days 4, 6 and 9 of storage at 25, 20 and 15 °C, respectively. The colour of the pulp, texture, taste and flavour of ripe fruits held at 25 and 20 °C were superior followed by fruits stored at 15 °C (Figure 3).

5.4.2 Effect of Chemicals

Dipping custard apple fruits in 100 ppm Naphthalene acetic acid (NAA) and 50 ppm GA_3 effectively reduced weight loss and slightly extended the marketability at ambient, whereas dipping fruit in 10^{-4} and 10^{-2} M indole acetic acid (IAA) accelerated ripening.

Custard apple fruits treated with $CaCl_2$ (6 per cent) and stored at 16 °C had reduced weight loss and respiration rates, less peroxidase activity, and higher firmness, and the biochemical processes of ripening were delayed. $CaCl_2$ dipping of custard apple fruits had a positive effect on reducing flesh browning.

Salicilic acid treatment lowered respiration, increased activities of antioxidant enzymes as compared to the control. TSS, total sugar content, softness and decay rate were significantly lowered in treated fruits, and in turn a delay in the fruit ripening process was achieved after 10 days of storage.

Coating cherimoya fruit with wax containing caranauba, morpholine fatty acid salt, sellac gum lac, microcrystalline wax, sodium polyphosphate, and orange oil, decreased respiration and ethylene production of the fruit, and extended the shelf life by 5 days, with less weight loss and minimal browning.

5.4.3 Ripening

In general, custard apple fruits are left to ripen naturally during transport to the consumer at ambient conditions. Being a climacteric fruit, it responds well to artificial ripening. At MPUAT, Udaipur custard apple ripening technology has been standardized by treating fruits with 100 ppm ethylene gas for 24 hours and keeping in a ripening chamber at 15 °C resulting in uniform ripening.

Atemoya fruit had acceptable eating quality for up to 4 days when ripened at 20 °C in ethylene free air or under propylene. Fruit ripened under ethylene free air had better fruit quality than fruit ripened under propylene. Ripening periods of 4-5, 5-6 and 8-9 days copuld be achieved in atemoya fruits at 24-25, 19-20 and 15-16 °C, respectively. The rate of ripening of 'Balanagar' custard apple fruits was delayed with decrease in storage temperature with more than 90 per cent of the fruits ripening on days 4, 6 and 9 of storage at 25, 20 and 15 °C, respectively (Vishnu Prasanna *et al.*,

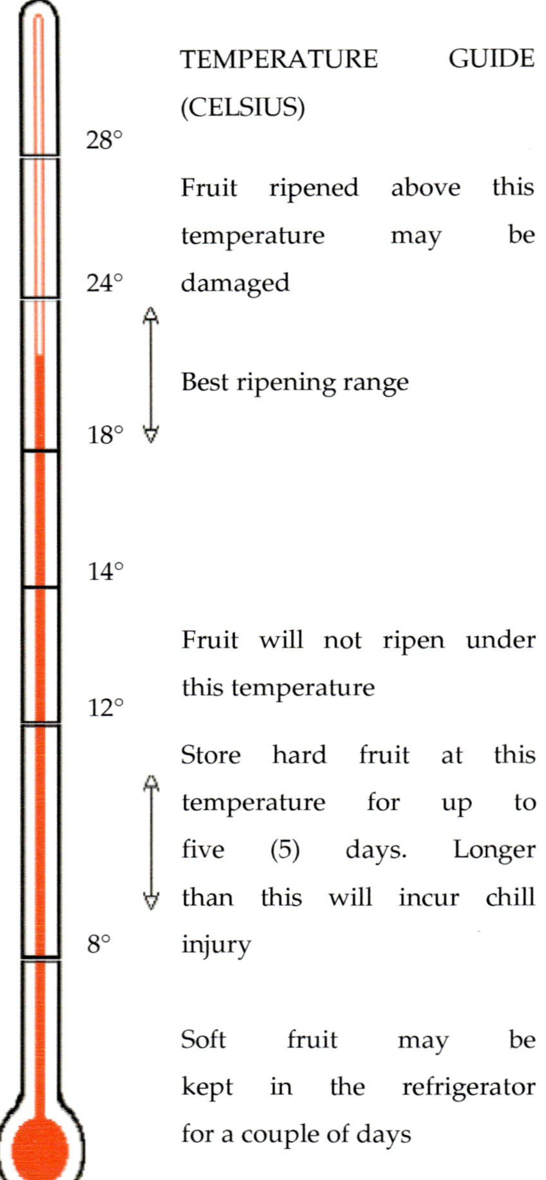

Custard apples are a delicate tropical fruit and are easily damaged if handled incorrectly. One can get a greater return if the following points are observed

1. Do not store green fruit under 8°C
2. Store at 8° to 12° for up to 5 days only, longer than this and chill injury occurs.
3. Fruit will not ripen under 14°C. Extended periods at this temp will lead to chill injury.
4. During winter store fruit in a warm area and cover to hold in their own warmth.
5. Best temperature range for ripening is 18° to 24°C.
6. Shelf life is short so buy small quantities more frequently.
7. Avoid small (under 300g) and dark green (immature) fruit.

Figure 3. Temperature Management in Harvested Custard Apple Fruit

2000). Custard apple fruits reached their edible quality at the beginning of the post-climacteric phase, with a postharvest life at 26 ± 2 °C and 60-70 per cent relative humidity of no longer than 4 days.

5.4.4 Physiological Disorders

a) Browning

Fruits are commonly susceptible to browning catalyzed by the enzyme PPO affecting its sensory and nutritional qualities. It has been reported that the PPO activity in soursop fruit was optimum at 7.0 for fully ripe and 7.5 for immature, mature and ripe fruits. It was optimum pH at 7.5 pH and 32 °C, respectively. Cherimoya fruit is also prone to browning, which is considered one the most important oxidative reactions leading to loss of sensory and nutritional quality.

Browning development is the most important factor limiting the quality of fresh-cut cherimoya. L-cysteine at 0.5 per cent was somewhat effective in reducing browning development, without affecting other quality attributes. PPO activity did not show differences between mature (at harvest) and ripe fruit (at processing) in both harvest times, but cherimoya fruits picked in November presented lower PPO activity and total phenolic content of L-cysteine-treated fruits did not show consistent differences with untreated fruit at 6 or 12 days at 0 °C. PPO activity was higher in the outer part of the cherimoya flesh compared to the middle or inner sector.

b) Fruit Splitting

Atemoya fruit splitting occurs frequently reducing marketable yield and quality, and increases disease attack. Splitting during atemoya fruit ripening first appears around the peduncle and radiates out from the fruit base. Fruit splitting can occur pre and postharvest and the intensity varies with the cultivar. Fruit early in the season with higher TSS are also more prone to splitting. It occurs during the respiration peak, at the start of ethylene production, with two to three fold increase in TSS in all the fruit tissue, increase in dry matter content of skin and receptacle, decrease in dry matter in the flesh, increased weight loss, increased receptacle diameter and a decrease in fruit circumference and peduncle diameter. This was due to the osmotic and subsequent turgor changes related to productions of neutral sugars during ripening led to the movement of water from the skin and receptacle to the flesh. The effect of water movement was compounded with weight loss during ripening. In addition, the receptacle, though it increased in dry matter content, also increased in diameter possibly imposing increased stress on the flesh and skin. Heating the fruit to 47.5 °C, with increasing levels of humidity increased split number and width. Shrink wrapped fruits had fewer and narrower splits. It starts at the respiratory peak when ethylene production increases and coincides with the decline to one fifth of the starch level in the fruit. The number and width of splits increased for two days, then reached a plateau. Weight loss increased continuously during ripening.

c) Chilling Injury (CI)

Custard apple fruit develops CI within 5 days at 4 °C, while other *Annona spp.* such as *A. cherimola* var. 'Concha Lisa' seems to be more resistant. Depending on the area of production and the month of harvest, 10 °C is the minimum storage temperature for local cultivars allowing ripening of cherimoya without impairing the normal organoleptic characteristics of the ripened fruit, and without the development of CI related symptoms. In cherimoya, CI was observed at 5 °C with the fruits becoming hard and black.

d) Pulp Hardening

Fungi often attack fruit leading the infested fruit to be mummified in which *C. gloeosporioides* can infect fruit of sugar apple, causing mummified fruit. When the fungus invades mature fruit, the resultant lesion is small with an expanding area of hardened tissues.

e) CO_2 Injury

Modified atmospheres would be beneficial for extending storage life of many fruits. In sugar apple, however, high levels of CO_2 in polyethylene (PE) bags at above 15 per cent for 10 days at 13 °C could cause peel darkening. Furthermore, the flesh colour turns from white to pink soon after peeled.

6.0 Domesting Marketing and Export

Being a minor tropical fruits, custard apple is exported in small volumes in the world market. In general, most fruit in the producing countries are supplied to domestic markets. Thailand and Brazil are major exporting countries. Importing countries in Asia include Hong Kong, Singapore, Brunei, China and Japan. Some European countries such as France, Germany and the Netherlands are also importers. Most of the Indian production are inferior in quality and therefore custard apple is not exporting from India.

7.0 Conclusion

Custard apple fruits do not ripen adequately when harvested at the immature stage, hence identification of the best stage for picking is essential (Figure 4). On the other hand, ripe fruits are particularly vulnerable to rot and decay, thus quickly losing their shelf life and quality. This raises the need for adequate on and off-farm facilities and technology to store and/or process the fruits.

Figure 4. Blackening in Custard Apple Fruit due to Immature Harvesting

Appropriate packaging techniques for long distance transport are necessary, especially for markets that demand better flavour and appearance, due to the fact that fruits are nearer maturity when picked.

Custard apple is such a tropical aggregate fruit with satisfied prospects both from attractive appearances and tastes. The fruit provides enriched sources of carbohydrates, vitamin C, and mono or sesquiterpenes. Fruit softening, fruit segment separation, and skin blackening during ripening are the main problems for the fruit quality at postharvest. Custard apple fruit biosynthesises low level of ethylene, but very high respiration rate during ripening. When pectatelyase (PL) plays an important role in the fruit splitting and softening, polyphenol oxidase (PPO) is mainly involved in the peel and pulp browning. The optimum temperature is between 13 and 15 °C that long storage of fruit under the temperature causes abnormal ripening and chilling injury symptoms. The combinations of modified atmosphere packaging and use of 1-MCP or plant regulators could be beneficial for extending the storage life and preventing postharvest disorders.

In some cultivars fruit splitting occurs with advanced ripening stages and increased ethylene production rates. It has been suggested that turgor changes related to production of neutral sugars during ripening lead to movement of water from the skin and possibly receptacle to the flesh. The increase in receptacle diameter increased the stress on the flesh and skin leading to fruit splitting. Important pathological disorders of custard apple fruits include anthracnose, which appears as dark lesions and may produce pink spore-masses under high humidity conditions, black canker which appears as purple spots on the fruit, becoming hard and cracked followed by development of small black bodies containing spores, and botryodiplodia rot which first appears purple, later pimpled with black pycnidia, and subsequently the flesh becomes brown and corky. Control of postharvest diseases includes good orchard sanitation to minimize sources of fungal spores, preharvest application of fungicides, careful handling to reduce physical damage, prompt cooling to 10 °C, and subsequent maintenance of optimum temperature and relative humidity during marketing.

Quality indices of these fruits include fruit size, colour, absence of defects and decay, and firmness. Optimum handling temperature is 8–12 °C depending on cultivar, ripeness stage, and duration, all at 90–95 per cent RH. Custard apple fruits are sensitive to temperatures below 8–12 °C, depending on cultivar and ripeness stage, resulting in chilling injury, causing darkening and hardening of skin, pitting, failure to develop full flavour, and mealy flesh. Exposure to ethylene (100 ppm for 1–2 days) accelerates ripening of mature-green fruits; they can ripen in about 2-4 days if kept at 15 °C to 20 °C. Ethylene removal can be helpful in retarding ripening of mature-green fruits. MA/CA can delay ripening, lower respiration and ethylene production rates, and firmness retention, and optimum atmosphere is 3–5 kPa O_2 and 5–10 kPa CO_2.

8.0 Important References

Aljque, R., Zamorano, J.P., Calvo, M.L., Carmen, M. and De La Plaza, J.L. 1994. Tolerance of cherimoya (*Annona cherimola* Mill.) to cold storage. *J. American Soc. Hortic. Sci.* **119**: 524-28.

Anonymous. 2013. National Horticulture Data Base 2012-13. *National Horticulture Board*, Gurgaon, Haryana.

APEDA, 2016. https: //agriexchange.apeda.gov.in/India%20Production/India_Productions.aspx?cat=fruit&hscode=1044

Babu, K.H., Zaheeruddin, M. and Prasad, P.K. 1990. Studies on postharvest storage of custard apple. *Acta Hortic.* **269**: 299-00.

Batten, D.J. 1990. Effect of temperature on ripening and postharvest life of fruit of atemoya (*Annona cherimola* Mill. x *A. squamosa* L.) cv. 'African Pride'. *Sci. Hortic.* **45**: 129-36.

Broughton, W.J. and Tan, G. 1979. Storage conditions and ripening of custard apple (*Annona squamosa* L.). *Sci. Hortic.* **10**: 73-82.

Brown, B.I., Wong, L.S., George, A.P. and Nissen, R.J. 1988. Comparative studies on the postharvest physiology of fruit from different species of *Annona* (custard apple). *J. Hortic. Sci.* **63**: 521-28.

Brown, B.I. and Scott, K.J. 1985. Cold storage conditions for custard apple fruit (*Annona atemoya*). *Singapore J. Primary Ind.* **13**: 23-31.

Campos-Vargas, R., Defilippi, B.G., Paula, Q.R., Hector, V.G., Paula, R.M. and Humberto, P.E. 2008. Effect of harvest time and L-cysteine as an antioxidant on flesh browning of fresh-cut cherimoya (*Annona cherimola* Mill.). *Chilean J. Agric. Res.* **68**: 217-27.

Chen, C.C. and Paull, R. 2008. *Annona squamosa*, sweetsop. **In**: The Encyclopedia of Fruit and Nuts. Janick, J. and Paull, R. (Eds.). *CAB International*, Wallingford, U.K. pp. 48-53.

Chunprasert, A., Uthairatanakij, A. and Wongs-Aree, C. 2006. Storage quality of 'Nang' sugar apple treated with chitosan coating and MAP. *Acta Hortic.* **712**: 857-64.

George, A.P., Nissen, R.J. and Brown, B.I. 1987. The custard apple. *Queensland Agric. J.* **113**: 287-97.

Gutierrez, M., Lahoz, J.M., Sola, M.M., Pascaul, L. and Vargas, A.M. 1994. Postharvest changes in total soluble solids and tissue pH of cherimoya fruit stored at chilling and non-chilling temperatures. *J. Hortic. Sci.* **69**: 459-63.

Hofman, P.J., Jobin-Décor, M., Meiburg, G.F., Macnish, A.J. and Joyce, D.C. 2001. Ripening and quality responses of avocado, custard apple, mango and papaya fruit to 1-methylcyclopropene. *Australian J. Exp. Agric.* **41**: 567-72.

Kamble, P.B. and Chavan, J.K. 2005. Effects of postharvest treatments and storage temperature on shelf-life of custard apple fruits. *J. Food Sci. Technol.* **42**: 253-55.

Ke, L.S., Yang, C.S., Yu, Y.B. and Tsi, P.L. 1983. Ripening and storage of custard apple fruits (*Annona squamosa* L.). *J. Chinese Soc. Hortic. Sci.* **29**: 257-68.

Lima, S., Barbosa Guerra, N., Sucupira Maciel, M.T. and Souza Livera, A.V. 1994. Polyphenol oxidase activity, polyphenols concentration and browning intensity during soursop (*Annona muricata* L.) maturation. *J. Food Sci.* **59**: 1050-52.

Martinez, G., Serrano, M., Pretel, M.T., Amoros, A., Riquelme, F. and Romojaro, F. 1993. Ethylene biosynthesis during the ripening of cherimoya (*Annona cherimola* Mill.). *Current Plant Sci. Biotechnol. Agric.* **16**: 148–49.

Mo, Y., Gong, D., Liang, G., Han, R., Xie, J. and Li, W. 2008. Enhanced preservation effects of sugar apple fruits by salicylic acid treatment during postharvest storage. *J. Sci. Food Agric.* **88**: 2693-99.

Monkong, J., Komkhuntod, R. and Wanichkul, K. 2010. Effect of bagging materials on fruit growth of *Annona* hybrid cv. 'Petch Pakchong'. *Proc. 48th National Conference of Kasetsart University*, 3–5 February, 2010, Bangkok, pp. 16-23.

Mosca, J.L., Asis, J.S., Alves, R.E., Filgueiras, H.A.C. and Batista, A.F. 1997. Physical, physical-chemical and chemical changes during growth and maturation of sugar apple (*Annona squamosa L.*). *Memorias, Congresso Internacional de Annonaceaes*. Universidad Autonoma Chapingo (UAC), Chapingo, Mexico. pp. 304-14.

Nakasone, H.Y. and Paull, R.E. 1998. *Annona*. **In**: Tropical Fruits. Nakasone, H.Y. and Paull, R.E. (Eds.). *CAB International*, London, U.K. pp. 45-75.

Nilam, P., Naik, A., Arbat, G. and Shakti, S. 2011. Response of postharvest chemical treatments on shelf-life and quality of custard apple cv. 'Balanagar'. *Indian J. Hortic.* **68**: 547-50.

Nissen, R.J., Smith, L.G., Broadley, R.H., George, A.P. and Sisson, C.S. 2002. Marketing of custard apple in Australia: from grower to consumer. *Acta Hortic.* **575**: 437-42.

Noichinda, S., Bodhipadma, K., Wongs-Aree, C., Komkhuntod, R., Sirisukchaitavorn, H., Rachakhome, P. and U-Nat, A. 2010. Evaluation of antioxidant properties in naturally and artificially ripening sugar apple (*Annona squamosa* L.) fruits. *Acta Hortic.* **875**: 377-81.

Nonta, P. 2009. Grading evaluation of sugar apple (*Annona squamosa* Linn.) fruit from Pakchong District, Nakhon Ratchasima Province. *M.Sc. thesis*, Kasetsart University, Bangkok.

Palma, T., Aguilera, J.M. and Stanley, D.W. 1993. A review of postharvest events in cherimoya. *Postharvest Biol. Technol.* **2**: 187-08.

Paull, R.E. 1996. Postharvest atemoya fruit splitting during ripening. *Postharvest Biol. Technol.* **8**: 329-34.

Pereira, M.C.T., Braz, L.C., Nietsche, S. and da Mota, W.F. 2010. Determining the harvesting maturity of the sugar apple fruits on northern Minas Gerais. *Acta Hortic.* **864**: 207-14.

Phillips, R.L. and Campbell, C.W. 1994. The sugar apple. fact sheet HS-38 of the Florida Cooperative Extension Service, University of Florida, p. 3.

Pinto, A.C.Q., Cordeiro, M.C.R., de Andrade, S.R.M., Ferreira, F.R., Filgueiras, H.A.C., Alves, R.E. and Kinpara, D.I. 2005. *Annona* species. *International Centre for Underutilized Crops*, University of Southampton, Southampton, U.K. pp. 263.

Preungvate, O. 1982. Starch, sucrose, glucose, fructose and titratable acidity in various Thai fruits. *J. Nat. Res. Coun. Thailand.* **14**: 29-53.

Salunkhe, D.K. and Desai, B.B. 1984. Custard apple. **In**: Postharvest Biotechnology of Fruits, Vol. 2. *CRC Press*, Boca Raton, Florida, USA, pp. 133.

TAS 21-2013. Thai Agricultural Standard for Sugar Apple; National Bureau of Agricultural Commodity and Food Standards: Ministry of Agriculture and Cooperatives, Bangkok, ISBN ICS 67.080.10

Torres, L.M.A.R., Silva, M.A., Guaglianoni, D.G. and Neves, V.A. 2009. Effects of heat treatment and calcium on postharvest storage of atemoya fruits. *Alim. Nutr. Araraquara.* **20**: 359-67.

Tsay, L.M. and Hong, Y.S. 1988. Effects of several physiologically active regulating substances on the ripening of sugar apple fruits. *J. Japan. Soc. Hortic. Sci.* **57**: 538-43.

Vishnu Prasanna, K.N., Sudhakar Rao, D.V. and Krishnamurthy, S. 2000. Effect of storage temperature on ripening and quality of custard apple (*Annona squamosa* L.) fruits. *J. Hortic. Sci. Biotechnol.* **75**: 546-50.

Wills, R.B.H., Poi, A., Greenfield, H. and Righey, C.J. 1984. Postharvest changes in fruit composition of *Annona atemoya* during ripening and effects of storage temperatures on ripening. *HortScience.* **19**: 96-97.

Chapter 16

Date Palm

Sunil Pareek

Associate Professor, Department of Agriculture and Environmental Science, NIFTEM, Plot No. 97, Sector 56, HSIIDC Industrial Estate, Kundli, District-Sonepat – 131 028, Haryana, India

E-mail: sunil.pareek@niftem.ac.in

1.0 Introduction

Date palm (*Phoenix dactylifera* L.) belongs to the family Araceae. Date palms grow in hot arid regions of the world, with most production concentrated in the Middle-East, western South-Asia, and Northern Africa.

Dates are considered nutritious and a high energy food (3000 calories kg^{-1}). Sugar contents range from about 40 per cent (fresh dates) to 80 per cent (dried dates) and mostly of inverted form (glucose and fructose). The ratio of glucose to fructose is almost equal. Dates contain 10.2 per cent dietary fiber. The highest protein content is observed during *kimri* stage (5.5-6.4 per cent), which gradually decreases to 2-2.5 per cent during the *tamar* stage. Dates contain all the essential amino acids. A consumption of 100 g of dates provides over 15 per cent of the daily Recommended Dietary Allowance (RDA) to adequate intake of selenium, copper, potassium, and magnesium. The pulps are rich in iron, calcium, cobalt, copper, fluorine, magnesium, manganese, potassium, phosphorus, sodium, copper, sulfur, boron, selenium, and zinc.

This chapter provides an overview of date production, cultivars, preharvest factors affecting postharvest quality, maturity indices, postharvest handling, packaging, storage, transportation and quality standards and trade.

2.0 Status of Cultivation

The area under date palm cultivation in the world has almost doubled between 1990 to 2012 (0.63 to 1.10 million hectares), while the production of dates during this period has increased almost 120 per cent.

Egypt is the top-most producer of dates (19.47 per cent of total world production) followed by Iran (1.06 mt), Saudi Arabia (1.05 mt), Algeria (0.79 mt) and Iraq (0.65 mt). Combined these top five countries contributed a 66.5 per cent share of total world production.

In India, grooves of seedling date palm are found in the coastal belt from Anjar to Mandavi in Kutch district of Gujarat. Presently, over 1.2 million date palm trees are in Kutch area. In Rajasthan, date cultivation was first introduced by the then ruler Ganga Singh of the earstwhile Bikaner state through the plantation of few suckers of 'Halawy', 'Khadrawy' and 'Zahidi' at Sri Ganganagr, a part of *Thar Desert* after development of Gang canal in the state. In north western Rajasthan, scattered plantation of few thousand date trees are confined mainly in the districts of Sri Ganganagr, Bikaner, Jodhpur and Jaisalmer.

Date palm is one of the important horticultural crops of the Kutch region which has most suitable climate for date palm plantations and thus enjoys the monopoly of commercial cultivation of this crop in India. The total area under cultivation has increased to 16,000 ha during 2010-2011, from 8,973 ha during 2000-2001, with production of 0.12 million tonnes. The Kutch is the largest producer of date palm in India. The date grooves of Kutch are probably about 500 years old and are believed to be developed from the seeds thrown by the army camps and partly from the seeds and off shoots planted by settlers. The settlers used to visit Middle East countries for *Hajj* or for trade from where they brought a lot of plant material. In due course of time, the initial plantations multiplied and developed into grooves.

Table 1. Potential Zones of Date Palm Cultivation in India

Zone	*Region*	*Edapho-Climate*	*Harvesting Stage*
I	Jaisalmer, western parts of Barmer, Bikaner and Jodhpur district	Extremely dry areas with lowest rainfall and highest heat units	*Dang* or early *Pind*
II	Arid coastal areas of Kutch and part of Saurashtra. Sub divided into 3 zones:		*Doka* or *Dang*
	Mundra, Dhrub and Zarpar	Deep coastal soil with very high water table (1-3 m)	
	Anjar and Khedoi	Deep sandy soil with low water table (10-30 m)	
	Kera, Tera, Vada	High water table (1-2 m)	
III	Jodhpur, Bikaner, eastern part of Barrmer, western part of Nagaur, Churu and Ganganagar	Dry areas with storm type rains	*Doka* or *Dang*
IV	Abohar, Sirsa, Ganganagar, eastern Churu and western Sikar districts	Dry areas with rains followed by cloudy weather	*Doka*

Dates has potential for successful cultivation in Sirsa, Mahendragarh, Hisar and Bhiwani district of Haryana state, Abohar and Ferozpur areas of Punjab state, Ganganagar, Bikaner, Churu, Jodhpur, Jaisalmer and Barmer districts of Rajasthan state, and Kutch region of Gujarat state (Table 1). It can be grown at such places where assured source of irrigation is available besides other climatic requirements.

3.0 Major Commercial Varieties

There are about 2000 date palm cultivars in the world. A large number of these cultivars are propagated by seeds. Very few cultivars are grown extensively in major producing countries, however, the date varieties are the common types showing Table 2.

Table 2. Common Types of Date Varieties

Variety/Type	*Speciality*
Amari	Dried date, soft, sweet, medium sized date
Barhee	Fresh consumption at *khalal* or early stage, sweet and juicy, yellow colour,
	available seasonally during harvest months
Delget Noor	Dried date, semi-soft, famous for flavour, light to dark brown colour,
	harvested semi-dry
Hadrawi	Dried date, sweet and fleshy date, dark brown colour
Hayani	Fresh consumption, black and shiny colour, long fruit, not very sweet
Medjool	Dried date, large fruit, soft and sweet, light brown to dark brown colour
Zahidi	Dried date, round, medium sized and not too sweet date, golden colour

The date cultivars that dominate the world market, based on consumption are 'Medjool', 'Delget Noor', and 'Barhee'.

Depending on the flesh consistency and moisture content at harvest when fully ripe, date palm cultivars are divided into three groups, namely soft (Figure 1), semi-dry and dry (Figure 2). However, fruit of any cultivar when left on the palm or exposed to excessive curing conditions will lose moisture and develop a hard texture. Other classification can be found within the same group based on fruit characteristics, size and sugar content.

In soft cultivars ('Hillawi', 'Abada', 'Amhat', 'Barhee', 'Bentaisha', 'Halawi', 'Hayany', 'Honey', 'Khadrawy' and 'Medjool'), almost all can sugar (sucrose) is converted into invert or reducing sugars (glucose and fructose) during ripening, with a moisture content >30 per cent.

Dry date cultivars (<20 per cent moisture) include cultivars such as 'Badrayah', 'Bartamoda', 'Delgate Beida', 'Horra', 'Sakoty' and 'Thoory'. Semi-dry date cultivars (20-30 per cent moisture) include cultivars such as 'Amry', 'Dayri', 'Delget Noor', 'Khalasa', 'Sewy' and 'Zahidi'.

Both dry and semi-dry dates retain a good amount of sucrose on full ripening, in addition to the reducing sugars.

The salient features of commercial cultivars of date palm in India (Figures 3–12) are described as under:

a) Barhee (Figure 3)

This variety is also suitable for cultivation in Rajasthan state. Trees are semi-vigorous and medium in height. Fruits are oval in shape and golden yellow in colour at *doka* stage. The average fruit weight is 7.5 g and size of 2.9 x 2.3 cm pulp thickness of 0.7 cm and TSS of 32.0 per cent at *doka* stage. Fruits of this variety are non-astringent at *gandora* stage also. This variety is most suitable for raw eating and is not good for soft dates. It is late in ripening as fruits of this variety reach at full *doka* stage by mid August and hence is moderately damaged by rains and high humidity. Average yield potential is about 120 kg *doka* fruits per plant per year.

b) Halawy (Figure 4)

It is one of the most suitable varieties for date palm cultivation in Rajasthan state. Trees are tall and vigorous. Fruit colour is light orange with yellow shade. The fruit is oblong elliptical with obtuse apex and almost horizontal base. The average fruit weight is about 9.5

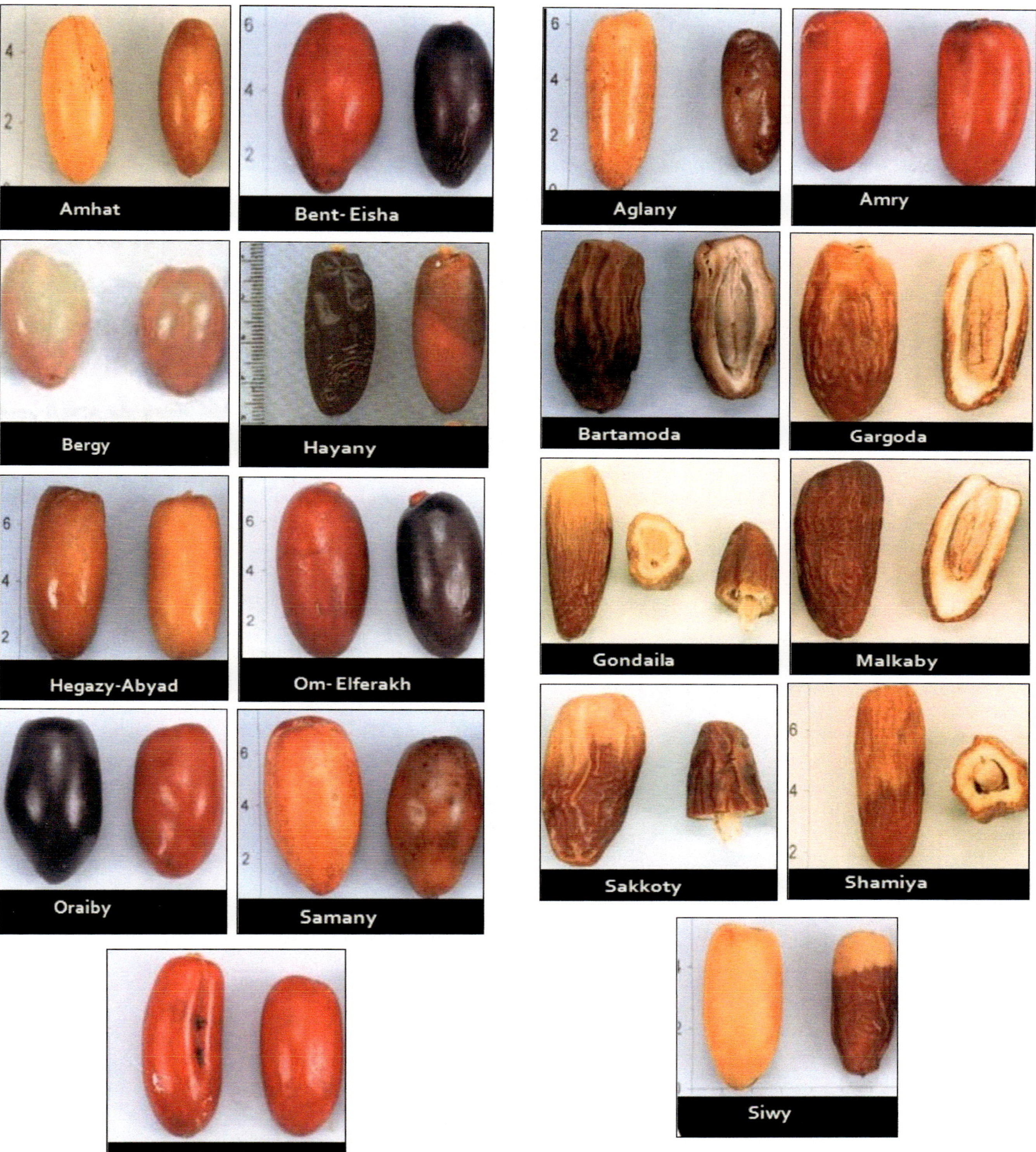

Figure 1. Variability in Fruit Morphology of the Best Known Soft Cultivars of Dates. (Photo courtesy. Prof. Jameel M. Al-Khayri, King Faisal University, Saudi Arabia).

Figure 2. Variability in Fruit Morphology of Semidry ('Siwy', 'Amry' and 'Aglany') and Best Known Dry Date Cultivars ('Sakkoty', 'Bartamoda', 'Gondaila', 'Malkaby', 'Shamiya' and 'Gargoda'). (Photo courtesy. Prof. Jameel M. Al-Khayri, King Faisal University, Saudi Arabia).

g, size of 3.5 x 2.0 cm, pulp thickness of 0.6 cm having 31.0 per cent TSS at *doka* stage. The fruits are sweet at full *doka* stage and may be eaten as raw. The fruits have a tendency to shrivel during the ripening stage particularly in case of insufficient irrigation. It is early in ripening as the fruit of this variety reach at full *doka* stage by second fortnight of July. Halawy has average yield potential of 100 kg *doka* fruits per plant per year.

Figure 3. Barhee

Figure 4. Halawy

Figure 5. Khadrawy

Figure 6. Khalas

Figure 7. Khunezi

Figure 8. Medjool

Figure 9. Sewi

Figure 10. Shamran

Figure 11. Zagloul

Figure 12. Zahidi

Popular Commercial Date Palm Cultivars

c) Khadrawy (Figure 5)

Trees of this variety are dwarf in growth. Fruits at *doka* stage are greenish yellow in colour and oblong in shape with broad apex and a bit slanting base. The average fruit weight is 7.7 g, size of 3.1 x 2.1 cm, pulp thickness of 0.66 cm with TSS of 30.0 per cent *doka* stage. The fruits are highly astringent at full *doka* stage and are badly spoiled due to rains at time of ripening. This variety is suitable for both soft dates as well as preparation of dry dates. It is medium in ripening as the fruits of this variety reach full *doka* stage by end of July. Average yield potential is 60 kg *doka* fruits per tree per year.

d) Khalas (Figure 6)

Trees are semi-vigorous in growth. Fruit colour is yellow at *doka* and golden brown at *pind* stage. Fruits are oblong in shape. The average fruit weight is 11.0 g, size of 3.2 x 2.4 cm, pulp thickness of 0.83 cm and TSS of 28 per cent. The fruit are sweet at full *doka* stage and suitable for raw eating as well as for soft dates. It ripens early as the fruits of this variety reach full *doka* stage in the second fortnight of July at Bikaner. Average yield potential is 75 kg *doka* fruits per tree per year.

e) Khunezi (Figure 7)

This is emerging as a potential date variety in India. Trees are semi-vigorous with spreading growth habit. Fruit is dark red in colour at *doka* stage with crispy pulp and dark brown at *pind* stage. Fruits are oval shaped with pointed apex. The average fruit weight is about 10.2 g, size of 3.5 x 2.2 cm, pulp thickness 0.75 cm and TSS of 43.0 per cent at full *doka* stage. The fruits are sweet at full *doka* and thus suitable for raw eating. These are not suitable for soft dates. The variety is also early as fruits reach full *doka* stage by the end of first fortnight of July. Average yield potential is 50 kg *doka* fruits per tree per year.

f) Medjool (Figure 8)

It is a popular variety and has potential of commercial cultivation. Trees are vigorous and tall with erect growth habit. The fruits are large but variable in size having attractive golden yellow colour and more firm than Barhee and Khadrawy. The irregularity in shape is common and there are ridges on the seed. The average fruit weight is 15g, size of 3.9 x 2.8 cm, pulp thickness of 0.8 cm with TSS of 29 per cent at *doka* stage. The fruits at *doka* stage are highly astringent. It is suitable for dry dates (*Chhuhara*) because of its large sized fruits and good pulp thickness. It is tolerant to rainfall and high humidity conditions. It is late in ripening as fruits of this variety reach full *doka* stage in mid August in Rajasthan. Average yield potential is 90 kg *doka* fruits per tree per year.

g) Sewi (Figure 9)

Trees are vigorous in growth, tall and erect with moderate canopy. Fruits are oblong in shape, yellowish green in colour at *doka* stage. The average fruit weight is 6 g, size of 2.9 x 1.9 cm, pulp thickness of 0.6 cm with TSS of 32 per cent TSS at full *doka* stage. The fruits are sweet at full *doka* stage therefore, suitable for both raw eating as well as soft dates. It is late in ripening as fruits of this variety reach full *doka* stage by the first fortnight of August. The fully grown tree of 'Sewi' has an average yield potential of 50 kg *doka* fruits per tree per year.

h) Shamran (Figure 10)

The trees of this variety are semi vigorous. The fruits are yellow in colour with slightly pink shaded base. The fruits are medium in size and oblong in shape with obtuse apex. The average fruit weight is 6.5 g and size of 3.5 x 2.18 cm. it has pulp thickness of 0.56 cm having 30 per cent TSS at full *doka* stage. The fruits are astringent at *doka* stage. It is suitable for dry and soft dates. Its soft dates have typical aroma. It is medium in ripening as fruits reach full *doka* by the end of July. It is a prolific bearer and has an average yield potential of 100 kg *doka* fruits per year per palm.

i) Zagloul (Figure 11)

Trees are vigorous with erect growth habit. This is also a promising date variety suitable for Thar desert. Fruits are red in colour and large in size. These are oblong, oblique and asymmetrical in shape. The calyx is small in size and slightly depressed at the base. The fruit bunches are bigger in size. The average fruit weight is 9.5 g, size is 3.9 x 2.2 cm, pulp thickness of 0.64 cm with TSS of 28.0 per cent at *doka* stage. The fruits are astringent at *doka*. The *dang* and *pind* fruits are relatively less sweet but have pleasant taste. It is suitable for soft dates and for preparation of fruit products such as date juice and preserved products. It is medium in ripening as fruits of this variety reach full *doka* stage by the beginning of second fortnight of July. Average yield potential is 120 kg *doka* fruits per year per tree.

j) Zahidi (Figure 12)

Trees are semi-vigorous. Fruits are yellow in colour at *doka* stage with smooth and hard surface. The fruits are ovate in shape with the broader end towards the apex. The average fruit weight is 10 g, size of 3.0 x 2.2 cm, pulp thickness of 0.68 cm with TSS of 30 per cent at *doka* stage. The fruits are astringent at *doka* stage. The partial *dang* fruits are eaten as raw dates. It is late in ripening as fruits of this variety reach full *doka* stage in mid August. It is a high yielding variety with average yield potential of 125 kg *doka* fruits per tree per year.

4.0Nature and Causes of Postharvest Losses

Date cultivars vary greatly in quality at harvest and during postharvest handling. A large number of date palm trees are grown from seeds and most of these trees produce low quality dates that end up as animal feed or waste. Improvement of date quality and value (marketability) can be achieved by selecting a few (no more than ten) cultivars of good quality that meet consumer preferences and by limiting future plantings to these cultivars, while gradually replacing the poor quality cultivars with the selected good quality cultivars.

Cultural practices (including irrigation and fertilization, pest management, fruit thinning, and harvest maturity) and climatic conditions (temperature, RH, rainfall, wind) influence the rate of development and quality at harvest of dates. For example, sub-optimal water supply to the palm tree reduces yield and quality of the dates. Managing the crop load by reducing the number of bunches on the date palm tree and fruit thinning improves fruit quality.

Only about 30-40 per cent of the fruit normally ripen on the tree. Bunch bagging of dates on the tree with black or blue polyethylene, white agrisafe (polyethylene fleece) or paper bags significantly increases the rate of fruit ripening and increases rutab yield per bunch. Black and blue polyethylene bags were the most effective followed by agrisafe and paper bags.

4.1 Darkening

Enzymatic browning of polyphenols proceeds at observable rates at room temperature, whereas at temperatures above 38 °C, sugar browning predominates. Insoluble tannins are responsible for non-enzymatic oxidative browning in dates. Vacuum packaging is a useful technique for reducing darkening of the date for long-term storage.

4.2 Skin Separation (Puffiness)

Skin is dry, hard and brittle, and is separated from the flesh. This disorder develops during ripening of soft date cultivars, which vary in susceptibility. High temperature and high humidity at a stage before the beginning of ripening may predispose the dates to skin separation.

4.3 Sugar Spotting (Sugaring)

Sugar spotting is characterized by light coloured spots under the skin, results from crystallization of sugars in the flesh of soft date cultivars. Although it does not influence taste it alters fruit texture and appearance. Incidence and severity of sugar spotting increase with storage temperature and time. Storage at recommended temperatures minimizes this disorder, which occurs mainly in cultivars in which glucose and fructose are the main sugars. Sugaring may be reduced by gentle heating of the affected dates.

5.0 Postharvest Management

5.1 Maturity Stages

Date fruit development progress through five maturity stages that take about 6-8 months; fruit: (1) grow rapidly, (2) turn their characteristic colour, (3) lose water, (4) accumulate sugars, and (5) ripen completely (Table 3). Ashraf and Hamidi-Esfahani (2011) described these stages as:

Table 3. Fruit Developmental Stages

Stage (India)	*Corresponding Arabic Term*	*Characteristics*
Hababouk	*Pre-Kimri*	Fruit round with whitish cream colour
Gandora	*Kimri*	Fruit is still hard and green
Doka	*Khalal*	Fruits are fully grown but remain hard, colour becomes yellow or red
Dang	*Rutab*	Softening of fruits start from tips and finally the whole fruit is soften
Pind	*Tamar*	Fruits fully ripened, fruit weight decreases as a result of dehydration

Source: Pareek, 2014.

a) Hababouk (Pre-Kimri)

The first stage that appears after pollination and continues for 4-5 weeks. The fruit has round shape, whitish-cream colour with green stripes.

b) Kimri

This stage appears in the first 17 weeks after pollination. The fruit is young, elongated, greenish in colour, hard in texture and with about 85 per cent moisture. Fruit weight increases significantly and the tannin concentration is high. Although the fruit is inedible for direct consumption at this stage, it can be used for making chutney or pickles.

c) Khalal

During the next 6 weeks, date fruit gains maximum size and weight, colour gradually becomes a typical yellow, purplish-pink, or red depending on the cultivar, with hard texture. At this stage, sugar increases slowly and becomes mainly sucrose. This is the stage at which dates are mainly consumed raw as fresh fruit or they can be used for jam, butter, or date-in-syrup.

d) Rutab

In the next 4 weeks, the dates loss water with half of the fruit becoming soft, sweeter and darker in colour (light brown), and less astringent. Sucrose converts to reducing sugars and protein, fat and ash percentage decrease. This stage is the start of ripening. Dates at rutab stage from many cultivars are eaten fresh or processed into jam, butter, date bars, and date paste.

e) Tamr or Tamar/Tamer

During this final stage, the typically lasts 2 weeks, the fruit gains maximum total solids, highest sweetness, lowest astringency, dark brown colour, soft texture, and a typical wrinkled shape. There is a high concentration of reducing sugars, especially glucose and fructose, with no or very low sucrose. The percentage of protein, fat and ash are less compared to *rutab* stage. Owing to low moisture and high sugar content, dates of this stage have good storage stability.

At Abohar, Ganganagar, Churu, Sirsa and Hisar, the fruits are usually harvested at *doka* stage due to early commencement of rains and also due to other climatic reasons whereas the late *doka* and *dang* stages can be obtained at Bikaner, Nagaur and Jodhpur areas of west Rajasthan and *dang* to early *pind*/full *pind* stages can be obtained in extremely dry areas like Jaisalmer and western parts of Barmer, Bikaner and Jodhpur districts. Therefore, it indicates the suitability of date palm cultivation in *Thar* desert of west Rajasthan where good quality of date palm berries may be harvested. At Abohar, *doka* stage is normally attained but afterwards most of the cultivars are seriously damaged due to rains and humidity during the months of July and August.

5.2 Harvesting

Dates are harvested and marketed generally at three stages of development: mature firm (*khalal*), full ripe (*rutab*) and dry (*tamar*). The decision for harvesting at one or other stage depends on cultivar characteristics; especially soluble tannins levels, climatic conditions, and market demand. Time of harvest is based on the date fruit's appearance and texture (related to moisture and sugar content). Proper timing of harvest reduces incidence and severity of cracking or splitting of dates, excessive dehydration, insect infestation, and attack by microorganisms.

Dang (Rutab)

Doka (Khalal)

Gandora (Kimri)

Pind (Tamar)

Figure 13. Developmental Stages of Dates in India

The time taken for fruit to ripen on the tree depends on the cultivar, the heat units received during the growing season and the stage at which the fruit are picked. For early ripening cultivars the fruit within the bunch may take as long as 3-4 weeks to complete ripening, while for late ripening cultivars, fruit within the bunch ripen in about 8-12 weeks. As ripening of dates is progressive on the bunch some fruits can be overripe while others are still at the *khalal* or *rutab* stages. Selective picking of individual dates or strands is often practiced for good quality at prime maturity. When this approach is adopted, a number of pickings are made before harvesting all fruits. Frequency of picking depends on several factors such as type of date (soft, dry, or semi-dry), climatic conditions, market demands, handling methods, cost of handling and availability and cost of pickers. On an average, however, when ripe fruit are picked from bunches, three pickings are required over several days.

Yellow, *khalal* 'Barhee' dates are marketed on branches (strands) or bunches. The whole bunches are harvested (when the dates are fully yellow) and lowered to ground level, then hung on a carrier for transportation to the packinghouse. Green to greenish-yellow and ripe (*rutab*) fruits are removed from the branches before packing in 5-kg fiberboard boxes for shipment to markets. These dates are hydro-cooled to 0 °C in 10 to 20 minutes, depending on initial temperature. However, it requires effective disinfection of the water and removal of excess surface moisture from the cooled dates before packing in the shipping containers. Hence, forced air cooling may be the better choice than hydro-cooling.

Harvest is generally by hand, with access to the crown of the tree being by way of climbing or mechanical lifts. As the palm tree grows taller, harvesting the dates becomes more difficult, dangerous and more costly. Hand harvesting of dates involves the use of aluminium ladders for short palms and picking platforms for taller palms. Ladders may be mounted on the palm tree to facilitate harvesting or various types of lifts, such as tree squirrel and self-propelled elevating platform, are used to elevate the harvesting labourers to facilitate harvesting (Figure 14). The fruits are placed within a bucket and lowered to the ground, and then packed in bulked bins and sent to the packinghouse. A usual procedure to climb the date palm trees is the use of a wide belt woven out of coir to support the climbers back and cutting of the whole bunch. Bunches may be lowered either by ropes or by passing the bunch hand-to-hand. However, hand harvesting is costly and may cost up to 45 per cent of total operational cost. Fruits are also harvested by shaking the bunch and all mature fruits which detach easily drop on to mats spread on the ground around the palm. Very soft fruits get damaged in this process.

5.3 Pack House Operations

In the packinghouse there are a number of processes, designed to improve or maintain fruit quality. These processes are: fumigation, washing, storage, refrigeration, hydration, dehydration and curing.

5.3.1 Fumigation

In order to store the fruit for a long period (several months to one year), it must be completely cleaned of any pests (eggs, pupas, larva or adults). This is done by fumigation, either in the field under various kinds of plastic sheets, or at the packinghouse in special sealed rooms.

i) Fumigation must not be carried out when the fruit is fresh, harvested at the *khalal* stage, ('Barhee', 'Khalas', 'Zaghlool' and 'Hayany') or when stored under refrigeration. The most frequently used treatment for insect disinfestations is Methyl bromide @ 30 g per cubic meter (30 ppm) for 12 to 24 hours at temperatures

Figure 14. Date Harvesting by Mechanical Ladders (*Source.* Glassner *et al.*, 2002)

above 16 °C. A potential substitute for methyl bromide is sulphuryl fluoride at 34 g per cubic metre for 24 hours at 20-25 °C. However, the U.S. Environmental Protection Agency has re-evaluated the current science on fluoride and is taking steps to begin a phased-down withdrawal of the pesticide sulfuryl fluoride primarily due to fluoride residue in foods. For dates grown and marketed by the bio-organic method, fumigation alternatives for methyl bromide are as follows.

ii) Phosphine is the principal alternative to methyl bromide for fumigation and is widely used in developing countries.

iii) Insecticidal Controlled Atmospheres high in carbon dioxide are in regular use in South East Asia for disinfecting bag-stacked durable commodities. Storage in N_2 or 100 per cent CO_2 can control insects for 1-2 days at ambient temperature, and low O_2 atmosphere can also inhibit enzymatic browning.

5.3.2 Drying (Dehydration)

Dates need to be dehydrated to the optimal moisture content for preserving their quality during subsequent handling and storage. Dehydration aims to achieve an appropriate sugar: water ratio which should be close to 2 for soft dates, greater than 2 for dry dates and lower than 2 for very soft dates. This ratio is a good indicator of date quality behavior in storage. Over drying to less than 20 per cent moisture should be avoided to keep the dates soft. The desired moisture content is 23 to 25 per cent. The temperature and duration required to reduce water content depend on the type of date, use and flesh consistency.

Dehydration may be done concurrently with ripening. If ambient conditions allow, dehydration can be done using solar energy by spreading the dates on trays that will be exposed to the sun or under plastic tunnels until drying is completed (Figure 15) to the desired moisture level. Drying in plastic houses protects the dates from dust, birds, rodents, and damaging factors. If solar or ambient air drying are not possible, heated air must be used to dry the dates to their desired moisture content.

Dehydration is also achieved by exposing dates to hot air (<70 °C) inside a solar or industrial oven to avoid browning of sugars. For drying soft dates 54 °C at 50 per cent relative humidity is recommended, while drying time depends on initial moisture content.

5.3.3 Hydration

If picked ripe and not over-dried, dates do not require hydration. However, sometimes hydration is used to soften the texture of some hard-type date cultivars. It is achieved by dipping dates in hot or cold water for a certain period of time. Dates are dipped in hot water or exposed to steam at 60 to 65 °C and 100 per cent RH for 4-8 hours. Hydration changes the dried dates into plump and glossy dates. Forced air circulation is used to improve uniformity of temperature and relative humidity throughout the hydration room.

Figure 15. Sun Drying of Dates (*Source*. Kader and Hussain, 2009)

5.3.4 Pasteurization

Dates may be pasteurized by exposure to 72 °C and 100 per cent RH air until their temperature reaches 66 °C, where it is kept for one hour. However, such conditions may induce colour darkening of the dates.

5.4 Preparation for Market

5.4.1 Transportation to the Packing House

The harvested fruit is transferred into large plastic or wooden bins for transport to the packing stations. Each container contains 200-450 kg of fruit and is suitable for dry fruit. Large wooden, plastic, or cardboard cases of various sizes are also used, as well as baskets, sacks (for very dry fruit), and trays, focusing on the need to prevent damage to the fruit (Figure 16).

When transporting the fruit also take into account its sensitivity, and the importance of every link in the chain in the treatment of the fruit. Dates harvested at the *khalal* stage must be transported as soon as possible to receive appropriate treatment. The fruit must be transported in the early hours of the morning to avoid the heat; if the distance is long. Refrigeration during transport is advisable. 'Deglet Noor', which is to be marketed on the branches must not be shaken during transportation in order to prevent the fruit from falling off the branches. Speedy transport will also prevent infection by pests which attack the fruit during the postharvesting period.

Figure 16. Transport of Bunches from Field to Packing House (*Source.* Glassner *et al.*, 2002)

5.4.2 Cleaning

Clean water must be used and care taken that all the fruit is washed. The fruit from 'Barhee' and 'Deglet Noor' are also cleaned by air pressure specially adapted for the removal of dust and sand, before they are packed on branches. While the fruit is still hanging, it can be cleaned by water spray, accompanied by the use of fine swivelling brushes, but they must be dried before being packed.

5.4.3 Drying

When the fruit is packed immediately after washing, it is important to dry it in drying cubicles or by means of large fans. Air-drying is designed to result in moisture content of 20 per cent or below to prevent growth of moulds and yeasts. Temperatures of 55 °C to 65 °C for drying of soft dates are generally used.

5.4.4 Sorting

In order to ensure appropriate sorting it is important to provide sorters with precise, unambiguous definitions of the defects of the fruit that can be identified in date fruits:

a) Defects Resulting from the Activity of Insects and Various Mites

The most common are the remains of various moths, sour bugs and mites. Some of these pests leave signs of nibbling inside the fruit; some spoil the look of the skin. Tolerance for these defects differs according to various standards, going up to 4 per cent; in all cases there must be no live insects inside the package. Defects caused by birds, mice, bats or other rodents (mainly signs of nibbling on the outside) are often found on fruit grown without being covered by a net or paper, or stored under inappropriate conditions. Such fruit must be removed. These pests may leave remains of feathers, excrement of mice or birds, which stick to the fruit and may cause microbiological contamination.

b) Mechanical Defects

As a result of the fruit being crushed while wet after harvesting, or grazed or scraped during the period of growth, leaving scars on the fruit. Sometimes the fruit is so badly soiled by earth and by mud that washing does not clean it.

c) Microbiological Defects

Defects stemming from microbiological processes, fermentation (alcoholic) resulting from the activity of yeast; souring resulting from lactobacilli, acetobacteri or *Aspergilus niger*, a fungus which creates a black promycelium which fills up the stone cavity. These types of defects cannot be tolerated. These defects may be due to inappropriate conditions during storage (for example wet fruit without refrigeration) or may arise while the fruit is still in the field.

d) Physiological Defects

The physiological defects are as under

i) Unpollinated fruit that reaches sorting in an unripe state (its colour depending on the variety);

ii) Shriveled and dry dates, usually dates which have been detached from the spikelet while still unripe;

iii) Defects caused by water stress (excess or shortage), which may lead to checking (in 'Barhee') or blacknose.

iv) Some defects will appear more frequently in certain species. Workers must become familiar with them and immediately sort such fruits out.

5.4.5 Sizing/Grading

This operation is done manually or mechanically to separate dates based on their size and weight. Uniformity of size in a package is one of the quality criteria for dates. Date size varies depending on cultivars. In postharvest operations date sorting and grading processes can greatly delay packaging and marketing of fruit. The reason is that they are repetitive, labour intensive, and time consuming processes and are carried out by humans manually through vision inspection. Colour, size, surface defects, and texture are external characteristics used in sorting and grading in date industry. Different size grades and their criteria is discussed under quality standards.

5.4.6 Metal Detection

This process is also carried out simultaneously with the grading or sorting. It is done for the sake of identifying any outside metal particle or item that could have been picked up with the crop. It is highly recommended for dates to pass through a metal detector since metals could prove to be highly injurious during chewing of dates.

5.4.7 Surface Coatings

The objective of this process is to reduce stickiness and improve appearance. Several materials have been recommended for this purpose including a 5 per cent or 6 per cent solution of soluble starch as a dip, 3 per cent methyl cellulose or a combination of 2 per cent butylated hydroxyanisole, 2 per cent butylated hydroxytoluene, 6 per cent vegetable oil, 90 per cent water and a wetting agent.

5.4.8 Packaging

Packing the fruit in various ways is the last stage of its preparation for the consumer (Figure 17). It is done to protect, contain and market the product. Methods of packing are of two kinds: in bulk and for retail sales.

a) Bulk Packaging

Dates are usually packed in cardboard boxes (sometimes in plastic bags for additional protection and preservation of moisture, before being placed in boxes). The usual weight is 5 kg. The quality of the fruit may differ according to customer requirements. The fruit is sold on the open market and intended for customers wanting to buy fruit in large quantities.

b) Retail Packing

It has been greatly developed in recent years. These packages have to be adapted to consumer demands at all levels, starting with the codes used by a certain network, to repackaging and to the surface on which they are to be placed, ending with the writing on the packages such as the nutritional composition, and the last date for sale or for use. Retail packages can be divided into two categories:

i) Packing according to some arrangement, usually 'fish bone', the traditional way, which was developed in Marseilles in France and is called 'glove box' or 'boite à gants' in French. There are usually 26 to 30 dates in this box, placed in two layers, separated by cellophane, weighing 220 g to 250 g. A natural or plastic spikelet is placed in the top layer. Packing is done manually and much time is invested in arranging the dates in the boxes. The fruit is usually covered with glucose (natural) to give it a shine appearance.

ii) Packing by automatic weighing (without any inner arrangement): Much packing is done in this way, starting with the 'window' type, where a cellophane window showing the fruit is part of the package design, which is usually made of cardboard.

Dates are also packed into tubs made of transparent plastic, showing the fruit as part of the package design. The information for the client is usually on the lid. This type of packing can be of varying sizes, according to the client demands. Bags, usually made of PET polyethylene, are the cheapest and most economical way of packing. Quality control of packed products is the last time the fruit is checked before reaching the customer. Documented checking of the packages entails:

i) Weight of the package;
ii) Weight of the fruit;
iii) Arrangement of the fruit (in glove boxes);
iv) Uniformity of the fruit;
v) Damage to the fruit;
vi) Defects; and
vii) Moisture content.

5.5 Ripening

A series of metabolic physiological changes occur in date fruits during growth, development and ripening. These changes are controlled by the genotype, branch size, temperature and heat units, light and water potential, carbohydrate supply and hormonal

Figure 17. Packing of *Tamar* and *Rutab* Dates
(Photo Courtsey. Dr. Elhadi M. Yahia, University of Quertaro, Mexico)

changes, especially gibberellin, auxin and ethylene levels. The physiological changes that occur in dates during ripening begin with the recognition of external signals including the ethylene spike that occurs before the respiratory rise.

Following complex signaling pathways, a number of transcriptional factors are activated inducing the transcription of ripening-related genes. Newly synthesized proteins are involved in many processes, including ethylene synthesis, which increases ethylene levels and accelerates ripening.

Dates may need to be ripened after harvest when picked early to avoid damage by rain, insects or other factors. For indoor ripening, rooms should be equipped with devices to control temperature and humidity and an adequate air circulation. The exact temperature and time of ripening depends on the type of date, stage of maturation and condition at harvest. A temperature of 40-43 °C is recommended for ripening 'Khdrawy', 'Hayani', 'Sayer', 'Khalasa', and 'Sphinks' dates. Temperatures of 45-46 °C and 70 per cent RH for a period of 2-4 days or longer are required ripen cultivars with thick flesh such as 'Iteema', 'Maktoom', and 'Saidy'. 'Delget Noor' dates should not be ripened at temperatures above 35 °C. Soft cultivars such as 'Halawi', 'Dayri', and 'Zahidi' can be ripened at slightly higher temperatures (35-38 °C). Ripening of these cultivars is complete in about 2-4 days when they have lost their translucency and little or no hard tissue remains. Use of higher temperatures is not recommended because it increases skin separation from the flesh of dates. Freezing at -35 °C to -50 °C for at least 24 hours can be used to accelerate ripening of *khalal* dates to *rutab* satge.

Postharvest dipping of fruit at the *khalal* stage in Ethrel at 4.2 ml L^{-1} and abscisic acid at 1.0 mM significantly enhanced ripening. Neither treatment showed any negative impact on the overall quality characteristics of ripe fruit, suggesting that they may be practical tools for increasing the ripening rate of 'Helali' dates at *khalal* stage.

5.6 Storage

5.6.1 Low Temperature Storage

Low temperature storage is the most effective method of maintaining high quality in dates. It minimizes loss of colour, flavour and textural quality, delays development of sugar spotting, incidence of moulds and yeasts, and insect infestation, and prevents development of syrupiness (due to conversion of sucrose into reducing sugars) and souring of excessively moist dates. Fruits must be dehydrated to remove excess moisture if successful storage is required both at room temperature and under refrigeration. In general, soft date cultivars require a lower relative humidity in the storage compared to semi-dry cultivars.

Fruits picked at the *khalal* stage should be stored at 0 °C and 85 to 90 per cent RH to reduce water loss, delay ripening to the *rutab* stage and maintain their textural quality and flavour. Packaging in plastic bags or use of a plastic liner in the box helps in reducing water loss. Date fruit at the *rutab* stage are highly perishable and require immediate refrigeration. Optimal temperature for *tamar* dates is 0 °C for 6-12 months, depending on the cultivar (semi-soft dates, such as 'Deglet Noor' and 'Halawi' have longer storage life than soft dates, such as 'Medjool' and 'Barhee'). For longer storage durations, use temperatures below the highest freezing temperature of -15.7 °C. Dates with 20 per cent moisture or lower can be kept at -18 °C for more than one year, or at 4 °C for 8 months, or at 20 °C for one month; relative humidity should be kept between 65 and 75 per cent in all cases (Yahia, 2004). Ripe dates at the *rutab* or *tamar* stages, commonly harvested and handled in the world market, are not sensitive to chilling and freezing temperatures. However, freezing temperatures can injure dates at the early stages of *khimri* or *khalal*.

Relative humidity can influence water loss, incidence of some physiological disorders, and fungal growth. Condensation of moisture on the commodity over long periods of time is probably more important in enhancing decay than is the RH of ambient air. Pathological and physiological deteriorations increase with increasing moisture content and storage temperature. Very soft and syrupy dates are subject to mould invasion and fermentation more than other types of dates. An appropriate RH range for dates is 65 to 75 per cent. Water activity of 0.65 to 0.85 corresponds with moisture contents of 15 to 35 per cent in dates (Figure 18).

Recommendations of FAO summarized the following storage conditions:

i) Storage under conditions of 26 per cent humidity or higher requires a temperature of 0 °C enabling a storage period of 6 - 8 months;

ii) The storage period can be more than 1 -year if humidity is less than 26 per cent;

iii) If humidity is less than 20 per cent, dates can be stored at 25 °C for up to 1- year; and

iv) High sugar content coupled to high humidity tends to aggravate the situation of fruit going bad.

v) Varieties sold at *khalal* stage, such as 'Barhee' and 'Zagloul', are stored at a temperature of 1 °C, which increases their shelf life from a few weeks up to 6 - 8 weeks.

5.6.2 Modified and Controlled Atmosphere Storage

The conventional method of date conservation after harvest is cold storage at -18 °C. This is the most suitable method for the sensitive soft-fruit cultivars, but it is highly energy consuming. In contrast to studies carried

Figure 18. Cold Storage of *Khalal* 'Barhee' Dates (Photo courtesy. Dr. Awad Hussain, United Arab Emirates)

out in modified atmospheres for insect disinfestations, only limited work has been carried out to determine the influence of modified atmospheres on date quality. CO_2 significantly delayed browning and sugar formation on dates, and extended shelf life. 'Barhee' dates can be stored at 0 °C under controlled atmosphere containing CO_2 to 20 kPa (balance air) is more effective for maintaining fruit colour, firmness, soluble solids, and total tannins for 26 weeks as against 7 weeks in air.

6.0 Domestic Marketing and Export

Major date exporting and importing countries, by quantity and value, are listed in Table 4. Tunisia led the list of exporting and India as importing country by value. Based on the quantity of dates exported, Iraq was the leader followed by Pakistan and Iran.

Table 4. Major Date Exporting and Importing Countries during 2012

Exporters			*Importers*		
Country	*Quantity (tonnes)*	*Value (1000 US $)*	*Country*	*Quantity (tonnes)*	*Value (1000 US $)*
Iraq	1,38,437	46851	India	2,56,295	1,41,713
Pakistan	1,13,358	64081	Morocco	33,140	75,208
Iran	1,12,030	160251	Yemen	30,364	14,071
Tunisia	86,910	211451	Indonesia	20,142	20,546
Saudi Arabia	77,795	86293	USA	21,398	30,360

Source: Adapted from FAO, 2014.

India is the largest importer of dates in the world. In 2012, total date imports were 2,56,295 tonnes with the value of 1,41,713 thousand US Dollars. Import quantity is more or less constant since 2001 and it is around 250 thousand tonnes (Table 5). India imports from more than 22 countries but ten countries (Pakistan, Iraq, Iran, UAE, Oman, Saudi Arabia, Afghanistan, Tunisia, Algeria and Indonesia) contribute more than 90 per cent. Pakistan tops the list (Table 6). In comparison to European countries, Indian imports were of inferior quality.

Table 5. Import of Dates in India Since 2001

Year	*Quantity (tonnes)*	*Value ('000 USD)*
2001	2,56,295	1,41,713
2002	1,93,467	95,042
2003	2,98,423	99,059
2004	2,30,926	81,318
2005	2,53,341	72,359
2006	2,86,317	74,760
2007	2,40,399	55,753
2008	2,47,875	46,407
2009	1,93,755	33,011
2010	1,71,523	27,798
2011	2,44,367	52,786

Source: FAO, 2014.

Table 6. Import of Fresh and Dried Dates in India in 2012-13 from Top Ten Countries

Rank	*Country*	*Quantity (thousand kg)*	*Value (million USD)*
1	Pakistan	115,758.3	85.6
2	Iraq	43,137.3	54.7
3	Iran	27,257.7	12.7
4	UAE	12,857.5	8.2
5	Oman	3,052.6	3.1
6	Saudi Arabia	2,331.1	1.9
7	Afghanistan	2,063.0	1.4
8	Tunisia	1,447.5	1.3
9	Algeria	589.0	0.4
10	Indonesia	412.8	0.1

Source: Ministry of Commerce, India, 2014.

6.1 Quality Standards

Quality characteristics depend on cultivar, type of date (soft, semi-dry or dry) and condition (whole, pitted, pieces or macerated dates). For fresh dates, high quality is attributed to dates with adequate size and colour, small pits, thick flesh, freedom from dirt, sand or leaf particles, no evidence of bird, insect or rodent damage, no fungal or mould infection, no sugar crystals formation and freedom from any other apparent alterations. The skin of dates should be smooth, with little or no shriveling, and golden brown, amber, green or black in colour, depending on the cultivar. The texture may be soft and syrupy, or firm or dry, depending on the cultivar. In general, texture and flavour are considered the most important indices of quality in dates. Colour is a good quality attribute in light coloured cultivars.

6.1.1 CODEX Standard

Quality factors in the CODEX Standards for dates include the following:

i) Dates should possess the characteristic colour and flavour for the variety, be at the proper stage of ripeness and be free of live insects and insect eggs and mites.

ii) Moisture content of 26 to 30 per cent, depending on the variety

iii) Minimum fruit size of 4.75 g (unpitted) to 4.0 g (pitted)

iv) Absence of defects, including blemishes, mechanical damage, unripe, unpollinated, embedded dirt or sand, damaged by insects and/ or mites, souring, mould, and decay.

v) Dates and their products should be free from objectionable matter and microorganisms that represent a hazard to human health.

The CODEX standard for dates includes three sizes based on the number of dates per 500 g:

i) Small: >110 dates without seeds or >90 dates with seeds

ii) Medium: 90-110 dates without seeds or 80-90 dates with seeds

iii) Large: <90 dates without seeds or <80 dates with seeds

6.1.2 US Standards

In the US Standards for Grades of dates, quality scores includes 20 points for colour, 10 points for uniformity of size, 30 points for absence of defects, and 40 for character (well developed, well fleshed, and soft). US grade A or US Fancy are given to whole or pitted dates of one cultivar that achieve a score of 90 or higher. Lesser grades include US Grade B or US Choice, and US Grade C or US Standard (Table 7).

In the US, Medjool growers use a Grading standard that differentiates four grades based on fruit size and freedom from defects as follows:

Table 7. U.S. Standards of Date Grades

Grade	*Dates per kg*	*Description*
Jumbo	35-42	No blemishes, skin separation, or dryness
Large	44-51	No blemishes, skin separation, or dryness
Extra Fancy	44-53	Minor blemishes, packed all sizes together
Fancy	44-57	Some dryness and skin separation, packed all sizes together

6.1.3 Quality Profile

Quality profile of dates involve an evaluation of four aspects:

i) Colour, shape, size, taste, texture, pit/flesh ratio and uniformity in colour and size of the fruit

ii) Moisture, sugar and fiber content

iii) Defects of the fruits, which may include discolouration, broken skin, sunburn, blemishes, shrivel deformity, *etc.*, and

iv) Insect infestation, foreign matter, pesticide residues, mould, and decay.

The most important physical quality attribute of dates is texture, which should be determined by instrumental analysis of date flesh. The physical quality attributes are hardness, cohesiveness, adhesiveness, springiness, resilience, and chewiness. Hardness, chewiness, and resilience usually increase exponentially with the decrease of moisture content and show a maximum value at critical moisture content 21.5 per cent.

A well-defined, standardized scoring system for evaluating the total quality of a date (*tamar*) based on consumer preferences was developed and tested as mentioned below. As a result, 11 defined quality attributes of the *tamar* samples using a 5-point scale (very poor, poor, satisfactory, good, and excellent) that later was transformed to quantitative scores were evaluated. For most of the attributes (colour, appearance, sweetness, fruit size, mouth feel, mouth shear, and pit size), there was a single preferred category by a majority of date consumers. With respect to chewiness, flesh thickness, solubility, and elasticity, there was not a single preferred category by most consumers. On the basis of preferences of consumers, each attribute was given a quality definition (Table 8), and a total quality score then was determined by adding its individual attribute scores.

7.0 Conclusion

Dates are an important fruit, especially in many African, Middle-Eastern, and Asian countries. In India,

Table 8. Quality Definition of Attributes Based on Consumer Preferences

Attribute	*Maximum Score*	*Quality Definition*
Colour	16.5	Good-quality dates tend to be light brown in colour
Appearance	13.8	Most preferable dates tend to be regular and long
Sweetness	13.7	Preferred dates tend to be moderately sweet
Fruit size	9.5	A good-quality date fruit tends to be moderately large
Chewiness	7.8	A date of good quality ranges from slightly to moderately chewy
Flesh thickness	7.0	The flesh thickness of a good-quality date fruit ranges from moderately thick to thick
Solubility	7.0	A good-quality date tends to be moderately soluble to very soluble when consumed
Elasticity	6.6	Dates of good preference tend to range between slightly elastic and moderately elastic
Texture and mouth feel	6.3	A good-quality date tends to have a smooth texture and mouth feel
Mouth shear	6.2	A slight force is needed to shear or tear a good-quality date fruit
Pit size	5.6	The pit of a good-quality date tends to be medium in size, *i.e.*, neither big nor small

Source: Ismail *et al.*, 2001.

it is now gaining popularity and converts dessert into oases in Rajasthan and Gujarat. Recently, Rajasthan Government launched a mega project for planting tissue cultured date palms in western Rajasthan. India is the largest importer of dates in the world. Postharvest handling in value chain of dates is most crucial issue. Many topics needs to be investigated and improved such as: selection of adequate cultivars for better quality fruits and smaller tree size, improvement of harvesting methods, ripening procedures, dehydration and hydration techniques, safe method for insect and pathogen control, practical methods for moisture determination, and adequate packaging and storage conditions.

Some of the important means to maintain quality after harvest include: developing adequate date harvest mechanization, adequate use of the cold chain, packaging and packages, food safety measures, methods of insect control and prevention of infestation during postharvest handling. Use of low temperature during storage and transport is the most important way of maintaining quality of dates because it minimizes loss of colour, flavour, and textural quality, delays development of sugar spotting, reduce incidence of moulds and yeasts, and insect infestation, and prevents development of syrupiness and souring of excessive moist dates.

In India, due to early rains, dates can be harvest at *doka* stage and then artificial ripening should be done. Fumigation is most important for dates and presently it is being done by methyl bromide. Alternatives to this such as insecticidal controlled atmospheres and irradiation need to be standardized. By adopting adequate postharvest technologies quality may be maintained and losses can be reduced.

8.0 Important References

Al-Farsi, M.A. and Lee, C.Y. 2008. Nutritional and functional properties of dates: a review. *Crit. Rev. Food Sci. Nutr.* **48**: 877-87.

Al-Redhaiman, K.N. 2005. Chemical changes during storage of 'Barhi' dates under controlled atmosphere conditions. *Hort Sci.* **40**: 1413-15.

Ashraf, Z. and Hamidi-Esfahani, Z. 2011. Date and date processing: a review. *Food Rev. Int.* **27**: 101-33.

Awad, M.A. 2007. Increasing the rate of ripening of date palm fruit (*Phoenix dactylifera* L.) cv. Helali by preharvest and postharvest treatments. *Postharvest Biol. Technol.* **43**: 121-27.

Chao, C.T. and Krueger, R.R. 2007. The date palm (*Phoenix dactylifera* L.): overview of biology, uses, and cultivation. *HortScience,* 42: 1077-82.

El-Hadrami, A. and Al-Khayri, J.M. 2012. Socioeconomic and traditional importance of date palm. *Emr. J. Food Agric.* **24**: 371-85.

FAO [Food and Agriculture Organization]. 2014. Crop production and trade data. Available at http: // faostat3.fao.org/home/index.htmal (assessed September 2, 2014).

Glasner, B., Botes, A., Zaid, A. and Emmens, J. 2002. Date harvesting, packinghouse management and marketing aspects. **In:** *Date Palm Cultivation*, Zaid, A. (Ed.). FAO Plant Production and Protection Paper 156. Rome, FAO. p. 237-67.

Ismail, B., Haffar, I., Baalbakic, R. and Henryd, J. 2001. Development of a total quality scoring system based on consumer preference weightings and sensory profiles: Application to fruit dates (*Tamr*). *Food. Qual. Pref.* **12**: 499-06.

Kader, A.A. and Hussein, A.M. 2009. Harvesting and postharvest handling of dates. ICARDA, Aleppo, Syria, 15p.

Pareek, S. 2014. Date palm status and perspectives in India. **In**: *Date Palm Genetic Resources, Cultivar Assessment, Cultivation Practices and Novel Products.* Khayri, J.M., Jain, S.M. and Johnson, V. (Eds.) Springer, The Netherlands.

Rygg, G.L. 1975. Date development, handling and packaging in the United States. ARS, USDA, Washington, D.C. *Agric. Handbook* 482. 56p.

Sardarkrushinagar Dantiwara Agricultural University, Date palm Resaerch Station, Mundra. 2014. http: // www.sdau.in/completed/datepalm-0/index.htm

Chapter 17

Grape

Ajay Kumar Sharma
Principal Scientist
R.G. Somkuwar
Principal Scientist
P.G. Adsule
Former Director
ICAR-National Research Centre for Grapes, P.B. No. 3, Manjari Farm, Pune – 412 307, Maharashtra, India
E-mails: sharmadrajay@gmail.com; rgsgrapes@gmail.com; adsulepg@gmail.com

1.0 Introduction

Grapes are believed to be originated from West Asia and Central Europe and from there have made the journey to all parts of the globe. Grapes found in the old world of Europe and near the East are botanically known as *Vitis vinifera*. The *Vitis vinifera* has more than 10,000 varieties. Resveratrol a type of polyphenol is found in widely varying amounts among grape varieties, primarily in their skins and seeds ranging from 50-100 micrograms per gram. Anthocyanins are the main polyphenols in purple grapes whereas flavan-3-ols (*i.e.* catechins) are more abundant in white varieties. These anthocyanins and phenolic compounds are beneficial to human health due to their high anti-oxidant activities that have been associated with the metabolism of lipids and inhibition of oxidation of low-density lipoproteins as well as aggregation of platelets. Although people in France eat higher levels of animal fat yet the incidence of heart disease remains low which is sometimes mentioned as French Paradox. This is probably due to the protective benefits of regular consumption of red wine having beneficial bioactive compounds

2.0 Status of Cultivation

In India, grapes are commercially cultivated for table purposes and have reasonably high level of productivity. An area of 118000 ha was covered under grape cultivation with a production of 2483000 MT during 2012-13. Grape is mainly cultivated in Maharashtra followed by Karnataka, Tamil Nadu and Andhra Pradesh. Some northern states *viz.* Punjab, Himachal Pradesh and Jammu and Kashmir are also producing grapes. Postharvest management of grapes is a major concern for maintaining the quality of fresh grapes as well as processed products. Table grapes undergo serious water loss during postharvest handling. Rachis browning, which occurs as a consequence of water loss, reduces postharvest quality of table grapes. Cumulative water losses occurring during postharvest handling may lead to stem browning, berry shattering, wilting and shrivelling of berries during marketing. Area, production and productivity of grapes in India are mentioned in Table 1.

3.0 Major Commercial Varieties

Major commercial varieties of grapes cultivated in India are described below (Figure 1).

a) Crimson Seedless

It was obtained as a result of a cross between Emperor x C33-199 at USDA, Fresno, California. Vines are very vigorous. Crimson Seedless berries are bright red, large, cylindrical-oval and seedless. They have thick,

Table 1. State-wise Area (000 ha) and Production (000 MT)

State	*2015-16*		*2016-17*		*2017-18*	
	Area	*Production*	*Area*	*Production*	*Area*	*Production*
Maharashtra	90.09	2048.11	104.08	2378.17	105.50	2286.44
Karnataka	23.35	429.7	24.29	450.79	26.61	524.20
Mizoram	2.47	22.55	2.45	18.00	2.45	18.00
Tamil Nadu	2.44	34.10	2.49	29.02	2.16	58.93
Andhra Pradesh	0.73	14.64	0.77	15.36	0.80	15.92
Telangana	0.59	13.54	0.29	3.47	0.35	5.27
Other states	2.08	27.40	2.27	26.89	1.04	11.33
Total	**121.65**	**2590.04**	**136.64**	**2921.70**	**138.91**	**2920.09**

Crimson Seedless

Manjari Naveen

Sharad Seedless

Thompson Seedless

Figure 1. Major commercial Varieties

tough skin, and firm crisp flesh with a neutral flavour. Bunches are medium in size, conical with a shoulder, and well filled to slightly compact. It is a late ripening variety and takes not less than 130 days from pruning. The fruits are acidic and usually harvested at a sugar: acid ratio of more than 25 for Indian markets. The fruits can be cold stored for up to 18-20 weeks.

b) Flame Seedless

It was obtained as a result of complex 3-generation hybridization made at USDA, Fresno, California, USA. The vines are very vigorous and the shoots tend to grow continuously under tropical conditions. The berries are bright red, large to medium in size, round and seedless. They have crisp skin and firm but juicy pulp with a distinctive muscat flavour. Bunches are medium to large, conical and well filled. It is an early variety and takes about 95-110 days to harvest from pruning.

c) Manjari Naveen

It is a clonal selection from Centennial Seedless made at NRCG, Pune and was released in December, 2009. This variety ripens about 25 days early as compared to Thompson Seedless. It is a white seedless, naturally bold berry with a mild muscat flavour table grape variety. It has uniform berries and clusters and does not require extensive thinning operations. Extended harvesting should be avoided, the ripe berries are less acidic and aromatic and hence should be quickly packed for precooling and cold storage. A yield of 20-25 tonnes of exportable quality fruits can be obtained per hectare.

d) Red Globe

This variety developed at UC Davis, California, USA was introduced to India in 1985 and after multi-locational evaluation, it was recommended for cultivation. Bunches are big having very bold berries (22-25 mm dia), red round seeded with meaty pulp. It is a late-ripening variety and takes more than 135 days from pruning. It has good keeping quality.

e) Sharad Seedless

It is a clonal selection from the Russian variety Kishmish Chernyi (Black Sultana). Recently, 4 clonal selections are made by farmers, which respond well to hormone applications with attractive berries. These selections are Jumbo (Nath Seedless) and Nana Purple for extra bold elliptic berries, Sarita and Krishna Seedless with oblong/elliptical shaped berries and attractive conical bunches. The berries are bluish-black with crisp pulp, oblong to elliptical. It has a medium maturity and takes about 125 days from pruning for harvesting.

f) Thompson Seedless and its Clones

Thompson Seedless is well adapted to Indian conditions and is used in commercial cultivation for fresh as well as raisin purpose for about the last five decades. Some clones of Thompson Seedless namely Tas-A-Ganesh, Sonaka, and Manik Chaman are also in cultivation. The berries of these cultivars continue to ripen even on warm summer days and these cultivars are suited for staggered production. Clusters are medium to large, cylindrical to conical shaped, heavily shouldered that ripen in about 130-140 days after pruning. The berries are seedless, small to medium sized, oval to ellipsoidal in shape, soft berry skin and greenish white to golden in colour with a firm, juicy pulp. The majority of exported grapes from India consist of this variety.

4.0 Nature and Causes of Postharvest Losses

The postharvest loss in grapes has been recorded by various workers within the range of 8.23 to 16 per cent in the country. As per the present estimate of 8.23 per cent, India is losing about 223 thousand tonnes of grapes annually. If the loss is calculated as economic loss, it is much higher than expected. Due to grapes' delicateness and extreme perishability, the losses suffered during the preparation, harvest, packing, storing, transport and distribution of table grapes can also be very high. Main factors which affect postharvest losses in grapes are discussed here.

4.1 Varieties

Each variety has its own shelf life in given conditions. Some varieties show deterioration in the quality of grape bunches in terms of rachis browning, shrivelling and shattering very early. TSS content of berries at the time of harvesting plays an important role. Skin browning in Princess variety is related to TSS content of more than 18 °B at the time of harvesting. Flame Seedless showed moderate to severe symptoms of rachis browning when water loss was 2.8 to 3.7 per cent while water loss of 3.6 per cent was reflected as light rachis browning in Thompson Seedless. The thickness of the skin and pulpiness of berries have an impact on the shelf life of grapes. Coloured seedless varieties (Nana Purple, Crimson Seedless) are widely demanded in domestic as well as export markets. But due to prevailing high temperature after veraison, bunches are become uneven with poor colour development. Berries with poor colour are discarded at the time of packing.

4.2 Improper Handling

Improper handling of bunches leads to more postharvest loss. From harvesting to consumption at every point proper care of bunches is required. Harvest wound, transit injury, or breakage at the cap stem attachment show symptoms of stem browning in forms of bleached colour, followed by sunken areas where accelerated water loss has occurred. Gentle handling, careful cluster cleaning, fast pre-cooling, grading and storage at low temperatures are helpful to maintain freshness in grape berries for a longer duration.

4.3 Weather Conditions

Prevailing weather conditions at the time of harvesting plays an important role. Water loss due to surrounding weather conditions especially higher temperature is responsible for the loss of grape quality very fast. There are three symptoms of water loss from grapes: (i) shrivelled stems (also known as dry stems) which usually become brittle and break easily; (ii) browning of the stem which occurs as stem dehydration becomes more severe; and (iii) berry softening which is followed by a wrinkle formation that starts to appear radiating out from the pedicel. Harvesting of grapes at higher temperatures (more than 20 °C) always leads to faster rachis browning and more berry shattering during storage and marketing. High humidity is also equally responsible to the reduce shelf life of grape berries. Higher humidity is responsible for more incidences of postharvest diseases. The rate of moisture loss from fresh grapes depend on the humidity and temperature of the surrounding, as well as the heat and mass transfer properties of the fruit such as thermal conductivity, thermal and moisture diffusivity, interface heat and mass transfer coefficients. In Indian conditions, cool chain is maintained in the market channel for export while this practice is neglected in domestic market. By adopting the cool chain in the domestic market postharvest loss can be minimized.

4.4 Packaging Materials

Packaging materials not only play important role in proper marketing, but it also helps in the longer shelf life of grape bunches. The suitable materials enhance the freshness of grapes during storage and passage through the market channel. Less postharvest loss is observed in the market channel of export than domestic. In the export market, the quality of packaging materials matches international standards while in the domestic market, due to easy availability and lower cost, substandard packaging materials are preferred. Apart from this grape guard was never used. Packing using bubble sheets along with grape guard improves the shelf life of grapes and reduces post harvest loss. The grapes remain fresh during transportation to distant markets when packed in thermocol boxes rather than CFB boxes. By using proper packaging materials a large quantity of grape produce can be saved.

5.0 Postharvest Management

5.1 Maturity Indices

As grape is a non-climacteric fruit, should be harvested when they are fully ripened since neither the colour nor the taste improves after harvest. Harvesting period is determined by the variety, climatic conditions, TSS, acidity and sugar acid ratio, depending on whether the grape is for the local or export market and also whether it is used for table or raisin purpose or for wine making. Maturity standards of grapes fixed under the AGMARK states that the minimum TSS of 16 °B and sugar acid ratio of 20:1 should be complied with. Although the sugar content of the berries is considered as the indicator of their level of ripening, the ratio of sugar/acid is the correct index of ripening, since this ratio indicates the taste of berries. Berries with same content of sugars, taste sour and less sweet when their acid content is more. A simple device called 'Refractometer' is used for determining the sugar content of berries. Refractometer measures the total soluble solids (TSS) as °B and in ripe grapes, the soluble solids are nothing but only sugars, therefore, the total soluble solids content in berries represents the sugar percentage.

Physical appearance is considered as one of the major criteria, so bunch and berry size with uniform colour is important. The characteristic uniform colour development is a reliable index of ripening in coloured varieties. In white varieties, uniform green colour is preferred in the export market. Change of green colour to straw or amber colour should not be taken to indicate the stage of ripening as exposed clusters show more colour change even at less maturity as compared to shaded clusters. With the maturity of berries, the colour of the peduncle also changes.

All grapes do not ripe at the same item in a vineyard. Variation also occurs in the level of ripening within a bunch. Normally the berries nearer the stalk of bunch are riper than those at the tip of the bunch. Variation in ripening of bunches in a vine is due to differences in the dates of bud burst, while variation in ripening of berries in different plots of a vineyard is attributed to the difference in the dates of pruning and bud break. Also bunches exposed to sunlight mature earlier than those in shade.

Because of these facts, the bunches which are ready for harvest have to be tagged a day before the proposed day of harvest and berries are sampled from these bunches randomly.

5.2 Harvesting

Only attractive bunches fulfilling minimum quality requirement should be harvested. Harvesting should be done by skilled workers wearing soft rubber gloves and using sharp secateurs/scissors for cutting.

Careful handling of grapes during harvesting, cleaning and packing is very essential to prevent injury and abrasion. The bunch should always be held by the stem/peduncle. Rough handling results in loss of bloom (thin wax coating on berry surface) making the berries susceptible to decay. Loss of water also causes the skin to pull away from the pedicle attachment. These weak pedicel attachment areas result in berry shatter, sulphur dioxide injury and fungal growth. Therefore, touching the berries by hand should be avoided as far as possible. Rubber gloves should be worn during cleaning operations and handling of bunches.

5.2.1 Time of Harvest

Bunches should be harvested during the early morning hours before the berry temperature rises above 20 °C. It is advisable to close harvest by 10 a.m. Otherwise, the berry's temperature can not be brought down to 4 °C by pre-cooling within the stipulated time of six hours. This will avoid extra pre-cooling period to remove field heat. Bunches harvested during high temperature leads to loss of more physiological weight plus pedicel desiccation.

In case there was heavy dew in the night, harvesting should be delayed till the dew has dried. The free moisture on the surface of the berries is very congenial for germination of fungal spores and will increase decay during storage.

If rainfall has occurred just before harvest, the fruit should not be picked for at least 3-4 days, as the free moisture present on the surface of the berries can lead to fungal infections. At harvest, the berries with visual symptoms of decay should be removed.

5.2.2 Method of Harvest

A day prior to picking, the broken, along with decayed, deformed, undersized, and discoloured berries are removed by cutting their pedicels from the selected bunches, using a long nosed scissors. Care must be exercised not to injure other berries with the scissors. Clusters should never be held against the naked palm while cleaning, harvesting or trimming. They should be held by their stalk, preferably by wearing rubber gloves. This care is taken not to erase the fine waxy coating called 'bloom' from the berry surface. Otherwise it leads to rapid loss of water through the skin of berries and they shrivel during storage.

At the present, hand harvest is the only harvesting system for table-grapes. The primary advantages of hand harvest are:

i) Human picker can handle bunches with care avoiding mechanical damage;

ii) Selection of fruits starts on the vine because the picker selects maturity and appearance;

iii) Multiple (generally no more than two) harvest for grading the bunches on the vine could be done;

iv) Minimum capital investment required.

The harvest rate depends on the vine growing system, the number and expertise of workers, and the packing system adopted. The main problems with hand harvest are encountered regarding labour management and picker skill. A short training period is necessary for harvesting and packaging grape according to the market request. At the present, table grapes harvesting is carried out by the picker who selects the bunch and cuts the stem using very sharp shears. Before releasing the bunch in the box, the picker generally removes brown and mouldy berries. Thus, the picker performs the following operations:

i) Bunch selection using colour and eventually dimension as selection criteria;

ii) Detachment of the bunch selected by keeping the bunch in one hand and then with sharp scissor cut the stem

iii) Bunch trimming and cleaning (removal of decayed and mouldy berries)

iv) Place gently the bunch in the box, without pressing or squeezing

5.3 Handling

5.3.1 Bunch Collection

Harvested bunches are placed gently in clean perforated plastic crates and left in the shade of the vines for subsequent transfer to pack-house. The crates should be lined with clean bubble sheets for cushioning and kept over newspapers spread on the ground to avoid contamination with vineyard dust. The bunches are kept in such a way that their stalks should not injure berries of other bunches.

5.4 Sorting and Grading

The sorting and grading of grape bunch is very much required for fetching the good value of produce in domestic market. The sorting in packhouse helps in improving the shape of bunch and helps in removing undersized, bruised, uneven coloured, diseased berry. In case of export, the grading of grapes as per AGMARK standards is primary requirement. So it is essential to grade the fresh table grapes according to the Agmark standards before issuance of the Phyto Sanitary Certificate (PSC). The grapes have three standards *viz.*; Extra Class, Class I and class II as per AGMARK. The details on these grades are available.

Some EU importers also classify grapes according to the uniform size of the berries as follows:

i) Grade one: 26mm - 28mm

ii) Grade two: 24mm - 26mm

iii) Grade three: 22mm - 24mm

5.5 Packaging

Table grapes shall be packed in such a way that the produce is suitably protected during transportation and handling. Packaging is normally done in corrugated or solid fiber board cartons. A layer of bubble pad or protective liner is placed at the bottom of the carton to protect the grapes from bruising and a polyethylene lining is placed over this. Grapes from field trays are weighed in 5 kg lots. Care should be taken not to erase the bloom of berries and not to injure them while trimming the bunches and weighing them and therefore, use of hand gloves is must while handling the bunches.

Bunches from these weighed lots are placed in small, thin and clean food grade polyethylene pouches. One or maximum of two bunches weighing neither less than 350 g now more than 650 g are placed in each pouch. No bunch weighing less than 150 g is placed in a pouch. The grapes are then pre-cooled to a temperature of 4 °C and then a sulphur dioxide generating pad enclosed in absorbent tissue paper is placed over the grapes. This is then covered with the polyethylene lining and the box is closed.

During packing care should be taken to avoid damage to the fruit due to shattering, splitting or bruising. Therefore, select only loose bunches and trim them before packing to make them further loose. While packing, the grapes should neither be compressed as to cause splitting nor so loose as to cause bruising due to rubbing of loose berries against the wall of the box or against each other. The cartons are then palletized strapped and shifted to the cold stores. While packing fresh table grapes for exports, the packing requirement of the importing countries should further be taken into account. The following essential factors should, however, be given due consideration.

Packing in multi-layers within a crate causes weakening of the pedicle attachment in the berries of the lower bunches due to pressure exerted by the bunches on top. A break in the skin at the pedicle attachment acts as the entry point for decay organisms and is also susceptible to sulfur dioxide injury. Apart from this, pressure on the lower bunches can also result in bursting of berries causing not only physical damage but the exuded juice creates an environment conducive to growth of decay causing organisms.

There are two main methods in use for packing grapes for extended storage and shipping. One is to pack the grapes with SO_2 generator pad as a perforated liner in a carton. The second method is to pack the grapes in a carton with a SO_2 generator pad, arranging the cartons on a pallet, rapidly cooling the grapes, and the next day wrapping the pallet with stretch film, leaving the bottom of the pallet unwrapped.

Table grapes must be packed in such a way as to protect the produce properly. In the case of 'Extra' Class, the bunches must be packed in a single layer. In Germany importers require grapes packed in 9kg, 8kg, 5kg and 4.5 kg corrugated cartons or wooden boxes. The retail packaging of 1 kg pack fresh-keeping plastic bags or transparent plastic trays, 10 packs per carton, are usually preferred.

5.6 Storage

5.6.1 Pre-cooling

The pre-cooling is practiced to reduce field heat. Prompt removal of field heat from harvested berries is the best way of retaining the freshness of grapes for a longer time. The temperature in the pack house should be maintained at 18-20 °C and the grapes should be transported to pre-cooling units with 4-6 hours of harvest. The temperature of harvested grapes should be brought down to less than 4 °C within six to eight hours in the pre-cooling chambers. The delay in bringing down this temperature will reduce the keeping quality of grapes. In situations, where the pre-cooling units are away from the production sites and packing houses, mobile pre-cooling units are to be used. The filled boxes without closing the polyethylene liner are placed in the pre-cooling chambers.

5.6.2 Cold Storage

After pre-cooling, the dual releasing sulphur dioxide pads (Grape guard) are placed with their coated surfaces facing downwards on the filled plastic pouches and covered with the plastic sheet lining. The boxes are closed and shifted to cold storage rooms where the temperature and humidity are maintained at 0 ± 0.5 °C and 93 ± 2 per cent respectively. Temperature of 0 °C and humidity of 95 per cent are the best for maintaining freshness and preventing decay. Freezing injury to berries, pedicels and rachis occur at –2 °C, while at slightly higher temperatures decay may occur. Care should be taken to maintain the temperature and humidity strictly during storage and transit.

Sodium bisulphite packets placed in the containers with grapes can substitute for the fumigation treatment during several weeks' transit. Packets are formulated to release SO_2 either immediately or several days later, and generally, both types of packets are placed in the container for long lasting effects. Pallets holding containers with SO_2 packets can be covered with polyethylene film to maintain high relative humidity. The rate of SO_2 released by these materials is greater at higher temperatures, but the concentration of SO_2 in the container is maintained at lower temperatures.

i) Temperature more than 0 °C and/or humidity less than 92 per cent: More moisture release from berries. Surge in the release of SO_2 from grape guard and dislocation of berries and SO_2 injury to the berries.

ii) Temperature less than 0 °C: Freezing injury to berries, pedicels and stem of cluster. Berries crack and decay.

6.0 Domestic Marketing and Export

6.1 Cultivars Suitable for Export

The grapes are exported mainly in two groups white and coloured varieties. The Thompson Seedless and its clones namely Tas-A-Ganesh, Sonaka, Manik Chaman are major white varieties good for export. Besides these varieties, Manjari Naveen, which is developed at ICAR-National Research Centre for Grapes also had potential in the export market. In case of coloured varieties, Sharad Seedless and its clones and introduced Red Globe have a prime position in past years. The details of the export of grapes from India are presented in Table 2. Similarly, the import of fresh grapes in India is presented in Table 3. In 2021-22 the total export of grapes from India is 2,63,075.66 tonnes worth US $ 305.66 million.

6.2 Criteria for Export

Importing countries have their specific requirements. Even in one country, various importers have their own requirements for the produce. The exporters must abide by international rules to make supply as per requirements of particular importing group. In general, the importing countries demand the grapes with following requirements:

i) Berry size- should be more than 16 mm in diameter.

ii) TSS- should be more than 17 °Brix.

iii) Bunch weight- should be between 300-750 g.

iv) Bunch colour- milky green.

v) The selected bunch should not be compact.

vi) All the berries should be of uniform colour and size in a bunch.

vii) Less than 2 per cent sunburnt or Sulphur bleached, bruised or crushed berries are allowed.

viii) Pedicels should be fresh and green.

EU countries have accepted the AGMARK standards for import from India. Beside the minimum physico-chemical parameters (AGMARK standards), the importing countries especially EU require a certificate of phyto sanitary congition assurance of safe produce (detection of agrochemicals, heavy metals and hormones not more than described MRLs), foolproof traceability system, HACCP (Hazard Analysis and Critical Control Point) certficiate for import. Some specific importing groups from EU are adopting private grades and requiring grapes without any detectable agrochemicals. The GrapeNet (a system for residue monitoring for EU countries) is a very unique system, which can trace exported grapes from a particular plot of vineyard. The plot size constitutes a maximum of 1.2 ha for this purpose.

All imported grapes should comply strictly to with EU standards on quality and food safety. Most of the importers give great importance to EUREPGAP or Global GAP Certification. Although EU consumers prefer seedless grapes, grapes with seeds still have a good proportion in EU market. German consumers prefer both red (mostly imported from France) and white

Table 2. Export of Fresh Grapes from India during 2019-2022

Country	2019-20			2020-21			2021-22		
	Qty in MT	Rs. Lacs	US$ Mill	Qty in MT	Rs. Lacs	US$ Mill	Qty in MT	Rs. Lacs	US$ Mill
Netherland	56,265.55	78,672.33	107.76	68,553.42	82,516.51	112.83	75,308.62	87,789.41	116.34
Bangladesh	32,742.75	13,784.62	19.04	55,103.12	23,677.45	32.45	79,968.32	35,356.32	47.03
Russia	23,736.47	26,558.79	36.53	24,228.01	23,670.85	32.19	20,947.19	22,348.16	29.87
U K	14,484.40	20,544.30	27.84	20,013.33	25,264.55	34.43	12,705.64	16,433.09	21.75
U Arab Emts	12,128.27	11,235.27	15.41	15,517.56	14,409.66	19.54	16,459.07	15,453.78	20.54
Germany	8,718.95	13,669.04	18.64	6,940.64	8,558.21	11.72	5,209.76	6,024.97	7.95
Nepal	7,249.50	2,001.33	2.73	12,535.51	3,016.51	4.11	17,683.51	5,573.02	7.4
Saudi Arab	8,709.26	7,729.59	10.53	11,942.35	10,182.64	13.77	5,782.28	5,457.68	7.25
Malaysia	2,673.13	3,832.78	5.24	3,378.35	4,299.57	5.83	3,121.60	4,146.18	5.53
Thailand	3,569.99	5,885.86	8.12	2,365.33	3,252.45	4.45	2,458.51	3,622.21	4.82
Canada	1,491.73	3,567.60	4.85	3,160.56	4,433.37	6.08	2,207.01	3,276.56	4.33
Oman	3,833.54	3,528.65	4.82	4,124.47	3,746.83	5.05	2,909.09	2,517.91	3.35
Sri Lanka Dsr	1,272.97	1,800.19	2.5	1,132.68	1,495.90	2.05	1,337.77	1,916.86	2.55
Hong Kong	2,381.19	4,410.32	6.03	3,036.95	4,547.32	6.2	1,068.49	1,567.74	2.09
Finland	1,050.77	1,872.83	2.56	1,131.04	1,879.85	2.58	803.36	1,454.09	1.93
Lithuania	777.92	882.8	1.21	823.13	809.39	1.11	1,109.94	1,173.40	1.56
Spain	725.51	1,005.18	1.37	761.6	1,048.82	1.43	927.98	1,166.93	1.55
Singapore	122.57	413.03	0.56	538.96	712.98	0.97	789.71	1,067.38	1.43
Norway	344.94	1,056.74	1.44	254	463.04	0.64	516	1,066.20	1.41
China P Rp	1,911.27	3,233.85	4.4	1,708.08	2,375.60	3.23	696.14	968.92	1.3
Switzerland	285.05	550.47	0.74	582.44	886.55	1.21	557.95	834.73	1.11
Denmark	1,546.19	2,508.62	3.45	422.59	468.56	0.64	792.99	824.39	1.09
Latvia	396.53	411.08	0.56	460.55	428.47	0.59	764.17	786.21	1.04
Sweden	217.81	322.46	0.45	383.6	470.18	0.65	619.58	713.73	0.94
Total	**1,93,690.54**	**2,17,686.82**	**298.05**	**2,46,107.38**	**2,29,845.04**	**313.57**	**2,63,075.66**	**2,30,216.45**	**305.66**

https://agriexchange.apeda.gov.in/indexp/Product_description_32headChart.aspx?gcode=0205

Table 3. Import of Fresh Grapes in India

Year	Quantity (MT)	Value (₹ Lacs)
2019-20	7,583.77	8,582.14
2018-19	7,006.13	8,615.67
2017-18	5,626.96	7,352.65

https://agriexchange.apeda.gov.in/importtoindia/Product_description_32head.aspx?gcode=0205

Table 4. Import of Fresh Grapes in India

Year	Quantity (MT)
2013	14,146
2012	10,813
2011	3,899
2010	3,238
2009	2,370
2008	2,942
2007	2,815

Source: APEDA.

varieties of grapes imported from Italy, which have good sweet flavour, rather than an acidic flavour. During the last few years consumers are demanding specific varieties (mostly patented) having unique colour, aroma or shape of berries.

6.3 Major Bottlenecks in Export

A study was conducted on the constraints analysis of grape exporting farmers of Maharashtra state revealed that the major constraints faced by the respondents were: difficulty to replace the varieties after plantation (production constraint), high rate of interest for the loan availed (financial), more incidence of pests and diseases, irregular supply of electricity (general constraints), high initial investment in establishment of vineyards (constraint related to cost), and difficulty to meet exports standards of grape (export constraint).

6.3.1 Quality

The quality of any product is decided by the preference of consumer. Besides appearance and physical parameters (size and shape of bunch, colour, size of berry) of bunches/berries, hidden attributes of berries like sugar content, acidity, ratio of sugar and acids etc also play important role in acceptability of particular lot in the market. Production of quality grapes is always a bottleneck to maintaining the position in the existing market as well as for entry into new markets. Peculiar cultivation practices in India make it difficult to supply quality produce in particular market. Besides targeted consumers in particular market making it more difficult. However, each country has its own quality standards for fresh grapes, though AgMark standards are adopted by EU countries. The residue levels of various agrochemicals including insecticides, fungicides, heavy metals, plant hormones etc in the grape berries affect the export. No one in society is ready to tolerate the menace of the residues of agrochemicals in the food intended for his/her nurture. On the other hand, being in the tropical belt we do not enjoy the luxury of avoiding the use the agrochemicals in pest management. However, implementation of the residue monitoring system through GrapeNet has indeed emerged as a model system in the whole country.

6.3.2 Infrastructure

A strong infrastructure is the first demand for export of grapes from India. Cold chain is not available from the vineyard to the pack house. Packhouses are not in sufficient number, during packing season, packhouses become overburdened and all available packhouses are not equipped with desired infrastructure.

6.3.3 High Production Cost

Grape production in India mainly prevails in the tropical region. Double pruning and single crop make it very peculiar. The vines remain always active and require maintenance. The maintenance of vineyards in form of water supply, nutrition, pruning, pinching, leaves thinning, management of insect-pest and diseases and other activities from March- April to September-October increases cost of production.

6.3.4 Market Intelligence

Marketing strategies and market intelligence always play important roles in domestic marketing as well as export of commodity. Various markets can be available for Indian grapes but our strategies as well as market intelligence must strengthen in this regard. APEDA is the nodal agency which is dealing with export of grapes and helping grape growers, exporters and other stakeholders in different ways. Besides APEDA, Grape Exporters Association of India is helping exporters in various ways and helping find out new markets.

6.3.5 Packaging

Packaging of table grapes is fancier now. Various packaging materials are coming into the market. The exporters are importing packaging materials including grape guards to add value to produce. Imported packaging materials are increasing total cost at arrival in market. Punnet packing is very popular in different markets. Punnets with three colour clusters are also in demand in a specific market.

6.4 Quality Control

Efficient quality control and food safety is the most important in improving the export potential in India. There have been many instances when commodities exported by India have been rejected by importing countries due to concerns about high pesticide residue levels in grapes. Thus, strict monitoring from production to transport including processing, handling, packing, *etc.* is necessary. WTO provided an International framework by adopting Sanitary and Phytosanitary Measures (SPS). Sanitary and Phytosanitary standards, as covered under the SPS agreement include health and hygiene standards or regulations to avoid the spread of animal and plant diseases and epidemics. These have been adopted by the Codex Alimentarious Commission (CAC) of Food and Agriculture Organization (F.A.O.) and World Health Organization (W.H.O.). GLOBALGAP (earlier known as EUREPGAP) is an internationally used management system for promoting Good Agricultural Practices (GAP) for fruits and vegetable trade. This system has become quite popular and acceptable (Figure 2). EUREPGAP (now GLOBALGAP) was initiated in 1997 with the aim to provide safe food products to consumers. GLOBALGAP is being followed by many leading European food retailers like Tesco, Sainsbury's, Marks and Spencers, Metro, Kesko, *etc.*

Codex Alimentarius Commission has also come forward with guidelines about Good Agricultural Practices (GAP's) and Good Manufacturing Practices (GMP's) that help to control microbial, chemical and physical hazards associated with all the stages of the production of fresh fruits and vegetables from primary production to packing. It is difficult to implement these guidelines where primary production is taken up in small holdings. All aspects like quality of water for irrigation, washing, contamination from unsterilized manures, biosolids, pesticide residues, safety of agricultural workers, personnel hygiene and sanitary facilities, cleanliness and sanitation of post-handling equipment, machinery, *etc.* have been taken into consideration while finalizing the standards.

International Organization for Standardization (ISO) has published Food Safety Management System and is classified as ISO 22000. This system promotes FARM TO FORK APPROACH and is the latest and most comprehensive food safety and quality management system. It demonstrates the ability to control food safety hazards and communicates food safety information throughout the supply chain (through traceability) by emphasizing on prerequisite programmes *i.e.* Good Agricultural Practices, Good Manufacturing Practices, and Good Hygienic Practices *etc.*

India being a member of W.T.O, has adopted sanitary and phytosanitary measures agreement. With

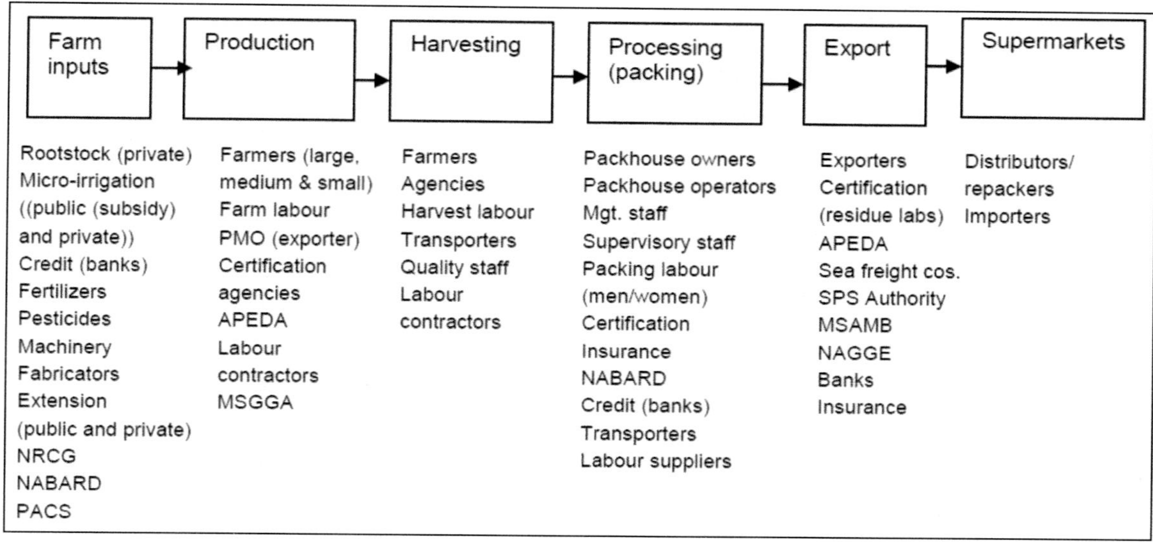

Figure 2. Grape Production Network for Export in India
***Source*. Singh, 2013**

Note: MSAMB = Maharashtra State Agricultural Marketing Board; NRCG = National Research Centre for Grapes; NABARD = National Bank for Agriculture and Rural Development; SPS = Sanitary and Phytosanitary; NAGGE = National Association of Grape Growers and Exporters; PACS = Primary Agricultural Cooperative Societies.

our experience of exports of fruits and vegetables especially to EU countries, India has realized that to stay in the competitive export market, our exporter will have to learn to grapple with existing food standards or else our farmers run the risk of missing the market access in other countries criteria. Thus, food safety and quality are inevitable for export market. India has accepted the importance of quality standards for different commodities among fresh fruits and vegetables and even Govt. of India has been participating in the meetings of Codex Alimentarius Commission for consideration of draft proposals of different commodities, so that various features and required parameters pertaining to Indian products are incorporated in the final version of quality standards. Recently, various standards applicable for exports of fresh fruits and vegetables have been notified under AGMARK Act and have been named under the title of GRADE DESIGNATION AND QUALITY. These standards under Agmark Act have been prepared keeping Codex Standards as a base. APEDA has also been taking various steps for promotion of quality and its control. APEDA provides assistance in the form of a subsidy up to 25 per cent for setting up laboratories for analyzing quality parameters like pesticide residue, metal content, *etc.* APEDA also promotes quality control by giving 50 per cent subsidy for installing quality management, quality assurance and quality control systems such as ISO series, HACCP and GAP Certification.

7.0 Conclusion

The area under grape production is increasing in country and grape quality from many areas is also at par with international standards. Large quantity of grapes can be saved which is lost between harvesting and delivery to consumers. Harvesting of grapes at proper maturity considering market, prevailing weather conditions at the time of harvesting, proper handling of bunches, timely precooling, sorting, grading, use of suitable packaging materials and maintaining cool chain till retailing have paramount importance in reducing postharvest losses and in maintaining freshness and quality of the grapes.

8.0 Important References

Anonymous 2007. The German Market for Fresh Grapes. *ftp: //ftp.moldova.cnfa.org/REPORTS/Marketing per cent 20Studies/The per cent 20German per cent 20Market per cent 20for per cent 20Fresh per cent 20Grapes per cent 20En.pdf*

Anonymous 2013 a. Regulation of export of table grapes to the European Union through control of residue of chemicals. *http: //www.apeda.gov.in/apedawebsite/Grapenet/rmpgrapesfor12-13.pdf*

Anonymous. 2013 b. Sanitary and phytosanitary requirements and quality standards. *http: //agriexchange.apeda.gov.in/Market per cent 20Profile/Market_Inteligence/IV.pdf.*

Banerjee, K.,Adsule, P. G. and Sharma, A. K. 2012. Food safety system in Indian table grapes and wine for export to the European countries. In *35th World Congress on Vine and Wine*, 18-22 June 2012 at Izmir, Turkey. Abstract Book pp. 149-50.

Crisoto, C. H., Smilanick, J. L., Dokoozlian, N. K. and Luvisi, D. A. 1994. Maintaining table grape postharvest quality for long distant market. International Symposium on Table Grape Production, 195-199.http: //ucanr.edu/datastoreFiles/234-35.pdf

Lichter, A., Zutahy, Y., Kaplunov, T.andLurie, S. 2008. Evaluation of table grape storagein boxes with sulfur dioxide releasing pads with either an internal plastic liner or external wrap. *Hortic. Technol.* **18**: 206–14.

Margaris, D. P. and Ghiaus, A. G. 2007. Experimental study of hot air dehydration of Sultana grapes. *J. Food Engg.* **79:** 1115-21.

Nelson, K. E. 1978. Pre-cooling its significance to the market quality of table grapes. *Int. J. Refrig.* **1**: 207-15.

Patil, A. B. 2008. A study on constraints analysis of grape exporting farmers of Maharashtra state. Ph D thesis submitted to UAS, Dharwad.,*http: //etd.uasd.edu/ft/th9696.pdf*

Ramteke, S. D., Sharma, A. K., Banerjee, K., Adsule, P. G and Khurana, S. C. 2012. *Manual for AGMARK in grapes.*, NRC for Grapes, Pune.

Sabir, F. K.,Sabir, A. and Kara, Z. 2010. Effects of modified atmosphere packaging and ethanol treatment on quality of minimally processed table grapes during cold storage. *Bulgarian J. Agril. Sci.* **16:** 678-86.

Sharma, A. K. and Adsule, P. G. 2013. Maturity indices, harvesting and postharvest quality management. In: *Good Agricultural Practices for production of quality table grapes.* Eds; Adsule, P. G.,Yadav, D. S., Upadhyay, A. Satisha, J. and Sharma, A. K., National Research Centre for Grapes, Pune. pp. 47-50.

Singh, S. 2013. Governance and upgrading in export grape global production networks in India. Capturing the Gains 2013, ISBN: 978-1-909336-86-5. *http: //workspace.unpan.org/sites/internet/documents/UNPAN20130726.pdf*

Zutahy, Y.; Lichter, A.; Kaplunov, T. and Lurie, S. 2008. Extended storage of Red Globe grapes in modified SO_2 generating pads. *Postharvest Biol. Technol.* **50**: 12–17.

Chapter 18

Guava

Neelima Garg *(Former Principal Scientist)*
A.K. Misra *(Former Project Coordinator (Fruits))*
A.K. Trivedi *(Principal Scientist)*

ICAR-Central Institute for Subtropical Horticulture, Rehmankhera, P.O. Kakori, Lucknow – 226 101, Uttar Pradesh, India

E-mail: neelimagargg@gmail.com; misra_a_k@yahoo.co.in; ajayakumartrivedi@gmail.com

1.0 Introduction

Guava belongs to the family Myrtaceae which include approximately 5,950 species in about 132 genera (Christenhusz and Byng, 2016; Govaerts *et al.*, 2008). Guava (*Psidium guajava*) is one of the important commercial fruits in India. It is the fourth most important fruit after mango, banana and citrus. It is native to tropical America where it occurs wild. Guavas originated from an area thought to extend from Mexico, Central America or northern South America throughout the Caribbean region. Archaeological sites in Peru yielded evidence of guava cultivation as early as 2500 BC. In India, it was introduced in the seventeen century. The important guava producing countries are India, Mexico, Brazil, Cuba, Venezuela, Australia, South Africa, Thailand, Malaysia, Indonesia, China, Sri Lanka, the Philippines, Bangladesh, Myanmar, Dominican Republic, USA (Hawaii, Florida and California) and Haiti.

The trees have the potential to bear fruits even thrice annually (ambey bahar, mrig bahar and hast bahar) with the adoption of crop regulation strategies. The crop bears heavily and gives handsome economic returns without many inputs. The shelf life of fruits is poor under ambient conditions; however, it can be increased with appropriate management strategy.

In India, a total of 265 000 ha area is under guava plantations with a production of 4054 000 Mt production and 15.3 MT/ha productivity (Govt. of India 2018). In the year 2015-16, total output of guava in India at current price was Rs. 741226 lakh. The area, production, productivity of guava during last 3 decades is given in Tables 1 and state wise area production and productivity during the year 2017-18 in Table 2.

Table 1. The Area, Production, Productivity of Guava during Last Three Decades

Sl. No.	*Year*	*Area ('000 ha)*	*Production ('000MT)*	*Productivity (MT/ha)*
1.	1991-92	94.0	1095.1	11.7
2.	2001-02	154.6	1715.5	11.1
3.	2010-11	205.0	2462.0	12.0
4.	2011-12	219.9	2510.4	11.4
5.	2012-13	235.6	3198.3	13.6
6.	2013-14	268.2	2667.9	13.7
7.	2014-15	246.2	3993.5	16.2
8.	2015-16	254.9	4047.8	15.9
9.	2016-17	260.1	3826.4	14.7
10.	2017-18	264.9	4053.5	15.3

Source: Horticulture Statistics at a Glance 2018.

Table 2. The Area, Production, Productivity of Guava in different States during 2017-18

Sl. No.	State/Uts	*Area ('000 ha)*	*Production ('000MT)*	*Productivity (MT/ha)*
1	Andhra Pradesh	9.53	229.78	24.12
2	Arunachal Pradesh	0.12	0.22	1.89
3	Assam	4.43	96.69	21.84
4	Bihar	27.61	427.61	15.49
5	Chhatisgarh	21.89	197.18	9.01
6	Gujarat	12.67	169.57	13.38

Sl. No.	State/Uts	*Area ('000 ha)*	*Production ('000MT)*	*Productivity (MT/ha)*
7	Haryana	12.09	137.02	11.33
8	Himachal Pradesh	2.32	2.61	1.12
9	Jammu and Kashmir	2.34	7.77	3.33
10	Jharkhand	8.32	89.31	10.73
11	Karnataka	7.18	140.23	19.52
12	Kerala	0.16	1.42	8.77
13	Madhya Pradesh	35.08	686.70	19.58
14	Maharashtra	9.07	122.83	13.55
15	Mizoram	0.42	2.55	6.07
16	Nagaland	0.58	4.75	8.16
17	Odisha	14.27	105.04	7.36
18	Punjab	8.69	195.60	22.50
19	Rajasthan	4.33	55.13	12.73
20	Sikkim	1.21	17.60	14.53
21	Tamil Nadu	9.69	155.06	16.00
22	Telangana	2.56	38.74	15.14
23	Tripura	0.70	3.40	4.85
24	Uttar Pradesh	49.53	928.44	18.75
25	Uttarakhand	3.62	20.37	5.63
26	West bengal	16.25	215.20	13.24
	Others	0.20	2.73	13.98
	Total	**264.85**	**4053.51**	**15.30**

2.0 Status of Cultivation

Guava as a commercial crop has captured the attention of many countries in the sub-tropical agro-climate. Worldwide, Mexico is considered the third largest guava producer after India and Pakistan. Brazil is another important guava producer in America. State-wise guava growing districts (belts) are mentioned in Table 3. Guava is credited with fourth ranking as regards the share of fruit production in the country. Uttar Pradesh is the largest producer of guava, followed by Madhya Pradesh and Biha. It covers a total area of 264850 ha with a production of 4053510 MT. In view of its uses, processing potential, popularity, health benefits, hardiness in cultivation and revenue generation potential, new areas are emerging in different states.

3.0 Major Commercial Varieties

Development of improved cultivars having comparative advantages in terms of traits, feasibility of adoption of high density planting systems, organic production and profitability prospects have been the driving force behind growing interest among orchardists to consider guava as one of the components of crop diversification strategies.

Table 3. Area and Production of Guava in Guava Growing Belts (Districts) in different States in India during 2016-17

Districts	*Area ('000 ha)*	*Production ('000MT)*
1. Uttar Pradesh		
Budaun	4.82	75.27
Kasganj	4.11	71.76
Aligarh	4.02	70.23
Farrukhabad	2.68	61.97
Etah	3.13	54.75
Hathras	2.94	51.34
Kanpur Nagar	2.13	38.02
Moradabad	1.19	35.05
Kaushambi	2.01	33.65
Bulandshahr	1.72	32.95
Rampur	1.41	30.90
Agra	1.17	30.25
Bareilly	1.00	30.03
Unnao	1.13	24.72
Sambhal	0.95	24.06
2. Madhya Pradesh		
Khargone	1.70	42.50
Sehore	1.30	41.86
Rewa	2.78	31.94
Vidisha	1.53	30.76
Katni	0.83	28.75
Indore	1.34	28.14
Bhopal	0.86	23.97
Singrauli	1.44	23.22
Sheopur	0.80	21.80
Morena	0.92	20.88
Chhindwara	0.81	17.75
Dhar	0.74	16.78
Ratlam	1.08	15.92
Jabalpur	1.26	15.09
Shajapur	0.53	13.38
Sidhi	0.63	12.42
Hoshangabad	0.55	12.10
Sagar	0.80	12.00
Shahdol	0.58	11.81
Tikamgarh	0.35	11.65
Datia	0.52	11.44
3. Bihar		
Nalanda	1.50	146.20
Rohtas	3.31	27.10
Bhojpur	1.93	16.50
Purbi Champaran	1.70	14.00
Pashchim Champaran	1.60	13.80
Buxar	1.60	12.50
Muzaffarpur	1.50	11.80
Vaishali	1.50	11.20
Kaimur (Bhabua)	1.40	10.60

Districts	Area ('000 ha)	Production ('000MT)
4. Andhra Pradesh		
Anantapur	2.29	187.41
Krishna	1.80	36.04
West Godavari	0.97	24.13
5. West Bengal		
24 Paraganas South	2.56	39.63
Birbhum	1.38	18.61
24 Paraganas North	1.13	16.52
Medinipur West	1.49	16.10
Nadia	1.32	13.63
Murshidabad	1.51	13.56
Purulia	0.90	10.78
Bankura	0.85	10.16
Purba Bardhaman	0.80	9.98
6. Punjab		
Patiala	0.97	22.08
Ludhiana	0.92	21.11
S.a.s Nagar	0.72	15.88
Sangrur	0.69	15.71
Muktsar	0.58	13.04
Jalandhar	0.56	12.65
Bathinda	0.47	10.59
Rupnagar	0.43	9.77
Fazilka	0.36	8.30
7. Chhattisgarh		
Korba	2.10	22.62
Bilaspur	2.46	17.89
Kanker	1.76	14.61
Kabirdham	0.98	14.26
Mahasamund	1.00	12.02
Janjgir-Champa	1.35	11.50
Surguja	1.05	9.45
Raigarh	1.16	9.06
Raipur	1.04	8.25
Korea	0.80	7.31
8. Haryana		
Sonipat	0.95	20.11
Karnal	0.66	16.68
Yamunanagar	0.86	15.81
Hisar	1.03	14.35
Jind	0.57	12.07
Jhajjar	0.80	11.96
Panipat	0.46	10.64
Fatehabad	0.52	9.63
Ambala	0.45	8.42
9. Gujarat		
Bhavnagar	3.96	47.25
Vadodara	1.68	28.06
Kachchh	0.60	10.17
Gandhinagar	0.62	9.01
Kheda	0.56	8.86
Chhotaudepur	0.51	8.56
Mahesana	0.82	8.12
10. Karnataka		
Dharwad	0.90	20.30
Chikballapur	0.64	15.88
Bangalore Rural	0.75	14.39
Haveri	0.58	13.84
Kolar	0.67	11.39
Bengaluru Urban	0.35	7.68
Mandya	0.37	7.11
Koppal	0.31	6.30
Belgaum	0.27	6.05
11. Maharashtra		
Solapur	3.55	31.81
Jalgaon	0.87	15.21
Nashik	1.18	15.17
Ahmednagar	1.26	12.60
Pune	1.05	12.60
Osmanabad	0.52	5.63
12. Tamil Nadu		
Dindigul	3.33	31.17
Madurai	1.19	15.34
Vellore	0.83	13.95
Virudhunagar	0.80	8.77
Cuddalore	0.73	5.86

Source: Horticulture Statistics at a Glance 2018.

The important commercial cultivars of guava in the world are listed in Table 4.

Table 4. Commercial Cultivars of Guava in the World

Country	Cultivars
Australia	Allahabad Safeda, Beaumont, Lucknow-49, Ka Hua Kula
Bangladesh	Kanchannagar, Kazi, Mukundapuri, Swarupkathi
Brazil	Cortibel, Kumagai, Ogawa, Paluma, Pedro Sato, Rica, Sassaoka, Seculo XXI, Yamamoto
Colombia	Extranjero, Puerto Rico, Rojo Africano, Trujillo
Costa Rica	Tai-kuo-bar
Cuba	EEA 1-23, Enana Roja Cubana
Egypt	Allahabad Safeda, Apple Colour, Basateen Edfina, Basateen El Sabahia, White Fleshed
Malaysia	Bentong Seedless (Malaysian S), Hong Kong Pink, Jambu Kapri Putih, Kampuchea (Vietnam GU 8), Maha 65, Taiwan Pear
Mexico	Acaponeta, China La labor, Coyame, Media China, Regional de Cavillo
Puerto Rico	Corazol Mixta, Corriente, Seedling 57-6-79
South Africa	Fan Retief, Frank Malherbe
Taiwan	Tai-kuo-bar
Thailand	Glom Sali, Glom Toon Klau, Khao Boon Soom
USA (Hawaii)	Beaumont, Pink Acid, Ka Hua Kula
Vietnam	Ruot hong da lang, Xa ly don, Xa ly nghe

Source: Pommer and Murakami, 2009.

In the Indian scenario, cultivars Allahabad Safeda and Sardar (*Synonym*: Lucknow-49) has dominated the commercial production scene country-wide apart from other cultivars with regional importance. Research on guava at Central Institute for Subtropical Horticulture, Lucknow, Indian Institute of Horticultural Research, Bangalore, Govind Ballabh Pant University of Agriculture and Technology, Pantnagar and AICRP on Sub-tropical Fruits has regulated in the development of trait-specific cultivars and improved production technologies. Following are the cultivar dynamics of guava in the country:

The important guava cultivars grown in different states of India are listed in Table 5.

Table 5. Commercial Cultivars of Guava in India

State	Cultivars
Andhra Pradesh	Allahabad Safeda, Anakapalli, Arka Mridula, Banarasi, Chittidar, Hafshi,Lucknow 49, Safed Jam, Smooth Green
Assam	Am Sophri, Madhuri Am, Safrior Payere
Bihar and Jharkhand	Allahabad Safeda, Apple Colour, Chittidar, Hafshi, Harijha, Selection-8
Jharkhand	Allahabad Safeda, Lucknow-49
Karnataka	Allahabad Safeda, Arka Amulya, Arka Mridula, Bangalore, Dharwa, Lucknow-49
Madhya Pradesh	Allahabad Safeda, Gwalior-27, Hafshi, L-49, Seedless Chittidar
Maharashtra and Gujarat	Dharwar, Dholka, Kothrud, L-24, Lucknow-49, Nagpur Seedless, Nasik,Sindh
North-eastern States	Allahabad Safeda, Lucknow-49, Red Fleshed
Tamil Nadu	Anakapalli, Banarasi, Bangalore, Chittidar, Green, Hafshi, Nagpur Seedless,Smooth
Uttar Pradesh	Allahabad Safeda, Allahabad Surkha, Apple Colour, Chittidar, CISH-G-1,CISH-G-2, CISH-G-3, Lucknow Safeda, Lucknow-49, Mirzapuri Seedless, Red Fleshed, Sardar
West Bengal	Allahabad Safeda, Baruipur, Chittidar, Dudhe Khaja, Gole Khaja, Harijha, Kabli, Lucknow-49, Sardar

3.1 Some Improved Guava Cultivars of India

a) Allahabad Safeda

It is a clonal material derived from the seedling population from Allahabad region of Uttar Pradesh and a popular variety of Allahabad. The fruits are medium, roundish with smooth skin, white pulp with few soft seeds and pleasant flavour. Keeping quality is good (Figure 1a).

b) Arka Amulya (Seedless x Allahabad Safeda)

This hybrid has been developed by Indian Institute of Horticultural Research, Bengaluru, from a cross between cv. Seedless and Allahabad Safeda (Figure 1b). Fruits are medium in size with white colour and sweet pulp having good keeping quality.

c) Arka Kiran (Kamsari x Purple Local)

A hybrid developed by IIHR, Bengaluru (Figure 1c). Fruit pulp pink in colour, mean fruit weight, 200-220 g, seed hardiness: 9 kg/cm^2, TSS: 12 °B;lycopene content: 5 to 6.0 mg/100 g.

d) Arka Mridula

This variety is a seedling selection from Allahabad Safeda and was released from Indian Institute of Horticultural Research, Bengaluru. Fruits are smooth, medium sized (180– 200 g) and pulp is white (Figure 1d). Fruits are soft seeded and have good keeping quality. It is good for processing due to high content of pectin (1.041 per cent).

e) Chittidar

It is a popular variety in Uttar Pradesh. The fruits are characterized by numerous red dots on skin, high sweetness and small and soft seeds. Fruits are sub-globose in shape, straw-yellow in colour. Keeping quality is excellent (Figure 1e).

f) Lalit

It is a chance seedling selection from red pulp guava identified at CISH, Lucknow (Figure 1f). Fruits are Saffron yellow coloured with red blush, pulp firm pink and weight 185-200g with good blend of sugar and acid. It has 250.63 mg/100g vitamin-C content. The pink colour in the beverage remains stable for more than a year in storage.

g) Lucknow-49

It was a selection made at Pune. Thi is also known as 'Sardar' guava. Trees are semi dwarf, and bear profusely with large fruits, roundish-ovate in shape, skin primrose-yellow pulp white, seeds soft and sweet. Keeping quality is excellent (Figure 1g).

h) Pant Prabhat

A seedling selection identified by GBPUAT, Pantnagar with upright growth habit. Fruits round, medium (150-175g) (Figure 1h). Peel, smooth and light yellow, pulp white with small and soft seeds as compared to Sardar, sweet,ascorbic acid content 125 mg (rainy season) – 300 mg (winter) per 100 g; TSS 10.5-13.5 °B.

i) Shweta

It is a half sib selection from Apple colour identified by CISH, Lucknow. Fruits are globose, medium size, weight 225g, creamy white epicarp, with red spots or blush snow-white pulp, high TSS (12.5-13.2 °Brix) and vitamin C (300 mg 100g1 pulp) with good keeping quality, high yield, good quality and attractive fruit appearance (Figure 1i).

(a) Allahabad Safeda

(b) Arka Amulya (Seedless x Allahabad Safeda)

(c) Arka Kiran (Kamsari x Purple Local)

(d) Arka Mridula

(e) Chittidar

(f) Lalit

(g) Lucknow-49

(h) Pant Prabhat

(i) Shweta

Figure 1a-i. Improved Guava Cultivars of India

In recent years, newer areas are being covered by Lalit (pink pulped) guava developed by CISH Lucknow especially in the states of Maharashtra, Andhra Pradesh, Gujarat, Bihar, Odisha; Shwetha (white pulped) varieties developed by CISH, Lucknow in the states of Haryana and Punjab; Arka Kiran (pink pulped), from IIHR, is gaining popularity in Karnataka and Tamil Nadu. A new variety named VNR-B which produces fruits of about 600 g fruit weight with better adaptability, good quality, fewer seeds and better shelf-life, transportability is also gaining popularity in Madhya Pradesh, Chhattisgarh and Karnataka.

4.0 Nature and Causes of Postharvest Losses

4.1.1 Losses Due to Fruit Fly

Fruit fly is one of the serious yield limiting factors in guava production. If not managed, it has the potential to cause nearly 50 per cent yield losses. Incidence of fruit fly (*Batocera zonata, B. correcta*) in guava orchards could be effectively managed by the use of pheromone traps which are based on male annihilation technique. These involve preparing the solution of pheromone with ethanol, methyl eugenol and malathion (6: 4:1). The 2″x 2″ water absorbable plywood pieces are soaked in the solution and fixed in a small plastic jar (Figure 2). Four holes of 1.5 cm diameter are made in the opposite directions on the upper side of the jar and installed in orchards @ 10 traps/ha. The fruit fly incidence by this approach could be reduced to 10 per cent, as compared to

50-60 per cent, especially during the *Ambe bahar* period. The imbalanced pesticide usage adopted for fruit fly control by growers could also be reduced by 80-90 per cent by this technique.

Figure 2. A View of Methyl Eugenol Pheromone Trap Set Up in the Guava Orchard.

4.2 Fruit Bagging

Some workers have recommended bagging of fruits for obtaining better quality. Newspaper or brown paper bags are slipped on the selected fruits at pre physiological mature stage, one month prior to harvesting. This not only protects the fruits from fruit fly and borer, but also gives a them a better look in terms of uniformity in colour and blemishless epicarp.

Fruit thinning is always accompanied by fruit bagging operation. Different types of bags that can be used to obtain high quality fruit. Bagging of fruits serves three functions:

i) Protection from insects, especially fruit fly damage.
ii) Prevents chemical contamination when pesticide applications are carried out in the field.
iii) Maintain a stable environmental condition for the fruit development.

Plastic bags with a small perforation at the end of the bags are utilized. The perforation serves as a drain outlet. Foam net is more frequently used to prevent fruits from sun burn in the dry season and bird damage. Waxed paper bags can also be used. Fruits are ready to be bagged when they are about 2.5-3 cm in diameter.

4.3 Losses Due to Diseases

Guava has the potential to produce crops throughout the year provided good orchard management practices including pruning, irrigation and pests and diseases management practices are considered. Majority of guava growing countries ubiquitously exhibit multi-nutrient deficiencies of P, K, Zn, Cu and B which need site-specific appraisal for amelioration for ensuring quality production.

About 177 pathogens are reported on various parts of guava plants or associated with guava fruits, which are due to fungal, bacterial, algal, nematodes and epiphytes *etc.* These pathogens cause various diseases pre and postharvest diseases. Besides this, fruit rots in the field (pre harvest rots) or during transit and storage (postharvest rots) are also the important and cause huge losses as the fruit, is perishable. Due to the perishable nature of fruit, it can be said that guava suffers badly by different rot pathogens (Misra, 2003).

4.3.1 Anthracnose

Disease causes die back, twig blight, wither tip and fruit spots/rot. In India disease is a serious problem in Uttar Pradesh, Punjab and Karnataka. Die back phase is caused by *Gloeosporium psidii* resulting in the death of plants.

Anthracnose in guava is also reported from Taiwan and Southern Nigeria. Eighty per cent of the guava plants in Nigeria are reported infected with anthracnose disease and fruit produced on those trees were severely infected.

Anthracnose of guava is caused by *Colletotrichum psidii* Curzi.

Pathogen produces setae and conidia in the acervuli. In moist weather, acervuli can be seen as black dots on twigs or fruits, and later produce pinkish spore mass (Figure 3). Spores are disseminated by wind or rain and initiate fresh infection where it fall.

Effective control of anthracnose can be achieved by sprays of Bordeaux mixture (3:3:50) at 7 days interval. Copper oxychloride and cuprous oxide also significantly control the disease (Tandon and Singh, 1969) but Bordeaux mixture and other copper fungicides cause russeting of fruits especially in cv. Allahabad Safeda and reduce their market value (Sohi and Sridhar, 1969). Difolatan (0.3 per cent), Dithane Z-78 (0.2 per cent), Thiabendazole, Aureofungin, Thiophanate methyle, Benomyl and Thiabendazole (5-50 ppm) are reported for the management of disease.

Storage of fruits at 10°C prevents decay but when guava fruits are again brought back to room temperature, the decay is enhanced fast. Higher temperature generally favours decay, so fruits should be stored in cooler temperature.

Figure 3. Anthracnose on Guava Fruit.

4.3.2 Fruit Canker

In India fruit canker caused by *Pestalotia psidii* was first recorded from Bombay and later from Mysore, Thane, Dharwar, Poona, Ponta Vally, Himachal Pradesh and Lucknow.

Fruit canker of guava is caused by *Pestalotia psidii*. The disease generally occurs on young green fruits and rarely on leaves. Initially infection on fruit is the appearance of minute, brown or rust coloured, unbroken, circular, necrotic areas. In advanced stage of infection it tears open the epidermis in a circinate manner. The margin of lesion is elevated and a depressed area is noticeable inside (Figure 4). The creater like appearance is prominantly noticeable on fruits than on leaves. The canker is normally confined to a very shallow area and does not penetrate deep into the flesh of the fruit. In older cankers, white mycelium consisting of numerous spores is noticeable. In severe cases, raised, cankerous spots develop in great numbers and the fruits break open to expose seeds. The infected fruits remain underdeveloped, become hard, malformed and mummified and drop in great numbers. In winter the cankerous spots are common but in rainy season minute red specks are formed.

Figure 4. Fruit Canker of Guava.

The spread of disease (in early stage of infection) is controlled by 3 to 4 spraying of 1 per cent Bordeaux mixture or lime sulphur (1 in 25) at 15 days interval.

4.3.3 Algal Leaf and Fruit Spot

In guava the alga causes spots on leaves and fruits. *Cephaleuros virescence* was reported on leaves, fruits, twigs and bark of guava from Florida (Ruehle, 1941).

Algal leaf and fruit spot is also called red rust because of the upper surface of the thallus produces erect, yellow to red filaments and fruiting bodies. The pathogen extends itself between cuticle and epidermis and penetrates the epidermal cells. The alga sporulates readily during the period of greatest rainfall.

Figure 5. Algal Leaf and Fruit Spot

During early spring flush *Cephaleuros* infects immature guava leaves. Small, shallow brown lesions appear on leaves. Later the disease progress and the lesions enlarge to 2-3 mm in diameter (Figure 5). Penetration of fruit is confined to several layers of cells beneath the epidermis. Fruit lesions are usually smaller than leaf spots. They are darkish green to brown or black in colour. Generally disease begins to appear from April and is more serious during May to August. During the period of highest rainfall (July-September) the pathogen sporulates readily. The disease incidence is greatest during September.

The disease can be managed by spray of copper oxychloride (0.3 per cent) 3-4 times at the interval of 15 days. Spray of copper oxychloride in rainy season is more effective compared to other fungicides.

4.3.4 Pseudocercospora Leaf Spot

Pseudocercospora leaf spot is prevalent in warm, humid and rainy areas of south Florida.

Figure 6. Pseudocercospora Leaf Spot.

Symptoms are seen primarily on leaves but also on stems and fruits. Small lesions (2-8 mm) of irregular to sub-circular in shape, with dark smokey brown colour and with a darker brown diffuse border develop as on the upper surface of leaves (Figure 6). Under high humidity, sporulation takes place in the centers of the lesion as greenish-gray, felty tufts of mycelium. Several lesions may coalesce to form large areas of necrotic tissue.

Copper based fungicides may be used for the management of the disease.

4.3.5 Phytophthora Fruit Rot

Fruit rot of guava is caused by *Phytophthora parasitica* Dastur (Mitra, 1929; Sohi and Sridhar, 1971) and *Phytophthora nicotianae* var. *parasitica* (Singh *et al.*, 1976) *Phytophthora citricola* (Ko *et al.*, 1982).

Figure 7. Phytophthora Fruit Rot of Guava.

The fruits especially affected are those, which either hang near the ground level or have fallen on the ground or (Figure 7). It is also found on fruits which have been placed in storage. The disease generally starts at styler end. Whitish cottony growth develops on fruits very fast on the ripening fruits and is able to cover almost the entire surface within a period of about 3-4 days during humid weather. The fruits which hang near the soil level or covered with dense foliage under high relative humidity are generally severely affected. Below the whitish cottony growth, the skin of the fruit becomes soft, turns light brown to dark brown and emits a characteristic unpleasant smell. The diseased fruits retain their normal shape unless they are invaded by saprophytes, which cause rotting. Sometimes the fruits drop due to infection. When the young and half grown fruits are infected, they shrink, turn dirty brown to dark brown, remain hard in texture, either remain intact as mummified fruit, or drop off (Singh, *et al.*, 1976). *Phytophthora citricola* affected fruits hanging close to the soil surface also rot during rainy season. The affected area becomes grayish brown and water soaked with grayish black centre.

Diathane Z-78 (0.2 per cent) or Aureofungin (10 ppm) are effective in controlling for the management of the disease. Copper oxychloride is reported toxic to the fruits (Sohi and Sridhar, 1971).

Cultivar Tehsildar is reported moderately susceptible and Chittidar and Apple Colour are quite resistant to Phytophthora fruit rot of guava (Gupta *et al.*, 1977).

4.3.6 Phomopsis Fruit Rot/Stylar End Rot (*Phomopsis destructum, Phomopsis psidii*)

Phomopsis fruit rot caused by *Phomopsis destructum* in India was reported from Saugar (Rao *et al.*, 1976). About 60 per cent of the fruits of a plant were found affected.

The infected fruits show disease symptoms near the stalk. Under favourable environmental conditions, infection centers are numerous. Lesions are dark brown, at first, small and increase in size to 2 cm diameter. The tissues soften and the entire fruit rots within 8-12 days (Figure 8). The rotten fruits fall from the parent plants causing heavy loss in the yield of the crop.

Figure 8. Phomopsis Fruit Rot of Guava.

First visible symptom of the disease is the discoloration in the region lying just below and adjoining the persistant calyx (Figure 12). The discolouration area gradually increases in size and turn dark brown. Later the affected area becomes soft. Along with the discolouration of epicarp, the mesocarp tissue also shows discolouration and the diseased area is marked by being pulpy and light brown in colour in contrast to the bright white colour of the healthy area of the mesocarp. At an advanced stage due to disorganisation of the inner affected tissues, size of the fruit shrinks and concentric wrinkles develop on the skin. Finally the whole fruit is affected and is covered with pycnidia. At all stages of development of the disease the affected tissue show abundant fungal hypahe which are mostly intercellular (Rai, 1956).

For the management of *Phomopsis destructum* griseofulvin, auriofungin, nystatin, Tetracycline and Chloramphenicol were found effective *in vitro* tests (Rao and Agarwal, 1976a). For the management of *Phomopsis psidii* Alpha-naphthol (50 ppm) and guaicol (250 ppm) cause complete inhibition of mycelial growth and fruit treated with this do not show disease symptom (Khare *et al.*, 1994).

4.3.7 Diplodia Dry Rot

The dry rot of guava fruits was reported from Vellayani (India) and infected trees had even more than 40 per cent of the fruits infected (Rajgopalan and Wilson, 1972 a).

Light brown spots mostly at the stalk end or at the calyx end of the fruit develop initially and spreads quickly and within 3-4 days entire fruit is affected. Completely infected young and mature fruits become dark brown or black in colour and finaly dry up. On the infected trees number of dry fruits hangs (Figure 9). Numerous pycnidia of the pathogen can be seen as pinhead like structures on the rind of the dried fruits (Rajgopalan and Wilson, 1972 a).

Figure 9. Diplodia Dry Rot of Guava Fruit.

Ziride 3000 ppm is effective for the management of Diplodia dry rot of guava fruits in orchards and does not cause any phytotoxic effect on flowers and fruits of guava when it is applied at 15 days interval (Rajgopalan and Wilson, 1972 b).

5.0 Postharvest Management

Guava (*Psidium guajava* L.) is one of the most popular fruits of the Indian sub continent. The fruit bears a pleasant aroma and taste besides high nutritional qualities. It is grouped under fruits of high ascorbic acid content and contains 2 to 5 times more ascorbic acid than oranges. Apart, the fruit is a rich source of pectin, calcium, iron and phosphorus. Guava is grown in almost all parts of the country and excels most other fruit crops in productivity, hardiness and adaptability. Owing to wide availability, excellent taste and flavour and high nutritional value, the fruit is often called 'common man's apple'. Nutritional value of the fruit is given in Table 6.

Table 6. Nutritional Value of Guava Fruit

Nutrient	*Proximates Value per 100g (Fresh fruit basis)*
Water	80.8 g
Energy	68 kcal
Protein	2.55g
Total lipid (fat)	0.95g

Nutrient	*Proximates Value per 100g (Fresh fruit basis)*
Carbohydrate, by difference	14.32g
Fiber, total dietary	5.4g
Sugars, total	8.92g
Minerals	
Calcium, Ca	18mg
Iron, Fe	0.26mg
Magnesium, Mg	22 mg
Phosphorus, P	40 mg
Potassium, K	417 mg
Sodium, Na	2 mg
Zinc, Zn	0.23 mg
Vitamins	
Vitamin C, total ascorbic acid	228.3 mg
Thiamin	0.067 mg
Riboflavin	0.04 mg
Niacin	1.084
Vitamin B-6	0.11 mg
Folate, DFE	49 µg
Vitamin A, RAE	31 µg
Vitamin A, IU	624IU
Vitamin E (alpha-tocopherol)	0.73mg
Vitamin K (phylloquinone)	2.6
Lipids	
Fatty acids, total saturated	0.272
Fatty acids, total monounsaturated	0.087
Fatty acids, total polyunsaturated	0.401

Adapted from USDA.

In India, guava is grown in almost all the states. However, its production seasons varies from region to region. In Northern India, there are two crops of guava- winter season and rainy season. The fruits of rainy season guava are inferior in quality as they are highly infested with insect attack. They also lack optimum flavour and taste and sometimes fetch a lesser price in the market. Many times, a rainy season crop is regulated to get a better yield in the winter season, which are characterized by rich organoleptic and nutritional qualities. It is only the winter guava that is utilized for processing purposes.

The crop bears heavily and gives handsome economic returns without much input. However, the shelf life of fruits is very poor under ambient conditions. Moreover, due to lack of proper infrastructure facilities and postharvest management measures, about 20-30 per cent of the produce gets spoiled, causing heavy economic losses. These losses mostly occur during the chain of postharvest practices like harvesting, handling, storage, transportation, processing, *etc.* Through the adoption of proper postharvest management practices, these losses can be minimized, thereby increasing the availability of fresh fruit for consumption as well as processing.

5.1 Maturity Indices

Maturity of fruit is an important factor for harvesting as it determines its postharvest quality. Early or late harvested fruits lack certain important attributes *viz.*, optimum colour, flavour, taste as well as nutritional qualities. Additionally, their storage life also gets impacted adversely. The colour changes of skin is the most common criteria for judging maturity. However, total soluble solids content, acidity and specific gravity are some other parameters for assessing maturity of the fruit. The mature guava fruits have a specific gravity less than one as they float on water. The fruits having specific gravity between 1.00 to 1.02 are better suited for long distance transportation as they have longer shelf life than mature fruits (floaters). The fruit turns pale green or yellowish green upon maturity.

Guava fruits usually mature after 90-150 days of flowering depending upon the variety. However, days from anthesis to harvesting can vary from 120 to 220 days depending upon temperature during fruit development. The fruits of 'Allahabad Safeda' and Lucknow-49 take 165 and 155 days, respectively from flowering to ripening under Bangalore, condition. The 'KaHua Kula' guava enters the ripening phase when the fruit is only half the maximum size, while in 'Allahabad Safeda' ripening changes occur towards the end of the growing period. Besides fruit age, detachment force, texture and skin colour are also useful maturity indices. In 'Paluma' guava the best maturity indices were reported as skin colour, pulp firmness, titratable acidity and ratio of TSS/TA. 'Paluma' on ripening showed 11.1 °Brix and 7.41 g 100 g^{-1} soluble sugar content.

5.2 Harvesting

The maturity of the fruit at the time of harvest determines its quality and storage life. In India two crops, *viz.* rainy and winter season are taken. The qualities of the winter season fruits are better than those of rainy season. The maturity of fruits has been worked out on the basis of colour and specific gravity. The fruits are usually ready to harvest after 4-5 months of flowering, when the fruit colour changes from dark green to pale green or yellowish green. The period of maturity. Fully mature, firm guava fruits are usually handpicked. Notched stick with a pouch is often used for harvesting of fruits. However, it leads to accidental falling of fruits causing injuries like bruising, cutting, cracking, *etc.* Guavas need care when being picked and handled. The rainy season fruits should be harvested at 2-3 days interval, while those of winter season should be harvested at 4-5 days intervals. It is desirable to harvest the fruits with the stalk along with one or two leaves attached. Harvesting should be done in the morning hours to avoid quality deterioration by solar heat. Fruits should be cut with a clean pruning shear from the branch. Harvested fruits should be carefully placed in crates with a soft inner sponge lining to protect fruits from physical damage on their way to the market or cold storage.

The harvested fruits should be assembled in baskets or plastic crates, since keeping them on ground may lead to latent infections from soil microbes causing spoilage of fruits during storage and ripening. The containers should be kept under shade to avoid sun heat, which may cause rise in the rate of respiration resulting in shorter shelf life of fruits.

5.3 Sorting and Grading

In guava, the infected and bruised fruits cause spoilage of other fruits in packing during storage and transportation and it is necessary that such fruits should be segregated from healthy ones. The fruits are generally graded according to their size, shape and weight. Freedom from blemishes and freshness of fruits are also used as parameters for grading. The fruits having less than 5 per cent blemishes are graded as grade A while those with more than 5 per cent are graded as grade B. Fruits are also graded on the basis of their weight, size and colour.

5.4 Postharvest Treatments

5.4.1 Hot Water Treatment

It was reported that immersion of fruits in hot water at 46.1 °C for 35 minutes controlled the infestation of Caribbean fruit flies. It was suggested that hot water treatment at 49± 1 °C could be adopted for fruit fly disinfestations while the disinfestations of Oriental fly is possible by vapor heat treatment at 46.5 °C for 15 or 35 minutes.

5.4.2 Calcium Treatment

Vacuum infiltration of fruits in 1-2 per cent $CaCl_2$ at ambient temperature retarded softening and suppressed increase in soluble pectin, however, it had no effect on incidence of disease as compared to untreated fruits. Fruits of guava cv. Lucknow-49 treated with 1 per cent $CaNO_3$ and stored at 6-8 °C remained marketable up to 30 days.

5.4.3 1-MCP

Studies have shown that the effectiveness of the use of 1-MCP in extending shelf-life varies with cultivar, ripening stages, concentration, exposure temperature and duration. It has been reported the extension of shelf-life of Allahabad Safeda by about 4-5 days when treated with 1-MCP at 300 nl L^{-1} for 12 and 24 h or 600 nl L^{-1} for 6 h and stored at 10 °C. A significant reduction in the decay incidence of 1-MCP-treated fruit was observed at ambient as well as in cold storage conditions. 1-MCP at 600 nL L^{-1} for 12 h, in combination with cold storage (10°C) seems a promising way to extend the storage life of guava cv. 'Allahabad Safeda' while 1-MCP at 300 nL L^{-1} for 12 and 24 h or 600 nL L^{-1} for 6 h, may be used to provide 4–5 days extended marketability of fruit under ambient conditions (Singh and Pal 2008)

5.4.4 Surface Coating

Surface coating of mature green guavas with cellulose or carnauba based emulsions delayed colour development and suppressed increase in TSS. Coating fruits with 2 or 4 per cent hydroxypropyl cellulose also was found to markedly reduce softening while Media China, a Mexican guava when coated with carnauba wax increased shelf-life under refrigeration.

Guava (*Psidium guajava)* fruits of cv. L-49 with individual shrink wrapping using 9 µ LLDPE film could be successfully stored up to 12 days at ambient and 18 days in the evaporative cool chamber with negligible loss in vitamin C content. The untreated fruits lose 25-30% of ascorbic acid within 1week after harvest. Delay in senescence and metabolic activities as supported by less changes in soluble solids, sugars, acidity, respiration, and ethylene evaluation rate was also observed in individual shrink wrapped fruits in cool chamber. The spoilage of fruits by Fusarium rots was significantly less in cool chamber in individually shrink wrapped fruits followed by Sta-fresh treatment. Sta-fresh was more effective in cool chamber than ambient. Individually stored wrapped fruits scored a better value of sensory score than Sta-fresh under both the storage condition (Pal et al. 2004)

Postharvest treatments of guava fruits of cv. Allahabad Safeda with 1 per cent chitosan delayed the ripening process and pro-longed storage life up to 7 days at ambient conditions (28-32 °C and 32-41 per cent RH). The fruits retained desirable texture and postharvest quality till the end of their storage life. Thus, chitosan 1 per cent can be integrated into the supply chain management of guava fruit to extend storage life, marketability and maintain quality during transport and storage under ambient conditions.

5.4.5 Irradiation

Irradiation of guava fruits has been found to enhance shelf-life and kill fruit fly. Irradiation with 100 Gy and application of 6 per cent waxol to the fruits after harvest caused minimum physiological loss in weight and maintained fruit quality in storage for 16 days at ambient. Fruits of Allahabad Safeda when irradiated with 1.0 kGy showed 8 days shelf-life as against 4 days in control. It was observed that fruit fly control is also possible through irradiation.

Gamma radiation treatment of guava fruit (cv. Allahabad Safeda and L-49) with 0.25 kGy dose increased the postharvest life by 3–4 days, maintained fruit quality, and reduced the decay incidence. The optimal dose (0.25 kGy) for postharvest life extension of guava fruit may be exploited to provide phytosanitary security against many insect pests including fruit flies (Singh and Pal 2009).

5.5 Packaging

Guava fruits are generally packed without washing or pretreatment. However, for export purposes, the fruits need to be first washed under running water in order to remove dirt and dry blossoms. Then, they are given hot water treatment at 48 °C for 5 minutes, surface dried and packed. This prevents postharvest diseases like anthracnose during packaging and transportation. Since guava is consumed without removing the peel, no fungicidal treatment is recommended.

The conventional packages, bamboo baskets, gunny bags and wooden boxes along with newspaper as lining material used for packaging guava fruits are unsuitable for long distance transportation as they may cause injury to fruits. Corrugated fiber board (CFB) boxes have been found very suitable for packaging of guava as they possess good flexibility and durability besides stacking quality. These boxes having 0.5 per cent ventilation have been found to enhance the shelf life of fruits up to 9 days under ambient conditions. Ventilation provides good circulation of air and helps in checking accumulation of moisture and ethylene within the box. Foam sleeves (polystyrene/polyurethane) are also used for individual wrapping of fruits. This prevents bruising during long distance transportation. However, due to higher cost of the netting, it is mostly used for export purposes.

Packaging of guava is done usually in Arhar baskets (Cajanus cajan), bamboo baskets, gunny bags and wooden boxes along with newspaper lining. Dry grass and leaves, paper cutting shreds are also used as cushioning material.

5.6 Storage

Guava fruits ripen rapidly and hence are highly perishable, with a very short shelf life ranging from 2-3 days at room temperature. This underscores the need to optimize technologies to increase shelf-life, access distant markets faster and improve market process. The fruit ripening in guava is characterized by loss of green colour, softening, shrinkage, loss of brightness and rot development. Retailing of guava fruits in India is usually carried out under non-refrigerated conditions. The fruits of cultivar Allahabad Safeda can be stored for 6 days while those of Sardar for 9 days under ambient conditions (20-25 °C). The shelf life can be extended up to 20 days by keeping the fruits at low temperature of 8 °C and 75-85 per cent relative humidity. It can be stored for about 10 days at room temperature (18-23 °C) in polyethylene bags having a ventilation of 0.25 per cent. Storage of fruits under ambient conditions is highly desirable in order to increase their shelf -life to facilitate long distance transportation, increase marketable period and thereby improving its marketability. Different storage methods have been tried with varying degree of success. A few of them are detailed below:

5.6.1 Low Temperature Storage

Low temperature storage of fruits is an effective means to extend the shelf life of guava. Guava can be stored for 2-3 weeks at 8-10 °C and 85-90 per cent RH. MAP in combination with low temperature (5 °C) storage can further enhance the shelf life for 4 weeks. However, symptoms of chilling injury appear on fruits causing total loss in marketability of fruits when stored below 5 °C. An optimum storage temperature of 8-10 °C for mature green fruit has been reported.

5.6.2 Controlled Atmosphere (CA) Storage

Controlled atmosphere (CA) storage is yet another effective measure to slow down the ripening process and increase the shelf life of fruits. Little information is available on CA storage of guava fruits. In CA storage, by increasing the level of CO_2 to more than 5 per cent and reducing the level of O_2 to less than 5 per cent, the shelf life of guava could be increased up to one month. The response of different varieties to CO_2 and O_2 levels varies and so also the shelf life. Storage at 8-10 °C with 2-5 per cent O_2 and 0-1 per cent CO_2 is recommended. Guava fruits exposed guava to CA containing 5 per cent CO_2 and 10 per cent O_2 for 24 hours and subsequent storage at 4 or 10 °C for two weeks found delayed colour development during ripening at 20-30 °C for 3 days and the fruits did not show chilling symptoms even after storage at 4 °C. It was concluded that Lucknow-49 fruits harvested at mature green stage could be stored for one month in unripe conditions under CA containing 5 per cent O_2 and 2.5 per cent CO_2 at 8 °C and fruits ripened normally at 20 °C with best sensory and nutritional quality. Guava cultivars, 'Lucknow-49', 'Allahabad Safeda', and 'Apple Colour' may be stored for 30 days at low temperature (8 C) supplemented with 5 kPa O2 + 2.5 kPa CO_2, 5 kPa O_2 + 5 kPa CO_2, and 8 kPa O_2 + 5 kPa CO_2, respectively.

5.6.3 Modified Atmosphere Storage

Modified atmosphere packaging (MAP) in the form of 0.5-2.0 per cent ventilated polythene bags of 200-600 gauge thickness enhances the shelf life of guava cvs. Allahabad Safeda and Lucknow-49 for 8 and 12 days, respectively. Packing of Kampuchea guava in low-density polyethylene films slowed down ripening and increased shelf-life. Wrapping of these fruits in low-density polyethylene bags showed less dehydrated, less acid, lower chilling injury intensity and disease severity (*Colletotrichum gloeosporioides* and *Pestalotia psidii*) after two weeks of storage at 8 °C. Paluma guava maintained marketable quality at 10 °C for up to 16 days when wrapped in transparent polyethylene bags.

5.7 Transportation

Road transportation is the most common method for transporting guava fruits across the country. Carts and trucks are the principal modes of road transportation. Fruits in this mode are often transported loose or packed in gunny bags or bamboo baskets. These practices cause huge losses in the form of bruises and damages during transportation, particularly in carts causing huge postharvest losses. Packing in CFB boxes has been found to be highly suitable for road transportation which however adds to the cost. Flexibility and stacking qualities of these boxes protect the fruits from damages caused by jerks. Reefer vans could also be introduced for transportation of guava as they can maintain a reasonable amount of freshness in the fruits.

5.8 Domestic Marketing and Export

Guavas are harvested throughout the year (except during May to June) in one region of the country or the other. However, the peak harvesting period in north India is August for the rainy season crop, November-December for the winter season crop and March-April for the spring season crop. For the local market, fully yellow but firm fruits are harvested, whereas half yellow fruits are picked for distant markets. The major wholesale markets for guava are Delhi, Bengaluru, Bhopal, Gangtok, Jaipur, Kolkata, Mumbai, Patna and Raipur. Maximum arrivals take place in Mumbai, Kolkata, Jaipur, Raipur and Delhi during the peak season. The growers normally sell their produce to contractors or middlemen since there are no organized market facilities.

5.9 Export Potential

In the international market, there is good demand for fresh fruits as well as for processed products. The major global exporters of guava are Brazil, Mexico, Thailand, Peru, Philippines while the main importers are USA, Netherlands, UK, Germany and France. India exports both fresh fruits and processed products, to Saudi Arabia, UAE, Sudan, Oman, Nepal, UK, Qatar, Bahrain, Indonesia and other Asian countries.

Codex Standards for Export

a) *Definition of Produce*

This Standard applies to commercial varieties of guavas grown from *Psidium guajava* L., of the Myrtaceae family, to be supplied fresh to the consumer, after preparation and packaging. Guavas for industrial processing are excluded.

b) *Provisions Concerning Quality*

In all classes, subject to the special provisions for each class and the tolerances allowed, the minimum requirement of guava should be

- Whole;
- Sound, produce affected by rotting or deterioration such as to make it unfit for consumption is excluded;
- Clean, practically free of any visible foreign matter;
- Practically free of pests affecting the general appearance of the produce;
- Practically free of damage caused by pests;
- Free of abnormal external moisture, excluding condensation following removal from cold storage;

STATE/UTS	JAN	FEB	MAR	APR	MAY	JUN	JUL	AUG	SEP	OCT	NOV	DEC
ANDHRA PRADESH	Round the year	Round the year	Round the year	Round the year	Round the year	Round the year	Round the year	Round the year	Round the year	Round the year	Round the year	Round the year
ASSAM					Peak	Peak	Peak	Lean				
BIHAR								Peak	Peak			
CHHATTISGARH	Peak	Peak	Lean	Lean	Lean	Lean	Lean			Peak	Peak	Peak
GUJARAT	Peak	Lean								Lean	Lean	Peak
HARYANA	Peak	Peak	Lean	Lean	Lean	Lean	Lean	Peak	Peak	Peak	Peak	Peak
HIMACHAL PRADESH	Lean							Peak	Peak			Lean
JAMMU & KASHMIR			Peak	Lean		Peak	Peak	Lean				
JHARKHAND							Lean	Lean				Peak
KARNATAKA		Lean	Lean	Lean			Peak	Peak	Peak			
MADHYA PRADESH	Lean	Lean								Peak	Peak	Peak
MAHARASHTRA	Lean	Lean	Peak	Peak	Lean						Lean	Peak
MIZORAM							Lean		Peak			
NAGALAND								Lean	Peak			
ODISHA	Peak		Lean	Lean	Lean	Lean	Peak	Peak	Lean			Peak
PUNJAB	Peak							Peak	Lean			Peak
RAJASTHAN	Peak	Lean										
SIKKIM										Peak	Peak	Lean
TAMIL NADU	Lean	Lean	Peak	Peak				Lean		Peak	Lean	
UTTAR PRADESH											Lean	Peak
UTTARAKHAND	Peak								Lean	Lean	Peak	Peak
WEST BENGAL						Lean	Peak	Peak	Lean			
ANDAMAN NICOBAR									Peak	Peak		
D&N HAVELI	Peak	Lean								Lean	Lean	Peak
DAMAN & DIU	Peak	Lean								Lean	Lean	Peak

INFORMATION AVAILABLE: Round the year; Peak season; Lean season

Source: Director of Horticulture/ Agriculture of respective State/ UT's

Figure 10. Harvesting Season of Guava across the Country.

Free of any foreign smell and/or taste;

Firm;

Practically free of bruising.

The guava must be carefully picked and reach an appropriate degree of development and ripeness in accordance with criteria proper to the variety and to the area in which they are grown. The development and condition of the guava must be such to enable them to withstand transport and handling; and to arrive in satisfactory condition at the place of destination

c) *Classification*

Guavas are classified into three classes as defined below:

i) "Extra" Class: Guavas in this class must be of superior quality. They must be characteristic of the variety and/or commercial type. They must be free of defects, except for very slight superficial defects, provided these do not affect the general appearance of the produce, the quality, the keeping quality and presentation in the package.

ii) Class I: Guavas in this class must be of good quality. They must be characteristic of the variety and/or commercial type. The following slight defects, however, may be allowed, provided these do not affect the general appearance of the produce, the quality, the keeping quality and presentation in the package:

Slight defects of colour or shape;

Slight defects on the skin due to rubbing and other superficial defects such as sunburns, blemishes and scabs not exceeding 5 per cent of the total surface area.

The defects must not, in any case, affect the pulp of the fruit.

iii) Class II: This class includes guavas which do not qualify for inclusion in the higher classes, but satisfy the minimum requirements above. The following defects, however, may be allowed, provided the guavas retain their essential characteristics as regards the quality, the keeping quality and presentation:

Defects in shape and colour;

Defects on the skin due to rubbing and other defects such as sunburns, blemishes and scabs not exceeding 10 per cent of the total surface area.

The defects must not, in any case, affect the pulp of the fruit.

d) *Provisions Concerning Size*

Size is determined by the weight or maximum diameter of the equatorial section of the fruit. The details of size codes are mentioned in Table 7.

Table 7. Size Codes of Guava

Size Code	*Weight (g)*	*Diameter (mm)*
1	> 450	> 100
2	351–450	96–100
3	251–350	86–95
4	201–250	76–85
5	151–200	66–75
6	101–150	54–65
7	61–100	43–53
8	35–60	30–42
9	< 35	< 30

6.0 Conclusion

India is one of the important countries for production and consumption of guava. A medium to tall plant, less seeded fruit, coloured variety with good keeping quality and resistance to multiple diseases is still needed for expanding acreage and production. So far, research on postharvest management of guava is meagre and scanty. Systematic and thorough study on postharvest management is needed to standardize different postharvest operations. Usually after harvesting fruits are packed in wooden boxes or bamboo baskets to transport to the market or processing units. Guava fruits should not be packed in hessian bags as it may cause considerable damage to the fruits even in short distance transport. For long distance transport, a layer of cushion material such as dry grass, paddy straw, paper wool etc. should be used at the bottom of the container. If refrigerated transport facility is available; it reduces the spoilage of fruits. Guava is a climacteric fruit, showing a typical increase in respiration and ethylene production during ripening. The loss of tissue firmness during ripening is accompanied by a decrease in the level of total pectin. In general, fruits with higher specific gravity can be stored longer than fruits with low specific gravity. Fruits positioned in the natural posture with the pedicel end vertically upward during transport show better keeping quality than fruits kept in the reverse or horizontal position. Coating of mature-green fruits in cellulose or carnauba-based emulsions delays colour development and suppresses increase in the level of total soluble solids. However, there is a need to manipulate ethylene biosynthesis pathway and ethylene sensitivity in guava for better shelf life.

References

Anonymous 2018. Horticultural Statistics at a Glance 2018.Horticulture Statistics Division Department of Agriculture, Cooperation and Farmers' Welfare, Ministry of Agriculture and Farmers' Welfare Government of India. P 490.

Ariosa, T. 1982. New diseases of guava (*Psidium guajava* L.) in Sancti Spiritus Province. *Centro Agricola*, 9, 3-7.

Chibber, H.M. 1911. A working list of diseases of vegetable pests of some of the economic plants, occurring in the Bombay Presidency. *Poona Agriculture Collage Magazine*, **2**, 180-98.

Christenhusz, M. J. M.; Byng, J. W. 2016. The number of known plants species in the world and its annual increase. *Phytotaxa*, **261**: 201–17.

Dey, P.K. 1948. Plant Pathology. Administrative Report of Agriculture Department, U. P. 1945-46, pp. 43-46.

Govaerts, R. *et al.* 2008. World Checklist of Myrtaceae. Royal Botanic Gardens, Kew. xv + 455 pp.

Gupta, J.H. 1979. Control of damping off of guava by seed treatment with systemic and non-systemic fungicides. *Prog. Hortic.* **10**: 53-55.

Gupta, P.C., Madaan, R.L and Suhag, L.S. 1977. Varietal reaction of guava fruits to *Phytophthora nicotianae* var. *parasitica*. *Indian J. Mycol. Pl. Pathol.*, **7**: 177.

Harnam, Mewada, Jain, M.K. and Barche, Swati 2013. Physiological and biochemical evaluation of guava papaya mixed fruit toffee. *Int. J. Hortic.* **3**: 11-15.

Khare, V., Mehta, P., Kachhwaha, M. and Mehta, A. 1994. Role of phenolic substances in pathogenesis of soft rot diseases. *J. Basic Microbiol.* **34**: 323-28.

Ko, W.H., Kunimoto, R.K. and Nishijima, W.T. 1982. Fruit rot of guava caused by *Phytophthora*. *Pl. Dis.*, **66**: 854-55.

Misra, A.K. 1987. Studies on diseases of fruit crops. Annual Report, CIHNP, Lucknow. pp.124-25.

Misra, A.K. 2003. Guava diseases-their symptoms, cause and management. In: *Diseases of Fruits and Vegetables: Diagnosis and Management*. Vol. 2 (ed. S.A.M.H. Naqvi). Pub. Kluwer Academic Publishers, The Netherlands. pp 81-119.

Mitra, M. (1929). *Phytophthora parasitica* Dast. causing damping off disease of cotton seedlings and fruit rot of guava in India. *Trams. Br. Mycol. Soc.*, **14**: 249-54.

Narsimhan, M.J. 1940. Annual Report, Mysore Agriculture Department for the year 1938-39. pp. 96-97.

Pal RK, Ahmed MS, Roy SK and Singh Manoj 2004. Influence of storage environment, surface coating and individual shrink-wrapping on quality assurance of guava (*Psidium guajava*) fruits. *Plant Foods for Human Nutrition* **59:** 67-72

Pommer, C.V. and Murakami, K.R.N. 2009. Breeding guava (*Psidium guajava* L.). 83-119 pp. **In:** *Breeding Plantation Tree crops: Tropical Species*, New York, Springer v.1

Poonam, Mall and Tandon, D.K. 2007. Development of Guava Aonla Blended Beverage Proc. Ist IS on Guava Eds. G. Singh *et al.*, *Acta Hortic.* **735**: 555-60.

Rai, J.N. (1956) Stylar end rot of the guava fruit. (*Psidium guajava* Linn.). *Proc. Indian Acad. Sci.*, Sec. B. **43**: 55-61.

Rajgopalan, B. and Wilson, K.I. 1972a. *Diplodia natalensis* causing dry rot of guava fruits in south India. *Pl. Dis. Rep.* **56**: 323-24.

Rao, D.P.C. and Agarwal, S.C. 1976a Efficacy of antibiotics against *Phomopsis destructum* causing fruit rot of guava. *Hindustan Antibiotic Bulletin*, **18**: 108-10.

Rawal, R.D. 1993. Yield loss in guava by different root rots. *Int. J. Trop. Pl. Dis.* **11**: 69-72.

Ruehle, G.D. (1941) Algal leaf and fruit spot of guava. *Phytopath.* **31**: 95-96.

Singh, G., Chohan, J.S. and Mann, S.K. 1976. Fruit rot of guava - a new disease problem in Punjab. *Indian J. Mycol. Pl. Pathol.* **6**: 77.

Singh S.P. and Pal R.K. 2009. Ionizing radiation treatment to improve postharvest life and maintain quality of fresh guava fruit. *Radiat. Phys. Chem.* **78:**135-40.

Singh S.P. and Pal R.K.2008. Controlled atmosphere storage of guava (*Psidium guajava* L.) fruit. *Postharvest Biol.Technol.* **47:** 296-306

Singh S.P. and Pal R.K. 2008.Response of climacteric-type guava (*Psidium guajava* L.) to postharvest treatment with 1-MCP. *Postharvest Biol. Technol.* **47:** 307-14

Sohi, H.S. and Sridhar, T.S. 1969. Injurious effect of copper sprays on guava fruits. *Indian J. Hortic.*, **26**: 155.

Sohi, H.S. and Sridhar, T.S. 1971. Controlling fruit rot of guava. *Indian Hortic.* **16**: 9-10.

Srivastava, M.P. and Tandon, R.N. 1969. Postharvest diseases of guava in India. *Pl. Dis. Rept.* **53**: 206-08.

Tandon, I.N. (1961) A new seedling blight of guava and it's control. *Indian Phytopath.*, **14**: 102-03.

Tandon, I.N. and Singh, B.B. 1969. Studies on anthracnose of guava and it's control. *Indian Phytopath.* **22**: 322-26.

Tandon, R.N. and Agarwala, R.K. 1954. Pathological studies of *Gleosporium psidii* causing die back of guava. *Proc. Indian Acad. Sci.*, Sec. B. **40**: 102-09.

Thirumalachar, M.J. 1945. An Ascomycetous parasite of *Cephaleuros*. *Proc. Indian Acad. Sci.*, Sec. B. **22**: 374-77.

Venkatakrishniah, N.S. 1952. *Glomerella psidii* (Del.) Sheld. and *Pestolitia psidii* Pat. associated with cankerous disease of guava. *Proc. Indian Acad. Sci.*, Sec. B., **36**: 129-34.

Yadav, A.S. 1953. Some new hosts of *Cephaleuros* from Bihar. *Curr. Sci.*, **22**: 280.

Chapter 19

Litchi

Vishal Nath

Former Director

ICAR-National Research Centre on Litchi, Mushahari, P.O. Ramna, Muzaffarpur – 842 002, Bihar, India

S.K. Purbey

Principal Scientist

ICAR-National Research Centre on Litchi, Mushahari, P.O. Ramna, Muzaffarpur – 842 002, Bihar, India

Alemwati Pongener

Scientist, ICAR-Indian Agricultural Research Institute, Gogamukh – 787 034, Dhemaji, Assam, India

E-mail: nrclitchi@yahoo.co.in; E-mail: skpurbey_nrcl@yahoo.com

1.0 Introduction

Litchi [*Litchi chinensis* Sonn. (*Nephelium litchi* Cambess)] is an important sub-tropical evergreen tree and a renowned edible fruit of the soapberry family, Sapindaceae. Besides litchi, the other important members of this family are longan, rambutan and pulasan. The fruit is native to southern China, where it flourishes especially along rivers and near the seacoast. It has been praised and pictured in Chinese literature from the earliest known record in 1059 A.D. The cultivation of litchi is spread over the years through neighbouring areas of South Eastern Asia and offshore islands. From china, litchi reached Eastern India first via Burma by the end of 17th century and thereafter in Bengal by the end of 18th century.

Litchi bears, showy fruits, in loose, pendent clusters of 2 to 50 which are usually strawberry-red, sometimes rose, pinkish or amber, and some types tinged with green. Most are aromatic, oval, heart-shaped or nearly round, about 2.5 cm wide and 4 cm long; have a thin, leathery, rough skin, flexible and are easily peeled when fresh. Immediately beneath the skin of some varieties is a small amount of clear, delicious juice. The glossy, succulent, thick, translucent-white to greyish or pinkish fleshy aril, usually separates readily from the seed. The flavour of the flesh is sub-acid and distinctive.

Litchi fruits are very rich in Vitamin C, and contain about 72 mg of Vitamin-C per 100 grams. Hence consumption of 100g of litchi would meet 119 per cent of the recommended daily requirement of vitamin-C in human diet. It is also rich in copper and phosphorus. It provides many B complex Vitamins and is a good source of fibre too. It contains a polyphenol "oligonol" that has antioxidant and anti-viral properties.

2.0 Status of Cultivation

Presently, India ranks second in production of litchi fruits in the world and the area under its cultivation is showing an increasing trend. During the last decade or so, the acreage and production of litchi has increased, stretching from 60,000 ha and 3, 68,600 MT in 2004-2005 to 827000 ha and 5,80,100 MT 2012-13 (Figure 1).

Litchi occupies an important place in the horticultural landscape of Bihar owing to its geographic confinement and 44 per cent of its share to the overall production in the country. The primary production zones of litchi in Bihar are Muzaffarur, Samastipur, Vaishali, Champaran-East and Champaran-West. West Bengal is the next important state with 16 per cent share of total

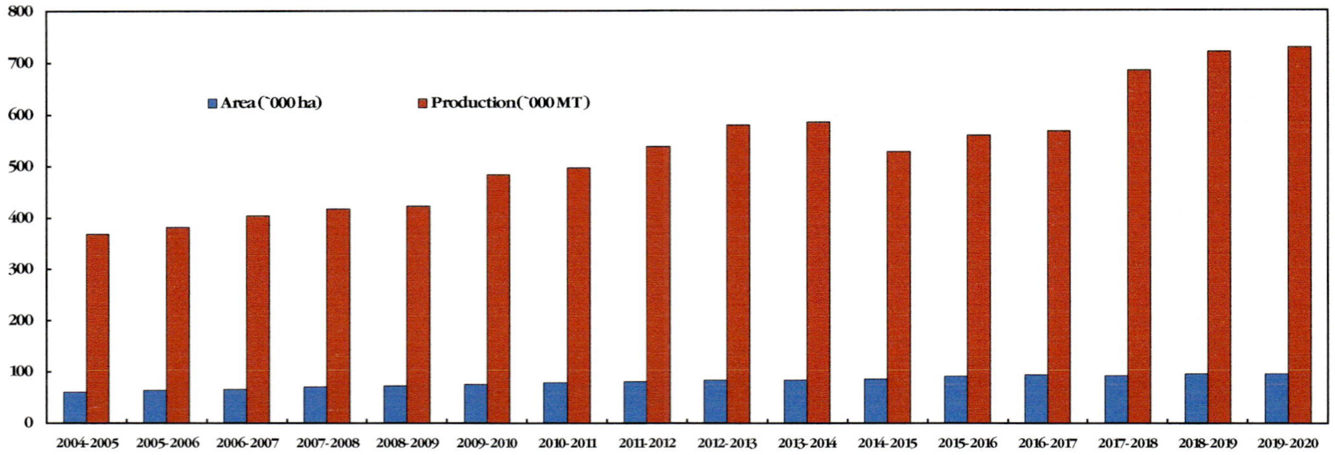

Figure 1. Growth of Area and Production of Litchi in India during the Last Decade

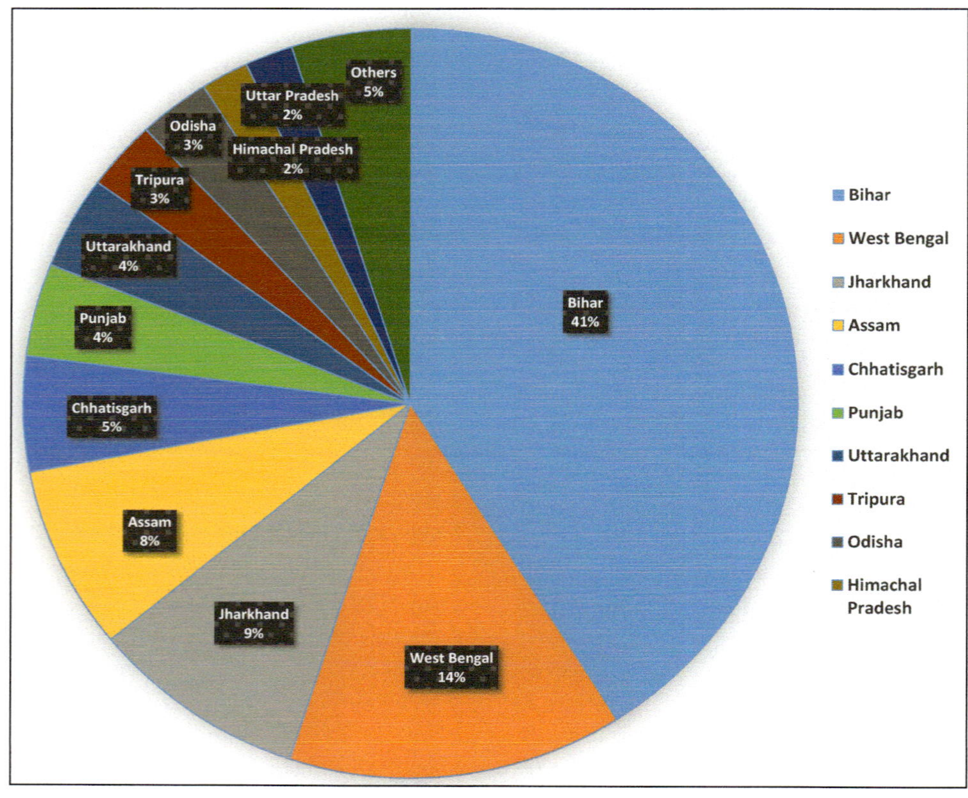

Figure 2. Share of different States in Litchi Production in India (*Source.* NHB 2020)

production followed by Jharkhand and Assam with 10 and 9 per cent respectively (Figure 2).

3.0 Major Commercial Varieties

The major cultivars of different litchi producing countries of the world are listed in Table 1.

Litchi cv. Shahi and cv. China are the major cultivars in India. Important varieties of litchi grown in different states are listed in Table 2.

4.0 Nature and Causes of Postharvest Losses

The major postharvest disorders associated with litchi is the pericarp browning. The bright red colour of litchi pericarp rapidly turns brown within 1-2 days at ambient condition which is due to desiccation, mechanical injury, chilling injury, heat stress and pathogen attack. Browning initiates from the distal end of the fruit and thereafter, extends over the entire pericarp surface, until the pericarp eventually becomes dry and brittle.

Table 1. Major Litchi Producing Countries, their Popular Varieties and Period of Harvesting

Countries	Varieties	Fruit Harvest Period
Australia	FayZee Siu (Feezixio), Kwai May Pink (Bosworth 3), Souey Tung, WaiChee, (Huaizhi), Tai So (Dazao), Huang (Mauritius)	October to March
China	Bah Lup, Baitang Ying, Feezixio, Heiye (Haak Yip), Kwai May Red, Lanzhu, No Mai Chee, Huaizhi (WaiChee)	2nd week of April-August
Florida (USA)	Brewster, Tai So, Haak Yip, Kwai May	May to June
India	Shahi, China, Bedana, Calcuttia, Longia, Bombai, Kasba, Rose Scented	3rd week of April to Mid-July
Indonesia	Local Selection	November to December
Madagascar	Madras, Mauritius	Late December to January
Philippines	Sinco, Tai So, ULPB Red	May to June
South Africa	Mauritius, McLean's Red	Late December to March
Thailand	Hong Huey, Chacapat, Haak Yip, Kom (Khom), Tai So, Wai Chee	May to June
Vietnam	Vaitheil, Local Selection	Late April to June

Table 2. Recommended Varieties for Commercial Cultivation of Litchi

Variety	States
Bedana	Bihar, Jharkhand, Uttar Pradesh, Uttarakhand, West Bengal
Bombai	Assam, West Bengal
Calcuttia	Himachal Pradesh, Jammu and Kashmir, Punjab, Tamil Nadu, Uttar Pradesh, Uttarakhand, West Bengal
China	Assam, Bihar, Jharkhand, Nagaland, Uttar Pradesh, Uttarakhand, West Bengal
Dehradun	Himachal Pradesh, Jammu and Kashmir, Jharkhand, Punjab, Tamil Nadu, Uttar Pradesh, Uttarakhand
Elaichi	West Bengal
Late large Red	Bihar, Chhattisgarh, Jharkhand, Madhya Pradesh, Manipur, Meghalaya, Nagaland, Odisha, Tamil Nadu, West Bengal
Late Seedless	Bihar, Jharkhand, Punjab, Tamil Nadu, Uttar Pradesh
Purbi	Bihar, Jharkhand, West Bengal
Rose Scented	Bihar, Himachal Pradesh, Jammu and Kashmir, Jharkhand, Punjab, Tamil Nadu, Uttar Pradesh, Uttarakhand
Shahi	Assam, Bihar, Chhattisgarh, Jharkhand, Madhya Pradesh, Sikkim, Uttar Pradesh, Uttarakhand

Source: Indian Horticulture Database, NHB 2013.

Table 3. New Varieties of Litchi Released by ICAR-NRC on Litchi

Variety	Plant Stature	Maturity	Fruit	Other Features
Gandaki Yogita	Dwarf, suitable for HDP	Late maturing	Pulp recovery > 75 per cent TSS 18-20°B	Free from fruit cracking and fruit borer, suitable for Indo-Gangetic plains
Gandaki Lalima	Robust	Mid late maturing	Fruit weight 28-32 g Pulp recovery > 60 per cent TSS 18-19°B	High yielder; Free from fruit cracking and fruit borer, suitable for Indo-Gangetic plains
Gandaki Sampada	Spreading	Mid late maturing	Fruit weight 35-42 g Pulp recovery > 83 per cent TSS 18-20°B	High yielder; Free from fruit cracking and fruit borer, suitable for Indo-Gangetic plains

4.1 Moisture Loss/Desiccation

Fruit moisture loss is encouraged by poor skin resistance to water vapour movement, air current, room temperature, low RH and temperature difference between air and fruit. Desiccation leads to increased conductivity of pericarp tissues, loss of membrane integrity, breakdown of vacuoles, leakage of anthocyanin, and destroys cellular compartmentalization and release of browning related enzymes and their substrate. The mechanism of litchi pericarp browning is attributed to the oxidation process of phenols and degradation of anthocyanin (red pigment) by the enzymes polyphenol oxidase (PPO) or peroxidase (POD) and the formation of polymeric browning pigments (α-quinones or melanin). Moisture loss and increased pH in pericarp tissues directly influence the PPO activity. The findings related to litchi pericarp browning have been schematically presented in Figure 3.

4.2 Pathological losses

Postharvest decay is one of the major factors contributing to postharvest losses of litchi fruits. A wide range of fungal pathogens have been found to cause pathological diseases in litchi. The list of micro-organisms reported to be associated with postharvest decay of litchi fruits are *Aspergillus* sp., *Botryodipoidia* sp., *Colletotrichum* sp., *Peronophythora*, *Geotrichum* sp. in Asian countries; and *Penicillium*, *Phomopsis*, *Trichoderma* and *Fusariums* sps. in South Africa and Madagascar. Control measures include treatments with fungicides, irradiation, heat, controlled atmosphere and use of biological agents.

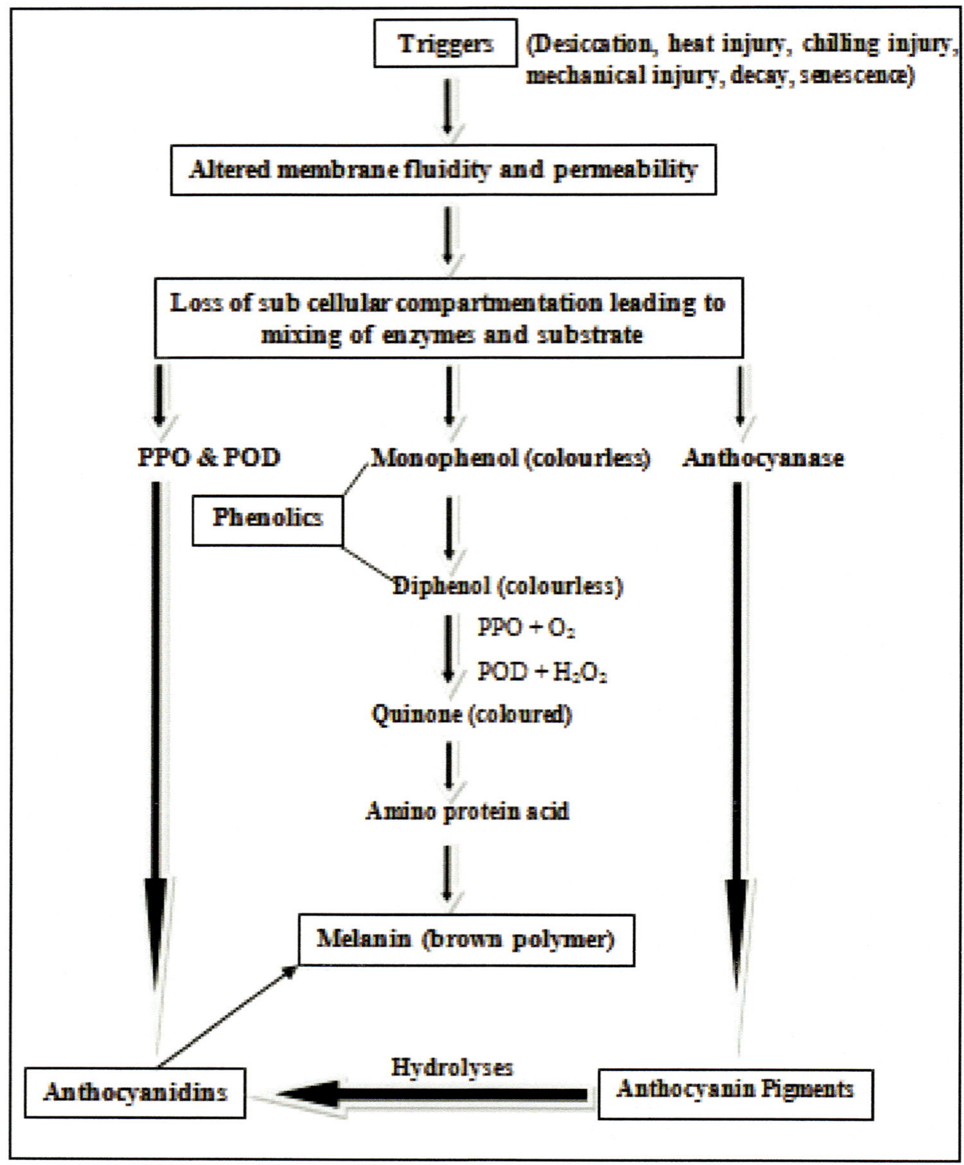

Figure 3. Enzymatic Browning of Litchi Pericarp
(Based on Jiang *et al.*, 2006 and Somboonkaew, 2010)

5.0 Bunch Bagging to Improve Fruit Quality

Bagging of litchi fruit bunch is a low cost, eco-friendly, sustainable, and economically viable technology that improves fruit quality. Fruit borer, sunburn and fruit cracking are major issues that limit production of quality fruits in litchi. Non-woven polypropylene bags have been found superior in terms of conferring physical protection against entry of insect pests and reduce sunburn and fruit cracking by creating a shading effect and favourable in-package micro-climate. For best results bagging operation is done one month before harvest. Almost complete freedom from fruit borer can be achieved if bunches are bagged at the right time, in addition to about 34 per cent reduction in sunburn, uniform size and colour of fruits and 30-36 per cent higher class-I fruit. Bagging also lowers production cost as growers can forgo chemical sprays.

Figure 4. Bunch Bagging in Litchi with Non-woven Polypropylene Bags.

Figure 5. Phases of Postharvest Deterioration of Litchi

6.0 Postharvest Management

Litchi is a delicate fruit, so minimal handling is preferred. In commercial production environments where fruits are to be transported to distant markets or the rate of consumption does not match the supply, appropriate postharvest management is critically required to minimize postharvest losses and successful marketing.

Litchi fruits mature in hot summer season (Figure 5) and are highly perishable in nature due to physiological and microbial spoilage. These last only for 2-3 days at ambient condition. This short postharvest life limits its long distance transportation and marketing. As per a study at NRC Litchi, Muzaffarpur, postharvest losses at various stages of fruit handling account for 20-25 per cent in case of cv. *Shahi*. Therefore, the postharvest management to prevent pericarp browning and enhancing shelf life at every stage of produce handling starting from harvesting to marketing is very much important.

Among various postharvest technologies that have been developed to retain peel colour and prevent fruit decay/rotting, the major breakthrough has been made by sulphur dioxide fumigation of litchi fruits which has been the common treatment for export and long distant marketing of litchi fruits. But due to growing concern over food safety awareness (sulphur residues on fumigated fruits) and restrictions imposed by many importing counties, sulphur-free postharvest treatment(s) which can maintain red pericarp colour and enhance shelf-life are increasingly important. In this endeavour, more emphasis is being given in recent years to develop active smart packaging techniques combined with non-hazardous chemicals and irradiation processing to make the litchi fruits available at every nook and corner of the world.

5.1 Maturity Indices

Litchi being a non-climacteric fruit does not improve in quality after harvest. Therefore, appropriate physiological maturity at harvest is crucial for proper quality and shelf-life. While pericarp colour is the most commonly used harvest index and has been suggested as picking maturity indicator it varies with cultivar, region of cultivation and cultural practices. A useful test of ripeness is red colour development of the inner surface of the skin in addition to development of surface colour. The concentration of sugar principally sucrose, fructose and D-glucose increases while the concentration of organic acid (mainly malic acid) decreases with maturity in litchi fruits. These attributes again seem to be highly variable according to cultivars, climate and cultural practices (Figure 6). Similarly, flattening of the fruit tubercles and the number of days from full bloom or fruit set to harvest are also considered to be important

Figure 6. Mature Fruits of different Litchi Cultivars

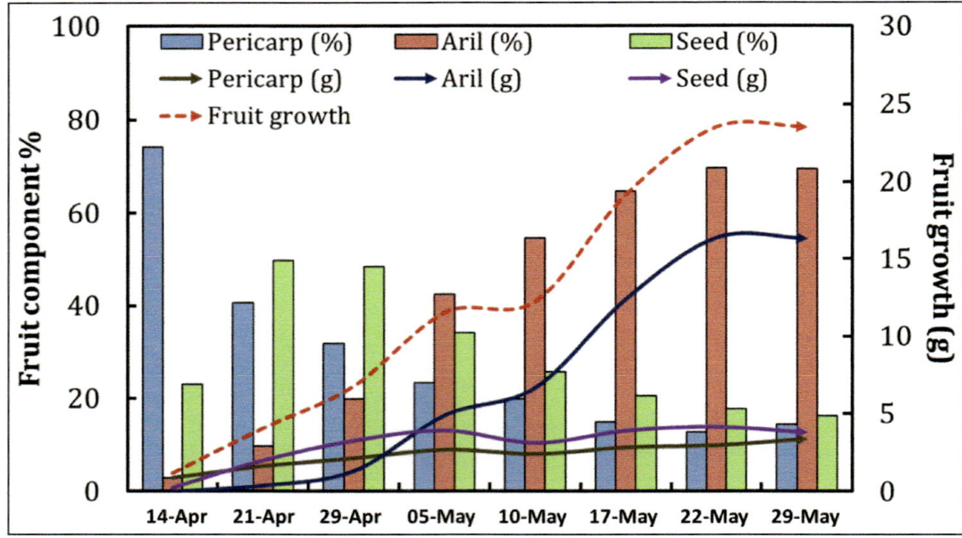

Figure 7. Fruit Growth and Development in Litchi cv. Shahi.

criteria for judging the maturity index of litchi fruit.But the correlation of these indices to mean eating quality has been often variable with the weather. The most guided and reliable index of maturity in litchi is the sugar: acid ratio (TSS/TA) which is highly correlated with the taste of litchi. A wide range of TSS/TA ratios from 15 to 65 have been recommended for mature litchi fruits from different areas. In practice, growers often rely on individual experience in terms of size, pericarp structure, colour, flavour, aroma, and taste of the fruit.

A common practice among litchi growers is early harvesting when fruits are harvested once the fruits attain colour-break stage, to get better prices early in the season. This stage occurs around the 10th of May under Muzaffarpur conditions. As depicted in Figure 7, litchi fruit at this stage weighs about 12-13 g, with low pulp content and high acidity of edible portion. Much of the fruit growth (25-40 per cent) takes place in the last 7-10 days after the colour break stage when fruit pulp expands and sugars accumulate with typical rosy flavour of litchi. Therefore, early harvesting not only results in quantum loss to growers but also qualitative loss. And since litchi is a non-climacteric fruit, where fruits do not ripen after harvest, fruits must be harvested only after attainment of optimum harvest maturity.

Recommendations vary, but a TSS/TA ratio of 40 or greater is recommended for commercial acceptability of fruit. The progressive farmers carry out the harvesting of litchi based on the development of surface colour and flattening of tubercles. The TSS in early harvested fruits range from 14-15 °Brix but the optimum mature fruits have TSS range of 19-22 °B.

5.2 Harvesting

Picking of litchi, for the local market is ideally done after attaining the full red colour and ripeness, whereas fruits intended for distant market are often picked when the pericarp partially turns red or at 75-80 per cent maturity. Harvesting is a very important operation to ensure good fruit quality after harvest. It should be done with care since mechanical damage leads to increased fruit desiccation and decay. The water content of litchi fruit on the tree fluctuates throughout the day. Maximum harvesting early in the morning ensures the highest available water content and minimizes moisture loss and rise in fruit temperature. Once fruits are harvested, exposure to the sun and air can increase the water loss by a factor of ten. Light spraying of fruit with cold water may help to maintain the fruit quality in hot and dry weather. Harvesting should never be done during rains since wet fruits are prone to rotting and decay.

Figure 8. Harvesting with a Pole Harvester

Litchi fruits are harvested by removing the whole panicle along with a portion of the branches (25-30 cm). Fruits are best harvested using clippers, pruning scissors or secateurs (Figure 8) to avoid the splitting of bark

Figure 9. (A) Hydraulic Platform Harvester (Developed at IIHR) at National Research Centre for Litchi, (B) Platform for Harvesting Big Litchi Tress in South Africa. (Photograph courtesy Alister Shaw) and (C) Litchi Harvesting in Florida

or breaking of peel at the point of attachment. Injury caused by rough handling contributes to fast rotting and infestation. Dropping of litchi fruit from more than 30 cm on the hard surface or 60 cm on the other surface can cause cracking if the fruits are fully ripe and turgid.

The litchi tree is not suitable for use of ladder while harvesting. The litchi fruit clusters at a higher level are usually harvested by metallic or bamboo pruning poles. Tractor mounted hydraulic platforms used for harvesting and picking have been found appropriate to harvest fruits from tall trees avoiding injury to the fruits. In the USA, South Africa, Madagascar and Australia ladders and such mechanized harvesting machines are used commercially (Figure 9).

Keeping the harvested fruits under shade, avoiding soil contact fruits, rapid transfer of fruit from the orchard to the pack house and minimum delay between harvesting and precooling are some of the important factors that determine postharvest shelf-life of litchi fruits. Fruits harvested before 10 a.m. should be sent separately for long distant marketing for better quality retention. Fruit harvesters (Figure 8) may be used for minimizing mechanical injury on the peel of fruits that leads to rapid browning of harvested produce.

5.3 Sorting and Grading

5.3.1 Sorting

Sorting of litchi fruits in different categories is a much required operation which can be done during the harvesting operation itself or at farm level. However, the growers rarely give sufficient attention to the logistics of this operation. Scarred, cracked, sun burnt, discoloured or otherwise affected fruits can be placed in separate harvest containers or discarded completely. More intensive sorting should be done at temperature and humidity controlled pack house (Figure 10). Panicles are split into sub-panicles and leaves and twigs are removed (leaving 10-12 cm part of twigs only). The entire surface of each

Figure 10. Sorting Operation in Temperature and Humidity Controlled Pack House

fruit must be observed to ensure that immature, overripe, misshapen, spotted and damaged fruit are not packed.

5.3.2 Grading

Grading separates fruits into different sizes and specifications to suit different markets. Fruits are graded according to their size, weight, colour and maturity. In India, grading is mostly done during sorting based on colour particularly, for domestic market. However, for export, a higher standard than the domestic market is required. Fully automatic grading machines have been developed and are in use at different places including NRC on Litchi, Bihar (India) which grades the fruit based on different sizes according to export/world standards.

5.4 Precooling

Precooling requirement of fruit is largely determined by its physiology concerning harvest maturity, temperature and humidity at harvest and ambient condition. Highly perishable litchi must be precooled to remove the field heat before packaging and storage, as the temperature at the time of harvest in India varies from 35-45 °C with 40-45 per cent RH at the field and 30-37 °C at ambient condition. Precooling is highly important to reduce the fruit pulp temperature which finally reduces fruit-respiration, prevents desiccation and maintains moisture level in fruits and keeps the peel fresh. This is the first step of good temperature management in cool chain system of litchi.

Rapid precooling is important immediately after arrival of fruits from fields or after sorting and grading. Room cooling, forced-air cooling, hydro-cooling and vacuum cooling are various precooling options that can be used for litchi fruit. However, hydro-cooling is more convenient and practical method being practiced abroad. Hydro-cooling utilizes cold water dips, sprays or ice slurries to rapidly chill the fruit. However, free water remaining on the fruit surface can increase postharvest pathogen growth during storage. Commercial hydro-cooling has been progressively adopted in Australia, Thailand, Florida and China. The water to be used for hydro-cooling may be acidized using citric acid (10g/l.) or chlorinated or fungicide mixed to ensure better colour retention and check fruit decay.

Forced-air cooling drives air flow in a single direction through a stack of fruits and has substantial benefits over room cooling. As per the speed of cooling, moist air or air with high Relative Humidity can be used for forced-air cooling, without causing significant water loss. Forced-air cooling may be a viable alternative for litchi providing the benefits of rapid cooling, without causing fruit wetness. However, hydro-cooling is a faster and inexpensive technique. Practically, on-farm hydro-cooling (0-2 °C) and packing in sealed polyethylene bags and shipping at same temperature can enhance the shelf life of produce. If cool-chain system does not exist throughout the postharvest cycle, a light precooling should be done at 7-10 °C for 10-20 minutes to minimize the fast respiration process.

Water is an efficient medium of heat exchange to achieve rapid removal of field heat. Pulp temperature of litchi aril may be as high as 31-33 °C at harvest which needs to be rapidly cooled to a safe pre-storage temperature to maintain fruit quality and increase shelf life. Studies at ICAR-NRCL have concluded that hydro-cooling of litchi fruits can be easily achieved by dipping harvested litchi fruit for 15 minutes in water maintained at 10 °C. Better results can be further achieved by acidifying the hydro-cooling medium with 4mM salicylic acid or any organic acid, which not only extends the storage life under refrigerated conditions, but also significantly reduces fruit rot and decay during storage.

5.5 Packaging

Proper packaging is important to maintain freshness, quality and prevent fruit decay during transit or storage. In ancient times packaging of litchi was done in woven grass bags, clay Jars, bamboos, baskets. In Asia, large bamboo baskets and wooden boxes cushioned with litchi leaves and lined with newspaper/brown paper are still used for local market. Early attempts at developing more effective packaging have made use of various types of paper cloth, PVC bags and perforated plastic bags. Results were highly variable with either heavy condensation promoting disease incidence leading to decay of litchi fruits or inadequate moisture control leading to desiccation of fruits resulting in colour loss and a high rate of physiological loss in weight (PLW). The fruits are then packed in plastic crates or Corrugated Fiber Board (CFB) or polythene bags (Thermocol boxes) with ice pads for long distance transportation (Figure 11).

Fresh, de-stemmed litchi fruits packed in shallow, ventilated cartons with shredded paper as cushioning material are shipped by air from Florida, USA to markets throughout USA and Canada. In Thailand, the fruits after hydro-cooling packed in plastic crates lined with polyethylene sheet and cushioned with foam are transported in ice lined Refrigerated Vans to China and Indonesia.

The packing of litchi fruits in ventilated punnet boxes and overwrapping with semi permeable membrane or PVC bags creates a modified atmospheric condition leading to reduced respiration rate of fruits besides providing a barrier to water loss. This practice has been found to enhance the shelf life of litchi by 5-7 days at ambient condition (*i.e.* 28±1.5 °C and 70 per cent RH) and by 2 weeks at low temperature (4 °C 90 per cent RH). In South Africa, sulphited fruits are packed in 2 kg capacity well ventilated cartons for export, whereas in Florida 10-45 lb heavy gauge polyethylene bags are generally used. In India, 2-6 kg ventilated CFB and 20 kg wooden boxes are used for domestic market.

Figure 11. Litchi Fruits Packed in Breathing Bags and Thermocol Boxes for Distant Transportation

A modified atmosphere (17 per cent O_2 and 6 per cent CO_2) created in biaxially oriented poly propylene (BOPP) films enable the retention of litchi fruit quality during storage by reducing pericarp browning. The successful implementation of MAP technology however depends on cultivar type, fruit maturity absence of decay as well as storage temperature. The maintenance of low temperature is vital for the success of MAP as it maintains the gaseous equilibrium required for a prolonged shelf life of litchi fruits. Litchi fruits cv. Shahi packed in LDPE/PP with modified atmosphere (13-15 per cent O_2 and 6 per cent CO_2) and stored at 6+2 °C (85-90 per cent RH) retained the colour and firmness up to 12 days as compared to 6 days at ambient storage. Applications of a "Protectant" such as biological control agent can also be used to reduce postharvest decay. A few recent technological developments on MAP of litchi reported beneficial from Mauritius are mentioned in Table 4.

On the basis of cited literature and work it can be inferred that litchi fruits after harvest could be successfully stored up to 3-5 weeks using MAP technology at low temperature with 90-95 per cent RH. Further, it has been reported that a combination of chitosan (1 g/l)+ MAP was effective in preventing decay, browning and retaining peel colour up to 21 days of storage time at 2 °C.

5.6 Storage

A proper storage technique is also essential for extending the shelf life, enhancing the availability and consumption period of the fruit, regulating their supply to the market and also suitability for long distance transportation.

Litchi fruits are harvested during high temperature and low relative humidity conditions leading to high moisture loss after harvest causing a decline in

Table 4. Effect of Modified atmosphere Packaging (MAP) on Quality of Litchi Fruit

Cultivars	*Type of Packing*	*MAP Condition*			*Treatments*	*Storage Time (days)*	*Effects*
		O_2 (kPa)	*CO_2 (kPa)*	*Temp. (°C)*			
Mauritius	Micro-macro-PE	15	5	1.5	Hot water brushing	28	Micro-PE: accumulation of CO_2, acetaldehyde and ethanol inhibited fungal growth and prolonged shelf life, better than macro-PE
Heiye	PE	15-19	2-4	3	CA	42	Colour, anthocyanin and phenols MAP and CA did not affect PPO and POD activities
McLean's Red	BOPP	16	6	3	Anti-microbial agents	18	Decay, PPO activity and browning changed red colour to be yellow-orange (h °)
Mauritius	BOPP	17	6	2 and 14	–	36	Respiration rate and weight loss Colour—and excellent eating quality
Mauritius and McLean's Red	BOPP	17-18	4-5	2	1-MCP	21	PPO, POD and browning Anthocyanin and colour
Mauritius	Punnetted in PFAM	–	–	13	–	9	Retained anthocyanin content, maintained sugar, acid, increase CO_2, moderate ethylene

BOPP: Bi axially oriented polypropylene; PE: Polyethylene; CA: Controlled atmosphere; 1-MCP: 1- methyl cyclopropane; PPO: Polyphenol oxide, POD: peroxide.

appearance and longevity, temperature and humidity management after harvest and during storage is highly effective means of increasing their shelf life.

5.6.1 Storage at Ambient Condition

Storage of litchi fruits at ambient temperature (25-35 °C) and RH (45-55 per cent) mainly rely on various pretreatments and packing with plastic film/bag. Under such condition litchi fruits may be stored for 5-8 days depending on cultivars and time of harvesting.

5.6.2 Storage at low temperature

Litchi fruits are more tolerant to low temperature and, thus can be stored in the low temperature range of 1-7 °C with 90-95 per cent RH. Although, low temperature storage increases storage life of fruits, yet rotting of fruits due to microbial infection, loss of peel colour, flavour and very short post cold storage shelf-life are some of the major problems. Thus, a temperature is 2-5 °C along with high humidity of 90-95 per cent is optimum for storage of litchi fruit for about 3-4 weeks.

5.7 Transportation

Litchi fruits for the local market are transported directly to retail outlets by road. However, proper and adequate transportation is required for distribution in the National and International markets. Transportation of litchi fruits is done by road, sea and air depending on the market, destination, quantity of produce to be transported, freight rates *etc.* Highway transport is the predominate method used to haul large volumes of produce across the nation mostly by non-refrigerated trucks. The fruits along with twigs are packed and transported to wholesalers. In China, ice is widely used in Litchi fruit for short distance transportation of 1-3 days in non-refrigerated vehicles. Fruits are packed into a PE bag then sealed and placed in a polystyrene box (Thermocol box) containing ice on top of the bag in ice: fruit ratio of 1: 3 or 1: 4. In China, Litchi fruits are usually transported from south to north by refrigerated trucks and exported to overseas markets usually by the sea in refrigerated containers. South Africa, Madagascar, Thailand and Australia export sulphited litchi through marine transport due to a lesser cost of shipment and higher loading capacity. However, Israel is exporting their litchi fruits by air transport, which is expensive but fast, and is highly required for produce having such a short shelf-life.

In India, about 90 per cent the litchi fruits are transported through surface route by trucks (non-refrigerated) mostly to Delhi, Mumbai, Bangalore, Varanasi, Kolkata, Amritsar and other markets. During the start of harvesting period rail transport is the most preferred means of transportation due to less quantity of produce and being the cheapest mode of transportation. Poor transportation and handling are the major cause of postharvest losses in Asia particularly in India. Adoption of the following recommendations during surface transportation as prescribed by the postharvest national consultant of FAO during the year 2011 could be of immense help to the litchi traders and growers to get their due share. The inner wall of the truck should be provided with *khas-khas* pads that should be kept wet by watering at certain intervals.

i) The boxes should be stacked as interlocking blocks leaving approximately 2.5-4.5 cm gap between boxes.
ii) About 5 cm gap from the side wall in alternate rows of boxes should be provided.
iii) Bottom layer of the boxes should rest on bamboo or wooden legs in order to ensure proper air circulation.
iv) White side of the tarpaulin covering the truck should face the sun.
v) An air gap of approximately 30 cm above the stacked boxes should be kept.
vi) A slight gap (5-7 cm) should be ensured between the top of truck's body wall and the top tarpaulin cover to escape the vital heat.
vii) Whenever the driver takes rest, watering should be done on the top cover of the tarpaulin for cleaning of dust for maximum reflection of solar radiation to keep the produce cool.
viii) As far as possible the truck drivers should be instructed to park the loaded vehicle near shed having source of water.
ix) A component of incentive scheme for the truck drivers and cleaners should be ensured for transporting litchi to the destination market in good condition.
x) Litchi fruits are to be protected against temperature variation, water loss, physical injuries and harmful gasses throughout the journey. Therefore, transport vehicle should be well designed and constructed, property loaded with well precooled produce and an optimal temperature maintained to retain the freshness of fruits throughout the distribution system.

5.8 Marketing

There are generally 3 modes of litchi supply chain in the world litchi industry (Figure 12). First type of supply chain is followed in India, whereas 2nd and 3rd type of supply chain are followed by Thailand, China, South Africa, Australia, Madagascar. In general loss encountered is more in mode-1 due to lack of appropriate handling.

The growing concern over food safety awareness, environment pollution and outbreak of new insect-pests has become a driving force for developing safe, environment friendly, economically viable and socially acceptable postharvest treatment of litchi fruits. Litchi fruits are good host of fruit borers that are subject

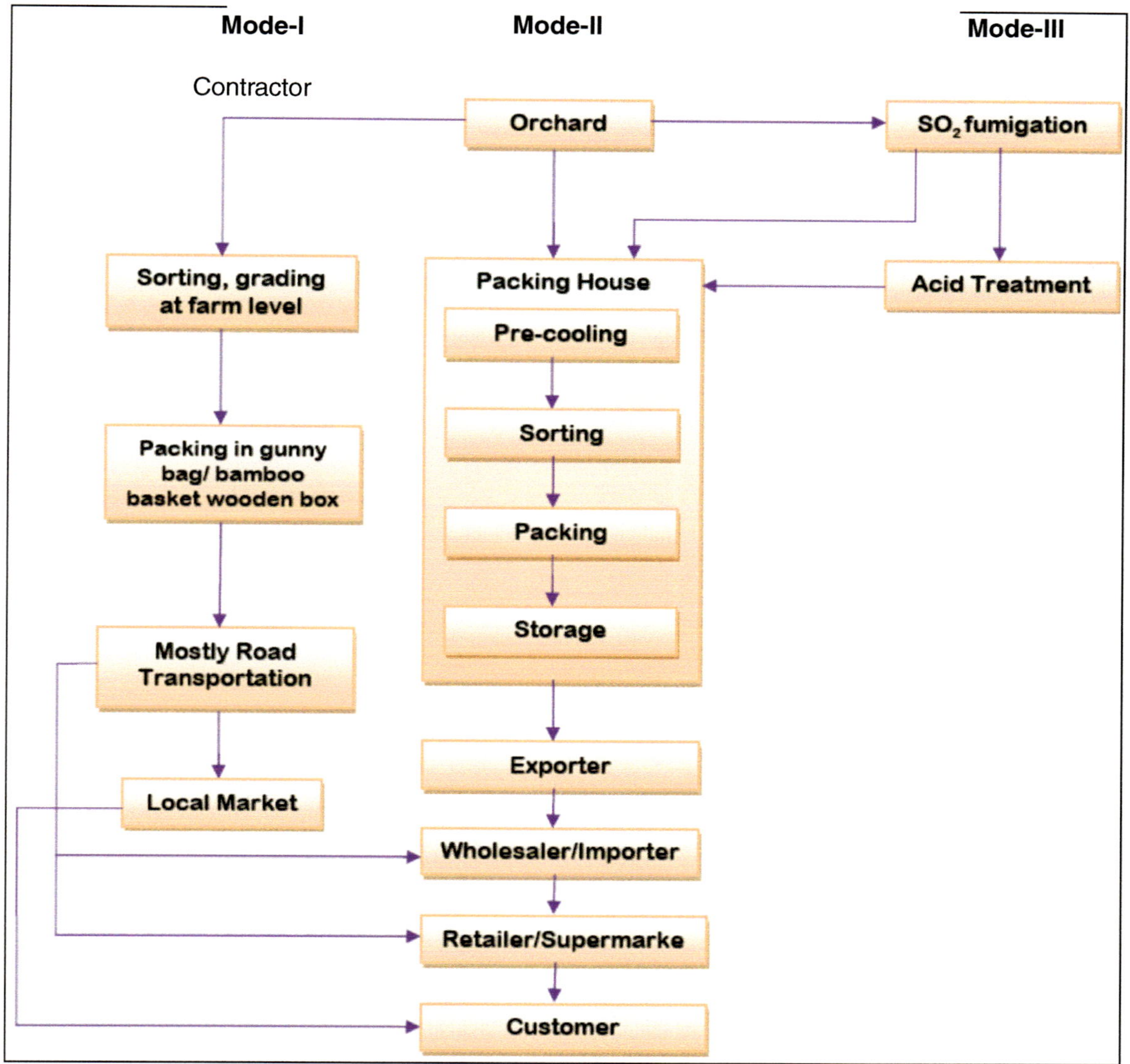

Figure 12. Supply Chain of Litchi

to specific prohibition by quarantine authorities of importing countries especially the USA, European Union and Japan. There are many methods for insect disinfestations, including heat, cold, irradiation and controlled atmosphere treatments. A proper quarantine treatment must meet the standard of probit 9, causing no heat injury to the commodity and be safe to humans. Japan requires Litchi fruits could be thermally treated until the pulp temperature of fruit reaches 46.2 °C and for 20 min., followed by cooling until the fruit reaches 2 °C for 42 hr in order to disinfest fruit fly. Cold storage of litchi fruit at 1 °C for 15 days or 1.39 °C for 18 days is mandatory requirement of USA as method for insect disinfestations.

Irradiation is a promising and effective disinfestations treatment for litchi fruits. Irradiation treatment of LDPE packed litchi fruits with gamma irradiation at doses of 0.5 KGY could be employed as a quarantine treatment. Gamma irradiation at doses of 75 or 300 GY kills eggs and larva of Queensland fruit fly without affecting the physical, chemical or organoleptic properties of the litchi. Permissible limit of SO_2 concentration in sulphitate litchi pulp should not be more than 11 ppm.

6.0 Conclusion

World production of litchi fruit is rapidly increasing from existing areas of production along with expansion in new areas like South Brazil, Mexico and India. The postharvest losses due to physiological disorders like pericarp browning, cracking and decaying in litchi fruit are also increasing and are of great concern. Proper postharvest handling of litchi fruit is still one of the main problems in litchi industry. Over the last 10 years, there has been a resurgence in litchi postharvest research aimed at inhibition of enzymatic degradation and control of postharvest decay. Many countries have adopted sulphur dioxide (SO_2) fumigation to prevent pericarp browning and decay. However, SO_2 fumigation leaves undesirable residues on the fruit, affecting fruit quality and health of sulphur-sensitive workers and consumers. Litchi fruits treated with SO_2 followed by acid dip treatments result in more intensified micro-cracks on the pericarp and contribute to the infestation by *Penicilium,* causing major losses to producers and industry. Besides, all the technologies mentioned are focused on maintaining the overall fruit quality up to 30 days. Since, sea shipment used as a convenient and economical mode of transportation to the export destination and sea export consignments can exceed 30 days, further research is needed to improve the currently developed technologies to retain overall quality for longer storage period.

In recent years, several alternative postharvest treatments have been applied to retain overall fruit quality. These include modified atmospheric packaging (active and smart) by combining treatments such as heat, gamma irradiation, bio control agents, coatings, 1-MCP, new plastic film along with cold chain system for litchi fruit during handling, storage, transportation and marketing remains the most promising tool for reducing postharvest losses. Furthermore, with the narrow spectrum of chemical applications currently available for use and likely impact on the environment, treatments should be minimized with the possible integration of new bio-control agents and nanotechnology. In Australia, new anti-browning pineapple cultivar is being developed using anti-sense RNA of PPO. Similarly, the potential of genetic manipulation to prevent the litchi browning and spoilage should be explored.

7.0 Important References

Archibald, A. and Bower, J.P. 2008. Effect of temperature management and packaging on retention of litchi colour and quality. *Acta Hortic.* **786**: 252-32.

Badilyala, S. D. 1993. Maturity standards for 'Muzaffarpur' litchi fruits. *South Indian Hortic.* **41**: 223-24.

Chen, X.J. 1998. Fresh-keeping effect of coating litchi with edible gluten. *Jour. Huan Agric. Univ.* **24**: 35-38.

Duan, Jinglian and Zhang, Shaoying. 2013. Application of chitosan based coating in fruit and vegetable preservation: A review. *Food Proc. Technol.* **4**: 227.

FAO/WHO, 2010. Codex Alimentarius: Codex standard for litchi, (Codex Standard 196-1995, amended 2005).

http: / /www.Codexalimentaxiug.net/download/ standards/320/cxs.1962.pdt.

Gaur, S.S., and Bajpai, P.N. 1977. Maturing Standards for Litchi fruit (*Litchi chinensis* Sonn.). *Prog. Hortic.* **9**: 11-17.

Ghosh, U., Bhattacherjee, A., Bose, R.K.,Choudhari, D.R. and Gangopadya, H. 1998. Physiological changes of litchi in modified atmospheric storage. *Indian J. Chem. Technol.* **59**: 309-14.

Holcroft, D.M. and Mitcham, E.J. 1996. Postharvest physiology and handling of litchi (*Litchi chinensis* Sonn.). *Postharvest Biol. Technol.* **9**: 265-81.

Holcroft, D.M., Lin, H.T. and Ketsa, S. 2003. Harvesting and Storage. In: *Litchi and Longan, Botamy, cultivation and uses.* Menzel, C. and Mira, S.K. (Eds), CAB International, Wallingford, U.K.

Huang, S., Hart, H., Le, H. and Wicker, L. 1990. Enzymatic and colour changes during postharvest storage of lychee fruit. *J. Food Sci.* **55**: 1762-63.

Jacobi, K.K., Wong, L. S. and Janet, E.G. 1993. Lychee (*Litchi chinensis* Sonn.) fruit quality following vapour heat treatment and cold storage. *Postharvest Biol. Technol.* **3**: 111-19.

Jiang, Y.M. and Chen, F. 1995. A study on polyamine change and browning of fruit during cold storage of litchi (*Litchi chinensis* Sonn.). *Posharvest Biol. Technol.* **5**: 245-50.

Jiang, Y. M., Li, J. and Jiang, W. 2005. Effect of chitosan coating on shelf life of cold-stored litchi fruit at ambient temperature. *Lebensnittel-Wissenschaft and Tech.*, **38**: 757-61.

Jing, G.J., Huang, H., Yang. B., Li. J., Zheng, X. and Jiang, Y. 2013. Effect of pyrogallol on the physiology and biochemistry of litchi fruit during storage. *Chem. Cent. J.* **7**: 1-11.

Kumar, Sanjeev, Mishra, B.B., Saxena, S., Bandopadhyay, N., More, V., Wadawan, S., Hajare, S.N., Gautam, S. and Shaume, A. 2012. Influence of pericarp browning and shelf-life extension of litchi by combination of dip treatment and radiation processing. *Food Chem.* **121**: 1223-32.

Kumar, V, Purbey, S. K., Pongener, A. Anal, AKD, and Nath, V. 2016. Effect of some fructoplane antagonists and postharvest dip treatments on litchi fruit rots and shelf life. *Int. J. Trop. Agric.* **34**: 333-43.

Kumar, V. and Purbey, S.K. 2011. Investigation of losses into postharvest disease and their management. *Annual Report,* 2011-12. *NRC on Litchi, Muzffarpur,* pp 33.

Mitra, S.K. 2002. Overview of lychee production in the Asia-Pacific region. *FAO/RAP Publication: 2001/09* 5-13.

Nagar, P.K. 1994. Physiological and biochemical studies during fruit ripening in litchi (*Litchi chinesis* Sonn.). *Postharvest Biol. Technol.* **4**: 255-34.

Nainwal, N.C., Vishal Nath, Das, B. and Joshi, N. 2003. Postharvest handling of fresh litchi for export. *Prog. Hortic.* **35**: 1-10.

Neog, M. and Saikia, L. 2010. Control of Postharvest pericarp browning of litchi (*Litchi chinensis* Sonn.). *J. Food Sci. Technol.* **47**: 100-04.

Olesen, T., Nacey, L., Wiltshire, N. and O'Brien, S. 2004. Hot water treatments for the control of rots on harvested litchi (*Litchi chinensis* Sonn.) fruit. *Postharvest Biol. Technol.* **32**: 135-46.

Paull, R.E., Chen, N.J., Deputy, J., Huang, H.B. Cheng, G.W and Gao, F.F. 1984. Litchi growth and compositional changes during fruit development. *J. Amer. Soc. Hortic. Sci.*, **109**: 817-21.

Paull, R.E. and Chen, J.N. 1987. Effect of storage temperature and wrapping on quality characteristics of litchi fruit. *Scientia Hortic.* **33**: 223-36.

Paull, R.E., Reyes, M.E.Q. and Reyes, M.V. 1995. Litchi and Rambutan insect disinfestation treatments to minimal induced pericarp browning. *Postharvest Biol. Technol.* **6**: 139-48.

Pathak, P.K. and Chakraborty, I. 2005. Studies on the effect of some organic acids, sulphitation and low cost storage technique on postharvest life of litchi fruits. *The Hortic. J.* **18**: 137-40.

Pesis, E., Dvir, O., Feygenberg, O, Ben-Arie, R., Akerman, M. and Lichter, A. 2002. Production of acetaldehyde and ethanol during maturation and modified atmospheric storage of litchi fruit. *Postharvest Biol. Technol.* **5**: 157-62.

Pal, R.K. 2011. *Final Report FAO Project on Improving Productivity and Quality of Litchi in Bihar* submitted by R.K. Pal, National Consultant on Postharvest Management, *FAO Project Code TCP/IND/3202*.

Pongener, A., Purbey, S.K., Nath, V., and Puja, K. 2018. Salicylic acid maintains membrane stability and reduces pericarp browning in litchi. *Acta Hortic.* **1211**: 45–52

Prasad, S. S. and Bilgrami, R. S. 1973. Investigation on diseases of litchi. III. Fruit rots and their control by postharvest treatments. *Indian Phytopathol.* **26**: 523-527.

Purbey S.K., Singh, S.K. and Vishal Nath. 2012. Postharvest management of litchi fruits. In: Workshop on Postharvest care and value addition of Horticultural crops, Bhubaneswar, Orissa, on 11th May, 2012 Directorate of *Hortic.* and Orissa Horticulture Development Society, pp. 42-52.

Purbey, S. K. 2014. Supply Chain Management in Litchi. In: Sivaramane, N. and Reddy, G.P (eds) *Supply Chain Management in Agriculture*. ICAR-NAARM, Hyderabad, pp. 39-53

Purbey, S. K. and Kumar, A. 2015. Effect of Pre-harvest bagging on quality and yield of litchi (Litchi chinensis Sonn) Fruits. The Ecoscan **VII** (Special issue). 197-201.

Purbey, S. K., Pongener, A. and Marboh, E. 2019. Effect of pre-harvest spray of chemicals on quality, yield and storage life of litchi fruits. *Int. J. Curr. Microbiol. App. Sci.* **8**: 207-214 doi: 10.20546/ijcmas.

Purbey, S. K., Singh, S. K. and Pongener, A. 2019. Management of light for quality production of litchi. *Int. J. Bio-resour. Stress Manag.* **10**: 529-38.

Purbey, S.K, 2010. Effect of different packaging and lining materials on shelf-life of litchi fruits. *Annual Report, (2010-11), NRC on Litchi*, Muzaffarpur, pp. 29.

Purbey, S.K, Pongener, A., Kumar, V., and Nath, V. 2018. Effect of time of harvest and packaging on quality and shelf life of litchi fruit. *Acta Hortic.* **1211**: 65–70.

Purbey, S.K. 201. Development of modified atmospheric packaging technologies for extending the shelf life of litchi fruits. *Annual Report (2012-13), NRC on Litchi*, Muzaffarpur, pp 27.

Purbey, S.K. 2011. Effect of different packaging and lining materials and other treatments on Shelf-life of litchi fruits. *Annual Report (2011-12), NRC on Litchi*, Muzaffarpurpp, 32.

Rai, M., Vishal Nath and Das, B. 2002. Heat unit submission- an index for predicting fruit maturity in litchi. *Indian J. Hortic.* **59**: 34-38.

Revathy, J. and Narasimham, P. 1999. Litchi (*Litchi chinensis*Sonn.) fruit, Influence of pre and postharvest factors on storage life and quality for export trade -a critical appraisal. *J. Food. Sci. Technol.* **34**: 1-19.

Ray, P.K. 1998. Postharvest handling of litchi fruits in relation to colour retention- a critical appraisal. *J. Food. Sci. Technol.* **35**: 103-16.

Ramesh C and Pal R.K.2006. Influence of active cushioning materials in packaging on shelf-life of litchi fruits. *Indian J. Hortic.* **63**: 31-35

Reichel, M., Carte, R.,Sruamsiri, P. and Neidhart, S. 2010. Influence of harvest maturity on quality and shelf-life of litchi fruit (*Litchi chinensis* Sonn.). *Postharvest Biol. Technol.* **57**: 162-75.

Reuck, K. de. 2010. Postharvest quality retention and decay control of South African Litchi in modified atmospheric packing. *M.Sc. thesis*, Faculty of Natural and Agricultural Sciences, University of Pretoria, S.A.

Rattanapahone, N. and Boonyakiat, D. 2001. Effect of packaging materials on shelf-life and physiochemical changes during storage of lychee fruits. *Acta Hortic.* **553**: 611-12.

Roth, G. 1963. Postharvest decay of litchi fruit. *Citrus and Subtropical Fruit Research Institute, Nelspurit, Technical Communications*. **11**: 1-16.

Sandhu, S.S. and Randhawa, J.S. 1992. Effect of postharvest application of methyl-2 benzimidzole carbonate and in-pack fumigant on cold storage life of litchi cultivars. *Acta Hortic.* **296**: 185-89.

Scott, K. J., Brown, B. L., Chaplin, G. R.,Wilcox, M. E. and Bain, J. M. 1982. The control of rotting and browning of litchi fruit by hot benomyl and plastic film. *Scientia Hortic.* **16**: 253-61.

Shi, J., Wang, C., An, X., Li, J. and Zhao, M. 2001. Postharvest physiology, storage, and transportation of litchi fruits-a review. *Acta Hortic.* **558**: 387-91.

Singh, A.,Abidi, A.B. and Srivastava, S. 1987. Variation in quality traits of Litchi (*Litchi chinensis*Sonn.) cultivars. *J. Food Sci. Technol.* **24**: 264-66.

Singh, H.S.,Vishal Nath., Singh, A. and Pandey S.D., 2012. *Litchi: Preventive Practices and Curative Measures*. Satish Serial Publishing House, Delhi, pp: 281-298.

Singh, R. 1957. Improvement of packaging and storage of litchi at room temperature. *Indian J. Hortic.* **14**: 250-12.

Singh G. and Vishal Nath, 2013. FAO study tour report to Thailand. Submitted to Ministry of Agriculture, Govt. of India, New Delhi. pp. 1-12.

Singh, G., Vishal Nath, Purbey, S.K., Singh, S.K. and Pal, R.K. 2012. Postharvest handling and valorizations of litchi fruits. *FAO Bulletin*, NRC on Litchi, p. 1-30.

Singh, G., Vishal Nath, Pandey, S.D., Ray, P.K. and Singh, H.S. 2012. The Litchi, FAO Rome: 1-207.

Sisler, E.C. and Serek, M. 1997. Inhibitors of ethylene response in plant at the receptor level- Recent developments. *Physiol. Plant.* **100**: 577-82.

Sivakumar, D.,Regnbier, T.,Demoz, B. and Korsten, L. 2005. Effect of postharvest treatments on overall quality retention in litchi fruit during low temperature storage. *J. Hortic. Sci. Biotechnol.* 80: 32-38.

Sivakumar, D.,Korsten, L. 2006(b). Influence of modified atmosphere packaging and postharvest treatments on quality retention of litchi cv. Mauritius. *Postharvest Biol. Technol.* **41**: 135-42.

Somboonkaew, N. and Terry, A.L. 2010. Physiological and biochemical profiles of improved litchi fruit under modified atmosphere packaging. *Post. Biol. Technol.* **56**: 246-253

Sun, D., Liang, G.,Xie, J., Lei, X. and Mo, Y. 2010. Improved preservation effects of litchi fruit by Combining Chitosan Coating with ascorbic acid treatment during postharvest storage. *Afric. J. of Biotechnol.* **9**: 272-79.

Swart, D. H. 1985. Sulphur content of fumigated South African litchi fruits. *Subtropica,* **6**: 18-20.

Tongdee, S.C., Scott, K.J. and Mc Glasson, W.B. 1982. Packaging and cold storage of litchi fruits. *CSIRO Food Res. Quart.* **42**: 25-28.

Underhill S.J.R. and Critchley, C, 1995. Cellular localization of polyphenol oxidase and peroxidase activity in *Litchi chinensis* Soon. pericarp. *Aust. J. Pl. Physiol.* **22**: 627-632.

Underhill, S.J.R., and Wong L.S. 1990. A maturity standard for Lychee (*Litchi chinensis* Sonn.). *Acta Hortic.* 181-187.

Chapter 20

Mango

W.S. Dhillon

Former Assistant Director General (Horticulture), ICAR, New Delhi – 110 012, India

E-mail: wasakhasingh@yahoo.com

S.K. Jawandha

Department of Fruit Science, PAU, Ludhiana – 141 004, Punjab, India

E-mail: skjawandha@pau.edu

1.0 Introduction

Mango (*Mangifera indica* Linn.) is the national fruit of India and is known as the "King of fruits". It is native to India from where it spread all over the world. While other *Mangifera* species (*e.g.* horse mango, *M. foetida*) are also grown on a more localized basis, *Mangifera indica* – the common mango or Indian mango – is the only mango tree commonly cultivated in many tropical and sub-tropical regions and its fruit is distributed essentially worldwide. It belongs to the family Anacardiaceae. Mango is essentially a tropical crop but it also grows well under sub-tropical conditions. It can be grown from sea level up to an altitude of 1500 MSL. It is one of the best fruit in the world market accepted by the masses owing to its excellent flavour, attractive fragrance, beautiful shade of colours and rich food value. The edible ripe fruit pulp is rich in a variety of phytonutrients *viz.* β-carotene - a precursor of Vitamin-A, Vitamin-C and Folate. A single mango can provide up to 40 per cent of the daily dietary fibre and 17 per cent carbohydrates. The fruit contains nearly 81 per cent moisture, 0.4 per cent fat, 0.6 per cent proteins, 0.8 per cent of fibre. The fruit contains important minerals like potassium, magnesium, sodium, phosphorus, and sulphur.

2.0 Status of Cultivation

Mango is well known to Indians for centuries and is under cultivation for probably over 4000 years. It spread from Indo Burma region to other parts of the world through Chinese travellers who visited India in 7th century. Approximately 50 per cent of all tropical fruits produced worldwide are mangoes. India is contributing 49.02 per cent of the world's area and 45.88 per cent of the total world production of mango. Mango is cultivated in about 111 countries of the world. Indonesia and China stood in second and third position, respectively, among mango producing countries. The other major mango producing countries are Pakistan, Mexico, Indonesia and Brazil. It is grown almost in all the states of India. Uttar Pradesh tops the list of mango producing states. Other major mango producing states are Andhra Pradesh, Maharashtra, Karnataka, Bihar Gujarat and West Bengal. Rest of the states also have substantial area and production of this fruit (Figure 1 and Table 1).

Though, India is the leading mango producing country in the world both in terms of area (2.57 million hectares) and fruit production (25.6 million tonnes), but its productivity is only 9.96 tonnes/ha as compared to 22.34 tonnes/ha in Brazil, followed by China and Indonesia with 13.17 and 13.14 tonnes/ ha, respectively (Table 2). When we compare to the scenario in India, Uttar Pradesh has the highest productivity (17.1 tonnes/ ha) and is comparable with Brazil. However, the area is the highest in Andhra Pradesh, but its productivity is quite low (Table 1).

a) Andhra Pradesh

Krishna Vizianagaram, West Godavari, Visakahapatnam, East Godavari, Srikakulam Khammam,

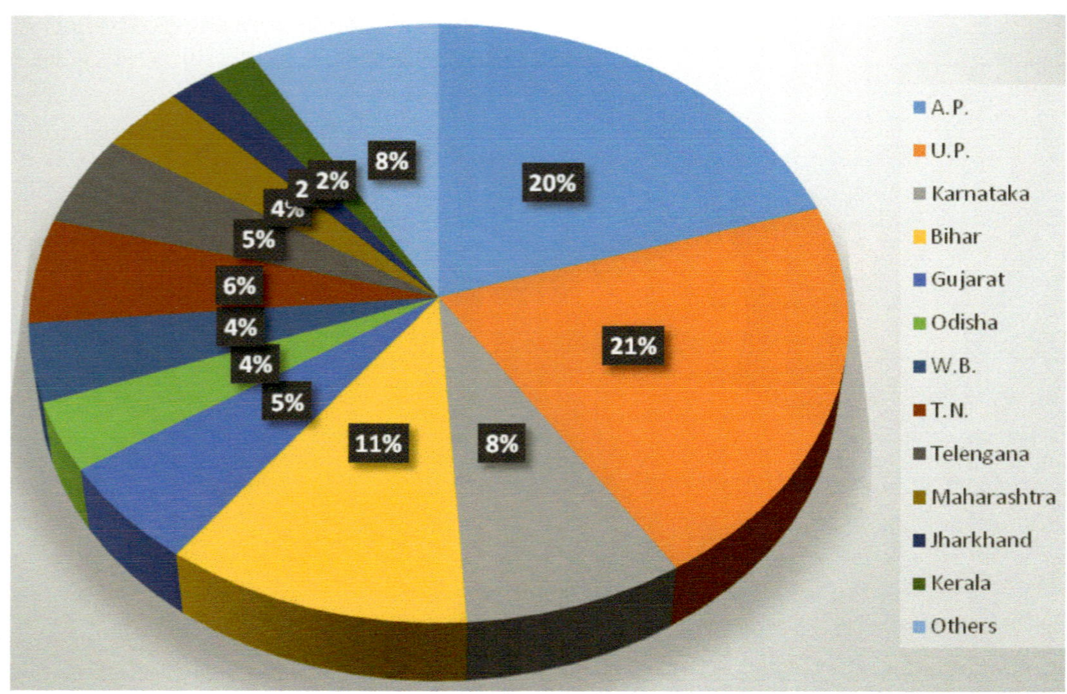

Figure 1. Production Share of Mango in India 2017-18
Source. Indian Horticulture Database, NHB 2018

Nalgonda, Karimnagar, Warangal, Mahaboobnagar, Chittoor, Cuddapah, Nellore, Prakasam.

Table 1: State-wise Area, Production and Productivity of Mango in India (2017-18)

State	*Area (000 ha)*	*Production (000 Mtonnes)*	*Productivity (tonnes/ha)*
Andhra Pradesh	363.00	4373.61	12.0
Uttar Pradesh	265.62	4551.83	17.1
Karnataka	183.23	1760.60	9.6
Bihar	149.28	2443.47	16.4
Gujarat	162.77	1207.78	7.4
Odisha	199.08	805.77	4.0
West Bengal	103.25	918.35	8.9
Tamil Nadu	152.57	1234.0	8.1
Telangana	115.99	1080.14	9.3
Maharashtra	166.76	791.36	4.7
Jharkhand	54.53	435.86	8.0
Kerala	83.12	439.20	5.3
Others	621.93	1780.35	2.9
Total	**2258.13**	**21822.32**	**9.7**

b) Bihar

Darbhanga, Bhagalpur, Muzaffarpur, East and West Chaparan, Sitamarhi, Sheohar, Vaishali, Saharsa.

Table 2: Global Scenario of Mango Production and Productivity in the Year 2019

Country	*Productivity (tons/ha)*	*Percentage Share in World Total Production*
Bangladesh	10.83	2.6
Brazil	22.34	3.57
China	13.17	4.62
India	9.96	45.88
Indonesia	13.14	5.89
Mexico	11.09	4.29
Nigeria	7.11	1.69
Other countries	8.29	23.15
Pakistan	10.56	4.06
Philippines	3.86	1.34
Thailand	7.78	2.91
World	9.99	

c) Gujarat

Valsad, Navsari, Surat, Vadodara, Bharuch Junagadh, Amreli, Bhavnagar.

d) Karnataka

Kolar, Bangalore, Tumkur, Chitradurga, Mysore, Hassan, Mandya, Chickmagalur.

e) Maharashtra

Ratnagiri, Sindhudurg, Raigarh, Satara, Sangli, Kolhapur, Latur, Nasik, Beed, Akola, Jalna, Ahmednagar, Buldhana, Osmanabad.

f) Tamil Nadu

Teni, Dharmapuri, Salem, Tirunelveli.

g) Uttar Pradesh

Lucknow, Sultanpur, Sitapur, Unnao, Hardoi, Barabanki, Faizabad, Saharanpur, Bulandshahar, Meerut, Muzaffarnagar, Bijnaur, Moradabad, Deoria, Basti, Maharajganj, Kabir Nagar, Gorakhpur, Kushi Nagar.

h) West Bengal

Malda, Murshidabad, 24 Parganas, Midnapur, Purulia, Bardhaman, Hoogly, Howrah, Nadia, Dinajpore.

3.0 Major Commercial Varieties

The major commercial varieties of mango are mentioned below.

a) Alphonso

This variety is medium in size, ovate oblique in shape and orange yellow. The pulp is yellow to orange, soft, firm and fibreless. It is a mid-season variety having great export potential.

b) Amrapali

This is a hybrid between Dashehari x Neelum. The trees are dwarf and regular bearing. Fruits bear in clusters, small size fruits, good keeping quality, suitable for high density planting.

c) Banganpalli

Fruit is large in size and obliquely oval in shape. The colour of fruit is golden yellow. The flesh is firm to meaty, fibreless. Good keeping quality and a mid season variety.

d) Bombay Green

It is one of the early varieties under north Indian conditions. Fruit size medium (250 g), pulp is soft and sweet. Fruits have strong and pleasant flavour.

e) Chausa

Fruit is large, ovate to oval oblique in shape and light yellow in colour. It is a late variety.

f) Dashehri

Fruit size is medium, shape is oblong to oblong-oblique and fruit colour is yellow.

The pulp is firm and fibreless and is a mid season variety.

g) Fazli

Fruit is very large, obliquely oval in shape. Fruit colour is light chrome; fruit quality and keeping quality are medium. This is a late season variety.

h) Kesar

Fruits are medium in size, flesh is sweet and fibreless. Colour of fruit is apricot yellow with red blush, an early season variety. This variety has a place in the export markets.

i) Langra

Fruit is of medium size, ovate shape and lettuce green in colour. The lemon- yellow flesh is juicy and flavourful. It is scarcely fibrous, a mid season variety.

j) Malda

Variety is an alternate bearer. Fruit is medium, ovate and yellow in colour. Keeping quality is medium

k) Mallika

A hybrid between Neelum x Dashehari. Regular in bearing, high TSS, good colour, uniform fruits, moderate keeping quality and is a mid season variety.

l) Neelam

A high yielding, late season mango variety of South India, proven regular bearer, medium size fruits, good flavour. Pulp is soft, yellow and fibreless. Good keeping quality.

m) Totapuri

Fruit size is medium to large, shape is oblong with a necked base and colour is golden yellow. Flesh is cadmium yellow and fibreless, a mid season variety.

4.0Nature and Causes of Postharvest Losses

In India about 25-35 per cent of mangoes after harvest are wasted due to lack of knowledge and faulty practices being followed during harvesting, handling, grading, packaging, storage, transportation and marketing. By employing scientific approaches these losses can be reduced to a considerable level. There are various reasons for postharvest losses of mango such as premature fruit harvesting, mechanical injury, injuries due to thermal shock, disease and pest attack, microbial attack, physiological and biochemical reasons. Some losses in process of grading take place due to improper handling of the fruits or due to over ripened fruits during marketing. Sometimes, fruits also get infected due to injuries that result in huge losses during grading. Some losses occurred during packaging due to improper handling, overloading and falling of fruits. Main losses are happening during transportation when it is not well planned or managed. This damage can lead to the non-acceptability of the produce by the consumers, resulting in overall economic losses. Losses are also occurring during faulty or inadequate storage conditions. Most of the losses during marketing are found at the wholesaler and commission agent levels. Improper handling, over ripening, falling of the fruits, rotting and shrinkage of the fruits are the main reasons for losses in preparation for the sale of the fruit. These losses are discussed under the following broader heads;

4.1 Preharvest Management

There are numerous preharvest factors that lead to postharvest losses and play role in deterioration of quality after harvest. The most important factor that leads to substantial losses both in quantity and quality is immature fruit harvesting. The immediately harvested fruits are forced to ripen with indiscriminate use of chemicals, but those fruits lose their weight quickly and quality never reaches near acceptable levels, ultimately also causing economic losses as such fruits do not fetch a good price. Other factors like mismanagement or no control of anthracnose disease attack on fruits, and field attack of insects like fruit flies or stone weevil. Mango fruit fly is a major pest of mango but can be well managed by a combination of preharvest and postharvest treatments. Preharvest treatments alone ensure 100 per cent control and it can be further subjected to postharvest treatments to satisfy international disinfestations requirement. These strategies are environment friendly and residue free. Due to the attack of such insects or their presence in the fruits, recently some countries have banned the export of mangoes from India. Stem end rot is another postharvest disease which is affecting significantly the produce, thereby consumers appeal is reduced and economic losses are heavy. To avoid such losses, the following points should be taken care of.

i) Produce from a healthy tree in a well managed orchard with good hygiene.

ii) Grow with adequate calcium and other elements.

iii) Do not subject to excessive N during fruit development and ripening.

iv) Harvest at the correct stage of maturity.

v) Do not subject to bruising or damage.

vi) Maintain fruit at ideal temperature after harvest.

vii) Pack and transport the fruits cleanly.

The quality of mango is decided by numerous pre and postharvest practices. The preharvest factors such as nutrient management, irrigation, use of growth hormones are the most important ones. Post flowering foliar application of potassium, in mango helped to increase fruit retention percentage, size, weight, fruit colour, flavour and taste. Preharvest putrisine spray exhibits better firmness, TSS and reduces fruit rot in Kensington Pride mango.

The spray of calcium is found useful in delaying ripening of mangoes, improves the storage life and increased marketability. Similarly, preharvest spray of fungicide like Dithane M-45 (0.2 per cent), Bavistin 0.1 per cent, Topsin-M (0.1 per cent) and Captan (0.2 per cent) is effective in controlling storage diseases due to anthracnose and stem end rot in mango varieties. The crop management practices adopted in the field for growing mangoes have a profound bearing on the product which is harvested. The stringent quality regime and international quality parameters which are to be followed for export should be carefully adopted. Some of the practices are:

i) Adoption of improved cultivars.

ii) Adoption of high density planting of cultivars to increase the productivity and ultimately to reduce the cost of production.

iii) Regulating flowering and increasing intensity of flowering.

iv) Pruning malformed mango shoots at bearing stage.

v) Alternate bearing needs to be managed by reducing fruiting during a year by adoption of pruning and thinning practices to regulate every year fruiting.

vi) Providing life saving irrigation at the critical stage of fruit growth.

vii) Integrated nutrient and water management.

viii) Management of spongy tissue through sod mulching.

ix) Integrated pest and disease management practices and timely control of pests and diseases, especially mango hopper, mango mealy bug, fruit fly, stone weevil *etc.*, and sprays of bavistin at fruit development stage especially after rains.

x) In warm and humid climate, the developing fruits suffer from fungal diseases and black spots appear on the fruits. In order to overcome the problem it is advised to spray 0.1 per cent carbendazim or hexaconlzol with any sticker or spreader @ 1 ml/litre for better effect.

5.0 Postharvest Management

Mango is a climacteric fruit and for its long shelf life and optimum quality, postharvest management in mango is extremely important. Postharvest management operations are quite diversified, consisting of collection, curing, pretreatment, grading, packaging, precooling, low temperature storage, pallet loading, transporting etc.

5.1 Postharvest Treatments

In general fruit requiring only short-term storage before retail sale are unlikely to suffer from fungal break down and postharvest fungicides may not be necessary. For longer storage of mangoes a fungicide treatment is usually required. However, effective organic treatment for postharvest fungal breakdown is yet to be established. The postharvest treatments like fungicidal dips, hot water dip, vapour heat treatment, irradiation *etc.* are found to increase the shelf life of mango fruits.

5.1.1 Hot Water Treatment

Mangoes may be immersed in hot water before storage or marketing to control postharvest diseases.

A common disease of fruit caused by the fungus *Colletotrichum gloeosporoides* can be successfully controlled by treating the mangoes at 53 °C for 5 minutes and or 51-55 °C for 30 minutes. The hot water treatment had little or no effect on the quality of marketable life of the fruits. Hot water treatment in 46.1 °C water for control of fruit flies is a mandatory requirement of the United States Department of Agriculture Animal and Plant Health Inspection Service (USDA APHIS). Hot water is an effective heat transfer medium and, when properly circulated through the load of fruit, quickly establishes a uniform temperature profile. Hot-water immersion also has the additional benefit of controlling postharvest microbial disease such as *Anthracnose* and stem end rot. It has been suggested that pathogenic fungi diseases such as anthracnose on mango fruit are latent and a main cause of serious decay. Hot water containing fungicide can also be effective in the control of stem end rot caused by the fungus *Dothiorella dominicana*. Hot-water immersion is currently used to successfully treat mangoes infested with the Mediterranean fruit fly and several different species of fruit fly before importation into the United States from Mexico. The application of quarantine hot water treatment is also known to induce chilling injury tolerance in mango fruits.

5.1.2 Vapour Heat Treatment

It was developed to control infections of fruit flies in fruit as postharvest treatment. Mangoes are subjected to 43 °C in saturated air for 8 hours and then holding the temperature for a further 6 hours. Fruit fly in Carabao mangoes can be controlled by exposing the mangoes to high humidity hot air *i.e.* 46 °C and 95 per cent RH and above. The exposure of Carabao mangoes to a core temperature of 46 °C for 10 minutes by vapour heat treatment also led to a significant reduction in both anthracnose and stem-end rot. Similar results were also observed in case of Baneshan and Alphonso mangoes in India.

5.1.3 Irradiation

Extensive studies have been conducted to determine the effect of gamma irradiation on the shelf life extension of mango. The optimum radiation dose for Alphonso mango is 25 Krad, giving an extension in shelf life of 6 to 8 days at ambient temperature. The response of Kensington Pride mango to g-irradiation which caused delay in ripening of less mature mangoes, characterised by inhibition of skin colour, de-greening and slow reduction of acidity, while the fruits at climacteric stage was unaffected. The lower doses of irradiation in combination with other treatments also used to induce delayed ripening and reduced postharvest spoilage. A combination of hot water treatment (55 °C for 5 minutes) followed by 30 Gy irradiation was found to be the best treatment in terms of shelf life extension and quality of mangoes. Irradiation can be used as a postharvest treatment to disinfest fruits from insects. It was observed that 300 Gy control mango seed weevil while 150 Gy was shown to control 11 species of Tephritid fruit fly and 75 Gy prevents the adults emerging from the fruits. However, irradiation followed immediately by hot benomyl treatment controlled anthracnose and stem-end rot during storage at 20 °C for 15 days. Irradiation at 300 Gy dose was found to be better in improving the storage life than 150 Gy and 225 Gy doses due to the concentrated level of rays, however the ray higher than 300 Gy level can be deleterious.

5.1.4 Insecticidal Controlled Atmosphere

Mangoes are infested with various species of fruit flies. Environmental concerns, detection of chemical residues in fruit and development of insect resistance to pesticides suggest using alternative strategies to combat these pests. Applications of insecticidal controlled atmospheres (ICA) alone or in combination with other treatments for postharvest insect pests have been tested. ICA involve the short-term exposure of a commodity to very low O_2 (<1 per cent) and/or very high CO_2 (50–80 per cent) at low or high temperature to achieve insect disinfestations. Low O_2 and/or high CO_2 injury along with off-flavour due to fermentation in tissue may result from ICA treatment. However, mango fruits can tolerate high CO_2 concentrations for a short period. Various tephritid fruit fly species have been successfully controlled in mango fruit with ICA.

5.1.5 Edible Coatings

These coating forms a transparent film on the fruit surface, acts as a barrier to water and gases (O_2 and CO_2), imparts gloss, better colour, extends storage life and minimise microbial spoilage. Edible coatings offer an alternative to film packaging due to their environment-friendly characteristics. The performance of different types of edible coatings is dependent on their composition and genotype. Various types of coatings such as carnauba wax, shellac, zein, cellulose derivatives, chitosan and its derivatives and other composite mixtures containing sucrose esters of fatty acids and a sodium salt of carboxymethylcellulose has been tested to extend the storage life of mango fruit with variable results. Nanoscale edible coatings as thin as 5nm have also been developed for mango which can provide a barrier to gas exchange and moisture and increase the shelf life of mango fruits more the conventional packings.

5.2 Maturity Standards

Mango fruits are generally harvested when shoulders are bulged and slightly turned light yellow. However, the maturity index for harvesting of mango varieties differed with cultivars. Time from flowering combined with fullness of mango cheeks is the most common criterion. The various criteria for judging mango maturity are:

i) Number of days (120-150 days) from flowering to maturity.

ii) Slight colour development on the shoulders.

iii) One or two ripe fruits fall from the plant naturally called *Tapka*.

iv) Change in fruit colour from dark green to light green or development of characteristic colour depending upon variety.

v) Lenticels on the fruit peel become distinctly clear.

vi) The fruits should be harvested at 85 per cent (14 Anna) maturity stage in cv. Alphonso for control of spongy tissue.

vii) Development of prominent beak (variety Pairi).

If picked immature, fruits develop white patches or air pockets and show lower amounts of brix, acid ratio, taste and flavour, whereas over mature fruits lose their storage life. Harvesting is traditionally done when a few semi-ripe fruits fall from the tree. However, it is not a scientific method. The accurate method of finding maturity is by immersing the fruits in water and when fruits fully sink in water, they are considered to have attained full maturity. Fruits sinking in water have specific gravity of less than 1.02. Harvest only matured fruits as frequently as possible in about 4-6 rounds. Injury to the fruits during harvesting brings down their quality and makes them prone to fungal attack.

The maturity indices which are generally followed for harvesting mango are: colour change from green to yellow; Development of shoulders on the stem end of the fruit; fruit flesh changes to yellow from white starting at the endocarp and progressing outward to the skin during maturation; Stone becomes hard and pulp colour changes from white to cream. Alphonso mango is considered mature and gives the best taste and flavour when the shoulder outgrows the stem and the external colour becomes light green with a yellowish red blush; In Dashehari and Langra cultivars, the fruits reach to maturity after 12 weeks of fruit set. However, Chausa and Mallika fruits take about 15 weeks to mature from fruit set. The best way to observe maturity in mango is the colour of the pulp, which turns cream to light yellow on maturity and hardening of stone. Mangoes are ready for harvest when the soluble solids content of mature fruit is at least 10 per cent.

5.3 Harvesting

The stage at which the fruit should be harvested has an important bearing on ripening and fruit quality. Unless the fruits are harvested at the correct stage, when they attain full size, the characteristic taste/flavour of the variety does not develop. Harvesting season varies with soil and climatic conditions. In India, mango-harvesting season differ in various parts of the country. In Southern parts of India (Kerala, Tamil Nadu) mango harvesting commence in December-January and completes in February- March, whereas in northern parts (UP, Bihar, Punjab, Haryana) it starts from the months of May-June and completes by August. In the Konkan region of Maharashtra mango harvesting begin from March to June, while in rest of Maharashtra it is from May to July. The harvesting time of mango generally varies with the distance to the market and local consumption.

In India, a common practice of harvesting fruits early in the season (premature stage) is practised to capture early market that leads to huge postharvest losses, besides, compromising on quality. Harvest the fruits, when yellow colour is developed on the fruits on the tree itself. For export market, harvest the fruits when firm and at the mature green colour stage. Such fruits ripe quite rapidly after harvest and begin to turn yellow within 3 to 5 days at ambient temperature, while fruits harvested at immature green stage will not ripen properly, have poor taste and shrivel soon.

Harvesting should be done in cooler part of day preferably in morning before 10.00 a.m. or late in the evening after 4.00 p.m. to keep fruits fresh, turgid for longer shelf life. Harvest fruits with sharp secateurs keeping 3-5 cm pedicel. All fruits should have pedicels intact to avoid oozing latex on the fruit surface spoiling appearance, development of black spots and paving way for entry of pathogens. Recently, mango harvesters or nippers have been developed at various Institutes. Dr Balasaheb Sawant Konkan Krishi Vidyapeeth, Dapoli (MS) has developed 'Nutan mango harvester' which is easy to operate light in weight and fruits can be harvested along with stalk. CISH, Lucknow has also developed an improved mango harvester by which 600-800 fruits can be harvested per hour from a tree of about 4 m height along with 50 per cent saving of energy. Fruits should never be harvested by hitting with stick, which may cause injury/bruises due to impact, resulting in decay and poor quality, and the produce will fetch a low price. The contact of fruits with the soil should be avoided and should be kept carefully in padded field crates, well-ventilated plastic containers, or picking bags. Harvested fruits should be kept in shade and not in the open sun and transported to the packhouse for postharvest operations.

5.4 De-sapping

If the stems are removed during harvesting, it will cause sap flow. The initial spurt of sap sometimes burns the fruit which lead to a blemish on the fruit. Always have short stems with the fruit while harvesting and hold the stem end down. Place those fruit on a de-sapping bench for 20 to 30 minutes to drain out the sap.

5.5 Curing

It is necessary when fruits are to be stored for long periods to extend the shelf life. The fruits are usually spread on the floor in the orchard's yard, having the cushion of paddy straw or simple straw for nearly 24 hrs and then washed to remove the dirt.

5.6 Sorting and Grading

Bruised and injured fruits must be separated, as they might cause damage to other fruits. It has also been observed that bigger size fruits take 2-4 days more time in ripening than smaller ones. Therefore, grading according to the size will play an important role in correct packaging of fruits and for their uniform ripening. Immature, overripe, damaged and diseased fruits should be discarded during this operation.

Most of the commercial mango varieties *viz.*, Alphonso, Totapuri and Dashehari mango are graded in 3 to 4 grades based on size. Fruits should be uniform in colour, size, shape, and free from pest or diseases, mechanical injuries, etc. Certain growers and traders use sieves having circular holes of 8 cm, 7.5 cm and 7 cm diameter for grading and distinguishing fruits by simple marks on the package. These days' mechanical graders for mango are available. Grading according to the specific gravity helps to separate fruits of the same maturity stage. In a general harvest, the percentage of fruits in different grades in cv. Alphonso is given in Table 3.

Table 3: Postharvest Distribution of Alphonso Mango Fruits in different Weight Grades

Weight Group	*Percentage of Fruits*
< 250 g	21.6
250-299 g	52.0
300-349 g	20.2
> 350 g	6.2

Generally, grading of fruits is done by farmers manually according to the size of the fruits. Grade will vary with variety, but weight is the best criterion to grade the fruits. The prevailing grading standards for Alphonso are given on weight basis are;

i) A + Grade - > 300 g

ii) A Grade - 250 –299 g

iii) B Grade - 200-249 g

iv) C Grade - 150-199g

v) D Grade - < 150 g

According to Agmark standards, mangoes are classified into the following classes, as per the details given in Table 4.

In Class I and Class II, scattered suberized rusty lenticels, as well as yellowing of green varieties due to exposure to direct sunlight, not exceeding 40 per cent of the surface and not showing any signs of necrosis are allowed. The size and grade of the mango fruit is determined by the weight, as mentioned in Table 5.

Due to the short postharvest life, precooling of mango is an important postharvest operation. The precooling process should be as fast as possible, but it should not result in any chilling injury to the fruit. The most appropriate precooling method for mango is forced air cooling with an air velocity of about 60-100 m/min. It is common, and very practical, to precool fruit right after packaging. Fruit is usually cooled to about 2 °C above its optimum temperature (about 15 °C). The time needed to achieve this temperature depends on the initial fruit temperature and velocity of precooling air. However, a period of 2 to 4 hours is commonly sufficient to reduce temperature from about 38 °C to 14-15 °C (temperature at which precooling should be terminated).

Table 4. Grades Standards of Mango

Grade Designation	*Grade Requirements*	*Grade Tolerance*
Extra class	Mangoes must be of superior quality. They must be characteristic of the variety. They must be free of defects, with the exception of very slight superficial defects, provided these do not affect the general appearances of the produce, the quality, the keeping quality and presentation in the package.	5% by number or weight of mangoes not satisfying the requirements for the grade, but meeting those of Class I or exceptionally, coming within the tolerances of that grade
Class I	Mangoes must be of good quality. They must be characteristic of the variety. Mangoes may have following slight defects, provided these do not affect the general appearance of the produce, the quality, the keeping quality and presentation in the package. - slight defects in shape; - slight skin defects due to rubbing or sunburn, suberized stains due to resin exudation (elongated trails included) and healed bruises not exceeding 2,3,4,5 sq. cm. for size groups A, B, C, D, respectively	10% by number or weight of mangoes not satisfying the requirements for the grade, but meeting those of Class II grade or, exceptionally coming within the tolerances of that grade.
Class II	This grade includes mangoes which do not qualify for inclusion in the higher grades, but satisfy the minimum requirements. Mangoes may have following defects, provided they retain their essential characteristics as regards the quality, keeping quality and presentation. - defects in shape, slight skin defects due to rubbing or sunburn, suberized stains due to resin exudation (elongated trails included) and healed bruises not exceeding 4,5,6,7 sq. cm. for size groups A, B, C, D, respectively	10% by number or weight of mangoes not satisfying the requirements of the grade, but meeting the minimum requirements.

Table 5: Size Grades of Mango

Grade	*Weight (g)*	*Permissible difference between Fruits within the Package (g)*
A	100-200	50
B	201-350	75
C	351-550	100
D	551-800	1255.7 Precooling

5.8 Packaging

Most popular means used for packing mangoes are baskets of various makes, crates and boxes. Previously, mango fruits were packed in bamboo baskets or wooden boxes. Packaging of mango fruits in bamboo basket of 50-100 fruit capacity is still a common practice in Western India.

Wooden boxes of 5-7 dozen capacity possess good stacking strength but are heavy, which adds more freight cost. The major demerit is extensive use of wooden boxes that depletes the forest wealth and harms the ecological balance. Corrugated Fibre Board (CFB) boxes improve the product image cause less damage to fruits and are easy to handle and print. CISH, Lucknow has developed telescopic type Corrugated Fibre Board (CFB) boxes of 2.0 and 4.0kg capacities having 0.5 per cent ventilation (Figure 2).

Figure 2: Packaging of Mango.

For domestic market, usually mangoes are packed in wooden boxes, details of which are given below (Table 6). However, Alphonso and Kesar are packed in CFB boxes.

Table 6: Specifications for Wooden Boxes for Packaging of Mango

Type of Carton	*Inner Dimensions (cm)*	*Capacity (kg)*
Wooden	45x30x30 (Ratnagiri)	16-18
	21.6x21.6x42 (Malihabad)	10-11

To overcome these problems, CFB Boxes of 5 kg and 10 kg capacity for packing and shipping of mango fruits are successful as an alternative to traditional nailed wooden boxes. Specifications of CFB boxes for packaging of mango are given in Table 7.

Table 7: Specifications for CFB Boxes for Packaging of Mango

Characteristics	*Requirements of different Packages*			
	Ring and Flap (4 kg)	*Full Tele-scopic (4 kg)*	*Ring Flap (8 kg)*	*Full Tele-scopic (4 kg)*
Material of construction	3 Ply CFB	3 Ply CFB	5 Ply CFB	5 Ply CFB
Grammage (g/m sq.), minimum (outer to inner)	*230/140/ 140	*230/140/ 140	*230/140/ 140	*230/140/ 140
Bursting strength kg/cm sq, minimum	6.50	6.50	10.50	6.50
Puncture resistance, ozs inches/tear inch, minimum	110	110	280	110
Compression strength, kgf, minimum	275	275	250	250
Cobb (30 minutes) g/m, sq, maxmium	130	130	130	130

* Outer ply of duplex board.

Source: Postharvest manual for export of mangoes, APEDA

5.8.1 Telescopic Tray Packaging

These types of trays in size of 45.0 x 29.0 x 10.5 cm are available in the market which are robust and protect the fruits well. A plastic cup insert, called a plix liner, is used, which acts as a packing guide and nests the fruit. For better look, fruits can be placed with stem end down and convex curve up. This position also prevents any sap that oozes after the fruit has been packed from spoiling appearance.

5.8.2 Modified Atmosphere Packaging (MAP)

Fruit are stored in sealed polythene film (40 μm thick low density polyethylene) bags at a suitable temperature. The fruits are removed from the bag to achieve normal ripening. MAP delays mango fruit ripening mainly by reducing respiration rate and inhibiting ethylene production. These films also minimise water loss from the fruit. MAP can be commercially used in postharvest handling and shipment of mangoes using a suitable film particularly to retard ripening during transportation. Other intelligent packaging using integrated sensor technologies that can indicate ripeness, maturity, spoilage and respiration rate are also evolving.

5.9 Ripening

Mango falls under the climacteric group of fruits based on respiratory patterns and ethylene biosynthesis during ripening. At present in India more than 90 per cent mangoes are ripened by using calcium carbide- a banned chemical, which is a toxic and extremely hazardous. The fruits ripened with this chemical do not ripe uniformly and quality of such fruits remains inferior. Many safe methods are available for ripening of mango.

The mango fruits are commonly harvested at full maturity but before ripening. Ethylene, either in the form of gas or an ethylene generator, is the most efficient and fastest means of ripening mango. It results in the most uniform ripening process. Ethylene generators are much safer, easier to use, and cheaper in the long term. For uniform and proper ripening, fruits should be exposed to 100 ppm ethylene gas for 24 hours in a ripening chamber maintained at 20-22 °C and 90-95 per cent RH. The fruits ripe in 3-4 days and attain uniform ripening and colour development.

The uniformity of the process of ripening and the quality of the ripe fruit depend on several factors such as maturity of fruit at harvest (fruit harvested immature will not properly ripen however; more mature fruit will ripen faster). Temperature and relative humidity during the ripening process are extremely critical and must be controlled very adequately. Optimum temperature during the ripening process is 20 to 22 °C, and optimum relative humidity is 90 to 95 per cent. Any deviation in these parameters will result in under ripening or over ripening of fruits. Lower temperatures, as 15.5 to 18.0 °C, can result in slower fruit ripening with attractive peel colour but flavour will not be adequate. Higher temperatures, as 27 to 30 °C, can result in a faster ripening process, a mottled skin and a strong flavour. But temperatures higher than 30 °C will retard ripening due to their inhibiting effect on ethylene production and action. However, at small scale the fruits can be ripened with postharvest dipping of fruits in aqueous solution of ethrel/ethephon (500 ppm) for 2-3 minutes followed by keeping under ambient condition in a well ventilated room. The fruits take 3-4 days for ready market.

Under tropical conditions the fruits ripen in 7-8 days after harvesting. The fruits are kept in single layers in paddy or wheat straw. This is usually done in a closed but well ventilated room. Fruits ripen uniformly at the temperature of 19.4 to 21.1 °C. Fruit coated with edible wax show slow ripening and give long shelf life. Ethrel treatment is the safest at desired concentration. Now days the ripening chamber concept is developing rapidly. The fruits are ripened in ripening chamber with controlled temperature, humidity and ethylene concentration. Dipping of 100 kg fruits in 100 litres of water containing 62.5 to 187.5 ml ethrel/ethephon at 52±2 °C for 5 minutes ripen fruits uniformly with attractive yellow colour without any spoilage within 4 to 5 days of treatment.

5.10 Storage

Storage is essential for extending the consumption period of fruits, regulating their supply to the market and also for transportation to long distances, and is essential during the period of glut. Mango is a climacteric fruit and unless the fruits are stored properly one can't be sure of the conditions in which the fruit will reach the designated market.

The mature-green fruits can be kept at room temperature for about 4 to10 days depending upon the variety. Shelf life of fruits could be extended by precooling, chemical treatments and storage at low temperature. The harvested fruits are precooled to 10 to 12 °C and then stored at an appropriate temperature. The fruits of Dashehari, Mallika and Amrapali should be stored at 12 °C, Langra at 14 °C and Chausa at 8 °C with 85 to 90 per cent RH. The fruits could be stored for 3 to 4 weeks in good condition at low temperature (12 to 14 °C) with 85-95 per cent RH. Fungicidal dip treatment helps in extending the storage life. Wax-coated fruits or fruits wrapped in tissue paper or packed in perforated plastic bags also showed better results in extending storage life. Before storage, the mangoes should be properly cleaned and dried. Damaged fruits should be discarded to avoid quality deterioration and pest attack. Some of the techniques which can be used to enhance the shelf life of mangoes are discussed below:

5.10.1 Low Temperature Storage

Being a tropical fruit low temperature storage has limited potential, in extending storage life of mango fruit, because it is susceptible to chilling injury. The most common visual symptoms of chilling injury in fruits are dark colouration of the skin, prominence of lenticels, uneven ripening, development of off flavour and poor fruit quality. Mango fruit can be safely stored at 12–13 °C for 2–3 weeks depending upon cultivar and harvest maturity. However, models have also been developed to predict the cooling time in Tommy Atkins mangoes based on the physico-chemical quality of the fruit. It is very necessary that no other fruit should be stored with mango under low temperature conditions. At the termination of low temperature storage gradually fruits are to be transferred to room temperature in order to minimize sweating. Only rigid containers that can withstand stacking without getting deformed due to high humidity should be used.

5.10.2 Controlled Atmosphere (CA) Storage

CA storage can also be alternative methods in extending shelf life, maintaining fruit quality of mango. CA storage is based on the principle of an atmosphere comprised of reduced oxygen (O_2) and enhanced carbon dioxide (CO_2) levels, usually mixed with nitrogen, at optimum temperature and relative humidity. But this technology has some limitations for mango such as poor colour development, alteration in texture, occurrence of undesirable flavour and odours, development of physiological disorders. However optimal limits of 5-8% O_2, 5-9% CO_2 and 86-91% N_2 gas mixtures have been found to be effective in maintaining quality attributes of 'Shelly' mango.

5.10.3 Hypobaric Storage

It is also known as low pressure storage. Produce is stored in a partial vacuum and chamber is vented continuously with water-saturated air to maintain oxygen levels and minimise water loss. The reduced partial pressure of oxygen in hypobaric storage retards ethylene action and production as well as ripening in some fruits. Hypobaric storage seems to be promising in extending shelf life of mango cultivars, ameliorating CI and reducing development of anthracnose and stem-end rot. It has also been reported to be an effective treatment in controlling tephritid fruit fly species. Hypobaric chambers are expensive to construct because of the low internal pressures requirement which tends to limit its application.

5.11 Transportation

Mango transportation needs careful handling since fruits on ripening become too soft, gets easily injured that reduces eating and keeping quality of fruits. In Delhi, up to 90 per cent of the total arrivals are transported by road. Similarly in Chennai city 83 per cent mangoes are transported by trucks. The loss during transit by trucks was very small (about 1 per cent) because the trucks run during night, thus avoids the excessive heat of the day. The road transport is preferred because of ready availability of trucks unlike rail wagons, their arrival to the destination on time, absence of pilferage *enroute* and facilities of delivery at the doors or to the market.

When the fruit is transported by rail, particularly over long distances, there is loss of 10-20 per cent in the consignment, mainly on account of frequent delays in the movement of the wagons. Hence despite concessional freight rates offered by the railways for fresh mango, the shift has continued towards road transport. In West Bengal and Assam, transport of mangoes by boats is a common practice. This method of transport is the cheapest. Bombay receives considerable quantities of mango by coastal steamers from Konkan tract during January to the end of May.

Generally, refrigeration facilities are not available during air shipment and on the contrary, low pressure environment with low RH are encountered at high altitude. This increases the rate of water loss of the produce. For air transportation, providing PE film liner with perforation within box or over-wrapping of unit load is necessary. Packed produce coming in for air transportation has to be precooled sufficiently to counterpoise the lack of refrigeration facility. As there are no cold storage facilities at airports in India, it is necessary to transport the precooled produce in insulated or refrigerated trucks or vans to airport to obviate possible delays in loading due to late arrivals or non-availability of required space on a particular flight.

6.0 Domestic Marketing and Export

India is a leading mango producing country in the world but its share in the international export markets is very limited owing to various reasons. However, India has a huge domestic market for mango. India exports mango to over 40 countries worldwide. The major importing countries of India's mangoes are UAE, Bangladesh, UK, Saudi Arabia, Kuwait and Bahrain etc. The arrival pattern of mango in domestic markets are summarised in Table 8.

Table 8: Arrival Pattern of Mango in Domestic Markets

State	*Season of Availability*	*Important Cultivars*
Andhra Pradesh	Mid Feb-mid July	Banganpalli, Neelum, Suvaranrekha, Totapuri
Gujarat	April-July	Alphonso, Kesar, Rajapuri
Karnataka	April-July	Alphonso, Banganpalli, Neelum, Pairi, Totapuri
Maharashtra	March-July	Alphonso, Kesar, Pairi
Uttar Pradesh	May-August	Amrapali, Bombay Green, Chausa, Dashehri, Langra

For maintaining highest quality standards, state-of-the-art pack-houses have been set up in major production zones. Keeping in view the requirement of different country requirements, internationally recognized treatment facilities like hot water treatment, vapour heat treatment and irradiation facilities have also been set up at various locations across the production belt.

Unique product identification system, compliant to the traceability networking and residue monitoring plan has been developed for the consumer safety and readiness to product recall in case of any emergency.

Fruits should be graded in each carton according to variety, size by weight or diameter giving a range of "counts" for each shipment and maturity (firm, green, full-mature and half- mature fruit will ripen at different rates and should not be packed in the same carton). Carton net weight is dependent on the importer, ranging from 3.5 to 5 kg (8 to 11 lbs) and must not be overfilled during packing. PMH facility for export of mango of MSAB shown in Figure 3. However, mixing of packaging of smaller fruits with larger ones should be avoided for adequate monitory benefit. The following precautions should be taken while packing for export markets.

i) Individual mango should be enclosed in a clean, white, soft, expandable and netted type polystyrene sleeve to prevent bruising before packing in a box.

ii) The mangoes must be packed in insect-proof boxes. If ventilated boxes are used, all the ventilator openings of the box should be covered with insect- proof screen and all the sides of box

Figure 3: PHM Facility for Export of Mango of MSAMB.

should be sealed with adhesive tape to prevent any entry of pests.

iii) The materials used inside the package must be new, clean, and of a quality such as to avoid causing any external or internal damage to the produce.

iv) The use of materials, particularly of paper or stamps bearing trade specifications is allowed, provided the printing or labelling has been done with non-toxic ink or glue.

v) The contents of each package must be uniform and contain only mangoes of the same origin, variety, quality and size. The visible part of the contents of the package must be representative of the entire lot.

vi) Mangoes shall be packed in each container in compliance with the Recommended International Code of Practice for Packaging and Transport of Fresh Fruits and Vegetables (CAC/RCP 44-1995, Amd. 1-2004).

India is a prominent exporter of fresh mangoes to the world. The country has exported 49,658 MT of fresh mangoes to the world for the worth of Rs. 400.21 crores during the year 2019-20. Major export destinations are UAE, UK, Saudi Arabia, Kuwait, Qatar and USA. The world trade in mango consists of an export of 9.29 lakh tonnes valued at US$ 6189.17 lakh and imports to the tune of 7.93 lakh tonnes estimated at US$7592.35 lakh.

Among internationally traded tropical fruits, mango ranks only second to pineapple in quantity and value. Major markets for fresh and dried mangoes are Malaysia, Japan, Singapore, Hong Kong and the Netherlands and canned mangoes are Netherlands, Australia, United Kingdom, Germany, France and USA. Southeast Asian buyers consume mangoes all year round. Their supplies come mainly from India, Pakistan, Indonesia, Thailand, Malaysia, Philippines, Australia and most recently from South Africa. The varieties in demand at the international market include Kent, Tomy Atkin, Alphonso and Kesar. Each exporting country has its own varieties, which differ in shape, colour and flavour. Prices are very low for varieties of Indonesia and Thailand, however, Indian mangoes fetch higher price in the International market. In the United States of America, the prices vary with the season. The year-round availability of mango in India is given in Figure 4.

Higher prices prevail during February and March, when mango availability in India is the lowest. The major chunk of international trade in fresh mangoes takes place within short distances. Mexico, Haiti and Brazil account for the majority of North America's imports.

India and Pakistan are the predominant suppliers to the West Asian market. Southeast Asian countries get most of their supplies from the Philippines and Thailand. European Union buyers source mangoes from South America and Asia. Although Asia accounts for 75 per cent of the world's production, its dominance does not translate into international trade. Asian producers find it easier to expand sales to the European Union. Europe's acceptance of different varieties is greater, because of a large demand from Asian immigrant groups. Transportation costs are not as big a factor in exporting mangoes to the European Union as in exporting to the United States market. India and Pakistan can compete with non-Asian suppliers to the European Union, whereas proximity gives Mexico and Haiti a clear advantage in supplying it to the United States market.

Fifty-four per cent of European Union imports enter during the periods May to July and November to December, with peak imports in June. French imports reach peak in April and May, whereas United Kingdom imports are concentrated during the May to July. German imports are spread more evenly throughout the year. Of the top suppliers, Brazil provides chiefly during the period November to December, the United States during June to October, South Africa during January to April and Venezuela during April to July. Pakistan supplies the majority of its exports to the European Union during June and July; Indian exports take place mainly during the month of May.

Although a lion's share of the Indian mangoes go to the Gulf countries, efforts have been made to exploit European, American and Asian markets. Alphonso variety is exported to Middle East, UK and Netherlands. Besides these, the fresh mangoes are being exported to Bangladesh, Bahrain, France, Kuwait, Malaysia, Nepal, Singapore and U.K. Varieties such as Alphonso, Dashehari, Kesar, Banganapalli and several others those are currently in demand in the international markets are produced and exported from India. Different quantities of mangoes exported from India are mentioned in Table 9.

Apart from being a leading mango producing country in the world, India has superior and leading varieties like Alphonso and Kesar. Beside, relatively long period of availability (February to August),

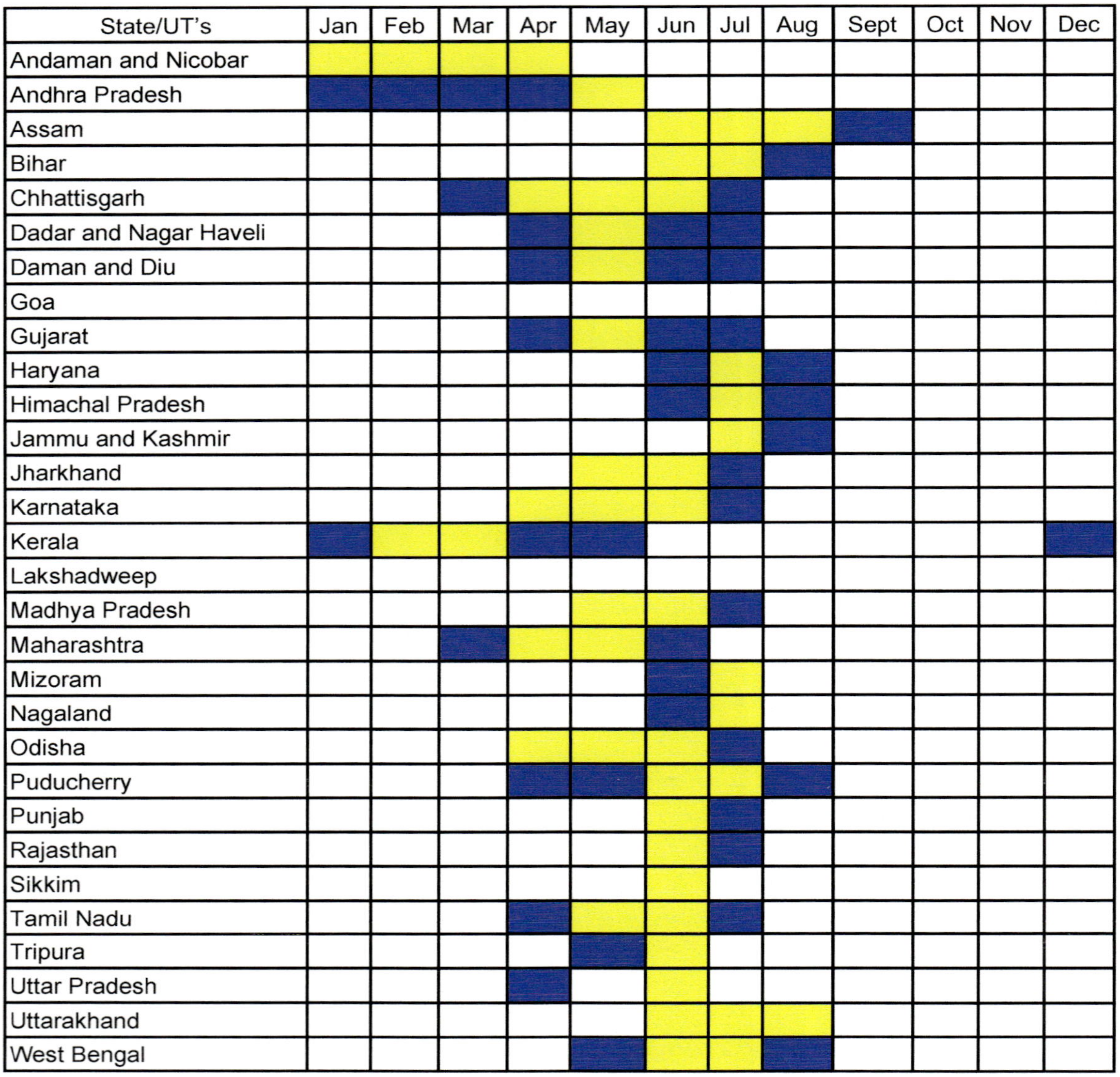

State/UT's	Jan	Feb	Mar	Apr	May	Jun	Jul	Aug	Sept	Oct	Nov	Dec
Andaman and Nicobar												
Andhra Pradesh												
Assam												
Bihar												
Chhattisgarh												
Dadar and Nagar Haveli												
Daman and Diu												
Goa												
Gujarat												
Haryana												
Himachal Pradesh												
Jammu and Kashmir												
Jharkhand												
Karnataka												
Kerala												
Lakshadweep												
Madhya Pradesh												
Maharashtra												
Mizoram												
Nagaland												
Odisha												
Puducherry												
Punjab												
Rajasthan												
Sikkim												
Tamil Nadu												
Tripura												
Uttar Pradesh												
Uttarakhand												
West Bengal												

Source: Indian Horticulture Database, NHB 2013.

Figure 4: Availability of Mango in India

diversity of the varieties, economic liberalisation and priority to export of fresh products have opened up the possibility of boosting mango exports from India. Therefore, India has a large opportunity to tap global trade in the export of Mangoes. The scenario of export of mango can be changed by efficient orchard management, improvements in infrastructure, use of advanced technologies, and meeting the food quality standards. India has the potential to become the leading exporter. The strategies for boosting up of mango export suggested by the Expert Group of the Planning Commission and APEDA are: Transhipment of mango by sea to make them competitive in price sensitive markets of Europe and Middle East; Promotion of non-traditional varieties like Chausa, Safeda and Badami;

Table 9: Export of Mango from India to different Countries (2019-20)

Country	*Quantity (MT)*	*Value (Rs in lakh)*
Bangladesh	3038	903
Kuwait	1170	1605
Nepal	11975	2451
Oman	3636	2756
Qatar	2744	2740
Saudi Arabia	1517	1299
Singapore	687	735
UAE	16567	14286
UK	4356	6761
USA	1095	3056
Others	2873	6185
Total	49658	40021

Source: APEDA, 2021.

Augmentation of supply during seasonal gaps (February/March in Middle East and March/April and July/August in Europe); and Development of postharvest management facilities/practices, if possible at every operational site, or at least at the cluster sites where maximum mangoes are produced. There is export specification to be followed while dispatching the consignments to the below mentioned countries: (Table 10).

Table 10: Export Requirement of Indian Mangoes

Variety	Countries				
	Middle East	Nether-land/ Germany	UK	Japan	USA
Alphonso (g)	200-250	250-300	250-300	250-300	250-300
Kesar (g)	200-250	225-250	225-250	250-300	250-300
Packing	1 Doz/2.5 kg	1 Doz/2.5 kg	1 Doz/2.5 kg	1 Doz/3.5 kg	1 Doz/3.5 kg
Storage temp. (°C)	13	13	13	13	13
Export	By sea	By air	By air	By air	By air

Source: Maharashtra State Agricultural Marketing Board Website.

In addition to the requirements of the Codex General Standard for the Labelling of Prepackaged Foods (CODEX STAN 1-1985, Rev. 1-1991), the following points should be followed; If the produce is not visible from the outside, each package shall be labelled as to the name of the produce and may be labelled as to the name of the variety; Each package must bear the following particulars, in letters grouped on the same side, legibly and indelibly marked, and visible from the outside, or in the documents accompanying the shipment. For produce transported in bulk these particulars must appear on a document accompanying the goods; Name and address of exporter, packer and/or dispatcher; Name of the produce if the contents are not visible from the outside; Country of origin and, optionally, district where grown or national, regional or local place name; Class; Size (size code or weight in grams); Number of units; Net weight etc.

For international exports it is pertinent to ensure that mangoes shall comply with those maximum levels for heavy metals and pesticides residues established by the Codex Alimentarius Commission. Similarly for hygiene, it is recommended that the produce covered by the provisions of this standard should be prepared. The produce should be handled in accordance with the appropriate sections of the Recommended International Code of Practice – General Principles of Food Hygiene (CAC/RCP 1-1969, Rev. 4-2003), Code of Hygienic Practice for Fresh Fruits and Vegetables (CAC/RCP 53-2003), and other relevant Codex texts such as Codes of Hygienic Practice, and Codes of Practice. The produce should comply with any microbiological criteria established in accordance with the Principles for the Establishment and Application of Microbiological Criteria for Foods (CAC/GL 21-1997).

7.0 Conclusion

Mango is the most important fruits that have been associated with large numbers of various stakeholders for their livelihood security right from production to consumption points. The losses of this crop after harvest is also very high primarily due to its season of production coincides with high summer temperature in India. Appropriate postharvest management of mango begins with proper care of preharvest spray schedule for control of latent infections followed by several postharvest operations *viz.* harvesting, trimming of peduncle, de-sapping, sorting and grading, pretreatments, packaging, precooling, transportation, short-term and long term holding, domestic and export marketing, ripening and ends with consumer satisfaction. This chapter discussed all the above major issues and suggested recommendations.

Although, India is the major mango producing country in the world but its productivity per hectare is quite low just half to Brazil and 30 per cent less than Pakistan. This adds additional cost on the fruit when traded in the domestic and international markets, thus becomes un-remunerative to producers, marketers and exporters. Some of the researchable issues flagged on postharvest management of mango are:

i) Development of environment friendly technologies for production of quality mangoes. This includes safe prophylactic sprays for disease and insect control in order to minimize the residue levels as per the International norms.

ii) Development of user friendly tools for adoption of fruit bagging technologies.

iii) Development appropriate crop management practices for synchronous maturity of mango in order to encourage more quantity of acceptable grades of fruits after harvest or in other words reducing the level of rejection of the harvested produce.

iv) Development and evaluation of novel safe molecules that could be applied as postharvest treatments in mango by replacing several fungicides that are either banned in many countries or are being phased out.

v) Commercial validation of many technologies *viz.* MAP, CA- storage, use of surface coating, use of 1-MCP etc.

8.0 Important References

Anonymous, 2008. *Mango production technology*. Published by Dr BS Konkan Krishi Vidyapeeth, Dapoli.

Anonymous, 1985. Effect of size and grading on storage behaviour of Alphonso mango fruits. *Horticulture subcommittee report, council meeting, October, 1985, held at Konkan Krishi Vidyapeeth, Dapoli,* India Pp: 228-232.

Davenport, T. L., Burg, S. P. and White, T. L. 2006. Optimal low-pressure conditions for long term storage of fresh commodities kill Caribbean fruit fly eggs and larvae. *Hortic. Technol.* **16**: 98–04.

de Mello Vasconcelos, O. C., Duarte, D., de Castro Silva, J., Mesa, N. F. O., Mederos, B. J. T. and de Freitas, S. T. 2019. Modeling 'Tommy Atkins' mango cooling time based on fruit physicochemical quality. *Sci. Hortic.* **244**: 413-20.

Din, A., Jahangir Chughtai, M. F., Mehmood, T., Khaliq, A., Nadeem, M., Ahsan, S., Amir, R.M., Abrar, M., Saeed, K. and Manzoor, M. S. 2021. Shelf life extension of mango fruit by using non-preservative technique. *Int. J. Fruit Sci.* 1-10.

FAOSTAT. 2019. Area and production of mango http: //faostat.fao.org

FAOSTAT. 2019. Top world mangoes exporters/ importers countries http: // faostat.fao.org Joshi, G.D. and Roy S. K. 1986. CFB box – an alternative to wooden box for transport and storage of mango cv. Alphonso. *Indian Food Packer,* **40**: 32-47.

Malik, A.U. and Singh, Z. 2005. Prestorage application of polyamines improves shelf- life and fruit quality of mango, *J. Hortic. Sci. Biotechnol.* **80**: 363–69.

Mukherjee, S.K. 1951. Botany and physiology of mango, *Indian J. Genetic.* **2**: 49. Mukherjee, S. K. 1953. Origin, distribution and phylogenetic affinities of the species of *Mangifera, J. Linnean Soc. Bot.* **55**: 65–83.

Ntsoane, M. L., Sivakumar, D. and Mahajan, P. V. 2020. Optimisation of O_2 and CO_2 concentrations to retain quality and prolong shelf life of 'shelly'mango fruit using a simplex lattice mixture design. *Biosyst Eng,* **192**: 14-23.

Rao, S.D.V. and G.K.P. Rao, 2008. Controlled atmosphere storage of mango cultivars 'Alphonso' and 'Banganapalli' to extend storage-life and maintain quality, *J. Hortic. Sci. Biotechnol.* **83**: 351–59.

Sadhu, M.K. 1999. Mango **In:** *Tropical Horticulture.* Eds. T.K. Bose, S.K. Mitra, A. A. Faroogi, M.K. Sadhu, Naya Prakash pp. 178-198.

Singh, Z., Singh, R.K., Sen, V.A. and Nath, P. 2013. Mango- postharvest biology and biotechnology, *Critical Rev. in Pl. Sci.* www.tandfonline.com/loi/bpts20

Singh, Z. and Singh, S. P. 2012. Mango. **In**: *Crop Postharvest: Science and Technology.* Volume 3: Perishables, pp. 108–142. Rees, D., Orchard, J., Eds., UK: Blackwell Pub.

Vega-Alvarez, M., Salazar-Salas, N. Y., López-Angulo, G., Pineda-Hidalgo, K. V., López-López, M. E., Vega-García, M. O., Delgado-Vargas, F. and López-Valenzuela, J. A. 2020. Metabolomic changes in mango fruit peel associated with chilling injury tolerance induced by quarantine hot water treatment. *Postharvest Biol. Technol.* **169**: 111-299.

Yahia, E.M. 2006. Effects of insect quarantine treatments on the quality of horticultural crops. *Stewart Postharvest Rev.* **1**: 1–18.

Yahia, E.M., Medina, F. and Rivera, M. 1989. The tolerance of mango and papaya to atmospheres containing very high levels of CO_2 and/or very low levels of O_2 as a possible insect control treatment. *Proc 5th Int CA Res Conf, Wenatchee,* WA. **2**: 77– 89.

Zahedi, S. M., Karimi, M. and Teixeira da Silva, J. A. 2020. The use of nanotechnology to increase quality and yield of fruit crops. *J. Sci. Food Agric.* **100**: 25-31.

Website References

www.apeda.org; www.ashiko.org; www.cisklko.org; www.dgcis.org; www.icar.org; www.ikisan.com; www.indianhorticulturedatabase.org; www. manarasoft.com; www.msamb.org; www.nhb.org; www.pau.org; www.phptc.org

Chapter 21

Papaya

V. Ponnuswamy

Former Dean

Tamil Nadu Agricultural University, Periyakulam – 641 020, Tamil Nadu, India

T.L. Preethi

Assistant Professor

KVK, Tiruvallur – 602 025, Tamil Nadu Agricultural University, Tamil Nadu, India

E-mail: swamyvp200259@gmail.com; E-mail: preethi.chari@rediffmail.com

1.0 Introduction

Papaya (*Carica papaya* L.) originated in southern Mexico and Costa Rica and was taken by the Spaniards to Manila in the mid-16th century from where it gradually spread to all tropical and subtropical countries. It is one of the most nutritious of tropical fruits. It is a wholesome fruit used mostly as dessert and is a rich source of vitamins. Unripe fruits are consumed as vegetable. It contains beta carotene equivalent to 2000-3000 I.U. of vitamin A. Among other vitamins, it also contains thiamin (15-63 µg), riboflavin (22-83µg), niacin (0.15-0.76 mg) and ascorbic acid (33-136 mg) per 100g of the fruit pulp. Papaya, therefore is of special interest and importance for human nutrition. A single 100g serving of papaya is sufficient to cover the daily requirement of vitamins and minerals for one person.

2.0 Status of Cultivation

Papayas are grown in about 60 countries. It is now widely cultivated in India, China, Sri Lanka, Malaya, Mexico, Brazil, Peru, Venezuela, Central and South Africa, Philippines, Australia and most of the pacific Islands. The estimated total world production for papayas in 2018 was 13,290,320 metric tonnes, up by 2.5% from 12,971,418 tonnes in 2017 as shown in Table 1. India is by far the largest producer, accounting for 45% of global production. Almost all of its production is intended for its domestic market.

Table 1. List of Countries Ranked According to Papaya Production

Rank	*Country/Region*	*2018*	*2017*
1	India	5,989,000	5,940,000
2	Brazil	1,060,392	1,058,487
3	Mexico	1,039,820	961,768
4	Dominican Republic	1,022,498	869,306
5	Indonesia	887,591	875,112
6	Nigeria	833,038	829,563
7	Democratic Republic of the Congo	213,769	214,405
8	Venezuela	188,636	178,164
9	Colombia	183,732	179,899
10	Cuba	176,630	189,086
11.	Others	16,93,196	16,73,611
	Total	**13,290,320**	**12,971,418**

During the year 2017-18, out of total area of 1.32 lakh ha, India produced 5381700 MT of papaya with average productivity of 40.7 tonnes/ha. The detail of state-wise area, production and productivity of papaya in India during the year 2012-13 are presented in Table 2 and Figure 1. The maximum share of production of papaya comes from Andhra Pradesh (28.19 per cent) followed by Gujarat (20.99) and Karnataka (9.92 per cent).

Source. https://the-biggest.net/other/which-country-is-the-largest-producer-of-papaya.html and https://en.wikipedia.org/wiki/List_of_countries_by_papaya_production

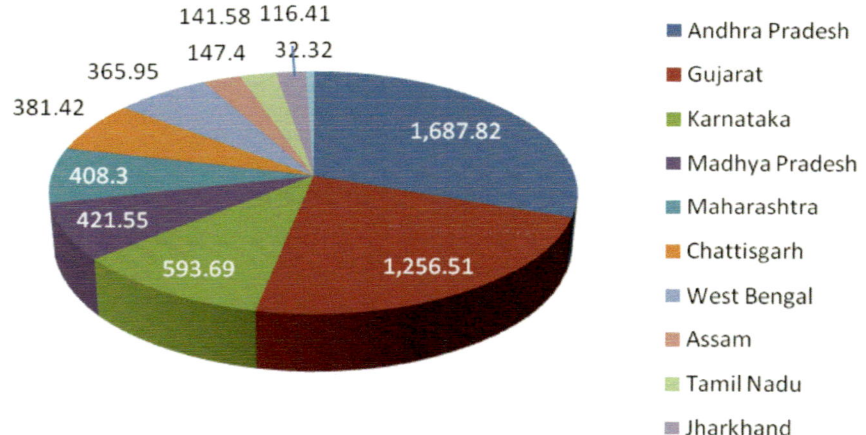

Figure 1. Per cent Share of Papaya Production among different States in India (*Source.* Indian Horticulture Database, NHB 2018).

Table 2. State-wise Area, Production and Productivity of Papaya in India

State	*2017-18*			
	Area(000 ha)	*Production (000 MT)*	*Productivity (tonnes/ha)*	*Share (%)*
Andhra Pradesh	18.01	1687.82	93.72	28.19
Assam	7.21	147.40	20.44	20.99
Chhattisgarh	14.40	381.42	26.49	9.92
Gujarat	20.31	1256.51	61.87	7.04
Jharkand	2.66	116.41	43.76	6.82
Karnataka	8.75	893.69	102.14	6.37
Madhya Pradesh	10.55	421.55	39.96	6.11
Maharashtra	10.28	408.30	39.72	2.46
Tamil Nadu	1.53	141.58	92.54	2.36
West Bengal	12.38	365.95	29.56	1.94
Others	125.31	6037.65	48.18	7.80
Total	**138.40**	**5988.93**	**43.27**	**100.00**

Papaya crop grown under net house recorded higher yield and quality attributes at Anantharajupet, Coimbatore and Gandevi for number of fruits (63.63, 77.20 and 46.38 respectively), fruit weight, marketable fruits, total yield (151.03, 236.71 and 77.79 t/ha respectively), TSS (12.13, 12.90 and14.64 °B respectively) as compared to crops grown in open conditions. Net house significantly protected the plants from PRSV incidence in the three test locations.

3.0 Major Commercial Varieties

There are two main types of papayas produced: the small-sized Solo-papayas (Hawaiian papayas), usually weighing between 500 g and 1 kg per fruit, and the large-sized (Mexican papayas), weighing up to 4.5 kg each.

The major commercial varieties of papaya grown in India are Coorg Honey Dew, Pink Flesh, Solo, Sunrise, Thailand and Washington. However, several varieties and hybrids developed in India have also been recommended for cultivation in different states. Papaya is virtually available all year round in India in almost all the states. Some of major cultivars *viz.* Croog Honey Dew, Pink Flesh Sweet, Taiwan are grown all over India. Nematode resistant variety of papaya is Pusa Majesty. Area-wise recommended varieties for commercial cultivation in India is mentioned in Table 3.

Table 3. Papaya Varieties Commercially Cultivate in India

Variety	*States*
Arka Prabhath	Andhra Pradesh, Karnataka, Kerala,Tamil Nadu
Arka Surya	Andhra Pradesh, Karnataka, Kerala, Tamil Nadu
Co-1	Andhra Pradesh, Bihar, Jharkhand, Karnataka, Odisha, Tamil Nadu, Uttarakhand
Co-2	Andhra Pradesh, Bihar, Chhattisgarh, Himachal Pradesh, Jharkhand, Madhya Pradesh, Odisha, Rajasthan, Tamil Nadu
Co-3, Co-4	Andhra Pradesh, Karnataka, Maharashtra, Tamil Nadu
Co-5	Andhra Pradesh, Tamil Nadu
Co-6	Andhra Pradesh, Bihar, Jharkhand, Tamil Nadu
Pant-1	Uttar Pradesh, Uttarakhand
Pusa Delicious	Bihar, Himachal Pradesh, Jharkhand, Rajasthan, Uttar Pradesh, Uttarakhand
Pusa Dwarf	Bihar, Himachal Pradesh Jharkhand, Uttar Pradesh, Uttarakhand
Pusa Nanha	Bihar, Jharkhand, Tripura

Source: Indian Horticulture Database, NHB 2013.

4.0 Nature and Causes of Postharvest Losses

The global papaya industry faces two major problems: the papaya ringspot virus (PRSV), a disease that drastically reduces fruit yield, fruit size and quality and in some cases results in total loss of production. The second major problem is the significant postharvest losses along the marketing chain. Factors such as fungal diseases, physiological disorders, mechanical damage or a combination of these are the leading causes of postharvest losses. While papaya suffers postharvest losses ranging from 30 to 60 per cent in the Southeast Asia region, simple technology and practices have helped to reduce losses and to extend storage life. At the field level, the losses are mainly due to immature and small size of fruits, malformation and harvesting injury. At the market level, bruises and pressing injury caused transit losses. Anthracnose and fruit rot due to *Alternaria* and *Phytophthora* are the main causes of loss during ripening. Rotting of fruits is the main reason for loss during retailing.

5.0 Postharvest Management

Size, shape, smooth skin, and absence of blemishes are major quality characteristics. Consumers in Western countries also prefer fruit without the heavy musky, sweaty found in some Southeast Asian cultivars. Small, dry brown-black "freckles" on the skin are non-pathogenic and do not detract from ripening (Reyes and Paull 1994).

5.1 Maturity Indices

In Hawaii, the minimum grade standard requires 11.5% SSC (Anon 1990). Therefore, fruit should have started ripening before harvest, as indicated by some skin yellowing (Akamine and Goo 1971). Less mature fruits are lower in sugar and ripen poorly.

Various non-destructive indices can be used to determine papaya harvest maturity, including the number of days from flowering, fruit size and external colour. It is important to harvest papaya fruit at proper maturity, because they do not increase in sugar content after picking. Papayas normally require about 3 months from flowering until fruit maturity. The pear shaped Hawaiian type papaya fruits generally weigh between 350 to 500g when mature. Native type fruits are significantly larger when mature, often weighing more than 1.5 kg when mature. The most obvious index of fruit maturity is external skin colour. As the fruit matures, the skin colour will change from green to yellow/orange.

5.2 Harvesting

Papayas are harvested manually depending on the size and age of the tree, using specialized tools, knives or by hand. When harvesting is done by hand or with knives, the peduncle is snapped or cut from next to the tree, then immediately trimmed flush against the top of the fruit. The specialized implement for harvesting of fruits which are inaccessible to hand due to tree height comprises of a long pole, a small circular hoop at the top, a small mesh bag attached to the hoop, and a horizontal blade above the hoop and the bag. The blade is positioned below the peduncle of the fruit and the pole moved upwards; the fruit is detached from the tree and then drops gently into the mesh bag below the hoop at the top of the pole. After harvest, the fruits are placed in single layers in shallow, light coloured field crates, preferably containing a foam layer for cushioning. Peduncle attached to the fruit should be trimmed after harvest to ensure that no stem to fruit rubbing occurs during transport to the packing facilities. Fruits should never be thrown or dropped. Field crates containing the fruit should be kept in shade for protection against sun and rain while awaiting collection for delivery to a packing facility. Mesh bags, sacks or baskets are unacceptable for papaya transport due to the high susceptibility to bruising. Fruits can be covered with paper to avoid bruising (Figure 2). Care should be taken during transport in field crates to minimize the movement of fruit. Papayas for export should be harvested between the one stripe stage (Figure 3) [one yellow stripe showing at the blossom end and the quarter-ripe stage [some yellow at the blossom end (Figure 4)]. Fruits harvested at these stages of maturity will withstand the rigors of shipping and transport to distant markets. Fruits harvested immature green will not ripen properly, will taste flat, and shrivel prematurely.

Figure 2. Papaya Covered with Paper in Order to Avoid Bruising

Figure 3. Papaya Fruits Harvested at the One-Stripe Stage

Figure 4. Quarter-Ripe Papaya Fruits

Flavour and edible quality generally improve with the advancement of ripening. Fruits intended for domestic markets should be harvested when the skin is partially yellow-orange (between one-quarter to half-ripe). A completely yellow-orange skin indicates full ripeness and generally a sweeter fruit. However, the postharvest life of this advanced maturity stage fruits will typically be less than a week (Figure 5).

Figure 5. Papaya Fruits Harvested at Full Ripe Stage

The fruits should be left on the tree until they mature fully. Usually, the fruits are harvested when they are full size, light green with a tinge of yellow at the apical end. On ripening, fruits of certain varieties turn yellow but some of them remain green. When the latex ceases to be milky and becomes watery, the fruits are considered suitable for harvesting. While picking the fruits from the trees, care must be taken that these are not scratched and are free from any blemishes, otherwise, there will be an attack of fungus and the fruits may start decaying during marketing.

5.3 Sorting and Grading

Suitable under grading must be done before packing. Pregrading and sorting should be carried out as soon as possible after harvest and fruits should be left under ambient conditions to continue ripening or placed at 10 ° to 12 °C for cooling and storage. Fruits that fail to meet market requirements should be removed before washing and decontamination stages and packed later or in a different packing line. The grading should be done for over-ripeness where the fruits exhibit localized softening or have more than three yellow stripes at the base and also for under-ripeness where the fruits have no yellow stripes. Sorting is also necessary for mechanically damaged fruits by knives, harvesting poles, rubbing in the crate *etc.* Graders remove any fruits that show signs of fresh mechanical damage or any of the conditions that qualify the fruit for rejection in the pre-grading stage.

Many factors contribute to papaya skin damage, including the harvesting tools used, dropping of fruit into crates, overfilling of crates, and jostling of fruit during field transportation. During ripening, bruised areas will develop into soft, dark regions that are susceptible to secondary microbial infection. Similar problems can result from poor handling during washing, grading and packing. Papaya handlers can minimize fruit damage by taking protective measures through all handling stages. Staff should be trained in proper harvesting techniques. At the packing house, crates should include foam in their base and contain only one layer of fruit. In automated operations, all machinery should be padded where possible. Palletized systems are preferable for moving produce in both field crates and final packages.

5.4 Washing

Pre-graded fruit should be placed in a water tank to remove latex and debris. In larger operations, fruit is washed with water in a separate tank or with overhead sprays before being passed to the main wash tank. Normally papayas require no additional cleaning, but some may benefit from careful wiping with a soft cloth. Fruit can also be cleaned with rotating soft brushes, but simple movements through a water tank for 10 to 15 minutes generally suffice (Figure 6). To help prevent disease and microbial build-up, sodium hypochlorite can be added to the water tanks at a rate of 100 parts per million. A timetable for re-administering the sodium hypochlorite and changing the water, based on the volume of fruit passing through the system is required to be developed. After pre-grading, washing and fungicide treatment, the papayas are left to dry and then are graded for packing.

5.5 Packaging

Standard packing house design and operations apply for papaya grading and packing. Operations can be carried out with basic equipment, including water tanks, field crates, and grading tables or with automated washing and weight-grading or separation systems. All tanks and grading tables should be covered with foam to protect the fruit from exposed edges; the skin of the papaya is delicate, and scratches will result in latex exudation and staining. Similarly, if the fruit is dropped, it will easily develop bruises as it ripens.

Figure 6. Washer for Papaya Fruits

The most common package size is a 4.5 kg (10 lb) carton, and larger, 10 kg (22 lb) cartons are also used. Cartons from areas requiring insect disinfestation are fully sealed to meet regulatory requirements, while fruit from other areas can be in open-topped cartons. Count size ranges from 6 to 18, depending on fruit and carton size. Fruits are marketed as "colour break" and , ½, and ripe, and it is normally ready to eat when there is 75% or more skin colour. Foam mesh sleeves, foam padding on the bottom of cartons, or paper wrapping prevent abrasion injury, which is a major problem in fruit still having green areas of skin (Quintana and Paull 1993; Sivakumar and Wall 2013).

Papayas fruits are classified for packing based on size, stage of ripeness, and, if applicable by shape. Fruits of different shapes may not be mixed in the same carton. While all the fruits in a carton must be similar in size, the counts across cartons can vary. A 4-kilogram carton, for example, can contain the following number of papayas, depending on fruit size (Table 5).

Table 5. No. of Fruits in different Grades of Papaya

Small	13 to 16 count	(308 to 250 grams)
Medium	9 to 12 count	(440 to 330 grams)
Large	6 to 8 count	(666 to 500 grams)

Papayas fruits are also separated according to degree of ripeness, normally as "one yellow stripe" (which is less than one-quarter ripe), "one-quarter ripe", and "one-half ripe". All fruit in a carton must be at the same stage of ripeness. In this way, supermarket produce managers, can easily identify which fruits are ready for display and which can be stored. Because papayas produce ethylene when they ripen, they should not be stored or shipped with ethylene-sensitive produce. Solo Sunrise papayas, the most popular variety, are packed in single-layer cartons, each with a net weight of 4 kilograms to 5 kilograms, depending on market requirement. Internal cushioning materials can include shredded paper in the base of the carton and individual tissue wraps for each fruit. Papaya may be individually labelled for appearance and easy recognition.

The fruit should be packed properly for dispatching to market. Since papaya fruits are easily perishable, care should be taken to wrap the individual fruit in paper and finally packed in wooden crates surrounded by soft material all-round and saw dust especially on the bottom. Papayas are best packed on their sides, in rows, with the stem ends at an angle; interlocking rows support one another. Papayas should not be packed on their bases because they soften from the base up. Packers should guard against overfilling the cartons. Stage of ripeness varies from 50 to 70 per cent yellow depending on market demand and time of year. Fruits can be prepared for packing in two ways *viz.* in single layer cartons with one-piece self-locking or two- piece full telescopic fibre board carton having a bursting strength of 14.06 to 17.57 kg/cm^2. Internal packaging material should include shredded paper in the base of the carton and individual tissue wraps for each or alternate fruit. Individual labels can be attached to the fruit for appearance and recognition. The internal dimensions of the cartons used for packing are: 10.9 x 34 x 26.9 cm or 10.2 x 43.2 x 27.9 cm.

Individual fruit packing using styrofoam sleeves and carefully placing the fruits with stem-end down in ventilated plastic crates in a single layer is helpful in minimizing mechanical damage. Spongy nature of styrofoam padding helps to absorb vibration damage during transport causing less skin damage.

5.6. Ripening

Little is known about papaya fruit ripening and its response to standard fruit storage techniques. The fruit ripening is probably regulated by naturally produced ethylene similar to that of apple, tomato and banana. Exposure to 100 ppm ethylene at 20 to 25 °C (68 to 77 °F) and 90-95 per cent R.H. for 24-48 h produces a rapid and uniform ripening of papayas harvested between colour break to 1/4 yellow. This treatment induces the yellowing of peel and pulp softening, but does not improve flavour. Experiments carried out at TNAU revealed the delay in ripening by 3 days modified atmospheric condition. These fruits when were exposed to 500 ppm of ethylene, at ambient temperature (24-26 °C and 85-95 per cent RH) showed a better texture, taste and colour compared to untreated ones.

When harvested, papayas are green with a stripe of yellow at their base, they ripen during transport or are commercially ripened on arrival. As papaya ripens, skin colour changes from green to yellow, and the fruit becomes softer and develops a fruity aroma.

Fruits should have smooth skin. Fruits with dark, sunken spots should be avoided as these may affect the flesh of the fruit and spoil the flavour. The fruit is sent to market at various degrees of ripeness - 1/4, 1/2 and 3/4 ripe - with skin colours ranging from green to yellow-green to yellow-orange. Papaya at its ideal stage of ripeness is 3/4 yellow or yellow-orange (depending on variety). Papayas are ripe when they emit a soft fruity aroma.

Room and forced-air cooling are commonly used. Hydro-cooling is possible. However, rapid cooling after insect disinfestation treatments can lead to skin scalding.

5.7 Storage

Fruits should be stored at room temperature unless they are completely ripe (100 per cent yellow). Weight losses greater than 8 per cent considerably diminish the postharvest quality of papaya. If they are completely ripe, fruits can be stored at 4.4 °C for 3 to 4 days. Storage of unripe papaya below 10 °C will result in chilling injury. The symptoms are indicated by surface pitting, discolouration of the peel and the flesh, incomplete ripening, poor flavour and increased susceptibility to disease incidence. Ripe papaya can be stored successfully at lower temperatures, but transport of ripe fruit (more than 50 per cent yellow colour) is not recommended due to the susceptibility to mechanical damage and bruising. The ideal storage temperature of papaya is 10- 13 °C with 80-90 per cent R.H. At this condition, the fruits could be stored up to 3 weeks depending on the stage of ripening. Papaya fruits, at colour turning, can be stored at 7 °C, to achieve normal ripening. Shelf life of fruit could be extended by storing at 13 °C with 1.0 - 1.5 per cent oxygen and 10 per cent carbon-dioxide.

Store from 7 to 13°C (45 to 55°F) with 90 to 95% relative humidity. At 7 to 10°C (45 to 50°F), storage-life is limited by chilling injury, while at 10 to 13°C (50 to 55°F) ripening occurs slowly (Chen and Paull, 1986). Papaya fruit at color-turning (break) stage can be stored at 7°C (45°F) for 14 days and will ripen normally when transferred to room temperature (Thompson and Lee 1971, Chen and Paull 1986). Ripe, fully coloured fruit can be held for more than 1 week at 1 to 3°C (33.8 to 37.4°F).

Shelf-life extension of 1 to 1.5 days was obtained when the fruits were stored at 12°C (54°F) in 1 to 1.5% O_2 for 6 days. Low O_2 (1 to 5%), with or without high CO_2 (2 to 10%), reduces decay and delays ripening (Chen and Paull 1986). High CO_2 (30%) adversely affects internal colour, aroma, and flavor, while there is no residual effect of 10% CO_2 on decay control, though skin de-greening is delayed. At 10°C (50°F), fruit can be stored for 36 days in 8% CO_2+3% O_2 and still have 5 days at 25°C (77°F) for retail (Cenci *et al.* 1997). Ethylene removal prior to storage has shown variable results. CA recommendations are in the range 2 to 5% O_2 plus 5 to 8% CO_2. However, no commercial use has so far been reported. Fruit stored at 10°C (50°F), 98% relative humidity and low pressure of 20 mm Hg, ripened more slowly than fruit at normal atmospheric pressure. Low-pressure storage appears to suppress disease development (Yahia 1998).

The optimum temperature for fruit ripening is 22.5 to 27.5°C (73 to 82°F), with fruit taking 10 to 16 days to reach full skin yellowing from the color-break stage. Severe weight loss and external abnormalities become significant at temperatures higher than 27.5°C (82°F). Display temperatures should not be less than 10°C (50°F) if fruit is not fully ripe. Fully ripe fruit at the edible stage can be held at 1 to 3°C (34 to 37°F). Fruit should not be stacked more than 2 or 3 deep in racks, and wicker baskets with uneven bottoms and sides should be avoided, or at least contain a layer of protection placed between racks and fruit. Loss of about 8% of weight from colour break produces rubbery, low-gloss, unmarketable fruit. Diseased and bruised fruit should be removed from display and used immediately, if possible, in salads or mixed fruit cocktails. Do not mist (Paull *et al.* 1997).

Figure 7. Chilling Damage to Ripe Fruit (Left) and Unripe Fruit (Right)

Chilling injury symptoms include skin scald, hard lumps in the pulp around vascular bundles, and water soaking of flesh. Fruit become progressively less susceptible to chilling stress as they ripen. Symptoms of chilling injury occur after 14 days at 5°C (41°F) for mature green fruit and 21 days for 60% yellow fruit. Skin scald can be induced in colour-break fruit after chilling at 1°C (34°F) for 24 h. At a 7°C (45°F) storage temperature for 14 days, storage decay is less than when fruits are held at 12 to 13°C (54 to 55°F).

Ethylene rates in ripening fruit are 6 to 10 ìL kg-1 h-1. Ethylene-treated papayas ripen faster and more uniformly in terms of skin degreening, softening and flesh color . Since papaya ripen from the inside outwards, the effect of ethylene treatment is to accelerate the rate of ripening of the mesocarp tissue nearer the skin that has not started to soften. Ethylene is not recommended commercially, as rapid softening severely limits available marketing time (Paull *et al.*, 1997)

Papaya are climacteric and begin to yellow from the blossom end (Akamine 1969). See Table 4 for respiration rates at specific temperatures. To get mL kg-1 h-1, divide the mg kg-1 h-1 rate by 2.0 at 0°C (32°F), 1.9 at 10°C (50°F), and 1.8 at 20°C (68°F). To calculate heat production, multiply mg kg-1 h-1 by 220 to get BTU per ton per day or by 61 to get kcal per metric ton per day.

5.8 Postharvest Diseases

Losses due to postharvest disease may occur at any time during postharvest handling, from harvest to consumption. Besides direct economic considerations, diseased produce poses a potential health risk. Common postharvest diseases resulting from wound infections include blue and green mould caused by *Penicillium* sp. and transit rot is caused by *Rhizopus stolonifer*. Bacteria such as *Erwnia caratovora* also invade the wound. Common postharvest diseases and their causal organisms are mentioned in Table 4.

Table 4. Common Postharvest Diseases of Papaya

Disease	Causal Organism
Anthracnose	*Colletotrichum* spp.
Black rot	*Phoma caricae papayae, Mycosphaerella caricae*
Phomopsis rot	*Phomopsis caricaepapayae*
Phytophthora fruit rot	*Phytophthora palmivora*
Rhizopus rot	*Rhizopus stolonifer*

Anthracnose is prevalent in papayas during long-term storage, particularly as a result of orchard humidity, inappropriate preharvest disease control or poor orchard management. The presence of the disease is characterized by small black or light- brown spots that gradually grow and may coalesce and sink (Figure 7). Anthracnose can be controlled or reduced with preharvest sprays, Postharvest cold-water dip or spray containing 0.05 per cent thiabendazole solution and surfactant. Before using any chemicals in the field or the pack house on papaya for export, growers and packers should consult legislation prevailing in the importing and exporting country.

Where injuries are present, the process of wound healing can be accelerated in some instances through manipulation of the postharvest environment (*e.g.* temperature and humidity) or by application of certain chemical treatments.

Papaya treated by vapour heat (VHT) or forced hot air must be picked before the fruit is suitably ripe. Many times this is a mandatory operation for export. In order to avoid premature shrinkage due to water loss growers need to harvest greener papayas that can tolerate heat treatment compared to semi-ripe or ripe stages. In

Figure 8. Postharvest Disease of Papaya

VHT, fruit temperature is raised by saturated water vapour at 44.4 °C until the centre of the fruit reaches that temperature, and then held for 8.5 hours. Forced hot air treatment is given for 2 hours at 43 °C + 2 hours at 45 °C + 2 hours at 46.5 °C + 2 hours at 49 °C.The single hot-water dip at 49 °C for 15 min is the optimum heat treatment for control of postharvest diseases of papaya with minimal detrimental impact on fruit quality. The additional single hot-water dip before or after forced air dry heat (FADH) or vapour heat (VHT) have been reported to achieve significant control of postharvest diseases of papaya. Waxing of fruit and storage under low pressure (LP) has also been successful in reducing the disease incidence and increasing the shelf life of papaya.

Chitosan (1 per cent) in combination with ammonium carbonate (3 per cent) has been reported to significantly reduce the incidence of anthracnose and the recovery of naturally-infected fruit. Treated fruits retain acceptable eating quality. It can also be controlled by use of fungicides *viz.* prochloraz or propiconazole. Furthermore, combined application of the biocontrol agent *Candida oleophila* with sodium bicarbonate-incorporated wax coating also resulted in significant and commercially acceptable control of anthracnose. A yeast isolate CEN63, *Cryptococcus magnus*, was found effective in controlling anthracnose in papaya.

During transportation, it was reported that damage due to abrasion and puncture injury was more important than impact injury for papaya fruit.

5.8.1 Physiological Disorders

There are several non-pathological disorders seen in marketed fruit (Paull *et al.* 1997; Sivakumar and Wall 2013). The disorders include the following:

Green, slightly sunken areas on ripe yellow fruit are due to abrasion injury that occurs when fruit is still green.

Unsightly skin freckles (small brown slightly raised areas) that are more common on the side of the fruit exposed to the sun and are seasonal, occurring when rainfall and low temperatures occur 2 months before harvest. This disorder is non-pathogenic and does not influence ripening or flavour.

Sunscald, a dark olive-brown discolouration, occurs on fruit developing on trees with very sparse foliage; on trees that are leaning over, with fruit directly exposed to the sun; and where harvested fruits are left exposed to the sun.

Blossom end defect can be severe at some times of the year. The disorder leaves an open channel at the blossom end to the fruit seed cavity. Fruits with this disorder are prone to bacterial diseases in the fruit seed cavity before harvest.

Hard lumps in the flesh occur infrequently in otherwise ripe, non-heat-treated fruit. The lumps are thought to be associated with insect injury, disease, or other foreign material in the flesh. The condition can also occur in heat-treated fruit.

Premature ripe fruit, referred to as "soft fruit," is related to low fruit calcium. The condition is more common following periods of heavy rainfall, 2 to 3 mo before harvest, at the start of the final phase of fruit growth.

Intra-ovarian ovaries are common in some strains of papaya. These ovaries occur as a proliferation of tissue in the seed cavity and can be a thread-like appendage, to round or elongated structures of various sizes and shapes. A few fill the entire seed cavity of the fruit and have their seed cavity.

Sunken dry brownish-grey areas caused by mites feeding on skin during early fruit growth. The red and black mite (Brevipalpus phoenicis [Geijskes]) generally causes this damage in Hawai'i.

Table 5. Respiration Rates for Papaya

Temperature	*mg $CO_2 kg^{-1} h^{-1}$*
5°C	4-6
15 °C	15 -22
20 °C –colour break	9-18
20 °C -ripe	**70 -90**

Postharvest diseases are effectively controlled by hot water at 49°C (120°F) for 20 min and fungicide treatment.

Figure 9. Abrasion injury

5.9 Transportation

Since many tropical products like papaya are high value and traded in small quantities, they are often shipped by air, even when surface transportation, could be used. Papaya does not respond well to shipping by sea. For domestic transportation, the use of road transport offers substantial advantages of convenience, availability, flexibility permitting door-to-door delivery, and reasonable cost. The use of road transportation for fresh produce is increasing and is likely to increase in countries all over the world. Produce may be transported by pick-up, enclosed truck, open truck or refrigerated vehicle.

6.0 Domestic Marketing and Export

Fruit fly infestation becomes a problem with papayas after fruits attain 25% skin yellowing (Seo *et al.* 1982). The damage caused by fruit flies includes small surface blemishes, destruction of the edible flesh, and spoilage from decay. Heat treatments and irradiation are used to achieve fruit fly disinfestation (Couey 1989, Paull 1990, Armstrong 1994). Occasionally, heat treatment can cause internal injury and scald (Paull 1995), limit postharvest life, and reduce quality (Paull 1990). Papaya can tolerate insecticidal atmospheres (0.17 to 0.35% O_2, balance N_2) at 20°C (6oF) for up to 5 days (Yahia *et al.* 1989), although its disinfestation potential has not been shown. Colour stages for air shipments can vary from 20 to 50 per cent, with selections generally being more rigid during the summer months because of accelerated ripening rates on arrival in importing countries. Fruits that are exported at less than 20 per cent yellow will fail to ripen adequately when temperatures in importing countries are low, particularly during winter.

In order for fruits to arrive in the importing country at the correct colour stage, exporters must be aware of the fruit's maximum and minimum colour stages upon its departure from the packing house, the length of the shipment period (for air shipment, 24 to 48 hours), and the temperature in the importing country. For fruits to arrive in the importing country at the correct colour stage, attention has to be paid to the maximum and minimum colour stage on departure, the length of transit, and the transit temperature. Many importers prefer fruits to arrive at the half-yellow colour stage.

On arrival in the packing facility, fruits should be washed in water to remove latex and debris and then treated in a 0.05 per cent Thiabendazole solution for anthracnose control. Washing, treatment and grading can be carried out using mechanized or manual systems. Grading in each carton is required in terms of size, shape and stage of ripeness. Female and hermaphrodite fruit cannot be mixed in the same carton; all fruits must be of a similar size in each carton resulting in a range of counts and separations must be made for the degree of ripeness. Carton net weight is dependent on the importer, ranging from 3.5 to 5 kg, and the carton must not be overfilled during packing.

White peach scale insect was seen on papaya and is considered as a quarantine pest on fruit for export. Accordingly, Papayas exported from Hawaii now receive a vapour heat quarantine treatment for fruit flies. Studies have revealed that vapour heat treatment poses no threat to quarantine security in export shipments. Most importers also require that papayas should be mostly yellow and have a uniform softness, smooth, unblemished skin and a minimum sugar content of 12 °Brix. Papayas are normally sold to consumers when they are at least one-quarter ripe. Optimal ripeness occurs at the three-quarter stage when the fruit bears a yellow-orange peel and an orange-red pulp. Peel colour may vary from yellow to reddish-orange, depending on the variety of papaya.

Figure 10. Anthracnose Disease

Figure 11. Mechanical Injury

Figure 12. Effects of Impact Bruising

Maintenance of papaya fruit quality during the supply chain depends on orchard management, harvesting practices, packing operations, postharvest treatments, temperature management, transportation and storage conditions. Postharvest losses are attributed to mechanical damage, rapid flesh softening, decay, physiological disorders, pest infestation, and improper temperature management. The aim of postharvest technology and management in the supply chain is to provide favourable conditions to extend storage life and retain quality and nutritional attributes. Fungal infection is mainly due to injury during storage and handling. Species of *Fusarium, Alternaria* and *Aspergillus* were found to be the disease causing organisms responsible for extensive damage to fruits in the markets.

7.0 Conclusion

The Papaya Supply Chain consists of a relatively large number of participants having significant negative implications in regard to quality management. The quality issues are serious and because there are so many different people handling the fruit from the farm to the market it is almost inevitable that physical damage will occur. Furthermore, because this physical damage does not show up until the fruits begin to ripen there is very little scope for the handlers who likely caused the damage to realize the result of their actions. With such a large number of players in the supply chain assuming responsibility, issues related to poor quality become a problem. Key issues of quality and consistency have to be addressed to maintain and expand the export market.

8.0 Important References

Abeywickrama, K.C., Wijerathna, N., Rajapaksha, K., Sarananda, K.H. and Kannangara, S., 2012. Disease control strategies for extending storage life of papaya (*Carica papaya*), cultivars 'Red Lady' and 'Rathna'. *Ceylon J. Sci. Biol. Sci.* **41**: 27-34.

Almora, K., Pino, J.A., Hernández, M., Duarte, C., González, J. and Roncal, E., 2004. Evaluation of volatiles from ripening papaya (*Carica papaya* L., var. Maradolroja). *Food Chem.* **86**: 127-30.

Akamine, E.K. and T. Goo. 1969. Effects of controlled atmosphere storage of fresh papaya (*Carica papaya*, L. var. Solo) with special reference to shelf-life extension of fumigated fruit. *Hawaii Ag. Sta. Res. Bull.* **144**.

Cenci, S.A., A.G. Soares, J.M.S. Bibino and M.L.M.Soiya. 1997. Study of the storage of 'Sunrise' and 'Solo' papaya fruits under control atmospheres. *Proc. 7th Intl. Contr. Atmos. Res. Conf.* July 1997, Univ. of Calif., Davis CA, Abstract No. 112.

Chen, N.M. and R.E. Paull. 1986. Development and prevention of chilling injury in papaya fruit. *J. Amer. Soc. Hortic.* Sci. **111**: 639–43.

Jung Chen, Manenoi, N.A. and Paull, R.E., 2007. Papaya Postharvest physiology and handling" Problems and Solutions. *Acta Hortic.* **740:** 285-93.

Lili Zhou, Robert E. Paull, and Nancy Jung Chen, 2014. Fruit, Nut, and Beverage Crops F_N-34

Maharaj, R. and Sankat, C.K. 1990. Storability of papayas under refrigerated and controlled atmosphere. *Acta Hortic.* **269**: 375-86.

Quintana, M.*E.G.* and R.E. Paull. 1993. Mechanical injury during postharvest handling of 'Solo' papaya fruit. *J. Amer. Soc. Hortic. Sci.*, **118**: 618–22.

Paull, R.E., W. Nishijima, M. Reyes, and C. Cavaletto. 1997. A review of postharvest handling and losses during marketing of papaya (*Carica papaya* L) *Postharvest Biol. Technol.* **11**: 165–79.

Sarananda, K.H., Balasuriya, S.T. and Ganeshalingam, K., 2004. Quality of papaya variety *Rathna* as affected by postharvest handling. *Trop. Agric. Res. Ext.* **7:** 72-78.

Sawant, S.G. and Gawai, D.U., 2011. Effect of fungal infections on nutritional value of papaya fruit. *Curr. Bot.*, **2:** 43-44.

Singh, Priyanka, Kumar Sanjay and Maji, Sutanu, 2012. Effect of ethrel on postharvest changes in papaya (*Carica papaya* L.) fruits. *Hort Flora Res. Spectrum*, **1**: July-Sept.

Sivakumar, D., Sultanbawa, Y., Ranasingh N. and Wijesundera, R.L.C., 2005. Effect of the combined application of chitosan and carbonate salts on the incidence of anthracnose and on the quality of papaya during storage. *J. Hortic. Sci. Biotechnol.* **80**: 447-52.

Sivakumar, D., and M.M. Wall. 2013. Papaya fruit quality management during the postharvest supply chain. *Food Rev. Int.* **29**: 24–48.

Thompson, A.K. and G.R. Lee. 1971. Factors affecting the storage behaviour of papaya fruit. *J. Hortic. Sci.* **46**: 511–16.

Yahia, E. 1998. Modified and controlled atmospheres for tropical fruits. *Hortic. Rev.* **22:** 123–83.

Websites

https: //the-biggest.net/other/which-country-is-the-largest-producer-of-papaya.html

https: //en.wikipedia.org/wiki/List_of_countries_by_papaya_production

Source: Annual Report 2019–20

http: //dare.nic.in/sites/default/files/AR-2019-20-English-Final.pdf

https: //eands.dacnet.nic.in/PDF/Pocket%20 2020-20Final%20web%20file.pdf

Chapter 22

Pineapple

Bidyut C. Deka

Vice-Chancellor,

Soumitra Goswami

Assistant Professor,

Assam Agricultural University, Jorhat – 785 013, Assam, India

E-mail: bidyutdeka@yahoo.com

1.0 Introduction

Pineapple (*Ananas comosus* L. Merr.) is believed to have originated in Southern Brazil and Paraguay and was spread by the American Indians to other parts of South and Central America. Pineapple grows and yields best in areas with warm and relatively uniform climates year-round; current production, however, remains restricted to the tropical regions of the world. The nutrient content of pineapple fruits varies within varieties and across maturity. Fresh pineapple is an excellent source of vitamin C (47.8 g/100g) and consumption of 100g of fresh pineapple meets 80 per cent of recommended daily dose of this vitamin. Pineapple fruits are rich in proteolytic enzyme bromelain that digests food by breaking down protein. Bromelain also has anti-inflammatory, anti-clotting and anti-cancer properties. It is derived from the stem and juice of the fruits (Gautam et al, 2010). It has been extensively used in the food industry for meat tenderization, baking processes, prevention of browning of apple juice (Azura et al, 2012), chill proofing beer; to increase the solubility of gelatin for drinking and in the leather-tanning process. Bromelain is also used as active ingredient to provide gentle peeling effects in cosmetic industries (Khan and Abourashed, 2010). The fruit is normally used as food whereas the stems are waste by-product and thus, it is a very cheap source for the production of bromelain (Tochi *et al.,* 2008). The highest protein contents and proteolytic activity were obtained from the extracts of the crowns while the lowest values were recorded from the stem of both cultivars. Studies revealed that regular consumption of pineapple helps fight against arthritis, indigestion and worm infestation. Moreover, this fruit is rich in B-complex group of vitamins like folates, thiamin, pyridoxine, riboflavin and minerals like manganese (0.93 mg/100 g) and potassium. Manganese is an important co-factor for the enzyme superoxide dismutase, which is a very powerful free radical scavenger.

2.0 Status of Cultivation

In India, pineapple is primarily cultivated in the north-eastern states *viz.* Assam, Manipur, Arunachal Pradesh, Tripura, Meghalaya and Nagaland. Other states, where it is cultivated widely are Kerala, Karnataka, West Bengal and Bihar. The total production of pineapple during 2020-21(final estimate) is 1798 thousand MT covering an area of 1.056 lakh ha with an average national productivity of 17.04 t/ha. Karnataka recorded the highest productivity of 57.36 t/ha followed by West Bengal (29.90 t/ha), Bihar (29.17 t/ha) and Assam (18.97 t/ha). As far as the production is concerned, 19.71 per cent of total produce of India comes from West Bengal, followed by Assam (17.52 per cent) and Karnatraka (9.15 per cent).

Table 1: Area, Production and Productivity of Pineapple in India in 2020-21 (final estimate)

State	Production (000 MT)	Area ('000,ha)	Productivity (t/ha)
Andhra Pradesh	67.18	4.20	16.00
Arunachal Pradesh	22.87	3.07	7.45
Assam	315.14	16.61	18.97
Bihar	113.76	3.90	29.17
Karnataka	164.61	2.87	57.36
Kerala	112.15	11.02	10.18
Manipur	134.82	12.12	11.12
Meghalaya	138.81	12.21	11.37
Nagaland	114.77	8.93	12.85
Tripura	144.60	9.92	14.58
West Bengal	354.64	11.86	29.90
Others	115.36	8.87	13.01
Total	**1798.71**	**105.58**	**17.04**

Source: Department of Agriculture Cooperation and Farmers Welfare, Government of India.

3.0 Major Commercial Varieties

The most popular commercial pineapple variety in India is Giant Kew. Other important verities are Queen, Kew, Mauritius, Charlotte, Rothchild, Jaldhup, Desi, Lakhat, *etc*. Qualitatively, Queen is the outstanding table variety. Internationally, however, numerous pineapple cultivars are grown which are grouped in four main classes: 'Smooth Cayenne', 'Red Spanish', 'Queen', and 'Abacaxi', despite much variation in the types within each class. There is always a very good demand of pineapple in the internal markets. It is in high demand from the processing industry as well. Of the four main commercial cultivars grown in India, Kew and Giant Kew is better suited for canning, while Queen and Mauritius cultivars are good for juice production.

4.0Nature and Causes of Postharvest Losses

The postharvest loss in pineapple is reported to be 5-10 per cent. The loss is about 50-60 per cent when ripe pineapples are transported in bulk without providing cushioning materials. Moreover, there is a loss of 34-35 per cent during transportation due to infection by the fungus *Thielaviopsis paradoxa* and other microbes during 4-5 days of road transportation. The loss was 10-15 per cent due to *T. paradoxa* in Karnataka.

Postharvest loss of pineapple in Assam is found to be 9.25 per cent of total production when the fruits are marketed within the state. However, the loss is higher when the pineapples are marketed in distant markets like Delhi. The loss at farmer's level accounts for 4.22 per cent while it is 5.03 per cent at middlemen level. Out of total postharvest losses at the farmer's level, the highest loss is during assembling (1.73 per cent) followed by losses in the market (1.22 per cent). Out of the losses at middlemen level the losses in the local market are the highest (1.65 per cent) followed by those in transportation (1.24 per cent) for a distance of 100-150 km.

In a study in Manipur, Ningombam *et al.* (2019) reported 32.12 per cent loss during various post-harvest stages in pineapple (Table 2).

Table 2: Overall Postharvest Losses of the Pineapple from Farm Level up to the Consumer

Sl.No.	Stages	Quantity loss (tones)	Losses (tones)	Losses (per cent)
1.	Farm level	370	25.4	6.82
2.	Wholesale market including transportation	338.5	10.46	3.09
3.	Retailer	5.95	3.25	5.44
4.	Storage	158.5	89.6	5.64
5.	Processing	2,50,000	8,000	4.31
6.	Consumers	0.07	0.052	6.82
	Total			**32.12**

4.1 Postharvest Diseases and Disorders

Microorganisms attack fresh pineapple easily and spread quickly, because the produce does not have much of a natural defense mechanism and has plenty of nutrients and moisture to support microbial growth. Postharvest decay control is becoming a more difficult task, because use of agro-chemicals now a days is discouraged is increasing as consumer concern for food safety increases.

The symptoms of the postharvest diseases of pineapple fruit may develop externally or internally. Diseases of pineapple fruit can be classified as:

1. Pre-flower infections by *Penicillium funiculosum* that begins in the floret before the flower opens and cause inter fruitlet corking, leathery pocket, and fruitlet core rot,
2. Flower infections beginning after the opening of flower and caused by bacterial complexes including pink and marbling disease,
3. Wound infections caused most frequently by *Ceratocystis paradoxa* that begin in wounds caused by harvesting, and
4. Physiological disorders, the most common of which is internal browning or black heart caused by chilling injury.

4.1.1 Black Rot

Black rot, also called Thielaviopsis fruit rot, water blister, soft rot, or water rot, caused by *Thielaviopsis paradoxa*, is the most common and well-known postharvest disease of the pineapple fruit. Infection takes place, usually, through wounds in the base of the fruit resulting from harvesting. The pathogen gains entrance also through natural growth cracks and wounds on the fruit surface generated during picking or packing. After penetration through the cut stem end the pathogen grows rapidly upwards and slowly into the flesh resulting, initially, in a dark yellow cone shaped soft rot. The severity of the problem is dependent on the degree of bruising during harvesting and packing, the level of inoculums on the fruit and storage temperature during transportation and marketing. As the disease develops, the flesh becomes watery and the juice exudes, resulting in an empty shell showing only dark fibres inside. Black rot does not occur in the field unless the fruit is overripe or injured.

The development of black rot in pineapple fruits is very fast at 25 °C and very slow at 8 °C. Due to that, severe rot problems may occur when untreated fruits are removed from refrigerated conditions and held at room temperature in retail market. For decades, benzoic acid, salicylanilide and sodium o-phenyl phenate tetrahydrate applied to the cut stem end or to the pineapple fruit surface, after harvest, to control the black rot. Of late, Benomyl, 0.12 to 0.24 per cent and Triadimefon, 0.05 per cent a.i., have found to be very effective postharvest treatments against this disease. In order to be effective, the treatment must be carried out within six to twelve hours after harvest.

4.1.2 Pink Disease

The pink disease is a bacterial infection of the pineapple fruit, characterized by the development of a brown colour in the flesh resulting from the infection of *Acetobacter aceti, Erwinia herbicola* and *Glucono bacteroxydans.* The bacteria are brought by insects to the open flowers and under favourable conditions they infect the ovary and reach the flesh of the fruit.

Pink disease can be controlled by harvesting the fruit before completely ripe, by promoting fruit production in season unfavorable to the disease development, or by dusting the fruits with Dissulfoton during the open flower stage. Smooth Cayenne variety is relatively resistant to this disease.

4.1.3 Yeasty Fermentation

It is caused by Saccharomyces spp. and is usually associated with overripe *fruit.* The yeast enters the fruit through wounds. Yeasts ferment sugar solution, producing alcohol and releasing carbon dioxide. The first symptom is a bubbling exudation of gas and juice through the crack or injury where infection occurred. The shell then turns brown and leathery and, as the juice escapes, the fruit becomes spongy. Internally, the decaying flesh turns bright yellow and develops large gas cavities.

Prompt cooling and maintenance of optimum temperature and relative humidity throughout the handling operations will reduce the intensity of the disease.

4.1.4 Fruitlet Core Rot (Green Eye)

This disease is present in most of the pineapple growing areas of the world. Fruitlet core rot is incited by either *Penicillium funiculosum* Thom., *Fusarium moniliforme* Sheld. or *F. subglutinans*. The pineapple fruit mite (*Steneotarsone musananas* Tyron), has been reported to be associated with fruitlet core rot and it seems to enhance the pathogenesis of the fungi that cause the disease.

A light to dark brown soft rot of the fruitlet core area characterizes the symptoms of fruitlet core rot. The disease extends from the base of the style, nectary ducts or placenta towards the core. Sometimes the infected tissue becomes corky, thus originating a dry rot. Low-acid cultivars being grown commercially are most susceptible (Rohrbach and Schmitt, 1994). No symptoms are observed in infected fruits of the cultivar Smooth Cayenne.

On the other hand, infected fruitlets of 'Perolera' and 'Queen' show a yellow to brownish colour and are borne in a lower level of plant in comparison with the surrounding healthy ones.

The incidence of fruitlet core rot is influenced by rainfall which helps in inoculum build up, and by the mean daily temperature ranging between 16 to 20 °C which favors mite population. The pathogen enters the developing inflorescence through unopened buds; it may also enter through the stylar canals and nectary ducts, as well as through insect injuries and natural growth cracks. This disease is believed to be caused by mite damage in the field allowing entry of the fungi. Preharvest spraying regimes are required to control the mite population.

The degrees of symptom depends on the time of infection, the pathogen or mixture of pathogens present, the cultivar and the environmental conditions. A global analysis of the data showed that average ascorbic acid content in fruits at harvest is negatively linked with the percentage of infected fruits. The nutritional status of the plants, especially low levels of calcium and magnesium, and/or high levels of nitrogen, are also a significant promoting factors for the development of the disease. It is concluded that Fruitlet Core Rot in Martinique is promoted by a combination of factors that, determine the sensitivity of fruits. Climatic conditions before harvest, acid content, and nutritional status are the main factors.

4.1.5 Sunburn

Sunburn is a problem resulting from excessive heating of one side of the pineapple fruit by the rays of the sun. This anomaly occurs usually when the pineapple plant leans or falls over exposing one side of the fruit directly to the rays of the sun, thus resulting in long exposure of the peel, and the flesh beneath it, to fairly high temperatures.

Affected fruits show a yellowish colour in the area exposed to the sun. As the anomaly progresses the yellow area changes to brown and the flesh beneath it becomes more translucent. The peel may crack between the fruitlets, and the flesh dries out and becomes spongy. The occurrence of sunburn during the early stages of fruit development reduces the growth rate in the area exposed to the sun thus resulting in an asymmetric fruit.

Control of pineapple sunburn depends entirely on cultural procedures. They may include promoting fruit production during periods unfavourable to sunburn development, providing adequate amount of fertilizers to reduce the possibility of the plant to lean over exposing one side of the developing fruit to the direct action of the rays of the sun; and providing mechanical protection of the developing fruits. In fact, mechanical protection seems to be the most important measure to control pineapple sunburn. Several methods for mechanical protection of the pineapple fruit are available. Materials such as straw, paper and polyethylene can be put on the fruits to protect them against the direct action of the sun. Another method consists of tying the pineapple leaves over the developing fruits.

4.1.6 Endogenous Brown Rot

The most economically important physiological disease of the pineapple fruit, endogenous brown rot, also known as internal browning or black heart, is usually a postharvest disorder even though it may occur under field conditions. The first symptom of endogenous brown rot is the appearance of a small translucent spot in the flesh of the fruit, close to the central cylinder, approximately 2 cm below the base of the crown. The affected area enlarges, and only a narrow band of the flesh between the peel and the rotted tissue remains healthy. Despite the internal alterations, no symptoms are observed externally and the infected fruit looks healthy. Fruit maturity significantly affects internal browning; immature and over mature fruits develop less browning than mature fruits. Ripe fruits are less susceptible than unripe or partially ripe fruits. The incidence is found in certain varieties and production areas and is generally enhanced during long-term storage.

Endogenous brown rot occurs when the fruit is removed from the low temperature (7°C) storage and shipping conditions to room temperature during commercialization. Besides the alternation effect of the temperature, fruits with low levels of ascorbic acid are more susceptible to endogenous brown rot. Adequate levels of ascorbic acid prevent the oxidation of phenolic compounds thus avoiding disease development. The incidence of endogenous brown rot can also be reduced by using cultural practices such as supplying high levels of potassium, known to increase fruit acidity. The field application of fused magnesium phosphate (FMP) @ 500 kg/ha to the soil reduces internal browning. Non-enzymatic browning and phenylalanine ammonia lyase (PAL) are not directly related to internal browning development. PPO and Peroxidae are correlated with the appearance of internal browning. Pre harvest application of potassium @ 16 g of K_2O per plant reduces IB in fruit of both maturities *i.e.*, colour break and half ripe stage. Partial to complete control of internal browning have been achieved by waxing, polyethylene bagging, heat treatments, controlled atmospheres. Application of 1-methylcyclopropene (1-MCP) can substantially reduce the internal browning of pineapple.

4.1.7 Flesh Translucency

Flesh translucency is a physiological disorder of pineapple (*Ananas comosus* L. Merr.) fruit whose flesh shows water soaking and has low porosity. It is also called as porosity that increases fruit sensitivity to mechanical injury. The intercellular free spaces in translucent fruit flesh are filled with liquid which reduces the porosity and light scattering ability of the tissue, making it translucent. Translucent fruit has higher relative water content and greater specific gravity than non-translucent fruit and this has been recommended as a nondestructive detection method of pineapple fruit translucency. Highly translucent fruit has flat and off-flavour. Translucent fruit are fragile and prone to mechanical damage during harvest and postharvest handling and therefore, difficult to ship fresh. Translucent fruits are comparatively more susceptible to diseases and pre-harvest sunburn. The incidence of fruit peduncle leakage is correlated with translucency severity that keeps the broken peduncle end wet and may lead to an unsightly dark bluish gray Penicillium growth postharvest.

The actual cause(s) of pineapple fruit translucency are unknown. It is observed that both crown weight and fruit translucency at harvest are correlated to the monthly average air temperature of 2 - 3 months before harvest, and there is a negative correlation between crown weight and translucency severity. These results suggested that higher temperature at the early stage of fruit development produced less translucent fruit. The crown may affect translucency occurrence by completing with the fruit for assimilates or shading the fruit to deduce fruit temperature during the daytime. Alternatively, both crown weight and fruit translucency could be affected by air temperature. Crown removal either at an early or late stage of fruit development do not have any significant effect on fruit weight or translucency occurrence.

5.0 Postharvest Management

5.1 Pre-Harvest Treatments

Pre harvest spray with 2-(chlorophenoxy) propionic acid at 40 days before harvest helps in inhibiting blackheart in pineapple during storage. Application of calcium fertilizer at 125 kg/ha as a basal dressing is the most effective treatment for the control of internal browning in 'Mauritius' pineapple. Calcium chloride (2 per cent) was found to be the most effective pre-harvest treatment for extending shelf life of 'Kew' pineapple. The edible coating with mucilaginous press-sap of *Neolitsea cassia* leaves significantly reduces weight loss during storage as well as internal browning of pineapple. Spraying whole fruits with K_2SO_4 solution (50 ml, 5 per cent w/v) plus 10 ppm alpha-NAA or Ethrel 4 weeks before harvesting significantly increases the potassium ion concentration of the fruits and decreases internal browning symptoms by 60-70 per cent in the core tissue and flesh of the fruits. Spraying @ 5 per cent K_2SO_4 and ethepon @ 100 mg/l, 4 weeks before harvest increases soluble solids, potassium content, flavour and reduces the harvest period by 1 week (Nanayakkara *et al*, 2005).

5.2 Maturity Standards

The maturity of fruits depends on the cultivars and season of growing (Figures 1 and 2). In addition, there are different optimum maturity levels for the same cultivars depending on season, intended use and storage life (Table 3). Harvesting too early results in off-flavour or lacking. In flavour, poorly coloured, small fruit, and subject to bitter pit and storage scald. Fruit maturity in pineapple is generally evaluated on the extent of fruit 'eye' flatness and skin colour. The fruit is ready to harvest when 30-50 per cent of eyes of from the base turn yellow. Pineapples are non-climacteric fruits and should be harvested when they are ready to eat.

Table 3: Maturity Indices of 'Kew' Pineapple

Indices	*Summer Season*	*Winter Season*
Days from flowering to harvesting	140-145	145-150
Specific gravity	> 0.93	> 0.93
TSS (° Brix)	> 12 ° Brix	> 12 ° Brix
TSS: Acid ratio	> 23	> 20
Sugar: acid ratio	> 20	14-15

Source: Deka *et al.*, 2008.

Pineapples are harvested when the colour changes from green to greenish yellow and smooth surface around the eyes and the tips of the bract dry up. 'Kew' pineapple should be harvested from141-150 days after flowering when the colour changes from green to yellow in 1/4-1/2 portion of the fruit from its base with a minimum TSS of 12° Brix and TSS: acid ratio of 14-20 irrespective of harvesting seasons. Delayed harvesting reduces the sensory quality of fruits (Table 4).

Figure 1. ¼th Ripe Stage

Figure 2. Half-Ripe Stage.

Table 4: Physico-chemical Changes of Pineapple Fruit at different Stages of Ripening during Summer Season

Different Stages of Ripening	*TSS (°Brix)*	*TSS : Acid Ratio*	*Ascorbic Acid (mg/100g)*	*Sensory Score*
Mature Green	14.53	18.87	13.70	4.23
1/8 Colour Development	15.66	23.37	13.88	5.51
1/4 Colour Development	16.24	25.38	14.40	6.70
1/2 Colour Development	17.40	34.12	14.92	7.50
2/3 Colour Development	16.51	30.57	15.32	8.26
Full Colour Development	15.22	26.70	14.50	7.45

Source: Deka *et al.*, 2008.

For distant market, harvesting should be done when the fruit is 10-20 per cent yellow stage or even 100 per cent green but mature stage just before striking colour changes begins (Adikaram and Abayasekara, 2012). The external colour and aroma of pineapple are important parameters to judge the quality of fruit by the consumer and minimum TSS of 12-14°B and acidity 0.9-1.3 per cent ensure the a minimum degree of consumer acceptance. The maturity index can also be estimated by relating it to the time after flowering or mid-flowering, but the number of days may vary from place to place. The period from flower induction to ripening varied from 140–221 days (Thompson, 2003).

Fleshy fruits undergo a natural stage of development known as ripening. This occurs when the fruit has ceased growing and is said to be mature. There are two characteristic types of fruit ripening that show different patterns of respiration. Based on this, fruits can be divided into two groups: (1) those that are incapable of continuing their ripening process once removed from the plant, known as non-climacteric fruit and (2) those that can be harvested at the mature stage and allowed to ripen off the plant, known as climacteric fruits. Pineapple belongs to group 1. Fruit of the Group 1 category, produce very small quantities of ethylene and do not respond to ethylene treatment except in terms of de-greening (removal of chlorophyll); these should be picked when fully-ripe, if good flavour quality is to be ensured. Fruit of the Group 2 category on the other hand, produce comparably larger quantities of ethylene which is associated with their ripening, and undergo more rapid and uniform ripening upon exposure to ethylene. However, Ethephon @ 25 ppm treatments responds well for uniform colour development in pineapple.

Pineapple starts ripening from the base of the fruits with an indication of colour change from green to yellow. The half-yellow stage is regarded as ripe and at this point the fruit is near the maximum mass if still on the plant. Fruit development and composition changes during growth have been reviewed. The most marked changes in flesh composition occur in the three to seven weeks prior to and at the half yellow shell colour stage. Chlorophyll levels show changes during maturation and decline very fast at ripening stages. Pineapple fruit produces around 22 ml CO_2/kg/h at 23 °C, with no dramatic respiratory or bio-chemical change during ripening. Ethylene production increases during ripening but has no pronounced peak.

5.3 Harvesting

Fruits are mainly harvested during July-August while a small crop is harvested during December to March. The sensory quality of the fruits harvested during summer season is better than the fruits harvested during winter seasons. If harvested early, they may develop woody pulp on ripening while delayed harvesting reduces the shelf life of the fruits.

Harvesting of pineapple for local markets is to be done at the full maturity stage and for distant markets at 75-80 per cent maturity stage. About 165 days are required to attain ripe stage in Kew pineapple. Defined pre-maturation stage at 0-120 days after flowering, early maturation after 120-150 days and late maturation after 150- 160 days in Kew pineapple.

Pineapple fruits are harvested by pickers who select fruit by size, colour, or both, and twist it from the stalk. Fruits after harvesting are packed either in the field or at a central packing shed. In small operations the harvested fruits may be placed in sacks and baskets that are carried to the end of the row for pickup Large-scale production however avoids hand carrying by using a harvesting aid consisting of an endless belt extending across a number of rows. Fruits picked by hand are placed on the belt and are carried to the machine where they accumulate in a bin or truck gondola. The machines and belts, usually mounted on a truck chassis, move slowly across the field at a speed determined by the pickers.

5.4 Postharvest Treatments

Quality of pineapple fruits cannot be improved after harvest, but can be maintained for certain period of time during storage with appropriate postharvest treatments. It is important to harvest pineapple at the proper stage and size and at peak quality as it is a non-climacteric fruit that does not continue ripening after harvest. Immature or over mature fruits may not last for long in storage as compared to the fruits that picked at proper maturity. Harvesting should be completed during the coolest time of the day, which is usually in the early morning, and produce should be kept shaded in the field and handled gently. Crops destined for storage should be as free as possible from skin breaks, bruises, spots, rots, decay, and other deterioration. Bruises and other mechanical damage not only affect appearance, but provide entrance to decay organisms as well.

5.5 Sorting and Grading

Fruits are graded on the basis of certain recognized parameters/characters. In countries like Philippines, Ivory Coast, Costa Rica, South Africa and Thailand grading is generally done on weight basis. The fruits however, need to be fully mature, firm and free from any defects with flat eyes and a minimum TSS of 12 per cent. The grading should be uniform not only in size but also in stage of ripeness. Usually, the following grades are made based on the weight of the fruits (Table 5).

Table 5: Size Codes of Pineapple

Size Code	*Size Specification*
Super quality size	1.75 kg
Above A	1.5 to 1.75kg
B	1.1 to 1.5 kg
C	0.8 to 1.1 kg
D	less than 0.8 kg
Baby	about 0.55 kg

Maximum preference is given for medium size fruits of 1.1-1.5 kg. Fruits less than 900 g are not preferred in the market. Super quality fruits weighing 1.75-2.0 kg are preferred and premium price is paid in the market. Crown size is also a crucial grade component, with a minimum size, and ratio of crown to fruit length (0.33-1.5), for the higher grades.

5.6 Pre-cooling

Good temperature management is the most effective way to reduce postharvest losses and preserve the quality of pineapple. Products harvested from hot fields often carry field heat and have high rates of respiration. Currently used pre-cooling methods include room cooling, forced-air cooling, water cooling, vacuum cooling and package icing. Pineapples are very sensitive to temperature. The fruits should be cooled within 8-10 hours of harvesting at a temperature of 7-10 °C. Pre-cooling is very important for successful sea shipping. There can be chilling injury if the fruits are pre-cooled at a temperature lower than 7 °C.

5.7 Waxing

Pineapples are commercially treated with a fungicide in a dip or spray application to control postharvest fruit rot, caused by the fungus (Robert et al, 2003). A food grade wax, usually containing polyethylene/paraffin or carnauba/paraffin based, may also be applied to the fruit with the fungicide. The major advantage of waxing is the reduction of the internal browning symptoms of chilling injury and it also reduces postharvest water loss and improves fruit appearance (Paull and Lobo, 2012).

Food grade waxes are commonly applied to reduce water loss and to improve appearance. It also provides protection against decay organisms. Waxing may be done after grading and fungicides may be added to the wax. Application of wax and postharvest fungicides must be indicated on each container where the refrigerated storage facilities are not available Protective skin coating with wax is one of the methods for increasing the storage life of fresh fruits (Table 6).

Table 6: Shelf Life of Pineapple (1/2 colour development stage) as Affected by different Treatments

Treatments	*Shelf Life (Days)*	*Remarks*
Control (No treatment)	12.22	All the fruits were having glossy and attractive look except those treated with untreated control.
Stafresh (1:2)	16.66	
Stayfresh (1:2)	17.00	
Semperfresh (1:4)	16.00	
Carbendazim (1000 ppm)	15.50	
Stafresh (1:2) + Carbendazim (1000 ppm)	17.00	
Stayfresh (1:2) + Carbendazim (1000 ppm)	17.00	
Semperfresh (1:4) + Carbendazim (1000 ppm)	16.33	

The selected waxes give a reduction in internal browning symptoms of chilling injury, reduce water loss, assure a more even application of the incorporated fungicide and improve fruit appearance. The shelf life of the fruits could be increased up to 17 days with minimum physiological loss in weight (PLW) of 6.34 per cent by treating the fruits with stay fresh as against 12 days in control with higher PLW of 13.01 per cent. The shelf life of the Stayfresh treated fruits could be further increased to 20 days by wrapping them with heat shrinkable polyethylene (25 µ). Ethephon can be successfully incorporated into the fruit wax. The ability to apply ethephon with the wax is useful, as waxing delays fruit degreening. Exposing waxed fruits to red light during storage at 8 °C accelerates the rate of shell degreening upon removal to ambient temperature. There is no worldwide uniformity in acceptance of wax components, so the restriction imposed by the importing countries need to be considered.

Wax treatment with Sta-Fresh 2952 at high concentrations (65 g L^{-1} and 90 g L^{-1}) delayed fruit color change and the fruits were found to be firmer than untreated ones (Li *et al.*, 2018). Waxing reduces rate of respiration and ethylene production resulting in delayed senescence (Wills *et al.*, 1998).

5.8 Packaging

Packaging of pineapple is not like other fruits, because of presence of crowns. Pineapples are packed along with crowns for prolonged shelf life and to avoid infection at the crown attachment point. Two methods are used *i.e.*, Horizontal packing and vertical packing keeping the crown on the topside. The fruit packed in bamboo or wooden boxes is more prone to physical injuries due to sharp edge of bamboo and nails present in wooden boxes. To minimize the physiological weight loss, fruit should be packed in corrugated fibre boxes (Singh, 2009). Pineapples will only fit into boxes in small numbers of around four or six packed vertically or horizontally with a crown of leaves cut to fit snugly into the box (Thompson, 2003).

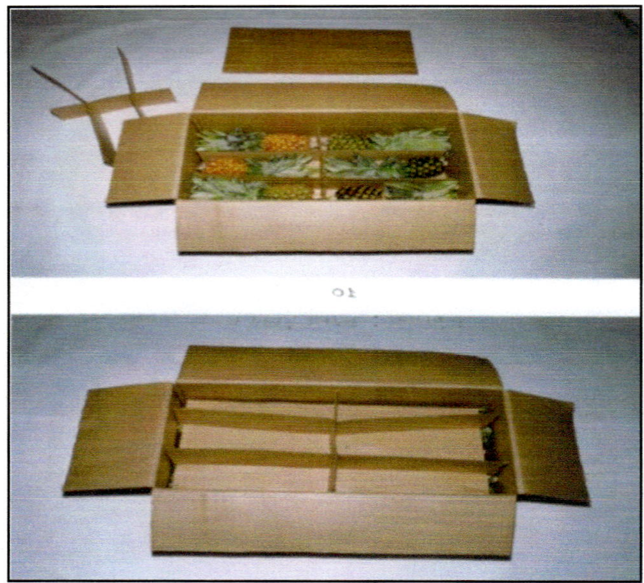

Figure 3: Packaging Methods

In India, packaging of pineapple is either not done at all or is inadequate, which is a major cause of postharvest loss. Straw packing in bamboo baskets for dispatching pineapple from Assam to Kolkata reduces the damage by 10-15 per cent during transport. A low cost packaging method was developed to bring down the heavy losses in the transportation of pineapple by road from North Eastern States to Calcutta during 1976 by CFTRI, Mysore. In this system a bamboo structure is created on the trucks to facilitate a tier system of loading the fruits.

Fruits are also packed into cartons of two different sizes: a large carton (18 kg) containing eight to ten fruits for surface shipment and a smaller container (9 kg) with five to six fruits for air-shipping. Tourist packs of two to four fruits are also prepared. Absorbent pads are used at the bottom of the carton and between layers if fruits are horizontally placed within the carton. There are two methods of packing, *viz.*, horizontal and vertical. The preferred method of packing is to place the fruit vertically on the base, and then to place dividers between the fruits to prevent rubbing and movement. In horizontal packing, higher number of fruits per unit of space can be accommodated. Twelve numbers of fruits can be accommodated in a CFB box having the dimension of 685 x 380 x 258 ((L x W x H in mm) in horizontal manner with a net weight of 15 to 18 kg.

5.8.1 Modified Atmosphere Packaging

Modified atmosphere packaging of fruits is a new dimension of extending the shelf life of pineapple. A modified atmosphere can be defined as one that is created by altering the normal composition of air (78 per cent nitrogen, 21 per cent oxygen, 0.03 per cent carbon dioxide and traces of noble gases) to provide an optimum atmosphere for increasing the storage length and quality of food/produce. This can be achieved by using controlled atmosphere storage (CAS) and/or active or passive modified atmosphere packaging (MAP). The recent developments in the packaging material particularly in the advent of selectively permeable plastic films have widened the scope of modified atmosphere packaging systems for the storage of the fruits.

Pineapple cv. Smooth Cayenne fruits of uniform size (1.5 kg average weight) harvested in the intermediate stage of maturation and packed in polyethylene without perforations reduces the black heart incidence significantly compared to unpacked fruits. Pineapple fruits (cv. Mauritius) sealed in polypropylene bags and stored at 8 or 10 °C show increased shelf life over control. The weight loss and intensity of internal browning are minimum in the fruits packed in polypropylene bags as compared to the fruits with no packaging. 'Kew' pineapple wrapped with heat shrinkable (25 µ) polyethylene can be stored for more than 20 days at ambient condition as against 10-12 days for the fruits without packing.

5.9. Transportation

Transportation is the most important factor in the marketing of fresh pineapple. Ideally, transport would take produce from the grower to the consumer, as happens in many developing countries. In more complex marketing systems (those serving towns, cities or distant countries) the cost of transport contributes significantly to the price paid by the consumer.

Fully ripe fruits are not suitable for transportation to distant markets. 1/4thcolour development stage from the base of the fruits is suitable for long distant transportation, while for local market, the fruits may be harvested at 1/2-2/3rd colour development stage. Immature fruits are not shipped since they do not develop good flavour, have low Brix and are more prone to chilling injury.

Transportation of pineapple is usually done either by rail or truck in most part of India. However, the major mode of transport of pineapple in Assam and some other parts of the country is bicycle (40.87 per cent) followed by hand load (39.05 per cent). The cost of transportation is found to be highest (Rs. 11.29 per km per 100 fruits) for Auto/Van. The middlemen however; transport the produce by truck.

Pineapple should be properly packaged and loaded on a suitable vehicle. No damage of pineapple fruits packed in CFB was observed after transportation by train to Delhi from Jorhat, Assam. Likewise, no fruit damage during and after transportation was recorded when pineapple packed in CFB boxes was transported for 1000 km by truck. Where sea-shipment is to be used, the fruit should be harvested on the day prior to shipment (Table 7).

5.10 Storage

Green fruit should be stored at 10 °C in 85 to 95 per cent relative humidity, and under these conditions, should have a storage life of two to three weeks. This will be dependent on the sugar content and the agronomic conditions during production, in addition to the handling and storage procedures. Where exports are made by air with fruit harvested at more advanced stages of maturity, pre-export storage can be used and the suitable storage temperature decreases to 7.5 °C in 85 to 95 per cent relative humidity.

Maturity at harvest is the most important determinant of storage-life and final fruit quality. Overripe fruits are likely to become soft and mealy with insipid flavour soon after harvest and can not be stored for long. Fruits for longer shelf life should be harvested either at 1/4 or 1/2 colour development stages. Low-cost storage structure based on evaporative cooling developed at Assam Agricultural University, Jorhat is suitable for short term storage of pineapple. Pineapple treated with stay fresh can be kept for 15 days in this low-cost storage structure.

In tropical areas, partially ripe, healthy and un-bruised pineapple could be stored for almost 20 days when refrigerated at 10-13 °C with RH 85-90 per cent. Fruits harvested at 1/4th colour development stage could be stored for 15-20 days at 12 °C with 80-85 per cent RH. The longer the fruit is kept out of refrigerated storage, the more likely, the fruit will be of poor eating quality. Exposure of pineapples to temperatures below 7°C results in chilling injury. Postharvest storage of pineapple at low temperature (8°C and waxing) can alter external and internal fruit characteristics. Shell colour increases slightly during storage at 8 °C, with little change in overall shell appearance.

Controlled atmospheres storage have shown minimal effectiveness in extending the shelf life of pineapple. Some beneficial effect is gained from controlled atmosphere (4 per cent oxygen) treatments in reducing the development of chilling injury symptoms. Controlled atmosphere storage (3-5 per cent O_2 and 5-8 per cent CO_2) delays senescence and reduces respiration. Exposure to O_2 levels below 2 per cent and/or CO_2 levels above 10 per cent should be avoided because of the potential for development of off- flavours.

In developing country like India; only the intermediaries (middlemen) practices storage. Pineapple fruits are stored mainly in open floor, gunny bags, bamboo basket or in thatch in Northeast India. Average cost of storage of pineapple is found to be Rs. 13.54 per 100 fruits. Wastage and building hiring charges are the major cost components of storage, which accounts for 39.65 and 25.78 per cent of the total storage cost. Both the farmers and intermediaries do not practice packaging in NE India.

6.0 Domestic Marketing and Export

In almost all the countries, most of the fruits and vegetables including pineapple are produced in such interior areas that the farmers cannot sell their marketable surplus in an easily accessible market. Such a situation is very common in the hilly areas of India where there is no well-developed road; transport and storage facilities. Due to underdeveloped marketing system, there has been great problem in channelizing the marketable surplus of pineapple in the developing countries like India. This situation compels the growers to sell their produce in throwaway prices. The total marketable surplus of pineapple in Assam is about 88.75 per cent; out of which only 8.76 per cent is marketed outside the state

Thailand, Philippines, Brazil, China, India, Costa Rica, Nigeria, Kenya, Mexico, and Indonesia are the top 10 pineapple producing countries of the world. Among these producers, Costa Rica, Indonesia, Philippines, Thailand, and Kenya send a significant proportion of their production towards international markets. About 60 percent of the world's fresh pineapple exports come from Costa Rica, the Ivory Coast, and the Philippines. Nearly 80 percent of world's canned pineapple exports come from Thailand, Philippines and Indonesia. Thailand and the Philippines also dominate world pineapple juice exports, accounting for more than half of total volume. Export of pineapple from India is primarily

Table 7. Condition of Boxes and Fruits after Transportation (1000km) by Truck

Type of Boxes	*No. of Boxes*	*No. of Boxes Damaged*	*Number of Fruits/ Box*	*Total Fruits*	*Total Damaged Fruits*	*Total Decayed Fruits*
CFB I	18	Nil	12.0	216	Nil	10.0
CFB II	18	Nil	12.0	216	Nil	09.0
CFB III	18	Nil	8.00	144	Nil	18.0
Wodden	18	2.0	12.0	216	3.0	23.0
Bamboo	18	9.0	12.0	216	10.0	21.0
Plastic crates	18	Nil	12.0	216	Nil	10.0

limited to SAARC and Gulf countries. During the year 2012-13 major quantity of pineapple (21.9 per cent) from India was exported to Qatar. The other importers of Indian pineapples were Nepal (17.8 per cent), Maldives (15.0 per cent), Bahrain (12.3 per cent), Saudi Arabia (11.2 per cent). Global sales for pineapples exports by country totaled US$2.2 billion in 2019. Among continents, Latin America excluding Mexico but including the Caribbean generated the highest dollar worth in international sales of pineapples during 2019 with shipments valued at $1.1 billion or roughly half (49.8 per cent) of the global total. In second place were European exporters at 22.2 per cent while 19.1 per cent of worldwide shipments of the sweet fruit originated from Asia. Smaller percentages came from North America (5.3 per cent), Africa (3.7 per cent) then Oceania (0.003 per cent) notably New Zealand and Australia.

The countries that exported the highest dollar value worth of pineapples during 2019.

1. Costa Rica: US$981 million (43.8 per cent of total exported pineapples)
2. Philippines: $323.8 million (14.5 per cent)
3. Netherlands: $250.6 million (11.2 per cent)
4. Belgium: $97.1 million (4.3 per cent)
5. United States: $93.3 million (4.2 per cent)
6. Taiwan: $65.4 million (2.9 per cent)
7. Honduras: $45.2 million (2 per cent)
8. Ecuador: $42.7 million (1.9 per cent)
9. Spain: $36.1 million (1.6 per cent)
10. Ghana: $31.6 million (1.4 per cent)
11. Ivory Coast: $30.8 million (1.4 per cent)
12. Portugal: $29.4 million (1.3 per cent)
13. Mexico: $24 million (1.1 per cent)
14. United Kingdom: $22.9 million (1 per cent)
15. Germany: $19.2 million (0.9 per cent)

Source: https://www.worldstopexports.com/pineapples-exports-by-country/

In terms of value, these countries shipped 93.5 per cent of global pineapples export in 2019. Among the top exporters, the fastest-growing pineapples exporters since 2015 were: Portugal (up 698 per cent), Philippines (up 207.3 per cent), Ivory Coast (up 154.2 per cent) and Taiwan (up 153.1 per cent). During the period, India earned $3,855,000 by exporting pineapples. Although there are about 99 countries and territories, which actively import Indian pineapple, the major countries importing pineapple from India are the Netherlands, China, UK, USA and Saudi Arabia.

6.1 Marketing Channels

Marketing channels are the paths through which the commodities are moved from the initial point of production until it reaches the hands of the ultimate consumer. The length of the channels varies from commodity to commodity, depending upon the quantity to be moved, the extent of consumer demand and degree of regional specialization in production. The flow chart presented below shows the flow of pineapple from the producer to the consumer level through different intermediaries. The following six major marketing channels of pineapple have been identified in Assam.

i) Producer → Retailers → Consumers

ii) Producer → Wholesalers → Retailers → Consumers

iii) Producer → Wholesalers via Commission agents → Retailers → consumers

iv) Producer → Wholesalers preharvest Contractors → Retailers → consumers

v) Producer → Preharvest Contractors → Wholesalers → Retailers → consumers

vi) Producer → Consumers

Producer→ Wholesalers (via Commission agents) →Retailers→ Consumers is the most efficient channel with an efficiency of 3.60. The most effective channel is Producer→ Wholesalers→ Retailers → Consumers through which about 60.0 per cent of the commodity are transacted. States that the highest cost is incurred by the channel V (Rs 233.92) while the lowest in the channel I (Rs 141.95).

6.2 Marketing Problems

In the existing marketing set up pineapple fruits are usually difficult to market than to produce in the country like India. Major marketing channels of pineapple in Assam are summarized in Figure 4. There are ready markets available daily or weekly for grain and livestock in almost all areas of the country; but very few similar markets for fruits, *etc*.

A grower has little chance of selling their produce directly to the wholesaler. Wholesale buyers have strict and specific guidelines in respect of product volume, quality and grade for each of the commodities.

A commodity may sell well and bring high prices for a long period, but demand and prices may drop drastically overnight. Supply and demand have a tremendous effect on marketability and prices of the produce like pineapple. There are no support prices for fruits and vegetables including pineapple to help the grower when market demand and prices of the produce drop. Daily market prices are now available for some commodities including pineapple in the internet. In such a situation, a producer gets to know what, when, and how much the buyer can use. The buyer gets to know what is available and when, as he has to keep the shelves stocked.

6.3 Recent Development in Marketing

Development of the marketing system is essential to

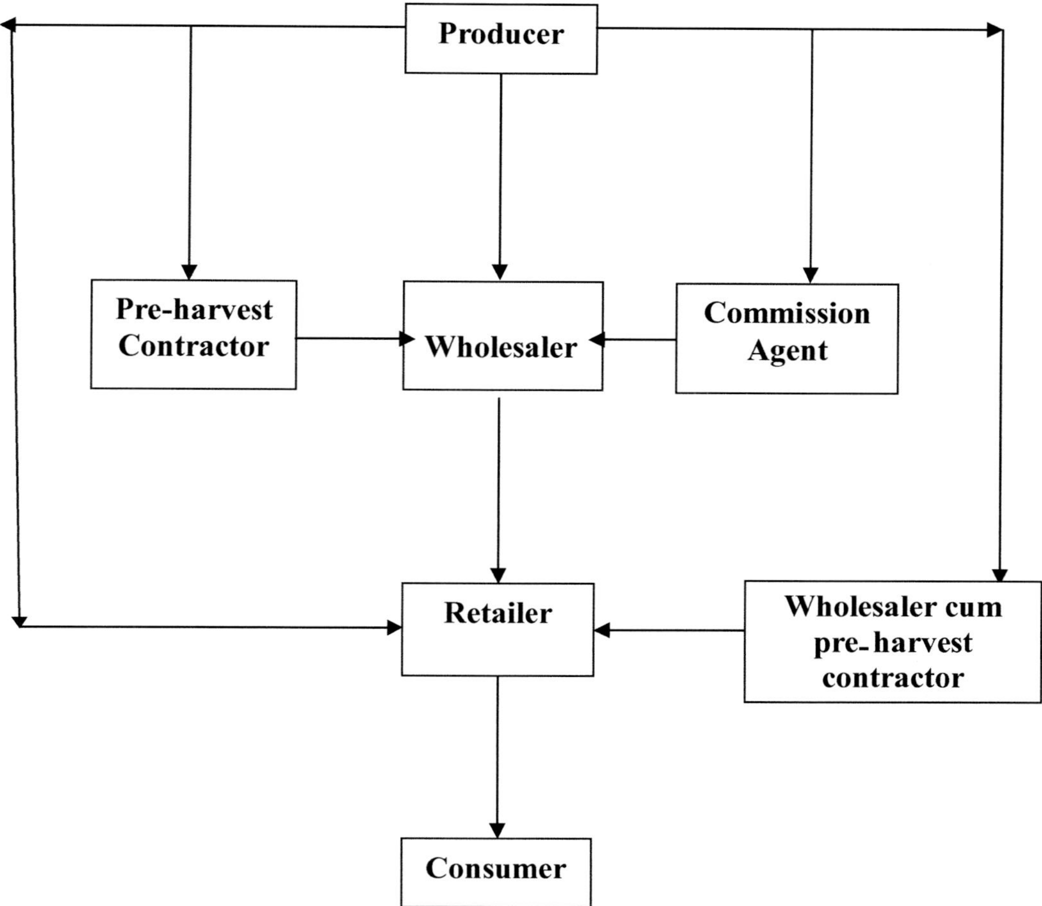

Figure 4. Major Marketing Channels of Pineapple in Assam (Hazarika *et al.*, 2003)

control the activities of the intermediaries and to help the regulation of trade. National Diary Development Board in Delhi, HOPCOM in Bangalore, FRESH in Hydarabad, Maxworth Harvest in Madras, Reliance Fresh, Field Fresh *etc.* have come up in a big way for retail marketing of the fresh and graded fruits including pineapple benefiting the grower as well as the consumer. APEDA is assisting the upcoming entrepreneurs for enhancing the exports by developing the market intelligence reports and survey of the potential export markets by providing infrastructural base at the commercial growing areas, growing centres and airports; by developing packages of the international standards and by providing airfreight subsidy.

7.0 Conclusion

Pineapple is one of the important fruit that needs appropriate postharvest management and handling although this fruit is not classified under highly perishable group. Major postharvest problem begins at harvesting of fruits at wrong maturity stage followed by rough and repeated postharvest handling during marketing. Slight bruising of the outer skin results in progressive development of microbial spoilage leading to huge postharvest losses. Further, packaging plays a crucial role in postharvest management of pineapple. In view of the huge potentiality of export of fresh pineapple from India, the role of postharvest management is very crucial for quality assurance and market expansion. Therefore, integrated postharvest management considering all the aspects of production to postharvest management *i.e.* end to end solution is the need of the hour.

8.0 Important References

Abdullah, H., Rohaya, M.A. and Zaipun, M.Z.1985. Effect of modified atmosphere on black heart development and ascorbic acid contents in ' Mauritius' pineapple (*Ananas comosus* cv. 'Mauritius') during storage at low temperature. *Asean Food J.* **1**: 15-18.

Adikaram, N., and Abayasekara, C. 2012. Pineapple, In: *Crop post-harvest: science and technology : perishables*, Debbie Rees, Graham Farrell And John Orchard (eds.), blackwell publishing ltd., West Sussex, UK, pp. 143-58.

Akamine, E.K. 1963. Fresh pineapple storage. *Hawaii Farm Sci.* **12**: 1-4.

Akamine, E.K. and Goo, T. 1997. Controlled atmosphere storage of fresh pineapples (*Ananas comosus* (L.) Merr. 'Smooth Cayenne'). Hawaii Agricultural Experiment Station Research Bulletin 152, 8 p.

Akamine, E.K., Goo, T., Steepy, T., Greidanus, T. and Iwaoka, N. 1975. Control of endogenous brown spot of pineapple in postharvest handling. *J. Am. Soc. Hortic. Sci.* **100**: 60-65.

Bowden, R.P. 1969. Further studies on ripeness in pineapple. *Food Tech. Australia,* **21**: 160-63.

Chadha, K.L., Mehta, K.R., Lodh, S.B. and Selvaraj, Y.1972. Biochemical changes associated with growth and development of pineapple var. Kew. I. Changes in physico-chemical constituents. *Indian J. Hortic.* **29**: 54-57.

Ching-Cheng Chen and Robert E. Paull 2001. Fruit temperature and crown removal on the occurrence of pineapple fruit translucency. *Sci. Hortic.* **88:** 85-95.

Dass, H. C., Randhawa G. S. and Negi S. P. 1975. Flowering in pineapple as influenced by ethephon and its combinations with urea and calcium carbonate. *Sci. Hortic.* **3:** 231-38.

Deka, Bidyut C., Saikia, J and S. Sharma 2005. Standardization of maturity indices of 'Kew' Pineapple. *Acta Hortic.* **682**, Vth International Postharvest Symposium, pp. 2215-2220.

Deka, Bidyut C., Saikia, J, Pal, R.K and Saikia, A. 2008. Physico-chemical changes of pineapple at different stages of maturity. *Indian J. Hortic.* **64:** 464-66.

Goswami, S., Borah, A., Saikia, A., Baishya, S. and Deka, Bidyut C. 2008. Low cost storage structures for shelf life extension of horticultural produces. *J. Food Sci. Technol.* **45:** 70-74.

Hine, R.B. 1976. Epidemiology of pink disease of pineapple fruit. *Phytopath.,* **66**: 323-27.

Hossain, M. and Bepary, R. H. 2015. Post Harvest Handling Of Pineapples: A Key Role To Minimize The Post Harvest Loss. *Int. J. Recent Sci. Res.* **6**: 6069-75.

Li, X., Zhu, X., Wang, H., Lin, X., Lin, H. and Chen, W. 2018. Postharvest application of wax controls pineapple fruit ripening and improves fruit quality. *Postharvest Biol. Technol.* **136**: 99–110.

Lodh, S.B., Selvaraj, Y., Chadha, K.L. and Mehta, K.R. 1972. Biochemical changes associated with growth and development of pineapple of pineapple fruit var. Kew. II. Changes in carbohydrates and mineral constituents. *Indian J. Hortic.* **29**: 287-91.

Mallikarjunaradhya, A.V., Bhatt, A.V., Crown, J.K., Vijayendra Rao A.R., Ramanna, K.V.R. and Narasimhan, P. 1979. Control of fungal stem end rot during the transport of pineapple. *J. Food Sci.* **16**: 232-34.

Nanayakkara, K.P.G.A., Herath, H.M.W., Senanayake, Y.D.A., 2005. Influence of ethephon (2-choloroethyl phosphonic acid) plus K_2SO_4 on the process of ripening and internal browning in pineapple (*Ananas comosus* L.) Merr. cv. Mauritius) under cold storage. *Acta Hortic.* **666**: 315-19.

Nanayakkara-KPGA, Herath-HMW and Senanayake-Y.D.A. 1997. Effects of some pre-harvest treatments of potassium and hormones on fruit core tissue deterioration and internal browning of pineapple under cold storage. *Trop. Agric. Res.* **9**: 90-95.

Ningombam, S., Noel, A.S. and Singh J. 2019. Post-harvest losses of pineapple at various stages of handling from the farm level up to the consumer in Manipur. *Int. J. Agric. Sci.* **11** : 9235-37.

Paull, R.E. and Rohrbach, K.G. 1985. Symptom development of chilling injury in pineapple fruit. *J. Am. Soc. Hortic. Sci.* **110**: 100- 05.

Paull, R.E. and Reyes, M.E. 1996. Pre-harvest weather conditions and pineapple fruit translucency. *Sci. Hortic.* **66**, 59–67.

Rohrbach, K.G. and Paull, R.E. 1982. Incidence and severity of chilling induced browning of waxed 'Smooth Cayenne' pineapple. *J. Amer. Soc. Hortic. Sci.* **107**: 453-57.

Rohrbach, K. G.and D. J. Phillips 1990. Postharvest diseases of pineapple. *Acta Hortic.* **269**: 503-08.

Rohrbach, K. and D.P. Schmitt. 1994. Pineapple. In: *Compendium of Tropical Fruit Diseases.* R.C. Ploetz, G. A. Zentmyer, W.T. Nishijima, K.G. Rohrbach, and H.D. Ohr. (eds) . *Am. Phytopath. Soc.* St. Paul MN, 45–55.

Selvarajah, S., Bauchot, A.D. and John, P. 2001. Internal browning in cold stored pineapples is suppressed by a postharvest application of 1-methylcyclopropene. *Postharvest Biol. Technol.* **23:** 167-70.

Selvarajah, S. and Herath, H.M.W.1997. Effect of an edible coating on some quality and physico-chemical parameters of pineapple during cold storage. *Trop. Agric. Res.* **9**: 77-89.

Selvarajah, S., Herath, H.M.W. and Bandara, D.C. 1999. Basal application of fused magnesium phosphate (FMP) on the incidence of postharvest internal browning of Mauritius pineapple. *Trop. Agric. Res.* **11**: 432-37.

Selvarajah, S; Herath, H.M.W., Bandara, D.C. andWicramasinghe, I.P. 2000. Physiological and enzymatic changes during postharvest low temperature storage periods and the manifestation of symptoms associated with Internal Browning in pineapples. *J. Food Sci. Technol.* **37:** 571-76.

Selvarajah-S; Herath, H.M.W, and Liyanage-K.K.1997. Effect of modified atmospheric storage on internal browning development in pineapple. *Trop. Agric. Res.* **9**: 388-91.

Shantha Krishnamurthy and Rao, D.V. S. 1998. Research and development in modified atmosphere packaging (MAP) and controlled atmosphere storage of fruits and vegetables. *and Food World,* **25:** 11-13.

Simth, L. G. 1983. Causes and development of black heart in pineapples. *Trop. Agric.* **60**: 31-35.

Smith, L.G. 1988b. Indices of physiological maturity and eating quality in Smooth Cayenne pineapples. 2. Indices of eating quality. *Queensland J. Agric. Animal Sci.* **45**: 219-28.

Soares, A.G., Trugo, L.C., Botrel, N. and Souza, L. F.S. 2005. Reduction of internal browning of pineapple (*Ananas comosus* L) by pre harvest soil application of potassium. *Postharvest Biol. Technol.* **35:** 201-07.

Sridhar, T.S. 1975. Black rot of pineapple-A new record from South India. *Curr. Sci.* **44**: 869.

Swete Kelly, D.E. andBragshaw, J. 1993. Effect of fruit handling and fruit coatings on black heart (internal brown spot) and other aspects of fresh pineapple quality. *Acta Hortic.* **334**: 305-16.

Talukdar, S and Ahmed F. 2002. Effect of pre-harvest chemical treatments on shelf life of pineapple fruit. *J. Food Sci. Technol.* **39:** 53-54.

Thompson, A.K., 2003. Fruit and vegetables: harvesting, handling, and storage. 2nd Ed. Blackwell Publishing: 306-08.

Wills, R., MeGlasson, B., Graham, D. and Yoyee, D. 1998. *Poatharvest- An Introduction to the Physiology and Handling of Fruit, Vegetables and Ornamentals.* Hyde Park Press, Adelaide, South Australia.

Zhou, Y., Daler. J.M. and Underhill, S.J.R. 2003. Enzymes associated with blackheart development in pineapple fruit. *Food Chem.* **80:** 565-72.

Chapter 23

Pomegranate

R.K. Pal
Former Director
Nilesh Gaikwad
Senior Scientist
K. Dhinesh Babu
Principal Scientist
ICAR-National Research Centre on Pomegranate,
NH-65, Solapur-Pune Highway, Kegaon, Solapur – 413 255, Maharashtra, India
E-mail: rkrishnapal@gmail.com, E-mail: nileshgaikwad98@gmail.com, E-mail: ckdhinesh@gmail.com

1.0 Introduction

Pomegranate (*Punica granatum* L.) is an important fruit crop in arid and semiarid regions of the world (Figure 1). It is a highly remunerative crop for replacing subsistence farming and alleviating poverty. The cultivation of pomegranate by mankind as a fruit crop dates back to antiquity. The usage of pomegranate is deeply embedded in human history with references in many ancient cultures to its use in food and medicine. It is associated with ancient civilizations of the Middle East. It grows naturally wild in western Himalayan regions of India (Himachal Pradesh, Jammu & Kashmir and Uttarkhand) and Pakistan. Pomegranate is one of the oldest known edible fruits.

Figure 1. Pomegranate in Flowering Stage

Exquisite fruit quality, enriched nutritional values, enormous medicinal values, and huge demand in domestic and international market, besides extended shelf life enables it to emerge as an eminent fruit crop of recent times (Pal *et al.*, 2014). As pomegranate is classified as difficult-to-eat fruit, minimal processing through the separation of edible parts (arils) for making convenient food is of great importance (Pal and Babu, 2014). Delayed harvesting often leads to internal breakdown of arils in pomegranate. At the same time, early harvesting would lead to poor quality. Being a non-climacteric fruit, pomegranate has to ripen on tree and must be picked at the right stage of maturity to ensure the best eating quality. The fruits when fully ripe have a waxy shining surface of red or yellow peel colour, depending on the cultivar (Rao and Narayana, 2014).

It has enormous nutritional value due to its rich content of anthocyanins, vitamins and minerals. It has immense medicinal value due to its healing property over numerous disorders and diseases that affect mankind, right from abdominal disorders (dysentery, diarrhoea), dermal defects (skin blemishes), cancer, cardiovascular diseases, etc. Owing to its immense nutritional and medicinal values, pomegranate is regarded as 'a treasure of health and nutrition' (Babu *et al.*, 2021).

It is believed to have originated in Iran (Primary centre of Origin) and is widely prevalent in Afghanistan, Pakistan and India, are the secondary centres of origin. From there, pomegranate cultivation spread to India, China and Mediterranean countries *viz.*, Turkey, Egypt, Tunisia, Morocco and Spain. Spanish missionaries brought the pomegranate to the America in 1500s.

Pomegranate arils provide 12 per cent of the Daily Value (DV) for vitamin C and 16 per cent DV for vitamin-K per 100 g serving, and contain polyphenols, such as ellagitannins and flavonoids. These are excellent sources of dietary fibre which is entirely contained in the edible seeds. People who choose to discard the seeds forfeit the nutritional benefits conveyed by the seed fibre and micronutrients. Pomegranate seed oil contains Conjugated Linoleic Acids (CLA) *viz.* punicic acid (65.3 per cent), palmitic acid (4.8 per cent), stearic acid (2.3 per cent), oleic acid (6.3 per cent) and linoleic acid (6.6 per cent). These CLA's are reported to be beneficial in delaying senescence or ageing of developing animal cells. Hence, pomegranate seed oil has a great potential in developing several pharmaceutical and cosmetic products. Further, CLA is also reported to reduce atherosclerosis thereby helping in preventing various heart ailments.

The antioxidants and hypoglycemic potential of six commercially grown pomegranate cultivars in India characterized by the author and his team revealed that the overall antioxidant composition of pomegranate especially total phenolics and total anthocyanin content could provide an attractive strategy to manage postprandial hyperglycemia. The most abundant polyphenols in pomegranate juice are the hydrolyzable tannins called ellagitannins also known as punicalgins. The red colour of the juice can be attributed to anthocyanins, such as delphinidin, cyanidin and pelargonidin glucosides. Generally, there is an increase in juice pigmentation during fruit ripening. Preliminary laboratory research and clinical trials have revealed that juice of the pomegranate might be effective in reducing heart disease risk factors, including reduction in LDL oxidation. In a limited study of hypertensive patients, consumption of pomegranate juice for two weeks was shown to reduce systolic blood pressure. It is also reported to inhibit viral infections.

Pomegranate fruit has a good consumer preference for its attractive, juicy, sweet acidic and refreshing arils. There is a growing demand for good quality fruits both for fresh use and processing into juice, syrup, squash, wine besides anardana, an acidulant.

For nutritional content, pomegranate is ranked better than the grape, mango, orange and papaya. It is one of the richest sources of Riboflavin. Rind of the fruit, bark of stem and root of pomegranate contain more than 28 per cent gallotannic acid and yellow dye that could be extensively used in tanning as natural bio-dye.

The nutritional parameters with mean and standard errors thereof for the three different cultivars of pomegranate are depicted in Table 1.

2.0 Status of Cultivation

Pomegranate grows under a variety of climatic conditions *viz.*, tropical and subtropical, temperate regions and in hilly areas up to 1800 m altitude. It is a deciduous shrub in the temperate region but evergreen or partially deciduous under tropical and subtropical conditions depending upon the variety/genotypes. It produces a better crop in the semi-arid climate where cold winter and hot summer prevails. The tree requires a warm and dry climate during fruit development and ripening.

Although it can withstand frost, it is injured by temperatures below -12 °C. Further, the crop can withstand heat, drought and moisture deficit. The threshold limit of pomegranate ranges from 44 °C to -12 °C. It can be grown successfully in arid regions where life saving irrigation facilities are available. It is therefore, highly remunerative crop to replace subsistence farming. Pomegranate is cultivated extensively in Mediterranean countries (Tunisia, Turkey, Israel, Egypt, Spain and Morocco), Iran, Afghanistan, India and to some extent in the U.S. (California), China, Japan and Russia.

In India, Maharashtra state is considered as the 'pomegranate basket of India' contributing more than 70 per cent of the total area.

Table 1. Nutritional Parameters of Three Pomegranate Cultivars

Cultivar	*TSS (°B)*	*TA (g CA/100 ml PJ)*	*pH*	*Maturity Index (TSS/TA)*	*Flavonoids (mg CE/100 ml PJ)*	*Anthocyanins (mg C3gE/100 ml PJ)*	*Free gallic acid (mg GAE/100 ml PJ)*
Arakta	15.10±0.12	0.28±0.01	3.43±0.02	53.41±1.91	46.38±1.36	16.53±0.57	47.03±3.80
Bhagwa	14.43±0.27	0.32±0.01	3.32±0.03	45.69±1.72	72.28±1.48	26.93±0.46	86.97±3.25
Ruby	14.07±0.28	0.22±0.01	3.64±0.04	63.13±2.73	57.25±3.11	18.14±1.31	54.59±2.40

Source: Fawole *et al.*, 2012.

This is followed by Gujarat (13.04 per cent) Karnataka (11.10 per cent), Madhya Pradesh (4.13 per cent) and Andhra Pradesh (4.04 per cent). In Maharashtra, the production of pomegranate is mainly concentrated in western Maharashtra and Marathwada region. Nashik, Solapur, Ahmednagar and Pune have a larger area under cultivation and production. Besides, pomegranate production is sizeable in Sangli, Dhule, Aurangabad, Satara, Osmanabad and Latur districts of Maharashtra state. The area under pomegranate cultivation is also expanding in Tamil Nadu. Largely, it is cultivated in marginal lands with fertigation system in India. Pomegranate has emerged as a commercial crop with better adaptability mainly due to technological manipulation (bahar treatment for regulation of flowering and fruiting) adopted by farmers. The state-wise area, production and productivity of pomegranate is given in Table 2a. During 2012-13, in India, Maharashtra produced 54.8 per cent of pomegranates followed by Karnataka (20.2 per cent), Gujarat (10.6 per cent), Andhra Pradesh (8.3 per cent), Madhya Pradesh (3.1 per cent), Tamil Nadu (1.6 per cent), Rajasthan (0.7 per cent) and others (0.7 per cent). The area under pomegranate during 2011-12 was 1.13 lakh ha with a total production of 7.45 lakh tonnes and productivity of 6.6 t/ha in India. The state-wise area, production and productivity of pomegranate are given in Table 2a. During 2017-18, in India, Maharashtra produced 62.90 per cent of pomegranates followed by Gujarat (16.23 per cent), Karnataka (9.42 per cent), Andhra Pradesh (4.88 per cent), Madhya Pradesh (4.01 per cent), Tamil Nadu (0.46 per cent), Rajasthan (0.93 per cent). The area under pomegranate during 2017-18 was 2.34 lakh ha with a total production of 28.45 lakh tonnes and productivity of 12.16 t/ha in India.

Table 2a. State-wise Area, Production and Productivity of Pomegranate in India 2017-18

Sl. No.	*State*	*Area ('000ha)*	*Production ('000t)*	*Productivity (t/ha)*
1	Andhra Pradesh	9.47	139.02	14.68
2	Chhattisgarh	0.76	4.34	5.71
3	Gujarat	30.51	461.75	15.13
4	Himachal Pradesh	2.77	3.15	1.14
5	Jharkhand	0.02	0.04	2.00
6	Karnataka	25.97	268.23	10.33
7	Kerala	0.10	0.33	3.33
8	Madhya Pradesh	9.68	114.27	11.80
9	Maharashtra	147.91	1789.46	12.10
10	Nagaland	0.09	0.50	5.56
11	Odisha	0.26	0.92	3.54
12	Rajasthan	4.44	26.58	5.99
13	Tamil Nadu	0.56	13.06	23.32
14	Telangana	1.49	19.87	13.34
15	Others	0.01	0.01	1.00
	Total	233.93	2844.52	12.16

Source: Horticultural Statistics at a Glance 2018.

3.0 Major Commercial Varieties

There are several types of edible pomegranate besides ornamental types with double flowers which are largely sterile (Figure 2). The edible type pomegranate which are grown commercially around the world include Wonderful in California and Israel; Mollar and Tendral in Spain; Schahvar and Robab in Iran; Hicaznar and Beynar in Turkey as well as Zehri and Gabsi in Tunisia.

Table 2b. Area, Production and Productivity of Pomegranate for Major Producing States/Districts (2016-17)

Sl. No.	*State*	*Districts*	*Area ('000ha)*	*Production ('000t)*	*Productivity (t/ha)*
1	Maharashtra	Nashik	38.80	628.11	16.19
	Maharashtra	Solapur	25.50	229.50	9.00
	Maharashtra	Ahmednagar	21.50	215.00	10.00
	Maharashtra	Pune	13.50	136.00	10.07
2	Gujarat	Kachchh	7.97	125.69	15.77
	Gujarat	Banaskantha	6.78	108.82	16.05
	Gujarat	Mahesana	1.04	16.44	15.81
3	Karnataka	Chitradurga	7.96	101.71	12.78
	Karnataka	Bellary	3.71	42.07	11.34
	Karnataka	Tumkur	3.33	35.40	10.63
	Karnataka	Bijapur	2.81	28.07	9.99
	Karnataka	Bagalkot	2.30	25.03	10.88
4	Andhra Pradesh	Anantapur	6.25	93.69	14.99
	Andhra Pradesh	Chittoor	0.80	12.00	15.00

Source: Horticultural Statistics at a Glance 2018.

There are more than 25 pomegranate varieties grown in different parts of India. Among the commercial cultivars, 'Bhagawa' occupies a major area under cultivation. Some of the other varieties cultivated on a commercial scale are listed in Table 3.

Table 3. Commercial Varieties of Pomegranate Grown in India

State	*Varieties*
Andhra Pradesh	Bhagawa, Ganesh
Gujarat	Bhagawa, Phule Bhagawa Super, Dholka, Ganesh
Himachal Pradesh	Bhagawa, Kandhari
Jammu & Kashmir	Bhagawa
Karnataka	Bhagawa, Phule Bhagawa Super, Ganesh
Madhya Pradesh	Bhagawa, Phule Bhagawa Super,
Maharashtra	Bhagawa, Phule Bhagawa Super, Ganesh, Mridula, Arakta, Solapur Lal
Rajasthan	Jalore Seedless, Bhagawa, Jodhpur Red
Telangana	Bhagawa, Ganesh
Tamil Nadu	Bhagawa, Ganesh

Figure 2. Major Commercial Varieties

4.0 Nature and Causes of Postharvest Losses

The juvenile period of pomegranate is relatively low compared to other tree fruit crops. Though the tree starts bearing fruits from the second year onward, it's not advisable to allow flowering and fruit set taking into consideration of tree vigour and orchard longevity. Hence, the flowers should be nibbed off till completion of 2 years. Flowering and fruit set have to be allowed only from 3rd year onwards to build up the vigour of the plants in orchard. The economic yield of the orchard could be obtained up to an age of 15-20 years and under good management condition up to 25 years. However, the new concept of high-density plantation/sequential planting keeping in view of commercial life of 4-5 years using tissue cultured planting materials either in open or under protected cultivation may come in practice in the forthcoming years. This may reduce the build-up of inoculum levels of various diseases particularly bacterial blight pathogen in the orchard for sustainable productivity.

Although the pomegranate industry has witnessed steady growth on area and production over the last one and a half decade, the growers are still facing problems like internal breakdown, sun-burning, fruit cracking, wilt and bacterial blight. The on-farm losses caused by these maladies are very high (12-18 per cent) depending on cropping season. By adopting fruit bagging, farmers can minimize these losses and could get an additional price of Rs. 8-10/kg over unbagged fruits.

Some of the problems that reduce the marketability of pomegranate fruits harvested during summer include sunscald, uneven colour development of fruits and aril paleness. This could be overcome by bagging the developing fruits 3 months before harvest. Butter paper, white colour cellulose based bags or zipped polyethylene or non-woven polypropylene bags (15cm x15cm) are suitable for bagging. Bagging is done on a bright sunny day when there is no moisture deposition around the fruits. White colour bags reflect sunlight and reduce the temperature around developing fruits. The optimal temperature helps in biosynthesis of anthocyanin pigments and prevents their denaturation during hot months. About 10-12 per cent higher anthocyanin is available in the arils of bagged fruits, compared to unbagged fruits. Recently many pomegranate growers are using UV stabilized non-woven fabric for whole tree covering to get rid of fruit sucking moth and other abiotic stresses.

Foliar application of plant biostimulant (cytozyme @ 4 ml/l) at 45 days after bud burst and again at10 days after fruit set is reported to improve fruit size (316.3 g/fruit) and reduce fruit cracking of pomegranate cv. Kandhari Kabuli.

5.0 Postharvest Management

India is the only country in the world, where pomegranate fruits are available throughout the year. In India, Maharashtra and Tamil Nadu have pomegranate fruits right from January – December (Pal and Babu, 2017). Although pomegranate is not a highly perishable fruit, yet appropriate postharvest management plays a crucial role in the retention of postharvest quality for fetching better returns of this high value produce. At any given time, approximately 50-60 per cent of harvested pomegranate meets the export requirement. Roughly 35-40 per cent of the produce is sold in the domestic market. However, approximately 10-15 per cent of the fruits do not fetch good market value due to poor colour, small size, deformity, cracking, sunburn *etc.* Hence, there is a need to adopt an integrated approach for postharvest handling, sorting and grading coupled with value addition that would fetch a better return.

5.1 Maturity Indices

Maturity indices may vary and are variety dependent which include external skin colour, juice colour, acidity, soluble solids content, *etc.* Pomegranate juice sugars consist of mainly glucose or fructose. At the time of maturity, fruits attain characteristic reddish colour for cv. Bhagawa and yellowish green for cv. Ganesh with waxy shining surface. The calyx at the anterior end of the fruit may get curved inside and becomes hard and dry at maturity. Fruit shape becomes compact and can be easily scratched with finger nails. On tapping, fruits give metallic sound. Arils become intense red in cv. Bhagawa and pink in Ganesh. Similarly, the colour of juice becomes red in Bhagawa, Arakta, Mridula, Ruby and pink in Ganesh. The TSS of the juice ranges from 13-16.5 ° Brix with titrable acidity below 0.8 per cent and brix acidity ratio in the range of 25-40.

The pomegranate fruits do not ripen after harvest; hence fruits are harvested only after attaining maturity in the tree. Total soluble solids/acid ratio is one of the most reliable maturity indices. The fruits should be harvested before they become overripe and crack (split) open, especially under rainy conditions. Early harvesting in order to avoid cracking (caused by sudden fluctuation in soil moisture, prolonged dry spell followed by a heavy rainfall) is one of the causes of poor quality of pomegranate in India.

From the pattern of CO_2 and ethylene production rates, pomegranate fruits are considered to be non-climacteric fruits. Hence, maturation and ripening should take place on the plant itself before harvesting to get quality fruits. The mature pomegranate may remain on the tree for a long period and picking time can be selected according to commercial needs. As pomegranate is a non-climacteric fruit, it has a relatively low respiratory rate. However, it was reported that the respiratory rate gets doubled with the rise in storage temperature from 5 °C to 10 °C.

The fruits become ready for harvest in 130-180 days after appearance of blossom *i.e.* Days After Anthesis (DAA). The maturity period for commercial cultivars during Mrig bahar is 135 DAA in Arakta and Mridula, 145-150 DAA in Jalore Seedless and Ganesh and 175-180 days in Ruby and Bhagawa and 180-185 days in Bhagawa.

5.2 Harvesting

Pomegranate tends to split open the fully ripe fruits as it is the natural way of seed release and dispersal. The practice of early harvest in order to mitigate fruit cracking at maturity stage is one of the reasons for poor quality of pomegranate in India. Late harvesting of fruits, on the other hand, results in a physiological disorder of Internal breakdown/aril browning. Hence, harvesting should be done at an appropriate time.

Shivashankar *et al.* (2017) considered aril browning to be a physiological phenomenon. They hypothesized that during the early stage of fruit development, seeds of variable sink strengths evolve due to differences in the levels of growth hormones like GA_3 and IAA. The formation of strong and weak sinks among arils of fruit leads to a reduction in the moisture content of weak sinks (seeds), resulting in either a reduction or complete loss of seed viability in them. The concurrent loss of moisture from the pulpy arils of such seeds results in the loss of membrane integrity followed by activation of PPO enzyme and the oxidation of phenols leading to AB. The present study has thus provided evidence for the direct role of seed viability in the development of aril browning in pomegranate. Fruits are generally clipped when they are ripe. Depending on the *Bahar* treatments, pomegranate becomes to harvest in installments. Usually, the harvest commences in December-January and extends up to June-July. Fruits ripening from April to June often get sunscald and may also crack if rains intervene or irrigation is irregular. In India, the flowering and harvesting period of pomegranate varies for different bahars (Table 4).

Table 4. Flowering and Harvesting Period of Pomegranate in India

Name of Bahar	*Period of Flowering*	*Period of Harvesting*
Ambe bahar	Jan-Feb	July-Aug
Mrig bahar	June-July	Dec-Jan
Hasth bahar	Sep-Oct	Mar-Apr

The following points have to be kept in mind while carrying out the harvesting operation. The mature fruits are harvested during the early morning or late evening using clippers. The scissors/clippers for picking should be sanitized using 1 per cent sodium hypochlorite. The fruits should be clipped at the base of stalk without inflicting any injury to the rind and keeping the crown intact. Ladders/tripod stands may be used to climb up and harvest the fruits from upright branches. Perforated plastic crates lined with 2mm foam at the base may be used to keep the harvested fruit in the field. The crates should be stacked under shade.

5.3 Sorting and Grading

Pre-cooling of the fruits at farm level is done preferably in cool chamber or in the shade. Sorting and grading of pomegranate fruits is important to obtain reasonable price for the produce from export and local market. Cracked, split, diseased and borer infested fruits should be separated and removed. The fruits are graded on the basis of their weight, size and external (rind) colour. The different weight grades for pomegranate are as follows (Table 5).

Table 5. Grades in Export of Pomegranate

Grades	*Fruit Characters*
Super Size	Fruits are attractive, very large, dark red in colour, free from blemish and weighing >750g
King Size	Fruits are attractive, large, free from blemish and weighing >500g and <750g
Queen Size	Fruits are attractive, large, free from blemish and weighing >400g and <500g
Prince	Fruits are attractive, large, free from blemish and weighing >300g and <400g
12 A	Fruits having no spots and weighing >250g and <300g
12 B	Fruits weighing >250 and <300g with spots

The fruits of 12A grade are generally preferred in south and north India whereas marketing of fruits of 12B grades are done in eastern part of India. However, due to establishment of more and more super markets and malls demand of high quality fruits in the Indian metropolitan cities is also increasing. After grading, the fruits are sometimes treated with ethyl oleate to provide luster. Surface coating with lac based wax emulsion is also beneficial.

5.4 Packaging

Packaging Varies for Export and Domestic Market

Usually for packing pomegranates for export purposes, a cardboard corrugated fibre board (CFB) box of 3.5 or 5.0 kg capacity is used (Tables 6 and 7, Figure 3).

Table 6. Packing Box Dimensions for Export

Fruit Weight Capacity	*Dimensions (l x b x h in mm)*
5 kg Box	380 x 270 x 125
3.5 kg Box	400 x 200 x 105

Table 7. Specification Details for CFB Boxes for Packing Pomegranates for Exports

	Ring and Flap Tuck in Type	*RSC (Regular slotted container)*	*Slide Type*
Material of Construction	5 Ply CFB	3 Ply CFB	5 Ply CFB
Bursting strength kg/cm^2	Min 10	Min 10	Min 10
Puncture resistance, inches/tear inch	Min 250	Min 250	Min 250
Compression strength, kgF	Min 350	Min 350	Min 350

Minimum Specification

Material of Construction	*Corrugated Fibre Board*			
External Dimensions (mm)	380X270X125			
*Style of Box***	*RSC-0201*	*Telescopic-0300*	*Telescopic-0306*	*Telescopic-0312*
No. of Plies	3	Lid-3, tray-3		
Type of Flutes	B (Narrow)	Lid B (Narrow), Tray B (Narrow)		
Direction of Flutes	Vertical			
Grammage (g/m^2) Outer to Inner (Indicative)	250/150/150	Lid-250/150/150 Tray-250/150/150		
Burst Factor of Paper (Kraft) (Minimum)	20			
Bursting Strength of Board (Kg/cm^2) (Minimum)	11	Lid-11, Tray-11		
No. of Pieces per Box	Not more than two			
Manufacturer's Joint	By Gluing			
Compression Strength of Box (Kgf) (Minimum)	350			
COBB (30 Minutes)*** (g/m^2) (Maximum)	60			
No. of Ventilation Holes	16			
Diameter (mm)	20			
Two Sides Each	3			
Two Ends Each	1			
Top & Bottom Each	4			
B. Recommended Specifications				
Edge Crush (Kgf) (Minimum)	34			

*Outer ply of white duplex board or bleached kraft; **Or any other suitable style which provides equal strength and performance; ***Outer ply to be laminated or coated for water proofing.

Notes: Dimensions of the box are modular to Euro pallets. However, any other dimensions or box styles to suit importers requirements could also be used. Grammage mentioned is indicative. However, any other grammage which provides equal strength and performance can be used. Diameter and no. of holes should be 20mm and 16 nos. respectively. However, sufficient ventilation depending upon the box style could also be used.

Figure 3. Packing of different Size of Fruits for Export

Figure 4. Pomegranate Packing for Local Markets

For domestic market, pomegranate is packed in CFB boxes having 4-5 kg of fruits, according to their weight (Figure 4). The graded fruits are placed on cushioning material followed by an attractive red colour paper on the boxes. Boxes made up of light wood, bamboo basket are also used for packaging. Dry grass, rice straws, or paper are used as cushioning material at the bottom and top of box or basket. Up to 6 dozen fruits are packed in each basket for transport by trucks.

5.5 Storage

Pomegranate fruits have better keeping qualities compared to other tropical fruits *viz.*, mango, grape and banana. The storage life of pomegranates is comparable to that of apples. Pomegranate fruits are susceptible to moisture loss and need to be stored at low temperature and high humidity. Under ambient conditions, the fruits can be stored up to 3-4 weeks depending upon the season. The safe temperature for cold storage is up to 10 weeks at 5°C and 85 per cent RH. For longer storage fruits should be kept at 10°C and relative humidity of 95 per cent to avoid chilling injury and weight loss (Figure 5). The treatment of packing material with sulphur compounds was found to give 50 per cent more fruit protection. The shelf life of pomegranate can be

Figure 5. Cold Storage of Pomegranate

extended up to 48 days in zero energy cool chamber when pretreated with waxing as against 3-4 weeks under room temperature.

5.5.1 Pre-storage Treatments

As the pomegranate fruits are highly susceptible to water loss resulting in shriveling of rind, the fruits should be pre-cooled to low temperature within 10 hours after harvesting. The shelf life of pomegranate fruit could be extended up to 30 days at room temperature (22.17-24.36 °C and 52-82 per cent RH), 48 days in cool chamber (16.14-18.22 °C and 85-95 RH) and 75 days in cool storage (8 °C and 90-95 per cent RH) when treated with waxol (12 per cent)+ carbendazim (0.1 per cent). The wax treatment coupled with fungicidal dip is considered beneficial in extending the shelf life of pomegranate in cool chamber and cool storage. Fruits of pomegranate cvs. 'Gyulosha Rozovaya', 'Uluchshennyl Kazake' and 'Bala Myursal' dipped in 4 per cent $CaCl_2$ solution for 3 minutes and stored at 2-4 °C and 85-90 per cent RH for 110 days showed low storage loss and retained quality.

Pre-conditioning of pomegranate fruits at moderate temperature (30-40 °C) and high R.H. (90-95 per cent) for a short period of time (1-4 days) is also applied before the conventional refrigerated storage. Pre-treatment at 35 °C and 90-95 per cent R.H. applied for three days prior to a refrigerated storage during 80 days at 5 or 2 °C and 90-95 per cent R.H. have shown to reduce the pitting and husk superficial scald (produced by the enzyme polyphenol oxidase) compared to untreated ones without the pre-conditioning treatment. The effect was more significant at 2 °C for retaining the quality during retail marketing for a period of one week at 15-20 °C and 70-75 per cent R.H. Pre-treatments of fruits at 30 °C and 33 per cent RH, or 20 °C and 47 per cent RH (for 1-4 days) have also been reported to retain good marketability up to 3 months or more during storage, depending on the cultivar.

5.5.2 Low Temperature Storage

The respiration rate of pomegranate cv. Wonderful remains low (less than 8 ml CO_2/kg/h) when stored at 0-10 °C for three months while it increases with increase in temperature. The rate of ethylene production from fruits stored below 10 °C is mostly below 0.1µl/kg/h. Storage at 10 °C is satisfactory, if a postharvest fungicide is used.

5.5.4 Relative Humidity

The control of relative humidity (RH) is critical in the storage of pomegranate fruits. Pomegranate can be best stored at low temperature and high humidity. At low relative humidity, the skin desiccates readily and the rind becomes dark and hard; the fruits become less attractive and have poor marketability. The ideal condition for storage of pomegranate is 5-7 °C and 90-95 per cent RH for storing up to 2-3 months. Fruits stored at 4-5 °C and 80-85 per cent relative humidity did not undergo any shrinkage or spoilage in a few months. A relative humidity of 85-90 per cent is recommended for storing pomegranate cv. Kandhari.

5.5.5 Controlled Atmosphere Storage

Storage under controlled atmospheres (10 per cent O_2 and 5 per cent CO_2; 5 per cent O_2 and 5 per cent CO_2; 5 per cent O_2 and 0 per cent CO_2) have been reported to reduce weight loss, fungal decay and chilling injuries. The storage in controlled atmospheres, however, reduces the vitamin C and sugars contents. Other studies on 'Wonderful' pomegranates stored in air enriched with 10 or 20 per cent CO_2, showed that the colour of the arils were better during refrigerated storage (10 °C, 6 weeks) in regular atmosphere than those stored with 10 per cent CO_2. It was even poor in those fruits stored in 20 per cent CO_2. Storage potential ranges from 3-4 months in air and from 4-6 months in a controlled atmosphere comprising of 5 per cent O_2 + 15 per cent CO_2 + 80 per cent N_2. Storage under controlled atmospheres (5 per cent CO_2, 5 per cent O_2 and 90 per cent N_2) at 5 °C and above 95 per cent relative humidity for two months have been reported to retain the quality attributes of the freshly harvested pomegranate with light to moderate rind scalding. Therefore, CA storage has been found little beneficial effect as compared to normal refrigerated storage primarily due to non-climacteric nature of the fruit.

5.5 Chilling Injury

Pomegranate fruits are susceptible to chilling injury which varies with maturity and variety. Fruits subjected to too low temperature (below 5 °C) develop brown discolouration of skin and pitting of the rind, internal browning of membranes, pale arils and increased susceptibility to decay. The severity of the symptoms increases with the exposure time.

Intermittent warming treatments have proved useful in the prevention of chilling injury symptoms in pomegranate cv. Molar de Elche stored at 0 °C and 5 °C and 95 per cent R.H., for 80 days. Intermittent warming treatment at 20 °C is applied in cycles, one day every six days storage, followed by a commercialization period of 7 days at 15 °C and 70 per cent R.H. Intermittent warming during the storage at 0 °C prevented fungal development, increased susceptibility to chilling injury. However, intermittent warming treatment followed by storage at 5 °C reduced considerably the chilling injury.

5.6 Transportation

For domestic market, the fruits are transported safely without any jerks through readily available means of transport and without loss of much time so as to maintain the glossiness and freshness of the fruits during the transit from the field to pack house. Fruits should be unloaded in the pack house without stacking too many crates over each other so that bruising of fruits is avoided. Tractor trolleys/mini vans and carts are used for transport from orchard/pack house to local markets after grading and packaging. Trucks/rail wagons are the means of transport from local assembly markets. The losses are very high (15-20 per cent) during transit through unvented rail wagons. Wooden boxes are also used for packaging and transportation. For export to an international market, the pomegranate fruits are transported through reefer containers by sea.

6.0 Domestic Marketing and Export

6.1 Domestic Marketing

In Maharashtra, pomegranate is marketed through a cooperative marketing society, locally called *sangha*. *Sangha* is a system in which a group of fruit growers join together on a cooperative basis to carry on some or all functions involved in transferring fruits from the farmers to the commission agents of the markets. Such registered cooperative societies are there at every taluka level. The growers finalize the picking and packing activity only after getting green signals from the *Sangha*. After receiving the boxes from the growers, the sangha decides at its discretion where to which commission agent and what number of boxes from assembled produce should be sent. Scheduling the dispatch of consignments in respective lots is the responsibility of the *Sangha*. Transport cost is in the range of Rs. 4-10 per box.

A second method of sale is where the growers send their produce to distant markets through their arrangements while well established farmers having their means of transport send the produce to the upcountry markets on their own. Sometimes, 5 to 10 growers of a particular region, with mutual understanding, come together and make common arrangements for sale of their produce. The third method is a contract system in which growers sell the standing crop to the trader/middleman on the farm itself. This system is locally known as 'Khoti'.

6.2 Exports

In Europe and other countries, pomegranate varieties such as Bhagawa (Kesar) and Mridula have good preference. Soft seeded, coloured varieties with high juice content and easy-to-remove arils are preferred. The fruits weighing more than 500 g with superior qualities have immediate and ready acceptance in the international market.

6.2.1 Codex Standards for Export

For export the following codex standards should be followed. The fruits should be 300-450 g, with red skin for cv. Bhagawa and Ganesh are destined for export to the Middle East, Netherlands, Germany and U.K. However, for Mridula and Arakta cultivars, the size should be 200-250 g with a deep red colour of the rind. Sea transport is primarily used for marketing of pomegranate for export by reefer containers maintaining a temperature of 5 °C.

Dark rose pink colour of the fruit with dark rose pink arils is preferred in the export market. Overripe fruits lead to severe cracking. Sunburn damage can be identified by the black colour on the surface of the fruit. The fruits affected by sunburn rot internally. The fruits should be free from the damage by thrips which leaves scars on the rind. Bracts (calyx) should be free from damage by mealy bugs and should appear fresh. Immediately after harvesting, pretreatment using Captan (50 per cent WP) @ 0.2 per cent could check the fungal spoilage during marketing. However, the removal of surface moisture before packing is essential.

6.2.2 Packing for Export

Requirement of packaging size for export marketing to the Middle East is a 5kg pack whereas in the Netherlands, Germany and U.K. customers prefer 3kg unit packaging of pomegranate (Figure 6). An orchard that can give an export quality fruits of around 40 per cent is desired. The entire export material is filled in cartons (4kg capacity) with different counts of 10, 12 and 15 fruits placed in a single layer. If the size of the fruit is big, only 10 may get into a carton. Usually, 200 cartons are required to make a pallet. A pallet usually contains 20 layers with each layer having 10 cartons. About 20 pallets can be loaded in a big container and 9 pallets in a small container. Therefore, 4000 cartons are needed to fill a big container. Pallets help in retaining the shape of the cartons and also protect the fruits from getting damaged while transportation.

Figure 6. Pomegranate Packaging for Export

The palletisation should be done in a proper manner by the carpenters which otherwise may create difficulty to fit these pallets into the container.

6.2.3 Export Potential

Considering the high keeping quality and availability throughout the year, India has great potential for export of pomegranate fruit. The commercial varieties *viz.*, Bhagawa, Ganesh and Arakta are suitable for the export market. Some gulf countries like Saudi Arabia, Qatar, Kuwait and Bangladesh are the major importing countries of Indian pomegranate. During the years 1992-93, a total quantity of 17,903 metric tonnes of fruit worth at Rs. 215.60 lakh was exported from India. However, the export of pomegranate increased to 25,000 metric tonnes in the year 2006. Afterwards, the export had crossed the mark of 35,000 tonnes per annum during the year 2012-13. India's share in world pomegranate export is 22.22 per cent and its only 2.53 per cent of the total pomegranate production for year 2019-20. During year 2020-21 Indian pomegranate export was pegged at 67,976 metric tonnes with revenue of Rs. 516.67 crores.

Export of pomegranate from Spain is mostly during October–December whereas pomegranate is available throughout the year in India. Pomegranates are available throughout the year in Maharashtra and Tamil Nadu states. This advantage of round-the-year availability has to be tapped for export promotion. The availability of pomegranate in different period in various states/ Union Territories of India indicates its great export potential. Year-round availability of pomegranate is given in Figure 7.

7.0 Conclusion

In the era of climate-resilient agriculture, cultivation of pomegranate is one of the important means for sustainable livelihood security of a large number of farmers in the dry arid and semi-arid agro-eco system. It is one of the most important crops that can give maximum returns on investment in horticulture based cropping system in the dry land area. However, quality assurance after harvesting plays a crucial role in reaping the maximum benefit from the harvested produce through adoption of smart and appropriate postharvest management practices for domestic as well as export marketing. Although pomegranate is not classified under highly perishable commodities among the fruit crops, yet the fruit quality and marketability are severely affected if left unattended primarily due to loss of moisture from the peel leading to a dull appearance. Hence, the postharvest quality assurance system should be integrated with the production system in order to achieve the desired goal. This should begin with adoption of *bahar-wise* management schedule including integrated disease and pest management.

The physical treatments used to prolong the commercial life of pomegranate should result in retaining the original quality of the fruit, avoiding fungal development, and the loss of quality characteristics (colour, flavour, texture) and nutritional properties (vitamins, antioxidants, health-promoting agents, *etc.*). These treatments modify the environmental conditions of pomegranate storage and have an effect on the fruit physiology and biochemistry and development of micro-organisms that contaminate the fruit surface. Thus, the possible application of the atmosphere during packaging or storage with gas composition different from the air (NO_2 enriched and/or reduced in O_2) need to be explored for postharvest quality retention. In addition, the use of thermal treatments for fruit conditioning and curing and intermittent warming during cold storage needs to explored commercially. Hence, harvesting at optimum maturity, sorting and grading as per market destination, pre-treatments, packaging and cool chain management are some of the key factors that govern the quality of the produce at different levels

8.0 Important References

Abubaker, A.R., Ashraf, N. and Ashraf, M. 2013. Effect of plant biostimulants on fruit cracking and quality attributes of pomegranate cv. Kandhari Kabuli. *Sci. Res. Ess.* **8**: 2171-75.

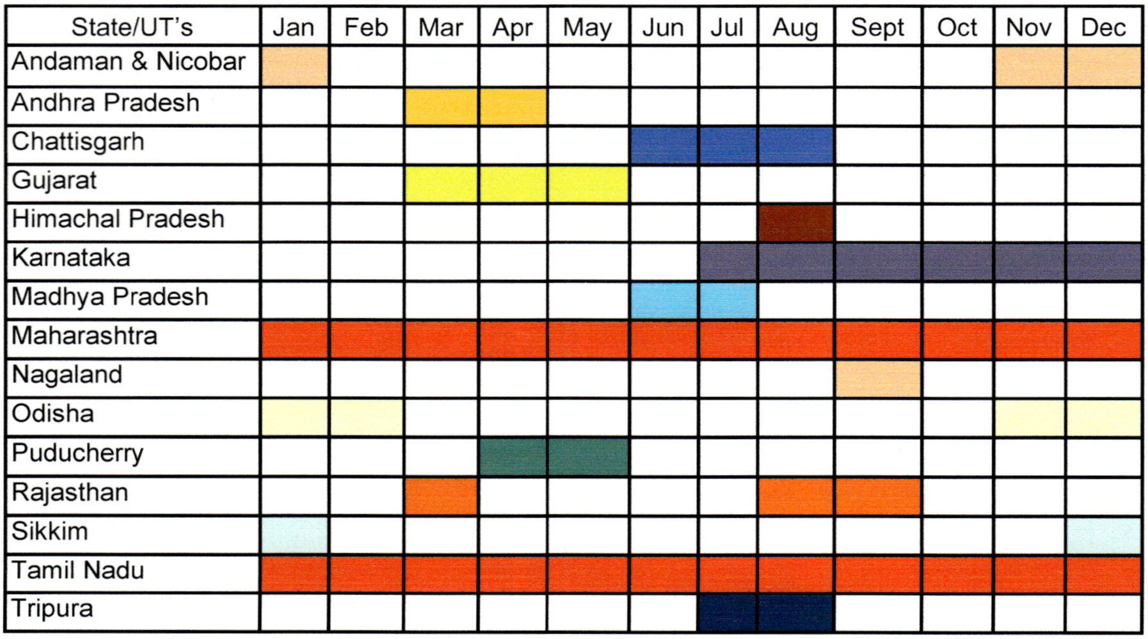

State/UT's	Jan	Feb	Mar	Apr	May	Jun	Jul	Aug	Sept	Oct	Nov	Dec
Andaman & Nicobar												
Andhra Pradesh												
Chattisgarh												
Gujarat												
Himachal Pradesh												
Karnataka												
Madhya Pradesh												
Maharashtra												
Nagaland												
Odisha												
Puducherry												
Rajasthan												
Sikkim												
Tamil Nadu												
Tripura												

Figure 7. Year-Round Availability of Pomegranate in India

Adsule, R.N. and Patil, N.B. 1995. Pomegrantae. **In:** *Handbook of Fruit Science and Technology*. Salunke, D.K. and Kadam, S.S. (Eds.) Marcel Dekker, Inc. New York, pp. 455-464.

Al-Khatani, 1992. Intercultivar differences in quality and postharvest life of pomegranates influenced by partial drying. *J. Amer. Soc. Hortic. Sci.* **117**: 100-04.

Artes, F., Tudela, J.A. and Gil, M.J. 1998. Improving the keeping quality of pomegranate fruit by intermittent warming. *Z. Lebersm. Unters. Forsch.* A **20**; 316-21.

Artes, F., Tudela, J.A. and Villaescusa, R. 2000. Thermal postharvest treatments for improving pomegranate quality and shelf life. *Postharvest Biol. Technol.* **18**: 245-51.

Asrey, R., Kumar, S. and Singh, S.K. 2007. Growing pomegranate in Punjab successfully. *Indian Hortic.* Jan-Feb: 9-11.

Asrey, R., Barman, K. and Kumar, A. 2011. Enjoying quality pomegranate for a longer period. *Indian Hortic.* July- Aug: 30-31

Awasthi, O.P., Saroj, P.L., Singh, I.S. and More, T.A. 2007. *Fruit based diversified cropping system for arid regions*. Technical Bulletin 25/2007, Central Institute of Arid Horticulture, Bikaner, Rajasthan. 18p.

Babu, K.D., Singh, N V., Gaikwad, N., Maity, A., Suryawanshi, S.K. and Pal, R.K. 2017. Determination of maturity indices for harvesting of Pomegranate (*Punica granatum* L.). *Indian J. Agril. Sci.* **87**: 1225-30.

Babu, K.D., Singh, N.V., Gaikwad, N, Maity, A., Marathe, R.A., Meshram, D.T., Patil, P.G., Shilpa P, Sharma,J. and Pal, R.K. 2019. Determination of maturity indices for harvesting of pomegranate cv. Bhagawa. 2019. *Indian J. Arid Hortic.* **1**: 69-72.

Babu, K.D, Sharma, J., Maity, A., Singh, NV., Patil, PG., Shilpa, P. and Marathe, R.A. 2021. Pomegranate: An ancient fruit for health and nutrition. *Prog. Hortic.* **53**: 1-11.

Barman, K., Asrey, R., Pal, R.K., Kaur, C. and Jha, S.K. 2014. Influence of putrescine and carnauba wax on functional and sensory quality of pomegranate fruits during storage. *J. Food Sci. Technol.* **51**: 111-17.

Ben-Arie, R. and Or, E. 1986. The development and control of husk scald on Wonderful pomegranate fruits during storage. *J. Amer. Soci. Hortic. Sci.* **111**: 395-399.

Ben-Arie, R., Segal, N and Guelfat-reich, S. 1984. The maturation and ripening of the Wonderful pomegranate. *J. Amer. Soci. Hortic. Sci.* **109**: 898-02.

Crisosto, C.H., Mitcham, E.J. and. Kader, A.A. 1996. Pomegranates. *Perishables Handling,* **85**: 17-18.

Darlington, D.C. and Janakiammal, E.K. 1945. *Chromosome Atlas of cultivated plants.*

De Candolle, A. 1967. *Origin of Cultivated Plants*, Hafner Publishers Co., New York, 468 p.

Du, C.T., Wang, P. L. and Francis, F.J. 1975. A research note on anthocyanins of pomegranate (*Punica granatum* L.). *J. Food Sci.* **40**: 417-18.

El-Kassas, S.E., El-Sese, A.M., El-Salhy and Abadia, A. A. 1998. Bearing habits in some pomegranate cultivars. *Assuit J. Agric. Sci.* **29**: 147-62.

Elyatem, S.M. and Kader, A.A. 1984. Postharvest physiology and storage behaviour of pomegranate fruits. *Sci. Hortic.* **24**: 287-98.

Fawole, O.A., Opara U.L. and Theron, K.I. 2012. Chemical and Phytochemical Properties and Antioxidant Activities of Three Pomegranate Cultivars Grown in South Africa. *Food and Bioprocess Technol.* **5**: 2934–40.

Gil, M.I., Artes, F., and Tomas-Barberan, F.A. 1996. Minimal processing and modified atmosphere packaging effects on pigmentation of pomegranate seeds. *J. Food Sci.* **61**: 161.

Hodgson, R.W. 1917. The pomegranate. *Calif. Agri. Expt. Sta. Bul.* 163-192.

Jadhav, V.T. and Sharma, J. 2007. Pomegranate cultivation is very promising. *Indian Hortic.*, pp. 30-31.

Jalikop, S.H. and Kumar, P.S. 2000. New fruit varieties for arid regions: Ruby (pomegranate) and Arka Sahan (custard apple). *Indian Hortic.* **45**: 19-20.

Kanwar, Z. and Thakur, D.P. 1972. Controlling postharvest soft rot of pomegranate fruits by treatment of packing straw. *Sci. Cult.* **38**: 450.

Kaur, C., Pal, R.K., Kar, A., Gadi, C., Sen, S., Kumar, P., Chandra, R., Jaiswal, A. and Khan, I. 2014. Characterization of antioxidants and hypoglycemic potential of pomegranate grown in India: a preliminary investigation. *J. Food Biochem.* doi: 10.111/fbc.12066

Khodade, M.S., Wavhal, K.N. and Kale, P.N. 1990. Physico-chemical changes during growth and development of pomegranate. *Indian J. Hort*ic. **47**: 21-27.

Kriventsov, V.I. and Arendt, N.K. 1981. Anthocyanins of pomegranate juice. *Trudy Gosudarstvennogo Nikitskogo bltanicheskogo Sada,* **83**: 110-16.

Kupper, W., Pekmezci, M. and Henze J. 1995. Studies on controlled atmosphere storage of pomegranate (*Punica granatum* L. cv Hicaz). *Acta Hortic.* **398**: 101-08.

Lee, S.W., Kim, K.S. and Kim, S.D. 1974. Studies on the compositional changes of pomegranate fruit during maturation. II. Changes in polyphenol compounds and anthocyanin pigments, *J. Korean Soc. Hortic. Sci.* **15**: 64.

Llacer, G., Martinex, R., Valero, Melgarego, P., Romero, M. and Toribio, F. 1994. Present status and future prospects of underutilized fruit tree crops in Spain, First Meeting CIHEAM Coop. Res. Network Underutilized Fruit Trees, Zaragoza, Spain, 63-75.

Mukherjee, P.K. 1958. Storage of Pomegranate (*Punica granatum*). *Sci. Cult.* **24**: 94.

Pal, R.K., Babu, K.D., Singh, N.V., Maity, A., Gaikwad, N. 2014. Pomegranate Research in India–Status and future challenges. *Prog. Hortic.* **46**: 184-01.

Pal, R.K., and Babu, K.D., 2014. Postharvest management and total utilization of pomegranate (*Punica granatum* L.). Souvenir – National Seminar cum exhibition on Pomegranate for nutrition, livelihood security and entrepreneurship development, Dec 5-7, 2014, ICAR-NRCP, Solapur, 252-261.

Pal, RK., and Babu, K.D. 2017. Postharvest management of pomegranate (*Punica granatum* L.). In : Pomegranate for nutrition, livelihood security and entrepreneurship development (Ed. Pal R.K, and Singh, N.V.), Daya Publishing House, New Delhi, pp. 235-242.

Pareek, O.P. and Godara, A.K. 1993. Crop regulation in pomegranate. **In**: *Advances in Horticulture, Vol. 3 Fruit Crops*. K.L. Chadha and O.P. Pareek (Ed.), Malhotra Publishing House, New Delhi, pp. 1229-35.

Patil A.V. and Karale, A.R. 1990. Pomegranate. **In**: Fruits: Tropical and subtropical (Bose, T.K. and Mitra, S.K.), Naya Prokash, Calcutta, 614-631.

Pota, S., Kesta, S. and Thongtham, M.L.C. 1987. Effect of packaging materials and temperature on quality and storage life of pomegranate fruits. *Kasetsart J.* **21**: 328-33.

Purohit, A.G. 1982. Flower induction in deciduous pomegranate in tropics. *Sci. Cult.*, **48**: 146-47.

Rao D.V.S, and Narayana, C.K.. 2014. Postharvest management of pomegranate fruits. Souvenir – National Seminar cum exhibition on Pomegranate for nutrition, livelihood security and entrepreneurship development, Dec 5-7, 2014, ICAR-NRCP, Solapur, 248-251p.

Roy, S.K. and Waskar, D.P. 1997. Pomegranate. **In**: *Postharvest physiology and storage of tropical and subtropical fruits*, S.K. Mitra (Ed.), p. 365-374.

Shantha, K. 1993. Effect of shrink wrapping on shelf life of pomegranate. *Abstract of IFCON-1993* held at CFTRI, Mysore, India, pp.150.

Sharma, S.D. and Sharma, V.K. 1990. Variation for chemical characters in some promising strains of wild pomegranate (*Punica granatum* L.). *Euphytica*, **49**: 131-33.

Sheikh, M.K. 2006. The pomegranate. International Book Distributing Co., Lucknow. 191p.

Shivshankar, S. Singh, H., Sumathi, M. 2012. Aril browning in pomegranate (*Punica granatum* L.) is caused by the seed. *Curr. Sci.* **103**: 26-28.

Shulman, Y., Fainberstein, L. and Lavee, S. 1984. Pomegranate fruit development and maturation. *J. Hortic. Sci.* **59**: 265-74.

Tregiazova, N.V. and Fetaliev, A.T. 1989. The effect of treatment with calcium chloride on pomegranate fruit storage. *Sadovodstro Vinogradastrvo*, **9**: 25-27.

Veres, M. 1976. A contribution to the study of the mechanical and chemical composition of cultivated pomegranate. *Hranai ishrana*, **17**: 426-32.

Waskar, D.P. and Khedkar, R.M. 1999. Effect of postharvest treatments on shelf life and quality of pomegranate in evaporative cool chamber and ambient conditions. *J. Food Sci. Technol.* **36**: 114-17.

Waskar, D.P. 1997. Postharvest management of pomegranate in Maharashtra. *Beverage and Food World*, **24**: 28-29.

Waskar, D.P. and Gaikwad, R.S. 2003. Studies on effect of postharvest treatments and storage conditions on shelf life and quality of pomegranate fruits. *Indian Food Packer*, **57**: 56-61.

Waskar, D.P., Kulkarni, S.R., Navale, A.M. and Shete, M.B. 2003. Bhagawa and Phule Arakta. MPKV Publication No. 41.

Westwood, M.N. 1978. *Temperate-zone Pomology*, W.H. Freeman and Co., San Francisco, 428 p.

Chapter 24

Sapota

Dev Raj

Professor and Head

Ashok K. Senapati

Assistant Professor

Department of Postharvest Technology, ASPEE College of Horticulture and Forestry, Navsari Agricultural University, Navsari – 396 450, Gujarat, India

E-mail: drdpandir@yahoo.co.in; E-mail: aksphtc@nau.in

1.0 Introduction

Sapota [*Manilkara achras* (Mill) Fosberg] belongs to Sapotaceae family. Sapota was first introduced to India during the late nineteenth century and its cultivation was first taken up in Maharashtra. Fully ripened sapota fruits are delicious and eaten as a dessert fruit. One hundred gram of edible portion of ripe sapota fruit contains carbohydrates (21.49 g), protein (0.7 g), fat (1.1 g), calcium (28 mg), phosphorus (27 mg), iron (2 mg), ascorbic acid (6 mg) and beta carotene (97 mg). It also has a lot of dietary fibre.

The sapota fruit is a fleshy berry and may be nearly round/oval/ellipsoidal. Though smooth-skinned, it is coated with sandy brown scurf until fully ripe. The colour of flesh ranges from yellowish to dark brown or sometimes reddish-brown. The texture of pulp is somewhat grainy but smooth; becomes soft and juicy on ripening with a sweet flavour resembling that of a pear. Some fruits of sapota are seedless. In seeded fruits, the number of seeds varies from 3 to 12 which are easily removable as they remain loosely held within a whorl of slots at the centre of the fruit. The average weight of fruit ranges from 70 to 300 g. The unripe fruit is hard to touch and contains high amounts of sapotin, which has astringent properties similar to tannin.

Sapota production is steadily increasing in India due to the expansion of the area under cultivation and improved production technology but the inadequacies of handling, storage, transportation and marketing pose problem during glut season which results in postharvest loss about 9.73 per cent (Anon., 2015).

2.0 Status of Cultivation

India is the largest producer of sapota followed by Mexico, Guatemala and Venezuela. The production of sapota in India is next only to mango, banana, citrus, apple and guava. Although, it originated in tropical South America, now it is one of the major tropical fruit crops grown in India. It has been so well acclimatized in the coastal tropics of India that it is often mistaken as an indigenous crop. Sapota fruits are grown mainly for table purpose besides making other value added products like jams, jellies, osmo-dehydrated slices and squash. The total area further declined from 107 thousand hectares in 2015-16 to 89 thousand hectares in 2018-19. Similarly, the production also decreased from 12.94 lakh tonnes to 10.59 lakh tonnes. Data on all India area and production of sapota are presented in Table 1.

In modern India, sapota is cultivated in Maharashtra, Gujarat, Andhra Pradesh, Karnataka, Tamil Nadu, Kerala, Uttar Pradesh, West Bengal, Punjab and Haryana with an annual production of 10.59 lakh tonnes from an area of about 8.9 lakh hectares in the year 2018-19 (Figure 1).

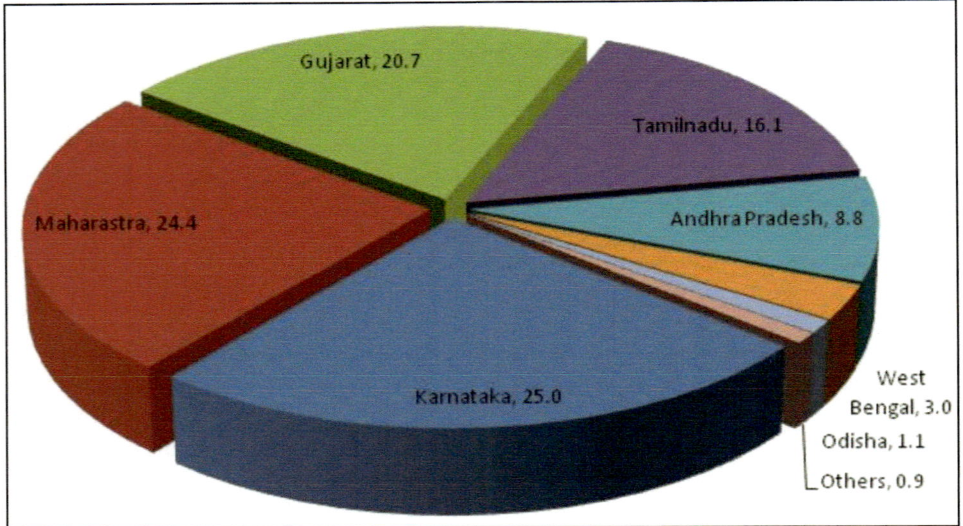

Figure 1. State-wise per cent Production Share of Sapota (NHB, 2012-13)

Table 1. Area and Production of Sapota in India

Year	*Area (in'000 Ha)*	*Production (in '000 MT)*
1991-92	27.2	396.2
2001-02	52.0	593.5
2005-06	139.7	1114.0
2011-12	163.4	1425.8
2015-16	107.1	1294.0
2016-17	99.2	1236.1
2017-18	97.0	1176.4
2018-19	89.0	1059.0

3.0 Major Commercial Varieties

The major commercially cultivated varieties of sapota in India are Kalipatti, Oval, Cricket Ball, Co-1, Co-2, Co-3, Kirtibharathi, Guthi, PKM-1, PKM-2, PKM-3, PKM-4, PKM-5 and Pilipatti. Important commercial varieties cultivated in different states of India are enlisted in Table 2. Salient features of these major varieties are discussed here:

Table 2. Important Commercial Varieties Cultivated in States of India

State	*Varieties Grown*
Andhra Pradesh	Cricket Ball, Kalipatti, Calcutta Round, Kirtibharathi, Dwarapudi, Pala, PKM-1, Jonnavalasa I and II, Bangalore, Vavivalasa
Bihar	Baramasi
Gujarat	Cricket Ball, Kalipatti, Pilipatti, PKM-1
Karnataka	Calcutta Round, Cricket Ball, DHS-1, DHS-2, Kalipatti
Maharashtra	Cricket Ball, Dhola Diwani, Kalipatti, Murabba
Orissa	Cricket Ball, Kalipatti
Tamil Nadu	Co 1, Co 2, Cricket Ball, Guthi, Pala, PKM-1
Uttar Pradesh	Baramasi
West Bengal	Baramasi, Baharu, Barada, Calcutta Round, Cricket Ball, Gandhevi

Source: Anon. 2014.

a) Bangalore

This variety is commercially grown in Andhra Pradesh. The fruits are large in size, oval with about nine ridges running from the base to the apex. The apex of the fruit is round and the pulp tastes medium in sweetness.

b) Co-1

This is a hybrid between Cricket Ball x Oval and is commercially grown in Tamil Nadu. Fruits are large, oblong, and sweet having reddish brown flesh and an average weight of 125 g. The TSS of fruit is around 18 °B.

c) Co-2

This is a selection of cv. 'Baramasi' from Tamil Nadu. The average weight of fruit is 125-150 g. Fruits are oblong-round in shape, medium-sized with soft juicy, sweet, gritty, light brown flesh having pleasant aroma and TSS around 23 °B.

d) Co-3

This is a hybrid between 'Cricket Ball' and 'Vavivalasa'. It is commercially grown in Tamil Nadu. Fruits are dull brown, oblong and sweet.

e) Cricket Ball

This variety is commercially grown in Tamil Nadu, Karnataka, Maharashtra, West Bengal and Andhra Pradesh. It is a table purpose variety. Fruits are large sized, round, and sweet with granular pulp.

f) Guthi

This variety is commercially grown in Andhra Pradesh and Tamil Nadu. Fruits are elliptical, small and brown. The pulp is less juicy, soft, and gritty, with a slight aroma.

g) Kalipatti

This variety is commercially grown for table purpose in Maharashtra, Gujarat and North Karnataka. The fruits are oblong/round in shape and mature in winter season.

h) Kirtibharathi

This variety is commercially grown in Andhra Pradesh and Tamil Nadu. It is suitable for both table as well processing purposes. The fruits are very small, egg-shaped with medium thick buff coloured skin having four to six ridges. The pulp is very sweet when fully ripe. This is ideal fruit for transporting to distant markets.

i) Pala

This is a popular and high yielding cultivar commercially grown in Andhra Pradesh and Tamil Nadu. The fruits are oval or egg shaped, thin-skinned, borne in clusters with apex broadly pointed. The pulp is very sweet and has a good flavour.

j) Pilipatti

This variety is commercially grown in Maharashtra and Gujarat. It is suitable for high density plantation. The fruits are oval, average fruit weight is 80-100 g with rough surface and gritty, medium sweet pulp.

k) PKM-1

This is a selection from Guthi made in Tamil Nadu. The tree is profuse bearing with medium sized, elliptical/oblong fruits. The average fruit weight is 80 g. The skin is thin and the TSS is around 23 °B.

4.0 Nature and Causes of Postharvest Losses

Since the tree bears flowers all year round, fruits in varying stages of maturity can be found on the tree at the same time, making it difficult to determine the optimum maturity for harvesting. The climacteric nature of sapota fruits further hinders the storage and distribution and necessitates careful postharvest handling to reduce losses. The fruits are classified under highly perishable group.

5.0 Postharvest Management

5.1 Maturity Indices

Fruits are harvested either when mature or when ripe. Fruits harvested earlier than physiological maturity may not soften, and such fruits are usually low in sweetness with high astringency. They also contain large amount of leuco-anthocyanin. Such prematurely harvested fruits provide an unappealing alcoholic after-taste. The formation of pockets of coagulated latex in these fruits makes them inferior in quality. Fruits harvested later than optimum maturity usually soften very rapidly and become very difficult to handle.

The maturity of sapota fruit is decided based on skin colour. A fruit with a smooth surface, shining potato colour and rounded styler end is considered to be mature. At this stage, practically no green tissue and milky latex are seen on fruits when scratched with nails. Fully mature fruit will have brown skin and will separate easily from the stem without leaking latex. The extent of scruffiness is also a good indicator of harvest maturity. The sucrose content and pulp-to-peel ratio increase during maturation. Sapota takes about 6-8 months from flowering to the maturity of fruits depending on variety and climate. At full maturity, the milky latex content of the fruit is reduced and the dried spine-like style at the top of the fruit drops off naturally or when gently touched.

5.2 Harvesting

Sapota is a seasonal fruit though it almost throughout the year is available in India. The fruit has an exceptionally sweet, malty flavour with a striking resemblance to caramel. The harvesting, season of sapota in different parts of India is illustrated in Figure 2.

State/UT's	Jan	Feb	Mar	Apr	May	Jun	Jul	Aug	Sept	Oct	Nov	Dec
Andhra Pradesh			Lean	Lean	Peak	Lean						
Gujarat	Round the year											
Haryana				Lean	Peak	Lean						
Karnataka	Lean	Peak	Lean									
Lakshadweep		Lean	Lean	Peak	Peak	Lean	Peak					
Maharashtra	Round the year											
Odisha				Lean	Peak							
Rajasthan							Peak					
Tamil Nadu			Lean	Lean	Peak				Lean	Peak	Lean	
Tripura				Lean	Lean	Lean	Peak	Peak	Lean	Lean	Lean	
West Bengal	Lean								Lean	Peak	Peak	Lean

Source. APEDA, 2012 — Lean season / Peak season

Figure 2. Harvesting Season of Sapota in the States of India

The fruits are handpicked or harvested with special harvester which has a round ring with a net bag fixed onto a long bamboo stick. Sapota starts bearing from the third year of planting but economic yields can be obtained from the 5th year onwards. There are two main seasons of flowering *viz.* October–November and February–March that correspond to two harvesting seasons, January–February and May–June respectively. Harvesting of mature sapota can be carried out in about 240 to 270 days after flowering. Sapota fruits are harvested with the stalk intact, individually by giving a twist and are collected without brushing.

Harvesting of fruits should be done in the early morning hours. Appropriate harvesting tools and equipment should be adopted to avoid external injury to the fruits. Fruits should be harvested with stem and kept in shade. In high-density plantations, the production increases from 10 tonnes/ha in the fifth year to 15 tonnes/ha in the 7th year followed by 20 tonnes/ha from 8th to 15th year.

5.2.1 Harvesting Method

A poor harvesting operation of sapota fruit would result in poor quality produce and a low selling price. The grower should plan the harvesting operation with great care, especially when the enterprise is on a commercial scale. Labour, equipment and transportation should be prearranged. Labour should be trained to pick the produce at the right stage (market harvest) and in a proper manner (*e.g.*, plucking and cutting) without bruising the produce. Mechanical injury is reduced by training workers to use appropriate containers in the best shape (not rough-edged, *etc.*), to avoid rough handling (throwing, dropping) and to pack properly. Harvesting equipment should be cleaned and kept ready for operation. When the crop is ready for harvesting, the decision as to when to start harvesting will depend largely on weather conditions and the demand for fruits in the market. The sapota fruits must be marketed as soon as they are ready or otherwise they will spoil.

a) Hand Harvesting

The manual method of harvesting sapota fruits has several advantages *e.g.*

i) Humans can accurately select for maturity, allowing grading and multiple harvests

ii) Fruits can be handled with minimum damage

iii) Hiring more workers can easily increase the rate of harvest

iv) Hand harvesting requires minimum capital investment.

The fruits thus harvested are spread on bamboo mats under shade for an hour or two. To avoid bruising fruits are collected in gunny bags and lowered to the ground carefully.

b) Mechanical Harvesting

Machines can be used to aid hand harvest *e.g.* clipper mounted on pole with bags, movable platforms and ladders. Alternatively, mechanical harvesting devices can be employed which make direct contact with fruits and harvest them by combing, cutting, pulling, snapping, twisting, stripping and compacting. Various types of harvesters used in sapota are: i) Clipper Mounted on a Pole with Bags (Khudi); ii) Bamboo with knife and net harvester (Khudi); iii) Bamboo cone fruit harvester; iv) Wire cone fruit harvester; v) Push and Twist type harvester; vi) Pull and cut type harvester; vii) Hold and twist type harvester; viii) Bamboo with sickle or hook harvester (Figures 3–5).

5.3 Sorting and Grading

Grading of sapota fruits is done according to the size, shape, colour, and volume/weight to fetch a high price in market. Fruits should be graded into big, medium and small sizes. Immature, diseased and badly bruised fruits are sorted out. The fruits are then dried in shade and graded according to their size. Most countries have their own set of standards for domestic and international trade. Grading is done manually or mechanically.

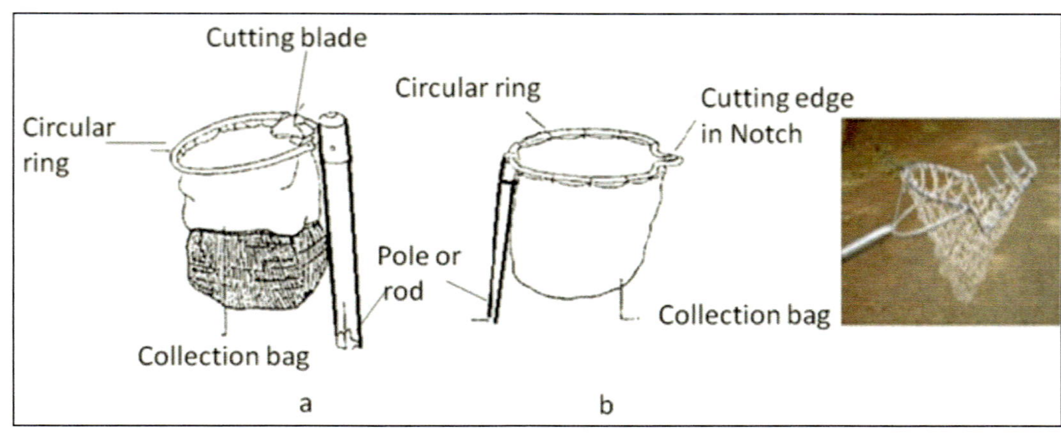

Figure 3. Clipper Mounted on Pole with Bags (Khudi) Used for Harvesting of Sapota (Developed by IIHR, Bangalore)

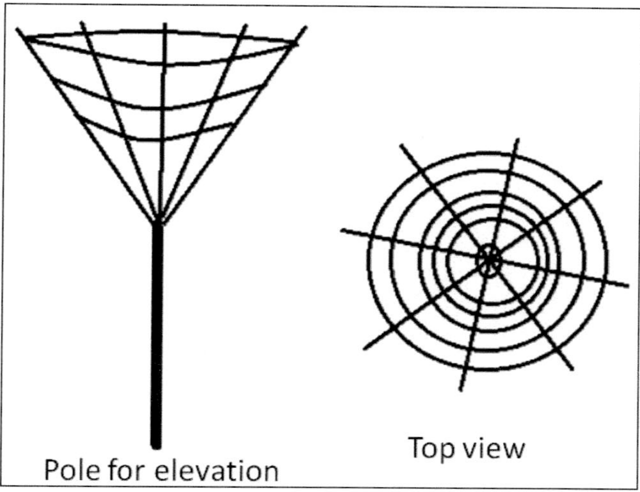

Figure 4. Bamboo Cone with Harvester

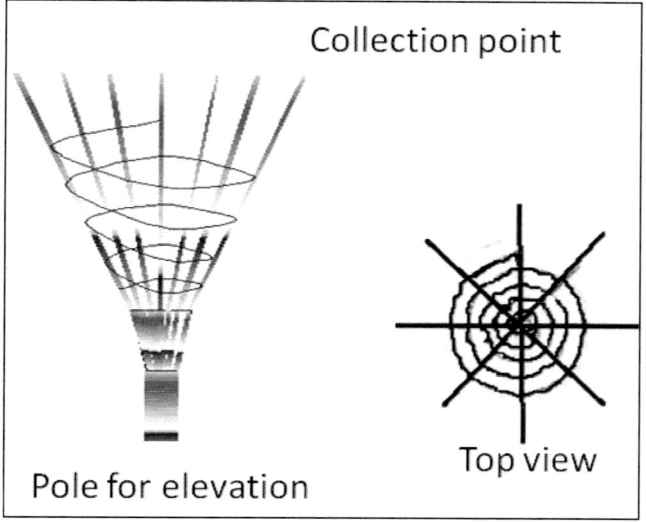

Figure 5. Wire Cone Fruit Harvester

Several graders have been designed and developed. A divergent roller type sapota fruit grader with roller speed of 300–450 rpm (gap between the rollers ranged from 3.0-7.5 cm) has maximum separation capacity of 1,800 kg/h with overall separation efficiency of 72 per cent. The best combination of roller speed, its inclination and roller gap was found to be 223 rpm, 4.5 °and 3.8 to 6.4 cm respectively, which provided the highest grading efficiency of 89.5 per cent.

5.4 Cleaning

It is performed to remove latex and scurf. For removing the latex oozing from the stalk, the fruits are washed immediately after harvesting. Washing improves the shelf-life of sapota by delaying their ripening. After washing, excess water should be removed which would otherwise encourage microbial spoilage.

6.0 Postharvest Treatments

Sapota fruits being climacteric are highly perishable and they undergo rapid ripening changes within 5–7 days during which the fruits become soft, sweet and develop excellent aroma with decline in tannins, latex, sapotin, aldehydes and acidity. These changes are associated with an increase in ethylene production, rate of respiration, catalase, peroxidase and PME activities. These changes can be regulated through proper postharvest treatments like chemicals, temperature and storage gas. Thus postharvest treatments may be given to slow down ripening, and for prevention of insect-pest attack and diseases.

Postharvest dip of sapota fruit in GA_3 (200 mg/l) reduced loss in weight, spoilage and shrivelling, induced more firmness and increased percentage of marketable fruits as well as shelf life. Storage of fruits in polyethylene bags with $KMnO_4$ + silica gel, gave encouraging results while unpacked fruits showed faster deterioration. For extending shelf-life and avoiding storage rots, fruits can be dipped in GA 300 ppm + Bavistin 1,000 ppm solution at the pre packing stage. Disinfestations may be done using either vapour heat treatment at 43 °C for 6–8 hr; or by ethylene di-bromide fumigation (18–22 g of EDB/cubic meter for 2–4 hr) (residues of inorganic bromide must not exceed 10 µg/g) or by cold treatment (exposure of fruits near to freezing temperature for a specified period).The fruits are dipped in GA (300 ppm) and Bavistin (1000 ppm.) solution at a pre-packing stage to extend the storage life of the fruits. Dipping treatment of sapota fruits in 1 per cent calcium hydroxide for 5 minutes followed by dipping in 2, 4-D, 4 ppm or GA3, 300 ppm or kinetin, 100 ppm extends the shelf life without affecting the quality of sapota fruits. The treatment of fruits with 1-methylcyclopropane (1-MCP) at 25 °C for 24 h in airtight chambers followed by cold storage at 16 °C significantly delays the ripening process and extends the postharvest life to 28 days. Thus, 1-MCP combined with refrigeration, is an adequate alternative for increasing the shelf life of sapodilla.

6.1 Disease and Pest Control

Important diseases like *Phytophthora palmivora* and species of *Pestalotiopsis* and *Phomopsis* can cause fruit rot. Some species of bacteria are associated with fruit latex. Anthracnose is a minor but serious disease of sapota fruit in some countries. Insects that infest sapodilla fruit include *Nephoptery xengraphella* Rag., fruit flies and borer. The most troublesome fruit flies are the Mediterranean fruit fly (*Ceratitis capitata*) and Mexican fruit fly (*Anastreph aludens*, Loew.). Postharvest application of Bavistin (0.1 per cent) and Topsin (0.1 per cent) controls storage diseases in Sapota. In Kalipatti, hot water treatment with Imazalil (0.1 per cent), Bavistin (0.1 per cent) and Benlate (0.1 per cent) is the most effective.

6.2 Ripening

Ripening transforms physically mature but inedible fruits into visually attractive and edible fruits. It marks the completion of the development and commencement of senescence of fruit and is normally an irreversible event. The treatment of fruits with 1-Methylcyclopropane (1-MCP) at 25 °C for 24 h in airtight chambers followed by cold storage at 16 °C significantly delays the ripening process and extends the postharvest life to 28 days. Treatment of sapodilla fruit with ethrel at 1 to 3 ml L^{-1} accelerated ripening, and reduced pectin content, phenolic content, SSC, sugar content and Vitamin C. Removal of ethylene delays ripening. In order to generate 1 cft of ethylene gas 198.8 ml ethrel is required and it should be made alkaline using caustic soda (3 g of caustic soda for 20 ml of ethrel) for its release. For uniform and rapid ripening ethephon (1,000 ppm) can be applied at 20 °–25 °C. Wrapping of individual fruits in tissue paper or newspaper causes synchronized ripening and reduces spoilage.

6.3 Packaging

Fruits should be tightly packed in cardboard boxes of 10kg capacity with rice straw as padding material and ethylene absorbents before being transported quickly to wholesale markets. In local markets, the graded fruits are packed in bamboo baskets containing straw as padding material. This helps in reducing bruising and promotes even ripening of fruit. For distant markets, fruits are packed in cardboard boxes. Corrugated trays are equally effective as packaging material while transporting the fruits. All the graded sapota fruits are wrapped in paper and packed in corrugated fibreboard (CFB) boxes. Wheat and paddy straw, banana leaves, dry grass, paper shreds *etc.*, which are easily available locally and less costly, are used mostly as cushioning material but are unhygienic and do not allow respiratory heat to escape from the packing box. In recent years, the use of moulded pulp trays, honeycomb partition, cell pack *etc.* have replaced the use of cushioning material in costly commodities and for export purposes.

The shelf life of sapota may be increased up to 49 days with minimum changes in physical, biochemical and sensory parameters at 6 °C temperature by packing them in 25 µ LDPE bags containing 5 per cent O_2 +10 per cent CO_2 gas concentration. Packaging of the precooled sapota fruits in 1 per cent vented LDPE bags along with CFBB extends the shelf life of the fruits for 4-5 days when compared to normal storage.

6.4 Pre-cooling

The shelf life of Sapota fruits can be increased by 2-3 days if they are precooled immediately after harvesting at 10 °C for 8 hours. Hydro-cooling at 12 °–15 °C with 500 ppm Bavistin increases shelf-life of sapota. Precooled sapota fruits should be packed in a perforated polythene bag (50 micron) and kept in CFB box and then stored at 12 °C in cold storage.

6.5 Storage

Sapota fruits are highly susceptible to chilling injury (CI). Storage of fruit at 6 to 10 °C causes irreversible damage and results in poor flavour. However, fruits waxed with a fatty acid of sucrose ester kept well for 40 days at 10 °C. Storing fruits in plastic liners and waxing can reduce water loss, especially under conditions of low relative humidity in the storage atmosphere. Postharvest life of sapota fruits may be 2 to 3 weeks at 12 to 16 °C with 85 to 90 per cent RH. Storage life is about 13 days at 25 °C, 15 days at 20 °C, and 22 days at 15 °C. Short-term holding of fruit at low temperature *i.e.* pre-cooling for less than 10 h at 4-8 °C before storage at 20 °C extended storage-life up to 24 days with satisfactory quality. Exposure of fruit to gamma irradiation at 0.1 kGy extended storage-life by 3 to 5 days at 26.7 °C (80 °F) and by 15 days at 10 °C (50 °F) without any effect on ascorbic acid content.

At storage temperature of 20 °C and RH of 85-90 per cent, the storage life can be increased for a period of 21-25 days by removing ethylene and adding 5-10 per cent CO_2 to storage atmosphere. Storage-life of sapota fruits at room temperature can be increased from 13 to 18 days with 5 per cent CO_2, and to 21 days with 10 per cent CO_2, 'Kalpatti' fruit treated with 6 per cent Waxol or 250 or 500 ppm Bavistin, or hot water at 50 °C (122 °F) for 10 min, and kept wrapped in 150 gauge polyethylene film with 1 per cent ventilation, showed shelf life extension. Fruit treated with 6 per cent wax emulsion and packed in 200-guage polyethylene containing ethylene and CO_2 absorbents had a shelf-life of 45 days at 12 °C. Fruits can be stored successfully using MAP for 4 weeks at 10 °C and 3 weeks at 15 °C, a week longer than fruit kept without MAP.

6.6 Transportation

Harvested fruits should not be transported in gunny bags. Ventilated wooden boxes using news paper as lining material can be used to avoid bruising for long distance transport. Junagadh Agricultural University assessed transportation losses for sapota with seven types of containers/bags. Minimum damage and loss of fruits were observed in foldable plastic boxes with cells while it was maximum in gunny bags. Refrigerated vans (12-23 °C) could be utilized for long distance transport and export.

NAU, Navsari has recommended to farmers and entrepreneurs to extend the shelf life of sapota fruits up to 12 days at 11°C including 3 days transportation by packing in CFB box (10 kg capacity) followed by pre-cooling at 10oC for 8 hours (Anon., 2018). Further, NAU developed a complete protocol for shelf life extension of sapota cv Kalipatti fruits for 21 days. The fruits should be disinfected by dipping them in 50ppm chlorinated water for 5 minutes, precooling at 10°C for 8 hours, packing in corrugated fibre board box (1kg) containing a sachet of 1 per cent potassium permanganate ($KMnO_4$) along with 0.5 per cent silica gel (SiO_2), shrink wrapping of CFBB using 20 micron polypropylene sheet followed by storage at 11°C (Dabhi 2021).

6.7 Domestic Marketing and Export

Cold chain transportation would extend the shelf life of sapota fruits up to 15 days without adverse effect on quality and thus aids in easy marketing. Brown coloured 3 ply CFB boxes are used for local market while white coloured 5 ply CFB boxes are used for distant markets.

India exports sapota to UAE, Bahrain, UK, Canada, Saudi Arabia, USA and other countries. The total volume of exports has increased in recent years. During 2006-07, India's sapota export was 2089.5 metric tonnes which decreased to 1603 metric tonnes during 2017-18 having a value of Rs. 8.91 crores. As whole, Asian countries are the major importers of sapota from India. Even though India is the largest producer of sapota in the world, the country exports a very minor fraction (0.4 per cent) of its annual harvest. This is due to large domestic consumption. Export data of sapota from India during 2017-18 are presented in Table 3. During 2021, Agricultural and Processed Food Products Export Development Authority (APEDA) helped to export GI certified sapota consignment of Dahanu Gholvad Sapota from the Palghar district of Maharashtra to the United Kingdom. GI certification of Ghovad Sapota is held by *Maharashtra Rajya Chikoo Utpadak Sangh* and the fruits are known for their sweet and unique taste. It is believed that the unique taste is derived from the calcium-rich soil of Gholvad village.

Table 3. Export of Sapota from India in 2017-18

Sl.No.	*Country*	*Quantity (000,MT)*	*Value (Rs. In Lakh)*
1.	Bahrain	269	145
2.	Canada	77	48
3.	Kwait	43	10
4.	Oman	121	74
5.	Qatar	99	62
6.	Saudi Arabia	70	40
7.	Singapore	29	20
18	UAE	824	425
9.	UK	134	40
10.	United States	58	38
11.	Others	17	11
	Total	1603	891

Source: APEDA Website September, 2018.

6.8 Marketing Standards

There are two grade standards for domestic market;

i) Fruits are graded into three sizes A (large), B (medium) and C (small);

ii) Fruits are graded into seven sizes. National Horticulture Mission, Govt. of India has given seven grading standards for the domestic market (Table 4).

Table 4. Grade Standards for Local Market According to NHB, GOI

No. of Fruits/Basket	*Fruit Weight (Grams)*
96	>160
120	125-160
180	85-135
240	65-85
300	50-65
360	45-50
420	<45

Source: Anonymous (2014), http://nhb.gov.in/sapota.

Specifications of sapota fruits for export to countries of the Middle East are presented in Table 5.

Table 5. Export Specification for Sapota for the Middle East

Specification	Middle East
Variety	Cricket Ball, Kalipatti
Packing	3 Kg
Storage	15-20 °C
Transport	By Sea

7.0 Conclusion

Sapota fruits are highly perishable due to their climacteric nature. The rapid softening of fruits is primarily due to the high activity of many oxidative enzymes and liberation of ethylene. Harvest maturity is a crucial deciding factor in the marketability of climacteric fruits. Postharvest life of sapota is very short and it is susceptible to chilling injury when stored for a long time at low temperature with symptoms including abnormal ripening, bronzing of the skin and decay. Therefore, harvesting at the appropriate stage of maturity with proper harvesting tools and grading of fruits as per the need of the market are essential to reduce postharvest losses. Use of ethylene scavengers and appropriate packaging for long distant transport would extend its marketability and would also retain quality for a long time. Adoption of recommended grades and standards and other package of practices would certainly boost the export of this important crop.

8.0 Important References

Anonymous, 2011. *Annual Report of AICRP on Application of Plastics in Agriculture at Central Institute of Postharvest Engineering and Technology*, Ludhiana

Anonymous, 2014. Postharvest Technology. **In**: *Sapota. http: //nhb.gov.in/fruits/sapota/sap008.pdf.* Retrieved 03/03/2014.

Anonymous, 2015. *Annual Report of AICRP on Application of Plastics in Agriculture* at Central Institute of Postharvest Engineering and Technology, Ludhiana.

Anonymous, 2018. Proceeding of the fourteenth meeting of Combined Joint Agricultural Research Council of SAUs and Kamdhenu University of Gujarat - 2017-18. Directorate of Research, Junagadh Agricultural University, Junagadh-362001 during 3-5 April 2018.

Anonymous, 2019. *Postharvest Technology*. In: Sapota. http: //nhb.gov.in/fruits/sapota/sap008.pdf. Retrieved 27/12/2021.

APEDA, 2018. https: //apeda.gov.in/apedawebsite

Bojappa, K.K.M. and Reddy, T.V. 1990. Postharvest treatments to extend the shelf-life of sapota fruit, *Acta Hortic.* **269**: 391.

Bose, T. K. and Mitra, S. K. 1990. *Fruits: Tropical and subtropical*. Naya Prakash, Calcutta, pp. 565-591.

Broughton, W.J. and Wong, H.C. 1979. Storage conditions and ripening of chiku fruits *Achras sapota* L. *Sci. Hortic.* **10**: 377-85.

Chundawat, B.S. 1991. Postharvest handling and marketing of sapota fruits. **In:** *National Seminar on Optimization of Production and Utilization of Sapota,* held during October 1991, Navsari, Gujarat, India.

Chundawat, B. S., 1998. Postharvest management. In: *Sapota (Manilkara achras* Mill.). Agrotech Publishing Academy, Udaipur, pp. 112–27.

Dabhi M.J. 2021. Studies on shelf life extension of sapota (*Manilkara achras* (Mill) Fosberg) fruits cv Kalipatti. *M.Sc Thesis*. Navsari Agricultural University, Navsari, Gujarat, India 396450, pp 91.

Ingle, G.S., Khedkar, D.M. and Dabhade, R.S. 1982. Ripening studies in sapota fruit. *Indian Food Packer,* **36**: 72-77.

Kadu, R.V. and Gajipara, N.N.2009. Studies on postharvest treatment of sapota fruit. *J. Life Sci.* **6:** 274-77.

Kute, L.S. and Shete, M.B. 1995. Sapota (Sapodilla). **In:** *Handbook of Fruit Science and Technology,* D.K. Salunkhe and S.S. Kadam (Eds.). Marcel Dekker, Inc., NY, pp. 475-84.

Lakshminarayana, S. 1980. Sapodilla and prickly pear. **In**: *Tropical and Subtropical Fruits. Composition, Properties and Uses*, S. Nagy and P.E. Shaw (Eds). AVI Pub, Westport CT, pp. 415-41.

Lakshminarayana, S. and Subramanyam, H. 1966. Physical, chemical and physiological changes in sapota fruit (*Achras sapota* L.) during development and ripening. *J. Sci. Food Sci. Technol.* **3**: 15-154.

Mohamed, S., Taufik, B. and Karim, M.N.A. 1996. Effects of modified atmosphere packaging on the physiological characteristics of chiku at various storage temperatures. *J. Sci. Food Agric.* **70:** 231-40.

Salunkhe, D.K. and Desai, B.B. 1984. *Postharvest biotechnology of fruits. Vol. II.* CRC Press, Boca Raton, Fl. pp. 387-396.

Shirol, A.M.,Hanamashetti, S.I.,Kanamadi, V.C., Thammaiah, N. and Patil, S. 2006. Performance of some sapota cultivars under Ghataprabha command area. *Karnataka J. Agric. Sci.* **19**: 366-70.

Ukey, P.D. and Unde, P. 2010. Design, development and performance evaluation of sapota fruit grader. *Int. J. Agril. Engg.* **3:** 35-39, 62-64.

Victor, M. 2013. Ripening of sapodilla fruit (*Manilkara zapota* (L.) treated with 1-methylcyclone propane after refrigeration. *Afr. J. Pl. Sci.* **12**: 561-70.

Chapter 25

Strawberry

Ram Asrey

Principal Scientist, Division of Food Science and Postharvest Technology, ICAR-Indian Agricultural Research Institute, New Delhi – 110 012, India

E-mail: ramu_211@yahoo.com

1.0 Introduction

Strawberry (*Fragaria ananassa* Dutch.) originated in France and it belongs to Rosaceae family. Strawberry is a temperate fruit, but, due to its short production cycle (100-120 days), it has now become the fastest growing crop in the tropical and sub-tropical regions of Asian nations. Initially, breeding efforts were made only to develop short-day cultivars, which were most successful in temperate climates. In India, strawberry was first introduced at NBPGR, Regional Research Station Shimla (H.P.) during the early sixties. But its cultivation could not get popularity on account of poor varietal adaptability and technical knowhow. With the development of day-neutral cultivars, its cultivation got a boost in the warmer region too. Later on, day neutral cultivars were introduced and accordingly different agro-techniques were also developed and modified. As a result, strawberry fruits are now available right from December to June (from the northern plains and hill regions of India). In recent years, farmers have shown interest in its cultivation due to very quick and high returns.

Strawberries are high in Vitamin C, pectin (0.55 per cent), in the form of calcium pectate and serve as an excellent ingredient for jelly making. It is also a rich source of potassium, calcium and phosphorus, The mature ripe fruit contains about 5 per cent total sugars (glucose and fructose are major sugars) and 0.90 to 1.85 per cent acids. The primary organic acid is citric acid, which comprises 88 per cent of total acid. Strawberry fruits contain significant levels of elegiac acid that possess anti-carcinogenic properties. It is also known for its characteristic aroma, which is mainly due to presence of volatile esters. However, concentration of these volatiles varies among cultivars. The red colour of the fruit is mainly due to the presence of anthocyanin, pelargonidin-3-monoglucoside and traces of cyanidine having antioxidant activities.

2.0 Status of Cultivation

The annual global strawberry production is around 3.0 million metric tons. USA is one of the leading strawberry producing countries with approximately 20 per cent of the world's production followed by Spain, Japan, Poland and Italy. The area under strawberry cultivation in India has not been properly estimated so far. However, due to introduction of day neutral varieties and increasing demand of consumers, its cultivation has gained momentum in the states of Himachal Pradesh, Jammu & Kashmir, Haryana, Punjab and Uttarakhand. Some places such as Ratlam, Jalgaon, Satara, Mahabaleshwar, Hisar, Gurgaon, Alwar, Solan, Shimla, Haldwani and Nainital are known for strawberry production. The strawberry mainly grown in Mahabaleshwar-Panchagani belt of western Maharashtra region has received the registration from the Ministry of Commerce (Geographical Indication status) and will now be branded as 'Mahabaleshwar Strawberry'. In addition, North eastern states are also growing strawberries on a substantial acreage.

As such strawberry is not grown under high rise polyhouses, but it gives a better response inside polyhouse production. In Northern and South Western plains strawberry fruits are grown mainly using drip and mulch system with low tunnel covering options during extreme winter conditions. Different cultivation systems had a great influence on postharvest quality parameters of the strawberry fruits. Open field condition grown strawberry fruits have higher firmness and storage life compared to green house conditions. However, soilless cultivation and green house grown strawberry fruits developed brilliant skin colour (scarlet) with golden achene embedded in.

3.0 Major Commercial Cultivars

a) Chandler

This is the oldest but still the ruling mid season variety in India. The vigorous, Chandler strawberry plants produce very attractive fruits which are large, firm, long and wedge-shaped to large and conical. Fruits are of brilliant red colour, glossy, and have a long lasting flavour. Chandler strawberries are good for eating fresh or shipping and very good for freezing but are being given only a fair choice for processing.

b) Fern

It is a day-neutral, early-ripening and over-bearing cultivar. Fruits large, medium, conical, solid internally, slightly hollow, skin red, flesh firm, flavour excellent, suitable for fresh market and processing. It tastes sweet to slightly acidic. Average berry weight 20-25g, TSS 11.2 per cent, acidity 0.98 per cent and sugars 6.1 per cent.

c) Sweet Charlie

This is an early-season short day cultivar. Plants are small and compact making picking relatively easy. Fruits are medium in size, but, as the name suggests, deliciously sweet and have a relatively low acid level. This variety is a great choice for the home gardener and excellent for fresh eating. It has a short shelf-life, and can tolerate gentle shipping over short distance. Generally, the variety has two weeks of early production over Chandler, however, after the initial two weeks; size tends to drop off drastically in the early midseason. It can be used for fresh-market and processing.

d) Selva

A day-neutral cultivar, it can produce off-season fruits. It is different from day-neutral or ever bearing types. Fruits are large, with firm flesh and skin, conic to blocky in shape, and dessert quality. It can be handled and shipped fairly well. Skin is bright red, attractive, flesh red, internally somewhat hollow, light in colour at core, average berry weighs 15-18g, TSS 11.1 per cent, acidity 1.0 per cent and sugar 5.5 per cent.

e) Winter Dawn

This is a mid season cultivar and now being popular in Himachal, Haryana Punjab and Delhi NCR region. This cultivar can produce relatively large fruit on small plants, and like 'Carmine' its fruit is resistant to botrytis and anthracnose fruit rots. 'Winter Dawn' produces large size fruits with mild TSS and acidity, suitable for fresh consumption.

f) Camarosa

Early-season short day cultivar. Fruit is larger and firmer than Chandler, and has a good flavor. The fruits can be used for fresh-market and processing.

4.0 Nature and Causes of Postharvest Losses

Estimates of the strawberry postharvest losses in India from mishandling and diseases are varying from 15-55 per cent; this means that one-quarter of what is produced never reaches the consumer for whom it was grown and the effort and money required to produce it are lost-forever. Strawberry is an extremely delicate fruit having ≈ 90 per cent water, and they continue their living processes after harvest. The high moisture content and soft texture of strawberry fruit makes them susceptible to mechanical injury, which can occur at any stage from production to retail marketing because of poor harvesting practices, unsuitable field containers and crates, which may have splintered wood, sharp edges, poor nailing or stapling. Over packing or under packing of field or marketing containers and careless handling, such as dropping or throwing off fruits and packed containers during the process of grading, transport or marketing. Abnormal physiological deterioration occurs when fresh produce is subjected to extremes of temperature, of atmospheric modification or of contamination. This may cause unpalatable flavours, dull colour, shrivelling, calyx browning and making it unfit for use.

Rhizopus rot or strawberry leak, a very common fungal postharvest disease and found in the field on ripe fruits as well as on transported and cold stored fruits. Affected fruit is characterized by distinctive brown spots of soft and watery rot, which causes the fruit to collapse and 'leak' juice. Botrytis fruit rot or grey mould rot (*Botrytis cinerea*) (Figure 1) is another major postharvest disease that loose 5 to 10 per cent of fruits at the field level itself. Infected berries may remain firm, yet become

Figure 1. Grey Mould Rot (*Botrytis cinerea*) Affected Fruits

covered with grey spores and mycelium. High humidity favors the moldy growth formation that is visible as a white to grey cottony mass. Sometimes these losses even vary up to 50-60 per cent if the rains are continuing for 2-3 days during the peak harvesting period. These enormous losses may be minimized to a negligible limit by adopting good harvesting and postharvest handling practices by the growers and traders.

Besides, postharvest losses, growers are also facing the problems like non fertilization of flowers during foggy weather and extreme cold resulting in catface and misshapen fruits. Attack of birds bats and rats on ripe fruits, higher incidence of red core disease, whitefly and sap beetle (*Stelidota geminate*) is also very high. Strawberry fruits require special packaging material and care during handling and transportation with assured cool chain right from harvesting to consumer's table. Browning of processed product within short span of storage and disappearance of flavour during hot processing and storage is also a major problem.

Figure 2. Different Fruit Maturity Stages
A. Full mature; B. 3/4th mature; C. Immature

5.0 Disorders

Albinism, cat face, misshapen fruit fasciation, dried calyx are major disorders in strawberry. Albinism affected fruits are mottled white in colour, insipid in taste having no flavor. There are several factors responsible for albinism such as excess use of nitrogen, warm weather and prolonged foggy conditions. The occurrence of misshapen fruit is a serious problem in strawberry cultivation in Northern India due to excess cold during January–February. Prolonged cold waves and foggy conditions obstruct pollination and results in the occurrence of misshapen fruit. The exact cause of dried calyx disorder is yet not known, however higher EC of irrigation water aggravates this problem.

6.0 Postharvest Management

6.1 Maturity indices

Skin colour, taste and T.S.S acid ratio (10:1) are taken into consideration to judge the maturity of fruits. But, the colour of fruits is one of the widely accepted maturity indicators by growers and consumers. Harvesting is the most crucial operation in strawberry production. Strawberry fruits rapidly get over ripe and delayed harvesting may cause unexpected fruit loss within a day or two due to their delicate nature. Fruits are usually harvested when a three-quarter (3/4) surface develops red colour. For medium distance and long distance (350 km and>400 km) market, fruits are harvested at 75 per cent and 50 per cent surface colour development stages. For the local market, fruits are harvested at the full skin colour development stage (Figure 2). Strawberry should not be picked over ripe, as colour, both externally and internally, darkens during storage and become more intense.

6.2 Harvesting

Fruits are preferably harvested in the morning hour to avoid the accumulation of field heat. Fruits picked during rainy days show poor keeping quality. However, a little dew does not seriously reduce the same. Fruits can be harvested just after rain if forced air cooling is available. Fruits should not be exposed to sun, dust and wind during harvesting and thereafter.

6.2.1 Hand Picking

In India, strawberry fruits are harvested manually and collected in trays. Harvesting with the help of fingers and nails, collecting in trays, grading according to size and packing in plastic punnets is common which is an undesirable practice. During this multi-hand handling process, delicate fruits get exposed to internal injuries and bruising. Such injured fruits cause maximum qualitative and quantitative loss during transit and storage (Figure 3).

Figure 3. Manual Harvesting

To avoid these losses, fruits are picked with the help of strawberry clippers and placed in different packing punnets according to their grades. In this improved method of harvesting, pickers carry tray with packing punnets in one hand and clipper in another hand. Thus, the harvesting, grading and packing are completed at a time in the field without touching the fruit. Hand picking is also prevalent in developed countries. While mostly strawberry producers pick fruits for local and retail markets, U- pick in consumers harvesting is popular particularly eastern and mid-western USA. The bulk of fruits destined for the fresh market is hand harvested on a 2-3 day cycle, except in extremely hot weather. The fresh market fruit is snap-picked with only a little stalk remaining. Mostly sorting and packing of fruits for fresh marketing is done directly in the field. Fruits meant for processing are often picked in large containers in the field with minimum care. These fruits are often left in the field for a longer period and their calyx is removed in the field itself.

6.2.2 Mechanical Harvesting

Several mechanical harvesters have been developed in western Europe and North America for strawberry harvesting. These machines are thought to have the potential to pick fresh quality fruit, but, to date these machines are solely used for fruit harvesting meant for supply to processing industries. Although mechanical harvesting system has shown high promise in many economic studies, they have not been widely utilized by growers (Figure 4). Total yield is reduced as all the fruits are stripped off (ripe and unripe) and a lower percentage of 'A' grade fruits is recovered than by hand harvesting (50 per cent vs 90 per cent). The non selectivity of harvest makes the timing of picking critical to obtaining a high proportion of firm and desired colour fruit. Mechanical harvesting is not yet practiced in India, virtually all strawberry fruits in India are harvested by hand due to limited cultivation of fruit in small areas.

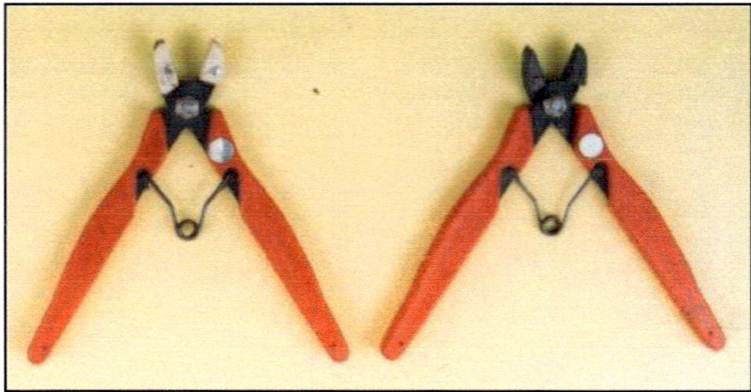

Figure 4. Harvesting Clippers with Cutting and Holding Action

6.3 Sorting and Grading

In India, harvested strawberry fruits are gathered at sorting and grading sites followed by manual grading and filling in packaging punnets. Fruits are generally graded into three groups based on their weight, size and colour (Figure 5). Fruits weighting > 20g are based in grade 'A' followed by grade 'B' (10 – 20g) and grade 'C' (< 10g). Workers engaged in sorting and grading must follow their hygienic practices, their nails should be properly trimmed and fingers free from metallic rings. In developed countries, colour vision mechanical sorters are also used for this purpose. The accuracy of these mechanical sorters varies between 85 to 98 per cent.

Figure 5. Manual Sorting and Grading

6.4 Pretreatments

Many postharvest treatments and microbial antagonists have been identified that can extend the postharvest life of fruit, but these are still in experimental stage. Pyrrolnitrin from *Penicillium cepacia* Burkh has been shown to prevent grey mould growth for 5 days on strawberries stored at 5 °C. Isolates from *Trichoderma viride Pers* and *Gliosporadium roseum* also inhibit grey mould sporulation growth. Fumigation with low concentrations of acetic acid can be used to reduce rot. The firmness of fruit after storage has been enhanced by postharvest dips in 0.5 - 1 per cent calcium nitrate. In addition, the polyamines spermine and spermidine, have been shown to increase the firmness of strawberry in storage. Brief heat treatment after harvest has been tested as a means of minimizing postharvest losses. Hot water treatment of 44 - 48 ° C for 15 minutes delayed the development of *Botrytis cineraria* in inoculated fruits with minimal loss 'of quality. However, the heat treatments diminished calyx colour and fruit skin brightness. In another experiment, it was concluded that Hot-Water Rinsing and Brushing (HWRB) treatments at 60 °C for 20 seconds significantly reduced fruit decay, did not affect its quality and could be suggested as a potential postharvest heat treatment for strawberry fruit.

Strawberry fruits treated with Carnauba and ash gourd wax remained fit for consumption for 5 days while the control fruits were rendered unfit for consumption at the second day of storage at 25 °C.

6.5 Pre-Cooling

Once strawberry fruit is picked, the removal of field heat is critical to preserve fruit quality. It is also desirable to remove field heat as early as possible (1–2 hours of harvest). Rapidly cooled fruit can maintain acceptable quality in storage for more than 10 days if stored below 5 °C at higher relative humidity (>90 per cent) to minimize fruit weight loss. For every 3 hour delay in cooling to 5 °C, decay loss is to be double. Fruits that were held at 30 °C for 6 hours before cooling, weight losses were found to be 4 per cent more and the fruits were 20 per cent softer than those chilled immediately. Strawberries are most commonly cooled either passively by placement in a refrigerated room *i.e.* room cooling or forced air cooling. Hydro cooling of fruits by immersion in chilled chlorinated water has also been shown to be a promising method as it reduces grey mould decay and causes less weight loss than forced air cooling. It is never sufficient to simply place the packaged strawberries inside a cooling room and allow them to cool gradually. For palletized loads, the cooling process should be allowed for a longer time. If the fruits in the centre of the pallet are not adequately cooled, they would begin to decay.

6.6 Packing

A wide variety of packing materials like polystyrene, foamed polythene, paper pulp basket, plastic punnets and plastic mesh baskets are used for packing strawberry fruits. Being a tender fruit, the unit packing size is kept small to avoid spoilage during transit. Owing delicacy of the fruit and small packing size, per kg packing cost varies from Rs. 15 to 20.

In India, fruits are packed in corrugated cardboard cartoons with ribbed punnets and paper liners. Punnets should be neither over filled nor under filled. For overseas trade, fruits are packed in polystyrene containers after cooling at 2-5 °C (Figure 6).

Figure 6. Filled Punnets Assembled in CFB Box

6.6.1 Regular Packing

Strawberries sent to the consumer market are typically packaged in individual punnets. Depending on the size of the package, 4 to 18 of the individual punnets are placed in a tray. The trays are made of either corrugated fiberboard or plastic. The plastic trays are called RPCs (returnable plastic containers), which are cleaned and returned to the strawberry growers for reuse. The design of the vents on both the individual container and the tray is critical for optimizing the rate of pre-cooling. Generally, every individual punnet has four vents two each at the top and bottom. Assembling trays also have sufficient vents to maximize air-to-product contact when the containers are palletized and forced-air cooled.

6.6.2 Commercial/Bulk Packing

Now consumer packing in India has changed and open top-type designed boxes with a 1kg filling capacity are being used for commercial packing. These boxes have a hinged lid with button hole closing provision. This design leads to less bruising and cutting of strawberries than the punnet basket. There has been an increased retail demand for a larger tray that fits five trays to a pallet layer rather than six trays per layer and these open top boxes are doing well.

6.7 Storage

Shelf-life of strawberry fruits is greatly influenced by the storage temperature. Firm ripe fruits stored at 1-5 °C and 95 per cent relative humidity retain their consumer acceptable quality for up to 12-15 days. The fruits stored after long transportation by road have shelf-life only for 4-6 days. Since high CO_2 concentrations (15-20 per cent) reduce fungal growth and help in maintaining quality, strawberries are frequently held in controlled atmosphere storage to increase shelf life. Typical gains in storage life compared to normal air are 3-5 days. Controlled atmosphere at elevated CO_2 (10-20 per cent) and decreased O_2 (5-10 per cent) can also lower decay and increase firmness. CO_2 injury, particularly when > 20 per cent CO_2 is used, is manifested as a bluing of the skin, whitening of inner fruit tissues and fermentative off-flavour.

6.8 Transportation

Packaging and transportation are the most important components of postharvest management and marketing of the strawberry fruits. Depending upon harvesting month, fruit maturity, transportation distance and method of packaging, 20-50 per cent fruit loss takes place during handling and transportation. Big size (A grade > 20g weight) fruits are more prone to transit damage as compared to small (B and C grade fruits).

The thickness of packaging punnets also affects the damaging rate. Therefore, punnets made up of > 0.5mm thickness should not be used for strawberry packaging meant for long distance transport. Right fruit harvesting stage is the key factor affecting the losses during transportation; 2/3rd right stage harvesting could save 50 per cent transportation damage over full ripe harvesting stage fruit during transportation.

For local and medium distance market (12 hour transportation) fruits may be transported through ordinary vans during winter months. For long and delayed transport, cool vans are always used. Before placing the CFB boxes inside the trucks/vans, the boxes should be pre-cooled till the core temperature reaches around 7 °C. CFB boxes having external dimensions of 400x300x65 mm with bursting strength of 11kg/cm^2, compression strength of 300kg/cm^2 with 19 holes should be preferred for palleting. The side wall and centre pallet should be separated with proper bracing. There should be temperature, humidity and CO_2 concentration monitoring devices fitted in the transportation van for regular monitoring of the internal environment of the pallets.

7.0 Postharvest Treatments

Strawberry is known for its unique flavour, high nutrient and digestible fibre content. It is a highly perishable fruit, prone to different postharvest losses due to the quick attack of fungal diseases. To minimize these losses mainly chemicals are used; but, their application is highly restricted due to food safety issues. Therefore, alternative green strategies are being explored to enhance storage life by increasing the natural resistance and anti-oxidative systems of the fruits. Some of the promising safe treatments are given in Table 1.

8.0 Domestic Marketing and Export

Over 85 per cent of the total production of strawberry in India is sold as table fruit in the domestic market, while about 400 ton goes for processing. Major domestic market for the fruit is Delhi, Mumbai, Pune, Bangalore, Hyderabad and Cochin.

Currently, Indian strawberries are being exported to Malaysia, Singapore, France and Belgium. As the strawberries are sent by air route (transportation cost 35 – 50 Rs./kg.), which brings down the profit margin of the growers and market agents. Maharashtra alone produce 80 per cent of the total Indian strawberry production. Strawberry availability in India commences in December and lasts up to March. Most European countries do not have strawberry crop during the winter months and they look for importing quality strawberries from Egypt and Turkey. Indian strawberry growers have a great potential to cater to the need of European countries. Maharashtra state government has recently set up the postharvest handling and packaging infrastructure in Bhilar, the village near Panchgani with the capacity to process 10 tons at a time. This will help increase the longevity of the fruit and prepare it for long distance transport and export.

Directorate of Marketing and Inspection (DMI), Ministry of Commerce, Govt. of India is in the process of formulating strawberry fruit standards for exporting to European Commission (EC) countries. Different countries have different fruit standards formulated by their agencies.

U.S. No. 1 grade consists of strawberries with calyx (cap) attached which are firm, not over-ripe or undeveloped, of uniform size and free of decay and damage. Each fruit must have at least 3/4 of its surface showing pink or red colour. U.S. No. 2 grade consists of strawberries free from decay or serious damage and with at least 1/2 of each fruit showing pink or red colour. U.S. Combination grade is a mixture of U.S. No. 1 and 2, with at least 80 per cent meeting U.S. No. 1 grade.

Canada No. 1 strawberries must be well formed with a minimum diameter of 3/4 of an inch. This means the fruit is not misshapen due to poor pollination, frost injury, insect injury, *etc.* Catfaces at the blossom end are poorly formed berries and are culls. Canada No. 1 strawberries must be of good colour, which means the berries are of the colour characteristic of the variety when mature. Although not defined in the standards,

Table 1. Effect of Postharvest Treatments on Quality and Shelf-life of Strawberries

Treatment	*Methodology*	*Effect on quality*	*References*
Hot water	Emersion of ripe fruit in hot water (45 °C for 3 h)	Inhibits PG activity and delays softening through hemicelluloses deprivation to enzymes. Anthocyanin and vitamin C were negatively affected by hot water, TSS increases while flavonoids, ellagic acid and phenol were least affected.	Panou, *et al.*, 2021
Hot water rinsing and brushing (HWRB)	Hot Water Rinsing and Brushing (HWRB) treatments at 60°C for 20 seconds.	Reduces fruit decay without affecting quality	Vicente *et al.*, 2005
UV-C Irradiation		Reduces calyx browning, increases anthocyanin, phenolics and easters. Increased shelf-life but negatively affected aroma.	Ortiz-Araque *et al.*, 2018
Gamma irradiation	Irradiation of berries @0.5-1.5kGy	Extends the shelf-life, reduce the severity of botrytis rot.	Panou *et al.*, 2020
Blue light illumination	Blue light application during storage at 5 °C.	Increased vitamin C, total sugar, acidity, phenolics and anti-oxidant activity.	Xu *et al.*, 2018
Ozone	Application gaseous ozone @ 0.5-1.5 ppm for 3 minutes	Ozonation at a concentration of 1.5 ppm retained higher firmness, reduced decay loss and extend shelf-life.	Panou, *et al.*, 2021 Maryam *et al.*, 2021

it is generally felt that at least 3/4 of the surface of individual berries should be of good colour. Berries that are hard and green and obviously will not ripen should be scored as grade defects. It is required that the calyxes be attached, berries with calyxes missing should be scored against the general tolerances. Canada No. 1 strawberries must be fairly clean. This means that the fruit may not be affected by dirt which is readily apparent.

The following shall be scored as damage:

i) **Wet bruises**: when the skin is broken or when the achene structure is more than slightly disturbed.

ii) **Soft bruises:** which are more than slight and are discoloured.

iii) **Bruises:** which aggregate more than 1/2 the diameter of the berry or exceed 1/8 inch in depth.

iv) **Flattened areas**: as a result of package or pressure bruises, found mostly around the shoulder should not be scored unless the appearance is materially affected. Materially affected means more than 1/3 of the surface of the berry is flat.

The tolerances for Canada No. 1 grade strawberries are applied by count. Any one defect - 5 per cent, decay - 2 per cent, undersize (under 3/4 inch min. diameter) - 5 per cent.

9.0 Conclusion

There is a need for screening germplasm and varieties for processing and export. Screening/breeding the varieties suitable for growing in summer prone areas. Design and development of fruit harvesting scissors provided with simultaneous cutting and holding action along with design and development of fruit harvesting trays meant for A,B and C grade fruit. Modification of the prevalent packaging punnets (200g filling capacity) in order to cost saving and simplifying the packaging operation is required. Optimization of CA storage conditions for enhancing the storage period. Processing technology for avoiding/delaying browning (pigment degradation) and retention coupled with flavour in processed products. Also, the assessing the demand and supply gap of strawberry fruits in South Europe, East Asia and Arabian countries are required to promote export.

10.0 Important References

Asrey Ram, Jain R.K. and Singh Rajbir 2004. Effect of preharvest chemical treatments on shelf life of Chandler strawberry. *Indian J. Agric. Sci.* **74**: 484-87.

Asrey Ram, Jain, R.K. and Singh, Rajbir 2008. Fruit packaging and transportation studies in strawberry. *J. Food Sci. Technol.* **45**: 145-47.

Bato, P.M., Nagata, M., Cao, Q.X., Hiyoshi, K.and Kitahara, T. 2000. Study on sorting system for strawberry using machine vision (part 2): Development of sorting system with direction and judgement functions for strawberry (Akihime variety)", *J. Jap. Soc. Agril. Mach.* **62**: 101-10.

Collins, J.K. and Perkins-Veazie, P. 1993. Postharvest changes in strawberry fruit stored under simulated retail display conditions. *J. Food Qual.* **16**: 133-43.

Ferriera; M. D; Bercht, J.K., Sargent, S.A, and Aracena, J.I. 1994. Physiological responses of strawberry to film wrapping and pre-cooling methods. *Proc. Fla. State Hortic. Soc.*, **107**: 265-68.

Garcia, J.M., Herrera, S: and Morilla, A. 1996b. Grey mold in and quality of strawberry' fruit 'following postharvest heat treatment. *HortScience* **31**: 255-57.

Jing, W., Tu, K., Shao, X. F., Su, Z. P. and Tang, J. 2010. Effect of postharvest short hot-water rinsing and brushing treatment on decay and quality of strawberry fruit. *J. Food Qual.*, **33**: 262–72.

Kader, A and Smith A. 1992. Modified atmospheres during transport and storage. In: Kader, A. (Ed.), *Postharvest Technology of Horticultural Crops*, University of California publication, **3311**: 85-92.

Kannaujia, P. K., Asrey, Ram and Jha, S. K. 2014. Cultivars and sequential harvesting influences physiological and functional quality of strawberry fruits. *Fruits*, **169**: 239-49.

Maryam *et al.*, 2021. Combined aqueous ozone and ultrasound application inhibits microbial spoilage, reduces pesticide residues and maintains storage quality of strawberry fruits. *J. Food Meas. Charact.* **15**: 1-15.

Mass, J.L., Wang, S.Y., and Galletta, G.J. 1996. Health enhancing properties of strawberry fruits. In: Pritts, M.P. Chandler C.K. and Crocker, T.E. (Eds) *Proceedings of the IV North American Strawberry Conference*, Orlando, Florida, pp. 11-18.

Mokkila, M., Randell,K., Sariola, J., Hagg, M. and Hakkinen, U. 1997. Improvement in postharvest qualtiy of strawberry. *Acta Hortic.* **439**: 553-57.

Moyls, A.L., Sholberg, P.L. and Gaunce, A.P. 1996. Modified Atmosphere packaging of grapes and strawberries fumigated with acetic acid. *HortScience* **31**: 414-16.

Nunes, M.C.N., Brecht, J.K., Morais, A.M.B. and Sargent, S.A. 1995. Physical and chemical quality characteristics of strawberries are reduced by a short delay to cooling. *Postharvest Biol. Technol.* **6**: 17-28.

Ortiz-Araque *et al.*, 2018. Cyclic low dose UV-C treatments retain strawberry fruit quality more effectively than conventional pre-storage single high fluence applications. LWT. *Food Sci. Technol.* **92**: 304-31.

Panou, *et al.*, 2021. Effect of Gaseous Ozone and Heat Treatment on Quality and Shelf Life of Fresh Strawberries during Cold Storage. *Int. J. Fruit Sci.* doi.org/10.1080/15538362.2020.1866735

Panou, *et al.*, 2020. Effect of gamma-irradiation on sensory characteristics, physicochemical parameters, and shelf life of strawberries stored under refrigeration. *Int. J. Fruit Sci.* **20**: 191-06.

Peng, G. and Sutton, J.C. 1991. Evaluation of microorganisms for bio-control of *Botrytis cinerea* in strawberry. *Canad. J. Pl. Pathol.* **13**: 247-57.

Sacks, E.J. and Shaw, D.V. 1993. Colour change in fresh Strawberry fruits of seven genotypes stored at 10°C. *HortScience* **28**: 209-10.

Smith, R.B. 1992. Controlled Atmosphere storage of 'Redcoat' strawberry fruit. *J. Americ. Soc. Hortic. Sci.* **117**: 260-64.

Sreenivas, K. M., Chaudhari, Kirtibala and Lele, S. S. 2011. Ash gourd peel wax: Extraction, characterization, and application as an edible coat for fruits. *Food Sci. Biotechnol.* **20**: 383-387.

Takeda. F., Janisiewicz, W.J., Roitman, I., Mahoney, N. and Abeles, F.B., 1990. Pyrrolnitrin delays postharvest rot in strawberries. *HortScience* **25**: 320-22.

Vicente *et al.*, 2005. Effect of heat treatments on cell wall degradation and softening in strawberry fruit. *Postharvest Biol. Technol.* **38**: 213-222.

Xu *et al.*, 2014. Effect of blue light treatment on fruit quality, antioxidant enzymes and radical scavenging activity in strawberry fruit. *Sci. Hortic.* **175**: 181-86.

Chapter 26

Walnut

Desh Beer Singh

Former-Director,

ICAR-Central Institute of Temperate Horticulture, Old Air Field, Rangreth, Srinagar – 5, Jammu & Kashmir, India

E-mail: deshbsingh@yahoo.co.in

Nazeer Ahmed

Former-Vice Chancellor

Sher-e-Kashmir University of Agricultural Sciences and Technology, Srinagar, J&K, India

E-mail: dnak59@rediffmail.com

1.0 Introduction

Persian (English) walnut (*Juglans regia*) is a member of the family Juglandaceae, and is by far the most important commercial walnut. The genus *Juglans* contains three other groups (the Butternut, East Asian species and Black Walnut species) that produce edible kernels. English walnut has its origins in Eastern Europe, Asia Minor, and points eastward to the Himalayan Mountains. However, there are native *Juglans* in the North, Central, and South America, Europe and Asia. Walnut is a high energy food, rich in oil including omega-3 fatty acids, vitamins and minerals, and valued as a healthy snack food and bakery ingredients. Walnuts are a nutrient-dense food and 100 grams of walnuts contain 15.2 grams of protein, 65.2 grams of fat, and 6.7 grams of dietary fibre. The protein in walnuts provides many essential amino acids. Its alpha linolenic acid has a substantial cardio-protective effect. It contains 'melatonin' an antioxidant produced by the pineal gland and responsible for inducing and regulating sleeps.

2.0 Status of Cultivation

Although native to Europe, it probably was not utilized there until improved forms were imported from Persia. The species came to the new world with English settlers and to California via missionaries. Walnuts were first introduced to California in the 1700s by the Spanish padres. The major walnut growing countries are China, the USA, Iran, Ukraine, Turkey and Mexico. World production of walnut is 3.28 Million MT (FAO 2012). Walnuts are produced commercially in 48 countries on 834.9 thousand hectares. In India, walnuts are grown in Jammu and Kashmir, Arunachal Pradesh, Himachal Pradesh and Uttarakhand. The total area under walnut cultivation in India is 149.5 thousand hectares with an annual production of 284.4 thousand MT in 2011-12 (APEDA 2012). Walnut is an important crop grown in Jammu and Kashmir and it contribute around 98 per cent of the country's output. The state produces about 86,263 tonnes from an area of 61,723 hectares.

The worldwide production of walnut seeds has been increasing rapidly in recent years, with the largest increase coming from Asia. The world produced a total of 3.28 million metric tonnes of walnut seeds in 2012; China was the world's largest producer of walnut seeds, with a total harvest of 1.70 million metric tonnes. The other major producers of walnut seeds are Iran, United States, Turkey, Ukraine, Mexico, Romania, India, France and Chile. Among the major producers, eastern European countries have the highest yield. According to the FAO, the most productive walnut seed farms in 2010 were in Romania, with yields above 23 metric tonnes per hectare (Table 1).

Although native to Europe, it probably was not utilized there until improved forms were imported from Persia. The species came to the new world with the English settlers and to California via missionaries. Walnuts were first introduced to California in the 1700s by the Spanish padres.

In 2019, the world production of walnuts (in shell) was 4.5 million tonnes, with China contributing 56 per cent of the total. Other major producers (in the order of decreasing harvest) were the United States, Iran, and Turkey (Table 1). Walnuts are produced commercially in 48 countries on 834.9 thousand hectares.

The worldwide production of walnut seeds has been increasing rapidly in recent years, with the largest increase coming from Asia. The world produced a total of 4.5 million metric tonnes of walnut seeds in 2019; China was the world's largest producer of walnut seeds, with a total harvest of 1.64 million metric tonnes. The other major producers of walnut seeds are Iran, United States, Turkey, Ukraine, Mexico, Romania, India, France and Chile. Among the major producers, eastern European countries have the highest yield.

Table 1. Top 10 Walnut Producing Countries – 2019

Country	Production (Metric Tons)
China	1,641,479.00
United States	628,255.00
Iran	418,871.00
Turkey	217,334.00
Mexico	162,848.00
Ukraine	125,387.00
Chile	87,323.00
Romania	54,016.00
Uzbekistan	37,228.00

Source: FAO, 2012.

In India, it is mainly grown in Jammu & Kashmir, Himachal Pradesh, Uttar Pradesh and Arunachal Pradesh. The total area under walnut is 67053 hectares with annual production of over 71758 metric tons. (APEDA 2019). J&K accounts for more than 98 per cent of the total walnut production in India.

Walnut cultivation in Jammu and Kashmir is mostly in Badrawah, Poonch, Kupwara, Baramulla, Bandipora, Gandarbal Budgam, Srinagar, Anantnag and other hilly areas. Walnut is also grown in Himachal Pradesh and Uttarakhand to a limited extent. In Himachal Pradesh, it is grown in Mandi (50 ha), Chamba (54 ha) and some areas in Shimla district. To promote walnut cultivation in Himachal Pradesh, a walnut cultivation development station has been established at Nohra in Sirmour district.

3.0 Major Commercial Varieties

Most of the production in India comes from the seedling trees scattered in the temperate region of the north-western Himalayas from 1200 m to about 3500 m above sea level. Many of these trees are as good as any improved cultivars. Therefore, known cultivars grown in the Himalayan region of India have resulted from local selections by the farmers based on their preferences. However, a good number of exotic germplasm has been maintained in the field gene bank at National Bureau of Plant Genetic Resources, Regional Station, Shimla in Himachal Pradesh. The promising selections from exotic collections were 'Lake English', 'Blackmore', 'Tutle-16' and 'Izvor'. Recently, a few indigenous selections were also made as 'Hamdan' and 'Sulaiman' from Jammu and Kashmir; 'Govind' and 'Partap' from Himachal Pradesh and 'Chakrata seedling 13' from Uttaranchal (Table 2).

Walnut (*Juglans* sp.) is the most important temperate nut fruit in the country. Walnuts in India are found in different sizes and shapes. The Indian walnuts are categorized into 4 categories *viz.*, paper-shelled, thin-shelled, medium-shelled and hard-shelled. Walnuts flourished at altitudes of 900 to 3000.

Table 2. Walnut Varieties Grown in India and its Production in 2017-18

State	Production (000 MT)	Varieties
Jammu and Kashmir	275.45	Lake English, Drainovsky and Opex Caulchry
Uttarakhand	21.17	Chakrata Selections
Himachal Pradesh	2.46	Gobind, Eureka, Placentia, Wilson, Franquetfe and Kashmir Budded
Arunachal Pradesh	0.36	Local

Source: National Horticulture Board (NHB).

However, a large number of cultivars have been evolved in the USA which is the main walnut exporting country of the world. Description of some important varieties is given here.

3.1 Exotic Varieties

a) Ashley

It is a high-yielding, early-bearing cultivar which requires heavy pruning to keep the tree vigorous. The kernel is of high quality, good in flavour and light tan in colour. This variety is unsuitable for high rainfall areas due to blight problem.

b) Chandler

It is a very promising cultivar which has evolved from a cross of Pedro X VC 56 -224. The nut is large, oval, and smooth with a good seal and excellent colour. The kernel percentage is 49. The tree is moderately vigorous and semi-upright.

c) Chico

It evolved from a cross of Sharkey and Marchetti. The variety is fruitful on lateral buds and the nut is small but excellent in quality with light-coloured kernel. Heavy pruning is required to maintain nut size.

d) Franquette

It is an old and leading cultivar of France. The nut is small, good shell seal and the kernel is light in colour. The tree is large and upright and is known for its late bud break thus escaping injury from frost during late spring.

e) Hartley

It is one of the most popular commercial cultivars of California. It is a selection from a seedling. The nuts are large with a broad flat base and pointed tip. The shell is light-coloured, thin and seals well. The variety is tolerant to codling moth and blight disease.

f) Howard

Howard is a lateral fruiting and high-yielding cultivar. Nuts are large, round and smooth with a good seal. The kernel is light in colour with a shelling percentage is 50. The tree is small, semi-upright and tolerant to blight.

g) Payne

It is the second leading cultivar of California which originated as a seedling. The nut is medium to small in size with a good shell seal. Trees are moderately vigorous, round and require heavy pruning to maintain vigour.

h) Pedro

It is an early bearing variety due to which it has become a popular cultivar in many walnut growing regions. The nuts are large shell heavy and seal fair. It sheds its pollen over a long period becoming a good pollinizer.

i) Serr

It evolved from a cross of Payne X PI 159568. It is heavy yielding and well adapted to warm conditions. The tree is very vigorous and gives poor yield on very fertile soils. The kernel is light in colour and good in quality. It is susceptible to codling moth and blight disease.

j) Sunland

It is a mid-season variety obtained from a cross of Lompoc X PI 159568 in California. The nut is very large, long, oval with a good seal and kernel colour. It is susceptible to blight and late spring frost.

k) Tehama

It is a heavy-yielding midseason and lateral fruiting cultivar evolved from a cross of Waterloo and Payne. The nut size is large but the shell seal is not so good because sometimes it cracks at the suture exposing the kernel to damage. The kernel colour is good with 50 per cent shelling. The tree is large and vigorous and mainly serves as a pollinizer.

l) Vina

It was evolved from a cross between Franquette and Payne. The kernels are light in colour, high quality and its shelling percentage is 48. It requires good pruning to maintain vigour and nut size.

3.2 Indigenous Varieties

Ten varieties of walnut are being developed at the Central Institute of Temperate Horticulture, Srinagar, J&K, India and released recently. The description of these is as:

a) CITH Walnut-1

High-yielding variety with the largest nuts (27.98g) kernel size (14.79g), good kernel recovery (52.6 per cent), light shell kernel colour. The nuts are long trapezoidal in shape with easy to remove kernel halves suitable for export market (Figure 1a).

b) CITH Walnut-2

Nuts are medium large, ovate, medium shell texture, medium shell colour, strong shell seal, intermediate shell strength, complete shell integrity, plumpy, easy to remove kernel halves and light kernel colour. It gives 16.51 g nut weight and 6.61 g kernel weight (Figure 1b).

c) CITH Walnut-3

Nuts are large, round, medium shell texture, medium shell colour, strong shell seal, strong shell strength, complete shell integrity, well-filled kernel, plumpy, difficult to remove kernel halves and light kernel colour. It gives a nut weight of 16.75 g and the kernel weight of 7.69 g (Figure 1c).

d) CITH Walnut-4

Nuts are medium-large, long trapezoidal, rough shell texture, light shell colour, strong shell seal, intermediate shell strength, complete shell integrity, thin, well-filled kernel, moderately plumy, very easy to remove kernel halves and light kernel colour. It gives nut weight of 14.24 and kernel weight of 6.92 g (Figure 1d).

e) CITH Walnut-5

High yielder, having extra light kernel colour, suitable for the export market, bigger nut (19g) and kernel (9.5g) size, good kernel recovery (48.9 per cent), light shell colour, ovate in shape, moderately easy to remove the full kernel halves (Figure 1e).

f) CITH Walnut-6

Nuts are large broad ovate in shape, shell colour light, intermediate shell seal, Intermediate shell strength, well-filled kernel, moderately plumy and easy to remove kernel halves. It gives in shell nut weight (25.6 g), kernel weight (12.4 g) with kernel recovery (58.8 per cent) (Figure 1f).

(a) CITH Walnut-1 (b) CITH Walnut-2

(c) CITH Walnut-3 (d) CITH Walnut-4

(e) CITH Walnut-5 (f) CITH Walnut-6

(g) CITH Walnut-7 (h) CITH Walnut-8

(i) CITH Walnut-9 (j) CITH Walnut-10

Figure 1. Walnut Varieties

g) CITH Walnut-7

Nuts are large in size, elliptic in shape, rough shell texture, medium coloured shell, strong shell seal, intermediate shell strength and well-filled kernel, plumpy, moderately easy for removal of kernel halves. It gives inshell nut weight of 24.7 g, kernel weight of 12.26 g with kernel recovery of 49.60 per cent (Figure 1g).

h) CITH Walnut-8

Nuts are having light kernel colour, nut weight (21.96g), and kernel weight (11.11g), good kernel recovery (54 per cent), light shell colour, long trapezoidal in shape, very easy to remove kernel halves, rough shell texture, strong shell seal and strong shell strength (Figure 1h).

i) CITH Walnut-9

Nuts are large in size, round in shape, light in colour, strong shell seal, intermediate shell strength and well-filled kernel, plumy, moderate to remove the kernel halves. It gives in shell nut weight of 21.23 g and kernel weight (10.65 g) with kernel recovery of 50.9 per cent (Figure 1i).

j) CITH Walnut-10

Heavy-bearing, Nuts are medium, long trapezoidal in shape, smooth shell texture, medium-coloured shell, intermediate shell seal, intermediate shell strength, well filled kernel, plumy and easy-to-remove kernel halves. It gives in shell nut weight (19.95 g) and kernel weight (11.08 g) with kernel recovery (55.5 per cent) (Figure 1j).

4.0. Nature and Causes of Postharvest Losses

Persian walnuts are best adapted to the Mediterranean climates, with dry, hot summers and mild winters. Cold hardiness is a major limiting factor of Persian walnut. Walnut is highly sensitive to the extremes of winter and summer temperatures as well as to its duration. It requires a climate which is free from frost in spring and from extreme heat in summer. A temperature of even 2 or 3 degrees below freezing point (0 °C) kills leaves, shoots and flowers and thus resulting in crop failure. A high temperature of about 38 °C occurred in the early season causes sun-burning of hulls and shrivelling of kernels resulting sometimes in empty nuts and if it occurs later in the season, the kernels may be partially shrivelled, dark in colour or they may be stuck to the shell.

5.0 Postharvest Management

5.1 Maturity Indices

Walnuts mature as soon as the husk cuts free from the nut, but they usually are not harvested until rains crack the husk, causing the nut to drop, usually in October (Figure 2). Husk denting is the method commonly used by growers to determine the optimum time of harvest. Kernels are considered mature when oil accumulation is complete. This is generally indicated by the browning of the internal packing tissue. However, harvest should not begin until the husk is well split and separated from the shell. In the hottest growing regions of California, kernels may mature 3 weeks before husk "maturation" (*i.e.*, dehiscence). Low temperatures and high RH at night advance dehiscence. Ethephon applications are used to advance the harvest and make nut maturation more uniform throughout the tree/ orchard.

Figure 2. Mature Walnut

5.2 Harvesting

Walnut farmers wait until late August when the green husks of the walnut begin to peel back due to their drying in the summer heat. Mature walnuts lose quality rapidly after they have fallen; hence fallen nuts should be frequently collected to prevent mould, discolouration and decay. To prepare for picking, the ground must be cleared of debris around the trees. Mechanical shakers are attached to the tree and shake clear all of the walnuts which can number in the thousands per tree. Nuts are collected from the ground during the months of September and October.

Put gloves on to prevent the hulls from staining. Step on the walnuts before picking them up to loosen their hulls. Use a pocket knife to pull the hulls off of English walnuts if you prefer not to step on the nuts. Rinse off the nuts to remove any remaining bits of hull.

5.3 Husk Removal and Drying

After removal of the husks, cleaning and drying, they should be stored and marketed separately to fetch a higher price. After collecting nuts, these are cleaned, washed and dried by spreading them on sheets or the floor. Sometimes to improve the appearance of nuts, these are bleached with either alkali or acid solution. Delay in drying causes rapid loss in nut quality and makes walnuts susceptible to mould. Drying nuts stabilises the product's weight and prolongs storage life. Once the hulls have been removed from either type of walnut, the nuts must be dried. There is a certain amount of natural moisture that is within the nut as it grows. This moisture must be removed before the nutmeats are stored (Figure 3).

Figure 3. Drying of Walnuts

The best method of drying the nuts is to place them in a single layer on a screen. An old screen door or window would be perfect for this task. The nuts should be kept in a dry and cool area that does not receive direct sunlight for a minimum of two weeks. After the drying time has been completed, crack the shell of one of the nuts open and remove the meat. If the nutmeat cracks crisply in half, the nuts are fully dried. Lay Walnuts may be dried in a single layer and covered on plastic netting or a screen. Screening the nuts keeps the wildlife from pilfering your harvest.

5.4 Sorting and Grading

Nuts which fall with their husks intact are generally second grade. Walnuts are usually sorted by size and colour, which can be done manually or through an electronic eye. A blower will remove any worm-infested or empty walnuts as their lack of weight lets them be blown away. Shelled walnuts are further graded by colour. Walnut kernels are screened and separated into different sizes. Then the walnuts are sorted by hand and then stored in large cargo containers

Following drying, sizing of the in-shell nut occurs. In-shell walnuts are sized as jumbo, large, medium, or baby according to USDA standards. Walnuts are mechanically cracked as needed. The shelled material is air-separated and screened for size and sent to electronic laser sorting units for kernel colour and shell removal. The product is then packed for shipment to the market place.

Walnuts produced in India have different sizes and shapes and are categorized into paper-shelled, thin shelled, medium-shelled and hard-shelled. These are classified based on their size and shell cracking rate also. The main grades are:

i) Indian special light half
ii) Indian light broken
iii) Indian light pieces
iv) Indian light crumbs
v) Indian light my-fire
vi) Indian light amber halves
vii) Indian light amber broken
viii) Indian light amber pieces.

5.5 Quality Standards

The primary quality criterion is a high oil content (55 to 65 per cent dry weight) that is free of off-flavours caused by the oxidation of polyunsaturated fatty acids. Thus, an important criterion is maintaining kernel water content below 4 per cent. Not only does this retard the progression of events that lead to rancidity, but it also prevents mould growth and maintains the kernel's crispness (Figure 4). If water content drops too low, however, damage to the kernel's covering can enhance O_2 penetration and rancidity.

Figure 4. Kernels

The skin covering the kernel contains chemicals that protect fatty acids in the kernel from becoming rancid. The light-coloured kernels earn a higher price because it indicates that the kernel still has a relatively long shelf-life. Retention of the light-colour is influenced by the integrity of the seal between the two halves of the shell because the shell is an important barrier to O_2 entry. Shell strength and seal integrity are also important in protecting kernels from insect damage and infections from fungi that often follow insect damage.

5.6 Packaging

Walnuts are packaged by weight in giant sacks for shipping. Generally, walnuts are packaged in, net bags (1.5 - 2.5 kg capacity), poly sacks (5 and 10 kg capacity), cartons (10 kg capacity) and flat jute fabric bags (55 kg capacity). For the export purpose, these are packed in double gunny bags. Walnuts are consumed in the winter season, so the problem of their shelf life is seldom felt. The quality of nut meat, however, deteriorates due to darkening and rancidity which are affected by air, moisture, heat and light.

Shelled nuts are packaged in, among other things, boxes lined with parchment paper, each containing 12.5 kg (net). The moisture content of the kernels should be

2-3 per cent. In-shell walnut kernels derive protection against oxidative changes from the intact shell and kernel skin. Shelled products should be packaged in airtight, moisture-proof, opaque or foil packages to maximize shelf life. Unroasted kernels are less likely to take up moisture than roasted kernels. FDA approved antioxidants such as butylated hydroxyanisole (BHA) and butylated hydroxytoluene (BHT) can be applied to the kernels in vegetable oil to enhance stability. Edible coatings may be used as O_2 barriers to retard kernel rancidity. Nitrogen flushing (replacing oxygen with nitrogen) can also extend the shelf life of the unopened sealed product.

5.7 Storage

The fresh taste of walnuts can be maintained by keeping them at low temperature. Heat causes the fat in walnuts to change its structure, which creates off odours and flavours. Fresh walnuts smell mildly nutty and taste sweet. The low water content and high-fat content of the kernel make it relatively metabolically stable and able to tolerate low temperatures. The primary objectives of storage are to maintain the low water content attained after preliminary drying (for enzyme activity suppression, retention of texture and reduction of microbial activity) and limit exposure to O_2 to minimize rancidity. The optimum temperature range for storage is 0 to 10 °C and 50 to 65 per cent RH. The best place to store walnuts is in the refrigerator or freezer, depending on when to use them. The following are some of the tips to maintain the high quality, freshness, flavour and nutritional value of California walnuts during storage.

Storage in air-tight packaging, away from foods with strong odours, like cabbage and onions.

Practice of rotational inventory management of FIFO – First In, First Out.

Storing at 0 °C to 5 °C at 65 per cent relative humidity up to six months.

Storing at -18 °C for more than six months.

Shelf-life can be extended by storage in < 1 per cent O_2. O_2 < 0.5 per cent (balance N_2) or CO_2 levels above 80 per cent can be effective in insect control.

6.0 Domestic Marketing and Export

The country has exported 7249 MT of walnuts mostly to France, Germany, Spain, Portugal, Austria, United Kingdom, Kuwait, Bahrain, Dubai and Saudi Arabia for the worth of Rs.199.80 crores during the year 2012-13. Major exporting countries are the UK, Egypt, Arab Republic, US, Germany, USA, Australia and Taiwan.

Walnut is traded all around the world. The data provided on the export analysis show that there are almost 58 countries and territories, which actively import Walnuts from India. The combined value of total export is 2.71 Million USD. "Connect2India" offers a complete guide on how to export Walnuts from India. From the perspective of the data on Walnut export, India's top 5 trade partners who import Walnuts from Indian exporters are mentioned in Table 3, although the total export value of the top 5 countries is 2.29 million USD yet it consists of 84.5 per cent of the total export value of Walnuts.

Table 3. Walnut Export from India

Country	*Value (USD Million)*
United Kingdom	0.74
New Zealand	0.47
USA	0.47
Germany	0.45
United Arab Emirates	0.16

6.1 Codex Standards for Export

This standard applies to walnut kernels from varieties (cultivars) grown from *Juglans regia L.*

The purpose of the standard is to define the quality requirements for walnut kernels at the export control stage, after preparation and packaging.

6.1.1 Minimum Requirements

In all classes, subject to the special provisions for each class and the tolerances allowed, walnut kernels must be: sufficiently dry to ensure keeping quality; sound; produce affected by rotting or deterioration such as to make it unfit for consumption is excluded; firm; sufficiently developed; shriveled kernels are to be excluded; clean, practically free from any visible foreign matter and from shell; free from insects or mites whatever their stage of development; free from damage caused by pests; free of any rancidity or oily appearance; free from mould; free of abnormal external moisture; free of foreign smell and/or taste.

The condition of the walnut kernels must be such as to enable them to withstand transport and handling, and to arrive in satisfactory condition at the place of destination. The walnut kernels shall have a moisture content of not greater than 5 per cent.

6.1.2 Classification

Walnut kernels are classified into the three classes as defined below according to their quality and colour.

a) "Extra" Class

Walnut kernels in this class must be of superior quality, uniformly light-coloured with practically no dark straw and/or lemon-yellow colour and with no dark brown patch. They must be characteristic of the variety and/or commercial type. They must be practically free from defects except with very slight superficial defects provided that these do not affect the general appearance of the produce, the quality, the keeping quality or its packaging. Scuffing is allowed on quarters, all pieces and halves, provided it covers *no more than 10 per cent* of the surface area of the skin.

b) Class I

Walnut kernels in this class must be of good quality, of colour not darker than light brown and/or lemon-yellow. They must be characteristic of the variety and/or commercial type. Slight defects may be allowed provided that these do not affect the general appearance of the produce, the quality, the keeping quality or its presentation in the package. Scuffing is allowed on quarters all pieces and halves, provided it covers no more than 20 per cent of the surface area of the skin.

c) Class II

This class includes kernels which do not qualify for inclusion in the higher classes, but satisfy the minimum requirements specified above. Walnut kernels in this class must be of colour not darker than dark brown. Defects may be allowed, provided that the walnut kernels retain their essential characteristics as regards general appearance, quality, keeping quality and presentation. This class also includes mixtures of kernels of different colours and is designated in the marking by the words 'mixed colours. Scuffing is not considered as a defect.

6.1.3 Sizing (styles)

Walnut kernels are classified by style as follows:

i) Halves: kernels separated into two more or less equal and intact parts;

ii) Quarters: kernels separated lengthways into four more or less equal pieces;

iii) Large pieces: portions smaller than a "chipped kernel" but larger than a "broken piece";

iv) Broken pieces: portions of kernels which can pass through an 8 mm sizing screen but not through a 3mm sizing screen;

v) Large pieces and halves: a mixture of kernels corresponding to the styles of large pieces and halves and of which the proportion of halves may be specified in the marking.

Tolerance in respect of quality, colour and type shall be allowed in each package for produce not satisfying the requirements of the class indicated (Table 4).

Table 4. Quality and Colour Tolerance

Defects Allowed	*Tolerances Allowed (per cent by weight of kernels)*		
	Extra	*Class I*	*Class II*
i) Kernels not satisfying the minimum requirements, which include not more than:	4	6	8
Rotten kernels 0.5 1 b 2 b	0.5	1[b]	2[b]
Mouldy kernels 0.5 1 b 2 b	0.5	1[b]	2[b]
Shell fragments or foreign matter	0.1	0.1	0.1
ii) Kernels darker in colour,	8	9	10
iii) Scuffing (halves only)	10	10	-

6.1.4 Mineral Impurities

Not greater than 1g/kg acid insoluble ash.

6.1.5 Size Tolerances (styles)

For all styles, a minimum percentage of kernels corresponding to the style indicated in the marking is required and a maximum percentage by weight of kernels different from the style indicated is tolerated (Table 5).

Table 5. Size Tolerance

Style	*Minimum Percentage and Tolerances Allowed (per cent by weight of kernels)*					
	Halves	*Chipped Kernels*	*Quar-ters*	*Large Pieces*	*Broken Pieces*	*Frag-ments*
Halves	85a	15b	5c	1c	1c	
Quarters			85a	15b	5c	1c
Large pieces				85a	15b	1c
Broken pieces				10b	90a	1d
Large pieces and halves	20 b			65a	15b	1c

a: Minimum percentage; b: Tolerances allowed; c: Included in 15 per cent tolerance; d: Included in 10 per cent tolerance.

6.1.6 Presentation

a) Uniformity

The contents of each package may be uniform and contain only kernels of the same origin, crop year, quality, style and when applicable of the same variety and commercial style. Uniformity of colour is compulsory for Extra Class and Class I. However, about shape, "halves" pass through a 15 mm mesh and "chipped kernels" may be included without limitation in consignments of "large pieces". The visible part of the contents of the package must be representative of the entire contents.

b) Packaging

Walnut kernels must be packed in such a way as to protect the produce properly.

If wooden packaging is used, the produce must be separated from the bottom, sides and lid by paper or suitable protective material.

The materials used inside the package must be new, clean and of a quality such as to avoid causing any external or internal damage to the produce. The use of materials and particularly of paper or stamps bearing trade specifications are allowed provided that the printing or labelling has been done with non-toxic ink or glue.

Kernels may be packed in airtight sealed containers, in a vacuum or an inert gas.

c) Marking

Each package must bear the following particulars in letters grouped on the same side, legibly and indelibly marked and visible from the outside:

i) Identification

ii) Nature of produce

iii) Origin of produce

iv) Commercial specifications

6.2 Definition of Defects for Walnut Kernels

Any defect adversely affecting the appearance or edibility of the kernel including:

i) Staining or discolouration: abnormal colouration which covers more than one-eighth of the surface of the kernel and which is of a colour in pronounced contrast with the colour of the rest of the kernel (dark blemishes or areas of discolouration);

ii) Embedded dirt: kernels or portions of kernels with dirt or other foreign material embedded into the flesh of the kernel.

iii) Crushing of more than 5 per cent of the volume of the kernel;

iv) Drying defect: the kernel is moist, soft or leathery.

6.3 Quarantine Issues

Insect infestation is a potentially important problem as are the fungal infections that often accompany insect damage. Fumigation with methyl bromide or phosphine were used for disinfestation but the use of the former is being restricted and insect resistance to phosphine has been reported. Initial disinfestation treatment of 0.4 per cent O_2 for 6 days followed by storage with 5 per cent O_2 at 10 °C and application of the Indian meal moth granulosis virus (as bioagent) for control of the navel orange worm and the Indian meal moth (*Plodia interpunctella* Hübner) has been reported as an alternative to fumigation treatments.

7.0 Conclusion

The walnut industry in India is going to grow rapidly. Because the fresh domestic market is limited due to the lack of proper marketing facilities as well as the demand for in-shell walnuts with cream-coloured kernels in the international market. Keeping in view defined constraints, efforts are required to place the walnut industry on sound footing. A strategic research and development program and ensured implementation are needed to popularize walnut cultivation in India to compete within the World Trade Organization regime. If the following measures are not implemented, exports will be negatively affected, resulting in loss of foreign exchange.

i) Identification and testing of superior genotypes based on inshell nut weight, creamish kernel colour and identification of superior trees from naturalized populations from the walnut growing areas need to be initiated to extend the new plantations. Many select trees are exceptionally good for nut size and kernel quality compared with exotic materials.

ii) Sound research programs on cultural techniques, propagation, pollination, nutrition and postharvest handling and storage of walnuts need to be strengthened and initiated for commercialization.

iii) Clonal rootstocks and dwarf cultivars of walnut need to be introduced for adaptive research and high-density plantation culture.

iv) Expansion of walnut growing in areas of northeastern hill region of India especially in the states of Meghalaya, Arunachal Pradesh, Sikkim and Darjeeling district of West Bengal needs to be commercialized as scattered plantings of walnuts in farmers' backyards were already established.

v) The existing walnut marketing system needs to be strengthened.

vi) Walnut growing may be diversified into other similar agro-climatic regions like the northeastern hill region states.

8.0 Important References

Anon. 2010. *Hortic.* Horizon: New Walnut Varieties released. *Ind. J. Hortic.* **67**: i-vii.

APEDA, 2012. *http: //agriexchange.apeda.gov.in/product_profile/prd_profile.aspx? categorycode=0203.*

Elaine, B and Feldman, M. D. 2002. The Scientific evidence for a beneficial health relationship between walnuts and coronary heart disease. *J. Nutr.* **132**: 1062S–1101S.

Beuchat, L.R. 1978. Relationship of water activity to moisture content in tree nuts. *J. Food Sci.* **43**: 754-55, 58.

Bhat, A.R. and Ahmad, F. 2002. Hamdan and Sulaiman–new walnut varieties for North-Western Himalayas. *Indian Hortic.* **47**: 33-34.

Diousse, L., Pankow, J.S., Eckfeldt, J.H., Folsom, A.R., Hopkins, P.N, Province, M.A., Hong. and Ellision,

R.C. 2001.Relation between dietary linoleic acid and coronary artery disease in the National Heart, Lung and Blood Institute Family Heart Study. *Am. J. Nutr.* **74**: 612-19.

ERS, 2005. http: //www.agmrc.org/commodities__ products/nuts/english-walnuts-profile/

FAO, 2012. http: //www.fao.org/statistics/en/.

FAO, 2004. http: //www.fao.org/statistics/en/.

FAS, 2012. http: //www.agmrc.org/commodities__ products/nuts/english-walnuts-profile/

FAO Stat. 2004. http: //faostat3.fao.org/faostat-gateway/go/to/download/Q/QC/E

FAO, 2020. http: //www.fao.org/statistics/en/.

Gangwar, R.P. and Kumar H. 1975. Studies on germination capacity of walnut (*J. Regia* L) in relataion to their size and position in the soil. *Haryana J. Hortic. Sci.* **4**: 142-44.

Gaur N.V.S. 1980. Effect of gibberellin, sulphuric acid and cracking on the germination and growth of seedlings of walnut. *Indian J. Hortic.* **37**: 16-19.

Johnson, J.A., Vail, P.V., Soderstrom, E.L., Curtis, C.E., Brandl, D.G., Tebbets, J.S. and Valero, K.A. 1998. Integration of nonchemical, postharvest treatments for control of Navel orange worm (*Lepidoptera: Pyralidae*) and Indian meal moth (*Lepidoptera: Pyralidae*) in walnuts. *J. Econ. Ent.* **91**: 1437-44.

Maté, J.I., Saltveit, M.E. and Krochta, J.M. 1996. Peanut and walnut rancidity: Effects of oxygen concentration and RH. *J. Food Sci.* **61**: 465-68, 72.

NASS, 2013. http: //www.agmrc.org/commodities__ products/nuts/english-walnuts-profile/

Olson, W.H., Labavitch, J.M., Martin, G.C. and Beede, R.H. 1998. Maturation, harvesting and nut quality. In: Walnut Production Manual, D.E. Ramos (ed) Univ. Calif., Div. *Agric. Nat.* Res. Pub. No. **3373,** pp. 273-276.

Puttoo, B.L., Razdan, V.K. and Baschadhary, K.C. 1990. Constraints in the production of walnuts in India. *Acta Hortic.* **284**: 327.

Qureshi, A.S. and Dalal, M.A. 1990. Status of nut crops in Jammu and Kashmir state. *Prog. Hortic.* **17**: 195-05.

Reiter, R.J., Manchester, L.C. and Tan, D.X. 2005. Melatonin in walnuts: influences on levels of Melatonin and total antioxidant capacity of blood. *Int. J. Appl. Basic Nutr. Sci.* **21**: 920-24.

Sharma, R. 2012. Area and production database of fruit crops. Directorate of Hortic., State Department of Horticulture, Jammu and Kashmir Government, India.

van Steenwyk, R.A. and Barnett, W.W. 1998. Insect and mite pests. In: *Walnut Production Manual*, D.E. Ramos (ed) Univ. Calif., Div. Agric. Nat. Res., Pub. No. **3373,** pp. 247-253.

Zettler, J.L., Halliday, W.R. and Arthur, F.H. 1990. Phosphine resistance in insects infesting stored peanuts in the southeastern United States. *J. Econ. Ent.* **82**: 1508-11.

Vegetables, Flowers, Plantation Crops and Seed Spices

Chapter 27

Onion

R.P. Gupta

Ex-Director, NHRDF

Chief Consultant MIDH, Dept. of Agriculture Co-operation and Farmers Welfare, Ministry of Agriculture and Farmers Welfare, Govt. of India, New Delhi

E-mail: rpgupta5@yahoo.com

H.P. Sharma

Joint Director, NHRDF

National Horticultural Research and Development Foundation, New Delhi – 110058, India

E-mail: hariprakash.sharma@gmail.com

1.0 Introduction

The onion (*Allium cepa* L.) is the most important commercial vegetable grown and consumed almost all over the world. Onion bulbs have been used as a food source since ancient times. Traces of onion remains were found alongside date stones and fig remains from Bronze Age that dates back to 5000 BC. However, it is not clear if these were cultivated onions. Although there is no clear-cut evidence it is believed to have originated in the central Asia as evidenced by onion traces being found in the eye sockets of Egyptian burials.

Onion is grown under a wide range of climatic conditions, *i.e.* temperate, tropical and subtropical. It grows well under mild climate without extreme heat or cold or excessive rainfall. Its growth, bulb development and bolting or flowerings are influenced by photoperiod and temperature. It does not thrive well when average rainfall exceeds 75-100 mm during the monsoon period. The ideal temperature for its vegetative growth is 12.8-23.0 °C. Warm temperature favours good development of bulbs. In India onion varieties grown in plains are of short-day type, requiring 10-11.5 hours day length, while long-day types require 13-14 hours day length. These long-day types are cultivated in hills. Sudden rise in temperature in *rabi* season results in the development of small sized bulbs that mature early. The long-day varieties do not develop bulb under short day conditions, whereas short day varieties if planted under long day conditions develop bulb. Onion has been reported to contain various organosulfur compounds that have antibacterial and anti-carcinogenic properties and are rich in flavonoids like quercetin a valuable natural source of antioxidant. Carbohydrates in onion constitute about 80 per cent of dry matter and the major non-structural carbohydrate of the onion bulb is fructo-oligosaccharides, well known as fructan, followed by glucose, fructose, and sucrose. Studies conducted at several places revealed that the postharvest losses of the onion range from 15-40 per cent. Among the different components of postharvest losses the decay losses contribute to around 8-10 per cent whereas the losses due to sprouting and desiccation ranges from 5-8 per cent and 20-25 per cent, respectively, depending upon the period of storage.

2.0 Status of Cultivation

2.1 Global Scenario

Onion is grown around the world (Table 1) covering an area of 51.93 lakh hectares with a total production of 999.68 lakh tons and productivity of 19.26 tons/ha.

China and India are leading onion growing countries followed by the U.S.A., Egypt, Turkey, Pakistan, Sudan, Bangladesh, Iran (Islamic Republic of), Russian Federation, Algeria, the Republic of Korea, Indonesia, Brazil, Mexico, Spain, Nigeria, Japan. The productivity of onion is maximum (73.22 tons/ha) in the Republic of Korea, followed by the USA (60.54 tons/ha), Spain (55.01 tons/ha), Japan (50.93 tons/ha), Iran (39.18 tons/ha) and the Netherlands (36.63 tons/ha) while productivity in India is quite low (18.70 tons/ha).

Table 1. Global Scenario of Major Onion Producing Countries (2019)

Countries	*Area (Lakh Ha)*	*Production (Lakh Tonnes)*	*Productivity (t/ha)*
Algeria	0.50	14.14	32.07
Bangladesh	1.72	18.03	10.45
Brazil	0.48	15.57	32.34
China, mainland	11.28	249.08	22.09
Egypt	0.88	30.81	35.03
India	12.20	228.19	18.70
Indonesia	1.59	15.80	9.93
Iran (Islamic Republic of)	0.45	17.79	39.18
Japan	0.26	13.19	50.93
Kazakhstan	0.33	9.16	27.93
Mexico	0.48	14.87	31.01
Myanmar	0.71	10.33	14.46
Netherlands	0.34	12.56	36.63
Niger	0.38	13.13	34.88
Nigeria	5.93	13.75	2.32
Pakistan	1.48	20.83	14.03
Republic of Korea	0.22	15.94	73.22
Russian Federation	0.58	16.70	28.68
Spain	0.27	14.59	55.01
Turkey	0.69	22.00	32.02
U.S.A.	0.52	31.70	60.54
Ukraine	0.54	9.98	18.52

Source: FAO Website.

2.2 National Scenario

Onion is grown widely in different parts of the country. It is an important foreign exchange earning crop to the country. India produces around 260.91 lakh tons of onion annually ranking second in the world. The production has increased more than twelve times during the past four and half decades. The reason for increase in production is due to an increase in area as well as productivity under onion cultivation from 1.94 lakh ha and 10.51 tons/ha during 1974-75 to 14.31 lakh ha and 18.23 tons/ha respectively in 2019-20. India possesses many innate advantages over other onion producing countries due to its large genetic base, varied soil and climatic conditions and skilled human resources. However, the productivity is still low. Poor soil fertility, use of fewer inputs like manures, fertilizers, crop protection chemicals, high labour cost and crop loss due to diseases, and lack of varieties resistant to pests and diseases are the major reasons for low productivity. Among the different onion growing states, Maharashtra ranks first in area and production, followed by Madhya Pradesh, Karnataka, and Gujarat. Maharashtra has the highest share in both (43.18 per cent) and production (40.94 per cent). However, the average productivity was found to be higher in Sikkim (56.45 tons/ha), Haryana (25.42 tons/ha), Madhya Pradesh (24.56 tons/ha) and Gujarat (24.52 tons/ha) (Table 2).

Table 2. Area, Production and Productivity of Onion in India (2019-20)

State	*Area (in '000 ha)*	*Production (in '000 MT)*	*Productivity (ton/ha)*
Andhra Pradesh	43.00	981.00	22.814
Assam	8.16	88.89	10.90
Bihar	58.00	1,313.00	22.64
Chhattisgarh	25.33	402.70	15.90
Gujarat	58.00	1,422.00	24.52
Haryana	24.00	610.00	25.42
Himachal Pradesh	3.06	65.11	21.27
Jammu and Kashmir	4.32	77.84	18.03
Jharkhand	19.77	293.27	14.83
Karnataka	160.00	2,275.00	14.22
Madhya Pradesh	173.89	4270.70	24.56
Maharashtra	618.00	10,683.00	17.29
Manipur	0.58	5.59	9.61
Meghalaya	0.56	5.04	8.98
Mizoram	0.27	1.80	6.67
Nagaland	0.61	5.72	9.40
Orissa	27.47	291.55	10.61
Punjab	10.34	236.49	22.88
Rajasthan	75.00	1,242.00	16.56
Sikkim	0.63	35.35	56.45
Tamil Nadu	40.00	435.00	10.88
Telangana	9.50	163.20	17.18
Tripura	0.16	1.05	6.43
Uttar Pradesh	28.00	454.00	16.21
Uttarakhand	4.46	45.47	10.20
West Bengal	37.39	669.80	17.91
Others	0.87	16.71	19.30
Total	**1,431.37**	**26,091.37**	18.23

Source: Horticulture Statistics Division, DAC and FW, MOA and FW, GOI, New Delhi

2.3 State-wise Status of Onion Production in India

Two to four crops of onions are now taken in various parts of the country. Fresh onion starts coming in July onwards in the markets from Kurnool in the State of

	Jan	Feb	Mar	Apr	May	Jun	Jul	Aug	Sep	Oct	Nov	Dec
Andhra Pradesh		■	■	■			■	■	■	■	■	
Bihar					■	■					■	■
Gujarat	■	■	■	■	■					■	■	■
Haryana					■	■					■	■
Himachal Pradesh						■						
Karnataka	■		■	■			■	■	■	■	■	
Madhya Pradesh	■	■	■	■						■	■	■
Maharashtra	■	■	■	■	■				■	■	■	■
Orissa	■		■	■								
Punjab					■	■					■	■
Rajasthan				■	■					■	■	■
Tamil Nadu			■	■			■	■	■	■	■	
Uttar Pradesh					■	■					■	■
Uttarakhand						■						
West Bengal		■	■	■								

■ Available season

Figure 1: Round the Year Availability of Onion in India

Andhra Pradesh, Tirpur in Tamilnadu, Chitradurga and Chellikere area of Karnataka and continues up to April-May. In Maharashtra State, fresh onion starts coming in market in September-October (Satara) and continues till April-May from Nasik, Ahmednagar, Pune and other district. Gujarat State produces three crops *i.e.* Kharif (November-January), late kharif (February-March) and rabi (April) in Bhavnagar, Amreli, Rajkot, Junagarh and Jamnagar districts. Kharif onion production although not common in north India is now taken up in some pockets of Rajasthan (Alwar, Dausa, Bharatpur and Ajmer), Haryana (Rewari, Mewat and Gurgaon districts) and Punjab (Patiala) states. In Northern, Eastern and Central India, *rabi* onion is commonly cultivated which is available in April-May. Multiplier onion (shallot) is grown in Tamilnadu and Odisha states. The quality of multiplier (shallot) onion produced in Odisha is better. Small pickling onion is produced in Andhra Pradesh near Mydukur in Cuddapah district during kharif season and all the three seasons in Karnataka near Chickballapur and Bangalore exclusively for export purpose. The quality of small onion from Karnataka is better due to the prevalence of mild climate almost round the year. The availability of onion in India from different locations are mentioned in Figure 1.

3.0 Major Commercial Varieties

A large number of onion varieties for cultivation in different regions and also in different seasons have been developed in India. The high-quality seed of all onion varieties developed by NHRDF are being produced and distributed and popularized by NHRDF all over the country. These varieties of onions have been widely accepted and commercially grown by the farmers in the country as well. As a result of which there is considerable improvement in production, productivity and availability of onion for domestic and export markets. The list of onion varieties developed by different agencies in the country is given in Table 3.

Table 3. The List of Onion Varieties Developed by Different Agencies in the Country

Name of Variety	*Developed by*	*Season*
COMMON BIG ONION (*Allium cepa* L.)		
Red Coloured Bulbs		
Agrifound Dark Red	NHRDF, Nashik	*Kharif*
Agrifound Light Red	NHRDF, Nashik	*Late Kharif and Rabi*
Arka Kalyan	IIHR, Bangalore	*Kharif*
Arka Niketan	IIHR, Bangalore	*Late Kharif and Rabi*
Arka Pragati	IIHR, Bangalore	*Rabi*
Arka Bheem(Syn-6)	IIHR, Bangalore	*Rabi*
Baswant -780	MPKV, Rahuri (MS)	*Kharif*
Bhima Raj	DOGR, Rajgurunagar	*Kharif and Late kharif*
Bhima Red	DOGR, Rajgurunagar	*Kharif*
Bhima Shakti	DOGR, Rajgurunagar	*Late kharif and Rabi*
Bhima Super	DOGR, Rajgurunagar	*Kharif and Late kharif*
Bhima Kiran	DOGR, Rajgurunagar	*Rabi*
Bhima Light Red	DOGR, Rajgurunagar	*Rabi*
Bhima Dark Red	DOGR, Rajgurunagar	*Kharif*
Bhima Shweta	DOGR, Rajgurunagar	*Rabi*
Bhima Shubra	DOGR, Rajgurunagar	*Kharif*
Bhima Safed	DOGR, Rajgurunagar	*Rabi*
Hissar-2	HAU, Hissar	*Rabi*
Kalyanpur Red Round	CSAUAT, Kanpur	*Rabi*

Name of Variety	*Developed by*	*Season*
N-2-4-1	Dept. of Agril.(MS)	*Late Kharif and Rabi*
N-53	Dept. of Agril. (MS)	*Kharif*
NHRDF Red (L-28)	NHRDF, Nashik	*Rabi*
NHRDF Red-2 (L-355)	NHRDF, Nashik	*Rabi*
NHRDF Red-3 (L-652)	NHRDF, Nashik	*Rabi*
NHRDF Red-4 (L-744)	NHRDF, Nashik	Late *Kharif and Rabi*
NHRDF FURSUNGI	NHRDF, Nashik	Rabi
Phule Samarth	MPKV, Rahuri (MS)	*Late Kharif*
Punjab Red Round	PAU, Ludhiana	*Rabi*
Punjab Selection	PAU, Ludhiana	*Rabi*
Pusa Red	IARI, New Delhi	*Late Kharif and Rabi*
Pusa Madhavi	IARI, New Delhi	*Rabi*
Pusa Riddhi	IARI, New Delhi	*Rabi*
Pusa Sona (Sel-153-1)	IARI, New Delhi	*Rabi*
Pusa Shobha (Sel-126)	IARI, New Delhi	*Rabi*
Pusa Ratnar	NBPGR, New Delhi	*Rabi*
Udaipur -101	RAU, Bikaner Campus, Udaipur	*Rabi*
Udaipur -103	RAU, Bikaner Campur, Udaipur	*Rabi*
VL – Piaz – 3	VPKAS, Almora	*Rabi*
Yellow Coloured Bulbs		
Arka Pitamber	IIHR, Bangalore	*Rabi*
Early Grano	IARI, New Delhi	*Late Kharif and Rabi*
Phule Suvarna	MPKV, Rahuri (M.S)	*Late Kharif and Rabi*
Spanish Brown	IARI, RRS, Katrain, Kullu (HP)	In Hills
White Coloured Bulbs		
Agrifound White	NHRDF, Nashik	*Rabi*
Akola Safed	PKV, Akola (M.S)	*Rabi*
Bhima Subra	DOGR, Rajgurunagar	*Rabi*
JNDWO	GAU, Junagadh	*Rabi*
GJWO-3	GAU, Junagadh	*Rabi*
GAWO-2	GAU, Junagadh	*Rabi*
N-257-9-1	Dept. of Agril.)M.S)	*Rabi*
Phule Safed	MPKV, Rahuri	*Rabi*
Punjab-48	PAU, Ludhiana	*Rabi*
Pusa White Flat	IARI, New Delhi	*Rabi*
Pusa White Round	IARI, New Delhi	*Rabi*
Udaipur-102	RAU, Bikaner Campur, Udaipur	*Rabi*
SMALL COMMON ONION		
Agrifound Rose	NHRDF, Nashik	*Kharif in Andhra and All* 3 seasons in Karnataka
Arka Bindu	IIHR, Bangalore	*All 3 seasons in* Karnataka

Name of Variety	*Developed by*	*Season*
Arka Swatista	IIHR, Bangalore	*White coloured – Kharif* and Rabi In Karnataka
MULTIPLIER ONION (*Allium cepa* var. *aggregatum*)		
Agrifound Red	NHRDF, Nashik	*Kharif and Rabi*
Arka Ujjwal	IIHR, Bangalore	*Dark Red Multiplier type* for Karnataka
Co-1	TNAU, Coimbatore	*Kharif and Rabi*
Co-2	TNAU, Coimbatore	*Kharif and Rabi*
Co-3	TNAU, Coimbatore	*Kharif and Rabi*
Co-4	TNAU, Coimbatore	*Kharif and Rabi*
Co-On-5	TNAU, Coimbatore	*Kharif and Rabi* (*Mutlore type*)
MDU-1	TNAU, Coimbatore	*Rabi*
ONION HYBRIDS		
Arka Kirtiman	IIHR, Bangalore	*Rabi*
Arka Lalime	IIHR, Bangalore	*Rabi*
Hybrid - 35	IARI, New Delhi	*Rabi*
Hybrid - 63	IARI, New Delhi	*Rabi*

4.0 Nature and Causes of Postharvest Losses

In onion, many operations are performed in preharvest and postharvest stages for quality retention and reduction of postharvest losses. Cumulative losses at different stages of handling of onion range from 35 to 50 per cent. The major loss in onion after the harvest is due to loss in weight since onion contains 80-85 per cent moisture. The factors causing postharvest losses are mechanical injuries, improper drying/curing, sprouting, decay and weight loss (shrivelling). Mechanical injuries are caused when too large bags are used, which are difficult to handle and cause mechanical damage to onion during loading and unloading for marketing. Drying/Curing of onion plays a vital role in reducing postharvest losses of onion. The purpose of drying is to remove excess moisture from the outer skin and neck region of the onions that prevents infection by disease causing organisms while minimizing shrinking caused by removal of excess moisture from the interiors. Sprouting of onion greatly reduces its, value and accelerates deterioration. Harvesting of bulbs at a premature stage increases chances of weight loss and sprouting during storage. Similarly, close neck cut of bulbs also aggravate postharvest decay. Various preharvest crop management practices *viz.*, irrigation, fertilizer and chemical application, time and stage of harvesting affect postharvest losses. Further, curing, sorting, grading, packing; storage and transportation are the major factors responsible for quality ensuring and reducing losses of the harvested onion.

4.1 Effect of Preharvest Factors on Postharvest Management

4.1.1 Varieties

Light red varieties *e.g.* Agrifound Light Red, Arka Niketan, N-2-4-1, Pusa Red and NHRDF Red-2, NHRDF Red-4, NHRDF FURSUNGI, Bhima Light Red, Bhima Kiran generally grown during the *rabi* season are good for storage. Dark red varieties having lower TSS and dry matter content as well as less of dried outer scales grown in the *kharif* season have poor storability compared to light red varieties.

4.1.2 Seed

Good quality seed from genuine sources should be used for getting good quality bulbs free from doubles, bolters, thick neck and long shelf-life.

4.1.3 Soil

It has been observed that the onion in sandy soil does not produce compact bulbs with high TSS as compared to the crop grown in clay loam soil. Hence clay loam soil should be preferred for raising the crop meant for long storage or long distance transportation.

4.1.4 Fertilizer

FYM @ 25 tonnes per hectare and Nitrogen @ 100 kg per hectare should be applied for bulbs with good storable quality. This will, however, depend upon the nutrient status of the soil. Potassium @ 50 kg per hectare is recommended for good quality storable bulbs. Top dressing with N should be completed before 60 days after transplanting to have thin necks. High dose of fertilizer (N: K: S) increases weight loss, rotting and sprouting percentage. Increased N doses lead to rotting when stored in natural conditions. Use of calcium ammonium nitrate as N source has been reported to improve the bulb shelf life of bulbs.

Supplementary application of organic fertilizers comprising of 3 per cent panchagavya + 50 per cent FYM + 50 per cent poultry manure recorded better postharvest quality of onion.

4.1.5 Irrigation

Light and frequent irrigations are necessary. Heavy watering must be avoided at all costs. Irrigation at 10-15 days interval should normally be given. It should be stopped 10-15 days before harvesting. Uprooting of bulbs immediately after irrigation reduces their storage life. Irrigation after the crop has reached maturity results in re-sprouting and doubles which increase losses in storage. Irrigation through drip has been reported to be better compared to flood irrigation in reducing the storage losses.

5.0 Postharvest Management

5.1 Preharvest Treatments

Bavistin @ 0.1 per cent sprayed 90 days after transplanting or Klorocin @ 25 kg per hectare as preharvest treatment reduces the spoilage of harvested produce.

5.2 Harvesting

The harvesting of onion bulbs should be done after one week of 50 per cent neck fall in *rabi* season. In *kharif* season when bulbs have fully developed and tops of the leaves have started yellowing, bulbs should be considered ready for harvesting. Bulbs harvested at a premature stage do not undergo the dormant stage and incur more losses due to weight loss and sprouting. Delay in harvesting results in splitting and bolting of bulbs. Such bulbs do not store well in storage.

No doubt production and productivity have increased substantially over the past 20 years; it is still quite low when compared to countries like the Netherlands, United States of America, China and Korean Republic. The constraints which have been identified for low production and productivity are enumerated as under:

i) Non-availability of high yielding storable varieties for all three seasons.

ii) Lack of strategy for expansion of area in kharif, late kharif and rabi seasons in non-traditional pockets.

iii) Non-availability of quality seed of improved varieties in adequate quantities.

iv) Onion varieties developed are not uniform in colour, shape and size. Bolting and doubles are observed in almost all varieties. The marketable recovery thus is low.

v) Non-availability of high yielding and storable F1 hybrids.

vi) Susceptibility to diseases and insect pests.

vii) Sub-optimal standards of cultivation adopted by farmers.

viii) Non-availability of high TSS white onion varieties suitable for dehydration and high TSS bigger sized bulb of yellow varieties for export.

ix) Inadequate storage facilities, transportation and marketing support.

5.3 Top Cutting

Cutting of tops close to bulbs causes entry of moisture and thereby more losses are caused due to decay. The NHRDF has standardized the technology and recommended to cut the tops of harvested onion at 2-2.5 cm above the bulbs so as to have closed neck.

5.4 Drying and Curing

The purpose of drying is to remove excess moisture from the outer skin and neck region. Curing is an additional process of drying to develop skin colour. Field curing for 3-5 days along with the tops by windrow method followed by 10-12 days of curing in shade before storage for *rabi* crop is necessary to minimize the losses. In *kharif* season if temperature at harvest is low, 10-15 days sun curing may be desirable, but it is not always possible.

The curing of kharif onion is very difficult in the rainy season due to cloudy conditions. The NHRDF has developed an artificial curing chamber for curing of kharif onion. The temperature and air velocity has been standardized as 35 °C and 3.2 ms^{-1} for perfect curing of bulbs in 16 hours for marketing using this chamber.

5.5 Sorting and Grading

The diseased, thick-necked, damaged, bolted and double bulbs should be removed at the time of sorting and grading. The medium sized bulbs of 4-6 cm size are good for storage. Sorting and grading are mostly done manually by the traders for domestic as well as export supplies. However, the NHRDF has developed electrically operated commercial graders, which are available for grading of onion in different sizes (Figure 2).

Figure 2. Onion Grader.

5.6 Packaging

Onions are often sent loose in tractor trolleys or bullock carts directly from farmer's fields to the market yard. In some markets, farmers bring onions in perforated 40 kg jute or plastic (9x10 mesh) woven bags having enough ventilation. Small pickling onions and multiplier onions are packed in 14-15 kg wooden baskets/bamboo baskets. For export, common big onions are packed in 5 to 25 kg size open mesh jute bags (Figure 3). However, packaging has been improved now, and plastic leno bags are gradually replacing the traditional jute bags. In some countries, plastic woven bags with good labelling and printed information about crop, variety, weight, the address of the supplier and other details are commonly shared. Use of netted leno bags has also been started on a small scale for domestic market as also for temporary storage. Plastic woven bags are not only attractive but uniform onions when graded, could also be easily seen from outside.

Figure 3. Packaging of Onion Systems.

5.7 Storage

Every agricultural commodity is required to be stored properly to prolong its availability with minimum qualitative and quantitative losses. Onion is not an exception. The onion bulb is a natural food store for the plant, and it undergoes a process of development towards sprouting, and is subjected to decay by various disease causing organisms. The objective of appropriate storage technology is to retain the quality of bulbs for as long as possible and for long distance marketing after removal from store, without much loss.

The onion bulbs are generally stored from April/ May to October/November for a period of four to six months. It has generally been noted that under normal conditions the total onion storage loss is around 15-40 per cent depending on the quality of stored material and duration of storage. The total storage losses comprise physiological loss in weight (PLW) showing shrinkage (10-25 per cent), rotting (3-8 per cent) and sprouting (2-10 per cent). The PLW can be minimized by harvesting at right time, proper curing, and subsequent storage at desired temperature and humidity conditions. Generally, the rotting losses are at a peak in initial months of storage, particularly in June and July, due to the prevalence of high temperature and high humidity. However, proper grading and selection of quality bulbs and good ventilation in storage condition can reduce rotting.

It is necessary to know the physiology of dormancy and epidemiology of storage diseases while deciding long term storage protocol.

Conditions unfavourable for disease development can be engineered using the physical principles of temperature and humidity control.

The physiological and pathological processes proceed within a store of onions interact with the physical process of heat and water vapour exchange and mutually influence the environment within the store. Main factors, which influence onion storage and bring change in the bulbs, are summarized as under.

i) Sprouting and internal root development change bulb shape, crack the skins and increase the water loss.

ii) Sprouting increases respiratory rates of onion bulb that leads to generation of heat and CO_2.

iii) Bulbs thus get deteriorated due to disease development.

iv) Bulb deterioration due to diseases increases respiratory output. The onion skins have vital role in the physical and physiological processes in the storage, as it is the main barrier to water loss and CO_2 exchange. Relative humidity of 65-70 per cent is desirable to maintain the skin fairly flexible and elastic. At lower RH, the skin becomes very brittle and gets easily cracked notably when skin moisture content falls below 20 per cent.

v) Ventilation is another key element for maintaining humidity in storage. Lack of ventilation often adversely affects the quality by increasing bulb respiration and heat accumulation.

vi) With the advancement in storage period requirement of ventilation increases

vii) Design of onion store should, therefore, match these requirements.

viii) High humidity with high temperature favours spread of pathogens within the store.

It is necessary to counter the above changes by proper monitoring internal environment. Heat and water vapour must be removed or introduced as necessary either by using heating or refrigeration or ventilation or a combination of all the mechanisms. However, under Indian conditions in onion growing states design to exploit natural ventilation is most economical.

5.8 Transportation

Onions are transported in bullock carts, tractor-trolleys to local markets. Trucks and railway wagons are used for long distance transport within the country. Ventilated containers are mostly used for export to neighbouring countries. A few exporters now started using refrigerated container for export of onion to different countries, especially to European countries.

6.0 Domestic Marketing and Export

Most of the farmers bring onion directly to the market after harvest, as proper storage facilities are not available with them. Fearing losses, farmers usually sell their entire stock within a month of harvest. As a result, during this period prices rule very low due to the glut situation in the market. Thereafter, the rise in prices is quite rapid and sometimes-wide fluctuations occur leading to dissatisfaction amongst the producers as well as consumers. There is a quantum jump in the export of onion during the last 3 decades (from 169771 tonnes during 1981-82 to 18.23 lakh tonnes during 2012-13) increasing the foreign exchange earnings from Rs. 29.43 to 2294.91 crores.

Farmers generally sell their produce through regulated assembly/wholesale markets in major producing areas in different states. Several agencies including retailers, wholesalers, commission agents and cooperatives are involved in marketing onions. Cooperatives have been playing a significant role in the marketing of onion, particularly after NAFED's entry into this trade. Cooperatives at district, talukas and market level act as commission agents for the sale of the produce of its members. Lasalgaon, Pimpalgaon, Chandwad, Manmad in Nasik, Rahuri in Ahmed Nagar, Lonand in Satara and Chakan in Pune district in Maharashtra are the major assembly markets in producing centers. Mahuva, Bhavnagar and Gondal in Gujarat, Bangalore, Hubli, Belgaon in Karnataka, Kurnool in A.P. are the major assembly markets in major producing pockets in respective states. From Rajasthan and Haryana onions are directly brought to Delhi or other places for local distribution.

Continued export fetches good returns for farmers. Open auction system, under cover, tender system and open agreement of sales are followed in different assembly and distributing markets of the onions in the country. The farmers send their produce to commission agents who arrange sale in the markets. After purchase, the traders and exporters take up sorting, grading and packing and dispatch to different destinations according to the demand.

It has been observed that in many areas regulated system of marketing does not exist. The informal group of farmers or cooperatives are being formed at village level for helping in marketing produce at reasonable price.

6.1 Production for Export

Continuity in onion export is a must to maintain the present status of production in the country and maintain the domestic prices at a remunerative level. Regularity in export also maintains the confidence of foreign buyers towards Indian onions.

Presently 90 per cent of export is comprised of mainly common big onion types grown in Maharashtra near Nasik, Pune, Ahmednagar, Satara *etc.*; Dindigul and Coimbatore in Tamil Nadu and Mahuva (Bhavnagar) in Gujarat and Patna and Biharsharif in Bihar. Small pickling onions grown near Cuddapah in A.P. near Bangalore and Chickballapur in Karnataka and multiplier onion gown near Dindigul are also exported.

Agrifound Dark Red, N-53, local Bellary Red, Patna Red, Agrifound Light Red and local Poona Red falls in this group. The big onions are exported to Gulf countries, Sri Lanka, Bangladesh, Singapore and Malaysia. Small onions such as Agrifound Rose and multiplier onions like CO_4, CO_3, $COOn_5$ and Agrifound Red contribute 10 per cent of Indian export. Small onions are exported to Bangladesh. Medium sized are exported to Singapore and Malaysia.

Yellow onion though are in demand in European and Japanese markets and are not being grown on commercial scale in India. Some yellow onion varieties have been tested by NHRDF and can be grown successfully for export purpose. There is wide scope for yellow onion export to European countries. It is however, necessary to ship in refrigerated or ventilated containers since these short day yellow onions have poor keeping quality. The NHRDF has tested yellow onion varieties *i.e.* Tana F1, Suprex, Granex-33, Granex-429, Mercedes, Excalibur, Equinex, Lexus, Cougar, Linda Vista, and Colina at its Regional Research Stations Nashik, Talaja (Bhavnagar) and Karnal. The varieties Excalibur, Mercedes, Cougur and Colina performed better by recording more than 70 per cent recovery of 6.50 cm and above size bulbs which are generally preferred in European countries.

NHRDF had also tested many white onion hybrid varieties *i.e.* Rio Jefe(H), White Creole, Perla, Estrella Blanco, Early Lockyer White, SSC-6008(H), Kristel, at its Regional Research Stations Nasik, Talaja and Karnal. The variety White Creole and Kristel performed better. White Creole recorded a high TSS of 16-18 per cent, while Kristel recorded TSS 9-11 per cent. Onion variety Kristal is suitable for fresh export to European countries and commonly grown in Gujarat under contract farming. The agro-techniques of above recommended varieties have also been developed by NHRDF.

White and yellow onions are not preferred for consumption in many areas in India, due to less pungency. These are mainly used exclusively for export or by dehydration industries. Therefore, the production of high TSS white onion varieties and yellow/red onion varieties could be increased by contract farming in Maharashtra and Gujarat. There is a need to increase the big onion production in Maharashtra and Gujarat and pickling type and small onion production in Andhra and Karnataka and multiplier onion production in Tamilnadu to increase the onion export from India.

6.2 Packaging for Exports

Packing size wholly depends on the requirement of importing country. Some of the netted poly bag sizes have been given below for different countries. Packaging and size requirement for export of onion to various countries are mentioned in Tables 5 and 6.

Table 5. Packaging Requirement of Export of Onion to different Countries

Importing Nation	*Pack Size*
Malaysia and Indonesia	10 kg
Srilanka	25 kg
Singapore	8 kg
Philippines	20 kg
Gulf countries	5 kg, 10 kg, 15 kg and 20 kg

Source: NIAM, Jaipur.

Table 6. Size of Onion Required by different Countries

Country	*Size (mm)*
Middle East and Gulf (Dubai, Sharjah, Doha,	40 – 60
Muscat, Bahrain, Dammam, Saudi Arabia, Kuwait)	
Malaysia, Singapore, Port Kelang and African Ports	25 – 30
Sri Lanka	25 – 30
Bangladesh, Pakistan and Nepal	25 – 30
Europe	60 – 70
Japan	60 – 70

Source: MSAMB, Pune.

6.3 Export Marketing

Global trade of onion export is 9.48 million tons in 2019. Netherlands is the leading exporter (1.60 million tons) of onion contributing to about 17 per cent of world export. India, China, Egypt, Spain and USA are at 2nd, 3rd, 4th, 5th and 6th position contributing to about 15 per cent, 10 per cent, 5 per cent, 4 per cent and 4 per cent share respectively in world export (FAO 2019). However, during the year 2018, India was the major exporting country in the world. USA, Malaysia, UAE, Netherlands, Saudi Arabia, Germany, Japan, Srilanka and Bangladesh are major onion importing countries in the world.

Even in 1951-52, the country exported more than 5000 tonnes of onion per annum. In 60 (s) the export was almost doubled and further in 80(s) it jumped manifold touching 2.74 lakh tonnes. Onion exports touched a new height during 2009-10 by exporting 18.73 lakh tonnes of value, Rs.2834 crores. However, the export to some extent came down to 13.41 lakhs tonnes worth Rs. 2159.06 crores during 2010-11 due to less production of *kharif* onion and ban on export. The onion export again affected during the year 2011-12 and reached only up to 15.52 lakhs tonnes with a value of Rs. 2141.43 crores due to less production of *kharif* onion and ban on onion export. Presently onion is exported to Malaysia, Sri Lanka, Bangladesh, United Arab Emirates, Nepal, Qatar, Philippines, Kuwait, Indonesia and Oman. Onion export in earlier days was mediated through the National Agricultural Cooperative Marketing Federation of India (NAFED) and other 13 state agencies including NAFED, which were facilitating the export of onion from India.

Presently onion export is free, however, at the time of domestic price rise because of crop failure due to unfavourable climatic conditions (drought/excess rains/hailstorm/flood), The government of India placed it on prohibited list of export to control the domestic price rise. The export is mainly from Maharashtra and Tamil Nadu ports. Export is also made by road to Bangladesh and Nepal from UP, Bihar and West Bengal.

6.4 Grade Designation and Quality of Onions for Export

The government of India has prescribed certain grades for exportable onions according to which the Onion should be:

i) Intact,

ii) Sound produce and free from rotting or deterioration which makes it unfit for consumption is excluded.

iii) Clean, free from any visible foreign matter,

iv) Free from damage caused by frost;

v) Sufficiently dry for the intended use (in the case of onions for storing, at least the two first outer skins and the stem must be fully dried).

vi) Without hollow or tough stems;

vii) Practically free from damage caused by pests; *viii*) Free of all abnormal external moisture;

ix) Free of any foreign smell and/or taste.

x) The stem must be twisted or clean-cut and must not exceed 6 cm in length (except for stringed onions).

xi) They shall comply with the residue level of heavy metals, pesticides and other food safety parameters as laid down by the Codex Alimentarius Commission for exports.

As per AGMARK onion is classified under three classes *viz.* Extra class, Class-I and Class-II, the details of which are mentioned in Table 7.

a) *Size Grading as per AGMARK*

Size is determined by the maximum diameter of the equatorial section as per Table 8.

Size tolerances: For all grades: 10 per cent by weight of onions not satisfying the size identified, but with a diameter of no more than 20 per cent below or above it.

Table 7. AGMARK Grades and Standards for Onion

Grade Designation	*Grade Requirements*	*Grade Tolerance*
Extra class	Onion shall be of superior quality. They shall be characteristic of the variety and/ or commercial type. The bulbs shall be: Firm and compact; Unsprouted (free from externally visible shoots); Properly cleaned; Free from swelling caused by abnormal development; Free of root tufts, however, in onions harvested before complete maturity, root tufts are allowed. They shall be free of defects, except very slight superficial blemishes, provided these do not affect the general appearance of the produce, the quality, the keeping quality and the presentation in the package.	5 per cent by number or weight of onion not satisfying the requirements of the grade, but meeting those of class I grade or, exceptionally, coming within the tolerances of that grade.
Class – I	*Onion shall be of good quality and characteristics of the variety and/or commercial type. The bulbs shall be: Firm and compact; Unsprouted (free from externally visible shoots); Properly cleaned; Free from swelling caused by abnormal development; Free of root tufts, however, in onions harvested before complete maturity, root tufts are allowed. The following slight defects, however, may be allowed, provided these do not affect the general appearance of the produce, the quality, the keeping quality and presentation in the package. A slight defect in shape. A slight defect in colouring; Light staining which does not affect the last dried skin protecting the flesh, provided it does not cover more than one-fifth of the bulb's surface. Superficial cracks and absence of part of the outer skin provided the flesh is protected.	10 per cent by number or weight of onion not satisfying the requirements of the grade, but meeting those of class II grade or, exceptionally, coming within the tolerances of that grade.

Grade Designation	*Grade Requirements*	*Grade Tolerance*
Class – II	Onions which do not qualify for inclusion in higher grade but satisfy the minimum requirements. They shall be reasonably firm. The following defects, however, may be allowed, provided the onion retain their essential characteristics as regard the quality, the keeping quality and presentation. Defects in shape; Defects in colouring; Early signs of shoot growth visible from outside (not more than 10% by number or weight per unit of presentation); Traces of rubbing; Slight marking caused by parasites or disease; Small healed cracks; Slight bruising, healed, unlikely to impair keeping qualities; Root tufts; Stains which do not affect the last dried skin protecting the flesh provided they do not cover more than half the bulb's surface; Cracks in the outer skins and the absence of a part of the outer skins from not more than one third of the bulb's surface, provided the flesh is not damaged.	10 per cent by number or weight of onion not satisfying the require-ments of the grade but meeting the minimum requirements.

Table 8. Size Grading of Onion

Size Code	*Diameter (mm)*	*Difference between the Diameter of the Smallest and Largest Onion in the Same Package (mm)*
A	10–20	5
B	21–40	15
C	41–70	20
D	71 and above	30

7.0 Conclusion

Onion is one of the most important commodities that play a critical role in the agri-based economy of many parts of India due to its great demand in India and abroad. However, awareness of postharvest management of this crop among the growers, wholesalers and retailers are very important for saving the huge losses after harvest. An integrated approach for reduction in postharvest losses is important. This should begin with development and promotion of varieties, refinement in agro-techniques with respect to production and crop protection and adoption of appropriate postharvest handling technologies.

The postharvest life of onion produced in any season largely depends upon proper curing of the bulbs, which involves field curing as well as shade curing. These two operations can be very well performed in late *kharif* and *rabi* crops, however, the kharif crop harvested in October-November hardly gets an opportunity for proper curing due to extended monsoon in western zones and due to low temperature in north zones. On-farm artificial curing facilities can resolve these problems. Similarly, large size godowns with forced ventilation and multi-tier system will be the future requirement. Standardization and designing godowns with 1000 tonnes capacity for traders and exporters are also the need of the hour. The marketable onions both for export and domestic markets suffer from poor grading. The real benefits of market price are masked due to improper/no grading and unattractive packing. The concept of mechanized grading is missing in the onion trade in the absence of proper graders at affordable prices. There is need for the development of on-farm motorized graders with a capacity of 2-5 tonnes/hr.

8.0 Important References

Bhonde, S.R., Srivastava, K.J., Sharma, H.K and Chougule, A.B., 1996. Effect of withholding irrigation before harvesting on storage life of onion. *NHRDF News Letter*, **16**: 1-14.

Bhonde,S.R., Ram, L. and Pandey, U.B. 1983. Effect of curing methods and stage of harvesting on storage quality of onion. **In:** *Proceedings of National Workshop on onion* organised by NAFED in collaboration with Min. of Agriculture, Govt. of India, Dec., 17-18th pp 13-22.

Chauhan, K.P.S., S.P. Singh and Chougule, A.B. 1995. Studies on the effect of windrow curing, neck cutting and shade curing on export quality of onion bulbs during storage. *NHRDF News Letter*, **15**: 5-7.

Cho Jungeun, Bare Rona, Lee Seungkoo, 2010. Current research status of postharvest technology of onion (*Allium cepa* L.). *Kor. J. Hortic. Sci. Tech.*, **28:** 522-27.

Chougule, R.R., Bhonde, S.R. and Pandey, U.B. 2004. Assessment of postharvest losses in onion. *NHRDF News Letter*, **24:** 11-16.

Deb, M., Pandit, M.K., Sarkar, A. and Maity, T.K. 2009. Response of irrigation and fertilizer on growth, yield and postharvest shelf-life of onion. *Env. Eco.*, **27:** 1025-27.

Gupta. R.P. and Srivastava, P.K. 1984. A study of storage diseases of onion. *Phytopathol.* **37:** 404.

Sankar, V.,Veeraragavathatham, D. and Kannan, M. 2009. Postharvest storage life of onion influenced by organic farming practices farming practices. *Asian J. Hortic.* **4:** 16-20.

Singh, D.K., Pandey, A.K., Pandey, U.B. and Bhonde, S.R. 2002. Effect of farm yard manure combined with foliar application of NPK mixture and micro nutrient on yield, quality and storage of onion variety Agrifound Light Red. **In:** *Role of Technology in improving the productivity of small farmer Agriculture-A8* NIRD Foundation Day Seminar on Rural Technology for poverty alleviation, Hyderabad, January 2-3, concept paper and Abstract pp. 28.

Zeka, N., Nasto, T.H. and Balliu, A. 2008. The influence of N sources and fertilization doses on dry onion yield and bulbs shelf life. **In**: *proceedings of the IV Balkan Symposium on vegetables and potatoes*, Plovdiv, Bulgaria, 9-12 September 2008.

Chapter 28

Potato

R. Ezekiel

Former NHB Chair Professor, National Institute of Food Technology Entrepreneurship and Management, Kundli – 131 028, Haryana, India and

Former National Coordinator, National Agricultural Innovation Project, ICAR, New Delhi – 110 012, India

E-mail: rajarathnamezekiel@yahoo.co.in

1.0 Introduction

The importance of potato (*Solanum tuberosum* L.) as food is well recognized throughout the world. It originated in South America, where it was cultivated by the Incas. It is believed that Potato was introduced into India by the Portuguese in the 17th Century. However, it was the English who were responsible for its wide-spread cultivation in the hills of the country. The potato is a staple food in Europe and North America. But, in India, potato is still considered a vegetable and is given the pride of place amongst vegetable crops. Earlier potato was considered to be a crop of developed nations in the world. This is no longer the case and in fact, developing countries now produce about 35 per cent of the world's potatoes and production in developing countries is growing rapidly. India has emerged as the second largest producer of potatoes in the world after China. Today potato is being grown almost throughout the country and under diverse agro-climatic conditions. The potato growing areas in the country can be broadly classified into three zones *i.e.* hills, plains and plateau. About 90 per cent of the potato crop is grown in the plains under sub-tropical conditions. Potato tuber is a wholesome food with plenty of carbohydrates and good quality protein of high biological value. Besides, potatoes contain vitamins, minerals and antioxidants. Potatoes are rich in vitamin C and potassium and low in sodium.

There are several success stories in Indian Agriculture, which has contributed to the present position of plenty and the success story of potato in India should easily rank among the best. During the last sixty years, there has been a revolution in potato production in India. The crop once considered best suited to temperate conditions has emerged as one of the major food crops in India and now is well suited to subtropical Indian conditions.

2.0 Status of Cultivation

In 1949-50, the potato production in the country was 1.54 million tons, the area 0.234 million ha and the average yield 6.58 t/ha. After seventy two years *i.e.* in 2021-22, the production has increased to 53.8 million tons, the area to 2.2 million ha and the average yield to 24.35 t/ha. Now, India is the second largest potato producer in the world. About 90 per cent of the potato crop in India is cultivated in the plains during the winter season. The Indo-Gangetic region comprising Uttar Pradesh, Bihar and West Bengal accounts for 80 per cent of the potatoes grown in the plains. Uttar Pradesh is the largest potato producing state in India followed by West Bengal and Bihar. The potatoes are harvested in this region during January-march. State-wise area, production and productivity of potato in India during 2018-19 are presented in Table 1.

3.0 Commercial Cultivars

In India, several commercial varieties of potato are grown in different states. Some of the most common varieties with their special characters are mentioned below (Figure 1).

a) Kufri Badshah

This is a hybrid between Kufri Jyoti and Kufri Alankar. The tubers are white, large, oval with smooth skin, fleet eyes and dull white flesh. Tubers turn

Table 1. Area, Production and Productivity of Potato in India 2018-19

Sl. No.	*State*	*Area (thousand ha)*	*Production (thousand t)*	*Productivity (t/ha)*
1.	**Assam**	102.87	720.97	7.01
2.	**Bihar**	304.78	7740.79	25.39
3.	**Gujarat**	133.29	3806.95	28.56
4.	**Haryana**	34.72	897.58	25.85
5.	**Jharkhand**	48.21	690.23	14.32
6.	**Karnataka**	35.53	509.48	14.34
7.	**Madhya Pradesh**	136.29	3144.64	23.07
8.	**Punjab**	98.52	2571.04	26.09
9.	**Uttar Pradesh**	614.78	15555.53	25.30
10.	**West Bengal**	427.50	12782.50	29.90
11.	**Others**	**205.23**	**2890.30**	**14.08**
	Total	**2141.72**	**51310.01**	**23.96**

Source: Horticulture Statistics Division, Department of Agriculture, Coopn and Farmers Welfare.

pale purple on exposure to light. The tubers mature between 100-110 days with very high yield potentiality of 50 t/ha. The tubers have medium dormancy. This is commercially cultivated in parts of Gujarat, Haryana, Jammu and Kashmir, Madhya Pradesh, Punjab and Uttar Pradesh.

b) Kufri Chandramukhi

The tubers are slightly elongated shape having medium dormancy of 6-8 weeks. It matures in about 90-100 days after planting having yield potentiality of 30 t/ha in plains. This is commercially cultivated in North and South Indian hills, parts of Bihar, Gujarat, Karnataka, Madhya Pradesh, Maharashtra, Punjab, Uttar Pradesh and West Bengal (Figure 1a).

c) Kufri Chipsona-I

Tubers are white, medium to large with smooth skin, fleet eyes, dull white flesh. The variety is medium in maturity (90-110 days), with very high yield potential

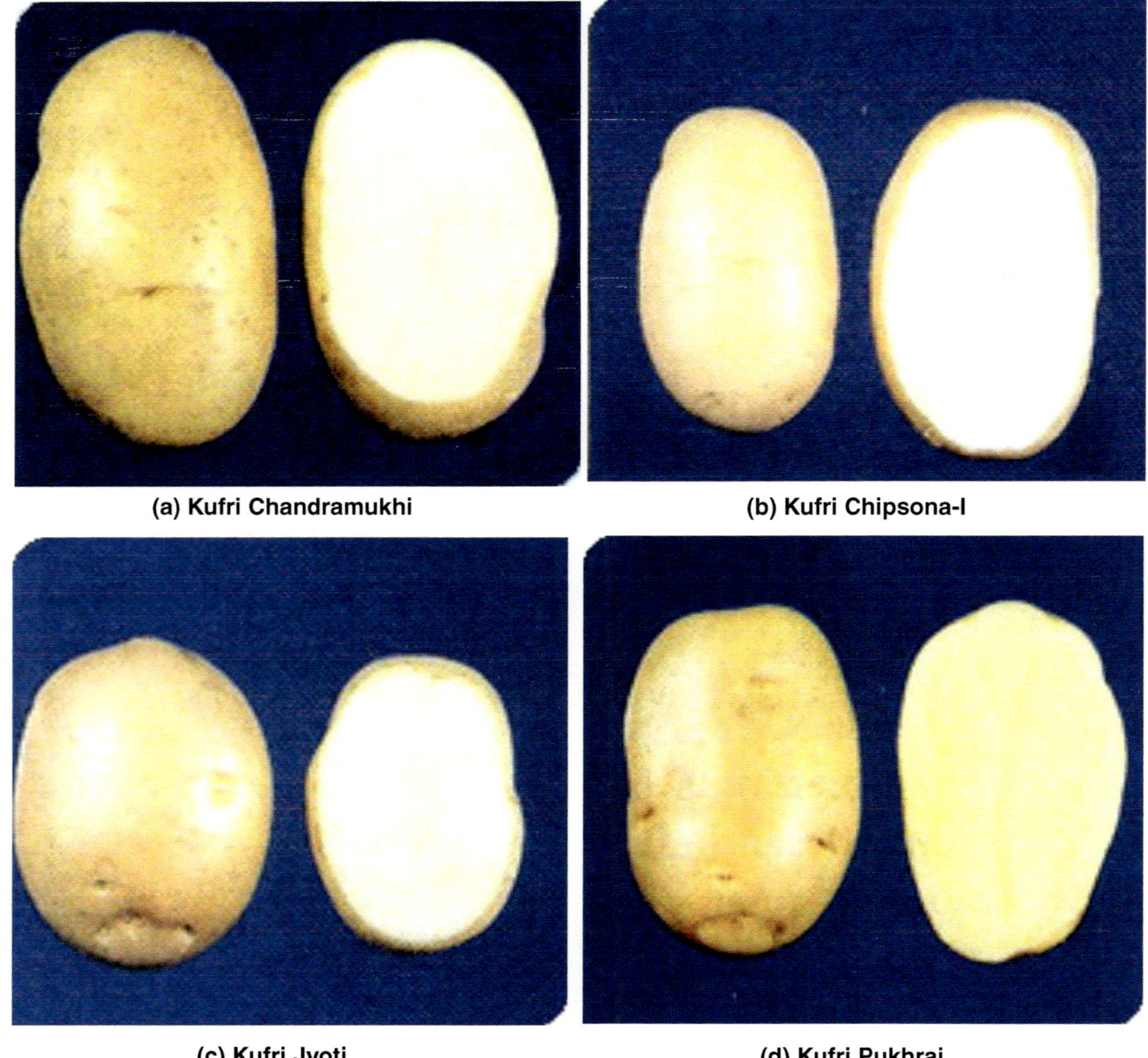

(a) Kufri Chandramukhi **(b) Kufri Chipsona-I**

(c) Kufri Jyoti **(d) Kufri Pukhraj**

Figure 1. Commercial Cultivars of Potato

of 40 t/ha and is resistant to late blight disease. This is commercially grown in Bihar and U.P. Due to high dry matter and low reducing sugars, the variety is suitable for making chips (Figure 1b).

d) Kufri Jyoti

Tubers are white, large, oval with smooth skin, fleet eyes, white flesh. There is a tendency to crack. The variety has high yield potential up to 30 t/ha in the plains with maturity at 90-100 days. The variety is having medium dormancy of 6-8 weeks. This is commercially cultivated in North and South Indian hills, parts of Bihar, Gujarat, Karnataka, Madhya Pradesh, Maharashtra, Punjab, Uttar Pradesh and West Bengal (Figure 1c).

e) Kufri Lauvkar

This is a hybrid of the cross between Adina and Sarkov. It can be grown both in Kharif and Rabi seasons in the plateau regions of Karnataka, Madhya Pradesh and Maharashtra. This is an early variety and matures in 75-80 days with a yield potential of 30 t/ha. The tubers have medium dormancy of 6-8 weeks.

f) Kufri Pukhraj

The tubers are white, large, and oval, slightly with smooth skin, fleet eyes, yellow flesh. This is an early maturing (75-90 days) with very high yield potential of 40 t/ha and moderately resistant to early and late blight diseases. Tubers have medium dormancy of 6-8 weeks and are not suitable for processing (Figure 1d).

Record levels of production during the last two decades have resulted in several postharvest problems and the major problem is that of storage. About 80 per cent of the total potato production comes from the Indo-Gangetic plains, which are harvested in February-March. Therefore, the arrivals in the market during this period are huge creating gluts resulting in distress sale by the farmer. Storage helps in regulating the arrival in the market and reduces transport bottlenecks at the peak period of production. Since fresh potatoes are available only for 2-3 months in a year, the requirement of table potatoes for the remaining period of 9-10 months has to be met from stored potatoes. Potatoes being semi-perishable in nature with 80 per cent water content cannot be stored for 9-10 months without refrigeration. Therefore, cold storage becomes essential.

The high water content in potatoes makes handling and storage difficult. It has been estimated that under tropical and sub-tropical conditions, losses due to poor handling and storage can amount to up to 40 per cent. Therefore, it is of utmost importance to minimize storage losses. It makes sense to minimize storage losses to increase the availability of potatoes as it costs less to store than to produce a given quantity.

4.0 Nature and Causes of Postharvest Losses

Good postharvest management begins at the production stage itself. The good quality of the tuber is greatly influenced by the cultural practices. Potatoes grown on improperly prepared fields are damaged while harvesting mechanically and have reduced sugar quality. Conditions for high yield and for good storage characteristics may be contradictory. Further, adequate pest and disease management is essential for producing potatoes with good storage quality. Therefore, proper management of pre-storage factors that affect the keeping quality of potatoes is the first step for good storage management. Whatever may be the storage method, its success depends on the quality of the tubers entering storage. If the tubers entering storage are not sound, the tubers that come out of storage will also be of poor quality.

4.1 Dehaulming

Removal of haulms at full maturity (*i.e.* 90-110 days after planting, depending upon the variety) is essential for proper curing of the skin of tubers. Dehaulming is done manually by cutting with a sickle close to the ground or by spraying non-selective herbicide paraquat @ 0.5kg/ha. If haulms are to be removed manually, it is preferred to keep the vines/haulms on the ridges to protect exposed tubers from high temperature and direct sunlight. Re-growth of leaves if any should also be cut after a week of dehaulming.

4.2 Time of Harvesting

Potatoes should be harvested in dry weather. Irrigation should be stopped two weeks before dehaulming. Potatoes harvested under wet soil conditions must be dried before storage because even little moisture on the surface of tubers could lead to infection and rotting during storage. Only mature tubers should be stored. Immature tubers have poor keeping quality due to lower dry matter content and a weak skin (Figures 2a and b). Harvesting is done 10-15 days after haulm cutting to facilitate proper skin curing. Despite of best efforts, cutting and bruising cannot be eliminated while harvesting and therefore, curing is essential to heal the wounds. Suberization is the process by which wounds are healed in potatoes and the optimum conditions for suberization are a temperature of around 25 °C and a relative humidity of 95 per cent. It is important to grade the produce into different grades before storage and grading should be done according to the requirement. Green, cut/cracked, damaged and undersized tubers (less than 2.5 inches) should be removed and only well-graded, defect-free tubers of desired size should be stored.

Figure 2. (a) Immature Tubers and (b) Well-Cured Mature Tubers

4.3 Dormancy

Dormancy is a state in which tubers will not sprout even when placed under conditions ideal for sprouting (18-20 °C, 90 per cent RH and darkness). All tubers remain dormant for some time and the duration of dormancy varies with the variety. Duration of dormancy is normally counted from the date of harvest. Another method of determining dormancy is from the date of detaching the tubers from the plant *i.e.* dehaulming. This is a better and more reliable method as the time between dehaulming and harvesting can vary. Dormancy is expressed as the number of days between the date of dehaulming and the date on which 80 per cent of tubers are sprouted. A tuber is counted as sprouted if it has at least one sprout longer than 2 mm.

4.3.1 Factors Affecting Dormancy

a) Cultivar

Based on the dormancy period from the date of dehaulming in the plains, Indian potato cultivars can be divided into three categories.

Short dormancy: < 8 weeks, *e.g.* Kufri Bahar.

Medium dormancy: 8-10 weeks, *e.g.* Kufri Jyoti.

Long dormancy: > 10 weeks, *e.g.* Kufri Sindhuri

Even within a cultivar, dormancy can be affected by growth conditions and tuber weight.

b) Growth Conditions

Soil and environmental conditions during crop growth has a strong influence on the duration of dormancy. Cold and wet weather is known to increase the dormancy while dry and warm weather reduces it. Potatoes raised under short-day conditions (plains) have shorter dormancy than those from long-day crop (hills). For example, tuber dormancy in Kufri Jyoti, Kufri Chipsona-1 and Kufri Chipsona-2 was 55,47 and 40 days, respectively, when grown in the northern plains under short days and 126, 84 and 91 days, respectively, when grown in the northern hills under long days. Season to season variation in the duration of dormancy can also be due to variation in the environmental conditions during the crop growth.

c) Storage Environment

Storage temperature has a strong influence on the dormancy. Higher storage temperature hastens dormancy release, while storage at a temperature of 4 °C or below prolongs dormancy by preventing sprout growth. Exposure to light during storage also affects dormancy but its effect is less as compared to the temperature effect. RH in the stores has very little effect on tuber dormancy. Concentration of CO_2 and O_2 in the storage atmosphere affects dormancy of stored potatoes. Lower concentration of oxygen (O_2) and higher concentration of carbon dioxide (CO_2) than the ambient hastens the release of dormancy and stimulates sprout growth. CO_2 concentration in the storage atmosphere is reported to affect dormancy duration and sprout emergence. It is also known to affect the number and length of sprouts in tubers stored at 4-10 °C. Work done at CPRI has shown that exposure of tubers of Kufri Chandramukhi and Kufri Jyoti to 5, 10, 15 and 20 per cent CO_2 concentrations for 7 days reduced the dormancy duration, and 20 per cent CO_2 concentration showed the maximum effect, reducing the dormancy duration by 20 days in tubers stored at 18 °C and 90-95 per cent RH. Sprout growth was also stimulated. Longer sprouts were observed when stored at 15 per cent than at 20 per cent CO_2 concentration.

Postharvest losses lead to reduction in the quantity as well as quality of potatoes. Quantitative losses are apparent and attempts are made to reduce these losses whereas qualitative losses are not apparent and their importance is underestimated. Since qualitative losses can greatly reduce the value of potatoes, adequate attention should be paid to prevent loss in quality as well. Postharvest losses result from physiological or pathological causes.

4.4 Physiological Losses

Storage losses are due to (i) respiration, (ii) transpiration, (iii) water loss, (iv) sprouting and (v) rotting. Potatoes being living material respire and excess of respiration results not only in the loss of dry matter but also in the generation of metabolic heat. Physiological losses occur through natural respiration and transpiration (evaporative loss of water). Major weight loss during storage is due to loss of water. Water loss occurs largely through periderm and to a smaller extent through lenticels. When the loss of water from tubers exceeds 10 per cent, the market value of potatoes is greatly reduced because of shrivelling. Freshly harvested immature potatoes lose more water than mature tubers because immature skin is more permeable to water. Sprouting results in increased water loss, as the surface of sprouts is 100 times more permeable to water vapour than the periderm of tuber per unit area and time. Sprouting also results in increased respiration leading to higher dry matter loss. The magnitude of these losses depends largely on the environmental conditions. Two important storage environmental factors that affect the storage behaviour of potatoes are temperature and relative humidity.

4.4.1 Effect of Temperature

Physiological damage can occur from exposure to high or low temperatures both before and during storage. Overheating of tubers either due to direct exposure to sun or high temperature and non-refrigerated storage could lead to black heart symptoms. At high temperature storage, the high rate of respiration leads to larger oxygen requirement, which results in asphyxiation. The discolouration and breakdown of the inner tissues of tubers is a result of asphyxiation. Tubers exposed to freezing temperatures also suffer internal damage. Tuber respiration during storage results in loss of dry matter. In addition, the heat generated during respiration called respiratory heat also influences the storage temperature and in turn affects the storage life of potatoes.

4.4.2 Effect of Relative Humidity

The relative humidity in the storage atmosphere affects the weight loss of tubers since water loss is directly proportional to the difference in water vapour pressure between the potatoes and the surrounding air. Generally, a higher relative humidity of 90-95 per cent is maintained in cold stores to minimize weight loss. The relative humidity in the storage atmosphere also affects the rate of sprout growth and the form of sprouts. The degree of branching in sprouts is greater under lower humidity levels. Relative humidity also affects wound healing. Wound periderm formation is rapid at higher humidity. For example, at a storage temperature of 10 °C, wound healing is faster when stored above 80 per cent RH.

4.5 Pathological Losses

Postharvest losses in potatoes caused by pathogens are greater than the losses due to physiological causes. Physical damage to tubers during harvesting and handling predisposes tubers to attack by bacteria and fungi leading to quantitative loss. The more common storage diseases caused by fungi are late blight, dry rot and pink rot. The most severe bacterial disease that causes rotting is soft rot. When infection occurs in the field, rotting begins in the field and continues during storage, for example, late blight, brown rot and pink rot. When infection occurs after harvesting, it is generally through mechanical injury as in the case of dry rot (Figure 6). High humidity and condensation of water on the tuber surface can lead to infection by soft rot causing bacteria *Erwinia* spp. Qualitative losses are caused by diseases such as common scab, powdery scab, black scurf and wart, which affect the appearance of tuber and thus reduce the market value of potatoes. Among the insect pests, tuber moth causes maximum damage during storage and is common in potatoes stored under higher temperatures, as is the case with non-refrigerated storage. The larval damage results in weight loss, and tuber moth infection greatly reduces the market value of tubers.

5.0 Postharvest Management

5.1 Storage Methods

Unlike in the temperate regions of the world where potatoes are harvested and stored during the cold, in India potatoes are harvested in February-March in the Indo-Gangetic plains when the temperatures begin to rise and have to be stored during the hot summer months. Potato storage methods being practised in India can be broadly divided into two categories, *viz.* Short-term storage (non-refrigerated) and Long-term storage (refrigerated).

5.1.1 Short-term Storage

When potato tubers are dormant it is relatively easy to store them under non-refrigerated conditions. But once dormancy is released and sprout growth begins, storage under non-refrigerated conditions becomes difficult as storage losses increase. Though refrigerated storage is the best option for long-term storage needs, considering the rising cost of construction, equipment and energy required for additional refrigerated storage capacity, alternate storage methods should be considered for meeting short-term storage requirements. Further seasonal production pattern, limited alternative market outlets and high cost of cold storage often result in gluts and the consequent reduction in price during the harvesting period. Potato prices rise rapidly in April-May and are almost double in July-August. Therefore, holding the produce for 3-4 months and then selling it in the market will be advantageous. The following non-refrigerated storage systems are effective for storing potatoes for 3-4 months under hot summer conditions in India.

For short-term storage, on-farm storage is economical because it requires only a small investment and no energy. Potatoes need to be stored on-farm for various reasons such as a farmer might decide to hold his produce for a few months, when the prices are low and are likely to increase in a few months. Storage is economical only if there is an increase in price during the storage period and the price increase is sufficient at least to cover storage costs. Crops have been stored on-farm for centuries and the storage methods have varied in different societies. In the first half of this century, potatoes were stored in West Bengal, Bihar and Uttar Pradesh in cool dry rooms or in pits. Potatoes were also stored in heaps and pits in Mumbai and Pune districts of Maharashtra. Potatoes were stored in pits in the erstwhile Bombay state from February to June. Tubers were examined periodically and rotted ones were removed. Table potatoes were disposed of before they started to sprout but seed potatoes were desprouted repeatedly and stored until the planting time. Some of these methods have survived and are practised even today in villages. The traditional methods of storing potatoes are very useful for storing small quantities of potatoes (15-20 tonnes) for 3-4 months due to the following advantages: (i) These methods are very cheap and no investment is needed on storage structures. (ii) The materials required are locally available and farmers can do it themselves thus saving on the cost of labour. (iii) They are readily acceptable to farmers. The traditional methods may be less efficient but if their efficiency is improved by minor modifications, the storage losses can be brought down to acceptable limits and they can also prove to be effective on-farm storage methods for meeting short-term storage requirements. Some of the common non-refrigerated methods used for on-farm storage of potatoes are discussed below.

a) *In-situ* Storage

In-situ storage means delaying the harvest of the crop until it is required. The disadvantage is that the crop may be exposed to pest and disease attack and the land remains occupied, and as a result a new crop cannot be planted there. *In-situ* storage of potatoes for up to a three months is practised in Meghalaya and Kumaon hills.

b) Sand Storage

Some of the earliest research attempts at proper storage of potatoes were made in the beginning of the 1900's and they were aimed at controlling tuber moth which threatened the potato industry in the country. It was discovered by Lafroy and Evans in 1910 that potatoes could be stored in the sand and this became a general practice in Bihar. It was also realized that rotting due to bacterial infection should be reduced and for this storage of potatoes free of injury and treatment with 2 per cent copper sulphate solution for 30 minutes before storage was recommended. Potatoes were stored in sand underground and above ground in Punjab during 1919-20. This kind of storage is practised by farmers in Bihar for the storage of small quantity (up to 200 kg) of potatoes meant for home consumption and the storage period can be extended up to few months. A layer of sand is spread on the katcha floor in a corner of the house or below the cot and potatoes are spread on them. They are turned over once in 15 days and the rotten tubers are removed. A total loss of 20 per cent has been observed.

c) Heap Storage

This is a common and popular traditional method of on-farm storage. Heaps have been used successfully for storing potatoes on-farm for 3-4 months. This method of storage is commonly practised in Madhya Pradesh, Uttar Pradesh, Gujarat, Maharashtra and Karnataka. In Ujjain district of MP, heaps are made under the shade of trees (Figure 3) and the quantity stored can vary from 5 to 33 t and the storage duration varies from 2 to 4 months. Smaller heaps are circular with a diameter of 5 to 7 m. The height varies from 1 to 1.5 m. Bigger heaps are 3 to 6 m wide with length varying depending upon the quantity of potatoes stored. The heaps are covered with 30 to 50 cm thick layer of straw or dried sugarcane leaves. The temperature inside the heap is around 20 °C at the time of making the heap and it increases to 30 °C by the end of the storage period. The temperature inside the heap is 11 to 14 °C lower than the ambient and the notable thing about heaps is that the decrease in temperature is achieved without any cooling mechanism. The RH inside the heap ranges from 46 to 63 per cent and this is 19 to 20 per cent higher than the ambient conditions. In Deesa district of Gujarat, 5 to 40 t of potatoes are stored for 2 to 3 months. While potatoes are stored in bigger heaps for 3 to 4 months in MP and Gujarat, in Karnataka smaller heaps of 1 to 20 t are made and potatoes are stored for 1 to 2 months in districts of Bangalore, Hassan and Dharwad. Heap is locally called *thamboo*. Storage losses vary from place to place (Table 2) and are dependent on temperature at the time of storage, duration of storage, variety, thickness of the straw covering layer and protection from rain and direct sunlight.

Table 2. Losses of Potatoes during on-farm Storage in Heaps at different Locations in the Country

Location	*Variety*	*Storage Period*	
		(Days)	*Total Loss (per cent)*
Deesa (Gujarat)	Kufri Jyoti	90	32.8
	Kufri Badshah	90	26.0 to 37.4
Hassan (Karnataka)	Kufri Jyoti	90	19.5 to 32.0
Jalandhar (Punjab)	Kufri Jyoti	105	9.9 to 12.1
	Kufri Badshah	105	12.7 to 19.5
Modipuram (UP)	Kufri Jyoti	90	7.4 to 14.7
	Kufri Bahar	90	9.5 to 11.2
Patna (Bihar)	Kufri Jyoti	75	8.4 to 31.3
	Kufri Sindhuri	75	34.3 to 36.7
Ujjain (MP)	Kufri Lauvkar	120	12.9

Weight loss forms a major part of total losses

Figure 3. Heap Storage of Potatoes.

in heaps. The storage losses of about 10 per cent are considered acceptable and more than 10 per cent is considered unacceptably high. Furthermore, when weight loss exceeds 10 per cent, it leads to shrivelling of potatoes resulting in drastic deterioration in quality. Therefore, it is necessary to improve the efficiency of heap method of storage and bring down the storage losses to less than 10 per cent. Work done at Central Potato Research Institute has shown that it is possible to achieve this by erecting a thatched roof over the heap and by advancing the date of storage from mid-March to end of February (Table 3). A thatched roof to cover heap helped in reducing the temperature by 2 °C besides giving protection from the rain. Advancing the date of loading of potatoes into heaps made the maximum difference. At Jalandhar and Modipuram, when potatoes were stored before 1st of March, storage losses were minimum. An additional advantage is that potatoes stored in heaps have lower reducing sugar content, which ranges from 75-131 mg/100g tuber (a reducing sugar level of up to 150 mg/100g is considered acceptable for chipping) and therefore, produce highly acceptable chips with colour scores of 1-3. (Chip colour was scored on a scale of 1-10 of increasing darkness and a chip colour score up to 4 was considered acceptable, and more than 4 was unacceptable).

Table 3. Reduction in Storage Losses of Potatoes (Kufri Bahar) Stored in a Heap under different Conditions for 90 Days

Storage Loss (per cent)	*Heap with Vent in the Sun*	*Heap in the Sun*	*Heap under the Shade of a Tree*	*Heap Under the Shade of a Tree and Protected with a Thatched Roof*
Sprout weight*	2.6	2.9	2.6	1.2
Rottage	6.0	5.4	4.6	0.5
Weight loss	15.6	9.4	9.2	8.1
Total losses	24.2	17.7	16.4	9.8

* Sprout weight in g/100g tuber FW.

d) Pit Storage

The pit is another traditional method of on-farm storage. Potatoes were stored in pits in the erstwhile Bombay state from February to June. In Meghalaya and Uttaranchal also, seed potatoes are stored in pits. Pits have been used for storage of potatoes for 3 to 4 months in Ujjain district of Madhya Pradesh. Two types of pits are used for the storage of potatoes and they are *katcha* pit and *pucca* pit. *Katcha* pit is a simple pit dug under the shade of trees. It is generally rectangular measuring 6 m in length, 4.5 m in width and 3 to 4 m in height. About 10-30 t of potatoes are stored in each *katcha* pit. Now-a-days use of *katcha* pit for storing potatoes has decreased mainly due to seepage of rainwater into pits leading to higher storage losses. *Pucca* pit is different from *katcha* pit in that it is lined with a layer of bricks made in the form of a circular well and is generally bigger (Figure 4).

Figure 4. Pit Storage of Potatoes.

Furthermore, it has a longer life, without recurring expenditure on digging and maintenance. A typical *pucca* pit measures 4.5 m in diameter and 4 m in depth and the quantity of potatoes, which can be stored varies from 10 to 40t. The temperature inside the pit at the time of storage is 19 °C and may increase to 29 °C by the end of the storage period. The temperature inside the pit is 11 to 15 °C lower than the ambient and the RH inside the pit ranges from 52 to 73 per cent and which is 25 to 33 per cent higher than the ambient. The storage losses of 10 to 40 per cent are unacceptably high in pits. It is necessary to improve the efficiency of pit method of storage to bring down the storage losses. A major part of total losses is due to rotting and the high incidence of rotting is caused by the pressure exerted by the top layers on the bottom layers of potatoes. By placing a platform made of bamboo placed 2 inches apart and 2 feet above the base of the pit and by reducing the quantity of potatoes stored, it is possible to bring down the storage losses to less than 10 per cent after 3 to 4 months of storage. Like in heap, potatoes stored in a pit have lower reducing sugar content, which ranges from 136-148 mg/100g tuber FW and, therefore, producing highly acceptable light colour chips with colour scores of 1-3. The storage cost comes to around Rs.18/q.

e) Room Storage

Traditional methods of potato storage were in vogue at the beginning of 1900s and storage in a cool dry room was one of the most common traditional methods of potato storage. Considerable quantities of seed potatoes were stored in cool dry rooms in the plains of Bihar, Uttar Pradesh and West Bengal during February-October. It is common to store potatoes in a dark, cool room in the basement or a corner of the house for 2 to 3 months. In Himachal Pradesh, Meghalaya and Uttarakhand, seed potatoes are either heaped or stored in gunny bags or bamboo baskets, in such rooms. In the plains, potatoes are stored in rooms in U.P. and M.P. Though the storage losses are as high as 27 to 46 per cent after 3 months due to excessive sprout growth, farmers in northern plains continue to store potatoes this way because of the higher price fetched in June-July. In the 1970's storage of potatoes in ordinary *katcha* farm store and double walled *katcha* farm store were tried at CPRS, Patna.

f) Ventilated Wooden/Bamboo Structures

Storage in diffused light helps in retarding sprout growth of potatoes. Accordingly seed potatoes were stored in diffused light in Assam during 1931-32. Small wooden/bamboo structures are built on a slightly raised site and potatoes are stored in these structures. Wooden planks are used in some places like Meghalaya while in Assam, bamboo is used. These structures are made above the ground level supported by wooden pillars and are well aerated. In Assam, bamboo structures are made in such a way that diffused light enters the store so that the tubers turn green and have small, stout sprouts (Figure 5). These bamboo structures have been found useful for on-farm storage of seed potatoes. The quantity of seed potatoes stored in each structure varies from 0.2 to 1 t. After 7-8 months of storage, 2 to 10 per cent losses occur due to rotting and the weight loss varies from 20 to 35 per cent. Work done at CPRS, Kufri has shown that seed potatoes can be stored successfully in a diffused light store in the hills of Himachal Pradesh.

g) Non-refrigerated Air-Cooled Store

Potatoes can be stored in non-refrigerated air-cooled stores in the hills. The cooler temperatures at night are utilized to cool the store. The ventilators of the insulated store are kept open during the night to draw in the fresh cool air and are closed during the day to prevent the entry of warm air. This has been tried at the Central Potato Research Station, Muthorai in the Nilgiri hills for the storage of potatoes.

h) Insulated Ventilated Cool Store

An insulated ventilated cool store was constructed at Daurala farm of Central Potato Research Institute Campus (CPRIC), Modipuram by the Central Potato Research Institute (CPRI) in collaboration with the Central Warehousing Corporation of India Ltd. in 1977. The store is located in a grove of trees. It is a masonry structure measuring 9.15m x 4.57m with 13 air-inlets and 22 ventilators, each 0.91 m x 0.30m on the walls. The air-inlets and ventilators each 0.91m x 0.30m and 2.13 m above floor level are provided with wooden shutters to exclude light from the store. A door 1.83m x 0.22m (with an L-shaped break-wall having an opening towards the east) is provided to serve as an ante-chamber. The ceiling walls, door, ventilators and air-inlets have been insulated with fibreglass. A coating of bitumen has been used on the wall to act as a vapour barrier. The roof consists of two parts, a shorter and steeper slope and a longer but less steep northern slope. The former is 1 m higher than the latter at the ridge. A north-facing row of sky-light ventilators are also provided. The parapet wall on the western side is raised to act as a screen. Store walls are white washed and the roof is painted with aluminium paint to reflect sunlight. To control ventilation, two exhaust fans are provided, one each on the eastern and the western walls. Thus the store

Figure 5. A Bamboo Structure for Potato Storage in Assam.

incorporates all the principles of durability, insulation and ventilation as well as other ancillary devices to reduce the heat load. In 1978, the daily minimum temperature in the cool store ranged between 8 °C at the beginning of March to 22 °C at the beginning of June, and the maximum temperature between was 14 °C to 24 °C. During this period, the corresponding range of maximum temperature was 8 °C to 33 °C in ordinary godown and of maximum temperatures were 20 °C and 40 °C. The relative humidity was between 60 and 90 per cent in the cool store and 15 to 65 per cent in the ordinary godown. Rotting percentage and driage in the store up to early June was generally lower in the former than in the latter. During storage, fewer eyes sprouted per tuber with Isopropyl-N-3-chlorophenyl carbamate (CIPC) treatment in the same period. Tetrachloro-nitrobenzene treatment had only a transitory effect on the number of eyes that sprouted.

i) Evaporatively Cooled Potato Store (ECPS)

Attempts were made in India in the 1920s to store potatoes in an evaporatively cooled potato store. A 20 t capacity insulated potato store equipped with passive evaporative cooling, which does not require electricity or any other source of power was evaluated at Central Potato Research Station, Jalandhar in 1981. It works on the principle of passive evaporation of water which results in the removal of heat from within the store lowering the inside temperature. It is a masonry structure measuring 9.1 x 4.6 x 3 m. The northern side has eight ventilators measuring 76 x 60 cm and the eastern wall has three ventilators, placed 15 cm above the ground level. The outer side of the ventilator is fitted with chicken wire mesh and 5 cm thick wood wool which is kept moist by water dripping from small holes made in a 1.25 cm water pipe placed above the ventilators and connected to a water tank. The ceiling is made of plywood mounted on wooden beams with 5 cm diameter holes at 10 x 10 cm spacing to facilitate the escape of hot air from within the store. The roof is made of galvanized iron sheet and is provided with four exhaust outlets (Figure 6). During the day the air inside the store becomes hot and rises through the ceiling and

Figure 6. An Evaporatively Cooled Potato Store.

this is replaced by the fresh cool air entering through the moist wood wool which lowers the inside temperature. The temperature inside the ECPS ranges from 18 to 30 °C (6 to 12 °C lower than the ambient) and the RH is 85 to 95 per cent (15 to 42 per cent higher than the ambient). Due to the low water vapor pressure deficit inside the ECPS, weight loss in potato tubers during storage is significantly reduced. Storage losses range from 10 to 15 per cent after 3 to 4 months of storage. Ware potatoes can be stored in good condition up to the beginning of June, when the prices in the market are much higher as compared to the prices at the time of harvest. Storage life of potatoes can be further extended up to the end of June by treating the tubers with a sprout suppressant like CIPC. Potatoes stored in ECPS have low reducing sugar content of around 120 mg/100g tuber and produce highly acceptable light colour chips with colour scores of 2-3. Potatoes produced in north-western plains are otherwise unsuitable for processing due to high levels of reducing sugars at harvest. This store of 20 t capacity costs around Rs. one lakh and is suitable for regions where the temperatures are high and humidity is low during the storage period. The disadvantage of ECPS is that it works only during dry and hot summer and it cannot be used in rainy season. This structure can be used for only 3 to 4 months for the storage of potatoes and remains unutilized during the remaining 8 to 9 months. To make storage in ECPS economical, it needs to be used throughout the year.

5.1.2 Long-term Storage

Potatoes harvested in February-March have to be stored until October and cold storage is essential for such long-term storage. The importance of potato storage in India was realized as early as 1925 and storing seed potatoes from one crop to the next crop was considered the greatest constraint. One of the first cold stores for the storage of seed potatoes in the country was set up in U.P. in the year 1940. Growth of the cold storage industry was slow in the 1940s and 1950s. Growth picked up in the 1960's with the availability of technology and a large number of cold stores were built. Easy availability of compressors and insulating materials also contributed to that growth. With the increase in potato production in the country, cold storage business became profitable. Consequently, competition started and large capacity cold stores (6-8 thousand metric tonnes) were constructed. This situation continued through, the 1970's. But in 1980's, there was little competition resulting in a monopoly like situation. Farmers found it difficult to get space in cold stores for the storage of potatoes. In the years of glut farmers faced very difficult situations and in some parts of the country, they had to reserve space in cold stores by paying the rent in advance. In the late 1990's, the cold storage industry went through hard times as electricity charges increased every year but rental rates did not increase proportionately. The situation changed in 2000 with the intervention of the government through subsidy schemes for multi-product, multi-chamber cold stores. These schemes were also instrumental in introducing limited modernization with newly constructed cold stores switching over from diffuser system to fin coil. Many state governments removed rental controls. From the year 2000, the introduction of high speed compressors resulted in reduced power consumption and easy maintenance. New entrepreneurs are now more interested in building modern cold stores with improved insulation, and air circulation for better and uniform maintenance of temperature. With the construction of multi-chamber cold stores and storage of potatoes at 10-12 °C with CIPC treatment, there is increased awareness among the cold store owners about the importance of monitoring and maintenance of uniform temperature throughout the store, high level of relative humidity and low level of CO_2 concentration. The number of cold stores in the private cooperative and public sectors are 4885 (90.8 per cent), 356 (6.6 per cent) and 140 (2.6 per cent), respectively. Thus a look at the sector-wise distribution of cold stores reveals that cold stores in India are predominantly in the private sector (90.8 per cent in terms of number and 95.7 per cent in terms of capacity). Only a small percentage (2.6 per cent) are in the public sector. Data on the commodity-wise distribution of cold stores show that most of the available cold storage space is used for the storage of potatoes (53.2 per cent in terms of number and 75.4 per cent in terms of capacity). By the end of 2009, there were 2862 cold stores in the country with a capacity of 18.42 million tonnes, used exclusively for the storage of potatoes. This figure does not include multipurpose cold stores and those used for the storage of fruits and vegetables. About 77 per cent of the cold storage space available for potatoes is located in two states only *viz.* Uttar Pradesh (47.3 per cent) and West Bengal (29.6 per cent). Long-term storage of potatoes is expensive. The cost of storage is very high due to high cost of erection and maintenance of refrigerated storage structures, electricity, and loading and unloading. During storage, potatoes are affected by the storage temperature, humidity, composition of the storage atmosphere, their physiological condition at the time of storage and degree of microbial infection. Deterioration during storage is through weight loss (moisture loss), respiration, pest and disease attack and physiological disorders. All these factors have to be considered before deciding on storage condition and duration. Storage of potatoes in cold store has become a common practice in the country.

Low temperature storage is the most common method of potato storage. Low temperature storage is effective because a storage temperature of 4 °C and below, sprout growth does not take place in most of the varieties. Further, the metabolic processes are slowed down therefore, losses caused by respiration are minimum. Pathogen metabolism is also slowed down at low temperatures hence rotting is checked. A normal cold store has a pre-cooling chamber where potatoes are kept at 15-16 °C for a few days and then shifted inside the cold store maintained at 2-4 °C and, 90-95 per cent RH. In India, cold stores were developed mainly for the storage of seed potatoes because non-refrigerated storage of seed potatoes in the first half of this century resulted in as high as 50 per cent storage losses. Storage requirements of seed potatoes are to minimize weight loss and suppress sprout growth without using a chemical sprout inhibitor so that the seed potatoes will be ready to sprout normally at the end of the storage period. But the present utilization pattern of cold storage space in different parts of the country shows that about 60 per cent of the cold storage space is being used for the storage of ware potatoes and 40 per cent for the storage of seed potatoes. Since the primary purpose of storage at 2-4 °C is to check sprout growth, storage of seed potatoes at 2-4 °C in cold store is ideal. But storage of ware potatoes at 2-4 °C is not desirable for the following reasons. i) Low temperatures induce accumulation of large amounts of sugars which make the tubers sweet and therefore less suitable for consumption. ii) Cold stored potatoes are unfit for processing as high level of reducing sugars causes browning in potato chips, iii) The keeping quality of cold stored potatoes deteriorates quickly once they are removed from the cold store as tubers begin to sprout, and iv) Marketing of cold stored potatoes is also a problem, as the trader cannot keep the potatoes at room temperature for long due to their poor keeping quality. The high sugar content of potatoes stored at 2-4 °C can be reduced to some extent by storing these potatoes at 20 °C for 6 weeks. This process is called reconditioning. However, reconditioning will not restore the sugar levels to those at the time of storage. Storage of processing potatoes requires a good technical knowledge of the crop.

Storage requirements of table or ware potatoes are to preserve their culinary properties, to prevent them from sprouting and to minimize storage losses. The problem of low temperature induced sweetness can be overcome by storing table and processing potatoes at 10-12 °C instead of 2-4 °C (Figure 7). However, the RH should be maintained at 90-95 per cent to minimize weight loss during storage. When stored at 10-12 °C, sugar accumulation is minimum therefore, potatoes do not taste sweet and hence are better than the cold stored potatoes for consumption. Since the sugar accumulation is minimum, they are quite suitable for processing, unlike potatoes stored at 2-4 °C. However, at10-12 °C, sprout growth takes place at the release of dormancy and therefore, CIPC a sprout suppressant is used to check sprout growth. Potatoes should be stored in leno bags to facilitate proper distribution of CIPC on the tubers during application.

As already mentioned, when potatoes are stored at 10-12 °C and 90 per cent RH sprout growth take place. If sprout growth is not checked, it would result in excessive

Figure 7. A Modern Cold Store with Chambers Used for Storage of Potatoes at 10-12 °C.

weight loss leading to shrivelling of tubers. When tubers of variety Kufri Jyoti were stored at 10 °C, the weight loss was 14.2 per cent after 145 days of storage (DOS). But when sprout growth was checked with CIPC, the weight loss was 3.4 per cent after 145 DOS. The weight loss at 12 °C with CIPC treatment was 4.3 per cent after 145 DOS. When potatoes are stored at 10-12 °C with CIPC treatment, reducing sugar level is within the acceptable limit and chips produced from these potatoes are light in colour as compared to dark brown chips produced from cold stored potatoes. Storage of potatoes at 10-12 °C with CIPC treatment, on a commercial scale was started in India in 1998, when CIPC was registered for the first time for commercial use in India. Since then it has become popular in Indore and Ujjain in Madhya Pradesh. It is also being used in Meerut, Agra and Sambhal in Uttar Pradesh, Deesa in Gujarat and Sainthian in West Bengal.

5.2 Quality of Stored Potatoes

As mentioned above, not all potato varieties can be stored at 10-12 °C with CIPC and used for processing. However, several of the varieties grown in the country can be used for table purpose after storage at 10-12 °C with CIPC. In a trial conducted at Central Potato Research Institute, Shimla, nine potato varieties grown at Modipuram *viz.* Kufri Jyoti, Kufri Chandramukhi, Kufri Sindhuri, Kufri Anand, Kufri Lalima, Kufri Bahar, Kufri Chipsona-1 and Kufri Chipsona-2, seven varieties grown at Jalandhar *viz.* Kufri Jyoti, Kufri Chandramukhi, Kufri Lauvkar, Kufri Badshah, Kufri Bahar, Kufri Sindhuri and Kufri Lalima, five varieties grown at Patna *viz.* Kufri Jyoti, Kufri Sindhuri, Kufri Ashoka, Kufri Chipsona-1 and Kufri Chipsona-2 and three varieties grown at Gwalior *viz.* Kufri Jyoti, Kufri Chandramukhi and Kufri Lauvkar were stored at 10-12 °C. CIPC treatment was given twice in the form of a fog. The formulation used was Oorja (Marketed by United Phosphorus Ltd., Mumbai) which contains 50 per cent a.i. It was applied @ 35 ml per t of potatoes. It was found that Kufri Jyoti, Kufri Lauvkar, Kufri Chandramukhi, Kufri Chipsona-1 and Kufri Chipsona-2 could be used for processing after storage at 10-12 °C with CIPC as the sugar level was within the acceptable limit. All the other varieties could be used for table purpose only. Though the sugar level in these varieties was higher than the acceptable level for processing, still they were much lower than that in potatoes stored at 2-4 °C. Thus, these potatoes were not sweet in taste and therefore, better than the cold stored potatoes for table purpose.

Besides variety, storage temperature and CO_2 concentration in the storage atmosphere play a major role in influencing the reducing sugar level in potatoes during storage and thus affect the processing quality of stored potatoes.

In a collaborative project between Central Scientific Instrumentation Organisation, Chandigarh, and Central Potato Research Institute, Shimla, the effect of storage temperature and carbon dioxide concentration on reducing sugar and sucrose level in stored potatoes was determined. Potatoes of four varieties *viz.* Kufri Chipsona-1, Kufri Chipsona-2, Kufri Jyoti, and Kufri Pukhraj were stored for 240 days at three temperatures *viz.* 8,10 and 12°C and RH was maintained at 95 per cent. Three carbon dioxide (CO_2) concentrations *viz.* 800ppm,1200ppm and 3000ppm were maintained.

Effect of Storage Temperature on Reducing Sugar Level in potatoes

The reducing sugar content was lower in Kufri Chipsona-1 (53 mg/100 g FW) , followed by Kufri Chipsona-2 (85 mg/100 g FW), Kufri Jyoti (187 mg/100 g FW) and it was highest in Kufri Pukhraj (193 mg/100 g FW), at the time of storage. During storage, irrespective of the storage temperature or CO_2 concentration , the reducing sugar content was consistently higher in Kufri Pukhraj followed by Kufri Jyoti and Kufri Chipsona -1. Kufri Chipsona-2 showed least amount of reducing sugar accumulation. The reducing sugar levels were higher at 8°C followed by 10°C and 12°C. Figure 8 shows the changes in reducing sugar levels in potatoes of variety Kufri Jyoti during storage at 8, 10 and 12°C with RH of 95 per cent and 800 ppm CO_2 concentration.

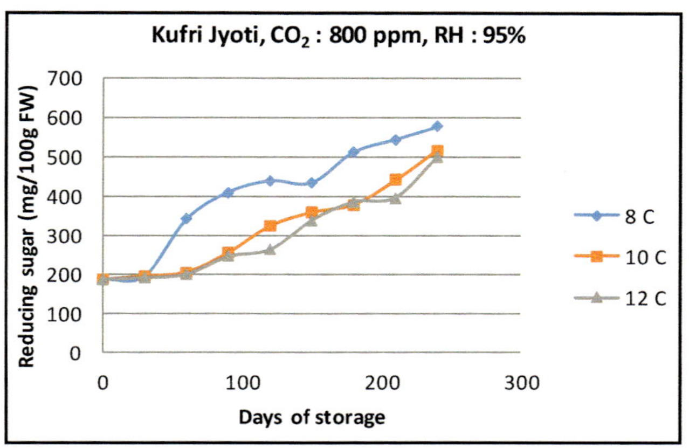

Figure 8. Effect of Storage Temperature on Reducing Sugar Level in Potatoes during Storage.

Effect of Carbon Dioxide Concentration on Reducing Sugar Level in Stored Potatoes

Reducing sugar level was generally lower at 800 ppm and 1200 ppm and the differences were not much between the two concentrations. However, at 3000 ppm CO_2 concentration, a clear increase in reducing sugar level could be observed. The reducing sugar level was even higher when the storage temperature was 8°C and CO_2 concentration was 3000 ppm. Figure 9 shows the changes in reducing sugar levels in potatoes of variety Kufri Jyoti during storage at 800, 1200 and 3000 ppm CO_2 concentrations. The storage temperature was 10°C and RH was 95 per cent.

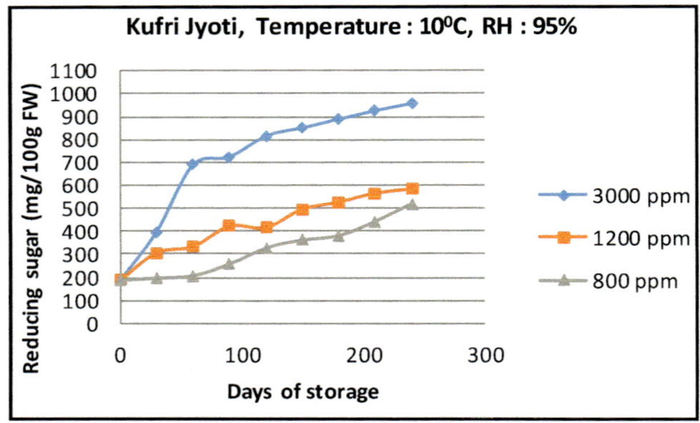

Figure 9. Effect of Carbon Dioxide Concentration on Reducing Sugar Level in Potatoes during Storage.

Effect of Carbon Dioxide Concentration on Sucrose Level in Stored Potatoes

Sucrose accumulation in potatoes stored at 3000 ppm was lower as compared to that at 1200 ppm and 800 ppm CO_2 concentrations. Figure 10 shows the changes in sucrose levels in potatoes of variety Kufri Jyoti during storage at 800, 1200 and 3000 ppm CO_2 concentrations. The storage temperature was 10°C and RH was 95 per cent.

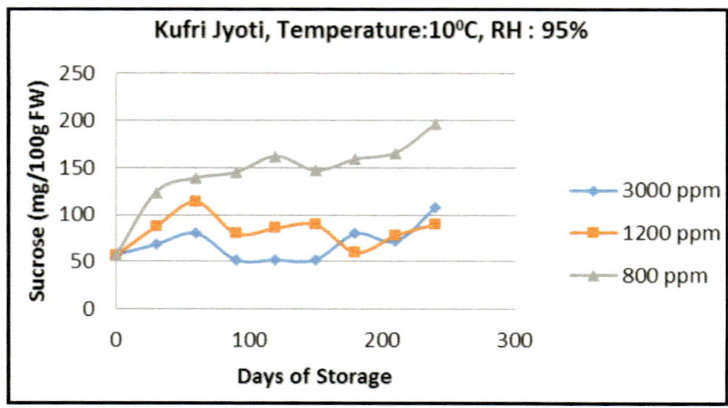

Figure 10. Effect of Carbon Dioxide Concentration on Sucrose Level in Potatoes during Storage.

There is a wrong notion prevailing in some places that potatoes stored at 10-12 °C with CIPC treatment contain low level of reducing sugar therefore, they are better for diabetic patients. For this reason, some people even call these potatoes "diet potatoes." It is wrong to call these potatoes "diet potatoes" and to claim that these potatoes are better than other potatoes for diabetic patients. The reducing sugar content in potatoes stored at 10-12 °C can vary from less than 100 to 300 mg/100g f. wt, whereas the reducing sugar content in cold stored potatoes can vary from less than 1000 to 3000 mg/100g f. wt. Because of the difference in storage temperatures, the sugar content is lower and starch content is higher in potatoes stored at 10-12 °C as compared to potatoes stored at 2-4 °C, which contain higher reducing sugar and lower level of starch. When the diet of diabetic patients is discussed, not only the free sugar level should be considered but also the total carbohydrate content of the food is to be considered. Therefore, if we consider the total carbohydrate content, there is little difference between cold stored potatoes stored at 2-4 °C and those stored at 10-12 °C with CIPC treatment. Further, the reducing sugar content in potatoes stored at 10-12 °C is comparable to freshly harvested potatoes. Therefore, they are as good as freshly harvested potatoes in taste. Nutrition wise barring the small loss in free sugars which are used up in respiration during storage there is not much difference in the carbohydrate level among fresh potatoes, potatoes stored at 10-12 °C with CIPC treatment and cold stored (2-4 °C) potatoes. All these potatoes can be consumed by diabetic patients provided the total carbohydrate intake does not exceed the level recommended by the doctor. Several studies carried out in Europe have clearly shown that potatoes do not cause or worsen diabetes.

5.3 Consumer Preference for Stored Potatoes

Surveys were carried out in five cities *viz.* Meerut (UP), Deesa (Gujarat), Indore (MP), Ujjain (MP) and Sainthian (WB), to see the preference of consumers to CIPC treated potatoes stored at 10-12 °C for table purpose in relation to cold stored (2-4 °C) potatoes which are normally available in the plains from May onwards. When asked which potatoes had better appearance, 61 per cent of the consumers felt cold stored potatoes had better appearance. When asked which potatoes they found sweet, 90 per cent of the consumers said cold stored potatoes were sweet. In Deesa and Sainthian, 100 per cent of the consumers felt cold stored potatoes were sweet. Majority of the consumers (58 per cent) preferred CIPC treated potatoes because they were not sweet in taste. Although CIPC treated potatoes were preferred for their better taste, not everyone who preferred them were willing to pay more for them. Only 42 per cent were willing to pay more for CIPC treated potatoes. Two cases *viz.* Meerut and Deesa make an interesting study. While 91 per cent of consumers in Meerut preferred potatoes stored at 10-12 °C with CIPC, 96 per cent of consumers in Deesa preferred potatoes stored in cold store at 2-4 °C (Table 4). This is because, consumers in Deesa like sweet taste in their vegetable preparations.

Table 4. Consumer Preference for CIPC Treated Potatoes

Particulars	*Meerut (UP)*	*Deesa (Gujarat)*	*Indore (MP)*	*Ujjain (MP)*	*Sainthian (WB)*	*Mean*
Sample Size	*57*	*69*	*54*	*73*	*29*	
Preference for appearance						
i) CIPC treated (10-12C)	59.6	11.6	35.2	43.8	44.5	38.9
ii) Cold stored (2-4C)	40.4	88.4	64.8	56.2	55.5	61.1
Which potatoes were sweet						
i) CIPC treated (10-12C)	1.8	0	25.9	21.9	0	9.9
ii) Cold stored (2-4C)	98.2	100	74.1	78.1	100	90.1
Preference for taste						
i) CIPC treated (10-12C)	91.2	4.4	51.8	75.3	65.5	57.6
ii) Cold stored (2-4C)	8.8	95.6	48.2	24.7	34.5	42.4
Willingness to pay more for CIPC treated potatoes						
i) Yes	54.4	4.4	40.7	46.5	62.1	41.6
ii) No	45.6	95.6	59.3	53.5	37.9	58.4

Values given are in percentage.

5.4 Biochemical Changes during Storage

The metabolic processes remain active in tubers even after their detachment from the plant. Storage conditions can alter the normal metabolic processes in a potato tuber. Major biochemical changes that occur in potatoes during storage are briefly discussed here.

5.4.1 Carbohydrates

Carbohydrates changes in potatoes during storage have been investigated in great detail. Only those changes that have a direct bearing on processing quality of potatoes are discussed here. Carbohydrates constitute 75 to 85 per cent of potato tuber dry matter, most of which is starch. The carbohydrate content and composition are important with respect to potato processing and its nutritive value. During storage, there is a gradual loss of carbohydrates from potatoes due to respiration. The rate of respiration is higher at temperatures below 4 °C and increases gradually up to 15 °C and then steeply with further increase in temperature. It has been observed that the respiratory peak corresponds with the peak of sucrose concentration. At low temperatures, therefore, besides increased respiration, free sugars also accumulate to cause low temperature-sweetening. Potatoes stored at low temperatures thus become unsuitable for processing into chips or French fries. The correlation between the rate of respiration and sucrose content can be explained as both results from alternative metabolic pathways of same substrate glucose-1 phosphate, produced by the phosphorylysis of starch, the ultimate respiratory substrate in potato tuber. The low temperature sweetening is negligible at around 10 °C. However, sugar content can increase at 10 °C also if sprouting takes place and the mobilized sugar is transported into the young sprouts. To prevent sugar accumulation at 10 °C, sprouting should be checked with a sprout suppressant like CIPC. Gamma irradiation has been reported to increase the sucrose content of potato tubers. Potato storage at higher temperatures ranging from 24.7 to 36.2 °C results in increase in total sugars but the increase is much smaller than that in cold storage. At higher storage temperatures, loss of dry matter and rotting increase sharply.

5.4.2 Nitrogen Fractions

The total nitrogen of potato tuber comprises soluble protein, insoluble protein and soluble non-protein nitrogenous compounds (NPN). The NPN compounds are free amino acids, the amides asparagines and glutamine and small amounts of nitrate nitrogen and basic nitrogen compounds including nucleic acids and alkaloids. Most of the storage studies have been carried out to investigate changes in free amino acids, soluble proteins and glycoalkaloids. The breakdown of proteins to free amino acids during prolonged storage of potatoes has been reported. Though reducing sugars are the major determinants of processing quality, anomalies in the relationship of reducing sugars to fry colour do occur and free amino acids may also act as determinants of processing quality, as illustrated by the greater browning per unit reducing sugar in tubers grown under high nitrogen conditions. Protein degradation has been associated with the end of tuber dormancy and the mobilization of nitrogen reserves for sprout formation and the breakdown of proteins is consistent with an increase of proteolytic enzyme activity. Protein synthesis may also occur during potato storage, although low temperature conditions reduce the rate of this process while enhancing protease activity. Storage studies at 5 and 10 °C up to 40 weeks have shown that free amino acids content of stored potato tubers are controlled by a combination of factors including proteinase activity and protein synthesis, the overall balance of nitrogen flow depending on the dormancy status of the tubers. Consequently free amino acid content changes with storage duration. Glycoalkaloids are another important nitrogen fraction and about 95 per cent of the total glycoalkaloids present in potato tubers are accounted for by two compounds, *viz.* alpha solanine and alpha chaconine. It has been suggested that at low concentrations the presence of glycoalkaloids may enhance the flavour of potatoes, but at levels exceeding 15 mg/100g FW, a bitter taste may be perceived. Potatoes containing more than 20 mg of glycoalkaloids per 100g FW of tuber are not safe for human consumption. When potatoes were stored at 4, 7 and 10 °C for 9 weeks it was observed that at all the three storage temperatures, the rate of glycoalkaloid accumulation was independent of the levels found at harvest and significant interactions were found between the cultivars and temperature.

5.4.3 Enzyme Systems

The potato tuber contains numerous enzyme systems that constitute a considerable proportion of the total protein, and some of which are important in relation to its processing and nutritional quality. Most of the studies carried out on this aspect deal with the enzymes associated with low temperature sweetening. Starch degrading enzymes decrease the degree of polymerization of starch, which contains predominantly alpha 1, 4-linked glucose units. The process of starch degradation during cold sweetening is believed to be principally phosphorylitic rather than hydrolytic in nature. In potato tubers two starch phosphorylase enzymes, types L and H, have been described and are believed to be responsible for the complete breakdown of starch. Some studies have suggested that an increase in starch phosphorylase activity acts as a triggering event in the sweetening of potato tubers during cold storage. Studies involving alpha amylase, beta amylase and debranching enzymes in potatoes stored at 4 and 10 °C showed that activities of these three enzymes and reducing sugar content increased at 4 °C while only a slight increase was observed at 10 °C. A cold inducible beta amylase has been purified and characterized from the variety Desiree. Its activity was low in tubers stored at 20 °C and was progressively increased when temperatures were decreased to 5 and 3 °C. The studies on cold resistant potato clone ND 860-2 suggest the ordered crystalline structure of starch granules as a key factor for its reduced sweetening potential in cold storage. Sucrose phosphate synthase, UDP glucose pyrophosphorylase (UGPase) and invertase are the three major enzymes directly involved in sucrose and reducing sugar formation. These enzymes have also been investigated in detail with respect to low temperature sweetening. In cold stored tubers, 70-80 per cent decrease in sucrose phosphate synthase expression resulted in a reproducible but non-proportional (10-40 per cent) decrease of soluble sugars. A relationship has been shown to exist between the allelic polymorphism of UGP ase exhibited among potato clones and their ability to accumulate sugars in cold storage. Genetic modification of acid invertase activity in potatoes has resulted in decreased concentrations of reducing sugars. Eight Indian potato varieties were stored at 3-5 °C and reducing sugars and invertase activity was monitored at periodic intervals. Reducing sugars and both basal and total invertase activity increased during storage in all the varieties, and there was a significant correlation between reducing sugar content and basal and total invertase activity. It was initially theorized that the cold liability of glycolytic enzymes led to an increase in hexose phosphate concentrations resulting in a stimulation of sucrose synthesis. But results of plant transformation studies have shown that there is no evidence to support the theory that a cold labile step in glycolysis is involved in triggering or regulating cold induced sugar accumulation in potatoes.

5.4.4 Other Biochemical Changes

Potatoes are a major and cheap source of ascorbic acid (vitamin C) in the human diet, but its levels decrease during storage. The most rapid loss occurs during the first months of refrigerated storage ranging from 40 to 60 per cent. Storage temperatures as well as time affect vitamin C losses and the losses are higher below 10 °C storage than between 10 and 20 °C. Polyphenolic compounds are present in potatoes and enzyme catalysed browning reactions involve the oxidation of phenolic compounds to quinnones, followed by transformation of the quinines to dark pigmens by the enzyme tyrosinase and polyphenol oxidase (PPO). These changes result in deterioration in flavour, colour and nutritional quality. The storage of potatoes at high temperatures did not affect the total phenolic and chlorogenic acid content, but both levels increased on storage at 10 °C for five weeks. PPO activity is higher at higher storage temperatures leading to greater discolouration. The principle organic acids in potatoes are citric acid and malic acid. Low temperature storage resulted in decrease in pH and an increase in malic and citric acid contents of tubers.

5.5 Economics of Storage

During the period of a study on the economics of potato storage, the normal cold storage rent varied from Rs. 0.60-0.65 per kg whereas, for CIPC treated potatoes, the rent charged varied from Rs. 1.0 –1.16 per kg. The most common modifications required for using an existing cold store for storing potatoes at 10-12 °C with CIPC treatment are as follows. (i) Partitioning is normally required as the existing chamber dimensions use to be too big for CIPC treatment (ii) Ducts may have to be installed to facilitate flushing out of accumulated carbon dioxide (iii) Humidifiers will be needed to maintain the R.H. The additional expenditure which cold store owners incur in making these modifications ranged from Rs. 0.01 to 0.11 per kg of potatoes. The cost of CIPC treatment varied from Rs. 0.11 to 0.36 per kg of potatoes. Such large variation in the cost of treatment was observed because some cold store owners used higher dosage of CIPC to account for leakage during treatment. While several cold store owners were able to get additional profit ranging from Rs. 0.04 to 0.37 per kg of potatoes, some cold store owners incurred losses due to various reasons like increased rottage and fall in potato prices at the time of selling. Farmers and traders who stored their potatoes at 10-12 °C with CIPC treatment were able to increase their profit by 37 to 58 per cent in some places but in other places farmers and traders incurred losses because there was a fall in potato prices at the time of selling and as a result, the cost of storage was higher than the rise in sale price of potatoes after storage. Only those, whose potatoes were acceptable to the processors, were able to make profit. But in many cases, the potatoes were not acceptable to processors, therefore, had to be sold as table potatoes resulting in either marginal profit or loss depending upon the market situation.

5.6 Future Storage Needs

The distribution of cold stores in the country is uneven. Most of the cold stores are concentrated in a few regions only. There is a need to identify areas that lack adequate cold storage facilities and promote the setting up of new cold stores in such areas in the future. Construction of new cold stores should be permitted based on potato production and seed requirement of a given area to increase accessibility to farmers. Cold stores were developed primarily to store seed potatoes at 2-4 °C. But at present cold stores maintained at 2-4 °C are used for the storage of seed as well as table potatoes. Storage at 2-4 °C requires considerable energy. Cold stores maintained at 2-4 °C should be used only for the storage of seed potatoes. If 60 per cent of cold storage (2-4 °C) space is converted into storage space for processing and ware potatoes to be stored at 10-12 °C, it will result in considerable saving of electricity. New cold stores should be encouraged to maintain two temperatures *i.e.* 40 per cent storage space at 2-4 °C to meet the storage requirements of seed potatoes and 60 per cent storage space at 10-12 °C to meet the storage requirements of ware and processing potatoes.

Majority of the existing cold stores have difficulty in maintaining uniform temperature, relative humidity and appropriate air circulation. The fresh air requirements of potatoes are not being met. However, awareness is increasing and many cold store owners are introducing new technologies to improve uniform maintenance of temperature through better air circulation, and to increase RH levels in the cold stores by using humidifiers. Cold store owners storing potatoes at 10-12 °C with CIPC treatment have also started monitoring CO_2 concentration in their cold stores and flush out accumulated CO_2 frequently. There is plenty of scope for improvement and modernization. Monitoring of temperature, relative humidity and CO_2 concentration on a continuous basis using computer controlled monitoring systems should become an integral part of every cold store. Arrangements should be made for improved availability of fresh air to the potatoes to meet their respiratory demand and for uniform air circulation to avoid development of hot pockets within the cold store. Humidifiers need to be used for maintaining high level of RH in the storage environment. Due attention should be paid to aspects like proper curing of potatoes before storage, and loading rate and pull down time at the time of storing the potatoes. Details on storage conditions required for optimum storage of Indian Potato varieties grown under different agro-climatic conditions need to be developed. The projected target for potato production in 2012-13 is about 40 million t. If storage of about 66 per cent of the total production is targeted, cold storage space will be required for about 26 million t of potatoes. As per a projection of National Horticulture Board, the additional cold storage capacity required is 8.280 million t and that will require an estimated investment of Rs. 66,240 million. It is hoped that this additional cold storage space to be created in the near future will make use of the latest technology available for the most efficient way of storing potatoes for up to 10 months with minimum storage losses, and with no deterioration in the quality of stored potatoes.

5.7 Use of Chemicals in Potato Storage

5.7.1 Sprout Inhibitors

Once the dormancy is released, sprouts begin to grow. Potatoes meant for ware and processing use are generally stored at 10-12 °C and at these storage temperatures sprout growth takes place which need to be checked with a sprout inhibitor. Sprout growth takes place in potatoes stored under non-refrigerated conditions also and use of a sprout inhibitor is needed to check sprout growth. Several sprout inhibitors have been used on potatoes. Some of the compounds used for checking sprout growth included several alcohols, acetaldehyde, and ethylene. Naphthalene acetic acid (NAA) and its derivative methyl ester of alpha naphthalene acetic acid (MENA), 2-4 dichlorophenoxy acetic acid (2, 4-D) and 2, 4, 5-trichlorophenoxy acetic acid (2, 4, 5-T) have also been reported to show sprout suppression effect. Although a number of chemical sprout inhibitors have been tried, only three have gained commercial acceptance. They are tetrachloro nitrobenzene, maleic hydrazide and isopropyl-N-chlorophenyl carbamate (CIPC). Of these, only CIPC is used commercially in India.

a) Maleic Hydrazide (MH)

Maleic Hydrazide is a sprout inhibitor that is applied as a pre-harvest spray to the foliage of potato crop. Work done at CPRI has shown that glyphosate can also be used as a sprout inhibitor that is applied as a preharvest spray to the foliage. Maleic Hydrazide is a sprout inhibitor that is applied as a pre-harvest spray to the foliage. MH applied to the foliage is translocated to the tubers and 10-15 days are needed for the accumulation of MH in tubers in sufficient concentration to inhibit sprout growth. MH was the only sprout suppressant registered for commercial use in India until 1998. 0.30 per cent MH applied 4 weeks before haulm cutting is effective in suppressing sprout growth without any adverse effect on tuber yield and quality. However, the time of MH application is crucial. If it is applied early yield is reduced, and if it is applied late, it is less effective. Because of the difficulty in applying at the right time, it is not being used on potatoes in India. Furthermore, often sprout suppression is not complete in MH treated tubers, and growth of cauliflower like sprouts has been observed. The tolerance limit to MH residues is 50 ppm. MH is not in use in European countries also. However, as per the latest notification of the Central Insecticide Board, Govt. of India dated 1st January, 2014 manufacture, import and use of Maleic Hydrazide (MH) is banned in India.

b) Glyphosate

Glyphosate [*N*-(phosphonomethyl) glycine] is another sprout inhibitor that can be applied as a pre-harvest spray to foliage. It is considered a non-residual and environmentally benign herbicide. Around 20 years back it was observed by researchers that when glyphosate is applied as a preharvest foliar spray, it affects the yield of potato. However, further work indicated that at lower doses it does not affect the yield but, has a strong effect on the sprout growth of the potatoes produced by the treated plants. Glyphosate applied to the foliage is translocated to the tubers where it affects sprout growth by inhibiting meristematic development of sprouts. EPSPS (5-enolpyruvylshikimate 3-phosphate synthase) is a key enzyme of the shikimate pathway which is required in plants for the synthesis of aromatic amino acids and other compounds including vitamins, plant growth substance and lignin and this enzyme is the target of glyphosate. In view of the versatile role of EPSPS enzyme in the plant metabolism, its inhibition by glyphosate affects many physicochemical and physiological processes in plant. Work done at CPRI, Shimla has indicated that single preharvest foliar application of glyphosate at sub-lethal dose of 0.694 kg (a.i.) in 600 litres of water ha^{-1} at 60 days after planting is effective in suppressing the sprout growth of the harvested potatoes during storage at 18 °C However, more research work is needed on glyphosate for establishing it as an effective sprout suppressant on potato.

5.7.2 Postharvest Application

CIPC and TCNB are the two sprout inhibitors applied postharvest to potatoes in storage

a) Isopropyl-N-3-Chlorophenyl Carbamate (CIPC)

CIPC is the most commonly used sprout suppressant on potatoes in many countries. It is very effective in checking sprout growth, when applied as a postharvest treatment on potatoes during storage at 10-12 °C. Details on this chemical, its application and effectiveness have already been mentioned above.

b) CIPC

CIPC (isopropyl N (3-chlorophenyl) carbamate) is the most common sprout suppressant used on potatoes. It is commonly called chlorpropham. CIPC is a mitotic inhibitor and checks sprout growth by inhibiting cell division. It inhibits spindle formation during cell division. CIPC was registered for use in India in 1998, and since then several cold stores in the country are using this chemical on potatoes stored at 10-12 °C. CIPC is available in three formulations, *viz.* solution form to be applied as an aerosol, emulsifiable concentrate for use as a direct spray and powder form for direct application. The liquid formulation is available at concentrations ranging from 25 to 50 per cent a.i. (active ingredient) and a fogger is used for its application. The rate and dose of application depend on the concentration of active ingredient, variety, duration of storage and temperature of storage. The emulsifiable concentrate is generally used on export consignment. The CIPC powder or dust is normally available as 1 per cent or 2 per cent active ingredient and is applied through applicators on potatoes going into store through conveyor belts. Powder applications could affect the natural skin colour of potatoes. A good storage can only maintain the quality of potatoes stored but it cannot improve it. Therefore, it is essential that potatoes for treatment with CIPC should be physically and chemically mature. The physical maturity refers to proper skin curing and wound healing. Potatoes treated with CIPC cannot form periderm layer as cell division is inhibited. Chemical maturity refers to low reducing sugars and sucrose levels. High sucrose content at the time of storage can lead to accumulation of reducing sugars during long term storage.

CIPC Treatment

Table potatoes can be treated with 25 mg CIPC dust a.i./t of tubers (commercial preparation of CIPC dust contains 1 per cent active ingredient) at the time of storage. For treating small quantity of potatoes, powder form of CIPC is convenient but for treating large quantity of potatoes, powder form is not convenient due to the practical problems in handling, since powder treatment is done manually in India. For large scale treatment of potatoes with CIPC, liquid formulation is preferred as it can be applied in the form of a fog using a fogging machine, without disturbing the stored potatoes. Chlorpropham (CIPC) is being marketed in India by United Phosphorus Ltd., Mumbai in the name of Oorja. This commercial preparation contains 50 per cent a.i. and 30-35 ml of this preparation is required for fogging one t of potatoes. However, this dose can vary depending upon the loss of fog during treatment due to leakage, the pore size of the bags used, the quantity of potatoes stored *etc.* After the CIPC treatment, the store should be kept air tight for 24-48 hours. The treated potatoes are safe for consumption after 3 to 4 weeks of treatment. First fogging is done at the first sign of sprout growth. Second fogging is done at about 45 days after the first fogging. The timing of first fogging is important. If it is done at the right time, CIPC is most effective in suppressing sprout growth (Figure 8a). Dealyed treatment of CIPC at inadequate concentration and after sprouting make it less effective and there is the risk of internal sprout growth. Figure 8b shows a tuber showing internal sprout growth after inadequate CIPC application. When internal sprout growth takes place, tubers become useless and have to be discarded. After CIPC application, ventilation should not be done for 24 hours. After this, regular flushing out of excess carbon dioxide is essential to maintain the quality of potatoes. Prolonged storage at 10-12 °C could result in sweetening, which is variety dependent. Senescent sweetening is irreversible and sugar level cannot be decreased through reconditioning. Contamination of seed potatoes with CIPC should be prevented as this can impair the emergence of the crop.

(a) Untreated Control

(b) CIPC treated

Figure 8. Effect of CIPC on Sprout Growth of Potatoes Stored at 10-12 °C.

CIPC Residue Level

Residue level of chemicals in treated foods has been an environmental and health concern. Therefore, we must know the residue level of CIPC in treated potatoes. Tuber samples were collected from two cold stores each at Indore, Ujjain and Meerut at regular intervals beginning from two days after treatment and CIPC residue levels in peels of treated potatoes and unpeeled potatoes were determined. The maximum residue level observed in peels was 14.7 ppm and the maximum residue level observed in unpeeled tuber was 2.5 ppm, which was within the tolerance limit prescribed at that time. Since the CIPC residue level did not exceed 2.5 ppm in unpeeled, unwashed tubers in samples collected from cold stores in India,it can be said that CIPC treated tubers are safe for human consumption. To be on the safer side, it is recommended that CIPC treated tubers should be consumed only after3- 4 weeks of treatment.

Around thirty years back, it was proposed that due to environmental concerns the use of CIPC may become restricted in future by several countries. During the last 20 years, declining trend in values for the levels of Maximum Residue Limit (MRL) of CIPC as recommended from time to time by different regulatory agencies is also an indication of potential harmful effects of CIPC. As per Federal Re-registration Eligibility Decision by the Environmental Protection Agency (EPA), the MRL (Maximum Residue Limit, also referred to as residue tolerance) for CIPC was fixed at 50 ppm. This limit was reduced to 30 ppm for fresh potatoes (meant for table purpose) in 2002 . Later, in 2009, European Commission and Health and Safety Executive (HSE) decided to keep the MRL at 10 ppm . The above decisions to reduce the limits of CIPC over the period are basically to put limits on the number of applications/overuse of CIPC so that desirable Acceptable Daily Intake Limit (ADIL) is maintained or achieved. The EU started phasing out CIPC from 2020.

Most of the EU countries have stopped using CIPC for sprout suppression in potatoes and have started using alternative sprout suppressants. Sweden has banned CIPC. Countries like Netherlands and Switzerland are using S-carvone, a naturally occurring substance as an alternative to CIPC for sprout suppression at commercial scale. In UK, Dimethylnaphthalene (DMN) is used as an alternative to CIPC. From 2021 the temporary MRL was 0.4 ppm, which means potatoes stored in EU countries are not be treated with CIPC. In view of these recent developments, India should also phase out CIPC and start using other sprout suppressants being used in EU.

c) Tetrachloro-Nitrobenzene (TCNB)

TCNB also known as Tecnazene (Commercial name 'Fusarex') is in commercial use in some countries but its use is limited. It is known to suppress sprout growth, when given as postharvest application on potatoes stored at 10 °C and does not inhibit wound healing. Since it is highly volatile, it can be used on seed potatoes also. It is applied at the rate of 135 mg/kg of potatoes.

d) Natural Substances as Sprout Inhibitors

The growing tendency to avoid consuming foods treated with chemicals has led to increasing research interest in natural substances with sprout inhibitory effect. The ultimate aim is to replace chemical sprout inhibitors with naturally occurring compounds. Several naturally occurring aromatic aldehydes (salicylaldehyde, benzaldehyde, cinnamaldehyde and cumminaldehyde) and aromatic alcohols (thymol) are reported to inhibit sprout growth for a short period of up to 14 days. Aromatic compounds like mono, di and trimethyl naphthalenes, benzothiophene and diphenylamine have also been found to have sprout inhibitory effect. Some volatile monoterpenes are also reported to inhibit sprout growth. Muna plants (family *Lamiacea*) contain essential oils that are comprised of over 98 per cent monoterpenes. Small scale experiments have shown that pulegium containing essential oils of *Minthostachys glabrescens* and *Mentha pulegium* and the carvone containing essential oils of *Anathem graveolens* and *Carum carvi* are efficient sprout inhibitors. 1, 8-cineole, available in large quantities (70-85 per cent) in oil of *Eucalyptus globules*, is a promising monoterpene with good sprout suppression effect. Vapour treatment (250 µl/l for ten days) with pure *E. globules* oil showed strong sprout suppression effect up to 80 days of storage at 24+ 2 °C. Carvone is the only monoterpene produced by *Carum carvi* (caraway), which has shown consistently good results in storage trials at 8-10 °C. It is found naturally in essential oils of *Cuminum cyminum, Mentha spicata, Mentha citrata, Mentha cardiac, Anathem graveolens* and *Anathem soya*. The S-carvone form is being marketed in the Netherlands under the commercial name "Talent". Since carvone is highly volatile, repeated applications are necessary to check sprout growth. This makes it very expensive and carvone treatment may be 10 times more costly than CIPC treatment.

5.7.3 Irradiation

Irradiation has also been used to check sprout growth in potatoes. Irradiation suppresses sprout growth by preventing cell division. In the early 1960's, gamma irradiation was tried to inhibit sprout growth in potatoes of old varieties like Phulwa and up-to-date. Periderm formation in stored potatoes is completely inhibited even when a low dose of 20-30 Gy gamma irradiation is used, as it affects the nuclei of potential periderm cells. Considerable amount of work on the use of gamma irradiation for sprout suppression of potatoes stored under various storage conditions has been done by BARC, Mumbai and CPRI, Shimla. These trials led to the conclusion that gamma irradiation cannot be used for suppressing sprout growth in potatoes stored under non-refrigerated conditions as higher temperatures during storage resulted in unacceptably high storage losses in irradiated potatoes due to increased rotting. Rotting generally increases in irradiated potatoes as irradiation interferes with wound healing by inhibiting cell division. Trials conducted at BARC have shown that under refrigerated conditions, irradiated potatoes could be stored successfully up to nine months at 10 °C. In a trial conducted at CPRI, Shimla during 2003-04, potatoes of three varieties *viz.* Kufri Chandramukhi, Kufri Jyoti and Kufri Chipsona-2 were exposed to 0.1 kGy and 0.5 kGy gamma irradiation and stored at 8, 12 and 16 °C. Potatoes could be stored for six months at these three temperatures but rotting was on the higher side in all the three varieties. While 0.1 kGy gamma irradiation was as good as the chemical sprout suppressant CIPC in suppressing sprout growth of potatoes (Figures 8a&b), irradiation induced accumulation of sugars made the stored potatoes less suitable for processing. Irradiation caused 18-35 per cent increase in reducing sugars content and 20-29 per cent increase in sucrose content. Higher concentration of reducing sugars resulted in browning of fried potato chips (Figures 8a&b). However, chemical sprout suppressants are preferred to irradiation, which has not found wide acceptance globally.

6.0 Conclusion

Potato production in the country has shown a phenomenal growth. The steady increase in the total production at the national level has not been matched by the developments in postharvest technology. This mismatch between potato production and postharvest technology and utilization has resulted in periodical gluts. Adoption of alternate storage methods for short-term storage of potatoes will reduce the demand for cold storage space in the country. There are a number of traditional, low-cost and non-refrigerated storage structures (essentially based on evaporative or passive evaporative cooling) in use in India with varying degrees of success. The traditional methods have been studied, validated and popularized for particular regions. In non-refrigerated storages, use of sprout suppressants has also been popularized to prevent excessive weight loss and shrinkage due to sprouting. Storage losses in traditional methods like heaps and pits could be quite high if proper care is not taken. For these methods of storage to become more widely acceptable throughout the country, it is necessary to improve their efficiency and to reduce the storage losses. Potatoes stored in heaps and pits are suitable for processing due to their higher dry matter and lower reducing sugars content. Storage of potatoes in cold store has become a common practice in the country and potatoes are generally stored in cold store at 2-4 °C and 90-95 per cent RH. When stored at this temperature potatoes do not sprout and weight loss is minimum. The conditions maintained in cold stores are ideal for storing seed potatoes but not for the storage of table and processing potatoes. Cold stored potatoes are especially unsuitable for processing since higher sugar levels result in unacceptable, dark brown chips. The problem of low temperature induced sweetness can be overcome by storing table and processing potatoes at 10-12 °C instead of 2-4 °C.Storage of potatoes at 10-12 °C with CIPC treatment is becoming popular in India. However, the storage technique being practised at present is not satisfactory. Greater awareness about suitability of a potato variety for storage at 10-12 °C, maximum duration of storage for a given variety at this temperature without affecting the processing quality and right time of application of CIPC, will help in getting the maximum benefit. Farmers, traders and cold store owners can gain from this storage technology if the processing quality of potatoes stored is acceptable to potato processors in the country. There is plenty of scope for improvement and modernization. Monitoring of temperature, relative humidity and CO_2 concentration on a continuous basis using computer controlled monitoring systems should become an integral part of every cold store. Potato cold storage industry has played a very significant role in sustaining the increased potato production in the country during the last sixty years. However, there are problems that need to be addressed. Unless immediate attention is paid to the problems associated with potato storage and utilization, it will not be possible to sustain the current trend of increase in potato production.

7.0 Important References

Burton, W.G. 1989. *The Potato*, 2nd Edn., Longman, Essex, 742 pp.

Burton, W.G., A Van Es and K.J. Hartmans, 1992. The physics and physiology of storage. In: *The Potato Crop*, 2nd Edn. (P.M. Harris, Ed.), pp. 6-8-727. Chapman and Hall, London.

Gottschalk, K.and Ezekiel, R. 2006. Storage. In: *Handbook of Potato Production, Improvement and Postharvest Management*. (J. Gopal and S.M.P. Khurana, Eds), pp 489-521, The Haworth Press, New York.

Sukumaran N.P. and S.C. Verma 1993. Storage and processing. In: *Advance in Horticulture*. Vol. 7-Potato (K.L. Chadha and J.S. Grewal, Eds), pp. 701-32, Malhotra Publishing House, New Delhi.

Talburt, W.F. and O. Smith. 1987. *Potato Processing*, 4th Edn. Van Nostrand Reinhold, New York. 796 pp.

http: //www.potaoworld.in

Chapter 29

Chilli (*Capsicum annum* L.)

Surya Kumari Surepeddi

Principal Scientist (Horticulture)

Giridhar Kalidasu

Principal Scientist (Horticulture)

AICRP on Spices, Dr. Y.S.R. Horticultural University, Horticultural Research Station, Lam, Guntur – 522 034, India

E-mail: surya.surepeddi@gmail.com; E-mail: gkalidasu@yahoo.com

1. Introduction

Chilli as a spice has spread to the entire globe in a few hundred years though it is popular since ancient times. It belongs to the genus *Capsicum*. The genus *Capsicum* represents a diverse plant group and includes twenty-seven species; five domesticated and twenty-two un-domesticated (Bosland, 1993). *Capsicum annum* L., *C. baccatum* L., *C. chinenses* Jacq., *C. frutescens* L. and *C. pubescens* Ruiz and Pav. are considered domesticated species of peppers (Carolina *et al.*, 2016). The centre of diversity for *Capsicum* is in South-central South America (Gonzalez and Bosland, 1991), with the majority of species having some range in Brazil and/or Bolivia. Amazonia (in the northern area) is considered to be the centre for *C. chinense* (habanero) (Toquica *et al.*, 2003), and western Amazonia is perhaps the centre for *C. frutescens* (tabasco), which is more domesticated in Central America (Hernandez-Verdugo *et al.*, 1999). Many members of the genus *Capsicum* are valued for their taste, colour and spice value. They are widely appreciated for their active principles and essential oil. It is known by different names *aji*, *chile*, *chili*, *chilli*, *paprika*, *red chile*, *chilli pepper*, *hot pepper*, *hot red pepper*, *tabasco*, and *cayenne*. Capsicum pepper or sweet bell pepper refers to a non-pungent, chunky, sweet chilli pepper type that is used as a vegetable. The botanical name for the genus Capsicum originated from a Greek based derivative of Latin "Kapto" meaning "to bite", a particular reference to heat or pungency. Chilli was consumed by the people of South India and Central America as early as 7000 B.C., but it was cultivated some 2000 years later. Intriguingly, *Capsicum* seems to be absent from the more recent Chiribaya culture of 1100- 600 B.P. in far southern Peru (Flamini *et al.*, 2003), even though earlier (2200-1400 B.P.) in southwestern Peru, *C. frutescens* was being used. Many civilizations were known to use this species, especially the Mayans and the Aztecs. Though the crop was domesticated between 7000-5000 years ago, it is considered a new crop since its utilization drastically increased only in the last few hundred years, particularly outside its origin. A major portion of the genus *Capsicum* arose in a nuclear area in south central Bolivia with subsequent migration to the Andes and lowland Amazonia accompanied by radiation and speciation. Five species *viz. C. annuum* var. *annuum*, *C. chinense*, C. *frutescens*, *C. baccatum* varieties *pendulum* and *umbilicatum*, and *C. pubescens* were domesticated by native Americans. Of the five domesticated species, initially *C. baccatum* and *C. chinense* were cultivated in northern South America while it is probable that *C. annuum*, *C. frutescens*, and *C. pubescens* were domesticated in Mexico or northern Central America. Following the initial visits of the Spanish to the Americas, the plant was brought to Europe and from there to most tropical, subtropical, and temperate zones. Many consider the popular name "chile" or "chilli" originated from the hot pepper species cultivated in the South American country of Chile. However, the name "chilli" seems to have nothing to do with the country name and, on the contrary, is believed to have originated from a region of Central America. Among the several species, the most economically important species in the world are *Capsicum annuum*, *Capsicum frutescens* and *Capsicum baccatum*. Pungent cultivars (*ajies, paprika, chilies, and hot peppers*) are generally used as a spice; whereas the sweet types (*sweet pepper, bell pepper, pimento*) as vegetables. Chilli pepper generally refers to a pungent chilli variety. Considering the trade, consumption and volume of production, red pepper products, pungent and non-pungent, represent some of the most important spice commodities in the world. Dry chilli is extensively used as a spice in the entire globe. Dry chilli along with other spices and condiments are used to prepare various spice mixtures and curry powders. It is also used for seasoning of egg, fish and meat preparations, sauces, chutneys, pickles, frankfurters, sausages *etc.* Bird chilli is used in making hot sauces such as pepper sauce, tabasco sauce and certain alcoholic beverages like Mandrani of West Indies. Chilli is a good source of β-carotene (Vitamin A), ascorbic acid (Vitamin C) and Vitamin E, with minor contents of Niacin, Vitamin B6, Folic acid and Vitamin K. It also contains minerals such as P, K, Ca, Mg, Fe, Zn, Cu, Co, Ni and Na. Other important phytochemicals include Tannins, Flavonoids, Alkaloids, Anthraquinones, Phenolics, Saponins, Glycosides, Terpenoids, Carotenoids and organic acids. The leaves, fruits and seeds have many nutraceutical, medicinal and other beneficial properties. These properties include *i.e.* analgesic, analgesic in neuropathy and neuralgia, anti-inflammatory effect, anti-bacterial, anti-viral, anti-protozoal, anthelminitic, anti-fungal, anti-pathogenic, anti-dermatomycoses, mitogenic activity, antioxidant, antinociceptive effect, reduction in plasma glucose level, cytoprotective activity in peptic ulcer, gastric acid secretion, gastroprotection, hyperplasia, hypoglycaemic effect, insulinotropic effect, insecticidal, acaricidal and larvicidal activity (Maji *et al.*, 2016; Saleh *et al.*, 2018). The hot and pungent fruits are employed in ethnic, traditional and modern medicine, capsaicin being the most important bioactive compound. Its pharmacological action as counter-irritant, stomachic, gastro-protective, neuropathy, diabetic neuropathy, beneficial effect on metabolic and arterial function and carcino-protective effects were widely appreciated. Capsaicin has been used in different allopathic creams to alleviate pains as it interferes with the sensory nerve terminals. Capsaicin incorporated creams are widely used in dermatologic therapy and have proven useful in preventing chronic pain associated with post-herpetic neuralgia, diabetic neuropathy, and other pain syndromes. Pepper spray is also used in defence and self defence applications to ward off enemies or attackers. High heat oleoresins are used for this purpose. Apart from its noted medicinal values, chilli recently attained status of a beverage component with its capsaicin being used as an active ingredient because of its pleasurable and euphoriant effects after ingestion. Value added products such as dehydrated chilli, pickle powder, paste, sauce *etc.*, can be prepared for higher returns. Almost all the chilli growers sell the produce directly. Through value addition at the farm, they can realize higher returns; enhance shelf life and quality of the end product.

2. Status of cultivation

2.1. Global Scenario

Chilli transcended as an excellent substitute for black pepper after the discovery of the New World by Columbus. Though, a native of warm, tropical America it found its way to Europe, Asia and Australia. Many local types are still consumed in the South American states. Now, these hot peppers or chillies are grown worldwide. The prominent producers are India, China, Pakistan, Korea, Mexico and Bangladesh. The global area, production, and productivity of dry and green chillies are presented in Figures 1 and 2 (FAOSTAT, 2014). India contributes to 41 per cent of the global production, with 45.1 per cent of the global area under dry chillies.

2.2. Indian Scenario

India is the largest producer, consumer and exporter of dry chilli in the world. Among the spices produced in the country, chilli (with 40 per cent production share in all the spices) is the most important one followed by garlic, turmeric, ginger and seed spices. India contributes about 41 per cent to the total world production, with 45.1 per cent of global area under dry chillies (FAOSTAT, 2019). It commands a lion's share of exports with 50 per cent of global trade by volume, with China being its closest competitor, but it is far behind in the second position. In 2019-20, India exported 4.84 lakh tonnes of chilli. The crop is cultivated in all the states and Union territories of the country. It is cultivated in 7,23,213 ha in the country with a production of (Directorate of Arecanut and Spices Development, 2019-20). No country in the world has so much area and production as much as India. Chilli has two important commercial qualities – pungency and colour. Some varieties are famous for their typical red colour and others are known for pungency. India is the only country rich in many varieties with both quality factors. Breeding work on chillies was carried out at different places and improved varieties were released for commercial cultivation. However, the postharvest handling and value addition of chilli was not given much importance. Consequently, Indian chilli could not compete in the international market for export purpose, particularly in certain value added products. Although Indian chilli is exported to over 90 countries, the percentage of chilli export in the international trade from India is very low in case of paprika and oleoresins. Pesticide residues and detection of aflatoxin are the issues affecting chilli exports from India. Moreover, during storage preservation of attractive typical red

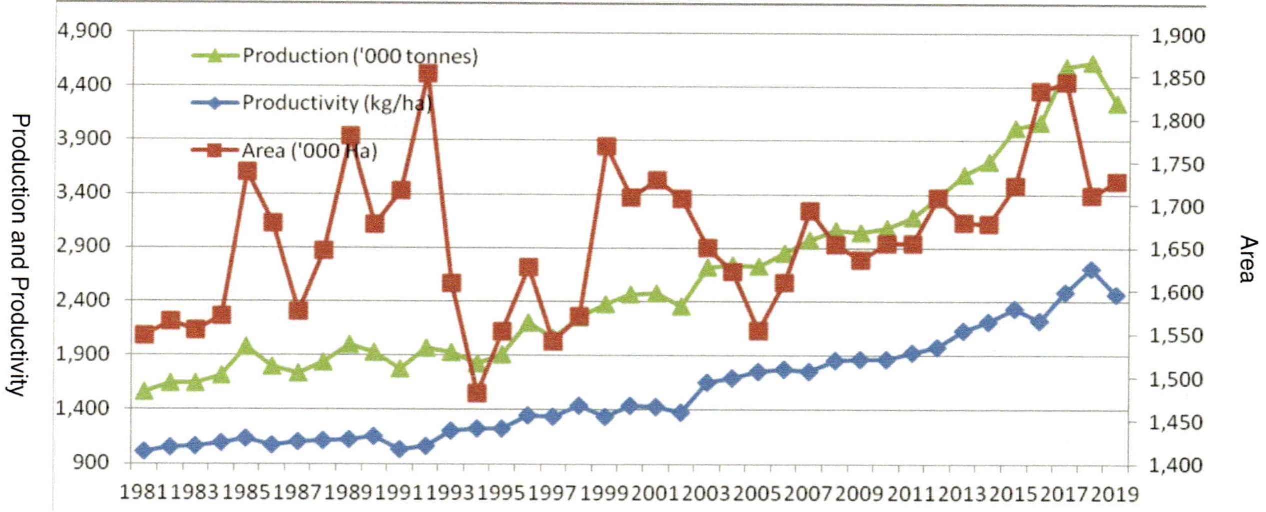

Figure 1. Global Area, Production and Productivity of Dry Chillies (1981-2019).

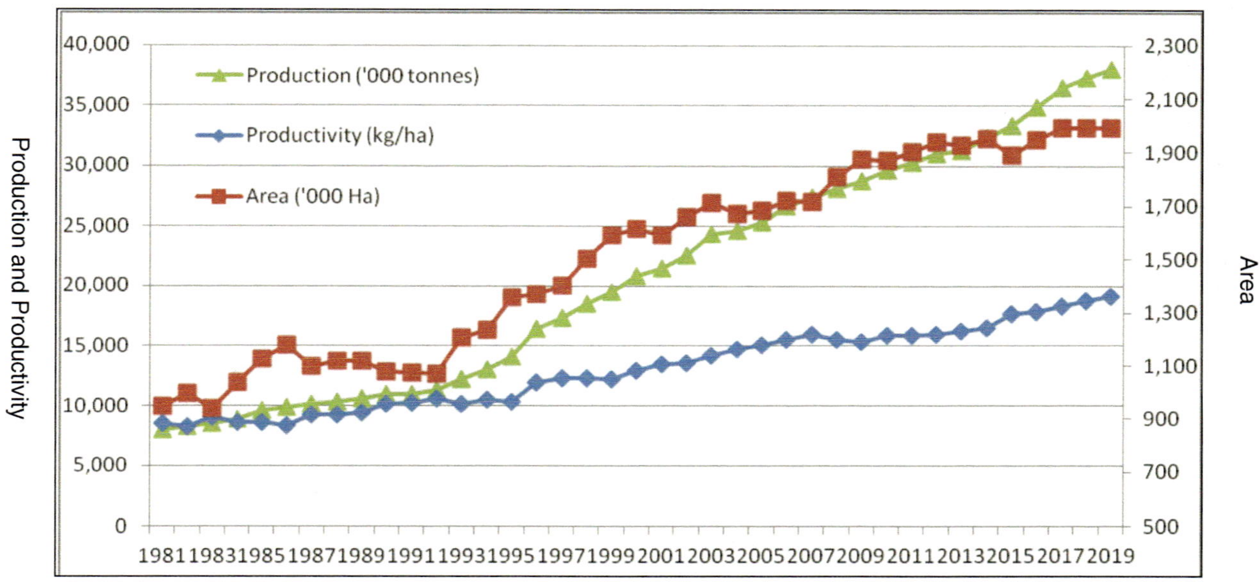

Figure 2. Global Area, Production and Productivity of Green Chillies (1981-2019).

colour and retention of the pungency have become a problem. The production of chilli in India is dominated by Andhra Pradesh which contributes 42 per cent to the total production. Telangana is the second largest producer contributing 20.5 per cent, followed by Madhya Pradesh (15 per cent), Karnataka (9.8 per cent), Odisha (3.5 per cent), Tamil Nadu (1.3 per cent), Maharashtra (1.3 per cent), Gujarat (1.1 per cent), and others about 5.5 per cent during 2020-21 (II Adv. Estimates, Directorate of Arecanut and Spices Development; Figure 3). The major chilli growing districts in Andhra Pradesh are Guntur, Warangal, Khammam, Krishna and Prakasam. Guntur is the biggest chilli producing region, contributing 30 per cent to the total production of Andhra Pradesh. Chilli production in India is moving northwards on increasing demand from diversified sectors and changing consumption patterns. Dry chilli production rose by nearly 291 per cent from 5.091 lakh tonnes in 1980 to 19.883 lakh tonnes in 2020-21. Rising export demand coupled with higher price realization in the domestic market has motivated farmers to bring more area under chilli cultivation. The average productivity of the crop was highest in Andhra Pradesh (4644 kg ha^{-1}) followed by Telangana (4510 kg ha^{-1}) in 2020-21.

3. Major Commercial Varieties

3.1. Globally Traded Types and Cultivars

In the genus *Capsicum*, enormous morphological variability exists for fruit colour, size, shape, pungency, flavour and aroma. Based on fruit size, shape and degree of pungency, a large number of horticultural types are recognized worldwide and at least 20 types are cultivated on a large scale in one or other parts of

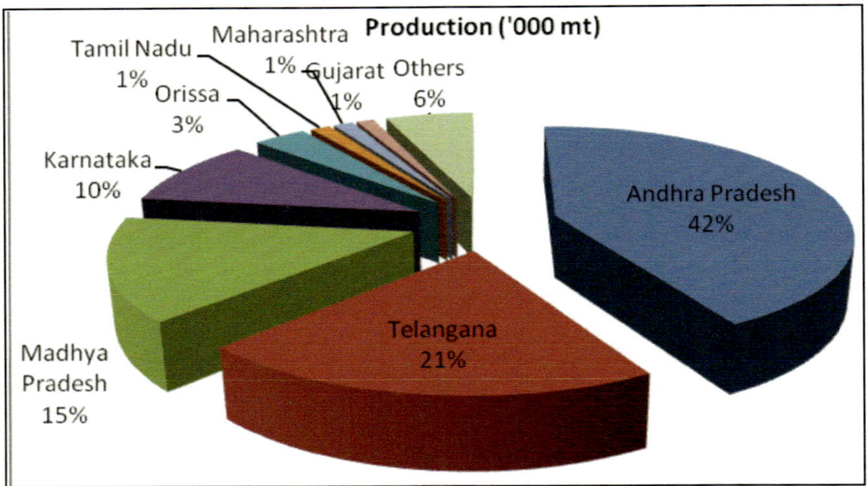

Figure 3. Dry Chillies Production and Production of Share of States in India, 2020-21.

the world. Some of these fruit types such as *ancho, bell, jalapeño, pasilla, new mexican, yellow wax* have specific traits for processing, fresh use, flavour and pungency. In India, several genotypes with specific fruit size, shape and quality attributes are commercially cultivated in different regions. The colour value is generally expressed as EOA or ASTA units. Codex standards prescribe pungency expressed in Scoville Heat Units (SHU) into four categories *i.e.* mild (900-1999 SHU), medium (2000-19999 SHU), hot (20000-100000) and extra hot (> 100000 SHU). Due to location specific consumer needs and nature of end use; colour, shape, size, pungency and flavour vary widely according to local preference. The fruit quality and type depend mostly on the target market. Considering the fresh market, low pungent pimentos are preferred in parts of Europe. In India, extremely hot Jwala and S4 types are preferred. Similarly, in dry chilli for best quality hot to mild paprika preparation, half a dozen types varying from long red wrinkled to round red cherry peppers are commonly used. The crop is generally distinguished by its pod type and several hundred pod types are grown worldwide. The important quality traits for different markets are presented in Table 1 and the commercial uses of chillies are presented in Table 2. Numerous subtle flavours of chillies exist and from experience one can know these flavours. For example *ancho* is sweetish, *mulatto* is chocolaty, *mirasol* is fruity, and *chipotle* is smokey (Bosland *et al.*, 2012). Grinding the pods produce one flavour, toasting before grinding produce another, and soaking the chillies in water produce yet another. Many pod types are central ingredients in ethnic dishes such as ancho and New Mexican in chile rellenos, pasilla in mole sauce, and serrano in pico de gallo. A detailed description and use of the various pod-types is presented by Dewitt and Bosland (2009) and Bosland *et al.* (2012). The details of important chilli varieties across the globe are presented in Table 3.

Table 1. Important Traits for different Markets

Market Type	*Important traits*
Fresh whole fruit market	Colour (light green, dark green, yellow, orange and multicolour), size (2-15 cm), shape (cherry or bell type or short stout to long slender), exocarp glossyness, exocarp thickness (both thin and thick), mesocarp thickness, seeds per pod, endocarp to seed ratio, seed colour, pungency, flavour, vitamin C and A.
Processing of fresh chillies	Colour, pungency, flavour, size, shape, pericarp thickness, endocarp to seed ratio, shelf life before processing, suitability for processing and shelf life after processing.
Dried chillies (whole and powdered)	Colour, pungency, flavour, dry weight, crude fibre, endocarp to seed ratio and seed content.
Oleoresin	Colour (high), pungency (nil to extra hot) and essential oil.
Capsaicin	Extra hot (> 100000 SHU) pungency and recovery.
Capsanthin	High EOA colour (above 50000 EOA) value and recovery.
Ornamental	Colour (multicolour and variegated), fruit shape, size, erect and clustered fruits.

3.2. Commercially Traded Types in India

India has emerged as a major player in chilli global trade. The major types grown in India are pungent ones and a few are low pungent types. The most important types traded are sannam, mundu, wrinkled and bird's eye chillies. There are also many other varieties like Ellachipur Sannam, Hindpur-S7, Kanthari White, Madras Pari, Kashmir chilli, Nalchetti, Ramnadu Mundu, Sattur-S, Byadagi, Warangal Chapata and Tomato Chilli. Each type is known for its intrinsic value, characteristics and specific market. The most important chilli types traded in India are presented in Figure 4. The important chilli growing areas are presented in Table 4.

Table 2. Commercial Uses of Chillies and Preferred Types

Capsicum Type	Purpose	Preferred Types	Products
All edible types (*Capsicum spp.*)	Fresh whole fruits	All types	—
C. annuum L.	Canned or bottled, frozen, pickled or brined, fermented fruits and sauces	Pimento, cherry peppers, jalapeño, serrano, yellow wax, pepperoncini, Thai chilli and New Mexican	Pimientos, Canned jalapeño, convenience food, pickled Pepperoncini, stuffed pepper, Ceviche and Sambal oelek
C. frutescens Roxb.	Fermented sauces	Tabasco pepper, Malagueta chilli and Thai pepper	Creole cooking and Tabasco Sauce
C. chinense Jacq.	Fermented sauces	Habanero, Scotch Bonnet, Yellow lantern chilli and King chilli	Puree, Salsa, Jerk dishes, Habanero Hot sauce, Scotch Bonnet Sweet Sauce
C. annuum L.,*C. frutescens* Roxb., *C. chinense* Jacq.	Spice extract, Paprika and chilli oleoresins	Hungarian paprikas, Pimento and Indian Sweet types	Oleoresin Capsicum (OC), Oleoresin Paprika (OP)
C. frutescens Roxb.,*C. chinense* Jacq.	Pharmacological and defence (Capsaicin)	Red Savina, Habanero and Naga jalokia (Tejpur)	Oleoresin Capsicum (OC)
C. annuum L.,*C. frutescens* Roxb.,*C. chinense* Jacq.	Whole pods, Paprika powder	Many types - New Mexican, cayenne, ancho, pasilla, mirasol, piquin, de arbol, and Indian hot and sweet types	Consumer unit packed pods, crushed pods, chilli powder, and spice mixtures.
	Pharmacological (Capsanthin)	Hungarian paprikas, Pimento and Indian Sweet types	Oleoresin Paprika (OP)
	Decoration and pot plants	Christmas peppers	Colourful strings (ristras), potted peppers and Yellow lantern plants

Table 3. Important Chilli Varieties of the Globe

Species	Type	Important varieties
C. annuum L.	Ancho or Mulato or Pablo Type (wide throated)	Ancho 101
	Asian types	Pusa Jwala and many Indian varieties, Takanotsume, Shishito, Santaka, Prik Kee Nu, Prik Chee Faa, Thai Chile, Chao Tian Jiao, Chi Chien, Sichuan chilli, Erjingtiao Chilli, Mantianxing, Chaotianjiao, Denglongjiao, Sanyingjiao, Zidantou
	Bell types (blocky fruits)	California Wonder, Dove, Golden Bell, Valencia, Lilac Belle and Mexi-Bell.
	Cascabel types (rattling pod, ball chilli)	Cascabel Local (Mexican)
	Cayenne types	Large Red Thick, NuMex varieties
	Cherry types (cherry pods)	Chilhuacles, Beslers Cherry, Hot Cherry Pepper, Cherry Pick and Cherry Bomb.
	Chiltepin (small round pods) or Chile piquin (small elongated pods) types	NuMex Bailey Piquin
	Chipotle types (smoked red ripe jalapeño)	Jalapenos
	Costeño types (used for sauses)	Costeño and Costeño Amarillo
	De Árbol types (treelike plant, mature plant resembles a small tree)	De Árbol Purple
	Ristra types (fruit wreaths)	NuMex Sunburst, NuMex Sunflare and NuMex Sunglo
	Jalapeño types (named after town of Jalapa, Mexico; thick walled conical fruits)	Early Jalapeno, NuMex Jalmundo, NuMex Piñata, NuMex Primavera, and NuMexVaquero.
	Mirasol types (looking at the sun, upright fruits)	NuMex Mirasol, Guajíllo and Pulla.
	New Mexican types (long green pods)	Chimayó Española, Improved New Mexico No. 6-4, NuMex Heritage 6-4, NuMex Big Jim and NuMex Heritage Big Jim, NuMex Conquistador, NuMex R, Naky NuMex, Sweet NuMex Sunrise, NuMex Eclipse, and NuMex Sunset, NuMex Joe E. Parker and Sandia.
	Onza types (short slender fruits)	Onza Roja and Onza Amarillo
	Ornamental types	Black CubanChilli, Chilli Jigsaw, NuMex Centennial, NuMex Twilight, NuMex Christmas, Peter Pepper and Poinsettia.
	Pasilla types (little raisin types; dark brown colour and raisin like aroma)	Apaseo, Salvatierra and Holy Mole.
	Piment D Espelette (heirloom chile pepper from Espelette inSouth-western France.)	Espelette
	Pimiento/Pimento (sweet heart shaped, thick-walled fruit)	Perfection Pimiento, Pimiento Select, Pimiento Sweet and Pimiento L

Species	Type	Important varieties
	Serrano types (meaning from the highland or mountain cylindrical, pendant, and bluntly pointed pods)	Balin and Tam Hidalgo and Serrano Tampiqueño.
	Wax/güero types (yellow waxy immature fruits)	Floral Gem, Mississippi Sport, and Santa Fe Grande
C. frutescens Roxb.	Borne erect, extremely hot fruits, fruit flesh often soft	Tabasco, Greenleaf Tabasco, Malagueta, Siling Labuyo, Prik Khee Nu, Xiao Mi La, Shuan La Jiao, Da Shu La Jiao, Sangpangmerutsu, Dhan Jolokia, Mem Jolokia and Sohmynken Khnai.
C. pubescens Ruiz and Pav.	Fruit flesh thick, firm	Canario, Manzano, Peron and Rocoto Longo.
C. chinense Jacq.	Fruits are generally wrinkled, slight to clearly segmented and very hot	Ají Limo, Ají Panca, Bhut Jolokia, Cajamarca, Charapita (wild), Chocolate Habanero, Datil, Fatalii, NuMex Suave Orange, NuMex Suave Red, Orange Habanero, Red Carribean, Rocotillo, and Scotch Bonnet, Bhut Jolokia, Huangdijiao
Capsicum baccatum L.	Pods with distinctive fruity flavour with low to high heat types	Ají amarillo, Ají ayucllo, Ají Limon, Christmas Bell, and Omincolor.

Source: Adapted and enhanced, Dewitt and Bosland, 2009.

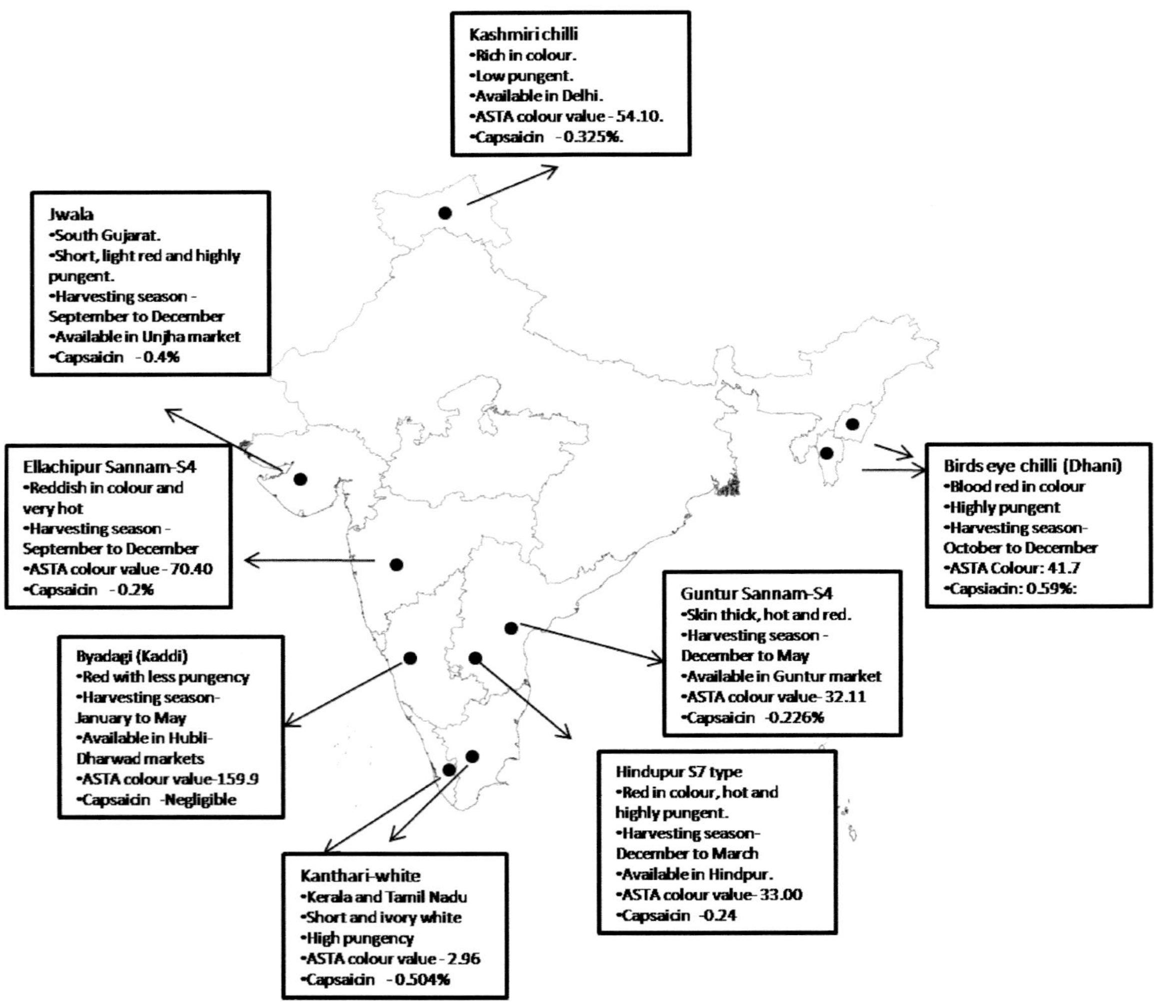

Figure 4. Important Chilli Types Traded in India.

The movement of chilli along the supply chain in India is depicted in Figure 5.

4. Nature and Causes of Postharvest Losses

Chilli is a crop of the old and new world with more than 400 varieties available all over the world with different pungency, size, shape, colour and usage. In India, more than 50 varieties are under cultivation and the production of the spice is erratic due to unpredictable monsoon. Drought plays a key role in the production of chillies in Africa. In India there is also a dearth of

Table 4. Important Chilli Growing Areas in India

State	*Important Chilli Growing Areas*
Andhra Pradesh	Guntur, Krishna, Prakasam, Kurnool, Kadapa, Anantapur, West Godavari
Telangana	Khammam, Warangal, Adilabad, Medak, Nizamabad
Karnataka	Belgaum, Dharwad, Gadag, Bijapur, Davangere, Raichur, Bangalore, Haveri
Gujarat	Banskantha, Jamnagar, Junagarh, Rajkot
Maharashtra	Nasik, Pune
Tamil Nadu	Coimbatore, Salem, Dindigul, T'hruvallar, Thiruchapalli, Ramanthpuram,, Sivanagar, Karur, Namakkal, Dharampuri, Virudhnagar
Assam	Nalbari, Darrang, Dhubri, Nagaon, Sonitpur
Manipur	Imphal, Churchandpur, Thoubal
Nagaland	Kohima, Dimapur, Phek, Wokha, Mokokchang, Mon Zuenhoboto
Meghalaya	West Garo Hills, South Garo Hills, North Garo Hills

analysis of various vital aspects of the chilli production and marketing such as latest prices and trends at major national and international markets, seasonal price fluctuations, demand and supply pattern, overall production constraints in the country, stocks available in cold storages, scientific forecasting, crop and weather information, its impact on agriculture and allied sectors and hedging among the various varieties of chillies. Many times low prices are of serious concern. Apart from this chilli crop is seriously affected by more than a dozen pest and diseases. The most important insect pests are thrips, mites, whitefly, borers and midge. Important diseases are fruit rot, powdery mildew and wilt. International trade is concerned with issues like aflatoxins, pesticide residues, microbial contaminations, capsaicin and colour values. Major importing countries in Europe and America have fixed quality requirement standards for chillies. Many times strict adherences to these are not followed and such exports were rejected

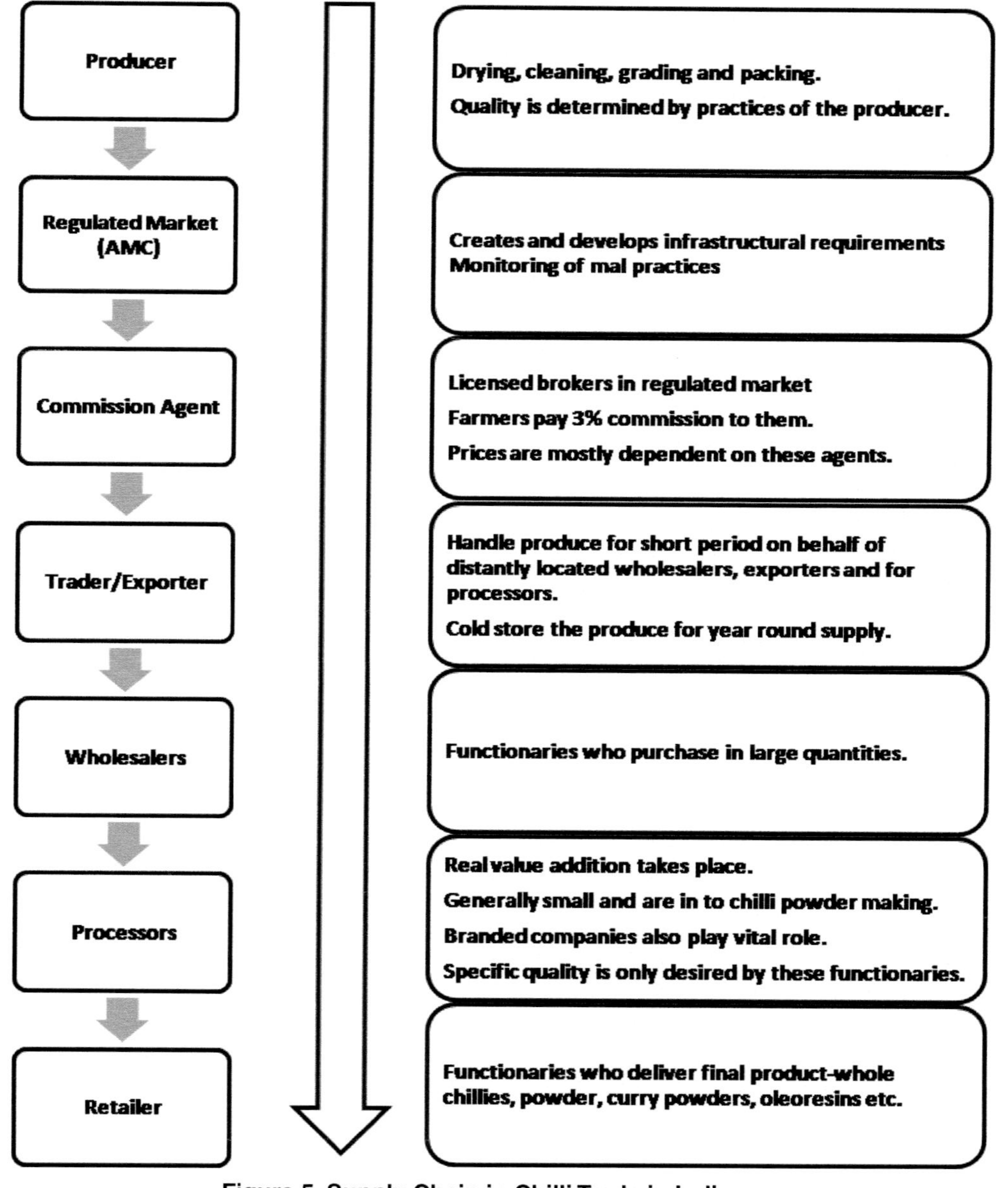

Figure 5. Supply Chain in Chilli Trade in India.

by the importing countries. Exports were also rejected due to the presence of carcinogenic dyes like Sudan IV. There is a need for robust extension support with new approaches that are different from those of the past. These include technical and organizational assistance for planning, storage, packing, grading and linking small farmer organizations with the organized private sector. Further data acquisition about the procedures involved in preparation of the crop for grading and marketing; the proportion of produce sold to different markets; transport facilities and cost of transportation; and the price spread in various marketing channels to develop decision support systems are needed. The focus should be on leveraging more profits for the farmers by minimizing role and charges by the intermediaries. Establishment and maintenance of an accurate spatial database using a Geographic Information System should also be given a high priority.

Chilli is unique among vegetables because the green form is used as a vegetable and the dry form is used as a spice. Green chilli, like other vegetables, is a perishable commodity. While the dry pods and ground chilli due to improper postharvest management lose the important quality traits of colour and pungency giving poor appeal to the consumer and low returns to stakeholders. Due to their perishable nature, quality of fresh fruits quickly changes during storage, transit and marketing. Spoilage was due to wilting, shrivelling, pathogenic disorders, water loss, *etc.* after harvest is common under improper postharvest management. The high perishability of the chilli fruits reduces their postharvest conservation and availability in the market, mainly for "in natura" consumption. Postharvest losses in fresh chillies go up to 20 per cent in market chain and maximum loss occurs at wholesale trader level. Transpiration losses, losses due to bacterial and fungal pathogens, physical damage during processing, losses due to sorting and storage, and retailing generally account for the losses rather than damage in transit. Transpirational losses in bell peppers are very high, limiting their storage life. New Mexican type peppers become flaccid in three to five days at 20°C (7–10 per cent weight loss) (Prakash and Eipeson, 2003). Apart from physical changes like bruises and wrinkles, loss of fruit weight is another serious damage to fresh chilli which distances the consumer away from buying the vegetable. Loss of glossiness, green or red colour and lowering of ascorbic acid content is usually associated with physical changes that occur due to ageing. The fungal and bacterial disorders also cause severe loss of shelf life and quality. Among these most important are Anthracnose and bacterial soft rot which are quite common causing damage, particularly in storage. Ethylene accumulation, presence of ethylene emitting fruits, accelerates senescence and causes deterioration during storage. Chillies are very sensitive and easily get damaged while handling. They also lose shelf life and quality quickly under improper storage conditions. Further, they are quickly damaged by too low or too high temperatures, during storage. Even storage temperatures affect the losses. Bell peppers are susceptible to chilling injury at temperatures below 7°C. Chilling injury also predisposes them to Altemaria rot and Botrytis rot. Food safety challenges such as microbial contamination, mycotoxins, adulteration, contaminants and pollutants affect postharvest management practices. Utilization of proper harvest and postharvest handling methods is essential for maintaining acceptable quality and maximizing shelf life.

4.1 Crop Management Practices Affecting Postharvest Losses

Proper pre harvest crop management solves many postharvest problems. Poorly managed crop ultimately gives low quality produce. Fresh fruits in bad condition deteriorate quickly after harvesting and cause severe loss to the stake holders. Hence, the crop should be managed with a holistic view for continuous supply of high quality chillies. Crop nutrition plays important role in getting good quality produce. Maintaining good, long-term soil health with an optimal supply of nutrients to plants in the growing season will help in achieving quality fruits throughout the season. Abu-Zahra (2011) had noted that inorganic fertilizer use increased the moisture content of pepper fruits with higher fruit size. The implication is that shelf life might be affected as the fruits would deteriorate fast due to microbial contamination. N application significantly increased the fat, protein, carbohydrate, and crude fibre and ash contents; and vitamin C and mineral nutrients of the fruits. The higher moisture content with N application emphasizes the need for proper postharvest handling and prompt marketing and use of the fruits to enable farmers to derive economic benefits from the higher yields (Ayodele *et al.*, 2015). Irrigation is one such factor that helps not only productivity of the crop but also desirable quality and shelf life of the produce. A fruit with optimum moisture content, tends to thrive long during storage, transit and marketing overcoming the moisture loss during these processes. Water stress during the growing season can affect the size of the fruit, and lead to soft or dehydrated fruit that is more prone to damage and decay during storage. Drought increased the oxidative membrane damage, indicated by the increased presence of Malondialdehyde, in the chilli pepper fruits, thus affecting the fruit quality. Further, there are stress-induced changes in metabolites in chilli pepper, namely capsaicinoids, ascorbic acid and total phenols, as well as antioxidant enzyme activities, in turn changing the quality of fruits (Kopta *et al.*, 2020). Incidence of blossom-end rot and total fresh mass of fruit was related to moisture deficiency, optimal irrigation schedule not only increases fresh fruit weight but also reduces the incidence of blossom-end rot (Dorji *et al.*, 2005). Excess moisture in the harvested fresh chillies increases the incidence of postharvest diseases and development of aflatoxins. Irrigation systems affect the micro-climate, the harvestable fruits and fruit quality. In one study, drip irrigation on raised bed with 100 micron Linear Low Density Poly Ethylene plastic mulch was observed to give maximum fruit yield with the lowest fruit rot (Bhardwaj *et al.,* 2018). High relative humidity, total rainfall and number of rainy days shared a positive significant influence on anthracnose – chilli patho-system, which affects the harvestable fruit quality (Saha and Bera, 2021). Harvesting during rainy periods encourages the postharvest decay on storage of fruits. Extended shelf life of fresh chillies produced under greenhouse conditions was observed compared to the produce from open field conditions. Maintaining ideal soil moisture content in the soil not only increases the total fruit biomass on the plant but also reduced the incidence of blossom end rot, which otherwise augments by unhealthy water relations. A series of experiments conducted on sweet pepper in the Netherlands revealed that increasing EC levels of the nutrient solution reduced the shelf life, while it improved flavour, measured in sensory assessments. Further, growing environment with 24 h optimum humidity of 80 per cent gave the best flavour and shelf life.

4.2 Management of Postharvest Losses

Chillies harvested at fully ripe stage are used as vegetable and dried spice. The dry chillies are mosty used for further processing to produce commercially traded forms. The most important operation in chilli processing is drying. The postharvest losses among tomato, brinjal, cauliflower and chillies vary from 9.5 per cent to 26.6 per cent at different stages of marketing and loss in chillies being the least (Singh *et al.,* 2013). Based on studies in Rwanda, overall postharvest losses and for green chillies were 7 per cent (Kitinoja *et al.,* 2018). It was estimated that in postharvest losses in chilli at different stages 5.66 ± 0.66 per cent *i.e.* at harvesting 1.64 ± 0.71 per cent, 0.71 ± 0.58 during collection, 0.67 ± 0.85 per cent during cleaning, 0.56 ± 0.29 per cent during drying, 0.12 ± 0.22 per cent during packaging, 0.13 ± 0.47 per cent during transportation, which is 3.95 ± 0.70 per cent at farm gate, while total loss in storage was 1.66 ± 0.55 per cent (Narayana *et al.,* 2014). In Indonesia, red chilli had total loss estimation of 29.32 per cent with details of losses estimation at farmer level of 2.33 per cent, 5 per cent losses at middleman level, while waste estimation at seller level was 10.80 per cent, and customer level was 11.18 per cent (Wigati and Darmawati, 2019). Different percentages of food losses and waste estimation between tomatoes and red chilli caused by different treatments, natural characteristics of products, handling by the actors, seasons and different economic value of the product. The prime requisite for chillies is harvesting them at the correct stage of maturity without much physical damage. Chilli like all other agricultural commodities invariably contains high moisture content in the range of 60-85 per cent at the time of harvest, which must be brought down to 8-12 per cent. The majority of chilli produce is dried on the ground in open space. The major change during drying is weight reduction which amounts to a reduction of 20-35 per cent of the total weight of the pods. Losses also occur during farm processing. These are spillage in field (1-10 per cent), wastage during farm assembling (5-10 per cent) and wastage during transportation and handling (2-5 per cent). Proper postharvest management not only helps to reduce postharvest losses but also improves the quality and shelf life of the final produce (Taufik, 2011).

5. Postharvest Management

5.1 Harvest Maturity

Size of the fruit is the most widely used harvest index in chilli. Apart from size, the calyx and stem should be fresh, firm and green. External colour is another widely used index of harvest maturity. Most chilli cultivars change their skin colour as they become fully mature. The colour of the fruits varies from dark to full green, pale green or whitish green finally turning to red in most cases. In general, fruits are harvested according to consumer preference. The nutrient composition also varies with the stage of harvest, and the highest capsaicin, ascorbic acid and total sugars were found when the fruits are 50 per cent red (Noichinda *et al.,* 2016).

5.2. Harvesting

Harvesting is a crucial operation in chilli which decides its postharvest quality, shelf life and marketability. Harvesting of chillies depends on the purpose for which it is grown. The fruit is non-climacteric in nature; and stratified harvesting is the norm as the flowering is asynchronous. The effect of harvesting at different maturity stages of the fruit indicated that the colour values were generally highest at the withering stage, moderate at the full ripe stage and lowest at the turning stage. At the turning stage, fruits were generally most pungent except in some cultivars which were most pungent at the full ripe stage or at withering stage.

5.2.1. Harvesting for Green Chillies

For vegetable purpose, fruits are harvested at green immature stage. Under well managed situation, 10-12 pickings can be harvested as green chilli. The quality of green chilli is decided by its size, colour and plumpness of fruit. Long, green, smooth, shiny and moderately plump fruits are preferred. For pickles, picking is done both at green and ripe stages. The crop takes about 40 to 50 days to flower after transplanting and a further 10 to 20 days to develop fruit suitable for green harvest. Green chillies of 20 days were proved to be excellent raw materials for making chilli powder in terms of colour, flavour, pungency and overall acceptability than chillies of other maturity stages (Tahmidul *et al.,* 2015).

5.2.2. Harvesting for Dry Chillies

Red mature fruits are harvested for dry chillies used for spice purpose. Stratified harvesting at different dates after transplanting had a significant influence on fruit yield and quality. Harvesting in chilli extends from three to seven months after transplanting or direct sowing as the case may be. Ripe fruits are handpicked at frequent interval, because retaining the fruits for a long period on plants causes wrinkles and fading of colour. Harvesting of chillies is a labour intensive operation and is carried out mostly by women labour. Chillies meant for making dried red chillies are ready for harvest after about 45-50 days from flowering when they are fully mature and red to nearly red ripe. *Capsicum annuum* required 35-46 days to reach maturity, and showed non-climacteric behaviour, unlike fruits of *C. Baccatum* (Mendez *et al.*, 2004). Generally, 6-8 pickings are taken under irrigated conditions and 3-4 pickings under rainfed conditions. Mature fruits after the colour break stage are harvested and heaped for 1-2 days for uniform ripening and colour development.

The suggested alternative to manual harvesting is mechanical harvesting. For cost effective machine harvesting the correct plant spacing and determinate cultivar with large pendulous fruit need to be selected. The performance of double-helix-type mechanical harvester was evaluated using four red chilli pepper cultivars (New Mexico 6-4, Sonora, B-18 and B-58). Ethephon was applied 3 weeks before harvest at 1.75 litres ha^{-1} to promote uniform ripening, however, it caused considerable fruit loss affecting yield. Harvest efficiencies of 73-83 per cent were observed among the cultivars. Trash content of the harvested chilli varied from 25 to 42 per cent of the dry weight. The trash was predominantly infected and had discoloured fruits, leaves, and small stems (Wall *et al.*, 2003).

5.3. Grading

Grading of fresh chillies is done mainly with manual labour. *E. coli* contamination is the most feared one in many countries and warrants minimum human contact and maintenance of hygiene in chilli processing particularly in grading. Grading fresh chillies is relatively easier when compared to dry chillies. Proper grading of green chillies facilitates efficient filling and canning in processing industry. The performance of an automatic Jalapeno chilli grader by three different width sizes was designed where the conveyor uses baby suckers to align each chilli during sensing. Chilli width was determined by using a photodiode scanner, which detects the incoming radiation sent by a laser line generator. Chillies with necrosis are detected with a radiometer and removed to increase the product quality. The accuracy of the necrosis detection and width classification was 96.3 and 87 per cent, respectively (Hahn, 2002).

Grading is done during and after drying. Colour, glossiness, size, moisture content, pest and disease infected fruits and whitened fruits are some of the important factors that traders consider when they buy the produce from farmers. But, it is advisable to do grading at various stages starting from harvesting to prevent microbial contamination and for quality end produce. Farmers grade the produce according to the buyer's preference. The Indian chillies are graded mostly by farmers based on colour and size before they are brought to the market. Grading at farmers' level involves separation of all visible foreign matter, discoloured pods (whitened), undersized and curly pods. The damaged, discoloured and immature pods are removed depending on market demand. Fungus infected pods are separated so that contamination in subsequent processing and storage could be avoided. Depending upon the market requirement grading is performed either during drying or after drying. However, at traders' level the most important specifications for trade are moisture and stalks. At the regulated markets, the pods are further graded based on size, colour, taste, *etc.* Produce is sold based on several grades depending on size and girth, seed content, pungency, colour, export and processing needs. Excess moisture adds weight to the pods but gives room to various fungi to grow and subsequent development of aflatoxins. Similarly, if the stalk of the pods is broken, exposing the seeds entirely, the seeds may fall out. On the other hand in absence of optimum moisture, the pods may break and let off the seeds. Thus the seed to pod ratio in a lot is also a valuable parameter of grade.

For different purposes, various varieties of chillies are chosen by the end user. End users are mainly of two types, such as domestic retail users and industrial wholesale users. Industrial users who prepare chilli

Table 5. Grade Specifications for Chillies/Capsicums/Paprika (AGMARK, India)

Grade	*Quality*							
	Special Characteristics							
	Organic Extraneous Matter per cent (m/m) (Max.)	*Inorganic Extraneous Matter per cent (m/m) (Max.)*	*Unripe and Marked Fruits per cent (m/m) (Max.)*	*Broken Fruits and Fragments per cent (m/m) (Max.)*	*Moisture per cent (m/m) (Max.)*	*Total Ash per cent (m/m) (Max.)*	*Acid Insoluble Ash per cent (m/m) (Max.)*	*Capsaicinoid Content per cent (m/m) (Max.)*
Special	0.8	0.2	2.0	5.0	10.0	7.0	1.25	0.3
Standard	4.0	1.0	5.0	15.0	12.0	8.0	1.25	Not specified

powder gives preference for colour, pungency, thickness of pericarp and seed content. Domestic users prefer all varieties for different occasions. There are several local and conventional grades followed by the farmers, village merchants and itinerant merchants. Traders usually depend on the visual assessment of lots/heaps by picking a handful of pods to decide upon the grades and assess the prices both in an open and closed auction.

Table 6. ASTA Grade Specifications (Physical and chemical) for Whole Chillies/Pods

Specification	*Suggested Limits*
ASTA Cleanliness specification	
Whole dead insects, by count	4.0
Maximum excreta, by mg lb^{-1}	1.0
Other excreta, by mg lb^{-1}	8.0
Mold, per cent by weight	3.00
Insect defiled, infested, per cent by weight	2.50
Extraneous, per cent by weight	0.50
FDA DALs	
Insect infested or mouldy pods by weight	Ave. of 3 per cent
or	
Mammalian excreta per pound indentified as to source when possible	Ave. of 1 mg
Volatile oil	N/A
Moisture	11.0 per cent max.
Ash	8.0 per cent max.
Acid-insoluble ash	1.0 per cent max.

5.3.1. Grade Specifications

Grade specifications vary widely from country to country, predominantly determined by their internal quality specifications and requirements. For trade in India, the Bureau of Indian Standards has outlined specified standards for dried chillies based on physical characteristics, as well as other factors such as total ash, acid insoluble ash, non-volatile ether extract and fibre content. Minimum purity standards were laid out under the Prevention of Food Adulteration Act, 1951. Agmark specifications for the grading of dry chillies for export take into account various physical, chemical, sensory and microbiological parameters (Agmark, 1962). Grade specifications for chillies/capsicums/paprika according to Spices Grading and Marking Rules, 2005 (Government of India) are presented in Table 5. American Spice Traders Association (ASTA) has its specifications which are presented in Table 6 and 7. The quality minima prescribed by European Spice Association (ESA) are presented in Table 8.

Table 7. ASTA Grade Specifications (Physical and chemical) for Ground Capsicum (Not including paprika)

Specification	*Suggested Limits*
FDA DALs (6 sub samples)	
Insects or fragments	Ave. of 50 or more/25 g
or	
Rodent hairs	Ave. of 6 or more/ 25 g
or	
Howard mold count	Ave. of 20 per cent
Volatile oil	N/A
Moisture	11.0 per cent max.
Total Ash	8.0 per cent max.
Acid-insoluble ash	1.0 per cent max.
Military specifications (EE-S-631J, 1981) for crushed red pepper	
Volatile oil	N/A
Moisture	10.0 per cent max.
Total Ash	8.0 per cent max.
Acid-insoluble ash	1.0 per cent max.
Granulation	98 per cent min through a U.S.S. 4
	85 per cent min through a U.S.S. 8
	98 per cent min on a U.S.S. 20
Scoville pungency	30,000-55,000
Bulk index (Average bulk index, granulation will affect the number).	
Ground (ml 100 g^{-1})	200
Crushed	235

5.4. Handling and Precooling of Green Chilli

Freshly harvested green chillies should be handled carefully while transferring to field containers to avoid damage by bruising and punctures. Containers or sacks used for picking should have smooth inner surfaces to avoid skin damage or fruit punctures. Wide, shallow, and stackable containers are advantageous as they cause less bruising of the fruit at the bottom of the container. Plastic crates or troughs with good ventilation are generally used for this purpose. These troughs can be easily cleaned and transported. All the containers used in the field should be periodically cleaned and sanitized to avoid the occurrence of disease spreading organisms during transit and storage. Field containers should be washed with water to remove previous debris and sanitized with a sanitising solution consisting of 200 ppm sodium hypochlorite (common bleach) at a water pH of 6.5 by soaking for two minutes. Troughs or baskets made up of local materials like wood or straw can also be used. In such cases, providing a newspaper lining inside minimizes damage due to abrasion while handling and transit. For local markets, medium sized sacks made of gunny or nylon may be used. Large sized sacks must be avoided as they hold large volume thus restricting aeration and fruits also get damaged due to heavy weight.

Table 8. European Spice Association (ESA) specifications of quality minima for chillies and paprika (ESA Quality Minima Document, Revision 5)

Subject	*Quality Minima*
Extraneous matter	1 per cent, by weight
Sampling	(For routine sampling) Square root of units/lots to a maximum of 10 samples.
	(For arbitration purposes) Square root of all containers*e.g.* 1 lot of pepper may = 400 bags, therefore square root = 20 samples.
Foreign matter	Maximum 2 per cent
Ash per cent w/w Max	10
Acid insoluble ash (AIA) per cent w/w max	1.6 (2.0 for paprika powder)
Moisture content (H_20) per cent w/w max	11
Packaging	Should be agreed between the buyer and seller, if made of jute and sisal, they should conform to the standards set by CAOBISCO Ref.C502-51-sj of 20.02.95. However, these materials are not favoured by the industry, as they are a source of product contamination, with loose fibres from the sacking entering the product.
Heavy metals	Shall comply with national/EU legislation.
Pesticides	Shall be utilised as per manufacturers' recommendations and good agricultural practice and comply with existing national and/or EU legislation.
Treatments	Use of any EC approved fumigants in accordance with manufacturers' instructions, to be indicated on accompanying documents. (Irradiation should not be used unless agreed between buyer and seller.)
Microbes	Salmonella absent in (at least) 25g. Yeast and moulds 105/g target, 106/g absolute maximum *E. coli.* 102/g target, 103/g absolute maximum. (Yeast moulds maximum target 100000). Other requirements to be agreed between buyer and seller.
Off odours	Shall be free from off odour or taste.
Infestation	Should be free in practical terms from live and/or dead insects, insect fragments and rodent contamination visible to the naked eye. (Corrected if necessary for abnormal vision).
Mycotoxins	Aflatoxins Aflatoxin B1: 5 µg/kg, maximum Aflatoxin B1 + B2 + G1 + G2: 10 µg/kg, maximum Should be grown, harvested, handled and stored in such a manner as to prevent the occurrence of aflatoxins or minimise the risk of occurrence.
	Ochratoxin A (chilli powder, cayenne and paprika) 20 µg/kg Mixtures of spices containing chilli 15 µg/kg Should be grown, harvested, handled and stored in such a manner as to prevent the occurrence of Ochratoxin A or minimise the risk of occurrence.
Volatile oil (V/O) per cent w/w min	-
Adulteration	Shall be free from.
Bulk density	Understanding between buyer and seller.
Water activity	Understanding between buyer and seller.
Species	Understanding between buyer and seller.
Documents	Should provide: details of any treatments the product has undergone; name of product; weight; country of origin; lot identification/batch number; year of harvest.

Green chillies should be kept in a cool dry place immediately after harvesting. Tight packing of fruits should be avoided as it encourages build up of temperatures inside the containers thus causing physiological damage to fruits. The fruit temperature at harvest, is usually in unison with ambient temperature, may rise to 10°C above the ambient temperatures even under shade if tightly packed and stored. A few hours of such exposure can make the fruits to soften and show signs of shrinkage, reducing the shelf life by half. Freshly harvested fruits are graded and usually cleaned with chlorine water (150 ppm Sodium Hypochlorite) maintained at a of pH 6.5. The cleaned fruits are shade dried before packing. Fruit should be firm, well-shaped, and free from damage caused by scars, sunburn, disease, insects, or mechanical injury before packing. Fruits that are not in shape, cracked, damaged, decayed, or wilted are discarded while grading. All fruit packed in the same carton should be similar in appearance. Then they are packed in units as desired for bulk packing. They may also be packed into consumer packs. The produce is then pre-cooled in a cooling unit for a few hours and transferred to cold storage ready for export. After harvest, pre-cooling can be employed to remove field heat from the fruit and slow down metabolism. It is an important step in cold chain management. For chilli produced in the tropics, rapid cooling to 10°C at high RH would be sufficient, as a lower end-point temperature could injure the fruit. Pre-cooling can be done using forced air, hydro cooling, or vacuum cooling. Hydro cooling is simpler and cheaper than the other methods, but care should be taken to prevent the development of decay from inadequate drying after treatment. Pre-cooling is advisable before cold storage. As the shelf life of fresh chillies is very short, they are usually transported by air. The packages are stored or transported as a pallet load. To maximize savings in distribution costs, pallet loads should be established as soon as possible in the distribution chain and remain unbroken as long as possible. For international trade, the ISO has specified two standard dimensions as the primary handling unit size *viz.* 1200 x 1000 mm, 1200 x 800 mm. The above two pallet sizes are used for distribution within Europe as well as for transportation of produce arriving unpalletised from European seaports and airports to its European destination. In an alternative to cleaning with chlorine, efforts were made to explore numerous alternative methods, *e.g.* the use of hydrogen peroxide, organic acids, and UV radiation were tried by various researchers. Aqueous ozone treatment at 30 µmol mol-1 for 10 min before storage reduced microbial counts on chilli peppers during subsequent storage at 8 °C (Chitravathi *et al.*, 2015). Continuous exposure to ozone at 0.9 µmol mol^{-1} stored at 10 °C extended the shelf-life, reduced weight loss, shrivelling, and firmness maintenance in storage. However, 2 µmol mol^{-1}, resulted in bleaching of skin colour and a reduction in antioxidant content (Glowacz and Rees, 2016).

5.5. Packing

Packing is one of the most important operations in chilli processing and transport. Both fresh and dry chillies are usually packed at the farm level to ensure easy transport, to protect the produce from any damage during storage, transportation and marketing. Good packing of chilli not only facilitates convenience in transportation and storage but also attracts consumer to pay more. The packing reduces the marketing cost and protects the quality. Packaging can prevent the produce from drying by creating an atmosphere with a high relative humidity. It can also modify the headspace of the atmosphere yielding high levels of CO_2 and low levels of O_2. Packaging slows maturation and delays

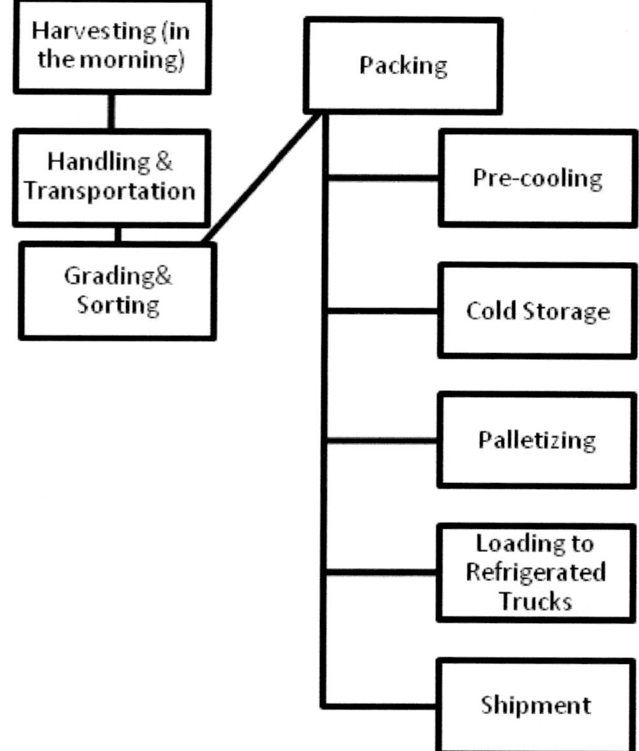

Figure 6. Flow Chart Showing Export of Green Chillies.

Table 9. Specification Details for CFB Boxes for Packing and Export of Green Chillies

Sl.No.	*Specification*	*Slide Type*	*Ring * Flap Tuck-In-Type*	*RSC (Regular slotted container)*	*Tray with LID*
		5 kg Box (Dimension: 450 x 265 x 110 mm)			
1	Material for construction	5-ply CF	5-ply CFB	5-ply CFB	5-ply CFB
2	Grammage (g/m²) (outer to inner)	*230X140 X140X140	*230X140 X140X140	*230X140 X140X140	*230X140 X140X140
3	Bursting strength kg/cm²	Min. 10.00	Min. 10.00	Min. 10.00	Min. 10.00
4	Compression strength Kg.	Min.350	Min.350	Min.350	Min.350
5	Cobb (30 minutes g/m²)	Max.130	Max.130	Max.130	Max. 130

*Outer ply of white duplex board.

Source: Postharvest Manuals on Exports of Fruits, APEDA, New Delhi.

changes in acidity, soluble solids, texture, colour and polygalacturonase and reduces water loss. Packaging can also be used to create brand names associated with freshness of a produce, besides being convenient to handle and maintain an attractive appearance of the product.

Freshly harvested green chillies are generally packed in gunny sacks, bamboo baskets, or polyethylene sacks and transported to nearby domestic markets (Figure 6). For long distance marketing within a country, they are packed in plastic trays, and transported. Generally, 3-4 inches length chillies are preferred for export. Green chillies from India are exported in Corrugated Fibre Board (CFB) boxes (Table 9). Fruits are packed in consumer packs of LDPE or PP (polypropylene). Twenty consumer packs of 250 gm capacity are placed in 2 x 2 x 5 pattern in a 5kg CFB box for export purpose.

In developed countries, they are packed in packing films in unit quantities, transported using a vegetable cold chain and marketed. The effect of modified atmosphere packaging (MAP) on the postharvest quality characteristics and biochemical properties of green chilli peppers (*Capsicum annuum* L. cv. Legon 18) resulted in significantly lower weight loss, better retention of green colour, less reduction in firmness at both 10 °C and 26-34 °C (ambient) temperatures, hence beneficial for extending the shelf-life (Edusei *et al.*, 2012).

Anti-fog film (RD45) packed samples maintained pigment stability for a longer duration and a shelf-life of 28 days was observed as compared to 15 days in the case of control samples at low temperature storage (8 ± 1 °C, RH 85–95 per cent) of green chillies. An increase in total phenolic content was observed, whereas a decrease in total antioxidant activity and capsaicin content was observed with progress in senescence (Chitravathi *et al.*, 2015). Active MAP having gas mixture (5 per cent O_2 + 5 per cent CO_2) filled in low-density polyethylene films and 5 g of $KMnO_4$ sachets as ethylene absorber at 10 ± 1 °C and 95 ± 5 per cent relative humidity, resulted in improved O_2 levels in active packages, with weight loss below 1 per cent and positive effects on the bioactive content of pepper (Cerit and Demirkol, 2020). Fresh green chillies treated with 3 µM gibberellic acid (GA_3) for 10min and packed in MAP with anti-fog film (RD 45 stored under low temperatures (8/±/2 °C) extended the shelf life up to 30 days, effective in preserving quality attributes (firmness, total phenolic content, total flavonoid content, antioxidant activity, total chlorophyll, total carotenoids, ascorbic acid, capsaicin content, pericarp color). Anti fog film RD 45 packed samples displayed optimal CO_2/O_2 gases composition and low respiration rate as compared to samples packed in LDPE film hence extending the shelf life of green chilli. Synergetic effect of GA_3 and MAP in improving the quality attributes of green chilli was observed in the samples (Maurya *et al.*, 2020).

5.6. Storage of Chillies

Both quantitative and qualitative losses at extremely variable magnitude occur at all stages of the postharvest system, including harvesting, handling, storage and marketing. Overall postharvest loss of 20 per cent was reported in chillies. Maximum losses were observed at the wholesale level. Losses caused by sorting and storage operations are higher than those during transport. Storage is a very important component of the supply chain and has a direct impact on the prices. Efficient and adequate storage facilities strengthen the supply chain over time and space. Storage function thus is responsible

The green chillies or ripe red chillies are perishable in nature and can be stored for 3 to 4 days under ambient condition without much appreciable damage to the appearance and quality. However, storing them beyond this period invites mould attack, physiological deterioration, quality loss and as a result quickly loose marketability. Due to its perishable nature, it is liable to fast quality changes and spoilage (wilting, shrivelling, pathogenic disorders, water loss, *etc.*) after harvest under improper postharvest management. Postharvest quality loss in stored chilli is often detectable as a change in texture, particularly when there is significant water loss. Fruits are also attacked by various types of fungi and bacteria leading to rots and fruit decay. Decaying fruits produce significant amounts of ethylene which may enhance the ripening and decay of other fruits and so, periodic inspection and culling out of decayed fruits should be done (Acedo and Thanth, 2006). Increase in the rate of loss because normal physiological changes is caused by conditions that increase the rate of natural deterioration, such as high temperature, low atmospheric humidity and physical injury (Food and Agriculture Organization, 1989)

The fruits of King chilli (*Capsicum chinense* Jacquin) grown under the foothills of Nagaland when treated with calcium chloride (0.5 per cent), packed in non-perforated bags and stored under the refrigeration at 5 °C recorded the lowest physiological loss in weight (2.55 per cent) change in fruit colour (10 per cent), highest value of ascorbic acid content (120.5 mg 100 g-1 of fruit) and longest shelf life (9 days) on 9th day after storage (Rongsennungla *et al.*, 2012). CFTRI, Mysore recommends storing fresh chillies at 10 ± 2 °C which extends shelf life for 10-12 days. Pre-packaging in polymeric film or paper bags improves the storage life and retains the freshness of chillies. Fresh chillies stored in polyethylene (PE) bags at 20 °C recorded a weight loss of up to 25-40 per cent accompanied by significant shrivel. The weight loss is associated with reduction in firmness of pericarp and firmness was better at lower temperatures (Jansasithorn *et al.*, 2010). Fresh green chillies can be stored up to three weeks without much postharvest losses when maintained at 10 °C (R.H. 80-90 per cent), however storing at less than 5 °C (R.H. 80-90 per cent) results in chilling injury to fruits. Green peppers in modified atmosphere packaging stored at 8°(±1°C) exhibited a shelf life of 7 days. Beyond this time, the microbial proliferation (total count>107– 108cfu/g) was over the safety limit for human consumption (Senesi *et al.*, 2000). Controlled atmosphere (CA) storage conditions (3 per cent O_2, 5 per cent CO_2) for green chillies at 10°C exhibited comparatively less weight loss per cent, decay per cent, wrinkling and ethylene production but more firmness as compared to all combinations and was effective in improving the quality and extending the shelf life of green chillies up to 6 weeks (Hameed *et al.*, 2009). Essential oil coating and *Aloe vera* gel coating were also found effective to a certain extent. Most of the decay in the storage of fresh chillies is caused by *Erwinia carotovora* and is hastened by high relative humidity.

Ethylene released by decaying or ripening fruits, generally, hasten the quality deterioration of fresh green chilli and cause turning of colour of the pods, resulting in poor market acceptability. Chillies are non-climacteric, although a rapid increase of C_2H_4 corresponds with color change. Corresponding with the rise in C_2H_4 is a sudden loss of chlorophyll, decrease in firmness, an increase in β-galactosidase activity, and the appearance of a new protein band at 35 kDa. However, colour development is a slow process (Biles *et al.*, 1993). Removal of ethylene from the storage and shipping environment retards the spoilage, reduces loss and increases profits. A study was conducted in South Korea, to know the effect of ethylene absorbents on fresh green chilli. Application with ethylene absorbents on chilli fruits gave many benefits to control the fruit quality and prolong the storage life when stored in LDPE bags. $KMnO_4$ based ethylene absorbent resulted in the reduction of the percentage of weight loss, higher retention in development of peel colour (PCI), fruit firmness, higher retention in TSS and TA values. The ethylene scrubber was prepared with one molecular weight (158 g) of $KMnO_4$, half molecular weight (49 cc) of H_2SO_4 and 2 kg of rice-husk ash were thoroughly mixed, and they were then put into the cloth bags as ethylene scrubbers for the treatment. The ethylene scrubber @ 8 per cent of fruit weight was found effective (Myint, 2020). In another study, gibberellic acid (GA3) coating with 2 ppm, by dipping for 30 seconds, then stored at 4 ± 1 °C, showed beneficial role in decelerating the mechanism of ripening throughout storage and provided expanded shelf-life of green chilli (Panigrahi *e al.*, 2017).

The dried chillies are stored in gunny bags for marketing. During this period, the deterioration in quality is more due to improper packing and storage. To slow down the colour loss during storage, it is important to keep the spice in cool store, away from light. In India, different methods of storage are followed at different places. The farm level storage capacity among the Chilli growers is not adequate in the country. Initially in most cases the chilli bags are stored in closed yards or shops for 5 to 30 days by commission agents. Chillies are mostly stored in gunny bags by the producers, wholesalers and exporters for a period extending from 2-12 months depending upon the market demand. In places like Orai, chillies are stored by producers in earthen pots even for one year. In cities like Murshidabad and Jalpaiguri of West Bengal chillies are stored in Bamboo basket by the farmers in their own houses. Well maintained storage units in the market yards with low and uniform storage charges would encourage more farmers to store chillies in the marketplaces and improve their bargaining capacity. Storage of chillies in cold stores is extensively followed worldwide. The temperature inside the cold store is maintained from 5-8°C and the relative humidity is usually about 55-65 per cent. The use of commercial cold store for chilli storage has now become almost a general practice among farmers. Deepa *et al.* (2013) evaluated the influence of vacuum packing and observed that the samples stored in vacuum packed bags maintained the quality with the least deterioration in all the quality parameters compared to samples stored in jute bags both at room temperature (25 ± 2 °C), cold storage (4 ± 1 °C) under both light and dark conditions for 24 months. Irrespective of the storage condition, all the cultivars show a declining trend in all the biochemical parameters with the progress of storage period. The cold storage of dried chilli was superior in the retention of biochemical parameters than storage under ambient conditions. Further, initial moisture content of fruits influence on changes in quality parameters in storage and chillies, initial moisture content of 12 per cent moisture is better for cold storage.

5.7. Postharvest Diseases of Chilli and Management

Chillies are susceptible to many postharvest diseases, most of which need entry point through mechanical damage or weakening of the tissue before they can penetrate the fruit. The waxy skin of chillies offer protection against pathogen infection, but it is easily damaged while handling. Bacterial and fungal pathogens may also enter the fruit through the severed stem tissue and natural openings around the calyx. The most common postharvest diseases of peppers are bacterial soft rot, grey mould, rhizopus rot, watery soft rot, black mould, anthracnose, and sour rot. The incidence of these diseases is always higher during wet, rainy weather. Proper pre harvest disease control practices along with careful handling, cleaning, and grading of the fruit, and appropriate postharvest temperature control will minimize the occurrence of disease. Dipping of fresh fruits in two per cent solution of sodium hypochlorite and sodium chloride and subsequent storage at 10 °C with 90-95 per cent humidity resulted in the lowest decay of fruits. Bicarbonate salts have broad spectrum antimicrobial properties and are Generally Recognised As Safe (GRAS) compounds. Dipping of harvested red peppers for 2 minutes in 1 to 2 per cent potassium bicarbonate reduced the decay to a commercially accepted level of 5 to 8 per cent where as higher concentrations reduced fruit quality significantly.

5.7.1. Bacterial Soft Rot

The disease, bacterial soft rot, is caused by the soil-borne bacteria *Erwinia carotovora*. It is the most destructive postharvest disease in chillies. The field infection occurs during rainy weather when soil containing the bacterium is splashed onto the susceptible fruit. It is also carried out, when contaminated water is used in postharvest cleaning operations. The pathogen usually enters the fruit through any wound, crack, opening or injury caused during insect feeding. Excess nitrogen fertilization is associated with increased susceptibility to soft rot. Initial symptoms begin as light coloured, water-soaked areas around the edge of wounds on the fruit or stem. The water-soaked areas quickly enlarge. Tissues around the infected site begin to soften and the fruit eventually turns into a mushy watery mass. Postharvest infections are particularly damaging, because if one fruit decays the disease may eventually spread to all fruit in the same container. Bacterial soft rot is easily recognized by a foul-smelling odour as the fruit decays. In later stages, bacterial ooze may develop from affected areas, and secondary organisms follow, often invading the rotten tissue. Both green and ripe fruit may be affected and the affected fruit hang from the plant like a water-filled bag. The optimal temperature for bacterial soft rot growth is 27°C to 30°C (81°F to 86°F) and the minimum temperature for growth is 3°C (37°F). However, the pathogen can effectively damage even at 10 °C if the conditions are conducive to pathogen development (Stommel *et al.*, 1996). Field infestation can be managed by controlling insects that create wounds for infection. Postharvest decay is reduced by picking fruit when it is sufficiently ripe and dry, avoiding injury during handling, and holding the fruit at 7°C (45°F). If the fruit is to be washed after harvest, the water should be properly chlorinated and the fruit should be dried as quickly as possible after washing. Pungent pepper cultivars usually have a defined abscission zone and do not suffer stem infections unless damaged during harvest.

5.7.2. Grey Mould

The disease is caused by the fungus *Botrytis cinerea*. It is the second most important postharvest disease of chillies after soft rot. Initial infection can be identified by small cream-coloured specks on wounded areas of the fruit surface. These spots enlarge into circular water-soaked lesions covered with a grey mould. The optimal temperatures for growth of the fungus are between 18°C to 23°C. The fungus may occasionally attack under cold storage of green chillies. Proper field sanitation and prevention of wounds on the fruit helps in reducing the incidence of grey mould.

5.7.3. Rhizopus Rot

The disease is caused by the fungus *Rhizopus stolonifer*. It is also a common postharvest disease of chillies. The pathogen primarily attacks wounded areas of the fruit tissue or enters through the broken stem. Small water-soaked lesions are seen initially which quickly enlarge without discolouration. In advanced stages, entire fruit is attacked by the pathogen. Greyish-white masses of mould develop over the wounded area, which eventually turn black. A clear liquid is released from diseased tissues. Nests of mould and decaying fruit may form within a carton of packed fruit. High temperature and humidity and fresh wounds promote disease development. The main ways to control Rhizopus are to sanitize all field containers and grading surfaces, prevent wounding of the fruit tissue, dry wet surfaces of the fruit, and cool the fruit to 7°C as soon as possible.

5.7.4. Scelorotium Watery Rot

The disease is caused by the fungus *Sclerotinia sclerotiorum*. It is seen under high humid conditions during rainy season. Air borne spores of this soil borne pathogen, anchors in wounded areas of the fruit and establishes there. Small lesions develop on the fruit surface; later develop in to a white mould. The fruit parts affected becomes soft and watery. The pathogen spreads easily among the fresh fruits stored in store houses or cartons. The disease is controlled by using good pre harvest sanitation practices. The disease attack can be mitigated by storing the fresh chillies at 7°C.

5.7.5. Black Mould

The disease is caused by the fungus *Alternaria alternata*. The pathogen attacks fruits that are having injuries. But the disease does not usually spread from fruit to fruit. Initially, small, circular, slightly sunken spots associated with cracks or other surface injuries are seen on the fruit. These spots enlarge into sharply sunken lesions, eventually becoming covered with a greyish to black mould. The lesions show alternating light and dark brown concentric zones typical of pathogen attack. The pathogen also attacks and grows on placenta and seeds. Storage of the fruit at low temperatures will significantly retard the development of black mould.

5.7.6. Anthracnose/Fruit Rot

The disease is caused by the fungus *Colletotrichum capsici*. The disease is a very important postharvest disease of chilli. Field infection is the primary reason for the damage. Infection usually occurs, when fungal spores are splashed onto developing fruits. The spores usually remain in a dormant state until conditions are right for sporulation and growth of the fungus. Ripe fruits are more susceptible to this disease. Initially small, water-soaked lesions appear on the fruit surface. These lesions rapidly grow and appear as sunken, black in colour. As the infection progresses, salmon-coloured spores appear either scattered or in concentric rings within the lesions. Field sanitation is the most important factor in the management of the disease. It is essential to remove infected crop debris and periodic applications of protective fungicides. Storing the fresh fruit at 7°C after harvest will also minimize decay by the pathogen. In dry chilli preparation, the disease is favoured by high moisture content in chillies and slow drying process especially by traditional method of drying resulting in deterioration and loss of quality of the chilli beside aflatoxin contamination. Drying chillies in solar tunnel dryer for more than 24 hours resulted in rapid removal of moisture and significantly low incidence of fruit rot.

5.7.7. Sour Rot

This is a minor disease occasionally seen in fresh chilli fruits. The disease is caused by the soil-borne fungus *Geotrichum candidum*. Initial symptoms appear as, the pale greasy water-soaked lesions. The fungus enters through wounds at cracks, cuts, or skin punctures. The affected tissues remain fairly firm until the decay is quite advanced. A typical sour odour is given off by the fruit as it rots. Control of this disease is obtained by storing the fruit at 7°C, preventing injury to the fruit during harvest and handling, and maintaining good ventilation during storage.

5.8. Storage Insect Pests and Management

The dry pods or powders when stored are, often attacked by several insect pests. The most important ones are the drugstore beetle (*Stegobium paniceum* Linn.) and the cigarette beetle (*Lasioderma serricorne*). Among these, cigarette beetle is known to cause severe damage under improper storage conditions. The pest has low preference for chillies when compared to other spices. The loss of weight due to infestation of cigarette beetle is comparatively less in chilli powder (1.67 and 4.32 per cent) at 3 and 6 months after storage compared to other spices. However, severe infestation of the chillies by the cigarette beetle is not uncommon in the local markets (Khare *et al.*, 1989). The Anthribidae species also feed on chillies, though the loss caused by them is negligible. *Paralycus chongqingensis* sp. *nov* and *P. longior* sp. *nov* were also reported to attack chillies during storage in China. Postharvest drying is crucial in curtailing the infestation of stored grain pests. It is always desirable to dry the produce to desirable moisture of 8 to 10 per cent before storage to ensure minimum incidence of stored pests. Further, the produce or product must be stored in well ventilated godowns or store houses. High humidity during the storage is favourable for the development of both insects and fungi and hence must be avoided. Spreading chillies in thin layers and exposing them to sunlight periodically will eliminate the infection if infested by storage pests. If large quantities are infested with these pests, fumigation with recommended chemical is the only remedy.

6. Marketing and Export of Chilli

The trade channel involves several members' *viz.* a village level trader, commission agent, wholesaler, retailer, agents for exporters and exporters. The commodity changes hands several times, exposing all these members to price risk. The commodity displays high volatility, with the prices heavily dependent on season, production in different producing tracts spread across the country, demand from exporters and the stock available at the cold storages. The chillies have a well established spot market in India at Guntur, Warangal, Khammam in Andhra Pradesh and Raichur, Bellary in Karnataka. Guntur is Asia's largest market for chillies. The marketing season here begins in the first week of February, peaks during April and closes by the middle of May. In addition to whole chilli, the marketing is also done in the form of stemless chilli and crushed chilli. For marketing, chillies are graded as first or second sort, mixture, *etc.* Grades such as special, medium and fair are also adopted. Important markets and feeder areas of chillies in India are presented in Table 10.

6.1. Whole Chilli

Dry whole chillies are graded for uniform size and colour. These are refilled into new bags having trader's logo and name for export purpose. Whole chillies are further processed in several ways.

6.2. Crushed Chilli

Chillies are crushed to produce pericarp and seed. Crushed chilli Pericarp is further processed to produce oleoresin, high quality chilli powder and other value added products such as chilli paste.

6.3. Deseeded Chilli

Deseeding is performed using chilli seed extractors or mechanical crushers. The product is separated by sieving to remove seeds and pericarp. Sometimes the chilli seeds are separated from pericarp by air classification.

6.4. Destemmed Chilli

Some buyers/exporters prefer stem-less chilli for both domestic and export market. Hence destemming is carried out at traders' level. Destemming involves removing the calyx and stalk portion of the chilli. During manual destemming, 15-16 per cent weight is lost in the form of stalk, calyx and seeds. There is a need to develop a mechanical destemmer as manual destemming is expensive. Stem-less chillies are packed into new bags by processors/traders for transport to other states or countries.

Table 10. Important Markets and Feeder Areas of Chillies in India

Sl.No.	State	Market	Feeder/Catchment Areas
1.	Assam	Nagaon	Kaliabor, Dhing, Samaguri, Raha, Jamunamukh, Murajhar, Lanka and Lumding
2.	Meghalaya	West Garo Hills	Phulbani, Rongamachokgiri, Rabbigiri, Kherapara, Rongohugui, Selsella, Tura, Batasing and Dalu.
3.	Andhra Pradesh	Guntur	Tangeda, Dachepalle, Piduguralla, Sattenapalli, Narasaraopet, Vinuknoda, Medikonduru, Krosuru, Tulluru, Pedakurapadu, Ponnekallu and Prattipadu (Krishna district) Vatsavai, Jaggayyapeta, Penuganchiprolu, Nandigama, Veerullapadu, Mylavaram, G.Konduru, Kanchikacherla, Chandarlapadu
		Kurnool	Nandyal, Owk, Gadivemula, Panyam, Jupadu Bunglow, Gospadu, Rudravaram, Allagadda, Midthur, Kowthalam, Pedda Kadubur, Yemmiganur, Adoni, Halaharvi, Chippagiri
4.	Gujarat	Jamnagar	Jodia, Kalavad, Jamjodhpur, Bhavad, Kalyanpur, Dwarka, Khambalia and Dhrol.
		Banskantha	Tharad, Dhabhar, Devgani,Vodgam, Danta and Vav
5.	Manipur	Thoubal	Yairipok, Wangjing, Wabagai and Kakchig.
		Charchandpur	Hanglep, Songsan, Tinaong, Senvon, Hanship, Molnom, Thinghat, Mulanil and Hangtam.
6.	Maharashtra	Nasik	Kalvan, Peint, Igatpuri, Sinnar, Niphad, Yeola, Nandgaon, Satara, Furgana, Dindori and Melgaon
		Pune	Junnar, Ambegaon, Ghod, Rajgurunagar, Wadgaonsirur, Mulshi, Welhe, Purandhar, Bhor, Baramati, Indapur, Daund and Saswad.
7.	Karnataka	Bijapur	Indi, Sindgi, Basavna Bagevadi, Muddebihal and Tikota.
		Belgaum	Athni, Arkali, Chikodi, Mukeri, Bailhongal, Ramdurg and Khauapur.
		Bangalore	Amekal, Sonnenahalti, Kannur, Bagalur, Nagarur, Marangondahalli, Haralur, Mantapa, Solurur and Chandapur.
8.	Tamil Nadu	Thiruchirapalli	Turaiyur,Thottiyam, Musiri, Manachanallur and Lalgudi
		Coimbatore	Muttuppalaiyam, Avinashi, Tiruppur, Palladam,Udumallaippettai, Pollachi and Valparai.
		Salem	Mettur, Yercaud, Idappadi, Omalpur, Attur, Gangavalli and Sankagiri.
9.	Telangana	Khammam	Ragunathapalem, Kamepelly, Enkoor, Tallada, Konijerla, Singareni, Kalluru and Mudigonda
		J. Gadwal	Manopad, Maldakal, Itikyala, Rajoli, Ieeja, Undavelly, Ghattu, Alampur
		Warangal	Kuravi, Garla, Dornakal, Mahabubabad, Nellikudur, Maripeda, Bayyaram, Narsimhulapet, Gudur, Danthalapalle, Kesamudram, Thorrur

6.5. Red Chilli Powder

Ground chilli is the most important value added product of chillies. This is either directly used or blended with other spices for further use and preparations. Grinding is usually done with electrical grinders to attain fine powder. Even though spice grinding is essentially a size reduction unit operation, the requirement of high quality products has prompted considerable engineering improvements. Different types of machines are used for spice grinding. They include attrition mills, impact mills, roller mills, vibro energy mills, fluid energy mills and bowl mills. Modern spice mills, which handle large volumes, have optimized material flow, closed circuit pneumatic conveying, dust collection aids and noise reduction fixtures. In order to reduce volatile loss and other quality deterioration due to the heat generated during the grinding operation; cryogenic grinding is used. The powder after grinding is passed through a mechanical sieve of size 300 micron for fine powder or 500 micron for coarse sized powder. Standard ground spices are usually ground to allow them to pass through US standard sieves ranging from No. 20 to No. 60 mesh. Spices are also ground to microscopic fineness with a particle diameter of 50 microns, which is one-twentieth of the size of 60-mesh particles.

6.6. Ground Paprika

Paprika is the ground form of fine chillies with low or nil pungency with very high colour value. Hungarian paprika is a globally appreciated, speciality product valued for its intense colour. Paprika is produced commercially in Spain, Portugal, Central Europe, Southern Africa, the US, but Hungary is by far the most famous paprika-producing country, with approximately 2,398 ha devoted to the paprika types with a production of 21864 metric tonnes (FAOSTAT, 2014). The fruits are freed from the calyx and peduncle which dilutes the colour and adds fibre (destemming). The destemmed pods are further processed to remove the placenta and ground to produce commercial paprika powder. The assumption is that the paprika is always mild (sweet). The more the colour in the product, the higher the price. The colour range is between 80 ASTA (light orange) to 160 ASTA (dark red).

Among the popular varieties of Nigeria, the most pungent chilli was *C. frutescens* cv. Ata Were with 0.49 and 0.37 per cent capsaicin in fresh and dried fruits, respectively (Gbolade *et al.*, 1997). The heat of different global and Indian types is presented in is presented in Tables 11 and 12.

Table 11. Scoville Heat Units of Major International Chilli Varieties

Chilli Type	*Species*	*Scoville Heat Units (SHU)*
Bell/sweet	*C. annuum* L.	0–100
New Mexican	*C. annuum* L.	500–1,000
Espanola	*C. annuum* L.	1,000–1,500
Ancho and Pasilla	*C. annuum* L.	1,000–2,000
Cascabel and Cherry	*C. annuum* L.	1,000–2,500
Jalapeno and Mirasol	*C. annuum* L.	2,500–5,000
Serrano	*C. annuum* L.	5,000–15,000
De Arbol	*C. annuum* L.	15,000–30,000
Guntur Sannam	*C. annuum* L.	30,000–40,000
Cayenne	*C. frutescens* Roxb.	30,000–50,000
TabascoChiltepin	*C. frutescens* Roxb.*C. annuum* var. *glabriusculum*	50,000–100,000
Scotch Bonnet and Thai	*C. chinense* Jacq.	100,000–350,000
Habanero	*C. chinense* Jacq.	200,000–350,000
Tezpur/Nagarhari (Assam)	*C. chinense* Jacq.	8,55,000-1 million
Trinidad Scorpion Butch T	*C. chinense* Jacq.	1 – 1.4637 million
Carolina Reaper	*C. chinense Jacq.*	2.2 milllion

6.7. Pesticide Residues

Excessive and indiscriminate use of pesticides not only increases the cost of production but also results in many human health problems and environmental pollution. Presence of banned pesticide residues as well as recommended ones were reported in fresh chillies. Recently Saudi Arabia imposed a ban on green chilli imports from India which indicates the severity of the problem in India. Ethion residue levels were in the range of less than 2 micro g kg^{-1} in all tested hot chilli pepper varieties and needed commodity monitoring for ethion in Kuwait (Bu-Abbas *et al.*, 2009). In India, 13-14 per cent of pesticides used in the country are applied on vegetables, maximum use is in chilli (5.13 per cent) followed by brinjal (4.6 per cent) (Kodandaram *et al.*, 2013). Among several vegetables studied, the frequency of contamination was highest (80 per cent) for chilli. Monitoring of pesticide residues in India revealed that main pesticide contaminants in chillies were mainly mancozeb, dithiocarbamates, chlorpyriphos and profenphos. It is estimated that the average pesticide use in chillies was 5.13 kg a.i. ha^{-1}, which was highest among the vegetables. An average of 10-25 sprayings is given during the crop season in many chilli growing areas in India. The pesticides recommended for chillies in India, formulation, dose and waiting period are presented in Table 13. The contamination of phosphamidon, HCH, carbofuran, monocrotophos, endosulfan and fenvalerate in green chilli was reported from villages near Jaipur, Rajasthan, India (Singh *et al.*, 1999). The chilli samples collected from farmers of Guntur district of Andhra Pradesh, India were evaluated for the residues of pesticides *viz.* chlorpyriphos, endosulfan, acephate, carbofuran, monocrotophos and quinalphos, and reported that residue values were below detectable levels in the whole and processed spice samples (Sanhera and Shobha, 2012). The dissipation of different pesticides applied on chillies is presented in Table 14.

Table 12. Colour and Capsaicin Content of Major Indian Chilli Varieties

Chilli type	*Location*	*ASTA Colour Value*	*Capsaicin (per cent)*	*Scoville Heat Units (SHU)*
Byadagi (Kaddi)	Hubli, Dharwar district of Karnataka and Goa	156.90	Negligible	8000-15000
Ramnad Mundu	Ramnad district of Tamil Nadu	32.95	0.166	24900
Tomato Chilli	Warangal, Khammam, East and West Godavari districts of Andhra Pradesh	125.26	0.170	25500
Ellachipur sannam	Amaravathi District of Maharashtra	70.40	0.200	30000
Guntur Sannam	Guntur, Warangal and Khammam, districts of Andhra Pradesh	32. 11	0.226	33900
Hindpur	Hindpur in Andhra Pradesh	33.00	0.240	36000
Kashmir chilli	Himachal Pradesh, Jammu and Kashmir and also in sub-tropical regions of North India	54.10	0.325	48750
Jwala	Kheda, Mehsana and in South Gujarat	-	0.400	60000
Kanthari-white	Kerala and some parts of Tamil Nadu	2.96	0.504	75600
Birds Eye Chilli	Mizoram and some locations of Manipur	41.70	0.589	88350

Source: Prasath and Ponnuswami, 2008 (After, modified).

Table 13. Pesticides for Use in Chillies, Dose and Waiting Period

Sl.No.	*Common Name*	*Strength and Formulation*	*Target Pests*	*Dose*			*Waiting Period/PHI* (days)*
				a.i (g)	*Formulation (g/ml)*	*Dilution in water(L)*	
	Insecticides						
1	Acetamiprid	20 per cent SC	Thrips	10-20	50-100	500-600	3
2	Buprofezin	25 per cent SC	Yellow mite	75-150	300-600	500-750	5
3	Carbofuran	3 per cent G	Thrips	1000	33300	-	-
4	Carbosulfan	25 per cent EC	Leaf folder, Aphids	200-250	800-1000	500-1000	8
5	Chlorantraniliprole	18.5 per cent SC	Fruit borer	30	150	500	3
6	Chlorfenpyre	10 per cent SC	Yellow mite	75-100	750-1000	500	5
7	Cyantraniliprole	10.26 per cent OD	Thrips, fruit borer, tobacco caterpillar	60	600	500	3
8	Cyenopyrafen	30.0 per cent SC	Mite	60-90	200-300	400-600	7
9	Deltamethrin	2.8 per cent EC	Fruit borer	10-12.5	400-500	400-600	5
10	Difenthiuron	50 per cent WP	Mites	300	600	500-750	3
11	Emamectin benzoate	5 per cent SG	Fruit borer, Thrips, mite	10	200	500	3
12	Emamectin benzoate	1.9 per cent SC	Fruit borer, Thrips,	7.13	375	500	3
11	Ethion	50 per cent EC	Mite, Thrips	750-1000	1500-2000	500-1000	05
12	Fenazaquin	10 per cent EC	Yellow mite	125	1250	400-600	10
13	Fenpropathrin	30 per cent EC	Thrips, Whitefly, Mites	75-100	250-340	750-1000	7
14	Fenpyroximate	5 per cent EC	Yellow mite	15-30	300-600	300-500	7
15	Fipronil	5 per cent SC	Fruit borer, Thrips, Aphids	40-50	800-100	500	7
16	Flubendamide	20 per cent SG	Fruit borer	50-60	250-300	500	5
17	Flubendamide	39.35 per cent SC	Fruit borer	48-60	100-125	500	7
18	Hexythiazox	5.45 per cent EC	Yellow mite	15-25	300-500	625	3
19	Imidacloprid	70 per cent WS	Jassid, Aphid, Thrips	700-1050	1000-1500	-	-
20	Imidacloprid	17.8 per cent SL	Jassid, Aphid, Thrips	25-50	125-250	500-700	40
21	Indoxacarb	14.5 per cent SC	Fruit borer	50-60	333-400	300-600	5
22	Lambda Cyhalothrin	4.9 per cent CS	Thrips, Pod borer	25	500	500	5
23	Lambda Cyhalothrin	5 per cent EC	Thrips, Mite, Pod borer	15	300	400-600	5
24	Lufenuron	5.4 per cent SC	Fruit borer	30	600	500	5
25	Methomyl	40 per cent SP	Pod borer, Thrips	300-400	750-1125	500-1000	5-6
26	Milebemectin	1 per cent EC	Mites	3.25	325	500	7
27	Novaluron	10 per cent EC	Fruit borer, Tobacco caterpillar	33.5	375	500	3
29	Oxydemeton methyl	25 per cent EC	Aphids, Mites, Thrips	1000	10000	-	-
30	Propargite	57 per cent EC	Mite	850	1500	500-625	7
31	Pyriproxyfen	10 per cent EC	Whitefly, aphids	50	500	300	7
32	Quinalphos	25 per cent Gel	Aphids, Mite	250	100	500-1000	-
		25 per cent EC	Aphids	250	1000	500-1000	-
			Mites	375	1500	500-1000	-
		1.5 per cent DP	Aphid	300	20000	-	-
33	Spinetoram	11.7 per cent SC	Thrips, fruit borer, tobacco caterpillar	56-60	470-500	500-1000	30
34	Spinosad	45 per cent SC	Fruit borer	73	160	500	3
35	Spiromesifen	22.9 per cent SC	Yellow mite	96	400	500-750	7
36	Spirotetramat	15.31 per cent OD	Thrips, Aphids	60	400	500	5
37	Thiacloprid	21.7 per cent SC	Thrips	54-72	225-300	500	5
38	Thiodicarb	75 per cent WP	Fruit borer	470-750	625-500	500	6
39	Thiamethoxam	30 per cent FS	Thrips	2.1	7	Seed dressing	

Sl.No.	Common Name	Strength and Formulation	Target Pests	Dose			Waiting Period/PHI* (days)
				a.i (g)	Formulation (g/ml)	Dilution in water(L)	
40	Diafenthiuron + Bifenthrin	47 per cent + 9.4 per cent SC	Thrips, aphids	293.75+58.7	625	500	7
41	Emamectin benzoate + Fipronil	1.5 per cent + 3.5 per cent SC	Thrips, fruit borer	07.5+17.5-11.25+26.25	500-750	500	03 (day) or 48 (Hrs) Re-entry period after each application
42	Flubendamide + Thiacloprid	19.92 per cent + 19.92 per cent SC	Thrips, fruit borer	48+48	200-250	500	5
43	Fipronil + Hexythiazox	7 per cent + 2 per cent SC	Mites, thrips	70+20	1000	500	7
44	Indoxacarb + Acetamiprid	14.5 per cent +7.7 per cent SC	Thrips, Fruit borer	88.8-111	400-500	500	5
45	Profenophos + Fenpyroximate	40+2.5 EC	Thrips, mites, fruit borer	4+.025	1000	500	7
46	Pyriproxyfen + Fenpropathrin	5 per cent + 15 per cent EC	Whitefly, fruit borer	25.0+75.0-37.5+112.5	500-750	500-750	7
47	Thiamethoxam + Lamda cyhalothrin	12.6 per cent + 9.5 per cent ZC	Thrips, Fruit borer	33	150	500	3

Source: CIB and RC, 2021) * PHI: Pre Harvest Interval ** as required.

Table 14. Dissipation of different Pesticides Applied on Chilli

Pesticide	Application and Dosage	Time to Reach Below Detection Level (BDL)
Insecticides and Miticides		
Acephate	1500 g a.i. ha^{-1}	15 days
Acetamiprid	20 g a.i. ha^{-1} and 40 g a.i. ha^{-1}	7 days
Chlorpyriphos 50 per cent + Cypermethrin 5 per cent	1 and 2 L ha^{-1}	10 days
Dicofol	250 and 500 g a.i. ha^{-1}	1 and 2 days
Dimethoate	150 and 300 g a.i. ha^{-1}	8 days
Fipronil	5 SC @ 0.01 per cent a.i.	5 days
Flubendiamide	60 g a.i. ha-1, 0.824 and 0.992 mg kg^{-1} and double dose	7-10 days
Fluvalinate	20 g a.i. ha^{-1}	10 days
Imidacloprid 17.8/SL	25, 50 and 100 g a.i. ha^{-1} and double dose	3 days for the first dose, 5 and 7 days for the next two.
Lindane	150 and 300 g a.i. ha^{-1}	30 days
Monocrotophos	0.45 and 1.05 kg a.i. ha^{-1}	10 and 21 days
Novaluron	37.5 or 75.0 g a.i. ha^{-1}	10 days
Permethrin	-	5 days
Profenofos	40 EC @ 0.1 per cent a.i.	19 days
Propargite 57 EC	570 g a.i. ha^{-1}	5 days
Spiromesifen 240 SC	96 g a.i. ha^{-1} and	5 days
Spirotetramat	60 and 120 g a.i. ha^{-1}	7 days
Triazophos	700 g a.i. ha^{-1}	15 days
Fungicides		
Mancozeb 75 WP	1.25 kg ha^{-1}	3 days
Propiconazole	125 and 250 g a.i. ha^{-1}	15 days for 125 g a.i. ha^{-1}; detectable even after 30 days @ 250 g a.i. ha^{-1}
Propineb 70 WP	1.75 kg ha^{-1}	3 days

6.8. Aflatoxins (AFs)

Aflatoxins are mycotoxins produced by moulds, specially *Aspergillus flavus* and *Aspergillus parasiticus* which are ubiquitous and all types of organic matters are suitable substrates for their growth. Besides *A. flavus*, several potential mycotoxin producing species including *A. westerdijkiae, A. melleus, A. terreus, A. awamori* and *A. niger* have also been identified on nuts and spices (chilli, red pepper, spice mixes) in Hungary. Chemically aflatoxins are difuranokoumarin derivatives. About 18 different types of aflatoxins have been identified and among these B1, B2, G1, G2, M1 and M2 are the most common. AFs B1 is considered as the most toxic in nature. *A. ochraceus* is associated with Ochratoxin A production for which there are no regulations for chillies currently in the EU. Furthermore, it is important to differentiate between natural contamination levels in crops and those which may be present from malicious intent in the current security conscious age (Paterson, 2007). Aflatoxins which are a natural contaminant in commodities require regular monitoring and effective control to prevent health hazards. Aflatoxins in chilli are a threat to the health of populations and a constraint for national development. A consistent screening of chillies for aflatoxin and ochratoxin content, strictly prohibiting the use of produce containing mycotoxin concentrations exceeding the maximum permitted level should be made a policy matter. As per the European Union (EU) regulations, only 5 µg kg^{-1} of aflatoxin B1 and 10 µg kg^{-1} of total aflatoxins are allowed in chillies. Many countries do not have any regulations for this commodity. However, the mycotoxin problem is most acute in developing countries where climate and lack of resources compromise production. The drive for exports to raise revenue and strict regulations in developed countries means that the best quality commodities are sold leaving the remainder for underdeveloped countries, which have higher toxin concentrations as a consequence. The perusal of aflatoxin contamination data indicates that individual localities have particular problems. Red chillies sun dried on ground leaving considerable scope for moisture accumulation at some stages, are conducive to the growth of aflatoxigenic fungi. Drying involves spreading of one layer thick pods on the soil. Sorting is sometimes undertaken to remove the damaged and diseased chillies, but it is rudimentary. Storage of chilli for longer a period leads to aflatoxin production irrespective of storage temperature (30±2 ° C and 4 ° C). Hence, it is advisable to consume spices such as chilli powder within a relatively short period to avoid build up of aflatoxins. The nutritional profile of chilli powder (*Capsicum annuum* var. *sannam* L.) showed highest share of total soluble sugars (32.89 per cent) and fibre content (21.05 per cent), followed by protein (18.01 per cent) and fat (13.32 per cent) making it an ideal solid-substrate for mould growth (Tripathi and Mishra, 2009). A gradual decrease in the levels of biochemical constituents, such as total protein, carbohydrates, reducing sugars, ascorbic acid and total phenols in cold-stored red chillies were observed. *Aspergillus flavus* and *Aspergillus niger* followed by *Mucor spp*., *Fusarium spp*. and *Alternaria alternata* were the prominent moulds associated with chillies in cold storage (Ravikiran *et al*., 2009). Among the mycoflora invading chillies in cold storage *Aspergillus spp*. is predominant. The Aspergillus incidence varied from 0 to 14 per cent in different cultivated areas of Karnataka and the isolates collected from Bellary were most toxigenic.

The pods and powders of chillies from India had high levels of AFB1, particularly in lowest grade pods (Ravikiran *et al*., 2009). After a survey of chillies traded in UK, objectionable levels of AFs were reported only in a few cases (UK Food Standards Agency, 2005). The market samples from Hyderabad, India were analyzed and none of the samples exceeded the maximum limit of 5000 ng kg^{-1} as prescribed in European Union regulations for aflatoxin B1 (Nejad *et al*., 2013). All samples from farmers in Pakistan contained aflatoxin B1 and high levels were reported in powdered samples (Paterson, 2007). Among the samples collected from rural, semi-rural and urban areas of Pakistan, 56 per cent contained total levels of AFs that exceeded European Union limits. Aflatoxins contamination in red chillies (whole, powdered and crushed) from all parts of Pakistan revealed that AFs levels in 62.4 per cent of whole, 26.1 per cent of powdered and 19.4 per cent of crushed chillies were found lower than the maximum limit (ML=10 micro g kg^{-1}) as assigned by the European Union limits. Among the spices traded in Italy, aflatoxin contamination remained below the legal limits but matrices most contaminated among them by the mycotoxins were chilli, paprika and pepper. Chilli powder was found to have the highest level of aflatoxins contamination followed by chilli pods in Thailand market samples. In Turkey, 47 per cent samples collected from the farmers of chilli exceeded the legal limits established by European Commission for AFB1 and total AFs (the sum of AFB1, AFB2, AFG1 and AFG2), respectively. In West Africa, market samples of dried chilli were found to be contaminated with AFs, hot chilli was naturally contaminated with aflatoxin B1 and aflatoxin B2, at concentrations of 3.2 micro g kg^{-1}on hot pepper.

A survey on commercial spice preparations with chilli from the Irish retail market reported that only 3.8 per cent of the samples was above the regulatory limits of the European Union (O'Riordan and Wilkinson, 2008). The contamination of AFs in baby food, peanut and corn products sold at retail outlets in Indonesia were studied and the Groundnut-chilli sauces were found to have the highest percentage (75 per cent) of AFs contamination (Razzazi-Fazeli *et al*., 2004). The market samples from Portugal were analysed for the presence of AFs (Martins *et al*., 2001), samples of paprika contained levels of aflatoxin B1 ranging from 1-20 µg kg^{-1} while chilli samples had levels ranging from 1-5 micro g kg^{-1}. Nearly 9 per cent of the chilli powders sold in supermarkets of India contained non-permissible aflatoxin levels and hence the use of grade 1 chilli pods to minimize aflatoxin contamination is recommended (Reddy *et al*., 2001). In conclusion, globally the concentrations were often greater than the statutory limits set by the European Union. Recently, gamma irradiation was used to minimise the toxicological effects of aflatoxins by inhibiting the growth as well as the proliferation of Aspergillus species. Physical, chemical and biological methods were tested for aflatoxin detoxification. In a study, direct oven heating at 120°C for 12 hours produced maximum (83.3 per cent) AFB1 degradation, UV exposure for 60 min produced 59.6 per cent AFB1 degradation, oxidation with H_2O_2 58.3 per cent degradation, 12 IU of partially purified peroxidase (POD) isolated from garlic had the biological detoxification of 66.2 per cent. The physical methods were more efficient over other methods in degrading AFB1, but produced significant nutritional losses. The ascorbic acid and carotenoids were the most sensitive quality attribute affected by the treatments (Tripathi and Mishra, 2009).

Usually, *A. flavus* contamination occurs in the field itself, which aggravates during early drying and long storage. In kitchen, aflatoxin contamination may be detoxified effectively by physical methods, but produced significant nutritional losses (Tripathi and Mishra, 2009). On the contrary, Macdonald and Castle (1996) found that aflatoxin levels in spiced sauces are not reduced by domestic cooking with either microwave or conventional oven heating. Hence, the management requires specific approaches from pre harvest period to drying and storage. Care should thus be taken in the choice of a suitable fungicide because their effectiveness may depend on intra-specific variation and temperature. Moreover, it is necessary to take into account that the most efficient fungicide in reducing the fungal growth is not always the best choice for pre harvest treatments because they may promote AFs production during postharvest period. Thus, the best fungicide is the one that can simultaneously prevent fungal growth and AFs production. In-vitro studies on the efficacy of certain fungicides inhibiting fungal growth and AFs production revealed that regardless of the strain tested tebuconazole 25 per cent and mancozeb 80 per cent applied at a concentration of 0.75 and 3.5 g l^{-1}, respectively were most effective. Pre harvest spray of NSKE (5 per cent) or mancozeb (0.3 per cent) or *P. fluorescens* (1 x 108 cfu/ml) 10 days before harvest of chilli is recommended for effective field level management of aflatoxin (Sudha *et al.*, 2013). Complete inhibition (100 per cent) of *A. flavus* was observed with spraying of neem seed kernel extract (NSKE), nimbicidin and pongamia oil at 5 per cent. Among fungicides, Mancozeb (91.1 per cent inhibition) and captan (85.2 per cent inhibition) at 0.3 per cent among non-systemic fungicides and carbendazim (73 per cent inhibition) at 0.15 per cent among systemic fungicides were found effective in inhibiting the growth of *A. flavus*. Biological methods to reduce pathogen infection in postharvest period were also explored by many. Postharvest treatment with *Bacillus amylo-liquefaciens* strain-L fermented liquid on chillies (green and red) reduced the decay rates and improved shelf life, sensory quality and nutritional value. Application of biological agents such as *P. fluorescens* and *Trichoderma viride* or *consortia of potent agents* lowered the soil inoculum level in the field lowering the aflatoxin contamination significantly. Gamma irradiation minimises the toxicological effects of aflatoxins by inhibiting the growth as well as the proliferation of Aspergillus species.

6.9. Postharvest Disorders of Chillies

6.9.1. Chilling Injury

Chilling injury include fruit softening, water-soaked spots, surface pitting, lack of colour development, discolouration of the seed cavity, and increased susceptibility to decay. Chilling also causes browning and decay of the stem. Ripe or coloured peppers are less chilling sensitive than green peppers. Fresh chillies for export markets are usually stored at low temperatures between 7 to 12 °C. However, chillies are subject to chilling injury when stored at temperatures below 7°C. Chillies will develop surface pitting in a few days when stored 2 °C. Fresh green chillies can be stored up to three weeks without much postharvest losses when maintained at 10 °C (R.H. 80-90 per cent). However, storing at temperatures lower than 5 °C (R.H. 80-90 per cent) promoted chilling injury in fruits. Fairly low levels of chilling injury can be maintained up to 28 days of storage when fruits are held at 10 °C (Edusei *et al.*, 2012). Severity of chilling injury increases with more exposure in low temperature and decreases with higher temperature. However, symptoms of chilling damage usually do not appear until peppers are moved from cold storage to ambient temperature.

6.9.2. Ethylene Injury

Fresh chillies are very sensitive to the natural ripening hormone ethylene. The presence of even minute quantities can cause a loss of chlorophyll and discolouration of the green pigment in chillies, resulting in a mosaic of yellowish, chocolate, and red colours. The combination of ethylene with high temperature accelerates the rate of loss in green pigment. Ethylene-producing crops such as bananas, plantains, and tomatoes should not be stored or shipped with green-coloured peppers. The chilli fruits particularly sweet peppers should never be exposed to ethylene.

7.0 Conclusion

Chilli, an important vegetable and spice, is used as a culinary supplement essentially for its colour, flavour and pungency. In recent years, its importance as a source of natural food colour and valuable pharmacological compounds has grown beyond imagination. There are specific varieties in Europe and South America suitable for different processing purposes, which are yet to be identified in Asia, the largest producing and exporting region. In order to realize benefits of trade and develop processing industry, development of newer varieties suitable for processing with higher levels of these compounds is needed. In chilli harvesting is highly labour intensive. Development of varieties suitable for mechanical harvesting and stripping will be more advantageous. Even though several options of drying for ripe chillies are available, sun drying is the most common one. Hence, more research is needed on development of energy efficient, cost effective, farmer friendly drying and sustainable technologies. Pharmacological application of capsaicin demands upgradation of technologies to produce end-product specific capsaicin formulations. Postharvest losses of vegetable chilli are very high and development of cold chain technology keeping in mind the sensitivity and chilling injury of the crop under low temperature storage is needed. Under ambient conditions, the storage life of fruits can be extended by packing them in perforated polythene film. Loss of colour and quality during storage is the major concern in dried red chilli. Drying temperature is another factor that affects the colour retention during storage; high temperature is harmful as it results in the degradation of antioxidants. Technologies to prevent degradation of pigments under storage are to be studied intensively for opening new avenues to retain quality during storage. Chilli seed lose viability in a few months of storage. Seeds with moisture content of 6-8 per cent and stored in airtight containers could maintain seed viability for two years, but under ambient conditions lose viability in a year. Seed viability can be extended to a longer period under cold storage in airtight containers; hence there is a need to study the seed storage behaviour under low temperatures. The ultra drying effects on seed longevity is not known and this aspect of drying requires further study. Global trade of chilli and its value added products is afflicted with trade rejections due to the presence of adulterants, pesticide residues and aflatoxins. The use of toxic dyes is reported both from developed and under developed countries. Similar is the case with pesticide residues and aflatoxins. Extension programmes to increase the awareness about the adulterants in producers, processers and exporters should be taken up with serious note. Public concern has grown about aflatoxins and pesticide residues due to active media and citizen groups, that too with valid reasons. All the stake holders of production, policy makers and research workers should draw their immediate attention to curtail practices that lead to the presence of these hazardous substances. Adoption of Good Agricultural Practices (GAP) must be promoted among the producers to deliver quality and hygienic produce to the consumers.

8.0 Important References

Ayob, O., Hussain, P.R., Suradkar, P., Naqash, F., Rather, S.A., Joshi, S. and Azad, Z.A.A., 2021. Evaluation of chemical composition and antioxidant activity of Himalayan Red chilli varieties. *LWT*, **146**: 111-413.

Ayodele, O.J., Alabi, E.O. and Aluko, M., 2015. Nitrogen fertilizer effects on growth, yield and chemical composition of hot pepper (Rodo). *Int. J. Agri. Crop Sci.* **8**: 666.

Bhardwaj, R.L., Sundria, M.M., Kumhar, S.R. and Kumar, N., 2018. Effect of irrigation methods and mulching on growth and yield parameters of chilli (*Capsicum annum* L.) in arid condition. *J. Spic. Arom. Crops*, **27**.

Biles, C.L., Wall, M.M. and Blackstone, K., 1993. Morphological and physiological changes during maturation of New Mexican type peppers. *J. Amer. Soc. Hortic. Sci.* **118**: 476-80.

Bosland, P. W., Votava, E. J. and Votava, E. M. 2012. Peppers: vegetable and spice capsicums (No. 22). CABI.

Bu-Abbas, A. H., Yusuf, A. A. A. E. S., Badawi, B. S. and Al-Shammari, H. F. S. 2009. Contamination of the organophosphorus insecticide ethion residues in commercial samples of red-hot chilli spice in Kuwait. *Int. J. Food Saf. Nutr. Public Health Technol.* **2**: 59-68.

Cerit, I. and Demirkol, O., 2020. Effects of modified atmosphere packaging conditions and ethylene absorber on the quality of red bell pepper. *J. Food Nutr. Res.* **59**.

Chitravathi, K., Chauhan, O.P. and Raju, P.S., 2015. Influence of modified atmosphere packaging on shelf-life of green chillies (*Capsicum annuum* L.). *Food Packag. Shelf Life*, **4**: 1-9.

Chitravathi, K., Chauhan, O.P., Raju, P.S. and Madhukar, N., 2015. Efficacy of aqueous ozone and chlorine in combination with passive modified atmosphere packaging on the postharvest shelf-life extension of green chillies (*Capsicum annuum* L.). *Food Bioproc. Technol.* **8**: 1386-1392.

CIB and RC. 2014. Major Uses of Pesticides. At: http: // www.cibrc.nic.in/rc.htm [Accessed on 6 May 2014].

Deepa, G. T., Chetti, M. B., Khetagoudar, M. C. and Adavirao, G. M. 2013. Influence of vacuum packaging on seed quality and mineral contents in chilli (*Capsicum annuum* L.). *J. Food Sci. Technol. (Mysore)*. **50**: 153-58.

Dewitt, D. and Bosland, P. W. 2009. The complete chile pepper book. Timber Press, Inc., London.

Edusei, V. O. Ofosu-Anim, J. Johnson, P. N. T. Cornelius, E. W. 2012. Extending postharvest life of green chilli pepper fruits with modified atmosphere packaging. *Ghana J. Hortic.* **10**: 131-40.

FAOSTAT. 2021. At: http: //faostat.fao.org/ [Accessed on 01 November 2021].

Gbolade, A. A., Omobuwajo, O. R and Soremekun, R. O. 1997. Evaluation of the quality of Nigerian chillies for pharmaceutical formulations. *J. Pharmac. Biomed. Anal.* **15**: 545-48.

Glowacz, M. and Rees, D., 2016. Exposure to ozone reduces postharvest quality loss in red and green chilli peppers. *Food Chem.* **210**: 305-10.

Govindarajan, V. S., Rajalakshmi, D., Chand, N and Salzer, U. J. 1987. Capsicum—production, technology, chemistry, and quality. Part IV. Evaluation of quality. *Crit. Rev. Food Sci. Nutri.* **25**: 185-282.

Hahn, F. 2002. Automatic Jalapeno chilli grading by width. *Biosyst Engg.* **83**: 433-40.

Hameed, R., Malik, A.U., Imran, M., Khan, A.S. and Schoten, S., 2009, April. Effects of controlled atmosphere on storability, shelf life and quality of green slender chilies (*Capsicum annuum* L.). In

X International Controlled and Modified Atmosphere Research Conference 876 (pp. 123-34).

Horvitz, S. and Cantalejo, M.J., 2008, November. Combined effects of gaseous O_3 and modified atmosphere packaging on quality and shelf-life of fresh-cut red bell pepper. In *III International Conference Postharvest Unlimited* 858 (pp. 335-40).

Howard, L.R. and HernandezBrenes, C., 1998. Antioxidant content and market quality of jalapeno pepper rings as affected by minimal processing and modified atmosphere packaging. *J. Food Qual.* **21**: 317-27.

Jansasithorn, R., East, A. R., Hewett, E. W., Molan, A. L., Heyes, J. A., and Mawson, A. J. 2010. Harvest Maturity Influences the Antioxidant Activity in Jalapeño Chili. In: *XXVIII International Horticultural Congress on Science and Horticulture for People (IHC2010): International Symposium, August, 2010.* pp. 379-84.

Kang, J.S. and Lee, D.S., 1997. Susceptibility of minimally processed green pepper and cucumber to chilling injury as observed by apparent respiration rate. *Int. J. Food Sci. Technol.* **32**: 421-26.

Kati Coulibaly, S., Coxam, V and Barlet, J. P. 1998. Physiological, pharmacological and toxicological interests of chillies and of their active principle: Capsaicin. *Médecine et Nutr.* **34**: 236-45.

Khare, B. P. Johari, R. K. and Saxena, R. P. 1989. Record of *Lasioderma serricorne* (F.) (Anobiidae: Coleoptera) infesting chilli (*Capsicum annuum* L.) seeds in Pantnagar. *Agril. Sci. Dig., (Karnal)*, **9**: 77-78.

Kirk, S. and Sawyer, R. 1991. Pearson's composition and analysis of foods (No. Ed. 9). Longman Group Ltd., pp. 236-328.

Kitinoja, L. and Tokala, V.Y., 2018. Brondy. A. 2018. A review of global postharvest loss assessments in plant-based food crops: Recent findings and measurement gaps. *J. Postharvest Technol.* **6**: 1-15.

Kodandaram, M.H., Sujoy Saha, Rai, A.B. and Prakash S. Naik. 2013. Compendium on pesticide use in vegetables. , IIVR, Varanasi, *IIVR Ext. Bull.* **50**: 133.

Koncsek, A., Kruppai, L., Helyes, L., Bori, Z. and Daood, H.G., 2016. Storage stability of carotenoids in paprika from conventional, organic and frost damaged spice red peppers as influenced by illumination and antioxidant supplementation. *J. Food Process. Preserv.* **40**: 453-62.

Kopta, T., Sekara, A., Pokluda, R., Ferby, V. and Caruso, G., 2020. Screening of chilli pepper genotypes as a source of capsaicinoids and antioxidants under conditions of simulated drought stress. *Plants*, **9**: 364.

Macdonald, S. and Castle, L. 1996. A UK retail survey of aflatoxins in herbs and spices and their fate during cooking. *Food Addit. Contam.* **13**: 121-28.

Maji, A.K. and Banerji, P., 2016. Phytochemistry and gastrointestinal benefits of the medicinal spice, Capsicum annuum L.(Chilli): a review. *J. Complement. Integr. Med.* **13**: 97-122.

Manolopoulou, H., Lambrinos, G. and Xanthopoulos, G., 2012. Active modified atmosphere packaging of fresh-cut bell peppers: effect on quality indices. *J. Food Res.* **1**: 148.

Martins, M. L. Martins, H. M. and Bernardo, F. 2001. Aflatoxins in spices marketed in Portugal. *Food Addit. Contam.* **18**: 315-19.

Maurya, V.K., Ranjan, V., Gothandam, K.M. and Pareek, S., 2020. Exogenous gibberellic acid treatment extends green chili shelf life and maintain quality under modified atmosphere packaging. *Scientia Hortic.* **269**: 108-934.

Mendez, M.A., Ligarreto, G.A., Hernandez, M.S. and Melgarejo, L.M. 2004. Evaluating growth and determining harvesting index in four types of hot chilli pepper (*Capsicum sp.*) grown in the Colombian Amazonian Region. [Spanish] *Agronomia Colombiana*. **22**: 7-17.

Myint, K.M.M., 2020. Effects of Ethylene Absorbents On The Storage of Fresh Chilli Fruits. *3rd Myanmar Korea Conf. Res. J.* **3**: 233-41.

Narayana, C.K., Pandey, B.K., Malhotra, S.K. and Pandey, V. 2014. Postharvest losses in selected fruits and vegetables in India (a compilation). Indian Institute of Horticultural Research, Bengaluru.

Narayanan, C. S., Sumathikutty, M. A., Sankarikutty, B., Rajaraman, K., Bhat, A. V. and Mathew, A. G. 1980. Studies on separation of high pungent oleoresin from Indian chilli. *J. Food Sci. Technol.* **17**: 136-38.

Nejad, A. S. M. Bayat, M. Ahmadi, A. A. 2013. Investigation of aflatoxin B1 in spices marketed in Hyderabad, India by ELISA method. *J. Pure Appl. Micr.* **7**: 3219-23.

Noichinda, S., Bodhipadma, K., Moungjomprang, D., Thongnurung, N. and Kasiolarn, H., 2016. Harvesting indices of Chi-fah Yai pepper (*Capsicum annuum* L.) fruit. *The J. Appl. Sci.* **15**: 1-19.

O'Riordan, M. J. and Wilkinson, M. G. 2008. A survey of the incidence and level of aflatoxin contamination in a range of imported spice preparations on the Irish retail market. *Food Chem.* **107**: 1429-35.

Othman, Z.A.A., Ahmed, Y.B.H., Habila, M.A. and Ghafar, A.A. 2011. Determination of capsaicin and dihydrocap-saicin in *Capsicum* fruit samples using high performance liquid chromatography. *Molecules*, **16**: 8919–29.

Panigrahi, J., Gheewala, B., Patel, M., Patel, N. and Gantait, S., 2017. Gibberellic acid coating: A novel approach to expand the shelf-life in green chilli (*Capsicum annuum* L.). *Scientia Hortic.* **225**L 581-88.

Pankar, D. S. and Magar, N. G. 1978. Capsaicin, total colouring matter and ascorbic acid contents in some selected varieties of chilli (*Capsicum annuum* L.). *J. Maha. Agril. Univ.* **3**: 116-19.

Paterson, R.R.M. 2007. Aflatoxins contamination in chilli samples from Pakistan. *Food Control*, **18**: 817-20.

Pradeepkumar, T. 1990. Interspecific hybridization in Capsicum. *M.Sc. (Ag.) Thesis,* Kerala Agricultural University, Vellanikkara, pp. 102.

Prakash, V. and Eipeson, W.E., 2003. Postharvest handling and processing of Capsicums. In *Capsicum* (pp. 183-194). CRC Press.

Prasath, D. and Ponnuswami, V. 2008. Breeding for extractable colour and pungency in capsicum – A review. *Veg. Sci.* **35**: 1-9.

Pruthi, J.S. 1993. Major spice of India - Crop management and postharvest technology. ICAR. pp. 221-22.

Purseglove, J.W., Brown, *e.g.*, Green, C.L. and Robbins, S.R.J. 1981. Species (vol-1). Tropical Agricultural Series, Longman Group Ltd., UK, pp. 356.

Raina, B.L. and Teotia, M.S. 1986. Evaluation of chillies (*Capsicum annuum* L.) grown in Jammu and Kashmir. *Indian Food Packer*, **40**: 6-10.

Ravikiran, D., Narayana, K. J. P. and Vijayalakshmi, M. 2009. Moldassociated biochemical changes and aflatoxin b1 production in coldstored chillies (*Capsicum annuum* L.). *J. Food Process. Preserv.* **33**: 203-13.

Razzazi-Fazeli, E. Noviandi, C. T. Porasuphatana, S. Agus, A. Bohm, J. 2004. A survey of aflatoxin B1 and total aflatoxin contamination in baby food, peanut and corn products sold at retail in Indonesia analysed by ELISA and HPLC. *Mycot. Res.* **20**: 51-58.

Reddy, S. V., Kiran Mayi, D., UmaReddy, U., Thitumala-Devi, K. and Reddy, D. V. R. 2001. Aflatoxins B1 in different grades of chillies (*Capsicum annuum* L.) in India as determined by indirect competitive-ELISA. *Food Addit. Contamin.* **18**: 553–58.

Rongsennungla, I., Kanaujia, S. P., Singh, V. B., and Maiti, C. S. 2012. Effect of pre- and postharvest treatments on yield, quality and postharvest shelf life of king chilli (*Capsicum chinense* Jacquin) under foothill conditions of Nagaland (India). *J. Spic. Arom. Crops,* **21**: 42-47.

Saha, P. and Bera, S., 2021. Weather impact assessment and development of prediction model for yield loss in chilli anthracnose patho-system. *J. Agromet.* **23**: 352-55.

Saleh, B.K., Omer, A. and Teweldemedhin, B., 2018. Medicinal uses and health benefits of chili pepper (*Capsicum* spp.): a review. *MOJ Food Proc. Tech.,* **6**: 325-28.

Sanhera, S. and Shobha, S. 2012. Pesticide residues in spices of Guntur district of Andhra Pradesh. *The J. Res., ANGRAU*, **40**: 99-100.

Sarojam, S., Mohan, N., Paul, S.S., Chacko, B.K., Nema, N.K. and Jacob, V., 2020. A Comparative Study on Cultivars of Capsicum: Critical Assessment based on Colour Values, Pungency and total Capsaicinoids Content. *The Open Food Sci. J.* **12**.

Senesi, E., Prinzivalli, C., Sala, M. and Gennari, M., 2000. Physicochemical and microbiological changes in fresh-cut green bell peppers as affected by packaging and storage. *It. J. Food Sci.* **12**: 55-64.

Shivhare, U.S., Narain, M., Agarwal, U.S., Saxena, R.P. and Singh, B.P.N. 1987. Some physico chemical properties of chillies. *J. Food Sci. Technol.* **24:** 98-99.

Singh, A. K., Neeraj Singh and Singh, B. B. 2013. Marketing and postharvest loss assessment of vegetables in Varanasi district (U.P.). *Int. Res. J. Agric. Economics Stat.* **4**: 47-50.

Singh, B., Gupta, A., Bhatnagar, A. and Parihar, N. S. 1999. Monitoring of pesticide residues in farm gate samples of chilli. *Pest. Res. J.* **11**: 207-09.

Srivastava, M., Nagamadir, M., Pandey, P.H. and Salokhe, V.M. 1990. Design and development of a low cost chilli dryer. **In**: *Proceedings of the International Agricultural Engineering conference and Exhibition*. 3-6 December, 1990. Bangkok, Thailand, pp. 505-515.

Stommel, J.R., Goth, R.W., Haynes, K.G. and Kim, S.H. 1996. Pepper (*Capsicum annuum*) soft rot caused by *Erwinia carotovora* subsp. *atroseptica*. *Plant Dis.* **80**: 1109-12.

Sudha, S., Naik, M. K. and Ajithkumar, K. 2013. An integrated approach for the reduction of aflatoxin contamination in chilli (*Capsicum annuum* L.). *J. Food Sci. Tech.,* **50**: 159-64.

Taufik, M. 2011. Farming analysis and postharvest handling of red chili. [Indonesian] *Jurnal Penelitian dan Pengembangan Pertanian*, **30**: 66-72.

Teotia, M.S. and Raina, B.L. 1987. Evaluation of chillies grown in Punjab. *Indian Food Packer*, **41**: 81-86.

Tewari, V.P. 1988. Genetic improvement of capsaicin content in hot pepper (*C. frutescens* L.). *Capsicum Newslett.* **7**: 41.

Tewari, V.P. 1990. Development of high capsaicin chillies (*Capsicum annuum*) and their implications for the manufacture of export products. *J. Plant. Crops*. **18**: 1-13.

Thain, E.M., Robbins, S.R.J. and Green, C.I. 1980. World trade in chillies. In: *Proceedings of All India Workshop on Chillies*. March 1980, SEPI, Hyderabad, pp. 143-57.

Tiwari, R.K. 2014. Indian Horticulture Database, 2013. Eds. Mistry, N.C., Singh, B. And Gandhi, C.P. National Horticulture Board, New Delhi, India.

Tripathi, S. and Mishra, H. N. 2009. Nutritional changes in powdered red pepper upon in vitro infection of *Aspergillus flavus*. *Braz. J. Microbiol.* **40**: 139-44.

Tripathi, S. and Mishra, H.N., 2009. Studies on the efficacy of physical, chemical and biological aflatoxin B1 detoxification approaches in red chilli powder. *Int. J. Food Saf. Nutr. Publ. Hlth.* **2**: 69-77.

UK Food Standards Agency. 2005. Food Survey Information Sheet 73/05.

Uma Jyothi, K. and Ravi Shankar, C. 2006. Effect of postharvest chemical treatments on storage of dried chilli cv. LCA 334. *Veg. Sci.* **33**: 38-44.

Uma Jyothi, K., Surya Kumari, S., Venkata Reddy, P. and Ravi Shankar, C. 2008. Influence of temperature and relative humidity on quality constituents of dried chilli cultivars in storage. **In**: *Proceedings of International symposium on Agrometeorology and Food Security*. 18-21 February, 2008, CRIDA, Hyderabad.

Wall, M. M. Walker, S. Wall, A. D. Hughs, E. Phillips, R. 2003. Yield and quality of machine harvested red chilli peppers. *HortTechnology,* **13**: 296-02.

Wigati, L.P. and Darmawati, E., 2019, February. Losses and waste of tomato and red chilli along the supply chain. In *IOP Conference Series: Earth and Environmental Science* (Vol. 230, No. 1, p. 012001). IOP Publishing.

Zaulia, O., Razali, M., Aminuddin, H., Omar, D.C., Ng, K.H. and Habsah, M., 2006. Effect of different packagings and storage temperatures on the quality of fresh-cut red chilli. *J. Trop. Agri. Food Sci.* **34**: 67.

Chapter 30

Ginger

Ananta Saikia

Professor

Sanjib Sharma

Assistant Professor

Department of Horticulture, Assam Agricultural University, Jorhat – 785 013, Assam, India

E-mail: anantasaikia@aau.ac.in; E-mail: sanjib.sharma@aau.ac.in

01. Introduction

Ginger (*Zingiber officinale* Rosco.) is one of the most widely used spices around the world and is used as a condiment in a large number of foods and beverages. Rhizomes of ginger plants are eaten fresh or cooked as vegetables and used for flavouring foods. Ginger originated in South East Asia/Indo-Malaysian/Indo-China region. It has approximately 1400 species in 47 genera, presently grown in different geographical regions of the world.

Gingers are perennial herbs producing aromatic rhizomes that are subterranean or above ground and are rich in nutrients, especially vitamins and minerals. Each rhizome can produce erect leafy shoots. Inflorescences are terminal, borne either on leafy shoots or on special erect shoots near the base of the plant. It belongs to the family *Zingiberaceae* and is botanically known as *Zingiber officinale* Rosco. The family has three tribes, *viz.*, *Zingibereae*, *Alpinieae* and *Hedychieae*. *Zingiber* is the only genus of the tribe *Zingibereae*. The tribe *Alpinieae* consists of the genera *Alpinia*, *Amomum*, *Elettaria*, *Etlingera*, *Elettariopsis*, *Geocharis*, *Geostachys*, *Hornstedtia* and *Plagiostachys*. The tribe *Hedychieae* includes *Boesenbergia*, *Camptandra*, *Curcuma*, *Haniffia*, *Hedychium*, *Kaempferia* and *Scaphochlamys*.

Ginger has a long and well-documented history of both culinary and medicinal use throughout the world, especially in Chinese, Indian and Japanese medicinal care. Ginger is considered one of the world's best medicines since time immemorial which was referred to as "Viswa Bhesaj" (meaning universal medicine) and "Mahoushadha" (meaning the great medicine) in Sanskrit, during ancient Gupta era in India. It is reported to warm and soften stomach, cure pains, cough, chest disorders and other bodily discomforts. Ginger's therapeutic applications include analgesic, anti-arthritic, anti-oxidant, anti-fungal, antibacterial, anti-viral, anti-ulcer, hypercholesteraemic, cardiotonic, stimulant, aphrodisiac and anti-helminthic, without adverse side-effects.

02. Status of Cultivation

The major ginger growing countries are India, China, Nigeria, Nepal, Indonesia, Thailand, Cameroon, Bangladesh, Japan, the Philippines, Peru, Taiwan, Mali, Sri Lanka, Ethiopia and Malaysia. India is a leading producer of ginger in the world and, in 2019, the country produced 1788 thousand tonnes of the rhizomes from an area of 164 thousand hectares. During the year, India contributed 38 per cent to the global production of 4686 thousand tonnes from 441 thousand hectares of cultivated land distributed in various countries. During the last decade, India continues to be the leading ginger producing country in the world continuously increasing contribution to world production from 18 per cent in 2010 to 38 per cent in 2019 (Figure 1).

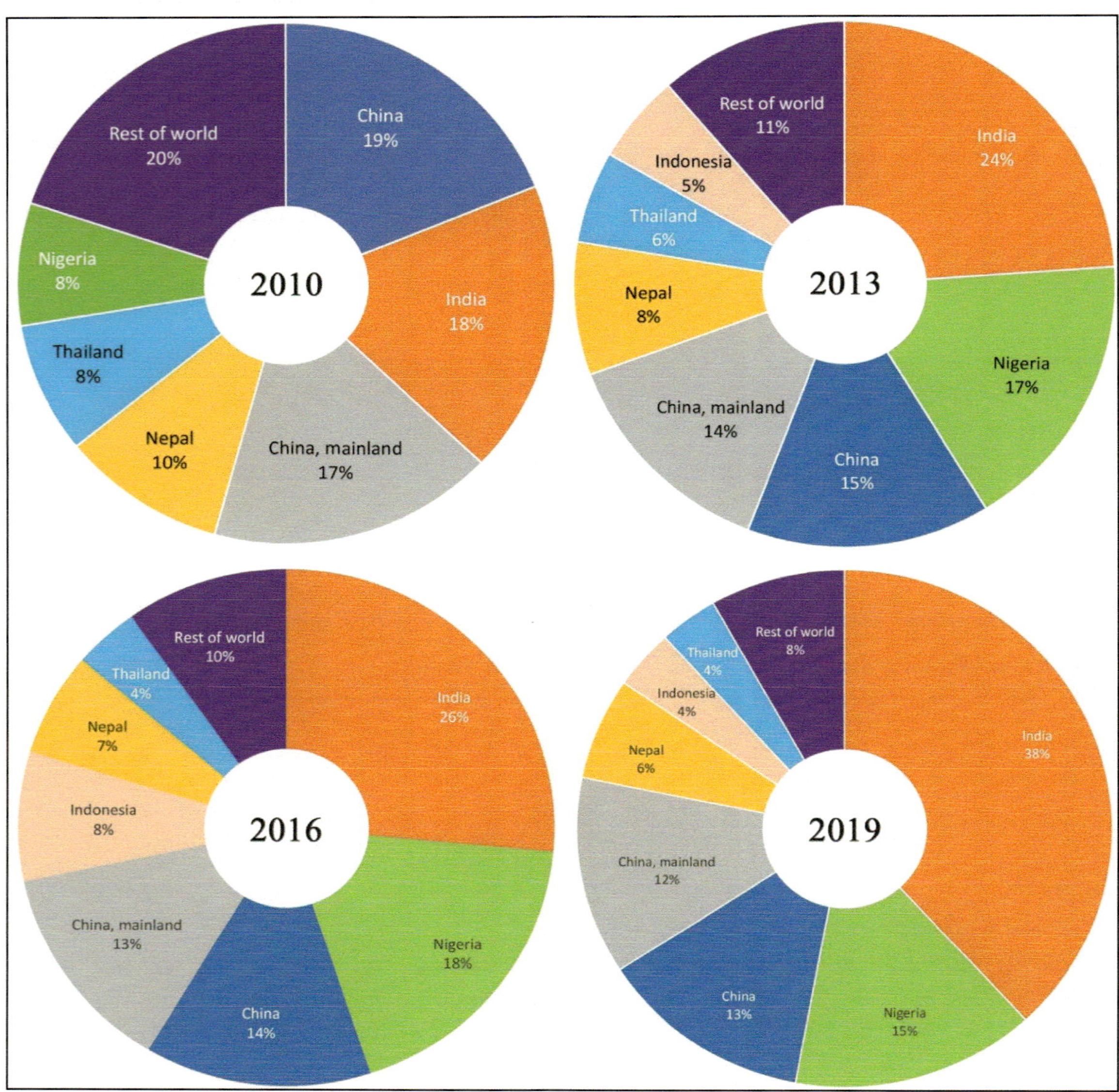

Figure 1. World Productivity of Ginger between 2010 to 2019 (FAO).

Ginger is cultivated in most of the states in India. According to the Department of Agriculture, Cooperation and Farmers' Welfare, Ministry of Agriculture and Farmers' Welfare, Government of India; the area under ginger in India is 160 thousand ha with a production of 1118 thousand tonnes with productivity of 6.99 t/ha during 2017-18. Assam is the leading state with a production of 167.39 thousand tonnes from 18.79 thousand ha and productivity of 8.91 t/ha. Assam, Maharashtra, West Bengal, Gujarat, Kerala, Meghalaya, Mizoram, Karnataka, Arunachal Pradesh and Sikkim are the major producing states and together contributed more than 87 per cent towards the country's total production during 2017-18. Gujarat is leading among all the states with a productivity of 22.23 t/ha (Table 1).

Table 1. Area, Production and Productivity of Ginger in India

States	*2017-18*		
	Area ('000 ha)	*Production ('000 t)*	*Productivity (t/ha)*
Andhra Pradesh	0.51	1.77	3.47
Arunachal Pradesh	7.65	56.58	7.40
Assam	18.79	167.39	8.91
Bihar	0.26	0.49	1.88
Chhattisgarh	1.93	2.36	1.22
Gujarat	4.87	108.25	22.23
Haryana	0.50	6.00	12.00
Himachal Pradesh	2.51	15.95	6.35
Jammu and Kashmir	0.02	0.03	1.50
Karnataka	23.09	58.39	2.53
Kerala	4.37	86.27	19.74

States	2017-18		
	Area ('000 ha)	Production ('000 t)	Productivity (t/ha)
Madhya Pradesh	12.50	20.20	1.62
Maharashtra	9.45	140.60	14.88
Manipur	2.40	3.84	1.60
Meghalaya	9.94	66.20	6.66
Mizoram	8.55	60.13	7.03
Nagaland	4.87	48.65	9.99
Odisha	16.57	31.99	1.93
Rajasthan	0.11	0.27	2.45
Sikkim	12.30	55.90	4.54
Tamil Nadu	0.39	2.64	6.77
Telangana	1.84	12.98	7.05
Tripura	1.88	15.74	8.37
Uttar Pradesh	0.88	4.17	4.74
Uttarakhand	1.81	19.07	10.54
West Bengal	11.99	130.40	10.88
Others	0.16	1.90	11.88
TOTAL	**16.14**	**1118.16**	-

Source: Horticultural Statistics at a Glance 2018, Department of Agriculture, Cooperation and Farmers' Welfare, Government of India.

Production area of ginger is mainly dominated by the North-Eastern and southern states in India. In NER, ginger is generally cultivated in the *jhum* fields in all the hilly states except for Sikkim and mainly as a rain-fed crop. Ginger is grown both as rainfed and irrigated crop in other parts of the country. The details of concentrated pockets of ginger in different states are given below (Table 2).

Table 2. Concentrated Pockets of Ginger in India

State	Districts
Arunachal Pradesh	Dibang valley, East Siang, Lohit, Tirap, Upper Siang
Assam	Borpeta, Darrang, Dibrugarh, Golaghat, Kamrup, Karbi Anglong, N.C.
	Hills, Sivasagar, Sonitpur, Tinsukia
Kerala	Alappuzha, Idukki, Palakkadu, Wayand
Meghalaya	East Garo Hills, Ri-Bhoi, West Garo Hills
Odisha	Kondhamal, Koraput
Sikkim	East Sikkim, South Sikkim

3.0 Major Commercial Varieties

3.1 Local Cultivars and Land Races

Cultivars are usually named in local languages in a geographical region where they are commonly grown and adapted by the farming community. Data presented in Tables 3–5 indicate certain important quality parameters with respect to yield, dry matter, crude fibre, oleoresin and essential oil content of both indigenous and exotic cultivars. The indigenous cultivars are Ernad, Wynad, Kuruppampadi, Thodupuzha, Narasapattam, Bhola, Jati, Maran, Thingpui, Thinglaidum, Tura, Himachal, Karakkal and Nadia. Few exotic varieties such

Table 3. Physico-chemical Characteristics of Local Cultivars/Land Races of Ginger

Cultivar	Fresh Mean Yield (t/ha)	Maturity (Days)	Dry Recovery (per cent)	Crude Fibre (per cent)	Oleoresin (per cent)	Essential Oil (per cent)
Assam	11.78	210	18.0	5.8	7.9	2.2
China	9.50	200	21.0	3.4	7.0	1.9
Himachal	7.27	200	22.1	3.8	5.3	0.5
Maran	25.21	200	20.0	6.1	10.0	1.9
Nadia	28.55	200	22.6	3.9	5.4	1.4
Rio-de-Janerio	17.65	190	20.0	5.6	10.5	2.3

Source: Sasikumar *et al.*, 2009.

Table 4. Physico-chemical Characteristics of some Improved Varieties of Ginger

Cultivar	Mean Yield t/ha	Dry Recovery (%)	Oil Content (%)	Oleoresin (%)	Crude Fibre (%)	Mean Days to Maturity
Himgiri Clonal selection from Himachal Local	14.0	20.6	1.6	4.3	6.0	230
IISR Mahima Selection from germplasm	23.2	23.0	1.7	4.5	3.3	200
IISR Rajitha Selection from germplasm	22.4	19.0	2.4	6.2	4.0	200
IISR Varada Selection from germplasm	22.6	19.5	1.7	6.7	3.3	200
Suprabha Selection from Kunduli local	16.6	20.5	1.9	8.9	4.4	230
Suravi Selection from germplasm	17.5	23.0	2.1	10.2	4.0	225
Suruchi Induced mutant of Rudrapur local	11.6	23.5	2.0	10.0	3.8	220

Source: Ravindran and Babu, 2005.

Table 5. Varietal Characteristics of some Commercially Grown Ginger Varieties

Cultivar	*Characteristics*				
	Rhizome Size	*Skin Colour*	*Pungency*	*Flavour*	*Fibre*
Maran	Medium	Buff	Pungent	High	Less
Nadia	Medium to Bold	Yellowish	Moderately pungent	Mild	Less
Rio-de-Janeiro	Bold	Buff	Pungent	High	Less
Thingpui	Medium	Buff	Pungent	High	Less
Wynad	Bold	Buff	Pungent	Moderate	Less

as Rio-de-Janeiro from Brazil, Taffingiva from Nigeria and the Peoples Republic of China had been introduced in India which later became popular in different parts of the country.

3.2 Improved Varieties

Distinct variations in plant morphology, rhizome characteristics and quality are also observed among the ginger cultivars grown in India. Continuous research efforts by the different organization have led to the development of many improved varieties suitable for growing in different agro-climatic zones of the country. Some of the prominent improved varieties are Suruchi, Surabhi, Suprabha, Himgiri, Manantoddy, Cochin ginger (NUGC), Calicut ginger (NUBK), IISR Varad, IISR Mahima and IISR Rajitha. Table 4 and 5 are show some popular ginger varieties with specific attributes are presented in Table 6.

Table 6. Ginger Cultivars with Specific Attributes

Attributes	*Cultivars*
High yield	Mahima, Rajatha, Suprabha, Suravi and Varda
High dry recovery	Jhorat local, Kuruppampadi, Mahima, Maran and Varda
High oil and oleoresin contents	Ernad, Kunnamangalam and Wynad
High gingerol and shogaol	Ambalavayal, Kunnamangalam, Rio de Janeiro, Sweetning Puri, and Wynad

Kerala is leading in exporting dry ginger 'Cochin ginger' and 'Calicut ginger' or 'Wynad ginger'. Malabar (Kerala) ginger is rich in volatile oil. Assamese ginger and Monantody gingers are rich in oleoresins. The local strains from Pouni and Choki are relatively better in terms of yield effectively resisting disease problems. The strains from Tanda Ganderwan are suitable for preparing '*sonth*' rhizomes.

4.0 Nature and Causes of Postharvest Losses

4.1 Postharvest Diseases

Loss from rot caused by pathogens begins in the field and continues in the packing house, cold storage and refrigerated containers on ships, causing significant losses for the ginger industry. Postharvest disease in ginger is normally due to improper handling during postharvest period which results in injury to the rhizome. Keeping ginger in a storage environment a favourable to pathogens accelerate postharvest decay. Careful harvesting and handling practices, sanitation of the cleaning water, curing of the rhizomes, application of a postharvest fungicide *etc.* can reduce postharvest decay of ginger. During storage of ginger, the dominant pathogenic fungi are *Fusarium oxysporum, Pythium deliense, P. myriotylum,* and other (saprophytic) fungi such as *Geotrichum candidum, Aspergillus flavus, Cladosporium lennissimum, Gliocladium roseum, Graphium album, Mucor racemosus, Stachybotrys sansevieriae, Thanatephorus cucumeris,* and *Verticillium chlamydosporium.*

a) Bacterial Soft Rot

Bacterial soft rot in ginger is caused by *Erwinia carotovora* and is considered the principal bacterial storage disease of ginger. These bacteria enter the ginger rhizome through wounds in the tissue. Symptoms include a soft wet rot of the tissue, which has a strong foul odour. The disease spreads rapidly under warm and humid conditions.

b) Blue Mould

Blue mould, caused by the fungus *Penicillium*, is a very common postharvest rot of ginger when the rhizome are stored at a high RH or in less ventilated store or when come into contact with another mouldy product. Moisture condensation on the rhizome surface encourages blue mould. This is a mild surface infection and is superficial only and does not lead to decay. But it imparts an unacceptable appearance to the rhizomes. However, if the mould develops on cut ends and injured areas, it will typically result in internal decay resulting in pale to dark brown, and firm or soft tissue. The decay is accelerated at temperature between 15 °C and 20 °C.

c) Dry Rot

Dry rot, caused by the soil-borne fungus *Fusarium oxysporum* Schl. is another common postharvest disease of ginger. The affected rhizomes do not show any external symptoms. Browning of internal tissues and yellow colouration occurs in the pith region of the rhizome. The buds on the rhizomes become black. This rot can cause serious problems invading rest of the rhizome that becomes brown, dry and shrivelled.

Its infection appears when rhizomes are wounded and damaged by insects or nematodes. External symptoms of dry rot include off-coloured dry sunken lesions, typically bordered by a brown margin, on the rhizome surface. Under high humid condition, the ginger surface is covered with a dense white mould.

d) Red Rot

The red rot of the rhizome is caused by *Verticillium chlamydosporium* Goddard and shows superficial mycelial growth with the internal tissues turning red.

e) *Sclerotium* Rot

Sclerotium rot, caused by the soil-borne fungus *Sclerotium rolfsii* Sacc. commonly appears on rhizomes harvested from stressed plants. The decay starts with a well-defined separation between rotten and healthy portions of the rhizome. Gradually, a white mould covers the ginger surface and the rhizomes start shrinking. Small spherical white fungal resting bodies, about 1 to 2 mm in diameter, develop on the mould which later turn brown.

f) Soft Rot

The soft rot or *Pythium* rot, caused by *Pythium ultimum* Trow, *P. aphanidermatum* (Edson) Fitzp., *P. vexans* de Bary, *P. deliense* Meurs and *P. myriotylum* Drech., has also been reported in which rhizomes become brown to black, soft and watery. This is a soil-borne fungus that can be severe on ginger harvested during the rainy season or on rhizomes grown in poorly drained soils. The buds become brown with mycelial growth on its surface. As the decay progresses, the tissue breaks down into a soft and watery mass. Fungicides, at this stage, are not recommended, but the disease can be reasonably controlled if the rhizomes are adequately cured and stored at 12-14 °C.

g) Watery Rot

Watery rot, caused by the fungus *Rhizopus*, is one of the most rapidly developing storage rots of ginger. A water-soaked lesion starts to develop a few hours after the root is wounded or bruised during harvest or handling. In 4 to 5 days, the entire inner tissues of the root become mottled, soft, slimy and watery as they liquefy and the infected portion is soon covered with large amounts of white mould which will eventually turn black. *Rhizopus* is a wound pathogen. Soft rot caused by *Rhizopus* has some similarities with bacterial soft rot, as both produce watery, flesh decaying rot and both attract fruit flies. However, bacterial rot produces an unpleasant smell while *Rhizopus* rot emits a pleasant fermented odour that attracts fruit flies. Also, bacterial rot does not produce whiskers, unlike *Rhizopus* soft rot.

h) Postharvest Treatments

Chemical treatments before storage of ginger rhizomes can prevent many of the above rots. Postharvest (prestorage) application of benomyl (750 ppm *a.i.*) and/or gibberellic acid (150 ppm) can be applied to control the rots. A combination of mancozeb and carbendazim treatment to rhizomes controls storage rot of ginger. *Trichoderma viridi* and *T. harzianum* have also been effectively used to control storage diseases of ginger.

4.2 Postharvest Disorders

a) Chilling Injury

Ginger rhizomes should not be stored at a low temperature below 12 °C, as it may cause chilling injury. This injury is manifested in the form of pitting and sunken lesions on the rhizome surface, shrivelling, softening, flesh darkening and decay. Those shorten the shelf life and lower the eating quality of ginger.

b) Sprouting

During storage, ginger rhizomes start sprouting if the temperature goes above 15.6 °C and the rate increases with the increase in temperature. Keeping the rhizomes at 12-14 °C will prevent sprouting. No effective chemical has been reported as sprout inhibitor for ginger. The use of gamma radiation has been so far effective in controlling sprouting. Dipping washed and dried rhizomes in *Trichoderma* suspension (10^8 spores ml^{-1}) followed by air-drying, packing in 250 gauge LDPE bags, and finally exposure to 50-60 Gy gamma radiation can inhibit spouting in the rhizomes.

4.3 Preharvest Crop Management

Various preharvest factors including management practices and environmental factors influence postharvest quality. The quality of the product during the postharvest phase can be ensured if proper preharvest management practices are followed. It is possible to get high market prices, create employment opportunities and sustainable supply only through the use of appropriate variety, improved cultural practices, irrigation, pest and disease management *etc.*

4.3.1 Cultural Practices

Selection of variety is crucial and determines the yield and quality of ginger crop. Thus, one should study all the available varieties critically, based on the anticipated end use of the produce, before selecting a variety for cultivation. It is, thus, essential to encourage producers to use appropriate varieties to get economically rewarding crop, increase income level and make more competition in the markets. The best time for planting ginger on the West Coast of India is during the first fortnight of May as soon as the premonsoon showers start. The usual time to start the crop in the North Eastern states is February-April. Under irrigated conditions, it can be planted well in advance during the middle of January or early February.

There is an increasing demand for organically-grown ginger both in fresh form as well as dried products. A few useful tips for organic ginger are as enumerated below.

i) A buffer zone of 20-25 feet all around from conventional farm should be maintained and a two year conversion period should be adopted.

ii) Tillage operations during land preparation should be kept to minimum.

iii) Soil should be solarised before planting of ginger rhizomes.

iv) Pest and disease free planting material should always be sourced from organically grown ginger belts which must be treated with Trichoderma (@ 2g/kg seed) both before storage and planting.

v) Application of FYM or quality compost (5-10 tonnes), groundnut cake (350 kg), neem cake (1700 kg), rock phosphate (250 kg), bio fertilizer (20 kg) and 1 tonne of wood ash in one hectare area at the time of planting should be practised.

vi) Mulching may be done with green leaves @10 tonnes/ha at the time of planting and @ 5 tonnes/ha each after 40 and 70 days of planting. Cow dung slurry or liquid manure could be poured on beds after mulching.

vii) Neem oil (0.5 per cent) should be sprayed at two week intervals to control insects. The caterpillars of shoot borer can be hand picked and destroyed.

4.3.2 Removal of Mother Rhizome

Traditionally, farmers of North East India in both plains and hills use bigger size rhizomes at the time of planting. They find it economical although this practice initially increases the seed rate to 2.0 – 2.5 t/ha against the normal rate of 1.5t/ha. Most of the farmers of the region remove the mother rhizome (Plate 2) when ginger crop attains 60 days of age or 3-4 leaves stage leaving the sprouted piece of rhizome in the soil. This practice provides the farmer with an additional yield of about 10 q/ha which is sold during off-season at much higher rates. After extraction of the mother rhizomes, FYM is applied and plants are earthed up. It is also believed that the removal of the mother rhizome gives sufficient space to the developing daughter rhizomes for growth.

5.0 Postharvest Management

5.1 Maturity Indices

Traditionally, farmers harvest gingers as soon as the foliage senesces and rhizomes attain considerable size. Gingers are harvested once the foliage begins to turn yellow and starts drying. This also indicates the maturity of rhizomes. This natural process of ageing, accompanied by foliage yellowing, typically sets in 7 to 9 months after planting. Delaying harvest till all of the leaves are dried up is not recommended as it reduces rhizome quality, and storage life, and increases fibre content and incidence of sprouting.

It is best to harvest rhizomes of ginger based on the end use. The rhizomes should be harvested, while they are still tender, medium in pungency and low in fibre content before they are fully mature for consumption as fresh or for making preserves. Thus, green ginger can be harvested right from five months onward. For making dried spices and extracting oils and oleoresins, it is better to harvest ginger at full maturity when the leaves turn yellow; however, harvesting beyond that stage may lead to reduce pungency and oil content with subsequent increase in fibre content in the rhizomes.

As per FAO's guidelines, the best harvest time for fresh consumption is 5 months after planting. For production of ginger preserve and making dried ginger or extraction of ginger oleoresins or essential oils, the rhizomes should be harvested between 5-7 months and 8- 9 months after planting when the leaves start yellowing.

When market demand for ginger is high with a rise in price, farmers occasionally harvest gingers at the premature stage as it seems to be cost effective. In this situation, one should sacrifice some randomly selected plants and dug up to assess the average rhizome size. However, it should be kept in mind that the early harvest of non-senescent plants can result in more weight loss and decay. The skin of early harvested ginger is very tender and fresh wound at the point of separation between the stem and rhizome will take more time to heal.

5.2 Harvesting

Harvesting of ginger in all the ginger producing states of India is carried out from November to January. The rhizomes are harvested by lifting the clumps carefully with a spade or digging fork or hoe by loosening the soil around the crown of the plant. Mechanical digging devices are also available for use on large plantations. Initially, a significant portion of the senescent foliage should be removed to make the rhizomes more accessible. The process should be done carefully to avoid damaging the rhizomes which have delicate skins and are easily wounded during harvest and handling. The rhizomes are then carefully separated from the plant. The rhizomes in a clump are highly irregular in shape and close to each other which favours the inclusion of soil and dirt in the interspaces. These should be removed by thorough washing. If needed, the interspaces are pricked clean using a pointed bamboo stick. The adhering roots are also to be trimmed. After washing, the ginger is lightly dried in the sun or under a hot air blower before being marketed in this form or being used for further processing. Young rhizomes harvested earlier will still have an actively growing stem which should be cut off slightly above the point of attachment to the rhizome. Harvesting during very wet or very dry conditions is not advisable as this will increase the amount of skinning and make digging rhizomes from the soil considerably difficult.

While field cleaning, one should wear cotton gloves for hygiene and to prevent infection. The rhizomes should never be thrown or dropped during postharvest operations. The rhizomes should be graded in the field itself and any unmarketable, damaged, or diseased rhizome should be discarded. The rhizomes should never be exposed to direct sunlight for more than 30 minutes to prevent desiccation. The harvested, cleaned and graded gingers should be placed directly into strong well-ventilated wooden or plastic field crates for transfer from the field. Transporting ginger in mesh sacks will increase the level of skin damage due to rubbing of the delicate skin against the mesh.

5.3 Pretreatments

Gingers are marketed as fresh rhizomes, whole dry form, bleached rhizomes, powder, flakes, paste, oils and oleoresins and other processed forms like preserves, pickles *etc.* Considering the above standards, one should adopt proper postharvest operations to improve the quality parameters of fresh and dry gingers to meet the above requirements.

5.3.1 Precooling

Generally, ginger should be harvested in the morning or the afternoon so that the operation is performed at the coolest possible temperature. The ginger rhizomes, immediately after harvest, should be precooled to remove the field heat. Preferably, forced-air or room-cooling to 12 to 14 °C should be used. In forced air cooling, refrigerated air is forced into the harvested rhizomes kept in boxes or crates. Several airflow systems are available, but tunnel cooler is common. Room cooling of the rhizomes can be done by keeping them in a refrigerated room. Alternatively, if these facilities are not available, ginger rhizomes can be precooled by keeping them spread over a clean concrete floor of an airy room.

5.3.2 Cleaning

Relatively clean rhizomes for sale in the domestic market may not require any further cleaning. However, thorough and proper cleaning is required for gingers intended for export. The rhizomes should also be properly cleaned for long-term storage immediately after harvest and then cured. Ginger should be scrubbed by hand or with a soft-bristled brush in clean water sanitized with 150 ppm hypochlorous acid. The rhizomes are to be washed as soon as possible after harvesting, as the adhering soil is more difficult to remove once dried (Plate 3). Pressure washing using a water jet is commonly advised as it efficiently cleans the rhizomes subsequently reducing the microbial load.

One should be careful enough to prevent rhizome breakage during handling. Broken, split, or bruised rhizomes can easily shrink due to moisture loss and are prone to decay. The pH of the cleaning water is to be maintained at 6.5 for maximising the sanitizing effect of the hypochlorous acid. The water needs to be changed after it builds up with soil particles and the strength of the hypochlorous acid goes down 150 ppm. A fungicide treatment, preferably with benomyl (500 ppm *a.i.*) or thiabendazole (1000 ppm), is also recommended and can be applied in the wash tank or as a separate overhead spray application after cleaning. After the cleaning process, the ginger should be air dried and sorted.

Specific equipment is suggested for optimum cleaning of the dried rhizomes. An air screen separator will help to remove dead insects, excreta and extraneous matter, while a rotary knife cutter with a screen separator will remove residual insects and other extraneous matter. The cleanliness specifications set by American Spice Traders Association are as followed in which the level of contaminants permitted must fall below the values shown (except for 'Whole insects, Dead' which cannot exceed the limit shown) in Table 7:

Table 7. Cleanliness Specifications for Ginger

Parameters	*Value*
Whole insects, Dead (by count)	4.0
Excreta, Mammalian (by mg/lb)	3.0
Excreta, Other (by mg/lb)	3.0
Mould (per cent by weight)	More than 3 per cent mouldy pieces and/or insect infested pieces by weight
Insect defiled/infested (per cent by weight)	More than 3 per cent mouldy pieces and/or insect infested pieces by weight
Extraneous/foreign matter (per cent by weight)	1.0

Source: APEDA 2014

The fresh gingers shall have to comply with the maximum levels of the Codex General Standard for Contaminants and Toxins in Food and Feed (CODEX STAN 193-1995) and the maximum residue limits for pesticides established by the Codex Alimentarius Commission. It is recommended that the produce covered by the provisions of this Standard be prepared and handled as per the appropriate sections of the Recommended International Code of Practice – General Principles of Food Hygiene (CAC/RCP 1-1969), Code of Hygienic Practice for Fresh Fruits and Vegetables (CAC/RCP 53-2003), and other relevant Codex texts such as Codes of Hygienic Practice and Codes of Practice.

The produce should comply with any microbiological criteria established in accordance with the Principles for the Establishment and Application of Microbiological Criteria for Foods (CAC/GL 21-1997). Quality specifications are imposed by the importing country on cleanliness specifications. Hence, suitable postharvest processing care must be taken to meet minimum requirements to avoid rejection. The rhizomes should be disinfected with ethylene oxide or irradiation before export. Before doing so, the exporter must be well-informed about the standards set by the importing country.

5.3.3 Curing

Ginger rhizomes, after cleaning, should be cured by air drying at ambient temperature (22 °C to 26 °C) and 70-75 per cent RH for several days to allow the skin to thicken and the cut surfaces to suberize. Curing helps in reducing postharvest weight loss and decay. After curing treatment, the ginger should be put in well-ventilated containers for long-term storage.

5.4 Sorting

After discarding damaged and injured rhizomes, the marketable portion should be sorted according to size and consumer acceptability. The fresh rhizomes for marketing should be clean, bright yellow brown in colour, and appear fresh without any symptoms of desiccation or sprouting, bacterial or fungal infection and objectionable skin blemishes.

As per the Codex Standards (CODEX STAN 218-1999) outlined by the Codex Alimentarius Commission, fresh whole ginger must be: "whole, firm, sound, produce affected by rotting or deterioration such as to make it unfit for consumption is excluded, clean, practically free of any visible foreign matter, practically free of damage caused by pests affecting the general appearance of the produce, free from abnormal external moisture, and properly dried if washed excluding condensation following removal from cold storage, free of any foreign smell and/or taste, free from abrasions provided light abrasions which have been dried properly are not regarded as a defect, sufficiently dry for the intended use; skin and stems and cuts due to harvesting must be fully dried".

Additionally, the development and condition of the ginger must be such as to enable it to withstand transport and handling and arriving in satisfactory condition at the place of destination.

There are no minimum or maximum size limits for domestic marketing of gingers. However, the minimum rhizome size for export is 150 g with no maximum size limit. The main stem of the rhizome should not be less than 3 cm thick and 12 cm long. Rhizomes should be large, well-formed with less than 4 side branches. The internal flesh should be firm and uniformly coloured, without any indication of darkening.

5.5 Grading

As per Codex Standards, the following classes of fresh gingers are defined as:

a) Extra Class

Ginger in this class must be of superior quality. It must be characteristic of the variety and/or commercial type. The roots must be cleaned, well-shaped and free of defects, except very slight superficial defects, provided these do not affect the general appearance of the produce, the quality, the keeping quality and presentation in the package. Quality tolerance limits to five per cent by number or weight of ginger not satisfying the requirements of the class, but meeting those of Class I or, exceptionally, coming within the tolerances of that class.

b) Class I

Ginger in this class must be of superior quality. It must be characteristic of the variety and/or commercial type. The roots must be cleaned, well-shaped and free of defects, except very slight superficial defects, provided these do not affect the general appearance of the produce, the quality, the keeping quality and presentation in the package. Slight skin defects due to rubbing provided they are healed and dry, and the total surface area affected does not exceed 10 per cent. Quality tolerance limits to ten per cent by number or weight of ginger not satisfying the requirements of the class, but meeting those of Class II or, exceptionally, coming within the tolerances of that class.

c) Class II

This class includes ginger which does not qualify for inclusion in the higher classes, but satisfy the minimum requirements specified above. The roots should be reasonably firm. The following defects, however, may be allowed, provided the ginger retains their essential characteristics as regards quality, the keeping quality and presentation. Skin defects due to rubbing are accepted in this class provided they are healed and dry and the total surface area affected does not exceed 15 per cent. Early signs of sprouting (not more than 10 per cent by weight by unit of presentation), slight markings caused by pests, healed suberized dry cracks, slight traces of soil are permitted in this class. Quality tolerance within this class limits to ten per cent by number or weight of ginger satisfying neither the requirements of the class nor the minimum requirements, with the exception of produce affected by rotting or any other deterioration rendering it unfit for consumption.

Codex Standards have also classified fresh ginger rhizomes into three grades based on size, *viz.*, A – 300g in weight; B – 200g in weight and C – 150g in weight.

According to Indian Standards (BIS, 14), there are six grades for dry gingers *viz.*, *i)* Garbled, non-bleached Calicut (NGK), *ii)* Ungarbled, non-bleached Calicut (NUGK), *iii)* Garbled, non-bleached Cochin (NGC), *iv)* Ungarbled, non-bleached Cochin (NUGC), *v)* Garbled, bleached Cochin (BGC) and *vi)* Ungarbled, bleached Cochin (BUGC).

5.6 Packaging

According to Codex Standards, "Ginger must be packed in such a way as to protect the produce properly. The materials used inside the package must be new, clean, and of a quality such as to avoid causing any external or internal damage to the produce. The use of materials, particularly paper or stamps bearing trade specifications are allowed, provided the printing or labelling has been done with non-toxic ink or glue. Ginger shall be packed in each container in compliance with the Recommended International Code of Practice for Packaging and Transport of Fresh Fruits and Vegetables (CAC/RCP 44-1995)" (Anonymous, 3).

Bulk fresh rhizomes can be packed in jute sacks, plastic crates, wooden boxes or lined corrugated cardboard boxes for domestic sale. According to Indian Standards (BIS, 14), the whole dried ginger should be packed in "sound double jute bags or in single jute bags with waterproof lining". The bags may be lined inside with cloth or with LDPE. The ground ginger should be packed in clean and sound tin plate or glass containers or paper cartons properly lined with waterproof paper or in moisture-proof bags made of a material which does not impart any smell to the ginger ground.

5.7 Storage

5.7.1 Fresh Consumption

Healthy undamaged fresh ginger rhizomes can be stored for several months in a cold and humid environment if the correct postharvest handling and storage procedures are followed. The ideal environment for storing mature ginger rhizomes is 12 to 14 °C with 85 to 90 per cent RH in cold storage. At this condition, the rhizomes remain in marketable condition for at least 3 months, depending on initial rhizome vigour and condition. Condensation over the rhizomes should not be allowed as it can encourage superficial mould growth. Lower RH in the storage environment leads to extensive dehydration and a wilted appearance.

Cold storage may not always be available in the production areas. In that case, a "Pusa zero energy cool chamber", designed by IARI, New Delhi, can be used to preserve fresh ginger, which can lower the storage temperature by 6 to 7 °C than the outside temperature. In the North Eastern states, the model developed by the Assam Agricultural University, a modified version of IARI's "Zero energy cool chamber", is more suitable for ginger storage.

Use of gamma irradiation at 50-60 Gy has been proved beneficial in inhibiting sprouts and extending shelf life of fresh ginger, although loss of volatile compounds have been reported after 5 months of storage. A combination of bio-control with *Trichoderma* sp. and storage in polyethylene bags at 25-30 °C has been reported to control the fungal rot during storage caused by *Sclerotium rolfsii* and prevented weight loss from dehydration.

5.7.2 Seed Purpose

Storage of seed rhizomes is also equally important for good germination in the next season. According to Spices Board of India, bold and healthy rhizomes from disease free plants are to be selected immediately after harvest for storing as seeds. For this purpose, healthy and disease-free clumps are marked in the field when the crop is 6-8 months old and still green. Those selected plants are harvested upon maturity and the seed rhizomes are treated with a solution containing quinalphos 0.075 per cent and mancozeb 0.3 per cent for 30 minutes and dried under shade. After treatment, the seed rhizomes are stored in pits of convenient size under shade. The walls of the pits should be plastered with cow dung paste. The seed rhizomes are placed in pits in layers covered with a 2 cm thick layer of dry sand/sawdust. In this manner, rhizomes are placed in alternating layers in the pit leaving sufficient space at the top for adequate aeration. The pits are then covered with wooden planks keeping one or two small openings for aeration. The seed rhizomes in the pits should be inspected once in three weeks to remove shrivelled and diseased rhizomes. The seed rhizomes can also be stored in pits dug in the ground under shade. Spices Board of India also recommends the incorporation of dried leaves of nux-vomica (*Strychnos nux-vomica*) into the rhizome and sand/sawdust layers to prevent infestation of rhizome scale.

5.8 Transportation

The fresh rhizomes should be transported promptly in clean and dry conditions. When available, refrigerated trailers and van containers should be used to transport gingers if the transit period is more than a week. The gingers are to be placed in clean baskets, dry sacks, trailers, hoppers or other well-aerated containers and carried to a central point for transport to the processing facility. All containers used at harvest should be cleaned regularly, especially before transporting a lot, to avoid contamination from previously-harvested plant products and other foreign matter. The containers should be kept in dry conditions, when not in use, in an area that is protected from insects, rodents, birds and other pests, and inaccessible to livestock and domestic animals. Vehicles used for transporting bulk plant materials from the place of production to storage for processing should be cleaned between loads.

6.0 Domestic Marketing and Export

Generally, producers manually transport ginger and deliver to the collectors. The collectors are responsible to transport gingers to the wholesalers and the wholesalers, in turn, are transporting ginger to the exporters. The exporters are responsible to transport ginger up to the Indian port of loading. Improper handling during loading and unloading of the packages at high temperature from the bus, truck and van adversely affects the quality of ginger. The bumpy roads during transportation aid to the loss of quality of ginger. The Codex Alimentarius Commission has recommended code of practice for packaging and transport of fresh fruits and vegetables (CAC/RCP 44-1995) which applies to ginger also.

India is presently leading ginger production in the world. More than 50 per cent of total ginger production takes place in North East India, Uttarakhand and West Bengal states. In many of these states, ginger is cultivated traditionally in organic environments. Several traditional varieties/cultivars have been used in cultivation in different states which are suitable for producing various products of ginger. In addition to these, ICAR-Indian Institute of Spices Research, ICAR-Institute of Horticultural Research and many state agricultural universities are continuously improving the existing varieties along with developing new, improved and high yielding varieties. Agricultural and Processed Food Products Exports Development Authority (APEDA) has earmarked several Agri Export Zones for fresh ginger in Assam, Orissa and Sikkim. Spices Board of India provides further support for the expansion of area under ginger. All the above facts contribute to strengthen India's position as a strong player in ginger export (Table 8).

The following measures can increase India's competitiveness in ginger exports:

i) India's main competitors in ginger export are China, Indonesia, Nepal and Thailand. India needs to percolate its superior and high yielding cultivars like IISR Varada *etc.* in states like Assam, Mizoram, Meghalaya, Sikkim and Orissa, to bring sizeable area under superior cultivars.

ii) With increased productivity, the cost of production will go down. This will help India to provide better quality ginger at competitive prices.

iii) In the international market, there is a demand for clean and bigger finger sized ginger. The gingers produced in North Eastern states are plump with less fibre and medium pungency having clump size above the Codex size standard for fresh ginger. Thus, India must establish postharvest handling facilities for cleaning, grading, postharvest treatment and appropriate packing of ginger for export.

7.0 Conclusion

Although, India is the leading ginger producing country in the world, the productivity is low compared to competing countries like China. Many constraints are limiting the production and productivity of ginger, chief among them being the farmers' adherence to traditional varieties despite the availability of high yielding varieties, pests and diseases, and improper postharvest management. The rhizome of ginger plant is used both as spice and medicine. India occupies a predominant position in ginger production, contributing 36.5 per cent of the total world production. This is because ginger is grown in almost all the states of India owing to versatility in adoption for climate and soil. The harvesting time of ginger varies depending on location, but largely, ginger is harvested between December to March. However, the demand for fresh ginger remains constant by the consumer throughout the year. The self-life of ginger is limited and moisture loss, sprouting, rotting, and rooting results in weight loss and shrivelling. Packaging of ginger plays a very crucial role in postharvest handling and storage. In India, the normal practice is to pack the ginger rhizomes in 55kg

Table 8: Country-wise Export of Ginger from India

Qty. in MT, Value in Lakhs

Countries	*2009–10*		*2010–11*		*2011–12*		*Per cent Share of Value*
	Qty.	*Value*	*Qty.*	*Value*	*Qty.*	*Value*	
Bangladesh	6276.74	1170.06	12647.51	2997.16	13062.24	2383.54	8.95
Egypt	50.00	48.43	218.02	311.54	728.02	843.88	3.17
Germany	146.03	161.37	297.21	437.14	1375.37	1964.74	7.38
Morocco	280.00	388.53	929.16	1569.42	1361.33	1684.86	6.33
Nepal	233.01	137.40	77.93	60.51	869.11	1083.47	4.07
Saudi Arab	727.09	767.75	997.56	1497.91	1675.60	2028.53	7.62
Spain	348.35	479.45	273.17	481.10	808.08	1032.44	3.88
UAE	414.50	301.32	741.42	790.48	2249.78	2524.77	9.48
United Kingdom	642.40	639.21	646.75	845.23	733.64	1120.72	4.21
USA	904.52	718.38	2302.42	2093.49	4303.35	3963.02	14.88
Other Countries	2152.27	2096.20	5498.94	5162.59	8449.88	7996.78	30.03
Total	**12174.91**	**6908.10**	**24630.09**	**16246.57**	**35616.40**	**26626.75**	**100.00**

Source: APEDA, 2014.

gunny bags during transportation. Hence, use of suitable packaging including MAP may have a great challenge in reduction of huge postharvest losses which sometimes goes up to 50 per cent due to lack of awareness, distance between production and market place, inaccessible road condition, poor transport vehicles *etc.*

8.0 Important References

Ali, B.H., Blunden, G., Tanira, M.O. and Nemmar, A. 2008. Some phytochemical, pharmacological and toxicological properties of ginger (*Zingiber officinale* Roscoe): A review of recent research. *Food Chem. Toxicol.* **46**: 409–20.

Al-Yahya, M.A., Rafatullah, S., Mossa, J.S. and Tariq, M. 1989. Gastroprotective activity of ginger (*Zinigiber officinale* Rosc.) in albino rats. *Am. J. Chin. Med.* **17**: 51-56.

Anonymous. 1995. *Recommended international code of practice for packaging and transport of tropical fresh fruit and vegetables (CAC/RCP 44-1995)*, Codex Alimentarius, World Health Organization, Geneva and, Food and Agriculture Organization of the United Nations, Rome.

Anonymous. 2004. *Ginger -postharvest care and market preparation*, Bull. No. 23. Ministry of Fisheries, Crops and Livestock; National Agricultural Research Institute and New Guyana Marketing Corporation, Georgetown, Guyana.

Anonymous. 2005. *Codex standard for ginger (CODEX STAN 218-1999)*, Codex Alimentarius, World Health Organization, Geneva and, Food and Agriculture Organization of the United Nations, Rome.

Anonymous. 2005. *Herbs, spices and essential oils: Postharvest operations in developing countries*, United Nations Industrial Development Organization, Vienna, Austria and, Food and Agriculture Organization of the United nations, Rome, Italy.

Anonymous. 2014. *Basic Report 11216, Ginger root, raw.* USDA National Nutrient Database, The National Agricultural Library.

Anonymous. 2014. Ginger. http: / / agriexchange.apeda. gov.in / Market per cent 20Profile / one / GINGER. aspx, APEDA AgriXchange, browsed on July 15.

Anonymous. 2014. State / spice wise area and production of spices in India. http: / / www.indianspices.com / pdf / state_prd.pdf, Spices Board of India, Cochin, browsed on July 26.

ASTA. 2007. *ASTA cleanliness specifications for spices, seeds, and herbs (Foreign and domestically produced)*, American Spice Trade Association, Washington, DC 20036, USA.

Azam-Ali, S. 2008. *Ginger processing*, Practical Action, The Schumacher Centre, Bourton-on-Dunsmore, Rugby, Warwickshire, CV23 9QZ, United Kingdom.

Balakrishnan, K.V. 2005. Postharvest and industrial processing of ginger. **In**: *Ginger The Genus Zingiber*, P.N. Ravindran and K.N. Babu (Eds.). CRC Press, Washington, D.C., pp. 391-34.

Baranowski, J.D. 1986. Changes in solids, oleoresin, and (6)-gingerol content of ginger during growth in Hawaii. *HortScience*, **21**: 140-46.

BIS. 2008. Indian Standard: Spices and condiments - Ginger, whole and ground - Specification. https: / / law.resource.org / pub / in / bis / S06 / is.1908.2008. html, Bureau of Indian Standards, New Delhi, India, browsed on July 22, 2014.

Chang, J.S., Wang, K.C., Yeh, C.F., Shieh, D.E. and Chiang, L.C. 2013. Fresh ginger (*Zingiber officinale*) has anti-viral activity against human respiratory syncytial virus in human respiratory tract cell lines. *J. Ethnopharcol.* **145**: 146-51.

DACFW. 2018. Horticultural Statistics at a Glance 2018. Horticulture Statistics Division, Department of Agriculture, Cooperation and Farmers' Welfare, Ministry of Agriculture and Farmers' Welfare, Govt. of India, New Delhi

Davis, C. and Wood, A. 2014. A literature review of the medicinal properties of ginger (*Zingiber officinale* Roscoe). **In**: *New Value-adding Opportunities for Natural Therapeutic Products in the Australian Ginger Industry*, C. Davis and J. Herse (Eds.). Horticulture Australia Ltd, Sydney. NSW 2000, Australia, pp. 59-113.

Dohroo, N.P. 2001. Etiology and management of storage rot of ginger in Himachal Pradesh. *Indian Phytopathol.* **54**: 49-54.

Dohroo, N.P. 2005. Diseases of ginger. **In**: *Ginger The Genus Zingiber*, P.N. Ravindran and K.N. Babu (Eds.). CRC Press, Washington, D.C., pp. 305-340.

Dohroo, N.P., Kansal, S. and Ahluwalia, N. 2012. *Status of soft rot of ginger (Zingiber officinale Roscoe)*, Dr. Y. S. Parmar University of Horticulture and Forestry, Nauni, Solan-173 230 (HP).

FAO. 2021. FAOSTAT. https: / / www.fao.org / faostat / en / #home,, Statistics Division, Food and Agriculture Organization of the United Nation, Rome 00153, Italy, browsed on November 5.

Gao, D. and Zhang, Y. 2010. Comparative antibacterial activities of extracts of dried ginger and processed ginger. *Pharmacog. J.* **2**: 41-44.

Govindarajan, V.S. 1982. Ginger - chemistry, technology, and quality evaluation: Part 1. *Am. J. Chin. Med.*, **17**: 1-96.

Govindarajan, V.S. 1982. Ginger - chemistry, technology, and quality evaluation: Part 2. *Am. J. Chin. Med.* **17**: 189-258.

Haware, M.P. and Joshi, L.K. 1974. Studies on soft rot of ginger from Madhya Pradesh. *Indian Phytopathol.* **27**: 158-61.

Kader, A.A. 2002. Pre and postharvest factors affecting fresh produce quality, nutritional value, and implications for human health. Proceedings of the International Congress on Food Production and the Quality of Life, Sassari, Italy, September 4-8, 2000, 1: 109-19.

Larsen, K., Ibrahim, H., Khaw, S.H. and Saw, L.G. 1999. *Gingers of Peninsular Malaysia and Singapore*, Natural History Publications (Borneo), Kota Kinabalu.

Mukherjee, P.K., Thomas, P. and Raghu, K. 2008. Shelf-life enhancement of fresh ginger rhizomes at ambient temperatures by combination of gamma-irradiation, biocontrol and closed polyethylene bag storage. *Ann. App. Biol.* **127**: 375–84.

Ogogi, T., Minami, M., Kuraishi, Y. and Satoh, M. 1992. Capsaicin-like effect of 6-shogol on substance P-containing primary afferents of rats, a possible mechanism of its analgesic action. *Neuropharm.*, **31**: 1165-69.

Okwuowulu, P.A. and Nnodu, E.C. 1988. Some effects of prestorage chemical treatments and age at harvesting on the storability of fresh ginger rhizomes (*Zingiber officinale* Roscoe). *Trop. Sci.* **28**: 123-25.

Onyenekwe, P.C. 2000. Assessment of oleoresin and gingerol contents in gamma irradiated ginger rhizomes. *Nahrung*, **44**: 130-32.

Paull, R.E. and Chen, C.C. 2014. Ginger. http: //www.ba.ars.usda.gov/hb66/067ginger.pdf, United States Department of Agriculture, Beltsville, MD 20705, USA, browsed on July 5.

Paull, R.E., Chen, N.J. and Goo, T.T.C. 1988. Compositional changes in ginger rhizomes during storage. *J. Am. Soc. Hortic. Sci.* **113**: 584-88.

Plotto, A. 2002. *Ginger - Postharvest operations*, Food and Agriculture Organization of the United Nations (FAO), AGST.

Premavalli, K.S. 2005. Ginger as a spice and flavourant. **In**: *Ginger The Genus Zingiber*, P.N. Ravindran and K.N. Babu (Eds.). CRC Press, Washington, D.C., pp. 509-26.

Pruthi, J.S. 1980. *Spices and Condiments: Chemistry, Microbiology and Technology*. Academic Press, New York.

Pruthi, J.S. 1992. Postharvest technology of spices: Pretreatments, curing, cleaning, grading and packing. *J. Spices Aromatic Crops*, **1**: 1-29.

Purseglove, J.W., Brown, *E.G.* and Green, C.L. 1981. *Spices*, **Vol. 1**. Longman, London and New York.

Rahman, H., Karuppaiyan, R., Kishore, K. and Denzongpa, R. 2009. Traditional practices of ginger cultivation in Northeast India. *Indian J. Tradit. Knowl.* **8**: 23-28.

Rai, S. and Anitha, G. 1999. Mother rhizome extraction of ginger (*Zingiber officinale* Roscoe) - an age old practice in Sikkim and Darjeeling hills. *Envir. Ecol.* **4**: 910-12.

Ravindran, P.N. and Babu, K.N. 2005. *Ginger The Genus Zingiber*. CRC Press, Washington, D.C.

Ravindran, P.N. and Babu, K.N. 2005. Introduction. **In**: *Ginger The Genus Zingiber*, P.N. Ravindran and K.N. Babu (Eds.). CRC Press, Washington, D.C., pp. 1-14.

Saikia, S. 1999. *Assessment of Growth Performance and Characterization of Products of Some Ginger (Zingiber officinale Rosc.) cultivars. Ph.D. Thesis* submitted to Assam Agricultural University, Jorhat, Assam.

Sasikumar, B., Thankamani, C.K., Srinivasan, V., Devasahayam, S., Santhosh, J., Eapen, A.K. and John, T.Z. 2009. *Ginger (Extension Pamphlet)*, Spices Board, Ministry of Commerce and Industry, Cochin - 682 025.

Sharma, S.K. and Dohroo, N.P. 1991. Postharvest management of rhizome rot *(Fusarium oxysporum* f.sp. *zingiberi* Trujillo) of ginger through chemical and antagonist. *Indian Cocoa Arec. Spices J.* **14**: 150-52

Shoji, N., Iwasa, A., Takemoto, Y.I. and Ohizumi, Y. 1982. Carditonic principles of ginger. *J. Pharm. Sci.* **71**: 1174-75.

Singh, H.B. 2013. *An Assessment of Pre and Postharvest Factors Affecting Quality of Ginger in The Export Chain - A Case Study of Salyan District and Nepalgunj City, Mid-Western Development Region, Nepal.* Master of Agricultural Production Chain Management thesis submitted to Van Hall Larenstein University of Applied Sciences, Wageningen, The Netherlands.

Soepadmo, E. 1976. Ginger plants. *Nat. Malaysia,* **1**: 32-39.

Srinivasan, K. and Sambaiah, K. 1991. The effect of spices on cholesterol 7-alpha-hydroxylase activity and on serum and hepatic cholesterol levels in the rats. *Int. J. Vitam. Nutr. Res.* **61**: 364-69.

Srivastava, K.C. and Mustafa, T. 1989. Ginger (*Zingiber officinale* Rosc.) and rheumatic disorders. *Med. Hypotheses,* **29**: 25-28.

Stoilova, I., Krastanov, A., Stoyanova, A., Denev, P. and Gargova, S. 2007. Antioxidant activity of a ginger extract (*Zingiber officinale*). *Food Chem.* **102**: 764-70.

Sutarno, H., Hadad, E.A. and Brink, M. 1999. *Zingiber officinale* Roscoe. **In**: *Plant Resources of South-East Asia, No 13 - Spices*, C.C. de-Guzman and J.S. Siemonsma (Eds.). Backhuys Publishers, Leiden, The Netherlands, pp. 239-44.

Teakle, D.S. 1965. Fusarium rhizome rot of ginger in Queensland. *Queensland J. Agric. Anim. Sci.* **22**: 263-72.

Trikarunasawat, C. 2008. Biological Control of Rhizome Rot of Ginger during Storage by Antagonistic Microorganisms and Medicinal Plant Extracts. Doctor of Philosophy (Plant Pathology). *Ph.D.* Thesis submitted to Kasetsart University, Bangkok 10900, Thailand.

Trujillo, E.E. 1963. Fusarium yellows and rhizome rot of common ginger. *Phytopathol.* **53**: 1370-71.

Wang, H. and Ng, T.B. 2005. An antifungal protein from ginger rhizomes. *Biochem. Biophys. Res. Comm.,* **336**: 100-04.

Weiss, E.A. 1997. *Essential Oil Crops.* CAB International Publishing, Oxon, UK.

Weiss, E.A. 2002. *Spice Crops.* CAB International Publishing, Oxon, UK.

Wu, J.J. and Yang, J.S. 1994. Effect of irradiation on the volatile compounds of ginger rhizome (*Zingiber officinale* Roscoe). *J. Agric. Food Chem.* **42**: 2574-77.

Chapter 31

Garlic

K.E. Lawande

Ex-Vice Chancellor, Dr. Balasaheb Sawant Konkan Krishi Vidyapeeth, Tal. Dapoli, Dist. Ratnagiri – 415 712, Maharashtra, India

P.C. Tripathi

Principal Scientist, Indian Institute of Horticultural Research, Bangalore – 560 089, Karnataka, India

E-mail: kelawande2011@gmail.com; E-mail: prakaashtripathii2000@yahoo.co.in

1.0 Introduction

Garlic (*Allium sativum* Linn.) is used around the world for its pungent flavour as a seasoning or condiment. It is a major constituent of many Ayurvedic medicines. Animal studies, and some early research studies in humans, have suggested possible cardiovascular benefits of garlic. Many studies found garlic supplementation reduced the accumulation of cholesterol on the vascular walls of animals and in humans. Garlic is produced only in rabi season in plains. It comes from MP in February and continues up to March-April. In Southern hills particularly Nilgiri hills garlic is cultivated in two seasons *i.e.* March-April to July-August and sometimes also in October to March-April. In Northern hills, it is cultivated in one season *i.e.* October to June.

2.0 Status of Cultivation

India ranks second in the area and production of garlic in the world after China. In India garlic is grown on over 3.45 lakh hectares with a production of 29.18 lakh tonnes (2018-19; *Source*: www.dasd.gov.in). Madhya Pradesh and Rajasthan are the leading states in India contributing more than 45 per cent of both area and production. The other major garlic growing states are Rajasthan, Uttar Pradesh and Maharashtra. Garlic is a traditional exportable crop of India, however, export from India is not stable. Garlic export from India is largely influenced by the production and the domestic demands of the country. Currently, small clove garlic is exported for consumption by ethnic Indian populations in the Middle East and South East Asian countries *viz.*, Bangladesh, Indonesia, Philippines and Gulf countries. A large quantity of big clove garlic is also imported from China and other countries when domestic production is low.

3.0 Major Commercial Varieties

More than 20 varieties of garlic have been released from different research organizations. The varieties Godavari and Sweta from at MPKV, Rahuri, HG-1 and HG-2 at HAU, Haryana, Pusa Selection from IARI, New Delhi, LCG-1 from Ludhiana, ARU-52, Almora, VL-6 and VL-7 from VPKAS, Almora, Agrifound white, Yamuna safed-1, 2,3, 4, 5 and Agrifound Pravati, Agrifound Parvarti-2 from NHRDF, Nasik, Bhima Omkar, Bhima Purple, DOGR, Rajgurunagar were released during last 30 years. Most of these varieties have smaller bulbs and numbers of cloves per bulb are more than 20 cloves. Some small clove varieties released have purple colour while others are white. These varieties are suitable for export purpose for those countries which demand small clove garlic. The export of selected bigger size bulbs (more than 40 mm) of small cloved landraces is done from Madhya Pradesh. There are a few bigger cloved cultivars released for cultivation. Yamuna Safed-3 (G-282) and Yamuna Safed-5 are released for plains and AgrifoundParvati (G-313), Agrifound Parvati -2 are released for Northern hills. Big clove garlic varieties are cultivated in Uttarakhand, Himachal Pradesh and

Jammu and Kashmir. Some of these varieties can also be cultivated in Kodaikanal and Mettupalayam in Tamil Nadu. The bulbs of these varieties are white and 5.0 -6.0 cm in diameter and the number of cloves per bulb is less than 20 and have a good shelf life. These are suitable for export.

4.0 Nature and Causes of Postharvest Losses

During the period of storage, 30-40 per cent losses occur mainly due to loss of moisture, sprouting and rotting. More than 95 per cent produce of garlic arrives in the market in March to May. But there is a constant demand for garlic in the market as it is required daily in small quantities in almost all houses. Further, there is glut in the market during the harvest season. As a result, farmers do not get the appropriate prices and they are forced to sell their produce at a lower price. Thus it becomes essential to store garlic for regular supply in the market and to stabilize the price. It is estimated that 8-9 lakh tonnes of garlic need to be stored to fulfil the domestic and export demand. Generally, management of storage losses is initiated after the harvest of the crops. There are many pre harvest aspects which affect the storage of garlic. These include genetic makeup of cultivars, type of manure and fertilizer and their time of application and quality, quantity, method and time of application of water, pre harvest treatments, harvesting time *etc.* Postharvest factors include field curing, drying, store environment, packaging, transport *etc.*

5.0 Postharvest Management

5.1 Harvest Maturity

The change in leaf colour is the sign of maturity in garlic. The stage of maturity of cloves at harvest has a significant influence on dormancy. As the bulbs reach maturity, the plants cease to produce new leaves and senescence starts. The tops just above the neck of the bulbs weaken and bend down still remain green. Further, there is development of bulbils in the pseudostem just above the ground. This is also an indication of maturity in tropical type of garlic. Bulbs continue to increase in size and weight for sometime after softening of the neck fall. This increase is due to continued accumulation of dry matter in the bulb. Bulbs harvested after neck fall have been reported to take less time for drying and curing. Garlic harvested when the leaves are still green, require more time to sprout than mature bulbs harvested when the foliage has dried. It has also been reported that harvesting garlic bulbs before the tops have fallen over indicating full maturity, results in poor storage whereas delayed harvesting results in splitting of bulbs and re-sprouting of bulbs.

5.2 Harvesting

Harvesting is done manually by uprooting. Some time the hand hoe is used in hard soil for harvesting of garlic but it increased the chances of damage of the bulbs. These bulbs get disease infection and exhibits more losses in storage. After harvesting the bulbs are is left in the field for 3-4 days for removal of excess surface moisture and drying of leaves. This allows the formation of strong outer protective scales and hardening of neck. Garlic bulbs, after harvest, are cured naturally in the field by windrowing in such a way that the bulbs of one row are protected from the sun by the foliage of the adjacent row. After field curing, the bulbs are cured in a ventilated shed for 7 to 15 days either with tops or after cutting the tops leaving about 2.5 cm neck intact and removing the roots. These bulbs are than graded as per size.

5.3 Curing

It is extremely essential that the garlic bulbs destined for storage should be well cured in the field. The garlic bulbs should be provided with adequate air circulation through the storage containers to remove the transpired moisture. Incomplete curing results in rapid decay particularly when the bulbs are held above 0 °C.

5.4 Sorting and Grading

After removal of leaves the damaged, sprouted and diseased bulbs are removed. Then the bulbs are graded as per size. The size of bulbs varies in different varieties or land races. The grading is mostly done manually but now size graders are available in India. The details of grade designation and definition of different qualities of garlic are mentioned in Table 1.

Table 1. Grade Designations and Definitions of Quality of Garlic

Grade	*Size (Minimum diameter in mm)*			
	Pune/Nasik Garlic	*Madurai Plain Garlic*	*Nilgiri Hill Garlic*	*Jamnagar Garlic*
Big	30	20	35	35
Medium	25	15	25	25
small	10	10	10	10

5.4.1 General Characteristics

i) The bulbs shall be reasonably uniform in colour, shape and pungency characteristics of the variety type.

ii) The bulbs shall be well cured and dried.

iii) Bulbs shall be compact with cloves well filled and fairly plump.

iv) Bulbs shall be free from damage caused by moulds, decay, split cloves, dirt, sun burn, dust, sprouting *etc.*

The tolerance for size is such that incidental error in sizing is not more than 5 per cent by weight of the next lower grade. The tolerance for requirement with respect of general characteristics is allowed for variations of other than size. Incidental error due to improper grading and handling not more than 10 per cent by weight of garlic in any containers may be below the requirement specified under general characteristics but not more than a total of 1/5th of this tolerance or 2 per cent shall be allowed for garlic which is affected by decay.

5.5 Packaging

5.5.1 Domestic Market

Garlic is packed in Hessian cloth bags/jute bags of varying capacity for domestic market. Recently nylon net bags or leno bags are used for domestic market. Garlic is packed in 90kg packing in Andhra Pradesh. But generally, 40-60 kg capacity jute bags are used for domestic marketing in Karnataka, Maharashtra, Gujarat, Madhya Pradesh and other garlic growing states. Packaging in small size bags 5 to 25 kg capacity is also gaining popularity in domestic market.

5.5.2 Export Market

Garlic grading and packing rules provide for packing size of 18 kg, 25 kg and 50 kg. For 50 kg packaging jute bags are preferred and 18 kg and 25 kg packing is done in perforated corrugated cardboard boxes for export. The packages are labelled with information about variety, weight, the address of the supplier and other details. The strength of the jute bags used is less. Due to this there is loss during various processes of handling many times. Preference for plastic woven bags is increasing due to their attractiveness and easy visibility of the quality of garlic in the bags. Although in Europe and other Western country's preference is given for jute bags because of disposal problem in plastic woven bags.

5.6 Storage

Garlic is stored for about 7 to 8 months with periodical release for domestic and export purposes. Since garlic is produced mostly in March and April, it has to be stored for augmenting continuous supply to the market. The storage of garlic besides genotypic factors and pre harvest management, largely depends on storage conditions. There are two distinct temperature regimes where storage losses are minimum. One is high temperature regime, where storage temperature is 25-30 °C and another is low temperature regime, where temperature is augmented maintained 0 to 2 °C. Best results under both the temperature regimes are obtained when humidity is maintained at 65 to 70 per cent. The storage losses in high temperature conditions (25-30 °C) are high (20-30 per cent) but storage cost is low, while in low temperature conditions (0-2 °C) losses are minimum (0.5 per cent) and storage period is longer. However, storage cost is high. High temperature storage (more than 30 °C) at ambient leads to high weight loss while storage at a lower temperature (less than 10 °C) enhances sprouting losses. High humidity (more than 70 per cent) coupled with high temperature enhances storage diseases, while low humidity enhances weight loss in storage.

Garlic should be stored with leaves hung by their tops or stored in circular heaps showing minimum losses at high temperature regime. This practice is adopted for garlic storage in most of the garlic growing areas of the country particularly for domestic consumption and seed purpose. The storage of garlic with dried foliage in bundles tightly heaped in ventilated storage structures is done at NRCOG and also by farmers in Nasik/Pune area which gives low storage losses. However, the presence of leaves for marketing and large scale storage is undesirables as it affects the grading, packing and marketing. Thus the leaves of garlic are removed after curing or sometime after on farm storage. The garlic storage structures should be designed in such a way that the optimum conditions of temperature and moisture maintained. In India, cold storage of garlic is not done commercially because of several reasons and almost all garlic is stored at room temperature conditions. Various types of storage structures have been designed and tested for garlic storage. Among these structures, bottom and side ventilation of 10, 25, and 50 tons capacity are recommended. Low volume structures are with single row, while high volume ones are with double rows with a gangway between two rows. Studies on width and height of heaps revealed that 120 cm staking width and 150 cm of height are most appropriate. Any increase beyond this recommendation increases storage losses.

Garlic can be stored in cold store at 0-2 °C temperature and 65-70 per cent relative humidity from May to January. The results revealed that total losses in cold storage were significantly lower (11.5 per cent) than at ambient condition (33.9 per cent) after 5 months storage. Physiological weight loss, diseases and drying of cloves were lower in cold store than at ambient. However, when these cold stored bulbs were taken out and kept at ambient storage conditions, rapid sprouting was observed and it was 13.3 per cent after 20 days and further increased. The post cold storage induced sprouting can be prevented with irradiation of garlic before storage. Irradiation with 2-6 krad of Cobalt 60 gamma rays have also been recommended for controlling sprouting in storage. The irradiation given to the bulbs within 8 weeks of harvest inhibited sprouting effectively.

6.0 Transportation

Harvested garlic is transported to the collection centre in open bullock carts or tractor trolleys. It is also brought in jute bags or nylon netted bags in trucks. While loading bags in trucks, care should be taken in loading up to 6-8 bags height of stacking as overloading will damage the garlic bulbs in the bottom layer of bags by splitting. The split bulbs do not get good rates in market and increase the chances of rot. Garlic bulbs are generally sold to local whole sellers after reaching collection centres. These transactions usually take place with the help of commission agents. The exporters either buy their requirements directly from collection centres through local commission agents or from whole sellers and then transport to ports by trucks after grading and packing.

7.0 Domestic Marketing and Export

Garlic is a traditional export item. In 1961-62, India exported about 2000 tonnes of garlic. Garlic is exported to Qatar, Saudi Arabia, United Arab Emirates, Bahrain, Mauritius, Kuwait, Bangladesh and Sri Lanka. The export has been ranging from 0.10-3.0 per cent of domestic production. Exporters of garlic purchase their requirement from important assembly markets in Gujarat, Madhya Pradesh, Tamil Nadu, Karnataka and Uttar Pradesh. The export of garlic, has also gone up from 2000 tonnes in 1961-62 to about 23,000 tonnes during 2012-13. The export of garlic, however, came down to 578 tonnes during 2002-03 and 182 tonnes in 2007-08 due to less availability. The import of garlic was very less till 1997 because of restricted imports of garlic with specific licenses. However, Govt. of India has lifted the restrictions. Thereafter import has substantially increased in recent years. Apart from fresh garlic, garlic products are also imported (Table 2).

Table 2. Export and Import of Garlic in India (1998–2018)

Year	*Export (Tonnes)*	*Import (Tonnes)*
1998-99	3978.00	5189
1999-00	8542.00	11636
2000-01	4442.55	2984
2001-02	657.00	36187
2002-03	578.00	40576
2003-04	1567.23	18039
2004-05	2236.47	19534
2005-06	32495.24	–
2006-07	12210.79	–
2007-08	182.47	–
2008-09	858.69	–
2009-10	19767.32	–
2010-11	21827.16	–
2011-12	1325.60	–
2012-13	22665.00	–
2013-14	23,869.27	–
2014-15	12550.56	-
2015-16	15440.59	-
2016-17	18910.01	-
2017-18	26527.72	594.0

7.1 Export Requirement

In earlier times the export demand for Bangladesh, Sri Lanka and Middle East countries were of small cloved types. However, bigger cloved garlic bulbs are now in demand from all over the world. Preference is for garlic bulbs having 40-60 mm diameter with 10-15 cloves in each bulb. Philippines still require smaller cloved varieties. Govt. of India has prescribed certain grade designation for different qualities of garlic for export.

7.2 Non-availability of Export Worthy Stocks

There is not much organised production of garlic for export. There is no uniformity in size, shape or colour. There is splitting of the bulbs due to delayed harvesting or other factors. Adequate care in harvesting and sorting at farmer's level is not taken. All these lead to the availability of poor quality stocks at the market yard. Most of the varieties of garlic were released during the last 30 years and still sizable area is under land races and variety wise material is not available for export

7.3 High Domestic Demand and High Ruling Prices

The domestic demand for garlic is high. Garlic crop is mainly produced in those areas which are prone to drought which affects production. The failure of crop in those years causes shortage in supply for the domestic market and prices of garlic in the domestic market goes up and usually little remain left for export. This creates problem in regular supply of garlic for export. Due to this, many times it is not possible to compete with other countries resulting in reduced export.

7.4 Lack of Proper Postharvest Practices and Quality

New exporters in garlic export business are not aware of proper postharvest practices and also quality standards for export. Many exporters do not have adequate facilities for handling the produce and sorting, grading and packing are done with the help of local packers where precautions for proper checking about quality, weight *etc.* are not taken properly by the packer as well as exporter. Appropriate sorting and grading is not practiced as per the requirement of the importers. This does not only affect export but hampers the image of the country. The stocks of garlic in many times kept in the open sun by traders and exporters which results in sun scalding. Many times farmers bring the stocks without proper curing. If such stocks are packed without removing field heat, the chances of rot increase during transit. The hooks used during loading and unloading on the trucks or in wagons also injure the bulbs. High temperature, high humidity and improper ventilation during transit increase respiration rate which results in rotting of the bulbs.

7.5 Inadequate Transport Facilities

Transportation from producing centres to port towns is done by rail and road both. Rail transport though is cheaper but takes a longer time. Sometimes wagons are also not available. Further, there is no ventilation in the wagons. Road transport though takes lesser time it is costlier. Adequate trucks at a time are also not available. In case of rail transport sometimes it is difficult to synchronise the movement of the stock by rail with the sailing schedule of vessel. Speedy rail transport with insulated wagons will lessen the cost of transportation.

7.6 Non-availability of Adequate Storage and Handling Facilities

At ports, many times storage is to be arranged before the vessel commences loading. Garlic bags are stacked in sheds without any ventilation and also stacking height is very high. While handling, the labourers sit on the bags or walk on the same. Further, there are no suitable vessels for carrying garlic from Indian ports. It has also been observed that even for loading, the vessels have to wait for getting berth like general cargo vessels resulting in inordinate delays in loading. Vessels do not directly go to the final destination but touch many other destinations as a result more time is involved thereby bulbs are damaged. Now containerised shipments are preferred for door to door delivery. The problem in these containers is that there is no system of ventilation. Due to this, many times losses are more. Modified containers with proper ventilation are required to decrease losses during shipment.

7.7 Inadequate Market Intelligence

Accurate information is not available at a time about price of different places internally or internationally so also the prospect of production within the country and abroad. Because of this many times, it is not possible to get the stocks of particular quality at a competitive price. Because of this uncertainty many times commitments made at a time are not possible to be fulfilled.

7.8 Sanitary and Phyto-sanitary Requirements for Export

The garlic grades and qualities are as follows:

a) General Characteristics

Garlic shall be obtained from varieties *Allium sativum* L. It may be fresh, semidry or dry. Garlic bulbs shall be (a) sound, fresh in appearance; (b) clean, free of any visible foreign matter; (c) firm; (d) free from damage caused by pests; (e) free from damage caused by frost or sun; (f) free from externally visible sprouts; (g) free of abnormal external moisture; (h) free of any foreign smell and/or taste. They shall comply with the residue level of heavy metals, pesticides and other food safety parameters as laid down by the Codex Alimentarius Commission for exports.

b) Criteria for Grade Designation

The grade designation requirement and grade tolerances are given in Table 3.

c) Other Requirements

The development and condition of the garlic must be such as to enable it to withstand transport and handling and arrive in satisfactory condition at the place of destination. Garlic must be presented as loose in the package, with cut stems. The length of the stem should not exceed10 cm. in the case of fresh and semi-dry garlic and 3 cm in case of dry garlic.When the garlic are presented in bunches it should indicate the number of

Table 3: Grade Designation and Requirements

Grade Designation	*Grade Requirements*	*Grade Tolerances*
Extra class	Garlic shall be of superior quality. They shall be characteristic of the variety and/or commercial type. The bulbs shall be intact; regular in shape; properly cleaned. They must be free of defects, except very slight superficial blemishes, provided these do not affect the general appearance of the produce, the quality, the keeping quality and presentation of the package. The cloves must be compact; The roots of dry garlic must be cut off flush with the bulb.	5 per cent by weight of Garlic not satisfying the requirements of the grade, but meeting those of Class I grade or, exceptionally, coming within the tolerances of that grade.
Class I	Garlic shall be of good quality. They shall be characteristics of the variety and/or commercial type. The bulbs shall be intact; of fairly regular shape The following slight defects, however, may be allowed provided these do not affect the general appearance of the produce, the quality, the keeping quality and presentation in the package. Slight tears in the outer skin of the bulb; cloves must be reasonably compact.	10 per cent by weight of Garlic not satisfying the requirements of the grade, but meeting those of Class II or, exceptionally, coming within the tolerances of that grade.
Class II	Garlic which does not qualify for inclusion in the higher grade, but satisfies the minimum requirements. The following defects may be allowed, provided the garlic retains its essential characteristics keeping quality and presentation. Tears in the outer skin or missing parts of the outer skin of the bulb; Healed injuries; Slight bruises; Irregular shape; Up to three cloves missing.	10 per cent by weight of Garlic not satisfying the requirements of the grade but meeting the minimum requirements.

bulbs, net weight and the stems must be evened off. In the case of dry and semi-dry produce the garlic may be presented in strings indicating number of bulbs. There should be at least six bulbs per string and mention of net weight. In the case of presentation in bunches or strings, each package must have uniform characteristics (number of bulbs or net weight). Irrespective of the type of presentation, the stems must be cut cleanly. Package shall be marked 'fresh garlic', 'semi-dry garlic' or 'dry garlic'.

Size is determined by the maximum diameter of the equatorial section. The minimum diameter shall be at 45 mm for garlic classified in the 'Extra' Class and 30 mm for garlic classified in Classes I and II. Where the diameter of the smallest bulb is less than 40 mm for garlic presented loose with cut stems or in bunches, the difference in diameter between the smallest and the largest bulb in the same package may not exceed15 mm. Where the diameter of the smallest bulb is than 40 mm or more this differences should not exceed 20mm. Maximum permissible limit for variation for all grades of produce is 10 per cent by weight of bulbs not satisfying the requirements as regards sizing and the size indicated but conforming to the size immediately above and/or below that specified. Within these tolerances, not more than 3 per cent of bulbs may have a diameter smaller than the specified minimum but not less than 25 mm. Garlic bulb without clove (single) may have minimum diameter of 20 mm for Extra class and 15 mm for Class I and Class II.

8.0 Conclusion

Appropriate postharvest management of garlic is essential for minimizing the losses during marketing and storage. Sprouting, rooting and decay are some of the major factors responsible for huge postharvest losses. Hence, care must be taken right from harvesting at appropriate maturity, curing for removal of moisture from outer skin, packaging and storing at 0 °C with 65-75 per cent RH for long duration. However, for short term storage garlic could be stored at 27-32 °C with low RH. Similar to onion gamma irradiation of well cured garlic after harvest is useful for minimizing the storage losses due to sprouting and rooting. These treatments are also effective in checking discolouration during storage. Proper grading and use of SPS protocol are necessary for fetching better price of the produce in the export market.

9.0 Important References

Anonymous 2019. Directorate of Areca nut and Spices Development (DASD) , Calicut.

Bhonde, S R. and Hari Prakash 2006. Garlic cultivation in India, NHRDF, Nashik. *Tech. Bull.* **17**L 50.

Bonginwar, D.R. and Shirsat, S.G. 2000. Demonstration Food irradiation facilities for prevention of losses due to sprouting in onion/garlic by use of gamma irradiation at Lasalgoan in Nasik district of Maharashtra. *National Symposium on Onion-Garlic: Production and Postharvest Management.* In: *Challenges and Strategies. Souvenir* pp. 33-42.

Kendler, Barry S. 1987. Garlic (*Allium sativum*) and onion (*Allium cepa*): A review of their relationship to cardiovascular disease. *Preventive Medicine*, **16**: 670–85. doi: 10.1016/0091-7435(87)90050-8. PMID 3317392.

Keys, Ancel1980. Wine, garlic, and CHD in seven countries. *The Lancet*, **315**: 145–46. doi: 10.1016/S0140-6736(80)90620-0. PMID 6101471.

Lawande K.E. and Tripathi, P.C.2006.Marketing of irradiated onion and garlic. *Proc. of National symposium on radiation processing of foods, food products and feeds. Hyderabad*, pp. 82-90.

Tripathi P.C. and Lawande, K.E. 2004. Pyajavam Lahsun Bhandaran. *Tech. Bull.*, **10** Published by NRC Onion and Garlic, Rajgurunagar pp. 20.

Tripathi, P. C. and Lawande, K. E. 2009. Effect of storage environment and packing material on storage losses in garlic. *Indian J. Hortic.* **66**: 511-15.

Tripathi, P. C. and Lawande, K.E. 2007. Effect of cold storage and gamma-irradiation on storage losses in garlic. *J. Spic. Aromatc. Pl.*, **17**: 23-26.

Chapter 32

Cut Flowers

Kushal Singh

Senior Plant Physiologist

Department of Floriculture and Landscaping, Punjab Agricultural University, Ludhiana – 141 004, Punjab, India

Ramesh Kumar

Former Director, Directorate of Floricultural Research,
ICAR-Indian Agricultural Research Institute, New Delhi – 110 012, India

E-mail: kushal.flori@pau.edu; E-mail: directordfr@gmail.com

1.0 Introduction

Cut flowers are an important component of domestic as well as international floriculture trade. Cut flowers are used for decoration in vases, flower arrangements, in bouquets, for decoration at auspicious occasions and as wreaths. The demand for cut flowers in the global market is increasing at the rate of 10-15 per cent each year. Important flowers which dominate the global trade include rose, carnation, chrysanthemum, orchids, tulip, gerbera, anthurium, lilies, gypsophila, bird of paradise, alstroemeria *etc.*

The cut flower trade is highly competitive and requires high quality products. Since flowers are highly specific in their temperature requirements, most cut flower crops are cultivated under protected conditions where the temperature conditions and nutrition is precisely controlled. Due to increase in labour costs and freezing temperatures during winter, the cultivation of cut flowers in Western Europe has become cost intensive. As a consequence, cut flower production has shifted to developing countries.

The varied climate of India is highly suitable for growing cut flowers throughout the year through the selection of the flower type and agro-climatic zone in the country.

2.0 Status of Cultivation

As per the NHB database on horticulture, the total area under flower crops in India was 3 lakh hectares with total production of 762 000 MT in the year 2019-20. The leading cut flower producing states are West Bengal (28 per cent), Karnataka (16.3 per cent), Chhatisgarh (10 per cent), Uttar Pradesh (9.09 per cent), and Assam (7.65 per cent). Maximum production of cut flowers in India comprises of rose followed by tuberose, gladiolus, gerbera, carnation, anthurium, orchids *etc.*

3.0 Major Commercial Varieties

The important floricultural crops in the international cut flower trade are rose, carnation, chrysanthemum, gerbera, gladiolus, gypsophila, liastris, nerine, orchids, achillea, anthurium, tulip, and lilies. Floriculture crops like gerbera, carnation, rose, lilly *etc.* are grown in green houses etc are open field crops while chrysanthemum, roses, gaillardia, lily marigold, aster, tuberose *etc.* Postharvest lasting quality of flower species and cultivars vary considerably. For longer postharvest life, selection of variety is considered very important. Differences in flower longevity and quality among the varieties may be due to differences in their structural and physiological make up. For instance, *Alstroemeria* cv. Rosaria exhibits a vase life of 17 days whereas cv. Pink Panther shows vase life of 8 days. Likewise, cv. Marleen of Gerbera shows vase life of 20.5 days in comparison to cv. Agnes with vase life of 8.3 days. Similar variations in vase life have also been reported in spray carnation, *Anthurium*, *Lilium*, rose and gladiolus and fronds of *Asparagus* and *Nephrolepis*. The flowers belonging to families Orchidaceae and Asteraceae possess long vase life whereas flowers like iris, tulip and narcissus have short vase life. Not only this, the varieties that are grown under protected conditions generally posses higher vase life than those grown under open field conditions because the climatic factors and nutritional requirements of the crops are better managed in the green house. Greater variations are also reported to occur in vase life due to variations in climatic conditions, texture and pH of the soil.

Some the varieties of important crops are tabulated as under:

Rose - Miyabi, Red Naomi, Red Eagle, Pink Avalanche, Avalanche+, Top Secret, Black Baccara, Wow, Prima Donna+, Ice Breaker, Pink Florence, Cofidential, BelleRose, Upper Class, Samourai, Metilda, Roseberry, Sweetness, Revival, Cotton Candy, White Chocolate, Miami, Good Times, Duett, Decora *etc.*

Chrysanthemum-Sombrella, Aurora, Vanilla Sorbet, Etrusko, Alto, Evora, Cornemuse, Deborah, Liora, Kena, Rosela, Nimbus, Vahina White, White Star, Yellow Star, Reagan White, Yellow Star, Reagan White, Reagan Emperor, Yellow Delight, Punjab Shyiamli, Pusa Centenary *etc.*

Gerbera- Intenzz, Stanza, Winter Queen, Cacharelle, Jaffa, Sangria, Diana, Thalsa, Sonsara, Paganini, Anneke, Nette, Rosetta, Gloria, Ginna, Ingrid, Pricilla, Alexias, Intense, Sunway, Zingaro, Havana, Julia, Balance, Rosaline, Sweet Frostring, Sweet Honey, Classic Sylvana *etc.*

Carnation- Pearl Lady, Pomarosa, Armonia, Viana, Master, Tanga, Hikran, Don Pedro, Americana, Francesco, Kaiser, Eskimo, Everest, Sicilia, Windsor, Melinda, Dracula, Sebas, Picasso, White unny, Zulu *etc.*

Gladiolus- Red Advance, White Prosperity, Priscilla, Punja Glad-2, Pusa Vidushi, Pusa Sirjana, Jester, Novalux, Jacksoville Gold, Advance, Big Business, Chinon, Costa, Jessica, Oscar, Traderhorn, Toscane, Sugar Plum *etc.*

Lilium - Brindisi, Litouwen, Pavia, Sulpice, Tresor, Eyeliner, Indian Diamond, Yellow Diamond, Little Yellow Kiss, Shine On, Corcovado, Casablanca, Haceinda *etc.*

Orchids – Sonia Red, Sonia White, Sara Elizabeth, Lone Jack, In Bud *etc.*

Anthurium- Angel, Arena, Calisto, Candy, Champagn, Carisa, Carnaval, Eterno, Moments, Saffron, Fire, Farao *etc.*

4.0 Nature and Causes of Postharvest Losses

Factors involved in the decline of vase life of flowers include loss of carbohydrates due to high respiration rates, water deficit conditions which are induced when the flowers are handled dry during postharvest handling, sensitivity to ethylene, incidence of pests and diseases, quality of irrigation water and pollutants present in the atmosphere or irrigation water. When detached from the plant the cut stem is deprived of its natural sources of food, water, minerals and hormones and carriers out its metabolic functions as an independent entity at the expense of its available resources. Postharvest management involves steps taken for the improvement of flower vase life so that the consumers can get the flower in fresh condition. Before reaching the consumer, the cut stem passes through many handling steps such as cooling, grading, treatment with preservatives, storage, packing and transit.

4.1 Water Relations of Cut Flowers

The wilting of petals of many cut flowers leads to the termination of vase life even though the stems are constantly placed in water. After cut stems are placed in water after harvesting, they undergo marked changes in fresh weight which initially increases due to bud development and then shows a decrease due to senescence of the foliage and the flower. The stems with large leaf area like those of rose and chrysanthemum are highly prone to water loss. In general, cut stem continues to exhibit a decline in water absorption with time. As water loss exceeds water absorption, the stem develops water deficit conditions leading to wilting of petals as well as leaves. The reduced uptake of water by the stem has been reported to be mainly due to plugging of xylem vessels by the growth of bacteria, fungi and other microorganisms in vase water or on the stem surface. Microbial growth causes an increase in the stem resistance to water transport via xylem channels and decrease in vase life. The metabolites produced by bacterial growth have been found to reduce longevity. The cut surface of a flower stem also releases proteins, amino acids, sugars and minerals into the vase water which provides an ideal food for the growth of bacteria, yeasts, and fungi. Decrease in water absorption by the stem and stomatal loss of water also leads to development of air-filled spaces (emboli) within the xylem, which causes cavitation and hinders the uptake of water by the stem. Emboli can, however, be reduced by re-cutting 2-3 cm bases of the stems under water, acidification of the solution to pH 3 to 4 or placing the stem neck deep in water. Xylem occlusion also occurs by ethylene produced at the cut end of the stem, wounding reaction that involves both peroxidase and catechol oxidase activity and also deposition of lignin, suberin and tannins. Water deficit conditions are known to cause physiological disorders in the cut stems such as stem break in gerbera and bent neck of rose. Adverse water relations also cause changes in hormonal balance.

Quality of water also considerably affects keeping quality of the flowers and the foliage. Hard water usually contains minerals that make water alkaline. High pH of water reduces water uptake by the stem. Total dissolved solutes (TDS) present in water also affect vase life of the flowers. Total dissolved solutes more than 700 ppm decrease the vase life of gladiolus whereas at low concentration *i.e.* at 200 ppm they adversely affect the vase life of carnation, chrysanthemum and rose. Tap water also contains fluorides which are toxic to many flowers such as rose, gladiolus, carnation and freesia. Cut roses are also sensitive to sodium ions. Deionized water, boiled tap water, distilled water or millipore-filtered water are considered appropriate for cut flowers. Acidification of water (pH 3 to 4) also helps to prolong longevity of flowers as low pH decreases the bacterial population and increases the rate of water flow in the stem.

5.0 Postharvest Management

5.1 Preharvest Factors

Postharvest behaviour of cut flowers is determined by the preharvest conditions under which the crop is grown. The preharvest conditions which have an important bearing on vase life of flowers are: selection of the variety, environmental factors *viz.* light, temperature and relative humidity, nutrition, irrigation, diseases and pests, pollutants *etc.*

The postharvest keeping quality of the flowers are determined by the preharvest growing conditions which account for 30-70 per cent of vase life.

5.1.1 Environmental Factors

Duration as well as intensity of light and temperature play a major role in determining the quality of the flowers. Most cut flower crops require well-lighted conditions *i.e.* about 3200 ft candle of light intensity for optimal growth and flower production. Low light conditions decrease in photosynthetic efficacy of leaves and result in flowers with short vase life. Reduced light intensity can also decrease anthocyanin content of petals and increase in bluing of petals in red cultivars of rose. Besides, low light intensities lead to weak stems and increased susceptibility to bentneck in rose. Very high light intensities accompanied by high temperatures and increased ethylene liberation cause abscission of flowers/petals. High temperatures also increase loss of water from the plants and also increase the respiratory activity which depletes the carbohydrate reserves of flowers. Flower crops require moderate and specific temperatures for growing. For instance, rose requires 20-25 °C day and 16 °C night temperature whereas chrysanthemum requires 20-28 °C day and 15-20 °C night temperature for flower bud initiation. Carnation, being a cool climate crop requires 20 °C day and 10 °C night temperature for optimum growth and flower production. Low temperatures below 10 °C cause retardation in plant growth and sub-zero temperatures can cause freezing injury to the buds. As a consequence, most flower crops are grown under protected conditions where optimum light and temperature conditions are precisely controlled. High relative humidity in the atmosphere is highly suitable for germination of fungal spores and hence, can cause incidence of diseases. *Botrytis* blight increases manifold when high relative humidity is accompanied by low temperatures. Therefore, relative humidity should be moderate and near 60-80 per cent for raising a good crop.

5.1.2 Nutrition

Flower crops require optimum supply of nutrients for sustained and quality production of flowers. Therefore, cut flower crops like rose, lilies, carnation, gerbera *etc.* are grown under green houses where the nutrients are provided through drip irrigation system. Excess of nutrients cause an increase in pH and adversely affect their uptake. Besides, high salt concentration in the root zone causes inhibition of root growth. Especially, application of high nitrogen doses should be avoided as they result in undesired longer and weaker stems and high incidence of diseases. The nitrogen deficient crop, on the other hand, produces pale green foliage and blooms of poor quality. Deficiency of micronutrients, especially those of iron, manganese and zinc is of common occurrence in many cut flower crops and should be corrected before they can cause any serious damage.

The supply of nitrogen and even the form, whether, ammonical or nitrate, also have been reported to influence the flower vase life. Ammonical nitrogen application, 50 mg/l in Ranunculus asiaticus is reported to influence the flower vase in positive ways but higher (100 mg/l and more) are reported to be detrimental for flower quality parameters (Bernstein *et al.*, 2005).

5.1.3 Irrigation

Irrigation schedule of flower crops varies with the species and prevailing climatic conditions. Excessive irrigations cause water-logged and anaerobic conditions which inhibit both root as well as shoot growth leading to production of poor quality flowers. Water-logged conditions also invite pathogenic fungi, especially grey mould (Botrytis cinerea). The irrigation regimes have a direct influence on postharvest vase life cut carnations as is evident in the vase life deterioration which was quicker in flowers grown at -75 mPa soil water potential (Driest soil conditions) as compared to -15 mPa (Taylor *et al.*, 2008).

5.1.4 Diseases and Pests

Flower crops are attacked by a variety of pathogens and pests. Especially, under polyhouse conditions, the incidence of spider mite is a major problem. The flowers damaged by pathogens, insects and pests show poor vase life and result in rejection in the market place. Damaged tissues are potential sources of ethylene which causes senescence of flowers and foliage on the stem. It is, therefore, essential to keep the crop free from diseases and pests. Sanitation of growing areas, especially greenhouses, grading rooms and cold store is very essential to keep flowers free from diseases during production as well as after harvest. Especially, green houses should be kept clean and ventilated to allow free circulation of fresh air.

5.1.5 Pollutants

The flower crops should not be grown close to the industries which release toxic gases like sulphur dioxide, nitric oxide and fluorides in the atmosphere. Sulphur dioxide and nitric oxide form weak acids when it comes in contact with moisture and cause discolouration of petals. Tip burn of leaves commonly observed in gladiolus is caused by high concentration of fluorides in the atmosphere and/or irrigation water. The fluoride accumulates in the tip regions of the leaves and causes burning of leaf tips.

5.2 Harvesting

Proper harvest maturity, time and techniques of harvesting the flowers determine postharvest vase life of flowers to a greater extent. Flowers should always be harvested at an optimum maturity stage. Too immature buds fail to open whereas mature flowers have short vase life. The stage of harvest varies with the species and even with the cultivar, prevailing temperature conditions, marketing system and requirement of the consumer. In general, for transport to the distant markets, the flowers should be harvested at bud stage whereas for the local market, the flowers are harvested at a relatively open stage.

5.2.1 Stage and Time

Flowers like rose, carnation, gladiolus, daffodils, lilies, iris, freesia and tulip can be harvested at bud stage as buds of these flowers possess the ability to open after storage, transport and distribution. On the other hand, flowers like orchids, dahlia, gerbera, calendula *etc.* are harvested when fully open. The optimum stages of harvest for some important flowers are given in Table 1.

Bud harvesting of flowers is always beneficial as the buds continue to open over a longer time and hence, possess long vase life. Besides, the buds are less sensitive to ethylene, easy to handle during storage and transport, occupy less space in packing boxes and are hence, economical to transport. Besides, the buds are less prone to damage by diseases and pests.

Table 1. Stages of Harvest for Important Cut Flowers

Flower Name/Species	*Stage of Harvest*
Alstroemeria (lily of the Incas)	4-5 florets open
Anthurium andreanum	spadix almost fully developed
Antirrhinum majus (snapdragon)	1/3 florets open
Bouvardia hybrids	Flowers start opening
Calendula officinalis	Fully open flowers
Callistephus chinensis	Fully open flowers
Cattleya spp. (orchid)	3-4 days after opening
Chrysanthemum morifolium (Florist's chrysanthemum)	
Standard	When outer florets fully expanded
Spray	Flowers open but before shedding of pollens
Pompons and decorative	Centre of the oldest flower fully open
Anemones	open but before central disc florets begin to elongate
Cyclamen persicum	Fully open flowers
Cymbidium species (orchid)	3-4 days after opening of all flowers
Dahlia variabilis	Fully open flowers
Dendrobium spp. (orchid)	Fully open flowers
Dianthus caryophyllus (carnation)	
Standard	Paint brush stage when flowers are half open
Spray	At least 2 flowers fully open
Freesia hybrids	Coloured buds
Gerbera jamesonii	Flowers open but outer 2 rows show shedding of pollens
Gladiolus cultivars	1-5 florets show colour
Gloriosa superba	Fully open flowers
Gypsophila paniculata	Flowers just open
Hemerocallis spp.	Coloured buds
Hippeastrum hybrids	Coloured buds
Iris germanica	Coloured buds
Lilium spp. (lily)	Coloured buds
Narcissus spp. (daffodil)	Goose neck stage
Nerine bowdenii	First bud almost open
Ornothogalum umbellatum	Coloured buds
Papaversominiferum	Coloured buds
Paphiopedilum spp.	3-4 days after opening
Polianthes tuberosa (tuberose)	
Single	Buds fully developed but unopened
Double	Maximum florets open
Rosa hyrida (Rose)	1-2 petals beginning to unfold
Sterlitzia reginae (bird of paradise)	First floret opens
Tulipa spp. (Tulip)	Half coloured buds
Zinnia elegans	Fully open flowers

Flowers should always be harvested when temperatures are mild because high temperatures increase respiration and also promote water loss. Flowers should, therefore, be harvested either in the morning or in the evening. In the morning, flowers are fully turgid due to low transpiration rate that occurs during night. Flowers such as rose, chrysanthemum and gerbera etc. are highly prone to water stress benefit when harvested in the morning. Evening harvest is also advocated because of high sugar level in the stem due to increased activity of photosynthesis during day time but due to high evening temperatures, these flowers are more prone to desiccation. After harvest, the stems should be immediately placed in water or preservative solutions to prevent water loss.

5.2.2 Method

Proper method should be adopted to harvest the flowers. The stems should always be cut with sharp tools (knives or secateurs). Specialized tools are also used to harvest the flowers. For example, rose shear which grips the stem after it has been harvested is used to harvest roses. While harvesting, care should be taken not to crush the stem. Hardwood stems should always be given slanting cut to expose the maximum surface area to ensure rapid water absorption.

Specific guidelines should be followed for harvesting of flowers. For instance, stems of chrysanthemum should be harvested at least 10 cm above the ground level. The stem close to the soil is hard and lignified. The lignified tissues hinder the uptake of water by the cut stem. The stems of rose should be harvested above at least two leaves (with five leaflets) leaves to ensure sufficient photosynthetic area on the plant for sustained production of flowers. Cutting the stem near the base may also result in blind shoots in the subsequent season.

Stems of some species like dahlia, poinsettia and papaver release latex upon cutting. Such stems should be given dip in hot water (80-90 °C) for few seconds to inactivate the latex-releasing tissues. Immediately after harvest, the flowers should be placed in water or preservative solution and shifted to cold rooms to remove field heat generated due to high rate of respiration. Under no circumstances, the flowers should be placed out of water for more than few minutes.

5.3 Cooling and Conditioning

Immediately after harvesting or when the flowers are received from far-off places, they are required to be brought to low temperature conditions. This will help to remove the field heat generated due to high rate of respiration, check the excessive opening of the flower bud, reduce water loss and production of ethylene from the flowers.

In many instances, the flowers and foliage after prolonged dry storage as well as transportation show signs of wilting due to excess loss of water and hence, require restoration of turgidity. Such stems are needed to be rehydrated by placing them in solution of biocide such as aluminium sulphate or citric acid (3 g/litre of water) or in water acidified to pH 3-4. This process of rehydrating the stem is referred to as "conditioning". In case of severe wilting, the stems should be placed neck deep in water. Before placing the stems in hydrating solution, should be given re-cut at the base to expose fresh tissues for improving water uptake. These should then be placed at cool temperatures (2-4 °C) to prevent further loss of water.

5.4 Floral Preservatives

Term "Floral Preservative" is used for any chemical formulation which is used to extend the vase life of flowers. The use of chemical formulations to extend vase life has been an integral part of postharvest management of cut flowers. Many commercial floral formulations are now available in several developed countries for improving the vase life of the flowers. The chemical treatments can be used in the form of pulsing, bud opening solutions or holding solutions and the composition of each varies in different species depending upon the mode of senescence.

Floral preservatives mainly contain sucrose (as a carbon source) and a biocidal compound (that curtails microbial growth in vase water as well as on the cut surface of the stem) as basic constituents. Besides, plant growth regulators or ethylene antagonists may also form a constituent part of the floral formulations. Floral preservatives can be applied as

i) Pulsing solutions;

ii) Bud opening solutions; and

iii) Holding or vase solutions.

5.4.1 Pulsing

The term "pulsing" refers to short duration (a few minutes to a few hours) pretransit or prestorage treatment of harvested flowers in a pulsing solution. Sugar (sucrose) is a main component of the pulsing solution. Since pulsing involves short duration treatment, high concentrations of sucrose are used. Pulsing treatment is given at moderate temperatures (20-25 °C) and well-lighted conditions to allow maximum solution uptake by the stem. Concentration of sucrose for pulsing varies with species and endogenous levels of carbohydrates in the bud. The suitable concentration of sucrose for pulsing treatment has been suggested to be 20 per cent for gladiolus and gerbera, 10 per cent for carnation, bird of paradise and gypsophila and 2 to 5 per cent for rose and chrysanthemum. Sucrose should be used in combination with a biocidal compound appropriate to the crop to prevent microbial growth. Buds in flowers like rose exhibit accelerated opening under ambient conditions. Such flowers should be given pulsing treatments at cool temperatures (2-4 °C). Since treatment with sucrose invites growth of pathogens, it is desired that the flowers which are to be exported or transported to distant markets should not be pretreated with sucrose. Instead, such flowers should be treated with biocidal solution to prevent microbial growth on the surface of the stem during transit.

Silver thiosulphate (STS), an inhibitor of ethylene action has found wide spread use in improving vase life of ethylene sensitive cut flowers. However, since silver in STS is seen as an environmental pollutant, its use is not desired in the developed countries. Ethylene-sensitive flowers *viz.* carnation can, however, be treated with 1-methylcyclopropene (1- MCP) before storage or transportation.

Stems of alstroemeria and lilium can be pulse-treated with GA_3 or those of chrysanthemum with thidiazuron (TDZ) to prevent premature yellowing of leaves. Short duration pulsing with NaOCl (1 per cent) plus tween 80 (0.01 per cent as a wetting agent) for one min or with $AgNO_3$ solution (1000 ppm) for 10 min has been recommended for gerbera stems. Suitable pulsing solutions for some important cut flowers are given in Table 2.

Table 2. Suitable Pulsing Treatments for some Important Cut Flowers

Name of the Flower	*Treatment*
Alstroemeria	1-MCP, $20nl^{-l}$ for 6 h
Anthuriummajus	BA, 150 mg/l
Antirrhinum majus L.	1-MCP, $20nl^{-l}$ for 6 h
Carnation	Sucrose, 10-20 per cent + Physan 20^R, 200 ppm for 16-20 h STS, 4mM for 10 min 1-MCP, $20nl^{-l}$ for 6 h Sucrose, 10 per cent +STS,1mM for 20 h 1-MCP, 100 ppb for 30 min
Chrysanthemum	Sucrose, 5 per cent + Physan 20^R, 200 ppm for 16-20 h Methanol, 4 per cent or BAP, 210 ppm for for 24 h
Gerbera jamesonii	NaOCl, 1 per cent +Tweeen 20, 0.1 per cent or $AgNO_3$, 1200 mg/l for 10 min
Gladiolus	Sucrose, 20 per cent + Aluminium sulphate, 400 for 20 h Sucrose, 20 per cent + Physan 20^R for 16-20 h
Gypsophila	Sucrose, 10 per cent + Physan 20^R, 200 ppm for 16-20 h.
Rose	Sucrose, 3 per cent + Citric acid, 320 ppm at 4 °C for 16-20 h Sucrose, 3 per cent + Aluminium sulphate, 300 for 16-20 h
Stock (*Mathiola incana* L.)	TDZ, 5mM for 24 h 1-MCP, $20nl^{-l}$ for 6 h
Sweet pea	4mM STS for 8 min

5.4.2 Bud Opening Solutions

The immature buds of many flowers which otherwise do not open in water can be made to open in chemical solutions before they are sold in the market and are referred to as "bud opening solutions". The components of bud opening solutions are essentially the same as those of pulsing solutions, but in case of bud opening solutions, lower concentrations (2 to 5 per cent) of sucrose are used. Buds are usually opened at moderate temperatures (20-25 °C). The bud opening solutions have been standardized in many flowers like standard chrysanthemum, carnation and gladiolus The immature buds of spray chrysanthemum could be perfectly opened in a solution containing 2 per cent sucrose plus silver nitrate (25 mg/l) plus citric acid (75 mg/l) plus 6-benzyl adenine (50 mg/l). Similarly, buds of standard chrysanthemum can also be opened in a solution containing sucrose and 8-HQC though the concentration of sucrose required for the opening of buds varies with the cultivar and ranges from 2 to 30 per cent. The immature florets of gladiolus spikes cut at tight bud stage can be made to open with a solution containing sucrose (2 to 4 per cent) and aluminium sulphate (400 ppm)

5.4.3 Vase Solutions

Holding or vase solutions are used to hold flowers continuously, till the termination of their vase life. The level of sucrose in vase solutions is also kept low (0.5 to 2 per cent) due to long durations for which flowers are placed in the solution. Suitable vase solutions for some important cut flowers are given in Table 3.

Water is the most important component of floral preservatives. High pH (alkalinity) and presence of total dissolved solutes (TDS) in tap water adversely affect flower vase life. High pH promotes the microbial growth and hence, reduces vase life. Therefore, acidification of water to low pH 3.0-3.5 is advantageous because it decreases microbial growth and improves water uptake by the stem. The threshold levels of TDS in water are toxic to some flowers (discussed under "water relations of cut flowers").

Sucrose is the most widely used sugar in floral preservatives though other sugars like glucose and fructose are also reported to improve vase life of flowers. This beneficial effect of sugars on flower senescence can be attributed to the supply of substrates for respiration, structural materials and osmoticum. Exogenous application of sugars supplies the flower with much needed substrate for respiration and not only prolongs the vase life but also enables cut flowers harvested at the bud stage to open, which otherwise would not occur naturally. Sugars also improve water balance of cut flowers. Besides, sugar decreases sensitivity to ethylene. The optimum concentration of sugar required, however, varies from species to species and type of treatment. Since sugar promotes microbial growth, it is usually combined with biocidal compounds before use. Important biocides used for treating cut flowers are 8-hydroxyquinoline citrate, silver nitrate, aluminium sulphate, citric acid, slow release chlorine compounds and quaternary ammonium compounds. HQC (8-hydroxyquinoline citrate) is very effective broad spectrum biocide. It acidifies water and also induces partial closure of stomata. This compound is used at concentration of 200-600 ppm. Silver nitrate ($AgNO_3$) completely inhibits the microbial growth at the concentration of 25 ppm.

Table 3. Suitable Vase Solutions for some Important Cut Flowers

Name of the Flower	*Treatment*
Carnation	STS, 4mM for 15 min f.b. Sucrose 5 per cent + Aluminium sulphate, 200
Chrysanthemum	Sucrose 1.5 per cent + 8-HQC 200 mg/l or Sucrose 1.5 per cent +
	$AgNO_3$ 25 mg/l + citric acid 75 mg/l
Dutch iris (*Iris hollandica*)	Sucrose 4 per cent + Aluminium sulphate 300 ppm
Eustoma (*Eustoma grandiflorum* (Raf.) Shinn	Sucrose 1.5 per cent + Aluminium sulphate 200 ppm
Gerbera (Gerbera jamesonii)	$AgNO_3$ 20 mg/l
Gladiolus	Sucrose 2 per cent + 8-HQC 200 ppm Sucrose 4 per cent + Aluminium sulphate 400 ppm Sucrose 4 per cent + Aluminium sulphate 300 ppm Sucrose 2-4 per cent + DICA 50 mg/l Sucrose 2-4 per cent + $AgNO_3$ 25 mg/ l+ciric acid 75 mg/l
Oncidium	Sucrose 4 per cent +8-HQS 100 mg/l
Rose	Sucrose 1.5 per cent + Aluminium sulphate 300 ppm Sucrose 1.5 per cent + Aluminium sulphate 300 ppm or chlorine solution 50 ppm prepared from $CaOCl_2$

Since silver, being a heavy metal is considered to be an environmental hazard, its use in cut flower industry is no longer desired. Aluminium sulphate (100-300 ppm) has also been found very effective in increasing keeping quality of many flowers and is a standard biocide used in commercial preservatives. Besides reducing pH of the solution, it inhibits bacterial growth and also causes partial closure of stomata. Since, aluminium sulphate is also reported to cause foliage wilting in chrysanthemum, it should be used with caution.

Citric acid (50-1000 ppm) has been reported to improve vase life of roses, chrysanthemum, gladiolus and carnation and is also a very effective compound for rehydration of cut stems.

Several slow release chlorine compounds such as sodium dichloroisocyanurate (DICA) and 1,3-dichloro-5,5-dimethyl hydantoin (DDMH) etc. have also been found useful for many flowers. Chlorine solution prepared from sodium hypochlorite or by dissolving bleaching powder (calcium hypochlorite) in water has also been reported to extend vase life of flowers like gladiolus, rose and carnation. The concentration of chlorine in solution should be kept at 50 ppm because high concentrations can cause damage to the foliage as well as flower buds. For roses, only bleaching powder is recommended as sodium hypochlorite causes necrosis of the leaves and abscission of buds. Quaternary ammonium compounds have also been found very useful biocides for cut flowers. Besides, some other compounds such as thiabendazol, dichlorophen, salts of nickel, cobalt and copper and benzoic acid have also been found to extend vase life of some flowers. The biocides should be used with caution because the suitability of biocidal compounds may vary with the species.

5.5 Role of Growth Regulators

Plant growth regulators (PGRs) play a central role in regulating the senescence process of petals. Petal senescence is probably controlled by an interaction of plant growth substances rather than by a single hormone. The role played by PGRs in senescence process is of considerable use in evolving techniques to improve the vase life of cut flowers.

5.5.1 Gibberellins

Gibberellic acid is known to extend vase life of excised petals of carnation and easter lily and accelerate opening of immature buds of gladiolus. Gibberellic acid reduces ethylene production and delay senescence of cut carnation flowers. GA_3 has also been reported to prevent chlorophyll degradation in leaves of *Zantedeschia*, lilum and alstroemeria.

5.5.2 Cytokinins

Cytokinins, often regarded as senescence retarding hormones, especially in leaves are also reported to delay petal senescence. The ageing of flower petals is often accompanied by a decline in the endogenous levels of cytokinins. Cytokinins have also been shown to reduce sensitivity of flowers to ethylene and also delay the ethylene climacteric. Cytokinins delay petal senescence by maintaining cellular integrity and proteins. There is an inverse relationship between cytokinin content and senescence in some flowers like petunia, roses and carnation There are also indications that cytokinins may also act as free radical scavengers in cells.

Cytoknins delay yellowing of leaves in miniature gladiolus, alstroemeria, some cultivars of chrysanthemum, asparagus and arum lily. Prestorage foliar sprays or foliar dipping with cytokinins as well as GA_3 are reported to delay yellowing of leaves though efficacy of these bioregulators varies with the species. Thidiazuron, a substituted phenyl urea compound with cytokinin like activity also retards postharvest senescence of leaves in cut flowering stems.

5.5.3 Polyamines

Polyamines which are endogenous growth regulators are known to possess anti-senescence properties. Polyamines have been reported to improve vase life of carnation and gerbera.

5.5.4 Ethylene

Senescence of many flowers is often associated with increased production of ethylene. In ethylene-sensitive flowers, ethylene is regarded as a signal, mediating a sequence of events that eventually lead to the death of petals. The typical effects of ethylene on flowers include inrolling of carnation and kalanchoe petals which indicates the first sign of senescence, fading and inrolling of corolla in *Ipomoea*, fading of sepal tips in orchids and abscission of flowers and petals. Ethylene causes premature senescence and wilting of petals in a wide range of flowers especially, those belonging to families Campanulaceae, Malvaceae, Caryophyllaceae, Scrophulareaceae and Orchidaceae. The sensitivity of flowers to ethylene depends upon the stage of bud development, concentration of ethylene and exposure time.The exposure of mature flowers to ethylene leads to autocatalytic production of ethylene associated with an increased respiration rate and consequent increase in fresh weight losses. In ethylene-sensitive flowers, pollination induces increased ethylene production and wilting or abscission of petals. Exogenous ethylene is also reported to accelerate chlorophyll degradation in some cultivars of chrysanthemum. Silver thiosulphate (STS), an inhibitor of ethylene action has frequently been used to extend vase life of ethylene-sensitive flowers. However, due to high toxicity of the silver ions included in STS which involves environmental concern, there is a great interest in 1-MCP, which prevents the increased production of this phytohormone through permanent bonding to the ethylene receptor. Moreover, 1-MCP prevents bud/flower desiccation and abscission as well as leaf yellowing at very low concentrations.

Since pollinated flowers release considerable amount of ethylene, they should immediately be removed from the greenhouses. The greenhouses, packaging areas and storage rooms should be kept clean by removing diseased and damaged plant tissues. Flowers should be handled with care so as to avoid any physical damage. Flowers should not be stored along with fruits and vegetables as the latter produce considerable quantities of ethylene. Flowers can also be protected from ethylene by removing ethylene from the atmosphere. The greenhouses, packaging areas and the storage rooms should be ventilated to remove ethylene that continues to accumulate in these areas.

5.5.5 Inhibitors and Retardants

Abscisic acid (ABA), a natural plant growth inhibitor when applied exogenously accelerates petal senescence in carnation and increases sensitivity towards ethylene. ABA content is reported to increase with petal senescence in carnation and rose Under well-lighted conditions ABA, however, prolongs vase life of cut roses apparently by closing stomata and preventing water loss. ABA increases vase life of cut roses in low pH 6 solution due to stomatal closure but at higher pH 8 it reduces vase life, probably due to the development of microbes in vase solution at high pH. Pulse treatment with sucrose plus abscisic acid suppressed leaf damage and improved the quality of cut *Eustoma* flowers

5.5.6 Growth Retardants

Growth retardants such as daminozide and chloromequat chloride have also been reported to extend flower vase life of some flowers, though their mode of action is not fully understood. Methyl jasmonate, another naturally occurring plant growth regulator has been reported to accelerate ethylene production and senescence in flowers of dendrobium and petunia.

Nitric oxide (NO), a diffusible natural growth regulator plays an important role in alleviation of biotic and abiotic stresses and programmed cell death. Nitric oxide donor compound 2,2–(hydroxynitrosohydrazino) bisethanamine (DETA/NO) extend vase life of flowers of snapdragon, delphinium, chrysanthemum, tulip, gerbera, oriental lily, rose and iris cut stems. NO seems to have commercial potential because solid NO releasing water soluble compounds are available.

5.6 Storage of Flowers

Storage of flowers is highly advantageous to regulate the supply of flowers in the market in view of increasing problems on the preservation of large volumes of flowers and their distribution to the consumers.

Different species as well as varieties exhibit significant variations to storage behaviour. Different storage methods have been suggested for the storage of flowers are Normal Refrigerated Storage, Controlled Atmosphere (CA) and Low Pressure Storage (LPS).

Important factors which determine the post-storage life of the flowers are quality of the cut stem to be stored, stage of harvest of the flower, temperature and relative humidity during storage, light, sensitivity of the flower to ethylene and pathogens.

The flowers meant for storage to be stored should be of high quality, free from physical injury and incidence of pests or diseases. Besides, the flowers must also be harvested at an optimum maturity for storage. Some flowers like gladioli, roses, lilies, narcissus and irises can be stored at bud stage as the buds of these flowers possess ability to open after storage or transport. The flowers harvested at bud stage also last longer in storage than those at open stage.

Temperature management during storage is of utmost importance to preserve the quality of the cut flowers. Respiration rates of rose and carnation flowers placed at 20 °C temperature are nearly 20 times higher than those placed at 0 °C. High respiration rates deplete the flowers of their food reserve due to over-utilization of stored carbohydrates. High temperatures also increase water loss. Low temperatures on the other hand, inhibit opening of the buds, ethylene production and growth of the pathogens.

Temperature in the cold room should be precisely controlled because temperature fluctuations lead to condensation of water on the stems which invite pathogens, especially *Botrytis cinerea*. Relative humidity inside the cold room should preferably be kept high (90-95 per cent) to decrease water loss from the stems and prevent desiccation. Only disease-free stems should be stored because high relative humidity in storage is highly favorable for growth of *Botrytis cinerea*. Cold rooms should be kept clean to prevent the spread of pathogens as well as accumulation of ethylene by the decaying plant tissues and microbes.

5.6.1 Refrigerated Storage

Normal refrigerated storage is the simplest and most frequently used method for the storage of cut flowers. It includes both wet and dry storage. In wet storage, the stems are stored with their bases dipping in water or solution of floral preservative, usually a biocidal compound to check the growth of microbes on the stem surface as well as in the vase water. Since during wet storage, the stems continue to absorb water/chemical solution, they exhibit higher metabolic rate as a consequence of which bud development and senescence processes occur more rapidly than in dry storage. Carnation stems stored in water respired 25-30 per cent faster than those stored dry. Wet storage is used to hold the cut flowers for short duration and their day to day handling. The temperature in wet storage is kept at 2-4 °C *i.e.* slightly higher than that in dry storage. Flowers like iris, lilies and gerbera are suitable for wet refrigerated storage.

In dry storage, the flowers are sealed in moisture-retentive plastic sleeves to prevent the loss of moisture. Dry storage is also referred to as Modified Atmosphere (MA) storage of because the level of O_2 and CO_2 in the storage packages gets modified *i.e.* the concentration of CO_2 increases and that of O_2 decreases due to respiratory activity of the tissues. The suitability of MA packages for storage depends upon the permeability properties of the plastic membrane for CO_2 and O_2. The plastic packages also involve hazards due to accumulation of ethylene, excessive increase in CO_2 or decrease in O_2 levels which create anaerobic conditions. In general, 5 to 7 per cent of CO_2 and 1 to 2 per cent O_2 are suitable for storage of flowers. Build up of very high levels of CO_2 (beyond 4-5 per cent) or extremely low levels of O_2 (lower than 0.4 per cent) in the packages may damage the flowers. Therefore, it is very important to identify suitable films for optimum retention of both CO_2 and O_2 levels inside the packages. Dry storage is more laborious but can be used to hold the flowers for longer duration. Proper pre-cooling of cut stems prior to storage may reduce the risk of damage to flowers caused by high accumulation of CO_2 or ethylene. MA storage of flowers is significant as flowers in these packages lose less water and also occupy less space in storage as well as in transit. Storage in MA packages also holds considerable scope for long term shipment or surface transport of cut flowers. Pulsing of flowers before storage significantly improves their post storage vase life. Pulsing of gladiolus spikes with a solution of sucrose, 20 per cent in combination with aluminium sulphate, 400 mg/l has been found to be very beneficial in extending the vase life of gladiolus spikes after storage in polypropylene packages or after wet storage. Addition of GA_3 to pulsing solution improved the post storage vase life of the spikes. It was reported that pulsing of gladiolus spikes with $AgNO_3$ (1000 ppm) for one hour followed by treatment with 20 per cent sucrose for 24 h before dry storage increased their vase life. Post storage treatment with vase solution containing sucrose (1.5 per cent) plus aluminium sulphate, 300 ppm also significantly improved vase life of cut rose flowers. Spraying the foliage with BA+GA_3 (25 mg/l each) before or after dry storage significantly increased vase life of flowers of Asiatic and Oriental lily. In Anthurium, $AgNO_3$ pulsing for 40 min before dry storage improved vase life of cut Anthurium flowers.

Most flowers are prone to freezing injury below -2 °C but since free water freezes at 0 °C, water droplets condensing on the petals can also cause injury to the tissues. Therefore, the flowers should be stored at 0.5-1 °C. The temperature range for tropical flowers like anthurium, *Cattleya* and *Poinsettia* is 10-15 °C and for sub- tropical flowers such as gladiolus, strelitzia and anemone is 2-8 °C. Response of some important cut flowers to dry and wet refrigerated storage is given in Table 4.

5.6.2 Controlled Atmosphere Storage (CAS)

Controlled atmosphere storage (CAS) refers to low temperature storage in gas tight chambers under decreased levels of O_2 and increased levels of CO_2. CO_2 levels higher than 4-5 per cent may produce anaerobic conditions. Major limitation in CA storage of flowers is that the optimum levels of O_2 and CO_2 required for storage may vary for different flowers and hence, different flowers cannot be stored at the same time in the same chamber.

Table 4. Storage Response of some Important Cut Flowers

Storage	*Crop*	*Storage Temperature*	*Maximal Storage Period*
Dry	Carnation	0 to 1 °C	16 to 24 weeks
	Chrysanthemum	1 °C	3 weeks
		4 °C	15 days
	Freesia	0-0.5 °C	I week
	Gerbera	2 °C	2 days
	Gladiolus	4 °C	2 to 3 weeks
	Rose	0.5 to 2 °C	2 weeks
Wet	Alstroemeria	4 °C	
	Anthurium	13 °C	2 to 4 weeks
	Carnation	4 °C	4 weeks
	Chrysanthemum	4 °C	2 weeks
	Dendrobium	5 to 7 °C	10 to 14 weeks
	Gerbera	4 °C	1 weeks
	Gladiolus	4 °C	9-12 days
	Phalenopsis spp.	7-10 °C	2 weeks
	Tuberose	7 to 10 °C	3 to 5 days
	Rose	2 to 5 °C	5 days
	Zantedeschia spp.	4 °C	1 week
	Zinnia	4 °C	1 week

5.6.3 Hypobaric or Low Pressure Storage (LPS)

Hypobaric or low Pressure storage (LPS) refers to storage at low atmosphere pressure *i.e.* 40-60mm Hg under continuous ventilation and high relative humidity. Major disadvantage of LPS is that it causes rapid loss of water and faster removal of gases from the tissues. LPS, however, holds considerable promise for prolonged storage of flowers but due to high cost of installation, it is of limited use.

The optimum storage temperature and duration of storage varies with each flower. In general, temperate flowers *e.g.* rose, carnation, lily, cymbidium and chrysanthemum are best stored at a temperature close to 0 °C *i.e.* above the freezing point of plant tissues. Flowers originating from tropical regions *e.g.* anthurium, vanda, cattleya and bird of paradise are sensitive to chilling injury and hence, have to be stored at higher temperatures.

Flowers exhibit decline in vase life and the buds also lose the ability to open after prolonged storage . Though precise mechanism underlying decrease in vase life of flowers following storage is not clearly understood, loss of carbohydrates, accumulation of ethylene in the atmosphere surrounding the flowers, loss of membrane phospholipids, changes in activity of membrane enzymes, transport systems and hormone binding systems as well as accumulation of toxic compounds like acetaldehyde and ethanol have been suggested to affect post storage vase life of flowers.

Many leafy stems *viz.* alstroemeria, asparagus, miniature gladiolus, lilium and chrysanthemum are prone to post storage yellowing of leaves. Dark conditions during prolonged storage hasten the yellowing of leaves in these species. The longevity of leaves can, however, be improved by the application of plant growth regulators like cytokinins, gibberellins although the efficacy of these growth regulators differs with the species. The effective treatments which can be employed to delay yellowing of leaves in some important species are presented in Table 5.

Table 5. Suitable Treatments to Prevent Premature Yellowing of Leaves in some Important Flower Species

Name of the Flower	*Treatment*
Alstroemeria pelegrina L.	GA_3, 10^{-5} M TDZ, 10mM for 24 h
Chrysanthemum	Methanol, 4 per cent continuous or BAP, 10 ppm as 24 h pulse TDZ, 10-100 mM for 24 h BAP, 100 ppm spray
Easter lily (*Lilium longi florum* Thunb.	Spray with GA_3+BA (500mg/l each)
Stock (*Mathiolain cana* L.)	TDZ, 5mM for 24 h
Tulip (*Tulipages neriana*)	TDZ, 10-100 mM for 24 h depending upon the variety

5.7 Postharvest Disorders and Diseases

Cut stems of some ornamental species are prone to physiological disorders which cause decline in their longevity and rejection in the market. These disorders include bent neck of cut roses, bluing and blackening of petals in red coloured cultivars of rose, topple of tulips, stem break in gerbera, yellowing and blackening of leaves and geotropic bending of stems. These are discussed below.

a) Bent Neck of Cut Roses

Bent neck of rose is a physiological disorder wherein the stem fails to hold the bud erect which droops in the peduncle region. Bent neck usually occurs in the stems which are harvested when the bud is immature or when cut stem is under severe water stress conditions. The bent neck in stems cut at too immature stage occurs due to lack of sufficient hardening (lignifications) in the peduncle region. The leafy stems of rose also lose considerable quantities of water through transpiration and are hence, prone to water stress conditions. Under severe water stress conditions, buds draw water from the peduncle which lies immediately below it. This leads to loss of turgidity in the peduncle cells which become flaccid. The incidence of bent neck also varies in different cultivars. Some cultivars like Cara Mia are highly susceptible to bent neck. In order to prevent bent neck, the stems should be harvested in the morning when they are fully turgid and at proper harvest maturity when the peduncle region is sufficiently lignified. The stems should not be placed out of water for prolonged duration. The wilted stems after prolonged storage or transportation should be placed neck deep in tepid water so that they regain turgidity.

b) Bluing and Blackening of Petals in Rose

Petal bluing in red varieties of rose occurs due to carbohydrate starvation and subsequent hydrolysis of proteins and respiratory loss of amino acids which lead to increase in free ammonia in the cells. Increased ammonia in the cells increases cytosolic pH which causes bluing of petals. Bluing of petals occurs to copigment complex of cyanidin 3,5 diglucoside, quercetin and kampferol glycosides. Treatment with preservatives containing sugars can be used to prevent bluing of petals. Petal blackening also occurs in some red cultivars of rose due to calcium and perhaps boron nutrition.

c) Stem Break of Gerbera

Breaking of the stem in the middle portion is commonly observed in many cultivars of gerbera. Stem break occurs due to development of water stress conditions in the stem. The cut stem of gerbera absorbs water through two different pathways: (a) direct one through the xylem vessels at the cut surface and (b) an indirect one through the cavity in the stem. Stem break occurs when the direct water uptake is inhibited by bacterial activity and the indirect uptake is hampered. The water imbalance caused by bacterial growth in xylem vessels are the major cause of stem break in gerbera. Pretreatment of the stems with one per cent sodium hypochlorite (NaOCl) for one min in combination with wetting agent such as Tween-20 (0.1 per cent) has been recommended to prevent stem break in gerbera. Likewise, impregnation of the stem with 1200 ppm $AgNO_3$ for 10 min or 20 ppm $AgNO_3$ in vase solution has also been found to completely inhibit microbial growth in vase water.

d) Topple of Tulips

Topple of tulips is associated with calcium deficiency and high atmospheric humidity. The stem collapses a few cm below the base of the flowers at the time of flowering or picking. The scape, generally at an upper internode becomes dark green and watery. The affected areas shrink and the scape collapses and topples under the weight of the flower. Under severe calcium deficiency, the symptoms also appear on the scape or on the leaves. Sprays of 2 per cent solution of calcium nitrate solutions can reduce the losses.

e) Yellowing and Blackening of Leaves

In general, the leaves last longer than the flowers. However, in certain species, leaves show yellowing or blackening much before senescence of the flower bud. Yellowing and blackening of leaves mark the end of display life in some flowers.

Yellowing of leaves is a major problem in certain cultivars of alstroemeria, chrysanthemum, statice, lilies *etc.* and occurs due to breakdown of chlorophyll promoted by high temperatures as well as dark conditions during storage. Yellowing of leaves can be delayed by the application of gibberellins, kinetin, benzyl adenine and thidiauron. The response of growth regulators, however, vary with the species.

Blackening of leaves is of common occurrence in certain species of *Protea* and results from polymerization of phenols and tannins. It is mainly induced by water stress, high temperature and dark conditions and competition for carbohydrates. Vase solutions containing sugar help to reduce this disorder. The concentration of sucrose higher than 0.5–1 per cent may, however, accelerate blackening, since proteas are also sensitive to exogenously-applied sugar.

f) Geotropic Bending

In certain flowers like gladiolus, antirrhinum, larkspur, lisianthus, stock, lupine *etc.* the flowering stems are sensitive to geotropic bending *i.e.* if placed horizontally, sub-terminal portion of the stem shows upward bending which is referred to as "geotropic bending". Geotropic bending occurs due to cell elongation in the sub-terminal region of the stem due to downward lateral movement of the auxin from the tip region. Geotropic bending leads to loss of quality and the affected stems are not preferred in the market. Low temperature conditions prevent geotropic bending but the stems regain the ability to bend when transferred to normal conditions. In order to prevent geotropic bending, the stems should always be held vertically during storage or transport.

g) Postharvest Diseases

Botrytis blight caused by grey mould fungus (*Botrytis cinerea*) is a major postharvest disease of cut flowers causing substantial losses. The disease develops very rapidly under high moisture and low temperature conditions. The flowers producing nectar are more susceptible to grey mould. The pathogen is usually carried along with the cut stem from the greenhouse or field conditions in which the crop is grown. Adequate measures are, therefore, required to control this disease in the field or greenhouse conditions. Especially, for storage only disease-free flowers or foliage should be used.

5.8 Grading

After cooling, the stems are shifted to air-conditioned grading room. In the grading room, defective stems with blemishes or symptoms of diseases or pest infestation are removed. The stems are sorted out in different grades of length manually or by colour coded board or automatic grading machines. Grading refers to categorization of flowers based on quality parameters. For any market, it is important that each bunch must be of same size, weight or quality of the bloom. The bud should be at right harvest maturity, free from blemishes or injuries due to attack of pathogens, insects or pests, straight and strong enough to stand the weight of the bud. The bud should be bright in colour. The foliage should be dark green and free from the dust or residue of fungicide or pesticide. The stem should be free from physiological disorders such as bent back (in roses), tip bending (in gladiolus and other flowers), stem break (in gerbera), calyx splitting (in carnation) *etc.* The Economic Commission for Europe (ECE) based at Geneva and the Society for American Florists have recommended grades for some species of cut flowers but at present, there are no uniform grading standards for cut flowers in the world. However, the growers can get guidelines from the above-mentioned categorization and accordingly, grade the flowers. It is, however, very important for the growers to first collect information from the flower market where they intend to market produce and judge for themselves as to what type of flowers fetch remunerative price in the market. Under the grading system suggested by the Society of American Florists, the flowers are categorized as blue (Fancy), red (Special), green (Standard) and yellow (Utility). The blue (Fancy) grade includes the best quality flowers. The U.S. standards for hybrid tea rose, gladiolus, carnation (standard) and chrysanthemum (standard) are given in Table 6.

Table 6: Standards for some Cut Flowers by the Society of American Florists

Gladiolus

Grade	*Spike Length (cm)*	*Minimum Number of Florets*
Blue (fancy)	107 and above	16
Red (special)	96-107	14
Green (standard)	81-96	12
Yellow (utility)	81 and above	10

Carnation (Standard)

Grade	Minimum Flower Diameter (mm)	Minimum Length Overall (cm)	
Blue (Fancy)	Tight	50	
	Fairly tight	62	55
	Open	75	
Red (Standard)	Tight	44	
	Fairly tight	56	43
	Open	69	
Green (Short)	Tight	none	
	Fairly tight	none	30

Chrysanthemum

Grade	*Minimum Flower Diameter (mm)*	*Minimum Length Overall (cm)*
Blue	140	76
Red	121	76
Green	102	61

Rose

Grade	*Minimum Length Overall (cm)*	
	Hybrid Teas	*Sweet Hearts*
Blue	56	36
Red	36	25
Green	25	15

Leaves from the lower 4-5 cm portion of the stem should be removed and stems put in buckets containing chlorine (50 ppm), aluminium sulphate or citric acid (300 ppm). These chemicals disinfect the stem and check microbial growth which otherwise could reduce shelf life of the stems. The graded stems are immediately shifted to cold rooms till packed.

5.9 Packaging

The flower should be graded and bunched before treating with chemicals or before storage. Flowers of orchids, standard chrysanthemum, anthurium, gerbera and some speciality flowers are, however, packed individually. The number of flowers in a bunch varies with the species and requirements of the market. Flowers are generally grouped into bunches of 10,12 or 20 stems and loosely tied with a rubber band to secure the buds in position. The bunches should be carefully put together to prevent any damage to the blooms and leaves. Individual flowers of gerbera, orchids, anthurium and standard chrysanthemum which are harvested at open stage, are packed individually to prevent any likely damage to the open buds due to friction. The individual flowers are protected by paper, cellophane or polyethylene sleeves. Cushioning material such as shredded paper is also used to further reduce damage due to friction. The bases of tropical flowers like dendrobium and anthurium are kept moist by placing them in specially designed vials filled with water or in moist wool. The bunches are held in polyethylene or cellophane sleeves or buds as in roses are wrapped in corrugated paper to protect them from mechanical damage. Different flowers require different methods of wrapping or sleeving. For instance, the buds of rose are wrapped with 2 ply soft corrugated paper. The wrapping paper should project at least 2-3 cm above the bunch to protect the buds from thorns. Long-stemmed varieties such as Grand Gala or First Red have bigger-sized buds. These buds can be packed in 5 different layers of 4 flowers in each layer at two different levels. Such staggered bunching is useful to avoid damage to the buds by mutual pressure.

Short-stemmed varieties have small-sized buds which can be packed keeping all the buds in the same plane. The bunches are packed in pre-cooled 4-5-ply telescopic (top completely overlapping the bottom) of fibreboard in such a way that lower heads face in opposite directions *i.e.* towards the width. The heads of the flowers should be placed 5-10 cm from the end of the box to allow effective pre-cooling and prevent damage due to bruising. The bundles should be placed in rows along the width. The second layer is placed in opposite direction of the first layer for proper utilization of the space. In the packing box, the layers of flowers are covered with soft paper. The quality board (usually 4-5 ply boxes should be strong enough to hold the weight of at least eight such filled boxes) should be used for packing the flowers. Some growers use newspaper to cover the stems but care should be taken to ensure that the newsprint does not leave marks on the stems. After packing the flowers, the box is closed with the lid and strapped properly. Mixing of varieties or stems of varying lengths in the same box should be avoided. The stems should be tightly held inside the boxes so that they are not subjected to jerk movements during transport. The boxes should be properly marked with symbol 'Fresh Flowers' on both sides so that they are handled with care at the airports. Besides, the boxes should have stickers indicating name of the variety, number and length of the stems it contains. For international market, the flowers of gerbera are packed in flat boxes containing paper inserts with holes for individual stems. The support is necessary to keep the stems in place. For domestic markets, however, before packing the individual flowers are inserted in polyethylene sleeves to protect them from bruising.

Polyethylene sleeves are generally employed for wrapping a wide range of flowers but such a wrap may tend to retain moisture and encourage fungal attack. Hence, many wholesalers prefer naked bunches or perforated polythene sleeves. Hot-needle sleeves which have small, perforated holes allow the bunches to breath and reduce the chances of fungal attack. The flowers are held tightly in the boxes so that they are not damaged by the jerk movements during transit. The dimensions of the packing boxes are worked out keeping in view the length of the stem, type of the flower to be packed and efficient utilization of the space in the cargo or in refrigerated trucks.

Over packing and under packing both causes excessive damage. A wooden batten should be placed across the centre of the box under light pressure and nailed to the sides to prevent the movements of bunches during transport. The dimension of the packing boxes, however, varies from grower to grower and depends upon the length of the stem and number of stems to be packed as well as type of the cargo space available in the aircraft so that the boxes can fit well into the available spaces.

5.10 Transportation

Flowers such as gladioli, snapdragon and lupine which show geotropic bending when placed in a horizontal condition during storage or transport should be transported in vertical hampers. A new packaging system, the 'Procona' system developed by Pagtar Innovations, Roosendaal, The Netherlands allows transport of flowers upright in water and is very useful to keep flowers hydrated and fresh during transport over long distances, especially when they are not transported under proper temperature conditions. 'Procona System' is a package made up of plastic container, a cardboard collar that fits the container and a lid with many ventilation holes. The corners of the lid are very strong so that the packages can be stacked without getting damaged. The lid also has ridges so that the packages do not shift when stacked on the lid. When stacked, the container does not shut off the lid of the package below. As a consequence, the water vapours can escape through the ventilation holes in the lid. During transportation the flower heads receive fresh, cold air through the ventilation holes. The consumer can also see the colour of the flowers through the ventilation holes in the lid. Since the flowers stand in water during transportation, they maintain their quality.

Flowers are usually lifted by air and by refrigerated vans/trucks. But when the distances are short, non-refrigerated insulated trucks are also put to use for transportation. Temperature management during transit is essential.

Therefore, the flowers and the boxes are essentially required to be cooled before transport. Once packed, the flowers are difficult to cool.

The simplest method to cool the flowers is to pack them in the cold room. This method is, however, labour intensive and takes longer to cool the flowers. Forced air cooling of flowers in boxes with closeable side holes is the most common and effective method for cooling flowers. Cold air is sucked or forced into the boxes to reduce their temperature quickly. Forced air cooling is efficient and the flowers can be cooled within 45 min to one hour.

Vacuum cooling is also the quick and cost-efficient cooling system for cut flowers. The technology is based on the principle that water boils at lower temperatures as the pressure decreases. At about 1/10th atmospheric pressure, water boils at 0 °C. In the vacuum cooler, the pressure is reduced to a level where water starts boiling at 1-2 °C. The boiling process removes heat from the product which can be cooled down to 1 or 2 °C within 15-45 min depending upon the type of the product and amount of heat required to be removed. Although this system cools leaves or stems efficiently, flower petals often take longer to cool as water does not escape readily from them.

Optimum temperature to which the flowers are cooled varies in different flower species. Flowers are generally cooled at 0-1 °C. Care should be taken to pack the flowers in such a way that cold air flows freely in the boxes. If the flowers are to be transported through refrigerated vans, the side holes should be kept open to remove heat generated through respiration and free circulation of cold air inside the boxes. The flowers originating in tropical and sub-tropical climates are prone to chilling injury and are required to be maintained at higher temperatures (as discussed under "Storage of Flowers").Before loading it is to be ensured that the truck or van is precooled to the ideal temperature. Before loading all boxes are also required to be precooled to the optimum transportation temperature. To avoid the box becoming warm, loading is to be carried out as quickly as possible.

6.0 Biochemical and Molecular Aspects of Petal Senescence

Considerable information has been generated on biochemical and genetic process involving petal senescence. Flowers show initial increase in fresh weight and then subsequent decline with the onset of senescence. Studies have indicated membrane deterioration with the onset of senescence which indicates loss of membrane permeability. During senescence, cell membrane undergoes loss of phospholipids and increase in micro viscosity. The loss of membrane integrity was found to be due to free radical production and concomitant increase in viscosity of the cell sap. Senescence is also accompanied by the loss of respirable substrates. At later stages of development, petals are reported to exhibit decline in the contents of starch, total soluble and reducing sugars and activity of hydrolytic enzymes. Kumar *et al.* (2007c) observed substantial increase in starch accumulation during early stages of flower bud development while starch degradation was observed in the later developmental stages. Decrease in soluble sugar content as well as invertase activity in senescing florets of gladiolus also takes place. In rose also, senescing petals exhibited significant decrease in invertase activity. (Kumar et al., 2008b) identified an invertase inhibitor in senescing petals of carnation.

Senescence in rose petals was found to be associated with increased production of reactive oxygen species and significant decline in SOD activity (Kumar *et al.*, 2007b). In chrysanthemum, lipid peroxidation and membrane damage were involved in flower deterioration. Lipid peroxidation and concentration of reactive oxygen species increased from young floret stage to senescence stage in chrysanthemum. Activities of superoxide dismutase, peroxidase and catalase also showed an increase from young to mature stage and then declined. It was suggested that capacity of antioxidative defence system increased during onset of senescence but imbalance between ROS production and antioxidant defences ultimately led to oxidative damage.

Increase in endogenous H_2O_2 levels and decrease in antioxidant enzyme activity may be partially responsible for initiating senescence in rose petals. It has also been reported that senescing petals retain continued synthesis of proteins and RNA until very late in senescence. Ethylene-induced senescence of carnation flowers were accompanied by changes in mRNA and protein profiles indicating the involvement of ethylene-induced gene expression both at the transcriptional and translational level however continued perception of ethylene was required for expression of genes. cDNAs encoding enzymes of ethylene biosynthesis in carnation petals have been isolated and cloned. Such studies are of considerable importance in understanding the mechanisms involving ethylene-induced gene expression and to devise biotechnological options for the production of flowers with low ethylene production and long vase life.

So far, considerable emphasis has been laid on the understanding of processes involved in ethylene-induced senescence of flowers. The process of petal senescence in ethylene-insensitive flowers is not completely understood. Many flowers of commercial significance are insensitive to ethylene and artificial increase in vase life of these flowers is still difficult. There are only a few known studies on the physiological and genetical mechanisms involved in petal senescence of such flowers. Rapid senescence of daylily flowers was accompanied by transient respiratory climacteric but not by any appreciable evolution of ethylene. Silver thiosulphate (STS), an inhibitor of ethylene action or aminooxyacetic acid (AOA), an inhibitor of ethylene biosynthesis, also did not delay senescence. Postharvest pulsing with STS also failed to improve vase life of gladiolus spikes. Flower senescence in chrysanthemum was accompanied by decline in total protein content and high activity of hydrolytic as well as proteolytic enzymes. Cycloheximide (CHI) extended vase life of gladiolus. iris and narcissus flowers. CHI -induced extension in vase life in gladiolus was accompanied by inhibition of protein synthesis in florets. CHI-induced inhibition of flower senescence was accompanied by maintenance of protein populations in the petals which indicated that *de novo* synthesis of proteins was required for flower senescence. Differential screening of cDNA library constructed from tepals of *Hemerocallis* flowers showing incipient senescence revealed 25 clones that were strongly unregulated. The derived amino acid sequence of full length cDNA (SEN12) had strong homology with cysteine proteases, reported from other plant tissues. It was suggested that the onset of flower senescence was associated with the upregulation of specific genes, and that at least one of the mechanisms for hydrolysis of corolla proteins. In *Iris* x *hollandica*, visible senescence was preceded by large scale protein degradation, which was mainly due to cysteine- and serine protease activity. Azeez et al., (2007) reported that proteolysis in floral organs of gladiolus was associated predominantly with serine protease of the total protease activity. Recent studies have shown that in chrysanthemum, an ethylene-insensitive flower, ethylene induced yellowing of leaves. Treatment with ethylene action inhibitors such as anionic silver thiosulphate and 1-methylcyclopropene (1-MCP) inhibited leaf yellowing. The release of cytochrome c from mitochondria, one commitment step in signaling of apoptosis in animal cells, was found to be dispensable in petal cell death. Apoptotic changes of nuclei, such as chromatin condensation and nuclear fragmentation correlated with cell death in gladiolus petals after full flower opening. Wu et al. (2017) found an ethylene-induced RhHB6-RhPR10.1 regulatory module that functions as a brake of ethylene promoted senescence by increasing the cytokinin content in rose. Silencing of RhPR10.1, a pathogenesis-related PR-10 family gene from rose and highly expressed in different floral organs during senescence, accelerated flower senescence accompanied by a reduction in cytokinin content and in the expression of three cytokinin signaling pathway genes. The enhanced rate of petal senescence in *RhPR10.1*-silenced plants was restored to normal levels by cytokinin treatment. Finally, the *TF RhHB6* was observed to bind to the *RhPR10.1* promoter, and silencing of its expression also promoted flower senescence. So, the effect of *RhHB6* and *RhPR10.1* silencing is that both accelerate flower senescence. The conclusion is that the ethylene-induced RhHB6-RhPR10.1 regulatory module inhibits ethylene-induced flower senescence through increasing the levels of cytokinins.

Significant achievement has been made in genetic transformation of many commercial ornamentals. Ethylene-induced accumulation of transcripts and their coded proteins speeded up abscission and reduced useful shelf life of cut flowers of *Delphinium*. Transgenic carnations showing anti-sense expression of ACC synthase and ACC oxidase were available. To inhibit the ethylene response expression of the *Arabidopsis thaliana etr1-1* gene in transgenic carnation plants was also attempted. Transgenic petunias which overproduced cytokinin exhibited extended flower longevity. Genetic transformation systems hold considerable promise in producing a diverse range of flowers with long vase life. Further, cultivation and commercialization of transgenic plants is a matter of great discussion and implementation is based on legislation of a country. Aswath and Hanur (2009) have reported the genes for improvement of vase life in flowers for various traits as under :

Candidate Gene and Pathway	*Traits*
ACC synthase	Inhibition with reduced ethylene production
ACC oxidase	Inhibition with reduced ethylene production
ACC deaminase	Over expression with reduced ethylene production
Etr 1	Expression of a defective gene with reduced ethylene production
ERS	Expression of a mutated gene with reduced ethylene production

Developing new ornamental cultivars with improved floral attributes are expected to be a major goal in floriculture. Ornamental plant attributes like floral color, fragrance, disease resistance, and vase life can be expectedly improved through genetic manipulation. Ethylene is a primary plant hormone involved in the senescence of many flowers. Inhibition of the synthesis or action of ethylene delays the onset of senescence symptoms and increases the vase life of the flowers. Generation of transgenic plants with suppressed production or action of ethylene is an excellent way to lengthen the longevity of ornamental plants (Netam, 2018).

7.0 Application of Nanotechnology

Pulse solution with 10 mg^{-1} silver nanoparticles (SNPs) increased the vase life in cut chrysanthemums significantly higher as compared to control (Kazemipour *et al.*, 2013). Nanotechnology based on nanoparticles, has also been applied in prolonging vase life of cut flowers. With the advancement in the use of various nanoparticles particularly silver nanoparticles (SNPs) in floriculture are widely used for their antimicrobial properties, high electrical and thermal conductivity, non-linear optical behaviour, catalytic activity and chemical stability. The various reason are already discussed which contributes to the depletion of vase life cut flowers. Thus, vase life can be extended by placing cut stems in preservative solution and providing a suitable environment containing suitable nano particles. SNPs not only kill the bacteria growing in the vase solution but also enter the cut flower stem through vascular tissues as these particles have very small size and inhibit bacterial colonization at the cut end. Furthermore, SNPs are translocated into other parts of the flower and limit ethylene biosynthesis by suppressing the action of ethylene biosynthesis genes (ACO1 and ACS1), thus increasing solution uptake, relative fresh weight, petal water content and ultimately enhancing the cut flower longevity (Naing and Kim, 2020). Copper nanoparticles (CuNPs) are reported to increase the vase life of cut carnation and chrysanthemum flowers by suppressing microbial growth along with morphophysiological improvement of the cut stems (Rashidiani *et al.*, 2020) Calcium oxide nanoparticles are reported to modify the stem firmness and decrease the stem bending considerably, which is the major cause of loss of vase life in gerbera. So the application of CaO-NP in vase solutions can be an alternative in postharvest treatments of cut gerbera flowers to maintain their quality (García-Gonz´ Alez *et al.*, 2022).

8.0 Domestic and Foreign Trade

Flowers from India are marketed both within India and are also exported to far places such as Europe and Japan by air transport. The boxes are taken to the airport in refrigerated vans to keep the flowers cool. The international airports also provide cooling facilities to the growers. In case of limited space in the cold room, air conditioning unit of the refrigerated truck can be plugged into the electric points provided by airport authorities to use refrigerated trucks as a makeshift cold store till the flowers are loaded into the aircraft. It should be ensured that the flowers are loaded into a special cabin meant for the transport of perishable commodities.

In the international flower trade, the flowers are traded through auction markets as well as directly through wholesale and retail outlets. In our country, the trade in cut flowers is rather of recent origin and hence, there is no organized marketing system of flowers. Mostly, the flowers are marketed by the growers directly to the retail outlets. Some efforts have been made in some states to develop auction centres for flowers.

Karnataka Agro Industries Corporation (KAIC) has started a flower auction centre at Hebbal, Bangalore which possesses a state-of-the-art electronic auctioning system and automatic cold storage system with a capacity to store 2.5 lakh stems. At present, the centre deals more with domestic trade but will be converted into an international auction centre. Maharashtra Agro Industries Corporation has also established a flower auction centre in Mumbai, which is yet to become operational. Similarly, flower market at Ghaziabad, Uttar Pradesh is also operational. Ghazipur Flower Trade Centre, Ghazipur on the outskirts Delhi gets hundreds of varieties of flowers from various parts of the country and even from some south Asian countries like Thailand, Indonesia, *etc.* In the international market, flowers are auctioned in the auction centres. There are seven auction centres in Holland. The largest flower auction centre in the world is Aalsmeer in the Netherlands. Flowers from different countries of the world are traded at these centres. Nearly 60 per cent of flowers are traded in Aalsmeer and Naaldwijk/Bleiswijk auction centres.

9.0 Conclusion

By observing the guidelines mentioned below, vase life of the flowers can be significantly improved.

1. Flowers should always be harvested at an appropriate stage of development. Too tight buds fail to open whereas mature flowers do not last long.
2. After harvesting, the stems should not be placed out of water for more than a few minutes.
3. The vases should be disinfected with chlorinated disinfectants before putting the flowering stems. The unclean vases harbour many microbes which multiply very rapidly when the stem is placed in the vase.
4. Water in the vase should be changed at 2-3 days interval to prevent, growth of microbes in vase water.
5. About 2-3 cm basal portion of the stem should be recut at 2-3 day intervals to expose fresh tissues for improving water uptake.
6. Lower leaves dipping in water/solution should be removed immediately as these leaves emit foul smell on rotting and also promote microbial growth which decreases vase life of the stem.
7. Vases should be placed in cool places.
8. Withered flowers should be periodically removed from the stem as they release considerable quantities of ethylene and cause senescence of other flowers.

9.0 Important References

Aswath C. and Hanur S.V. 2009. For choicest coloured flowers. Genes to fulfil consumers' demand. *Indian Hortic.* **54**: 30-33.

Azeez, A., Sane, A. P., Bhatnagar, D. and Nath, P. 2007. Enhanced expression of serine proteases during floral senescence in Gladiolus. *Phytochem.* **68**: 1352-57.

Bala, M., Kumar, R. and Singh, K. 2008a. Effect of pulsing and holding solutions on keeping quality of cut flowers of rose. *J. Orna. Hortic.* **11**: 54-57.

Bernstein, N., Ioffe, M., Bruner, M., Nishri, Y., Luria, G., Dori, I., Matan, E., Philosoph-Hadas, S., Umiel, N. and Hagiladi A. 2005. Effects of Supplied Nitrogen Form and Quantity on Growth and Postharvest Quality of *Ranunculus asiaticus* Flowers. *Hortic. Sci.* **40**: 1879–86.

Bovy, A.G., Van Altvorst, A.C., Angenent, G.C. and Dons, J.J.M. 1995. Genetic modification of the vase-life of carnation. *Acta Hortic.* **405**: 179-18.

Cho, M., Celikel, F.G., Dodge, L. and reid, M.S. 2001. Sucrose enhances the postharvest quality of cut flowers of *Eustoma grandiflorum* (Raf.) Shinn. *Acta Hortic.* **543.**

García-Gonz'alez A., Soriano-Melgar L.A.A., Cid-Lopez M.A., Cortez-Mazatan G.Y., Mendoza-Mendoza E., Valdez-Aguilar L.A., Peralta-Rodríguez R.D. 2021. Effects of calcium oxide nanoparticles on vase life of gerbera cut flowers. *Sci. Hortic.* **291**.DOI10.1016/j.scienta.2021.110532

Han, S.S. 2001. Benzyladenine and gibberellins improve postharvest quality of cutAsiatic and Oriental lilies. *Hortic. Sci.* **36**: 741-45.

Kazemipour, S., Hashemabadi, D and Kaviani B. 2013. Effect of silver nanoparticles on the vase life and quality of cut chrysanthemum (*Chrysanthemum morifolium* L.) flower. *Europ. J. Exp. Biol.* **3**: 298-02.

Karki, K., Kumar, S., Srivastava, R. and Jauhari, S. 2004. Effect of floral preservatives on vase life of cut roses grown under polyhouse conditions. *J. Orna. Hortic.* **7**: 121-23.

Khunmuang, S., Kanlayanarat, S., Wongs-Aree, C., Meir, S., Philosoph-Hadas S., Oren-Shamir, S., Ovadia, R. and Buanong, M. 2019. Ethylene induces a rapid degradation of petal anthocyanins in cut *Vanda* 'Sansai Blue' orchid flowers. *Front. Plant Sci.* **10**: 1-13.

Kumar, N., Srivastava, G.C. and Dixit, K. 2007b. Role of superoxide dismutase during petal senescence in rose (*Rosa hybrida* L.). *J. Hortic. Sci. Biotechnol.* **82**: 673-78.

Kumar, N., Srivastava, G.C. and Dixit, K. 2008b. Role of sucrose synthase and invertase during petal senescence in rose (*Rosa hybrida* L.). *J. Hortic. Sci. Biotechnol.* **83**: 520-24.

Kumar, N., Srivastava, G.C. and Dixit, K. 2008c. Senescence in rose (*Rosa hybrida* L.): role of the endogenous anti-oxidant system. *J. Hortic. Sci. Biotechnol.* **83**: 125-31.

Meeteren, U. van. 1978. Water relations and keeping quality of cut gerbera flowers. I. The cause of stem break. *Scientia Hortic.* **8**: 65-74.

Mor., Y., Reid, M.S. and Kofranek, A.M. 1984. Pulse treatment with silver thiosulphate and sucrose improve the vase life of sweet peas. *J. Amer. Soc. Hortic. Sci.* **109**: 866-68.

Namita, Kumar, R. and Singh, K. 2006b. Response of vase solutions on keeping quality of cut spikes of gladiolus. *J. Orna. Hortic.* **9**: 296-97.

Netam, N.2018. Improving ornamental's vase life through molecular approaches: A review. *J. Pharmacog. Phytochem.* **7**: 1687-91.

Nilofar and Paul, T.M. 2008. Postharvest management of gladiolus. *J. Orna. Hortic.* **11**: 69-71.

Naing, A. H. and Kim, C. K. 2020. Application of nano-silver particles to control the postharvest biology of cut flowers: A review. *Sci. Hortic.* **270**: 109-463 .

Pal, A., Kumar, S. and Srivastava, R. 2003. Effect of floral preservatives on postharvest management in gladiolus spikes. *J. Orna. Hortic.* **6**: 367-71.

Rahidiani, N, Nazari F, Javadi T and Samadi S. 2020. Copper nanoparticles (CuNPs) increase the vase life of cut carnation and chrysanthemum flowers: antimicrobial ability and morpho physiological improvements. *Ornam. Hortic.* **26**. 225-35.

Reid, M.S. and Kofranek, A.M. 1980. Postharvest physiology of cut flowers. *Chronica Hortic.* **20**: 25-27.

Serek, M., Sisler, E.C. and Reid, M.S. 1995a. Effects of 1-MCP on the vase life and ethylene response of cut flowers. *Plant Growth Regul.* **16**: 97-98.

Singh, K., Arora, J.S. and Bhattacharjee, S.K. 2001a. *Postharvest Management of Cut Flowers.*Tech. Bull. No. 10, All India Coordinated Research Project on Floriculture, IARI., New Delhi, pp. 39.

Singh, K., Singh, P. and Kapoor, M. 2007. Effect of vase and pulsing solutions on keeping quality of standard carnation (*Dianthus caryophyllus* Linn.) cut flowers. *J. Orna. Hortic.* **10**: 20-24.

Singh, K., Singh, P.J. and Arora, J.S. 2003. Effect of sucrose on vase life of cut roses. *J. Orna. Hortic.* **6**: 145-46.

Singh, K., Singh, P.J. and Kumar, R. 2004a. Effect of chemicals on keeping quality of cut roses. *Adv. Hortic. Sci.* **18**: 161-67.

Singh, K., Arora, J.S. and Bhattacharjee, S.K. 2001a. Postharvest Management of Cut Flowers. Tech. Bull. No. 10, All India Coordinated Research Project on Floriculture, IARI., New Delhi, pp. 39.

Taylor, R.D., Hill, J. and Grout B.W.W.2008. Does Irrigation Regime Affect the Postharvest Quality and Vase Life of Cut Flower Carnations (cv. Santorini) *Acta Hortic.* **792**: 663-38.

Wu, L., Ma, N., Jia, Y., Zhang, Y., Feng, M., Jiang, C., Z., Ma, C. and Gao, J. 2017. An ethylene-induced regulatory module delays flower senescence by regulating cytokinin content. *Pl. Physiol.* **173:** 853-62.

Chapter 33

Orchids

L.C. De

Principal Scientist, ICAR-National Research Centre for Orchids, Pakyong – 737 106, Sikkim, India

N. Pathak

Sikkim Himalayan Orchids Ltd., Assam Lingzey – 737 135, East Sikkim, India
Present address: Kangra, Himachal Pradesh

E-mail: lcde_horti@rediffmail.com; E-mail: narender.pathak@gmail.com

1.0 Introduction

Orchids comprise the largest family of flowering plants with 25,000 to 35000 species belonging to 600-800 genera. They are prized for their incredible diversity in size, shape, form and colour, attractiveness of their flowers and high keeping qualities even up to 10 weeks. Most of the orchids have originated from tropical humid forests of Central and South America, India, Sri Lanka, Burma, South China, Thailand, Malayasia, Philippines, New Guinea and Australia. Brazilian *Cattleya*, Mexican *Laelia* and Indian *Cymbidium*, *Vanda* and *Dendrobium* have played a major role in developing present day beautiful hybrid orchids which number more than 2,00,000. In international trade, among the top ten cut flowers, orchids rank the sixth position and occupies 3 per cent of the total cut flower production. *Cymbidium* holds the first rank among the orchids.

2.0 Status of Cultivation

Globally, trade on artificially propagated live plants are dominated by orchidaceae hybrids (28.7 per cent), Cymbidium species (26.9 per cent), orchidaceae species (18.9), Phalaenopsis hybrids (10.1 per cent), Phalaenopsis species (4.4 per cent), Dendrobium species (3.4 per cent), Cymbidium hybrids (3.3 per cent), Dendrobium hybrids (2.3 per cent), Cattleya species (0.4 per cent) and Oncidium species (0.2 per cent) (UNEP, 2017).

In India, orchids comprising 158 genera and 1331 species grow up to an elevation of 5000m. Indian terrestrials are naturally distributed in humus rich moist earth under tree shades in North Western India. Western Ghats harbour small flowered orchids. Epiphytic orchids are common in North eastern India which grow up to an elevation of 2000m from sea level. Indian orchid species having high ornamental values are *Aerides multiflorum, Aerides odoratum, Arundina graminifolia, Arachnis, Bulbophyllum, Calanthe masuca, Coelogyne elata, Coelogyne flavida, C. corymbosa; Cymbidium aloifolium, C. lowianum, C. devonianum, C. hookerianum, C. lancifolium; Dendrobium aphyllum, D. nobile, D. chrysanthum, D. farmeri, D. chrysanthum, D. densiflorum, D. moschatum, D. fimbriatum, D. jenkinsii; Paphiopedilum venustum, P. spicerianum, P. hirsutissimum, P. insigne, Phaius wallichii, Pleione praecox, Renanthera imschootiana, Rhyncostylis retusa, Thunia alba, Vanda cristata, Vanda coerulea* and *Vanda coerulescens*. Varied agroclimatic conditions in India are highly favourable for cultivation of commercial orchids like *Cattleyas, Vanda, Dendrobium, Arachnis* and *Aerides* in tropical areas, *Cymbidium, Paphiopedilum, Dendrobium* and *Phalaenopsis* hybrids in subtropical areas and *Cypripediums, Cymbidiums* etc in temperate areas. Among orchids, *Cymbidium, Dendrobium, Phalaenopsis, Cattleya, Vanda, Aranda, Arachnis* etc used as cut flowers whereas *Aerides, Arachnanthe, Bulbophyllum, Calanthe, Coelogyne, Eria, Phaius, Phalaenopsis, Pleione, Rhyncostylis* are used as potted orchids. The *Cymbidium* is mainly grown in NEH Region, Sikkim, Darjeeling hills and Arunachal Pradesh. Tropical orchids are cultivated in Kerala and some parts of Tamil Nadu. We should prefer those species which flower during winter and spring months to export flowers to temperate regions from December to May.

India accounts for nearly 7 per cent of the worlds orchid biodiversity which is distributed in five major phyto-geographical regions *viz.* North eastern Himalayas, Peninsular region, Western Himalayas, Western Ghat and Andaman and Nicobar Islands (Table 1).

Table 1. Status of Orchid Distribution in India

Region	*States*	*Genera*	*Species*
Andaman & Nicobar Islands	319 Islands and islets in the Bay of Bengal	53	115
Eastern Himalayas and North Eastern India	Arunachal Pradesh, Assam, Manipur, Meghalaya, Mizoram, Nagaland, Sikkim and Tripura	159	870
North -Western Himalayas	J&K, Himachal Pradesh, Uttarakhand	75	288
Peninsular India	Andhra Pradesh, Central India, Eastern Ghats, Gujarat, M.P., Orissa and Western Ghats	89	379

Source: Hajra and De, 2010.

3.0 Commercial Varieties

The commercial varieties of orchids grown belong to the following types.

Reported Convention on the International Trade in Endangered Species of Wild Fauna and Flora (CITES) trade in live artificially propagated plants is dominated by a small number of genera, with a large proportion of trade in hybrids (*e.g. Cymbidium* Sw., *Dendrobium* Sw. and *Phalaenopsis* Blume;). Orchids are consistently ranked among the best sellers in the global potted plant trade and also comprise c. 10 per cent of all fresh cut flowers traded internationally (De, 2015). This represents an economically significant global trade, with exports of potted orchids from the Netherlands alone valued at almost €500 million in 2015 (Flora Holland, 2015). The largest areas of production are in Thailand, Taiwan, The Netherlands and Japan, with demand for both potted and cut flowers growing in economic value annually (Hanks, 2015).

a) Cymbidium

Cymbidiums are among the most popular winter and spring blooming semi-terrestrial orchids originated from tropical and subtropical Asia covering North Eastern India, China, Japan, Malayasia, the Philippines, Borneo islands and North Australia, usually grown in cooler climates at high elevations. Cymbidiums are highly valued for cut flowers, hanging baskets, potted plants and herbal medicines. Cymbidium has been considered as top commercial orchids in Europe since many years. They fetch the highest price in the international markets of which major Asian markets of Singapore and Japan or the Dutch market.In India, Arunachal hills, Sikkim and Darjeeling hills with cool summer night and monsoon summer rains are ideal for cymbidium cultivation. In Sikkim, more than 350 hybrids of cymbidium orchids are commercially cultivated around 28 ha of land and about 15-20 million spikes are produced annually.

b) Dendrobium

Dendrobiums are popular flowering potted plants and cut flowers around the world due to their floriferous nature, wide range in flower colour, size and shape, year round availability and long vase life. Hawaii, California and Florida are major potted *Dendrobium* growing regions in the United States. In the Netherlands, production of potted orchids is now 40 to 50 million units with *Dendrobium* increasing in popularity. Imports from Thailand, the world's largest exporter of tropical cut orchids, and the second largest supplier to the EU, accounted for 22 per cent of supplies to the EU. Thailand holds a particularly strong position in *Dendrobium* orchids.

c) Phalaenopsis

Phalaenopsis is the second most valuable and popular flowering potted plant and cut flower around the world due to their easy cultural practices, diversity in flower colour, size and shape, year round availability, delicacy and long vase life. It is commercially grown in Germany, Japan, The Netherlands, Taiwan and United States. In the United states, 75 per cent of all orchids purchased are *phalaenopsis.*

3.1 Other Tropical Orchids

Vanda is widely distributed throughout Australasia from China through the Philippines, Indonesia, Malaysia, New Guinea and Australia, Myanmar, Thailand, India and Sri Lanka. In the world tropical orchid trade, *Dendrobium* is the most dominant crop in addition to *Mokara, Oncidium, Aranthera, Aranda, Vanda, Arachnis, Renanthera, Ascocenda, Phalaenopsis, Cattleya* and *Paphiopedilum* which are grown as cut flowers and potted plants.

3.2 Orchid Hybrids

Among the cut flower crops, *Cymbidium, Dendrobium, Phalaenopsis, Odontoglossum, Oncidium, Cattleya, Paphiopedilum, Vanda, Aeridovanda, Aranda, Mokara, Arachnis, Vascostylis, Renanthera, Rhyncicentrum,* and *Rhyncovanda etc.* are important.

A good quality cut flower of an orchid should have the following characteristics.

i) Minimum eight standard blooms per stem

ii) Flowers must be cleaned, evenly coloured and free from physiological disorders

iii) Stem must have flowers evenly arranged and around the stem.

iv) Two third of the stem should be covered with the flowers.

v) Flowers must have a firm texture and a luminescent sheen

vi) Stems must be firm when held up

vii) The minimum base diameter of the stem should be of 10mm

Important varieties and hybrids under different genera of orchids for cut flower purposes are listed in Table 2.

Table 2. Important Hybrids under different Genera of Orchids

Genera	*Hybrids*
Aeridovanda	Doctor Poyck, Vieng Ping, 'Bensiri', 'Noreen', 'Early Bird', 'Shiv Sidhu', 'New Dawn', 'Harrison Luke Somsri Sunlight'
Aranda	'City of Singapore', 'Hilda Galistan', 'Urmila Nandey', 'Christine', 'Thailand Sunspot', 'Millenium Dawn', 'Broga Giant', 'Salaya Red', 'Propine White', 'Propin Spot', 'Lueng Cholburi', 'Ishbel Manisaki', 'Baytown', 'Chao Praya Blue', Chao Praya Dot Com', 'Chao Praya Beauty', ' Ethan Pride', 'Taksari Chandrabir', 'Thailand Sunspot'
Arachnis	'Ishabel', 'Maggie Oei', 'Maroon Maggie', 'Bartha Braga'
Ascocenda	'Apinantat Red Berry', 'Pralor Tuyen', 'Pak-Kred', 'Bangkok', 'Surin', 'Karnada', 'Crownfox', 'Sundancer', Laksi 'Red Ruby',Guo Chia Long 'Spotty', 'Fuchs Angel Frost', 'Carol Belk', 'Renuka Angle', 'Joyce Bevins', 'Adisak Blue', 'Renu Gold', 'Tipi Blue Boy', 'Bobs Fortune', 'Rubychai', 'Shah Rukh Khan', 'Yang Sophia Firuz', 'Abdul Ghani Othman', 'Chunika', 'Fuch's Star'
Cattleya and allied	'Lovely Bangkok', 'Admiration', 'Bob Belts', 'General Patton', 'Joyce Hannington', 'Little Angel', 'Margaret Stewart', 'Nillie Roberts', 'Pearl Harbour', 'Primma Donna', 'Queen Sirkhit' 'Diamond Crown', 'Secret Love', Ladda Belle 'Pink Pearl', 'Maikai', 'Pastoral', 'Robert', Prism Palette 'Tricolour Magic', 'Ahmad Seikhi', 'Hsinging Catherine', 'Chia Lin New City', 'Chinese Beauty Orchid Queen'
Cymbidium	'Levis Duke Bella Vista', 'Madrid Forest King', 'Sparkle Late Green', 'Angelica December Gold', 'Sleeping Nymph', 'Pine Clash Moon Venus', 'Soul Hunt', 'Dr. H. C. Aurora', 'Susan Highes', 'Tia Gaig Suther Land', 'Miss Sanders', 'Amesbury', 'Kenny Wine', 'Red Star', 'Red Princess', 'Show Girl', Jungfrau 'Snow Queen', Jungfrau 'Dos Pueblos', Lilian Stewart 'Coronation', Lilian Stewart 'Party Dress', Orkney 'Pink Heather', Ensikhan 'Alpha Orient', 'Fire Storm Ruby', 'Fire Storm Blaze', Bob Marlin 'Lucky'
Dendrobium	'Emma White', 'Thongchai Gold', Julie', 'Erika, Sonia-17', 'Sonia-28', 'Kasem White', 'Madam Pompadour', 'Bangkok Blue', 'Ann', 'Gold Twist', 'Candy Stripe Pink', 'Genting Blue', 'Bengal Beauty', 'Sakura Pink', 'Burana Charming', 'Blue Fairy', 'Channel', 'Nette White', 'Dang Saard', 'Big White 4N', 'Ear Sakul', 'Erika', 'Lervia'
Mokara	Walter Oumae 'Seksan', Thailand, Sayan, Walter Oumae 'Royal Sapphire', Susan 'Orange', Walter Oumae 'Calypso', Eng Ling, Madame Panne, Mak Chin On, Bangok Gold, Bibi, Chao Praya Gold, Chark Kuan Orange, Chark Kuan Pink, Chark Kuan Rose, Chark Kuan Super, Dinah Shore, Kelvin Red, Kelvin Orange, Luenberger Gold, Margaret Thatcher, Pink Star, Sayan, Sayang Pink, Walter Oumae, WTO, Jiti, Happy Beauty, Salaya Gold
Odontoglossum	Carroll, Ismene, Palnina, 'Italian Job', 'Joyce Stewart', 'Pepe Gerald', 'Purple Rain', 'Katherine Jenkins', 'Roy Wittwer', 'Laguna Blanca', 'Precocious', 'Snow Fall', 'Pesky', 'Queen of Mars', 'Stam Point', 'Holiday Yellow', 'Ronald Norman', 'Bridget Ring Lawless', 'John Hanry Hanson', 'Point Pesky'
Oncidium	Aloha Iwanga Dogasima, Goldiana, Gower Ramsey, Golden Shower, Sum Lai Who Jungle Queen, Taka H and R, Sharry Baby Sweet Fragrance AM/AOS, Golden Glow, Popki Red, Irine Gleason Red, Vision Brownish Red, Catherine Wilson x New Calidonia Brownish Red, Robson Orchid Glad
Paphiopedilum	Niveum, Concolour, *P. rothschildianum* (3 to 5 flowers), *P. sanderianum* (3 to 5 flowers), Prince Edward of York, Michel Koopwitz, Saint Swithin, Mount Toro, Sorcerers Apprentice, Grande, Don Wimber, Elizabeth March, Hanne Popow, Jason Fischer, Living Fire
Phalaenopsis	Taisuco Crane, Taisuco Kochdian, Cygnus, Yukimai, Sogo Musadian, White Dream, Florida Snow, Nobby's Pink Lady, Minho Valentine, Minho King Beauty, New Cinderella, Taisuco Firebird, Sogo Smith, Carol Campbell, Emil Giles, Brother Lawrence, Taipei Gold, Golden Bells, Sogo Managers, Brother Passat, Be Glad, Cassandra, Vilind, Carmelas Pixie, Zuma's Pixie, Timothy Christopher, Be Tris, Quevedo, Strawberry, Detroit, Maki Watnabe, Kaleiodoscope
Renanthera	Brookie Chandler, Manila T-Orchids, Kilauea, Mok Yark-Seng, Poipu, Tom Thumb, 'Red Leopard', 'Scarlet Belle', 'Chanachae', 'Serdang', 'Brady Crocker', '20th WOC Singapore-2011', 'Bart Motes'
Renantanda	'Forever Yvonne', 'Inspiration Ng Teng Fong', 'Ladda Glow'; 'Polyetheramine Singapore', 'Momon Shija', 'Paul Gripp', 'Science Arts', 'Memoria Charles Darwin', 'Prof. G.J. Sharma', 'Kebisana Shija', 'Mary Motes', 'Kofi Annan'
Rhynchovanda	'Wilton Hill', 'Jammie Harper', 'Apichart', 'Noo Noi', 'Peter Draper', ' Brighton's Albino', 'Prairie Lady'
Vanda	Annette Jones, Antonio Real, Golamcos Blue Magic, Fuch's Charmer, Jimmy Millers RF Orchids, Keree Delight, Memoria Lyle Swanson, Motes Indigo x Merrillii, Motes Honeybun, Motes Primerose, Miss Joaquim,V. Rothschidiana, VTMA –Red, Pink, White, Vasco, Johnny Miller,Veerawan, Roberts Delight, Rasriprai, Pat Delight, Pakchong Blue, Mimi Plammer, Manuvade, Lumpini Red, Kultana Gold x Thongchai Gold, Fuchs Delight, Charles Goodfellow, Pine River,Adisak, Doctor Anek, John Club, Bill Sutton, Ellen Noa, Emily Notley, Evening Glow, Honomu, Honolulu, Hilo Blue
Vascostylis	Paragon Joy x Kasems Delight, Precious, Veeraphool, Crown Fox 'Red Yen', Aroon Fairy, Viboon Velvet, Chao Praya Lime', 'Lanna Rosy', 'Jeans Delight', 'Bay Sapphire', 'Spring Hill'

Source: Bhattacharjee and De, 2005; De *et al.*, 2013

***Cymbidium*, 'Fire Storm Ruby'** **Dendrobium*, 'Dang Saard'** ***Oncidium*, 'Sharry Baby Sweet Fragrance'**

***Vanda*, 'Roberts Delight Blue'** ***Phalaenopsis*, 'Strawberry'**

Figure 1. Commercial Orchids Varieties

4.0 Nature and Causes of Postharvest Losses

Opening of flower buds and rate of senescence accelerate at higher temperatures. At lower temperature, the respiration comes down and the flowers produce less amount of ethylene. Temperature plays an important role for flowers harvested at the immature stage for full expansion of buds and the flower buds are kept at temperatures as low as 0.5 to 4.0 °C in *Cymbidium* and *Paphiopedilum*, 5-7 °C in *Dendrobium* and 7-10 °C in *Cattleya*.

Light is essential for long distance transport or prolonged storage of cut flowers. Similarly, high light intensity is essential for opening of tight bud cut flowers. Florists had to maintain a light intensity of 2000-3000 lux for 12-24 hours in their shops for illuminations for most of cut flowers.

Table 3. Vase Life of Orchids for different Species and Varieties

Name of Species/Hybrids	*Vase Life (Days)*
Aerides odoratum, Aerides multiflorum, Cymbidium iridioides, Dendrobium nobile, Renanthera imschootiana	28-56 days
Aranda	18-28 days
Cattleya hybrids	10- 20 days
Cymbidium hybrids	20-55 days
Dendrobium hybrids	14-21 days
Paphiopedilum hirtussimum, P. wardianum	56 days
Phalaenopsis hybrids	25-30 days
Phaius tankervillae	28-42 days
Vanda coerulea, Vanda teres, Zygopetalum intermedium	14-21 days
Vanda, Mokara hybrids	14-30 days

Source: Bhattacharjee and De, 2005

Cut flowers are kept at 90-95 per cent relative humidity for maintaining turgidity. Flowers start showing wilting symptoms when they have lost 10-15 per cent of their fresh weight. The rate of transpiration from leaves is reduced with the increase of high relative humidity (Figure 1).

5.0 Postharvest Management

Postharvest life of orchid cut flowers is influenced by preharvest factor like varietal differences, light intensity, sugar level of flowers, temperature, nutrition and water loss; harvest factors like time, method and stage of harvest and postharvest factors including ethylene production, precooling, water quality, pulsing, use of preservatives, bud opening, packaging and storage. It is estimated that a third of the postharvest life of flowers is programmed by preharvest handling. Varietal differences in cut flowers had been reported due to variations in water uptake, fresh weight, flower diameter, stem lignification, vase life and senescence behaviour. The vase life of cut flower species and hybrids ranges from 10 to 56 days (Table 3).

5.1 Harvesting

Flowers should be harvested at mild temperature because high temperature causes rapid respiration rates and excessive water loss. Flowers should be harvested in the early morning or in the evening. In the early morning, flowers remain turgid due to transpiration at night and higher sugar levels. Similarly, flowering stems retain a higher amount of stored carbohydrates if cut in the afternoon and retain more vase life.

Table 4. Stage of Harvest, Spike Length and No. of Flowers of some Commercial Orchids

Orchid Hybrid	*Stage of Harvest*	*Spike Length (cm)*	*Flowers (No.)*
Aranda	50 per cent bloom	45-60cm	8-10
Cattleya	2-4 days before bud open	25-40 cm	1 or more
Cymbidium	75 per cent bloom or two buds open stage	60-90 cm	10-15
Dendrobium	All flowers except top bud	40-60cm	8-12
Oncidium	80 per cent bloom	60 cm	Many
Paphiopedilum	3 to 4 days after opening of flowers	25-40 cm	1-5
Phalaenopsis	Fully open flowers	40-60 cm	8-10
Vanda	Fully open flowers	50-75 cm	8-15

Sharp tools or secateurs are always used to detach the stem of flowers from the mother plant. The angle of the cut is given in a slanting position and the stem is not crushed during harvesting, especially hard wood stems. The spikes are dipped in a bucket containing water immediately after harvest.

The optimum harvesting stage of the commercial orchids is fully open and mature flowers. The stage of harvest, spike length and number of flowers of some commercial orchids are given in Table 4.

5.2 Temperature

Generally, higher temperature results in higher level of respiration. Cooling is essential to reduce other metabolic changes such as enzymatic activity and to slow the maturation of flowers. Cooling prior to packaging and transport reduces ethylene production and improved longevity.

5.3 Light Intensity

Light is found to determine the carbohydrate levels before harvest which in turn influences the keeping quality. Flowers containing relatively higher amounts of carbohydrates, especially mobile sugars lasted longer in the vase. Plants having few leaves, or leatherlike leaves (like most cattleyas and oncidiums), require a high-light environment. If the leaves are soft and limp (like some phalaenopsis and most paphiopedilum), the plants are probably very light-sensitive, and should not be placed in a sunny south-facing window. Most orchids prefer indirect or filtered light and 50 per cent shading. Some examples of light intensity requirements of orchid types are mentioned below:

i) Low light orchids (1200-2000 f.c.): *Phalaenopsis, Calanthe*

ii) Medium light orchids (2000-3000 f.c.): *Cattleya, Laelia, Brassovola*

iii) High light orchids (3000 f. c. or more): *Cymbidium*, Vandaceous groups

5.4 Precooling

It is an important operation in postharvest handling and transport of cut flowers for the fast removal of field heat wherever flowers are held dry pack. All flowers are precooled immediately after harvest by placing them in cold storage without packing or in open boxes until they reach the desired temperature. The temperature varies with the species and cultivars *e.g. Cattleya* (7-10 °C), *Cymbidium* and *Paphiopedilum* (0.5 to 4 °C), *Dendrobium* (5-7 °C). Precooling lowers respiration rate and decreases the break down of nutritional and other stored materials in the stems, leaves and petals; delays bud opening and flower senescence. It also prevents rapid water loss and decreases flower sensitivity to ethylene. Several precooling techniques such as room cooling, forced air cooling, hydro-cooling, vacuum cooling and ice bar cooling *etc.* are available. The resistance to water flow through stem segments increase predominantly in the lower most one to two centimeter of the cut stems. Re-cutting stems under water improves longevity due to elimination of air from the conducting vessels.

5.5 Conditioning

Conditioning or hardening is a simple process where flowers are kept standing loosely in a big container so that air can circulate. The purpose of the treatment is to restore the turgidity of cut flowers with water stress during storage and transport. Conditioning is achieved by treating the flowers with de-mineralized water supplemented with germicides and acidified with citric acid to pH 4.5 to 5.0 but with or without sugar. Hydration is improved when water is de-aerated or acidified or when a wetting agent like Tween 20 at the rate of 0.01 to 0.1 per cent is added. Flower stems are placed in warm water or a preservative solution in plastic jars at a depth of 2-4 cm and held at room temperature or in cold storage for several hours.

5.6 Impregnation

Sometimes, the cut ends of the flower stems were impregnated for a short time with chemicals. This treatment protects the blockage of the water vessel in the stem by microbial growth and stem decay. Impregnation of cut bases of flowers with high concentration (1000 ppm-1500 ppm) of silver nitrate, nickel chloride or cobalt chloride for 10-15 minutes improves the longevity of several flowers such as aster, gerbera, gladiolus, carnation, chrysanthemum, phalaenopsis, cymbidium and snapdragon.

5.7 Pulsing

The absorption of chemical solutions containing sugars and germicides through the lower cut bases of flower stems is known as pulsing. Pulsing could be used by growers, wholesalers or retail florists in order to enhance the cut flower vase life in water. Pulsing is employed with higher concentrations of sugar, mainly sucrose, the percentage of which varies with species and cultivars. Other chemicals used in the pulsing treatments were STS, $AgNO_3$, HQ, MH, AOA, $CaCl_2$, $CoCl_2$, nickel sulphate, aluminium sulphate and benzyladenine. Pulsing is found to be of great value in prolonging life, promoting opening and improving the colour and petal size of petals through osmo-regulation.In *Cymbidium* hybrid 'Red Princess' pulsing with 5 per cent sucrose increased vase life (56 days) followed by sucrose @ 8 per cent (54.78 days). In *Phalaenopsis*, pulsing with 0.5mM STS for 24 hours blocked the deleterious effect of ethylene. In *Dendrobium* cv. 'Sonia' pulsing with 4 per cent sucrose + 400 ppm HQ recorded the highest vase life of 21.33 days. Inflorescences pulsed with 6 per cent sucrose + 400 ppm HQ recorded the highest sugar content in the flowers (27.64 per cent).

5.8 Bud Opening

It is a procedure of harvesting flowers at a stage earlier than normally considered as the cutting stage and then opening the buds off the plant. Such types of postharvest handling could be applied by growers or wholesalers. Bud opening of flowers increases longevity of cut flowers by reducing the sensitivity of flowers to extreme temperatures, low humidity and ethylene, saving space during shipment and extending the useful storage life. The sugar concentration used is lower than the concentration of pulsing and the optimum temperature is kept lower.

In *Dendrobium* hybrids, HQS or $AgNO_3$ (50ppm) was effective for opening of tight bud cut flowers. In *Cymbidium* hyb. 'PCMV', sugar(4 per cent) + salicylic acid (200 ppm) showed maximum per cent of flower opening (75 per cent) and vase life (45 days) followed by sugar (4 per cent) + $Al_2(SO_4)_3$(100 ppm) (57 per cent and 44 days) and sugar (4 per cent) + 8-HQS (200 ppm) (53.8 per cent and 44 days) over control (bud drop and senescence on 27th days), respectively.

5.9 Preservatives

Preservatives are used in the holding solutions in the form of tablets containing a mixture of chemicals such as sugars, germicides, salts, growth regulators *etc.* Besides, the chemicals are employed during conditioning, pulsing and making bud opening solutions to improve flower shape, size and opening and colour of the flowers. Sugar, biocide, anti-ethylene compounds and hydrated compounds are used for conditioning. The sugar and biocide solutions are effective for opening of bud cut flowers. The vase solution should contain sugars, acidifying agent and a biocide. Citric acids are mainly used for acidifying agent and hydroxy quinoline as biocide. Metallic salts like silver nitrate, cobalt chloride, aluminium sulphate, zinc sulphate, calcium nitrate and nickel chloride have been found for prolonging postharvest life of various cut flowers. Among several growth regulators used to increase vase life of cut flowers, BA, IAA, NAA, 2,4,5-T, GA3, B9, CCC are

Table 5. Holding Solutions for different Types of Orchids

Name of Orchid	Holding Solution
Arachnis, Aranda, Aranthera, Cattleya	STS (1mM) + 1 per cent sucrose
Cymbidium	8-HQC 200 ppm + sucrose 2 per cent
	1-MCP (500ppb)
	1 per cent Sucrose + STS (1 mM)
Dendrobium	8-HQC (200 ppm) + sucrose (2 per cent),
	0.5 mM AOA + 4 per cent sucrose,
	$AgNO_3$ (30 ppm) + 4 per cent sucrose
	400ppm HQ + 30ppm $AgNO_3$ + 2 per cent sucrose
	200 ppm 8-HQS + 50ppm $AgNO_3$ + 8 per cent sucrose
Oncidium	8-HQC (100-200 ppm) + 4 per cent sucrose
	kinetin (50ppm) + 4 per cent sucrose
Paphiopedilum	8-HQC (200 ppm) +2 per cent sucrose
Vanda	$AgNO_3$ (30 ppm) + 1.5 per cent sucrose

Source: De *et al.*, 2013.

Table 6. Grading of Cymbidium and Dendrobium Cut Flowers

Cymbidium	*Grade*	*Flower Count*	*Spike Length*	*Other Consideration*
Cymbidium Standard	AAA	>12	1.25m	Strong straight stems, uniform length, no marks on flowers.
	AA	8	90 cm	
Cymbidium Miniature	XL	>15	65+	Bent crooked spikes but with perfect flowers.
	L	12-14	55-64	
	M	8-11	40-54	
	S	< 5	30-39	
Dendrobium	Small (S)	4-5	30cm	
	Medium (M)	6-8	40cm	
	Large (L)	8-10	45cm	
	Extra Large	>10	50cm	

common. New chemicals that had been found promising as floral preservatives are ethylene inhibitors like amino-oxyacetic acid, 1-amino cyclopropane, aminotriazole, aminoethoxy vinyl glycine, alpha aminoisobutyric acid, diazocyclopentadiene and phenidone. Different chemicals used in holding solution for improving vase life of orchids are listed in Table 5.

5.10 Grading and Packing

5.10.1 Grading

Grading is done based on the parameters like appearance, stage of maturity, blemishes or injuries due to diseases, infestations caused by insect pests, colour and size of the bud, straightness, strength as well as length of the stem. Flowers are generally grouped into bunches of 5, 10, 12 or 20 stems and loosely tied with rubber bands. Before placing them in the package, individual flower bunches are wrapped with suitable packing materials like cellophane paper, kraft paper, newspaper, tissue paper or corrugated cardboard sheet. For local markets, the bunches are held in buckets containing water or preservative solution. It is advisable that for long distance transport and storage, flower bunches are held in dry cardboard boxes. The minimum length of boxes should be about twice the width. Use of telescope type boxes made of CFB are ideal. Grading of cut flowers of orchids is done based on flower count and spike length of different genera (Table 6).

5.10.2 Packaging in orchids

Cut flowers are inserted in a tube containing water or water with preservatives or simply wrapped in wet cotton swab while covered with a piece of plastic and tied with rubber band to keep them in place. Bunches containing 5 or 10 spikes or individual spikes are placed inside the box in alternate fashion. Ethylene scrubbers with $KMnO_4$ or Purafil are also be kept in the box. For export purpose, packing of flowers in two piece box is a better option. In *Cymbidium*, single flowers backed by a fern leaf are inserted in small flasks containing preservative solution; the flasks are then packed in 3 sided box with a display window. In *Dendrobium* hybrid 'Sonia-17' a 100 gauge polyfilm and cotton dipped in 8-HQS (25ppm) covering the base of the spike rended maximum vase life and flower quality. Number of cut spikes packed per box for standard Cymbidium orchids are given in Table 7.

Table 7. Packing Requirement for Cymbidium Orchids

Stems/Box	*Standard Box*	*Super Box*
6	15+	—
8	11-14	14+
10	9-10	11-13
12	7-8	9-10
15	5-6	—

The benefit cost ratio of packaging of *Cymbidium* ranges from 1.0 in spikes to 1.5 in florets whereas in Dendrobium benefit cost ratio of packaging of spikes is 0.67.

Postharvest Management of Orchid Florets

In elite *Cymbidium* hybrids, longevity of cellophane paper packing of fully open florets ranges from 15 days (Soul Hunt 6) to 39 days (Caripepper Peachy Keen) over unpacked florets *i.e.* 9 days and 13 days, respectively.

In *Cattleya* hybrids, longevity of cellophane paper packing of fully open florets ranges from 14 to 45 days over unpacked florets *i.e.* 7 to 11 days.

In *Phalaenopsis* hybrids, longevity of cellophane paper packing of fully open florets ranges from 14 days (Phal. 'Boston') to 33 days (Phal. 'Ox Plum Rose x Black Jack').

5.11 Storage

Low temperature treatment during storage or shipment period reduces the entire metabolism in the tissues, slows down respiration, transpiration and ethylene action and retards the multiplication of bacteria and fungi. Various packaging of cut flowers are presented in Figure 2.

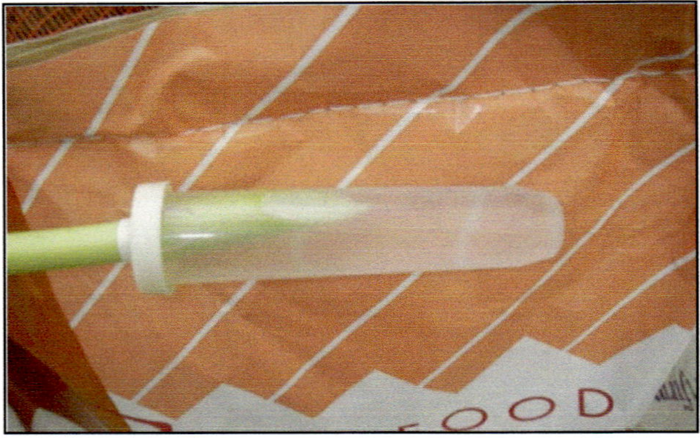

(a) Cut Flowers are Inserted in Tube with Preservatives

(b) Individual Spikes Inside the Box in Alternate Fashion for Domestic Transport

(c) Polyethylene Wrapping of Cut Cymbidium

(d) Packing of Cut Cymbidiums with Local Packaging Materials

(e) Packing of Cut Cymbidiums for Export Markets

Figure 2. Packaging of Cut Flowers

In general, temperate orchids are stored at lower temperature even at 5 °C in cold chambers whereas tropical orchids are stored at 7-10 °C. A 90-95 per cent relative humidity is necessary during storage to minimize moisture loss and prevent wilting.

There are two types of cold storage methods, namely 'Wet storage' and 'Dry storage'. In wet storage, flowers are stored with their bases dipped in water or preservative solution for a short time. Dry storage methods are used for long term storage. In this method, fresh flowers are harvested in the morning, graded and sealed in plastic bags or boxes to prevent the loss of moisture.

In Controlled Atmosphere (CA) storage, cut flowers are kept in gas tight cool chambers equipped with cooling systems with a higher level of CO_2 and lower level of O_2 to reduce the respiration rate and production and action of ethylene. Generally, the concentration of CO_2 is maintained higher than 4 per cent and not below 0.4 per cent in CA storage (Table 8).

Table 8. Storage of Orchid Cut Flowers

Name of Orchid	*Storage Temperature (°C)*	*Storage Period (days)*
Arachnis, Aranda, Aranthera, Ascocenda, Epidendrum	8-13 °C	10-14
Dendrobium	5-7 °C	10-14
Cymbidium	1-4 °C	14
Oncidium, Phalaenopsis, Odontoglossum, Cattleya	7-10	14
Paphiopedilum	–0.5-3. °C	20

5.12 Transport

Being short lived and perishable flowers are to be delivered to destination as early as possible immediately after harvest. For long distance markets, cut flowers are transported by cargo planes, merchant ships and trucks. Other modes of transportation are head loads, bicycles, two-three wheelers, cars, vans *etc*. Hence, for long distance transportation, advanced methods of postharvest handling like cooling, conditioning, impregnation, pulsing, bud opening and packaging were followed.

Short time pulsing of flowers with optimal concentration of sucrose, $AgNO_3$, STS and growth regulators etc. was important for long distance road and sea shipments. Refrigerated transport of cut flowers including temperate orchids from the growers to the consumers is the best method.

6.0 Domestic Marketing and Export

India is blessed with varied agro-climatic zones congenial for production of sensitive and delicate floriculture products.

Export Scenario in Orchids

Thailand is the largest exporter of orchids as cut flowers to India with 80.67 per cent of total import followed by Netherlands 15.54 per cent, New Zealand 2.29 per cent and China 1.5 per cent, respectively. Countries that export Dendrobium to India are Thailand and Netherlands. Maximum export of orchids was found in 2016-17 (Rs. 5.23 lacs) followed by 2017-18 (Rs. 4.89 lacs) (Table 9).

Table 9. Export of Orchids from India

Year	*Quantity (MT)*	*Value (Rs.in Lakhs)*
2013-14	0.72	2.72
2014-15	2.09	3.24
2015-16	0.99	2.11
2016-17	0.74	5.23
2017-18	2.54	4.89
2018-19	0	0

Source: DGCIS Annual report.

Import Scenario in Orchids

Data on the import scenario of orchids from India indicate the increased trend of import value of orchids as compared export value since 2013-14 to 2018-19 (Table 10). Highest import of orchids was recorded in 2013-14 (Rs. 3425.76 lacs) followed by 2015-16 (Rs. 2985.19 lacs) and 2018-19 (Rs. 2321.84 lacs) .

Table 10. Import of Orchids in India

Export of Orchids from India	*Quantity (MT)*	*Value (Rs. in Lakhs)*
2013-14	1248.77	3425.76
2014-15	640.49	1901.12
2015-16	1532.09	2985.19
2016-17	84.61	146.15
2017-18	832.66	1193.80
2018-19	1044.36	2321.84

Source: DGCIS Annual report

India's total export of floriculture was Rs. 541.61 crores 2019-2020. The major importing countries were USA, Netherlands, Germany, United Kingdom, Canada and Japan. There are more than 300 export-oriented units in India.

The value of fresh cut orchids and buds trade during 2007-2012 with the average trade value was US $ 483 million. In 2012, there were more than 40 exporting orchid countries and 60 importing orchid countries around the world, and the total size of the global trade was US $ 504 million (Table 11).

The Netherlands is the top most orchid exporting country (39. 67 per cent) followed by Thailand (28.41 per cent), Taiwan (10 per cent), Singapore (10 per cent) and New Zealand (6 per cent), respectively. Importing countries are mainly Japan (30 per cent), UK (12 per cent), Italy (10 per cent), France (7 per cent) and the USA (6 per cent), respectively. The total orchid cut flower trade of the world mostly consists of 85 per cent Dendrobium species and 15 per cent Phalaenopsis and Cymbidium species and Asia is the main source of orchids to enter the world. Major markets in Asia are occupied by Japan and Singapore. Total imports of orchids by Japan accounted for US$ 57.4mn in 2008 making it the largest importer of the orchids in the world. The main sources of imports include Thailand, Taiwan, New Zealand and Malaysia together accounting for as much as 96.5 per cent of the total imports of the orchids by Japan in 2008. Imports by Singapore of fresh orchids are mainly for Malaysia, Thailand and Taiwan. Indian exports of orchids are insignificant in global trade and this clearly indicates that there is vast scope of increasing India's exports to Singapore particularly considering the proximity of the country and India's East Policy.

Thailand is the largest exporter of tropical orchids. China is the largest consumer of orchid as cut flowers from Thailand followed by Japan, USA, Italy, India, Taiwan, Vietnam and the Netherlands. Other significant orchid genera being exported from Thailand were *Mokara,Aranthes, Aranda, Oncidium, Vanda, Arachnis* ans *Ascocenda* with very small market.

In India, it has taken almost a century for Sikkim and the northeastern region to realize the potential of *Cymbidium*, with *Dendrobium*, *Vanda*, *Cattleya*, *Phalaenopsis* and *Oncidium* orchids as revenue earners. While a late start it is turning out to be a promising sector. Sikkim now annually produces Rs 20 million worth 150-200 million spikes of orchids on an average. A cut stalk (spike) costs between Rs.20 and Rs.150 depending on the quality. The state government has introduced a special package in 18 clusters under which 50 flori-farmers have been given 500 orchid plantings (saplings) each with necessary support for cultivation, training and marketing. The state has to overcome several challenges to send the produce out to the national market. The first hurdle is transportation to the nearest market place. An effective cold chain is another area of concern.

Orchids and other horticultural products are marketed by SIMFED - the government's national sales network. Veteran flower trader Andy Warren, managing director of the New Zealand-based company Bloomz, proposed that Sikkim and the northeastern states - the country's traditional orchid hothouses - should concentrate on the domestic market and explore its "optimal capacity" for high returns instead of eyeing bigger export share globally. This region lacks postharvest facilities like good packaging houses and cold storages. During hot summer logistics to move the harvested products are inadequate. Flowers like orchids wilt. The region needs to look at expanding markets in New Delhi, Mumbai, Bangalore and Chennai to fetch good prices. The critical thing is to ensure top quality in the market every day.

Strategies to promote orchid export and to reduce the import of plastic flowers from other countries

(i) Exploration and identification of unique orchid germplasms having aesthetic values in terms of genetic resources as parental materials, cut flowers, potted ornamentals, hanging baskets, medicinal and aromatic plants.

(ii) Development of market driven hybrids of Cymbidium, Dendrobium, Vanda, Phalaenopsis, Cattleya, Oncidium and Paphiopedilum with attractive blooms of various colours, forms, shapes, sizes and long keeping quality.

(iii) Production of planting materials of commercial hybrids and valuable species at mass scale through micropropagation and other propagation techniques.

(iv) Round the year production technology of orchids through growing of selected hybrids and species.

(v) Area expansion of production in Cymbidium and Paphiopedilum in temperate regions, Oncidium, Cattleya, Zygopetalum in subtropical regions and Dendrobium, Vanda, Phalaenopsis and Mokara in tropical region.

(vi) Improved Postharvest technologies and value addition through dry flower production of unmarketable species and hybrids, single Floret packaging and wealth from waste.

(vii) Domestication, multiplication and popularization of lesser known but high value orchids such as *Dendrobium nobile, Renanthera imschootiana, Rhyncostylis retusa, Aerides spp.,* native *Paphiopedilum species, Vanda coerulea, Cymbidium whiteae, Malaxis spp., Habenaria spp.*

Table 11. Global Trade Value of Fresh Cut Orchids and Buds

(Unit. Million US$)

Year	*2007*	*2008*	*2009*	*2010*	*2011*	*2012*
Import	233,734,023	252,647,645	232,568,129	251,445,523	265,702,077	267,196,847
Export	230,470,421	238,702,950	217, 781, 745	227,389,789	244,996,271	237,543,797
Total	464,204,444	491,350,595	450,349,874	478,835,312	510,698,348	504,740,644

Source: Department of Foreign Trade, Thailand (2013).

(viii) Vertical gardening with suitable orchid genera such as Phalaenopsis, Coelogyne, Dendrobium, Bulbophyllum, Aerides *etc*

(ix) Development of epiphytic orchid based farming system including other ornamentals like Anthurium, Alstroemeria, Gerbera and terrestrial orchids.

7.0 Conclusion

The orchids have taken a significant position in the cut flower industry due to this attractiveness, long shelf life, high productivity, right season of bloom, easy packing and transportation. These account for a large share of global floriculture trade both as cut flowers and as potted plants and are estimated around 10 per cent of international fresh cut flower trade. Keeping in view of the export potential of Orchids, and based on the recommendations of the planning commission the Indian Council of Agricultural Research established a National Research Centre for Orchids in 1996. Orchids are very high value floriculture produce. Integration of crop management with postharvest management is essential for obtaining success for this industry. Selection of appropriate variety/species suitable for the location along with establishment of appropriate postharvest management facilities for handling, packaging, storage and transport etc. are the key requirement for any successful venture into orchid industry. For orchid trade to be successful attention needs to be given on the availability of adequate planting materials and quality flower spikes to compete in the international markets. There is also immediate need to upscale production strategies. The qualitative aspects need to be given due emphasis for value realization in the market. For this, maximum growing standards need to meet requirements, market standards and consumer preferences. However, erect stems, freedom from pests and mechanical injury, flower freshness are some of the indispensable criteria.

8.0 Important References

Bhattacharjee, S.K. 1997. Packaging fresh cut flowers. *Indian Hortic.* **41**: 23-27.

Bhattacharjee, S.K. and De, L. C. 2005. *Postharvest Technology of Flowers and Ornamental Plants.* Pointer Publishers, Jaipur, Rajasthan (India), Pp. 440.

Chowdhery, H.J. 2001. Orchid diversity in north-east India. *J. Orchid Soc. India,* **15**: 1-17.

Chowdhery, H.J. 2009.Orchid diversity in northeastern states of India. *J. Orchid Soc. India*, **23**: 19-42.

De, L.C., Deb, P. Chhetri, Geetamani, Medhi, R.P. and Pokhrel, H. 2013. *Postharvest management in orchids.*Technical Bulletin No.16, NRC for Orchids, Pakyong, Sikkim, pp. 37.

De, L.C. and Debnath, N.G., 2011. NRC for Orchids, *Vision 2030.* Pakyong, Sikkim, India, Pp.28.

De, L.C. and Bhattacharjee, S.K. 2000. Methods for prolonging vase life of cut flowers-a review. *Orissa J. Hortic.* **28**: 73-78.

De, L.C. and Bhattacharjee, S.K. 2002. Vase life of cut rose 'Queen Elizabeth' as affected by aquatic fungi and bacteria. *Indian Rose Annual*, **18**: 86-90.

De L.C..2015. Commercial orchid. Berlin De Gruyter Open. Available at: http: //www.degruyter.com/ view/product/456245

Flora Holland.2015. Facts & figures. Flora Holland facts and figures 2015. Available at https: // www.royalfloraholland.com/media/5685262/ RoyalFloraHolland_Annual_Report_2015_ENG_ facts_and_figures.pdf

Hajra, P.K. and De, Aparajita. 2010. Phyto-geographic analysis of orchid flora in India, *J. Orchid Soc. India,* **24**: 43-46.

Hanks F. 2015. Orchid cut flower industry. Report for the UK Agriculture and Horticulture Development Board and National Cut Flower Centre. Available at: https: //horticulture.ahdb.org.uk/sites/ default/files/u3089/A per cent 20review per cent 20of per cent 20cut-flower per cent 20and per cent 20foliage per cent 20production per cent 20statistics per cent 202015_0.pdf

Jawaharlal, M., Dinesh Babu, M.K. and Indhumathi, 2006. Postharvest packaging techniques for Dendrobium Hybrid Sonia -17. *J. Orn. Hortic.* **9**: 16-19.

Jomy, T.G. and Sabina,G.T. 2002. Effect of conditioning and pulsing on vase life of Dendrobium Sonia inflorescences. *J. Ornam. Hortic.* **5**: 80-81.

Sarkar, Indrajit, Mandal,T. Naveen Kumar, P., Kumar, Rajiv, Misra, Sanyat, Misra, R.L. and Singh, Kishan Pal,2009. *Temperate orchids.* AICRP on Floriculture, Technical Bulletin No.**28**, IARI, Pusa, New Delhi-12, Pp.77.

Singh, Foja 1990. Indian orchids. *Indian Hortic.* **35**: 14-15.

Chapter 34

Cashew

D. Balasubramanian
Principal Scientist
P.L. Saroj
Former Director
ICAR-Directorate of Cashew Research, Puttur – 574 202, Dakshina Kannada, Karnataka, India
E-mail: bavika13@email.com, E-mail: plsaroj@yahoo.co.in

1.0 Introduction

Cashew (*Anacardium occidentale*) is originally native to northeastern Brazil. It was introduced to the Malabar Coast of India in the sixteenth century by the Spaniards and probably served as a locus of dispersal to other centers in India and South East Asia. It is an important plantation crop in India due to its utility in soil and water conservation and to build up of balanced ecosystem. Among horticultural crops, cashew has been known to provide very high economic returns. The cashew nut, popular dry fruit is sweet and delightful in taste. It is a valuable food for both physical and mental health. It is considered a completely nourishing diet and functional food for various ailments. The analysis of the nut shows that it contains 21.2 per cent protein, 2.4 per cent mineral matter, 22.3 per cent carbohydrate, 46.9 per cent fat, 1.3 per cent fibre, 0.05 per cent, calcium, 0.45 phosphorous and 5.0 mg per 100 gm iron. It also contains carotene as vitamin A 100 IU, thiamine 630 mg, niacin 2.1 mg and riboflavin 190 mg per 100 gm of kernel. It is free from cholesterol and its calorific value is more than the egg. The presence of vitamin E in cashew takes care of all reproductive problems and prevents the development of oxidative rancidity in fats. Recently discovered vitamin-PP also develops while roasting cashew and it can exert a sparing action on the B group vitamins and assist in metabolism of lactose and thiamine.

Popularity of Cashew nut is increasing rapidly all around the globe due to its inherent taste and high nutraceutical properties in the human diet. India, the United States of America and Germany are the top three countries in the world with respect to consumption of cashew nuts in the world. It occupied the third most popular tree nuts after almonds and walnuts with 17 per cent of world tree nut production in 2019-20. (International Nut and Dried Fruit Council Foundation, 2020).

Recent reports reveal that the monounsaturated and polyunsaturated fatty acids found in cashews can help decrease LDL cholesterol and triglyceride levels. This reduces the risk of cardiovascular disease, stroke and heart attack (https://www.medicalnewstoday.com/articles/309369#benefits). However, consumption of moderate quantity of mixed nuts in the diet is a better option.

2.0 Status of Cultivation

In India, cashew is cultivated in a total area of 11.05 lakh ha with an annual production of 7.82 lakh MT during the year 2011-12. The average productivity in this country is around 707 kg ha^{-1}. Significant increase in the production of cashew at the global level registered over the past 60 years (Table 1). Africa was the largest producer of cashew till the 70s, accounting for about 68 per cent of the world's cashew production. During this

Table 1: Cashew Production in Major Cashew Growing Countries (1961-2020)

(Quantity in MT)

Country	*Year*						
	1961	*1971*	*1981*	*1991*	*2001*	*2011*	*2020*
India	85000	127223	185250	294590	450000	674600	772779
Brazil	9670	28602	74000	185965	124073	230785	139921
Mozambique	107000	202000	71100	31134	58000	72263	128225
Nigeria	7000	25000	25000	45000	485000	813023	98809
Philippines	7500	5014	3786	98283	111153	133388	255915
Vietnam	700	3500	6500	160000	292800	1272000	348504
Tanzania	50000	126409	64000	29850	122000	75000	232681
World	287535	558846	494157	1003094	2064375	4201010	

Source: FAOSTAT, 2020.

period, Mozambique and Tanzania were the dominant players in the market. In the last three decades, however, there has been a transformation in the global production scenario with Asia overtaking Africa. However, the production of cashews in India surpassed all the leading countries as per the FAOSTAT of 2020.

Table 2: Production Scenario of Cashew (2019-20)

State	*Area (000 ha)*	*Per cent Share*	*Production (000 MT)*	*Per cent Share*	*Productivity (Kg/ha)*
Kerala	98.821	8.78	69.624	9.91	762
Karnataka	132.433	11.77	70.624	10.05	565
Goa	58.350	5.19	27.048	3.85	481
Maharashtra	191.450	17.02	181.138	25.77	982
Tamil Nadu	150.297	15.36	70.11	9.97	499
Andhra Pradesh	192.446	17.11	115.39	16.42	625
Odisha	213.956	19.02	110.00	15.65	595
West Bengal	11.360	1.01	10.964	1.56	965
Jharkhand	15.580	1.38	5.762	0.82	346
Chhattisgarh	31.139	2.77	19.448	2.77	655
Gujarat	8.746	0.78	6.110	0.87	789
Pondicherry	5.000	0.44	2.030	0.29	221
Assam	1.050	0.09	1.062	0.15	1011
Tripura	4.250	0.38	3.243	0.46	729
Meghalaya	80780	0.78	9.50	1.35	1056
Manipur	0.900	0.08	0.305	0.04	610
Nagaland	0.500	0.04	0.508	0.07	564
Total	**1122.058**	**100.00**	**702.866**	**100.00**	**655**

Source: Directorate of Cashew and Cocoa Development, Cochin, Kerala.

In India, cashew is largely grown in different states of the coastal regions *viz.*, Kerala, Karnataka, Goa, Maharashtra, Tamil Nadu, Andhra Pradesh, Orissa, West Bengal and North East Hilly region (NEH) (Table 2). The Directorate of Cashew and Cocoa Development, the Directorate of Cashew Research, the Cashew Export Promotion Council and the Agricultural Universities across the states have been continuously intervening for the development of high yielding varieties, production of grafts and high density planting systems for augmenting the production of quality cashew nuts in this country. Current production data reveals that Maharashtra had emerged as the dominant player in the market contributing almost one-third of the national production but accounting for only one-fifth of the land area under cashew cultivation, showing a higher efficiency in production. Productivity is found to be nearly one and half times more than the national average.

3.0 Commercial Varieties

a) Bhaskara

This variety was released in March 2006 for coastal the region of Karnataka. This is having midseason flowering habit (Dec-Mar) with a flowering duration of 60 days and has the potential to escape from the attack of the tea mosquito bug (TMB) under low to moderate outbreak situations. But the regular insecticidal spray against TMB is essential under severe out break situation. The number of fruits per panicle (bunch) ranged from 4 -13. The average yield on the 13th year was 10.7 kg/tree with the highest yield of 19 kg/tree. The nut and kernel weights are 7.4 g and 2.2 g respectively. The shelling percentage is 30.6 and the kernel grade conforms to export grade W240. The apple colour is pinkish orange and the juice content is 67.5 per cent.

b) BPP 8 (H 2/16)

It is a hybrid (H2/16) derived from the cross Tree No.1 x Tree No.39 and released in 1993 for general cultivation in Andhra Pradesh. It has been performing well in Orissa and West Bengal also. This variety is superior to all the other six varieties developed from Bapatla. The variety has a mean yield of 14 kg/tree with better nut size (8.2g). Shelling percentage (29 per cent) of this variety is also better than the rest of the varieties released from Bapatla so far. Kernel grade is W 210 (export grade).

c) NDR 2-1 (Madakkathara-2)

This is a selection from germplasm collection made from Neduvellur in Kerala maintained at CRS, Anakkayam. This variety was also released in 1987. The mean yield is 17 kg/ tree. The nuts are bold (7.3 g nut weight) with shelling percentage of 26.2 per cent. Kernel weight is 2g having a count of W 240 export grade. Apple colour is red and with weight of apple 63.3g. Reducing sugar 7.8 per cent.

d) Priyanka (H 1591)

This is a new hybrid with parentage of BLA 139-1 X K 30-1 with jumbo nut size developed and released from Cashew Research Station, Madakkathara in 1995. The yield of nuts is 16.9 kg/ tree. The nut weight is 10.8g with kernel weight of 2.87g. Shelling percentage is 26.5. The export grade of kernels conforms to W 180. Colour of apple is yellowish red weighing 135g with juice content of 57.4 per cent.

e) Selection-2

This is a selection from the segregating seedling progeny of 2/9 Dicherla (VTH 40/1) originally a collection made from Andhra Pradesh. This variety was released in 1989. It has early season flowering habit (Nov - Jan) with a flowering duration of 74 days. The number of fruits per bunch is three with the yield of 9 kg/tree. The nut weight is 9.2g with shelling percentage of 28.6 and kernel quality conforms to export grade W 210. Colour of apple is pink.

f) Ullal-3

It is a selection from 5/37 Manjeri and released in 1993. It is early in flowering (November - January) and fruiting period is very short (50-60 days). The fruiting is from January to March and sometimes starts from last the week of December. It is a high yielding variety with an average yield of 14.7 kg/ tree. The nut size is medium with nut weight of 7g. The shelling percentage is 30 per cent and the kernel grade conforming to W 210 grade. The colour of apple is red.

g) Vengurla-4

The hybrid Vengulala-4 (Midnapur Red x Vetore 56) released in 1982 by Regional Fruit Research Station, Vengurala under Konkan Krishi Vidyapeeth, Dapoli. This is a popular variety with red colour apple. The variety is recommended for Maharashtra, Goa and coastal Karnataka. The average nut weight is 7.6 g, average shelling percentage is 31 and kernel is export grade (W 210). This is a cluster bearing variety and the average yield is 19.08 kg/tree. The variety is becoming popular in other parts of the country also.

h) Vengurla-7

Hybrid 255 evolved at Regional Fruit Research Station, Vengurla under Konkan Krishi Vidyapeeth, Dapoli was recommended for release under the name "Vengurla-7" in the XIII Biennial Workshop of AICRP on Cashew held in November 1997. Vengurla-7 is a hybrid developed from the cross Vengurla-3 x M 10/4 (VRI-1). The percentage of perfect flowers is very high (40 per cent). The average yield is 18.5 kg/tree. It is a bold nut type with nut weight of 10g and shelling percentage of 30.5. Kernel count is W 180. The colour of apple is yellow with apple weight of 60g and with juice content of 75 per cent. Average weight of kernel is 2.9g. This variety is recommended for the Konkan region of Maharashtra and adjoining cashew growing regions of Goa and Karnataka.

i) VRI-3 (M 26/2)

This is a selection from seedling progeny of a high yielding tree collected from a village Edayanchavadi in South Arcot District of Tamil Nadu and was released in 1991. It has 12.1 per cent perfect flowers. The average yield of this variety is about 10 kg/tree, thus the increase over VRI-2 and VRI-1 being 35 to 39 per cent respectively. The nut size is medium with 7.2g nut weight and shelling percentage of 29.1 per cent. The kernel grade conforms to W 210 export grade. This variety is picking up fast among farmers of not only of Tamil Nadu but also of other states.

j) Nethra Jumbo-1

A massive hybridization programme was undertaken at ICAR-Directorate of Cashew Research, Puttur to improve the nut size in released varieties of cashew. As a result of hybridization, followed by evaluation under both unreplicated and replicated trials, a jumbo nut hybrid from a cross combination of NRCC Selection-2 x Bhedasi was identified with superior traits for both yield and quality and released as Nethra Jumbo-1. The variety produces 12 g nut with a high kernel recovery of 29 per cent and the kernels fall under grade larger (W130) than presently available highest export grade (W180).

4.0 Nature and Causes of Postharvest Losses

In general, preharvest loss of raw cashew nuts is very minimal and mostly happens due to improper handling of the nuts while transporting them from field to processing site. Hard outer shell withstands impact load preventing physical damage and corrosive oil which is present within the shell structure known as cashew nut shell liquid prevents microbial entry and protect against deterioration. But moisture plays vital role to enhance shelf life of the harvested nuts and hence nuts are dried under sun to reduce its moisture to safer level and stored in gunny bags of 80 kg capacity. Nuts, at times gets spoiled due to high pressure when stacked one over the other and change in the environment, but it is estimated that it will be within 1 per cent only. As far as cashew kernels are concerned, it is handled properly being a costly commodity and only low grade kernels *viz.*, LW, SWP or BB will mix with testa during peeling and will be lost. Cashew apple produced is underutilized primarily due to astringency and lesser shelf life. Around 10-15 per cent of the total cashew apple produced in the country is utilized for the preparation of alcoholic beverage called 'Feni' in the Goa state. Although many cashew apple products are developed, it could not attract consumers.

5.0 Postharvest Management

5.1 Maturity Standards

Cashew fruit has two parts namely cashew nut (true fruit) and cashew apple (pseudo fruit). Developments of these are not parallel as the cashew nut develops first and the pseudo fruit develops later. The whole process of development of both nuts and apples will be completed within 60 days from anthesis. Cashew nut develop till 30 days from anthesis and it gets hardened during subsequent 10 days. Afterwards the development of cashew nut ceases and cashew apple starts developing only after the cessation of growth of cashew nut and loses moisture during subsequent 10 days.

Raw cashew nut has pericarp which is hard in nature with a spongy mesocarp containing caustic oil commercially known as Cashew nut shell liquid (CNSL). The edible kernel is wrapped up by leathery testa, which is the embryo comprised of two large cotyledons, a radical and a plumule. The nut and apple develop in three well defined stages and are delineated by the colour of the nut which generally begins as pink after fertilization, turns green and finally grey at maturity. The embryo and pericarp grow to maximum size when the nut is green. The nut is considered as mature when starts turning grey. The colour of the nuts changes from brownish green at fruit set to light green at ¼th growth of apple and thereafter turns grey irrespective of varieties. In the third stage of development, the growth rate of peduncle continues to increase. The size of the pericarp decreases slightly with the loss of moisture and the embyo remains constant in size. The total weight of the nut at high in the beginning of the third stage.

Development of cashew apple and nut has been studied by number of researchers in different cashew varieties. According to report, specific gravity could be considered as physical index of maturity of nut in cashew varieties. According to a report from Thailand the nut reached maximum size, approximately four weeks after fruit set, hardened in the subsequent week and declined in size by 10 per cent at harvest.

The growth of nut is faster during early stage of development than that of fruit. Changes in the composition of cashew nut and apple during development have been investigated by various workers. The amino acid content in kernel decrease with increased maturity. Cashew kernel lipid, free amino acids and CNSL have been shown to increase with increased maturity. Changes in the kernel sugars and starch were not uniform.

5.2 Harvesting

Raw cashew nuts are harvested in four different ways *viz.*, (i) collection after natural drop; (ii) thrashing by stick and collection; (iii) climbing on the tree and thrashing and (iv) shaking the tree. Studies indicated that about 72 per cent of the farms follow harvesting the nuts by thrashing with stick and collection method. Ultimately nut harvest by this method increased the quantum of immature nuts supplied for processing. Only 16 per cent of farms follow the collection of nuts after natural drop. Besides climbing on the tree and thrashing and shaking the tree to harvest the nuts amounts to 10 per cent and 2 per cent, respectively. These two methods increase the harvest of immature nuts and that lead to inferior quality of the end product. Harvesting by thrashing or climbing also results in the dropping of young flowers and ultimately reduces production. Besides, the nuts have to be harvested at right stage. Normally fallen nuts are collected from the ground, but cultivators tend plucking the nuts before the pseudo fruit develops completely because of pilferage. Fully mature apples are plucked from the tree with the help of a long pole with a *sickle* tied to it at one end. The best method of harvesting the nuts is to wait till it falls, but this damages the pseudo fruit. If one intends to utilize cashew apple, it has to be plucked to avoid damage to the fruit.

5.3 Drying and Storage

Once the cashew fruits are either harvested or collected from the field, the nuts are to be separated from the cashew apple and sun dried for 2 to 3 days (Figure 1). Raw nuts at the time of harvest will have moisture content ranging between 16 to 25 per cent d.b. Sun dried nuts are expected to have a moisture content of 5 to 6 per cent d.b. If the nuts are not dried and stored, it leads to fungal spoilage, resulting in poor quality of kernels. Fungi such as *Gonatobotryum*, *Helminthosporium* sp., *Corynespora* sp., *Alternaria* sp., *Verticillium* sp. and several species of *Aspergillus* have been isolated from stored cashew nuts.

The maximum permissible moisture content of raw cashew nuts is 8.7 to 9.1 per cent. It has been reported that moisture content of whole nut and kernel could be predicted from that of shell with more than 99 per cent accuracy. A higher kernel-shell weight ratio is inversely proportional to moisture content. It has been reported that nuts stored at 27 °C and at a relative humidity (RH) of 70 per cent had a moisture content of 9.2 per cent. Nuts exposed to higher RH (>75 per cent) lead to mould infection within a few weeks. Under high RH, nuts pick up moisture from the atmosphere till they attain equilibrium.

Figure 1. Sun Drying of Raw Cashewnuts

Gunny bags of 80 kg capacity are predominantly used for storage of craw cashew nuts in India. It has been observed that storage of raw nuts with a moisture content of 5 to 6 per cent for a period of 12 months at ambient temperature (25 to 30 °C) did not affect either processing or biochemical quality. Immature nuts and floaters contribute to 2-3% rejects during processing of cashew by steam method. Recent studies on the effect of maturity on the processing quality has clearly indicated that shelling percentage, peeling outturn, whole kernel recovery increases while kernel rejects decrease with maturity. Application of insecticides or pesticides are prohibited to improve shelf life of raw cashew nuts in warehouses.

Raw cashew nut is kidney shaped one with 3.5 mm thick soft leathery outer skin (Epicarp) and thin hard inner skin (Endocarp). Cashew nut shell liquid (CNSL) is present in the honeycomb structure located between these two walls of the shell. Kernel is inside the shell wrapped in a thin brown skin known as the testa (Figure 2).

5.4 Physical Properties

The size distribution of cashew nuts is given in Table 3. In the sample, about 52 per cent of the nuts had lengths ranging from 30 to 35 mm termed the medium fraction, whereas 11.5 per cent and 36.5 per cent had lengths greater than 35 mm (larger fraction) and less than 30 mm (small fraction) respectively. The mean values of 100 measurements at moisture content 8.46 per cent for the length, width, thickness and unit mass of the raw cashew nut were found to be 31.00 mm, 22.86 mm, 16.91mm and 6.04g, respectively. The corresponding values of the kernel were 24.67 mm, 12.06 mm and 1.89g, respectively. A higher mass ratio of 3.20 of nut to kernel indicates a relatively lower yield of kernel per unit weight of raw cashew nut. Shell thickness followed a linear relationship with moisture gain. The absorbed moisture, only after saturating the inter space available in honeycomb structure of shells move to the inner skin and kernel area and this is the reason for increase in its volume at once after soaking.

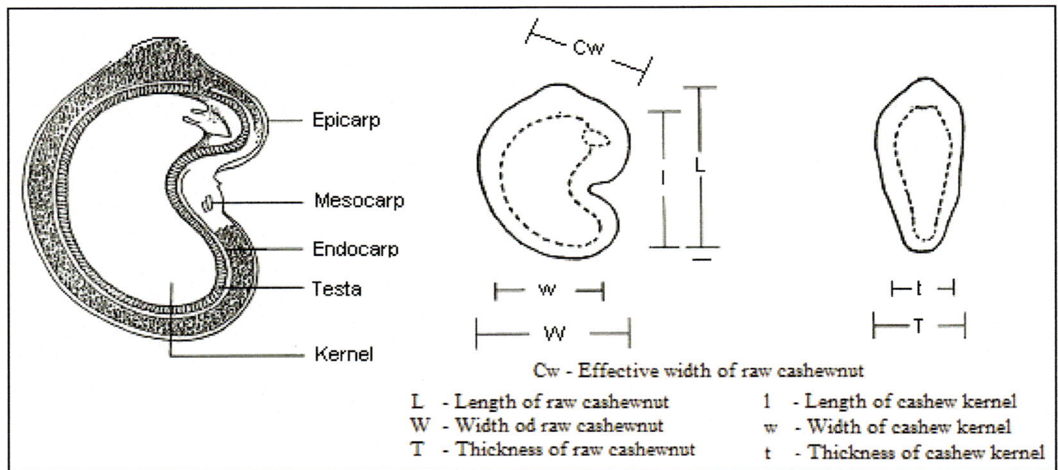

Figure 2. Geometry and Characteristics Dimension of Cashewnut

Table 3. Size Distribution of Cashew Nuts and Kernels at Moisture Content 8.46 per cent db

Particulars	*Ungraded*	*Size Category*		
		Large	*Medium*	*Small*
Length of Nut	23.06–40.06	> 35mm	30-35mm	< 30mm
Percentage of sample				
By number	200	23 (11.5)	104 (52.0)	73 (36.5)
By mass	1193	162.80 (13.7)	668.86 (56.1)	361.34 (30.3)
Average dimensions				
NUT				
(a) Length (L) mm		36.889 ± 1.655	31.868 ± 1.132	27.971± 1.454
(b) Width (W) mm		24.681 ± 1.629	23.270 ± 1.597	21.726± 2.068
(c) Thickness (T) mm		17.623 ± 2.046	17.297 ± 1.638	16.157± 1.655
(d) Mass (M) g		7.209 ± 2.245	6.447 ± 1.001	4.904 ± 0.842
KERNEL				
(a) Length (l) mm		28.169 ± 1.585	25.424 ± 1.473	22.548 ± 1.438
(b) Width (w) mm		12.491 ± 1.830	12.873 ± 2.662	13.300 ± 2.832
(c) Thickness (t) mm		12.556 ± 1.893	12.301 ± 1.350	11.578 ± 1.330
(d) Mass (m) g		2.353 ± 0.375	1.994 ± 0.356	1.611 ± 0.296
SHELL				
(a) Shell thickness		3.810 ± 1.117	4.041 ± 1.134	3.733 ± 0.423

Bulk density decreased with increase in moisture content, shows that the increase in mass owing to moisture gain in the sample is lower than the accompanying volumetric expansion of the bulk. The true density of nut found to increase linearly from 1201 to 1240 kg m^{-3} with increase in moisture content of 3.15 per cent db to 20.06 per cent db. This may be due to higher weight increase of nut in comparison to its volume expansion on moisture gain.

The storage life of raw nuts in the godown also depends on the quality of nut and its environmental condition during storage. Fluctuating temperature is delirious during storage due to condensation of moisture in the godowns at certain pockets and consequent microbial growth. Atmospheric relative humidity (RH) plays a key role in safe storage of raw nuts as the nuts absorb or desorbs moisture to maintain equilibrium with the atmospheric relative humidity. Equilibrium moisture content of raw cashew nut was determined by static method at constant temperature (30 °C) and different relative humidity and presented in Table 4. The increase in the moisture content with the relative humidity exhibits desorption isotherm up to 74.12 per cent, but at 81.33 per cent it followed adsorption isotherm. Mould growth and deterioration took place after 28 days at 81.33 per cent after the commencement of experiment. The data obtained in the Henderson's equation and by solving simultaneous equations the values of the constants were found to be 7.09 x 10^{-4} and 0.865 for c and n, respectively.

$$1 - R.H. = 7.09 \quad 10^{-4} \quad T \quad M$$

Where R.H is the relative humidity expressed in decimal, T is the absolute temperature in °K and Me is the equilibrium moisture content in per cent db.

Table 4. Equilibrium Moisture Content of Raw Cashew Nuts at Various Relative Humidities

Saturated Salt Solutions	*Relative Humidity (per cent)*	*EMC (Per cent db)*
Potassium acetate	26.83	1.669
Potassium carbonate	43.91	3.371
Sodium nitrite	61.17	5.167
Sodium chloride	74.17	7.148
Potassium chromate	81.33	10.786*

* Mild fungus attack started after 28 days of commencement of experiment.

5.5 Quality Assessment

Grade standards have been developed for cashew nut, based on a composite index. It has been found that the number of nuts per kg, moisture percentage and cutting test reduce the prices and these factors could explain 98 per cent variations in price. Among the quality variables considered, only three variables *viz.*, numbers of nuts, unit weight, floating test and nut length have been observed significant in price determination. These results could be used as a guide by the farmers for fetching better prices in the market and by the buyers for getting product of ascertained quality.

5.6 Grading

Kernels are graded based on the specifications prescribed by Govt. of India under the export (quality control and inspection) Act 1963, which recognizes 23 different export grades of kernels. The kernels are conditioned before packing in sealed tins. If the kernels are too dried at the time of packing, they are liable to breakage during transport by land and sea. If the moisture exceeds limit of 5 per cent kernels become susceptible to microbial and oxidative spoilage. Kernel grades of various cashew varieties released from cashew research centre in India is given in Table 5.

Table 5. Kernel Grades in different Cashew Varieties

Grade	*Varieties*
W-180	Priyanka, Vengurla -7, Ullal-4 and UN-50
W-210	NDR-2/1, Dhana, Vengurla -6. VRI-3. Selection-I and Selection-2
W-240	Ullal-I and Ullal-3
W-240	Chintamani -1
W-280	BLA-139/4, Kanaka, Vengurla -3 and Vengurla -4 and Ullal-2
W-320	Jhargram-I, BLA-39/1,K22, Bhubaneswar-1.Vengurla -1, Vengurla – 2 and Vengurla -5, VRI-I and VRI-2

Normally cashew kernels are sorted out manually based on size, surface colour and breakage (Table 6) and the average capacity of grading ranges from 80 to 120 kg and 35 to 60 kg for whole and broken kernels respectively depending on the degree of the skill. High degree of cleanliness is maintained in this section and workers are provided with apron and hair net. Minimum quality requirements are specified as processed kernels should be free from any deterioration; clean, practically free from any visible foreign matter; free from mould or rancidity; free from adhering testa and shell liquid and free of any foreign smell or taste. The moisture content of cashew kernels at the time of packaging is prescribed as 5 per cent d.b.

Table 6. Primary Criteria in Grading Cashew Kernels

Extent of breakage	Wholes (No breakage of kernel)
	Butts (2/3rd of kernel portion chipped off)
	Splits (Whole kernel separated into cotyledons)
	Pieces (All other broken)
Colour	White
	Scorched (Slightly brownish)
	Dessert (Deeply scorched with speckles)
Size	Wholes are graded according to size expressed as count per pound weight
	Pieces are graded according to the sieve size

5.7 Packaging

Cashew nuts are subjected to rancidity and become stale quickly due to its hygroscopic nature and oxidation. Therefore packing needs to have low permeability of moisture and oxygen. Method of packing involves either vacuum or inert gas inside packing. At present cashew nuts are packed by following two different techniques *viz.*, vita packing and molded vacuum packaging (MVP).

a) Mould Vacuum Packaging and Comparison

Introduction of an alternative method of flexible packaging (Mould vacuum packaging) with nitrogen as inert gas is gradually gaining importance. After securing ISO 9000 and introducing HACCP (Hazard analysis and critical control points), importing countries are following stringent quality standards for cashew kernels and the present system of tin packaging is bulky and difficult to dispose of these tin containers. MVP system produces consistent rectangular blocks ranging in size from 500 g to 25 Kg. This gives a large and obvious improvement in the quality production with benefits of minimum movement in transit, handling, display, stock count *etc.*

b) Vita Packing

The graded kernels are packed in 11.34 kg capacity tins on weight basis which are subsequently evacuated and infused with carbon dioxide gas with the help of packing unit called "vita pack" to suppress the possible insect attack. The packed tins are then labelled as per the grades across the lid using special tamper proof adhesive. Two tins each containing 25 lbs of cashew kernels are packed in corrugated carton box, which is bound by nylon strapping for export. The standard markings including brief description, name of packer, gross and net weight *etc.* are printed on the carton.

6.0 Domestic Marketing and Export

6.1 Import and Export Scenario

Over some time, a phenomenal growth is evidenced in the total quantity of raw cashew nuts processed in India. The pioneering spirit of the early entrepreneurs and the unsurpassed skills of Indian women extracting whole kernels from the nuts were the major factors influencing the birth and growth of cashew industry in India and for her preeminence as the world's largest processor and exporter of cashew kernel. A record quantity of 15.01 lakh MT of raw cashew nuts were processed during the year 2011-12, equivalent to 36 per cent of the global production. Although, a steady growth in the quantum of raw cashew nuts processed in this country, but the contribution of domestic nuts in the processing declined from 83.2 per cent during the year 1990-91 to 57.0 per cent during the year 2011-12 (Saroj and Balasubramanian, 2013). This indicates that sustainable production is not attained to operate the industry throughout the year to make it economically viable. Thus, there is a dependency on Indian cashew processors on the import of raw cashew nuts from African country.

Cashew kernels prepared in India are highly preferred by the importing countries and therefore, it stands in the first position as far as export of cashew kernel is concerned. Around 70,087 MT of cashew kernels were exported during 2020-21 earning a foreign exchange to the tune of 3112 crores (Agriexchange APEDA, 2021). The major markets for CNSL are the United States, United Kingdom, Japan and South Korea. Total world supply of CNSL is estimated at about 0.45 lakh MT with an average price of about US$ 300 per MT.

In India, only a small fraction of this potential CNSL is extracted from cashew shell obtained from industries following steam boiling mode of processing using expeller technique. However, these processors could change their processing methods provided the price of CNSL increases significantly.

6.2 Kernel Market

India was originally the dominant supplier of cashew kernel to the world market, but in the last 15 years, Brazil has also become a major exporter especially to the U.S. market. Tanzania and Mozambique were previously important suppliers but are now relatively small. In recent years, Vietnam has become a major source of kernel, as the government has assisted the development of the local processing industry by imposition taxes on the export of raw nuts. As far as the world market is concerned, there are two distinguished kernel markets *i.e.* International market, comprising 99.5 per cent of world trade of cashew kernels which allows the use of inorganic material during production. Another market is the organic market, (0.5 per cent world trade), relatively new development currently offers a 50 per cent price premium. The requirements in this industry are official certification that no chemicals or other banned substances are used during production and processing.

Major kernel markets in the world trade are i) United States Market, New York which is the world's largest cashew market; ii) European Market which includes Holland, Germany, UK and Rotterdam and iii) Australian Market, importing virtually 100 per cent of its cashew requirements as domestic production is still negligible.

In previous years the dominant source of supply was India, however in the past few years Vietnam has become the major supplier because the trader believes Vietnamese crop is better graded and shipping is quicker and cheaper. Although major portion of the processed kernels in India were exported, consumption of cashew kernels at the domestic level increased rapidly after 2000 A D, encouraging the cashew processors in India (Table 7). About 57 per cent of the totally processed cashew kernel is consumed by the Indians during the year 2010-11 indicating more or less 100 per cent in internal consumption over a period of 10 years. Strong processing potential with high demand for cashew kernels within India confirms sustainability cashew nut processing in India.

6.3 Raw Cashew Nut Price

Price information is essential as it provides cashew business to determine the trade operations economically viable either immediately or in a long term. Variation in price of domestic and imported raw cashew nuts are primarily due to the seasonality of indigenous nuts, increasing global demand for kernels and higher share of raw nuts in the processing cost.

The raw nut marketing in Kerala and Karnataka indicated that the nuts are sold immediately after harvest primarily due to the large processing facility in these states. In Goa, nuts are accumulated over a period of time and brought to the market during a particular period in a month. Besides, Agricultural Produce Marketing Centre (APMC) regulates the marketing of raw cashews in the yard and brings it to the marketing stream (Balasubramanian, 2003). Large growers in Andhra Pradesh tend to hold the nut and sell it in large volume. The factors influencing price of domestic raw nuts are i) Quantity of domestic production; ii) Kernel price in the international market; iii) Quantity of kernel exported; iv) Quantity of raw nut imported and v) Imported raw nut price.

The wide gap between domestic production and the demand of the finished goods at premium price

Table 7: Domestic Consumption of Cashew Kernels Processed from India

Quantity in lakh MT

Year	Raw Nuts Processed	Expected Kernel	Kernel Export	Domestic Consumption	Per cent
2000-01	7.21	1.51	0.92	0.59	39.07
2001-02	6.99	1.47	0.89	0.58	39.46
2002-03	8.27	1.74	0.98	0.76	43.68
2003-04	9.01	1.89	1.04	0.85	44.97
2004-05	9.88	2.07	1.01	1.06	51.21
2005-06	11.23	2.36	1.27	1.09	46.19
2006-07	11.38	2.39	1.14	1.25	52.30
2007-08	12.13	2.55	1.19	1.36	53.33
2008-09	13.01	2.73	1.09	1.64	60.10
2009-10	13.65	2.87	1.08	1.79	62.32
2010-11	11.68	2.45	1.06	1.39	56.78

Source: Directorate of Cashew and Cocoa Development, 2011.

and to utilize the labour in processing throughout year led to import of raw nuts into India. Processors execute forward trading with international kernel market at the beginning of the year and forced to import raw nuts during the season to fulfil the contract. Monthly imports are fairly even but decline in May and June when the factories are occupied with the local crop. The different harvesting period of cashew in the world ensures the supply of nuts around the year. This helps to have control over imported nut prices and the Indian processors are assisted by not carrying long-term stock. The nut supply from East African countries is improving due to closure of the processing facility in that country (Balsubramanian[2], 2002). Although, India has a strong domestic market for the kernels produced, the major advantages of exporting kernels over domestic trade are I) export promotion in terms of incentives *i.e.* 7 per cent of value of total exported kernel would be imported without tariff), ii) rapidly increasing equivalent currency rate and iii) kernel demand in bulk quantity.

7.0 Conclusion

Cashew has major economic, social and environmental benefits, not only for rural populations but also for the countries as a whole. In order to maintain India's position in the global market, certain research and development strategies have to be planned carefully and need to be implemented in time. Common quality standards for raw cashew nuts and cashew kernels need to be drafted and after assessment at various levels brought into practice to benefit all stake holders in the cashew value chain. Price stabilization mechanisms for cashew need to be worked out to explore the possibilities of further motivating the farming community. A major challenge to increasing the sales of low grade or broken kernels is to find industrial food processors likely to develop and commercialize cashew-based products or products using cashews as an ingredient on a large scale. Value addition to plain cashew kernels, improvement in packaging technique, branding of cashew, development and promotion of cashew by-products *i.e.* cashew apple and cashew nut shell liquid needs to be given utmost importance. National laboratories need to be strengthened and accredited for certification to develop confidence and assure quality at the consumer's level.

8.0 Important References

Acland, J.D. 1975. East African crops. FAO and Longman, **5**: 29–32.

Antarkar, M.V and Joshi, G.D. 1987. Studies on physical indices of maturity of some cashew nut varieties. *The Cashew,* **1**: 18-21.

APEDA Agriexchange. 2021. https: //agriexchange. apeda.gov.in/indexp/ Product_description_32 headChart.aspx?gcode=0701 visited on 13 jan,2022

Augustin, A and Unnithan, V. K.G. 1981. An attempt on maturities of cashew apple. *Indian Cashew J.* **14**: 9-11.

Balasubramanian, D and Joycy, R.L.K. 2012. Performance Evaluation of Mechanized Shelling Machine for steam treated raw cashew nuts. In: Abstract of PLACROSYM XX. Mechanization of sustainable productivity. Eds. Radhakrishnan, B, Rajkumar, R and Premkumar, R. Coimbatore, Tamil Nadu, India, P: 127.

Balasubramanian, D. 1997. Equilibrium moisture content of raw cashew nuts. *Cashew Bull.* **34**: 2-3.

Balsubramanian, D. 2002. Speculation of raw cashew nut Price – An over view. In: Proceedings of PLACROSYM XV. Sustainability of Plantation Crops through Integrated Approaches for Crop Production and Product Diversification, Mysore, Karnataka, India, P: 205

Balasubramanian, D. 2000. Status report of cashew industries in Dakshina Kannada. *The Cashew*, **14**: 12-25.

Balasubramanian, D. 2000. Physical properties of raw cashew nuts. *J. Agril. Engg. Res.* **78**: 291-97.

Balasubramanian, D. 2010. Design and development of radial arm type cashew kernel extracting machine" *Agricultural Mechanization in Asia, Africa and Latin America*, **42**: 49-55.

Balasubramanian, D. 2012. Comparative performance of mechanized peeling machines for unpeeled cashew kernels – A case study. In: Abstract of PLACROSYM XX. Mechanization of sustainable productivity. Eds. Radhakrishnan, B, Rajkumar, R and Premkumar,R. Coimbatore, Tamil Nadu, India, P: 130.

Hariharan, M and Unnikrishnan, K. 1985. Effect of gibberellic acid and total protein contents in developing kernel of cashew. *Acta Hortic.* **108**: 25-30.

Joseph, K.T. 1981. Seed-borne fungi on cashew (*Anacardium occidentale* L). *Cashew Causerie*, **3**: 12.

International Nut and Dried Fruit Council Foundation. 2020. Nuts and Dried Fruits Statistical Yearbook 2019/20. Reus, Spain.

Mathew, A.G. 1995. Improved shelling machine. *Cashew Bull.* **32.**

Mohankumar, G. N, Narayanaswamy, P and Mukashi, A. N. 1984. Fruit growth and development of cashew. *J. Plantation Crops*, **12**: 81-85.

Nagaraja, K.V. 1998. Quality of cashew kernels in relation to export. *The Cashew*, **12**: 143-48.

Okwelogu, T. N and Mackay, P.J. 1969. Cashew moisture relations. *J. Sci. Food Agric.*, **20**: 697-02.

Rao, C and Joshi, G.D.F. 1987. Studies on physical indices of maturity of some cashew nut varieties. *The Cashew*, **1**: 180-21.

Saroj, P. L. and Balasubramanian, D. 2013. Cashew Industry in India –a sustainable road map. *Indian Hortic.* **58**: 9-15.

Sumrit Feungehan, Taweekiat Yimaswat, Sorot Chindaprasert 1989. Physiological development of cashew nut on canopy flower and fruit. *Khon Kaen Agric J.* **17**: 239-53.

Thivavarnvongs, T.O. and Tand, K.O. 1995. Design and Development of Compact Sized Cashew Nut Shelling Machinery- Testing and Evaluation of Manual and Semi-Automatic Shellers. *J. Japan. Soci. Agril. Machi.* **57**: 57-65.

Chapter 35

Oil Palm

S. Arulraj

Former Director, ICAR-Directorate of Oil Palm Research, Pedavegi – 534 450, Andhra Pradesh, India

E-mail: aruldopr@gmail.com

1.0 Introduction

Oil palm (*Elaeis guineensis* Jacq.) is a native of West Africa and is grown extensively in South-East Asian countries (Malaysia, Indonesia and Papua New Guinea), African countries (Nigeria, Cote de'Ivoire, Ghana, Liberia, Sierra Leone, Cameroon, Republic of Congo and Zaire) and South American countries (Costa Rica, Panama, Columbia, British Guyana, Peru, Ecuador, Venezuela and Brazil). Malaysia, Indonesia and Nigeria are the leading producers of oil palm in the world. Oil palm is a humid crop which requires evenly distributed rainfall of 150mm/ month or 1800 mm/annum. As rainfall distribution is not even and adequate in India, the crop could be grown only with assured irrigation source.

Oil palm is known to be the highest edible oil yielding perennial crop that produces two distinct oils, *i.e.*, palm oil and palm kernel oil, which have culinary and industrial uses. Palm oil is derived from the fleshy mesocarp of the fruit, which contains about 45-55 per cent of oil. The palm kernel oil, obtained from the kernel of stony seed, is a potential source of lauric oil. Oil palm has greater advantage in terms of productivity that is much higher than that other oil yielding crops. It produces 4 to 6 tonnes of crude palm oil per ha annually and 0.4 to 0.6 tonnes of palm kernel oil from 4th to 30th year of its productive life span.

Oil palm is a crop of the present and future vegetable oil economy of the world as well as for India. Palm oil has good consumer acceptance as a cooking medium because of its price advantage. It is a good raw material for manufacturing oleo-chemicals used in the preparation of cosmetics, pharmaceuticals, nutraceuticals *etc.* Broadly, it could be mentioned that palm oil is a source for improving health and nutrition, value addition, waste utilization, eco-friendly, source of diversification, import substitution, co-generation and sustainability.

2.0 Status of Cultivation in India

Oil palm was first introduced to India at National Royal Botanical Gardens, Kolkata during the year 1886. The Maharashtra Association for Cultivation of Sciences (MACS), Pune later introduced African dura palms along canal bunds, home gardens and, to some extent, in forest lands near Pune during 1947 to 1959. Large scale planting of oil palm was launched from 1971 to 1984 in Kerala State by Plantation Corporation of Kerala Ltd. and Andaman Forest and Plantation Development Corporation Ltd., in Little Andaman Islands of Andaman and Nicobar Islands during 1976 to 1985. Oil palm has established as a successful crop in several states in the country and productivity levels of 6-8 tonnes of oil per ha could be achieved. The Technology Mission on Oilseeds and Pulses (TMOP) implemented by the Government of India looks after development of oil palm in the country through Oil Palm Development Programme (OPDP) along with other nine annual oilseed crops.

Though Oil Palm Development Programme in the country is progressing well, but area coverage is not taking place as per targets envisaged. As on 2020, various Expert Committees constituted by Ministry of Agriculture, Government of India have identified a total number of 22 states with 27,99,085 ha as potential area for oil palm cultivation. Out of them, 14 states (including A & N and Goa) have been found suitable for immediate allotment in the first phase and remaining eight states in the second phase of development. For the first phase of development, which could be combined with the on-going oil palm development, a total of 19,65,869 ha has been recommended. For the second phase of development, a total of 8,33,216 ha has been recommended as potential area.

An area of 3.23 lakh ha only has been covered under oil palm up to March, 2019 (Table 1). Production of palm oil in India continues to be at a meagre level with respect to its actual requirement. The FFB yield obtained by progressive farmers of Andhra Pradesh and Karnataka, under optimum.

Table 1. Area under Oil Palm cultivation and Production of Fresh Fruit Bunches (FFBs) and Crude Palm Oil (CPO) (March 2019).

Sl. No.	*State*	*Area (ha)*	*FFB Production (tonnes)*	*CPO Production (tonnes)*
1	Andhra Pradesh	168176.71	1379000.00	232938.00
2	Arunachal Pradesh	1595.00		
3	Assam	1699.16		
4	Bihar	0.00		
5	Chhattisgarh	4731.15	6.05	
6	Gujarat	5959.27	1053.01	
7	Goa	863.24		
8	Karnataka	41380.00	13238.00	2280.00
9	Kerala	5502.70	30269.00	4791.00
10	Maharashtra	474.20		
11	Manipur	0.00		
12	Meghalaya	0.00		
13	Mizoram	31597.07	5298.00	625.00
14	Nagaland	2640.00		
15	Odisha	19141.67		
16	Sikkim	0.00		
17	Tamil Nadu	20705.03	7014.00	1017.00
18	Telangana	19422.11	197231.22	37138.15
19	Tripura	31.00		
20	West Bengal	0.00		
	Total	323918.31	1633109.28	278789.15

Source: DAC&FW.

3.0 Major Commercial Varieties

Oil palm belongs to Palmae family of the order palmales. It is monoecious- male and female inflorescences are produced separately on the same palm in cycles. Oil palm fruit is a sessile drupe. The pericarp of the fruit consists of the outer exocarp (skin), mesocarp (or pulp rich in oil) and endocarp (hard stony shell surrounding kernel). The kernel has testa (skin), a solid endosperm (oil rich) and embryo. The fruit development takes place from about the 15th to 19th day after anthesis while oil formation takes place towards the end of maturity during which shell hardens and the embryo becomes viable. There are three types of oil palm fruits which impart characteristic feature to a palm as:

a) Dura

Dura palms have fruits with thick shell (2-8mm in thickness), low to medium mesocarp content (33-55 per cent) and are not commercially grown. These are used as female parent (mother palms) in developing tenera hybrid.

b) Pisifera

Pisifera palms are female sterile, shell-less and used as male (pollen) parent in development of tenera hybrids.

c) Tenera

This is the commercial planting material which is the cross between dura (female parent) and pisifera (male parent). Tenera fruits have thin shell (0.5-4.0mm) with medium to high mesocarp content (60-95 per cent). There is prominent fibre ring in its mesocarp.

Three oil palm varieties identified for release at CVRC: During 2020, three oil palm hybrids with desirable characters (high FFB yield, high oil yield and high sex ratio) were identified for State release:

1. Godavari Swarna (NRCOP-4): It has special characters like high fresh fruit bunch yield (26.87 t/ha/yr) and high mesocarp oil yield (5.71 t/ha/year). Suitable for coastal ecosystem with moderate humidity and temperature under assured micro-irrigation system; recommended for cultivation in Andhra Pradesh.
2. Godavari Ratna (NRCOP-2): It has special characters such as fresh fruit bunch yield of 22.44 t/ha/yr, average bunch weight of 18.31kg, high sex ratio (0.70), high mesocarp oil yield (5.36 t/ha/year) and high oil/bunch ratio (24.1). Suitable for high rainfall and assured irrigated conditions of coastal region of Maharashtra and Goa States.
3. Godavari Gold (NRCOP-17): It has special traits such as high fresh fruit bunch yield (27.23 t/ha/yr), more bunches/palm (11.74), high sex ratio (0.72), mesocarp oil yield (5.79t/ha/year) and oil/bunch ratio (21.28); suitable for coastal Tamil Nadu region under assured irrigated conditions.

4.0 Nature and Causes of Postharvest Losses

4.1 Preharvest factors

The notion that "palm oil is formed in the field not in the mill; the mills merely process to recover it" is significant that bunches harvested are ripe and at their optimal oil formation stage. The best bet for palm oil mills to achieve high oil extraction rate (OER) is to target for high degree of FFB ripeness of more than 90 per cent. Tight management and control on FFB harvesting, collection and handling is necessary.

Harvesting is among the most important activities of the whole production process of the oil palm because of its high cost (18 per cent of the total cost of the farm). Proper and timely harvesting of fruit bunches is an important operation which determines the quality and quantity of oil to a great extent. Harvest should be aimed at recovering the whole harvestable produce without loss of oil or loose fruits. The yield is commonly recorded in metric tonnes of fresh fruit bunches (FFB) per hectare per year, or metric tonnes of oil per hectare per year.

Harvest and postharvest operations in oil palm have considerable effect on the quality of oil. Presence of an enzyme lipase embedded in the membrane of fat globules in the fruit mesocarp causes accumulation of Free Fatty Acid (FFA) content of palm oil. Over-ripeness, improper handling, delayed processing *etc.* damage the delicate membrane, which triggers the enzyme activity and releases FFA. High FFA content makes the palm oil non-edible, enhances the rate of rancidity, fixes colour and increases refining losses. Processing the fruits within 24 hours is recommended to retain edible quality of raw palm oil. Harvesting at right stage, preventing excessive bruising of fruits during harvest and transportation and prompt processing are the key factors to overcome this deficiency. However, those who consume crude palm oil directly, like the Africans, the high FFA content is often ignored. Indian consumers prefer Refined, Bleached and De-odorized (RBD) palm oil and hence the build up of FFA has to be minimized during harvesting, handling and processing of the fruits.

4.2 Degree of Ripeness

The oil content in the mesocarp which is the major economic produce of oil palm is strongly related to fruit ripeness. In the early stages, the mesocarp has high water and carbohydrate content, and very little oil. As the fruit ripens, the oil content increases and then formation of free fatty acids (FFA) increase. As the fruit matures, the percentage of extractable oil increases to a certain degree, but at the same time, its quality decreases due to its acidity. The highest concentration of oil occurs during the final weeks of development of the fruit. A week before harvest, the oil content can increase to 80 per cent under irrigated conditions.

Oil is stored in the mesocarp cells in little vacuoles separated from the enzymes (Lipoxidases) by a delicate membrane, which when torn by overripeness or handling injury allows the two components (oil and enzyme) to mix leading to the formation of acids. The fruit adhered to the bunch has a FFA content of only 0.2-0.9 per cent, while in the oil that comes out after proper extraction in a mill usually reach up to 2 per cent. Internationally, the present permissible limit of FFA in food grade palm oil is 5 per cent. FFA content in the oil is affected by natural content of FFA in the fruit, action of the lipase in the fruit, action of microbial lipase and autocatalytic activity after processing.

4.3 Factors Affecting Ripeness

Optimum ripeness, measured as a function of fruit detachment, varies according to plant age, and environmental conditions. In some genetic material large amount of oil was found when 30 per cent of the fruits had detached from the bunch. While in other case increase in oil content in the mesocarp was found in detachments greater than 6 per cent (after cut and gathered). A faster rate of ripeness has been observed in the tenera types in the smaller bunches and in the bunches of younger palms. Besides, those bunches which develop during summer (those pollinated at the end of the rainy season) ripe faster (162 days) than those bunches that develop during rainy periods (178 days).

Changes in fatty acid composition, oil content and moisture content were analyzed in four adult oil palm tenera hybrids *viz.*, Malaysia, Palode, Deli x Ghana and Deli x Nigeria grown under irrigated conditions at ICAR-IIOPR. Fruit weight increased from 12th to 20th week (5.1 to 0.6g) in all the hybrids. Oil content (oil/dry mesocarp) increased from 22 to 79 per cent. Six fatty acid profiles *viz.*, myristic (0.5-4.3 per cent), palmitic (34.1-50.9 per cent), stearic (2.8-6.4 per cent), oleic (3.0-6.4 per cent), linoleic (7.8-11.5 per cent) and linolenic (0.3-3.3 per cent) were identified by standardizing the GC parameters.

An increase in oil or fresh mesocarp content was found when the number of fruits per bunch was less. According to this observation, bunches with more fruit (largest bunches) could be cut with a lower degree of ripeness. It is believed that bunches with less fruit are always associated with higher oil or fresh mesocarp content regardless of their degree of ripeness.

A report which studied the oil accumulation rate and oil weight per bunch of palms of different ages found a linear increase in both. However, the increase in the rate of oil accumulation was higher than the oil weight per bunch because of water loss. According to this information, an increase in oil percentage in stored bunches is a false indicator, since it is due to moisture loss rather than oil synthesis.

4.4 Fruit Ripeness and OER

The oil extraction rate (OER) is the actual oil production in the mill and this depends on the inherent oil/bunch ratio, harvesting standards and practices in the plantation and the efficiency of the mill. Modern mills do not vary much in efficiency and hence the major contribution to this variation must be originating from harvest practices. It was reported in 2003 that Malaysian industry was losing US$ 250 million per year because of the low OER which in turn was attributed to the prevailing harvesting standards and procedures and suggested that the loose fruit collection should be given high priority. While 20 per cent OER was considered normal in Malaysia. In Indonesia, many mills are obtaining OERs of around 24 per cent. One of the important causes suggested is low standards of harvesting, long intervals between harvesting rounds, harvesting of under-ripe fruits by harvesting too early or loss of high oil containing fruits because of too late and careless harvesting. With careless harvesting, many black bunches (unripe bunches) will be harvested leading to low OER. The oil extraction rate of the bunch is low during the first two or three years, but it reaches maximum values when the average bunch weighs more than 4 kg.

In India, the oil recovery rate (OER) is reported as 15 to17 per cent at present. This results in a recovery gap of at least 4 per cent which is equivalent to 11000 tonnes of crude palm oil (CPO), based on present statistics. Hence, correct maturity and grading are important in deciding the OER and thereby the profitability of the crop. By following proper harvesting standards and grading, more oil could be produced (recovered) from the existing plantations.

5.0 Postharvest Management

5.1 Harvest Time

The commercial oil palm variety tenera usually starts producing fruit bunches when the plant is 2-3 year old. To ensure better growth of palm and also have good yield during the first harvest, it is suggested to have ablation *i.e.* removal of early bunches. Flowering starts from 14th to 18th month after planting. Ablation should be started immediately after the appearance of inflorescences on the palms up to six months before the expected start of harvest. Ablation increases the storage of carbohydrate in the trunk which acts as reserve and gives more uniform yield. Some studies show that root development was encouraged by ablation thus giving greater drought tolerance to the plant.

Ablation (removal of inflorescence during juvenile phase) tool developed by ICAR-IIOPR eases the task of ablation, which is necessarily to be carried out in all juvenile oil palm plantations (1-3 years age oil palm plantation) for proper growth. This tool will have less drudgery, no damage to leaves, safe and manoeuvrable during intercrop period. On an average, in one hour a person can ablate 125 inflorescences.

5.2 Maturity Standards

The oil palm farmers have to be fully aware of the maturity standards and proper handling of the fruit bunches to ensure the quality of palm oil. The most common type of fruit is deep violet to black at the apex and pale greenish yellow at the base before ripening. Apex of the fruit turns deep red and golden yellow at the base as the fruit ripens. All the fruits in a bunch do not ripen simultaneously. As and when an individual fruit fully ripens, it becomes easily detachable from the bunch. The FFA content of palm oil increases as the ratio of detached fruit to total fruit on the bunch increases. The main aim of timely harvesting is therefore to obtain FFB, which will give good quality oil. Therefore the frequency of harvesting in an area has to be kept sufficiently high to eliminate over-ripening.

Bunch ripeness is identified by detachment of the fruit, change in colour and change in fruit texture. Granting that the fruits of a bunch do not ripe all at once, the way to obtain the maximum amount of oil is harvesting more frequently those bunches with an optimum ripeness degree; However, since this is not economically feasible, the harvest could be done at regular intervals. The bunches harvested in every round are the ones that contain a greater number of fruits with optimum ripeness and fewer under-ripe or over-ripe fruits. To determine the optimum degree of ripeness of the bunch based on fruit detachment, the following criteria are followed:

i) Number of fruits that have fallen before cutting the bunch: 10 or more fruits detached or easily removable for young palms, and 5 for adult palms.

ii) Number of fruits detached according to bunch weight counted after the bunch has been cut: 5 or more fruits/kg of bunch.

iii) Amount of fruit detachment per bunch: fruit detachment is 25 per cent or more of the bunch visible surface. Any of the above criteria recommended must be flexibly applied. Aspects such as harvesting frequency, climatic conditions and type of fruit must be considered in an efficient harvesting system.

Any of the above criteria recommended must be flexibly applied. Aspects such as harvesting frequency, climatic conditions and type of fruit must be considered in an efficient harvesting system.

5.3 Harvesting Frequency

Oil synthesis and free fatty acid formation occur during the ripening process and hence harvesting rounds should be carried out as frequently as possible to reduce the number of over-ripe bunches to be harvested, which otherwise leads to a high degree of fruit detachment and increased acidity of the oil.

Harvesting must take place in such a way that a bunch which is almost ready for one cycle will not be overripe for the following cycle. In low production periods, harvesting rounds can be made with less frequency, but the ripeness degree with which the bunches are to be cut should be less. This means that the longer the interval, the less is the fruit detachment accepted to cut the bunch (lower cutting order). According to current practices, harvesting rounds every 7-14 days have given good results. Factors determining harvest intervals refer to season, age of palms, labour availability, workers' skill and knowledge, transportation capability and processing capacity of the mill.

5.4 Harvesting Tools

In young palms, a 6-9 cm wide chisel attached to the tip of a 1.2-1.5 m long stick of wood or light hollow metal pipe with a handle is used for harvesting the bunches. With this kind of tool, the bunches can be cut without damaging the petiole of the leaves, and since the bunch is cut from a certain distance, workers are not hurt by the spines of the palm frond. When the harvest is started, it is advisable to cut down the bunch trying not to eliminate the leaf that supports it. Leaves are cut only to facilitate access to the bunch. The general rule for young palms is to avoid cutting down any green fronds unless it is unavoidable. During this task, it is important not to injure the trunk and the petiole of adjacent leaves. At the same time, rotten bunches (badly pollinated, diseased or failed) as and when found, must be eliminated in order to clean the plant. Use of the narrow chisel is usually continued until the palm reaches a height of 1m above the ground. The narrow chisel is substituted by a wider one (14 cm wide) attached to a longer stick or metal pipe which should be longer as the palm grows (1-3 m long). Wide chisels used for cutting bunches were found suitable for palms not higher than 3-4 m in height. Even though the bunches are cut from a distance, it is always advisable to use long leather gloves and cap, to give the harvesters protection for their hands and forearms.

In taller plants, bunches are cut by using pole and sickle/knife. A curved knife is attached to the end of a long thin and light pole which could be made of bamboo, aluminium or any other light and strong material. The knife can be fixed at the end of the pole with steel wire or with screws. There are different sickles available for the purpose *viz.* Malaysian sickle, modified Malaysian sickle, Kerala sickle, Andhra Pradesh sickle, Tamil Nadu sickle *etc.* DOPR1 and DOPR5 sickels are found to be ideal. Cutting bunches with pole and sickle requires skilled workers. A good cut depends not so much on strength but on the position/angle of the knife blade. There are ripe bunches whose base is difficult to find space to insert the tip of the knife to make the cut. Under these circumstances, the worker must find the right standing location to cut the bunch without hurting the young leaves of the palm. Cutting of subtending leaf is acceptable in case of adult palms since; the leaf area available is still adequate after cutting a few additional fronds.

In those fields where the height of the palm is uneven, short poles cannot be used with tall palms, and if a long pole is used for harvesting a short palm, the worker has to stand far from the palm and the task becomes more difficult. In these particular cases, one pole is used inside another (telescopic type). The inside pole can be used as an extension when the height of the palm requires it. When these instruments are transported, the blades must be covered properly to avoid accidents.

5.5 Harvest Management

Once the harvest interval has been decided, the area to be harvested every day must be determined. For this purpose, the total area is divided among the number of days of the planned interval. Then the area to be harvested is divided by the workers' harvesting capacity. In this way, number of workers needed to harvest each section within the desired interval can be determined. The aspects determining area, size, or number of rows to be harvested are i. Height of the palm: The taller the palm, the greater skill required to handle the cutting tools. ii. Number of bunches per area: When there are few bunches per area unit, the harvester must walk longer distances (more rows) to complete his task. iii. Average bunch weight: More effort is required to carry and transport heavier bunches. iv. Carrying distance: When collection centres are close to each other, bunch carrying is easier. v. Field conditions: Variations in topography, weed growth, poor drainage etc could affect the workers' efficiency.

Improved sickles and motorized harvester for harvesting oil palm: Two sickles DOPR-1 and DOPR-5 manufactured with stainless steel were designed and developed at ICAR-IIOPR to harvest bunches from oil palm plantations. Two models of oil palm harvesting tools *viz.*, back pack mounted and trolley mounted were also developed and tested in oil palm plantations. Based on the initial trials, it was observed that using the back pack model the operator could harvest comfortably up to a height of 5-6m and it has better adaptability and flexibility.

Height adjustable hydraulic lift platform for harvesting in tall palms: Height adjustable hydraulic lift platform was designed and fabricated at ICAR-IIOPR for harvesting bunches from tall palms. A person with average skill can reach up to a height of 4.54 m and with 6.1m pole attached to sickle, he can reach a height up to 12 m to harvest bunches.

5.6 Sorting and Grading

The primary objectives of oil palm bunch grading are:

i. To improve the quality of fresh fruit bunches (FFB) received at mills.
ii. To improve the quality of palm oil produced.
iii. To improve the efficiency of oil and kernel extraction rates.
iv. To ensure that farmers and processors obtain a fair deal from transactions and in India, as of date, there is no grading practiced for oil palm

FFB. Research work on this line is in progress considering the local features. However, owing to its importance in achieving better productivity and quality of oil, details of the grading system, as developed in Malaysia, are presented in Table 2.

Table 2. Palm Oil Registration and Licensing Authority (PORLA) Classification of Oil Palm FFB (at the time of inspection at mill)

Bunch Grades	*Description*
Dirty Bunch	With more than half of its surface covered with mud, sand, other dirt particles and mixed with stone or other foreign matters.
Diseased Bunch	Has more than fifty per cent (50 per cent) parthenocarpic fruits and is not normal in terms of its size or its density.
Dura Bunch	With shell thickness of 2-8 mm, Ratio of shell to fruit of 25-50 per cent, Ratio of mesocarp to fruit of 20-60 per cent, Ratio of kernel to fruit of 4-20 per cent and with no fibre ring around the shell
Empty Bunch	Has more than ninety per cent (90 per cent) of detached fruitlets.
Long Stalk Bunch	Has a stalk of more than 5 cm in length (measured from the lowest level of the bunch stalk).
Loose Fruit	Fruit detached from a fresh fruit bunch because of ripeness and is reddish orange in colour.
Old Bunch	Has been harvested and left at the field before being sent to the mill. The fruits still remaining on the bunch are dry and brownish black in colour. The stalk is also dry, soft, fibrous and blackish in colour.
Overripe Bunch	Fruits are darkish red coloured and has more than fifty per cent (50 per cent) of detached fruitlets but with at least ten per cent (10 per cent) of the fruits still attached to the bunch.
Pest Damaged	With more than thirty per cent (30 per cent) of its fruits damaged by pest attack
Bunch	such as rats, birds etc.
Ripe Bunch	Reddish orange in colour, outer layer fruit mesocarp is orange in colour, has at least 10 fresh sockets of detached fruitlets and more than fifty per cent (50 per cent) of the fruits still attached to the bunch.
Rotten Bunch	Partly or wholly and together with its loose fruits, has turned blackish in colour, rotten and mouldy.
Small Bunch	Has small fruits and weighs less than 2.3 kg.
Under ripe Bunch	Reddish orange or purplish red in colour, outer layer fruits mesocarp yellowish orange in colour and has less than 10 fresh sockets of detached fruitlets.
Unfresh bunch	Has been harvested and left at the field for more than 48 hours before being sent to the mill. The whole fruit or part of it together with its stalk has dried out. Normally, this type of bunch is dry and blackish in colour.
Unripe Bunch	Black or purplish black in colour, outer layer fruits mesocarp is yellowish in colour, does not have any fresh sockets of detached fruitlets and sockets (if any) on the bunch are not due to normal ripening process.
Wet Bunch	Consignment of fresh fruit bunches (FFB) which has excessive free water.

Palm Oil Registration and Licensing Authority of Malaysia under the Ministry of Primary Industries, has developed a Fresh Fruit Bunch Grading Manual during the year 1995 which has been given hereunder. The manual serves as a practical guide for grading of oil palm bunches based on various quality parameters.

Since the basic factor determining Oil Extraction Rate (OER) is the quality of FFB, any step to improve the OER should begin from the plantations. In order to ensure quality control in plantations, there should be a mechanism in the processing units to check the FFB quality. Quality control in processing units would act as a corrective measure for the farmers to improve their produce. Fresh Fruit Bunch grading is such a process whereby FFBs are assessed and classified according to the degree of ripeness and quality. Palm oil mills in Malaysia follows such grading systems.

Successful implementation of grading requires well-defined grades according to quality. Different FFB quality grades defined by Palm Oil Registration and Licensing Authority (PORLA), suitable for Malaysian conditions.

Trained grading staff, having the capability and experience in the grading of fresh fruit bunches (FFB), are appointed by the mills to ensure bunch quality. Well-defined sampling methods and step by step grading procedure should also in force to facilitate impartial grading. The minimum sample size of each consignment to be graded is based on the net weight of the consignment. In practice, it is not possible to obtain a hundred per cent good quality bunches and hence a reasonably good quality consignment constitutes the combination of bunch quality (Table 3).

Table 3. Bunch Quality in Oilpalm

Bunch Category	*Limit*
Dirty Bunch	0 per cent
Dura bunch	0 per cent
Empty Bunch	0 per cent
Long stalk Bunch	≤ 5 per cent
Old Bunch	0 per cent
Ripe Bunch	≥ 90 per cent
Rotten Bunch	0 per cent
Under ripe Bunch	≤ 10 per cent
Unripe Bunch	0 per cent
Wet Bunch	0 per cent

In order to facilitate payment according to basic oil extraction rate, standard extraction rates based on age of palm and bunch weight are established for different climatic regions. The graded oil and kernel extraction rate that can be considered for payment which is calculated by subtracting penalty rate from the basic extraction rate. The graded extraction rate is recorded in the monthly grading summary for the purpose of payment to the suppliers.

Case studies on the factors affecting OER in palm oil mills of Malaysia report that OER of above 19 per cent is possible by improving the FFB quality by observing the grading and quality control procedures, keeping other factors constant. One of the mills reported a high percentage of unripe (5.23 per cent) and under ripe bunches (34.42 per cent) was capable to give only 17.10 per cent OER and could improve the OER and above 18 per cent by following 100 per cent FFB grading, rejecting black and rotten bunches and imposing penalty for long stalks by weight deduction.

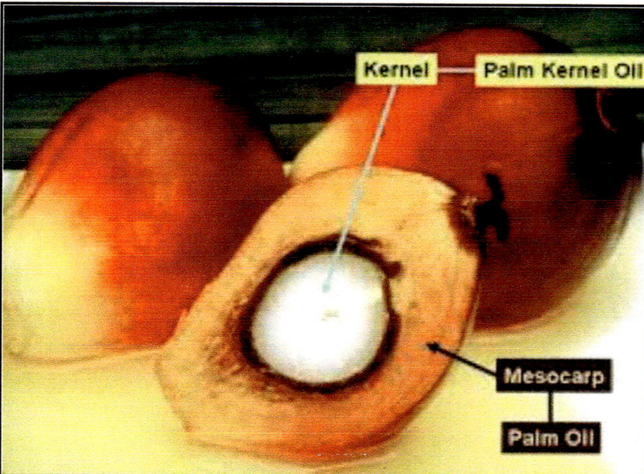

Figure 1. Kernel of Oil Palm

In India, as far as grading and quality control is concerned oil palm cultivation is in the infant stage, the Government and industry have not given much attention on the grading system and its implementation by the Indian palm oil mills. The kernel of oil palm is depicted in Figure 1. However, recently, after the CACP recommendation for Price Fixation Formula for oil palm recommended by Government of India, attempts were made by a few Processing Units for informal control on the quality of bunches and both farmers and Processors realized the benefits within a few months. Keeping area expansion in view, as the crop is gaining momentum, it was felt that rigid grading would create adverse impact on area expansion. Accordingly development officials appointed by processors as well as Scientists of Directorate of Oil Palm Research are making efforts to educate farmers about the importance of quality control and the ways to achieve good quality bunches. But harvesting of under-ripe and unripe bunches due to reasons other than the knowledge level of farmers could not be prevented and has resulted in less OER in the past years. Grading of FFB in the processing mills would act as a corrective measure to improve OER and subsequently for the FFB price realization by the farmers.

The Malaysian (PORLA) grading procedure and classification of bunches may not be suitable for Indian conditions due to obvious variation in maturity standards, bunch characteristics, FFB procurement methods *etc.* Hence a modified procedure suitable for Indian conditions is needed for successful implementation of such initiatives.

Fresh Fruit Bunches (FFB) are to be processed within 24 hours after harvesting to retain edible quality of crude palm oil. At present, India has 23 palm oil processing mills and the capacity ranges from 4 tonnes/hr to 60 tonnes/hr. Overall processing capacity of the mills is 266.3 tonnes/hr.

5.7 Storage

Oil palm fresh fruit bunches (FFB) cannot be stored after harvesting, but should be processed within 24 hours. Studies conducted at DOPR on the effect of low temperature on FFA content in the oil after harvesting of FFB indicated that the FFA content increase was steady up to seven days and hence there is no possibility of storing FFB at lower temperature before processing.

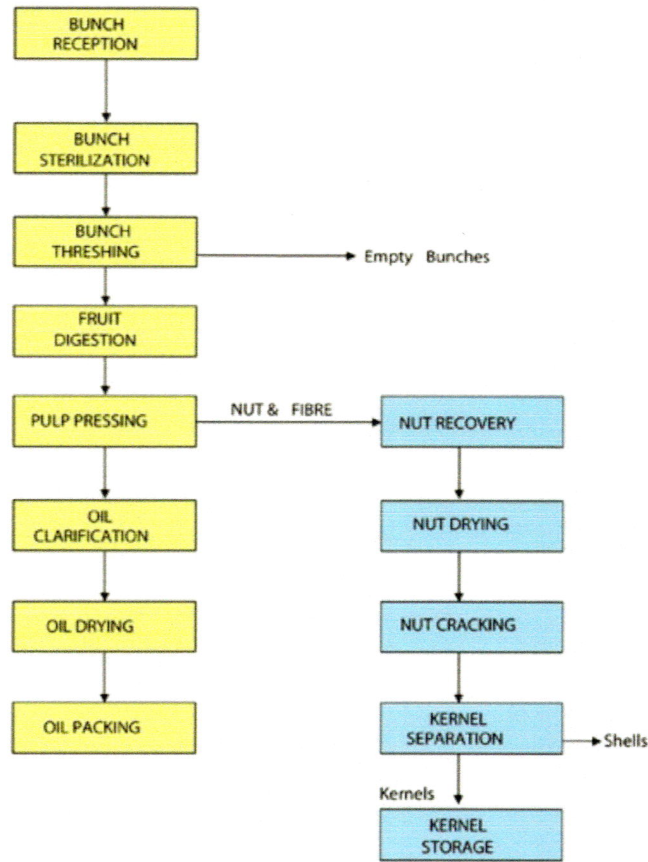

Figure 2. Unit Operations from Palm Oil Processing.

5.8 Transportation

Technology Mission on Oilseeds and Pulses constituted by the Government of India was looking after the National Mission on Oil seeds and Oil palm (NMODP) for improving vegetable oil production in India. The State Governments of respective states have identified entrepreneurs for establishing oil palm plantations and creating processing facilities in the states. The entrepreneurs are involved in the oil palm development programme from raising of the seedlings to providing inputs and guidance for collection of oil palm fresh fruit bunches and processing. Oil palm fresh fruit bunches are transported either by tractors or bullock carts to the collection centres established by the oil palm processing units nearer to the farmers fields or to the respective processing mill.

6.0 Domestic Marketing

The Oil Palm Act, implemented by various state governments facilitates the buy back of the oil palm FFB by the Oil palm processors. Presently, the State Governments are deciding the prices FFB as per the recommendation of the price fixation committee under the Chairmanship of Secretary (Agriculture/ Horticulture) of the concerned State Government based on the price fixation formula developed by CACP and recommended by Government of India. The price of FFB are influenced by the international and national price of palm oil.

7.0 Conclusion

With the present level of a rapid increase in the per capita consumption of vegetable oils in the country, it is expected that the estimated population of 1620 million in 2050 may need 40.5 million tonnes of vegetable oils at the estimated per capita consumption of 25.00 kg per year. Though it would be ideal, if the country could become self- reliant in the vegetable oil production sector, it would not be possible for us to spare the required land in view of demand for additional area for different crops as well as enterprises including urbanization.

Expert Committee recently constituted by Ministry of Agriculture, Government of India has identified, in 2020, a total number of 22 states with 27,99,085 ha as potential area for oil palm cultivation. Out of them, 14 states (including A & N and Goa) have been found suitable for immediate allotment in the first phase and remaining eight states in the second phase of development. For the first phase of development, which could be combined with the ongoing oil palm development programme, a total of 19,65,869 ha has been recommended. For the second phase of development, a total of 8,33,216 ha has been identified as potential area.

With intensive research efforts and active implementation of Oil Palm Development Programme, it is estimated that the oil palm productivity levels in the country could be enhanced to 7 tonnes of oil per ha, thus producing a total quantity of about 13.51 million tonnes of palm oil by 2050. Thus, the oil palm could contribute 33.35 per cent of the total requirement of vegetable oil by 2050.

The Indian Agriculture is already undergoing a paradigm a shift with focus from cereal production to diversified farming. Oil palm besides improving biological productivity and nutritional standards also has enormous scope for enhancing profitability. Past investment in oil palm research has been rewarding in terms of increased production and productivity of oil palm. However, challenges confronting oil palm sector are still many. While the edible oil demand is increasing, the production sector has to be competitive both in terms of quality and price. Improving the efficiency of oil extraction process and reducing the losses are also equally important from the point of view of boosting the oil production. Thus, the development of improved hybrids with high productivity, quality characteristics, resistance to pests and diseases and tolerant to abiotic stresses need prime attention. The technologies must improve the water and nutrient use efficiency and reduce variability in yield and quality and also reduce postharvest losses. Efforts will also be needed to ensure self sufficiency in production of quality hybrid seed and planting material. Development and popularization of Good Agricultural Practices on preharvest and postharvest management of oil palm would contribute significantly towards improving the system efficiency in oil palm sector.

8.0 Important References

Arulraj, S., Kalidas, P. and Suresh, K. 2014. Way forward for oil palm research and development in India. *Curr. Sci.* **107**: 1371-72.

Carrere, R. 2010. Oil palm in Africa; Past, present and future scenario. World rain forest movement http: //wrm.org.uy/countries/Africa/Oil-Palm-inAfrica.pdf.

Idris, O., Mohd, A.H., Mohd, H.H., and Mohd, B.W. 2003. Colour Meter for measuring fruit ripeness. MPOB Information Series TT No. 182. , June.

Jelani, A.R., Hitam, A., Shuib, A.R., Jamak, J., Noor, M. and Gono, Y. 2005. Motorised Mechanical Cutter for Efficient Oil Palm Harvesting. Proceedings of the PIPOC 2005 International Palm Oil Congress (Agriculture, Biotechnology and Sustainability). pp. 1189-1193.

Malike, F.A, N. Abdullah and Ahmad, N., 2011. Evaluation of Harvest Index in the MPOB germplasm collections. pp 172-175. In: Proceedings of Agriculture, Biotechnology and Sustainability Conference, PIPOC, 2011, 15-17, November, KLCC, Malaysia.

Murugesan, P., Bijimol, G., and Gopakumar, S. 2009. Nut Component Analysis of Exotic and Indigenous Sources of Oil Palm (*Elaeis guineensis*, Jacq.) Planting Materials. *Int. J. Oil Palm*, **6**: 19-21.

Prasad, M.V., Mary Rani, K.L., Arulraj, S. and Jameena, J. 2015. Estimation of cost of production of oil palm in Andhra Pradesh. *J. Plant. Crops*, **43**: 83-87.

Ravindra Naik, Annamalai, S.J.K., Prabinkumar, S. Prasad, M.V. and Vidhan Singh, T. 2013. Design

Concept for developing a back-pack power mounted harvesting equipment for oil palm. *J. Plant. Crops*, **41**: 98-100.

Shuib, A.R., Jelani, A.R., Jahis, S., Deraman, M.S., Khalid, M.R., Hitam, A., Basiron, Y. and Wahid, M.B. 2004. Development of a Machine for Harvesting Tall Palms, MPOB TT No. 217.

Sunilkumar, K. and Sparjan Babu, D.S. 2010. Studies on fruit maturity standards of oil palm in Andhra Pradesh under irrigated conditions. *Int. J. Oil Palm*, **7**: 5-8.

Webb, B.A. 1976. The development of suitable fruit harvesting system to improve the efficiency of oil palm production. Proceedings of Agriculture Oil Palm Conference, Kuala Lumpur: p. 741-752.

Chapter 36

Seed Spices

Gopal Lal

Member (Ag.), Cauvery Water Management Authority, New Delhi, India and

Former Director, ICAR-NRC on Seed Spices, Ajmer, Rajasthan, India

Shiv Lal

Senior Scientist (Hort.), ICAR-National Research Centre, Ajmer, Rajasthan, India

E-mail: glal67@yahoo.co.in

1. Introduction

India is 'The Land of Spices' and the glories of Indian spices are known throughout the world. Total 109 spices are listed by ISO and 63 spices are grown in India and out of which 20 are being classified as seed spices. Out of 20 seed spices, ICAR-NRCSS working of 10 most important seed spices namely cumin, coriander, fennel, fenugreek, ajwain, nigella, dill, anise, celery and caraway). Seed spices are high value export oriented crops and India is a major exporter of seed spices and spices products (Tables 1 and 2). It is highly felt at the moment to enhance the productivity level of our spices to grow varieties which yield good quality species and their products to bestow our maximum attention in on-farm processing for quality up gradation and greater emphasis on product diversification to the newer requirements of domestic as well as global market (Pruthi, 1992; Lal and Maheria, 2013). It is largely used for flavouring liquors, bread, fish, salads, soups, cheese and in the manufacturing of pickles, perfumes, soaps, cosmetics, cough drops, and other medicinal purposes and as a good source of antioxidants value (Garcia-Jimenez *et al.*, 2000; Choudhary *et al.*, 2017) 80 to 85% of the volume of spices is traded in whole ungrounded form. Quality requirements of the buying countries and price competitiveness are most important for the international market. Improved productivity and quality will increase farmers' income and country's foreign exchange earnings. India despite being the largest producer of seed spices could not exploit their value addition potential as majority of the export is in the form of raw produces (DASD, 2018-19). Of total spices, spice oil and oleoresins worth Rs. 219300 lakh was exported during 2018-2019. There is a great potential for increasing export of Indian spices in the form of value added products. To realize this potential there is need to enhance the quality as per international standard through post harvest technology and value addition at different level of post production system (Malhotra, 2009). In raw forms these commodities have certain disadvantages a whole or ground spice dose not impart there total flavour readily, and moreover, on being finely ground they lose their aromatic constituents partially (Rathore *et al.*, 2013). Some of these defects can be reduced by extracting oil and oleoresins (Kachru, 2006; Meena *et al.*, 2013). World over the people are becoming more and more conscious about the health, sanitary requirements, microbiological levels and pesticide residues. Emphasis should therefore be on production of more value added spice with quality products free from contaminants *etc.* (Malhotra, 2016). Since the seed spices have many therapeutic properties and to reduce the losses of seed spices, processing becomes very imperative.

Table 1. Country-wise Export in Terms of Quantity (descending order) of Major Seed Spices from India

Crop	*Country*
Coriander	Malaysia followed by UAE, Saudi Arabia, Nepal and UK
Cumin	Vietnam followed by Bangladesh, UAE, USA, and Nepal
Celery	USA, South Africa, China, Canada, Germany and Netherlands
Fennel	Vietnam, USA, Malaysia, Saudi Arabia, UAE and UK
Fenugreek	UAE, YAR, Srilanka, Saudi Arabia and USA

Source: DGCI and S Kolkata/Exporters returns/DLE from custom 2018-19.

Table 2. Country-wise Export in Terms of Value (Descending order) of Major Seed Spices from India

Crop	*Country*
Coriander	Malaysia, UAE, UK, USA, Saudi Arabia, South Africa, Nepal, Oman, Qatar, Australia, Singapore, Kuwait
Cumin	Vietnam, Bangladesh, USA, UAE, Nepal, Pakistan, UK, Brazil, Malaysia, Egypt, Saudi Arabia, Sri Lanka, Spain, Morocco, Mexico, Peru, Japan
Celery	USA, South Africa, China, Canada, Germany, Netherlands
Fennel	Vietnam, USA, Malaysia, Saudi Arabia, UK, UAE
Fenugreek	USA, UAE, YAR, Sri Lanka, Saudi Arabia, Korea (South), South Africa, UK, Sudan

Source: DGCI and S Kolkata/Exporters returns/DLE from custom, DASD, Calicut 2018-19.

2.0 Status of Cultivation

The area under seed spices is about 1.77 million hectares and production is about 1.68 million tonnes (DASD, 2019-20, Figures 1-3). India is a consistent source of seed spices for importing countries worldwide. There has been ever increasing demand or seed spices and importing countries look at India for quality products of seed spices. India is exporting only 15 per cent of its production annually and fulfils the 50-60 per cent of world demand. The total export of seed spice crops is more than Rs. 3685 crore, out of which cumin alone contributes more than Rs. 2884 crore annually (DASD, 2018-19). Seed spices are majorly cultivated in semiarid and arid parts of India. Out of 20 seed spice crops cumin, coriander, fennel, fenugreek, dill, and Ajwain contribute more than 95 per cent towards area and production. Cumin and fennel are dominant seed spices of Gujarat while coriander and fenugreek are important

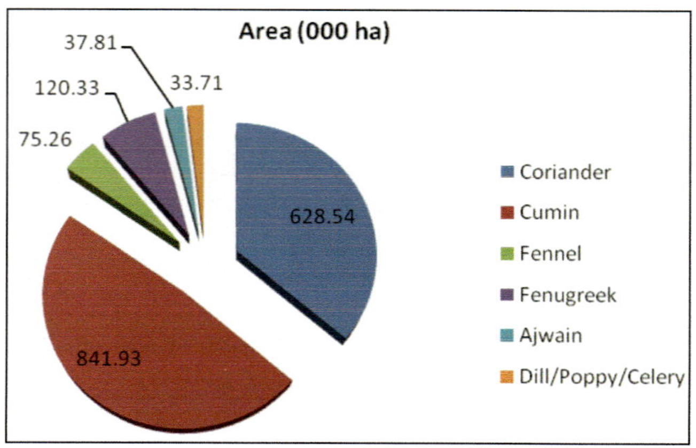

Figure 1. Crop-wise Cultivated Area of Major Seed Spices in India during 2019-2020.

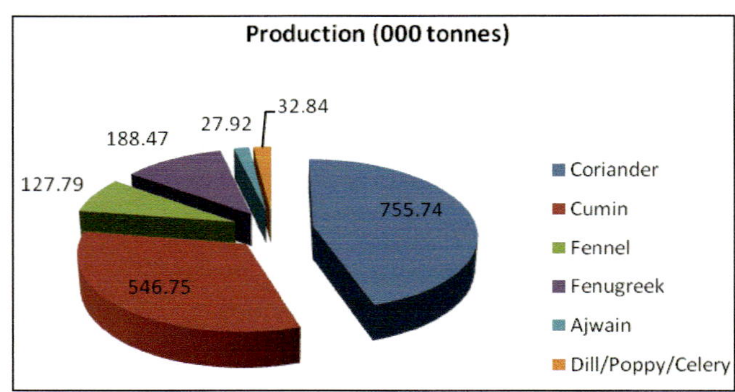

Figure 2. Crop-wise Productivity of Major Seed Spices in India during 2019-2020.

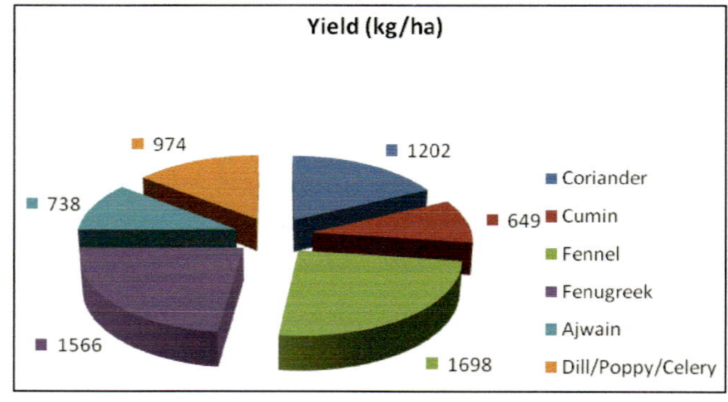

Figure 3. Crop-wise Yield of Major Seed Spices in India during 2019-2020.

Table 3. Status of Germplasm Wealth of Seed Spices in India

Crop	*NRCSS Collection*			*NAGS holding*
	Indigenous	*Exotic*	*Present Available*	
Cumin	100	7	107	247
Coriander	200	3	176	594
Fenugreek	82	59	135	745
Fennel	118	3	46	292
Ajwain	99	1	91	119
Dill	106	5	108	111
Nigella	21	3	24	37
Celery	36	-	36	36
Anise	18	-	18	18
Caraway	8	2	2	14
Total	**790**	**83**	**745**	**2220**

Source: NRCSS Annual Report, 2018-19.

in Rajasthan. Seed spices also grown in some other states in considerable areas such as Uttar Pradesh, Madhya Pradesh, Bihar, West Bengal, Orissa, Punjab, Karnataka, and Tamil Nadu.

In Indian context, ICAR-National Agriculture Research System for Seed Spices, Ajmer is the National Active Germplasm Site (NAGS) for seed spices and presently maintaining 2220 germplasm lines comprising exotic and native lines of different seed spice crops. The maintained lines are utilized in breeding for evolving improved varieties of mandate spice crops in India (Table 4).

Table 4. Improved Varieties of Seed Spices Devolved at ICAR-NRCSS

Sl.No.	*Name of Variety*	*Salient Features*	*Crop and Seed Photographs*
1.	Ajmer Coriander-1	It contains essential oil ~0.49 per cent; linalool (77.7 per cent), flavonoid (17.93 per cent) and phenols (34.12 per cent). Impart flavour in the essential oil. This coriander variety has been utilized to prepare squash and ready to serve drinks.	 
2.	Ajmer Coriander-2	It is a variety suitable for leaves and seed production, performs well under irrigation, essential oil content up to 0.50-0.60 per cent and average yield 12-13 q seeds per hectare. The plants are resistant to stem gall and have tolerance to powdery mildew. This variety is suitable for cultivation for green leaves during off-season also.	 
3.	Ajmer Coriander-3	It has 0.55 per cent essential oil and yield 16.89 q/ha. It is moderately resistant to powdery mildew.	 
4	Ajmer Fenugreek-1	The seeds are bold and large. The number of seeds per pod ranges from 17-20 with 17-20 g. test weight. The crop takes 137 days to mature and gives seed yield of 27.2 q/ha. The crop is grown exclusively for leave purpose, yield 76 q/ha, (green leaves) from three cuttings.	 
5	Ajmer Fenugreek-2	The seeds are small in size. The number of seeds per pod ranges from 16-18. The crop takes 138 days to mature and gives seed yield of 18.1 q/ ha. The crop is grown exclusively for green leaves.	 

Sl.No.	*Name of Variety*	*Salient Features*	*Crop and Seed Photographs*
6	Ajmer Fenugreek-3	Seeds contain 1.79 per cent diosgenin and 0.97 per cent Free amino acid (4-hydroxyisoleucine), thus useful for treating type II diabetes.	 
7	Ajmer Fenugreek-4	The seeds contain 0.94 per cent 4-hydroxyisoleucine and 1.74 per cent diosgenin. Seeds are rich in crude fibre (21.7 per cent). It has excellent anti diabetic properties. It has been fruitfully utilised in preparation of fenugreek biscuits. This variety is also suitable for green leaf production under shade net condition in summer/off season.	 
8	Ajmer Fenugreek-5	The seeds are bold and large. The crop take 106-141 days to mature with seed yield of 17.21q/ha. Seeds have higher antioxidant properties (66.4mg/BHTE/ppm). It has moderate resistance to powdery mildew and *Alternation* blight.	 
9	Ajmer Fennel-1	The variety is from the late maturity group and produces an average yield of 1950 kg/ha during *rabi* crop and 2510 kg/ha when grown as *rabi* crop. This variety is tolerant against *Ramularia* and *Alternaria* blight.	 
10	Ajmer Fennel-2	Seeds of this variety contain 1.9 per cent essential oil which has 57.5 per cent anethol and estragol together. It is useful for preparation of fennel based RTS and squash.	 
11	Ajmer Fennel-3	This is suitable for growing both for early sowing and as rabi crop. It yield 21.4q/ha during rabi. Its seeds contain 1.9 per cent essential oil and 45 per cent anethole+estyragol. This variety is moderately resistant to *Ramularia* blight	 

Sl.No.	Name of Variety	Salient Features	Crop and Seed Photographs
12	Ajmer Dill-1	The variety is suitable for cultivation under irrigated conditions and seeds are dillapiols less, a toxic substance as compared to Indian Dill varieties. It is susceptible to powdery mildew but possesses field resistance to blight. The variety is suitable for export purpose and seeds contain about 3.5 per cent essential oil. The average seed yield of 14.7q/ha can be obtained under irrigated conditions.	 
13	Ajmer Dill-2	It is an Indian dill type variety, grows to a height of 90 cm and is suitable for cultivation under both irrigated and rainfed conditions. This variety is also suitable for cultivation under both rainy and *rabi* reason. It takes about 135 days to reach maturity. The seeds are compact and require pressure to split. The seeds contain about 3.2 per cent essential oil content. An average yield of about 14.6 q/ha under irrigated conditions and 5.8 q/ha under rainfed conditions can be obtained.	 
14	Ajmer Ajwain-93	Early flowering (47 days) and matures in 130-138 days which is approximately 40-45 days early than identified check varieties, seed yield 14.15 q/ha and Essential oil content in this variety has been reported to be ~3.8 per cent, the variety is thermo- insensitive and contains an appreciable amount of thymol and p-cymene.	  AA-93 Ajmer Ajwain-93
15	Ajmer Ajwain-1	It is suitable for cultivation in both irrigated and rainfed situations. The average height of plant is 112 cm and bears an average 219 umbels/plant. It is a long duration variety and takes about 165 days to maturity. The variety is high yielding with a potential average yield of 14.26 q/ha under irrigated condition and 5.8 q/ha under rainfed condition. The essential oil content in the seeds is ~3.5 per cent.	 
16	Ajmer Ajwain-2	This variety has also been developed at ICAR-NRCSS, Ajmer through selection from local germplasm for earliness and matures in 147 days. It is suitable for cultivation under both irrigated as well rainfed conditions. The average height of plant is 80 cm and bears ~185 umbels/plant. The average yield is 12.83 q/ha under irrigated condition and 5.2 q/ha under rainfed condition. It possesses resistance to powdery mildew and contains approximately 3 per cent essential oil.	  NRCSS AA-2
17	Ajmer Ajwain-73	It is medium duration (165-170 days) and high yielding variety. It shows high tolerance to root rot and *sclerotium* rot. AA-73 gave 16-17 q/ha seed yield and contains about 3.15 per cent total oil and6.38 per cent essential oil	 

Sl.No.	Name of Variety	Salient Features	Crop and Seed Photographs
18	Ajmer Nigella-20	It matures in 140-150 days with 10-12 q/ha yield. Seeds of this variety contain 28.08 per cent essential oil.  	
19	Ajmer Nigella-1	It is suitable for cultivation in semi-arid region under irrigated conditions. The plants are 30-35 cm in height. This variety takes 135 days to reach seed maturity and has resistance to root rot. The average seed yield of Ajmer Nigella-1 is 800 kg/ha.  	
20	Ajmer Celery-1	This variety is suitable for cultivation in semi-arid regions under irrigated conditions. It produces an average yield of 8.01 q/ha with a potential yield of 12.1 q/ha. The variety yields essential oil content of 2.4 per cent from seeds.  	
21	Ajmer Celery-2	It yields 8.74q/ha. Its seed content essential oil 6.74 per cent. 	
22	Ajmer Green Coriander -1	Ajmer coriander-1 is suitable for green leaves, grown under 50 per cent green shade net  	
23	Ajmer Caraway-1	The ICAR-National Research Centre on Seed Spices at Ajmer has developed one promising variety Ajmer Caraway-1 with an average yield of 4-6q/ha for cultivation under semiarid conditions.  	

2.0 Major Commercial Varieties of seed Spices

In past two decades ICAR-NRCSS, Ajmer have developed 23 varieties (Coriander-4, Fenugreek-5, Fennel-3, Dill-2, Nigella-2, Ajwain-3, Anise-1, Celery-2 and Caraway-1) of 10 seed spice crops and identified unique promising genotypes by non-traditional breeding methods having desirable traits (Table 5).

Table 5. Other Promising Varieties of Seed Spices Released by different R&D Institutions of India

Crop	*Variety*
Cumin	RZ – 19, RZ – 209, RZ – 223, RZ-345, GC- 1, GC – 2, GC – 3, GC – 4
Coriander	RCr – 41, RCr – 20, RCr – 435, RCr – 436, RCr – 684, RCr – 446, RCr-728, GCr – 1, GCr – 2, CO- 1, CO- 2, CO- 3, CS- 287, RajendraSwathi, Rajendra Sonia, Sadhna, Swathi (CS – 6), Sindhu, GCr – 1, GCr – 2, CO- 1
Fennel	RF – 101, RF – 125, RF – 143, RF-205, PF – 35, GF – 1 (Gujarat Fennel – 1), GF – 2, GF – 11, Guj. F.-12, Co – 1, Hisar Swarup
Fenugreek	RMt – 1, RMt – 143, RMt – 305, RMt-361, GM – 1, CO – 1, Rajandra Kranti, Andhra Pradesh, Lam Selection–1, APHU Methi-1, Hisar Sonali, Hisar Suvarna, Hisar Mukta, Hisar Madhavi (HM – 350), HM-219, Pant Ragini, Pusa Early Bunching, Pusa Kasuri
Dill (Sowa)	GD-1, GD-2, RSP-11
Ajwain	Pratap Ajwain 1, GA – 1, Lam Selelction–1, Lam Selection–2, Pant Ruchika, RPA-68,
Nigella	Azad Kalaungi (1998), Pant Krishna (2001), RajendraShyama, Rajendra Mani

3.0 Nature and Causes of Postharvest Loss

A lot of energy, time, money and several inputs are invested in the production of seed spices. After the harvesting of a bountiful crop, if for some reason proper care is not taken of the ultimately harvested produce, it may be infested with insects, moulds, rodents *etc.* resulting in deterioration of the quality of final product. Thus post harvest management is very important.

4.1 Pre harvest Treatments

For producing high quality produce farm level operations or practices from seed sowing to harvesting should be managed very precisely following all standard scientific package of production (Singh and Malhotra, 2005). Without following the standard practices viz., selection of proper variety (Kalra *et al.*, 1999; Yildirim, and Gok, 2012), sowing at proper time with appropriate plant population (Yildirim, and Gok, 2012; Sharma and Israel, 1991), crop nutrition management (Baboo and Rana, 1995), water management (Singh and Rao, 1994), weed management (Choudhary and Gupta, 1991), crop health management (Khare *et al.*, 2014 a and b), *etc.*, one can harvest the quality produce with good shelf life (Malhotra, 2005; Kinyuru *et al.*, 2011). Although post harvest quality and shelf life generally depend upon the quality achieved at the time of harvest (Weston and Barth, 1997). A value chain of major seed spices from field to plate was developed under NAIP which included all the parameters of production, protection, post-harvest handling, processing, value addition, product development, marketing, trade, and export (Aglodiya, 2014).

4.2 Postharvest Factors

Very limited post harvest operations are carried out by farmers. Proper drying of the freshly harvested crop, threshing, seed drying, cleaning, grading, packing and transportation of whole seed to market, processing plant, or storage place is utmost requirement for quality produce. Value addition in seed spices is restricted to ground spices and curry powders which need to utilize new cryogenic grinding technology for retention of flavour and medicinal properties of spices. Value addition throws open ample opportunities in seed spice export. India can now boast as the monopoly supplier of spices oils and oleoresins to the world in addition to grounded spices or other value added products. Development of spice parks in seed spice belt has to be taken care of in years to come.

There are many constraints and challenges faced by the processing sector by which the expected growth is not achieved. Some most important constraints are:

1. Erratic and inadequate supply of quality raw materials.
2. Inadequate infrastructure development.
3. Inadequate investment in organized sector.
4. Fragmented research and development and lack of coordination.
5. Lack of adequately trained and dedicated human resource.
6. Lack of accredited quality testing and certification laboratories.
7. Long chain of intermediaries contributing to high costs and inefficiencies.
8. High costs of carrying inventories.
9. Working capital and taxation.
10. Less demand for processed products in the domestic market

5.0. Postharvest Management

5.1. Harvest Maturity and Harvesting

It is hardly possible to produce a good spice product from low quality harvested material (Weston and Barth, 1997). Most of the spices should be harvested when they mature fully (Vinod *et al.*, 2011). The immature and over mature/ripened are not good as far as quality is concerned. The colour turning stage from green to yellow is a good indicator for proper time of harvest (Pruthi, 1992; Singh and Malhotra, 2005). Harvesting stags impacted fatty acid, essential oil content and composition in different seed spices (Rebey *et al.*, 2013, Msaada *et al.*, 2008, Msaada *et al.*, 2009, Saharkhiz and Tarakeme, 2011). Harvesting is one of the major important factors that determine the quality of the produce (Mattheis and Fellman, 1999). Improved method of harvesting with the help of improved/new implements and harvesting machines at proper stage of maturity and proper drying of harvested material offers quality produce as well as overcome the problem of contamination (Vinod *et al.*, 2011) (Fig 4 and 5). Adoption of mechanized threshing preferably by seed spice threshers can further improve the quality and reduce the post harvest losses (Singh and Malhotra, 2005).

Figure 4. Manual Harvesting

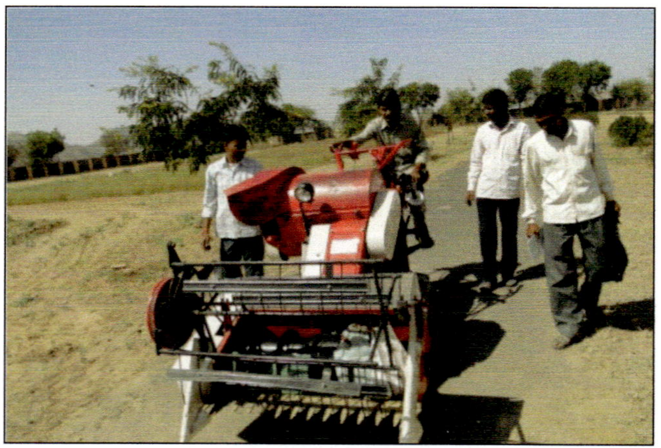

Figure 5. Combine Harvester

5.2 Sorting and Grading

Farm level processing includes major unit operations like cleaning, drying, and grading (Fig 6). This preliminary processing is required to provide a better income to the farmer, fewer losses in transport to urban areas for processing, creation of gainful employment at rural level besides enhancing the processable character of farm produce and to achieve good quality (Vinod *et al.*, 2011; Meena *et al.*, 2013).

Figure 6. Mobile Agro Processing Unit Suggested for Primary Processing of Seed Spices

5.2.1 Sorting

Any agricultural commodity including seed spices should be cleaned properly before processing. The dust and dirt be removed manually or mechanically (Vinod *et al.*, 2011). Small machines of different capacities are available for cleaning the spices and are of cost-effective. Cleaned produce fetch high price in the market always (Singh and Malhotra, 2005; Meena *et al.*, 2013).

5.2.2 Drying

This is the most important factor in the process to ensure good quality spices. Inadequately dried produce will lead to mould growth and spoilage in the storage (Vinod *et al.*, 2011). In addition, the growth of food poisoning bacteria on some spices is a real danger if proper drying is not done (Thankamani *et al.*, 2013). After threshing and cleaning the produce should be dried below 10% moisture level to avoid the pathogenic infections, insect infestations, *etc.* in the storage. Sun drying, solar drying are very much effective in arid areas and/or dry season and artificial drying with the help of electric ovens, or cabinet dryers with controlled temperature are equally effective in humid areas and/ or wet season (Singh and Malhotra, 2005; Meena *et al.*, 2013, Figures 7, 8, 9).

5.2.3 Grading

The Grade designations indicate the quality of spices. Spices can be graded by size, density, colour, shape, and flavour (Thankamani *et al.*, 2013). Machines like gravity separators or de-stoners are available for larger scale production units (Fig 10). Spices can be graded into 3-4 grades as per the quality. Top-quality produce fetches the high price in the market. *Coriander* is classified into different grades based on quality of the produce. In general, coriander has been classified into grades like Badami, Eagle, Scooter, single parrot, double parrot, green medium, green extra, and green special.

Figure 7. Traditional Drying

Figure 8. Solar Tunnel Drier

Figure 9. Seeds of Fenugreek Dried under different Drying Methods

After the harvesting of coriander, it is dried in sunlight; the excess drying leads to brown color and fetches a low price (https://www.indiamart.com/proddetail/eagle-coriander-16481159333.html). These brown colour seeds are called the Badami grade. The finer quality is called green and it is traded at a premium to all other grades. However, the Badami grade has the highest market share, constituting 50% of the total produce. It is also graded in the categories like standard and special

Figure 10. Seed Grader Used for Seed Spices

(http://www.rawther.co.in/images/wholespices/coriander.pdf).

5.3 Packaging

After grading packaging plays a major role in maintaining the quality of seed spices and in increasing shelf life (Siribel *et al.*, 2009; Lal *et al.*, 2015, Figure 11). The packaging requirements depend on: 1) the type of spice, 2) whether it is ground or intact and 3) the humidity of storage. Most intact spices will store adequately in sacks/boxes if the humidity of the air is not too high. Ground spices can also be stored without special packaging if humidity is low but over long periods there is a loss of flavour and risk of contamination and spillage. It is therefore better to store spices in a barrier film such as polypropylene (essential in areas of high

Figure 11. Packing of Seed Spices under different Packaging Materials

humidity) to provide an attractive package, retain spice quality and prevent contamination and losses (Babu *et al.*, 2013; Thankamani *et al.*, 2013; Lal *et al.*, 2015). Heat sealable cellulose film is adequate if polypropylene is not available. Polythene is a poor substitute and should only be used for short term storage as it allows the flavour/ aroma of the spices to escape (Lal *et al.*, 2015).

5.4 Storage

Cleaned, dried seeds must be filled in bags and stored in damp-free aerated storehouses. Drainage should be provided to stack the packed bags to prevent moisture ingress from the floor. Care should be taken to stack the bags 50 to 60 cm, away from the wall. No insecticide should, under any circumstances be used directly on the dried material. Stored material should be subjected to periodic fumigation for which only authorized persons should be engaged. Insects, rodents and other animals should be effectively prevented from getting access to the premises where the material is stored. Stored product should be periodically exposed to the sun. If care is taken in all stages of cultivation, harvesting, post harvest handling, processing, packing, storage and transportation by following sound methods and practices we will be able to prevent contamination in any farm produce and ensure consumer satisfaction. The dried seeds are usually packed into sacks and stored in cool dry room. It should be stored at room temperature (25°-28°C) where the critical moisture level and relative humidity should not be more than 13-19 per cent and 81 per cent respectively. During prolonged storage of spices, free fatty acids are increased by lipolytic action on fixed oil. This value is a good indicator of the period of storage. No insecticide should be used, under any circumstances, be used directly on spices. Stored material should be subjected to periodic fumigation for which only authorized persons should be engaged.

Post harvest management is the key to filling the major economic losses happening in seed spice produce management. Post harvest loss minimization is well addressed by the research going on at the centre for testing the affectivity of modified atmospheric packaging (Anonymous, 2013-14). Controlled Atmosphere Packaging or Modified Atmosphere Packaging (CAP or MAP) is fast emerging as the leading packaging option for preservation of fresh and processed foods which are required to be stored over longer periods and under diverse storage temperatures and conditions (Lal *et al.*, 2015). In the CAP system, the food product is packaged in a High Barrier Film or laminate, following which the atmosphere inside the pack is controlled by various gas flushing and vacuum options (Philips, 1996). After the pack is sealed or vacuum packed, the high barrier film prevents further transmission of gasses in or out of the pack, thus extending the shelf life and ensuring that the product reaches the customer in a fresher condition (Meena *et al.*, 2012). CAP films and pouches are based on Polyamide (Nylon) and EVOH High Barrier polymers, the contents of which are adjusted to meet the shelf life requirements of individual products. They are an ideal cost effective replacement to the expensive tin packing and aluminium foil based laminates, plus they offer the added advantage that the packed product can be visually inspected by the consumer before purchase (Lal and Mehta, 2013). Aluminium pouch packaging is found most effective and reliable method of packing materials for longer and safe storage. It is good moisture resistance stable coefficient of friction to ensure use in high speed machine suitable for shelf display convenient re-closable zip lock size and material according to client requests. These are available in desired quantity and specification of the client in the market (Lal *et al.*, 2015).

5.4.1 Consumer Packed Spices

The spices exported to developed countries are consumed in three main segments viz., industrial, retail, and institutional. The weight of retail packs generally range between 30 g to 200 g. However institutional packs range from 500g to 1.0 kg in weight and are also retained under some brands.

6.0 Domestic Marketing and Export

6.1 Marketing Channel Adopted by the Farmers in Sale of Major Seed Spices in Domestic Marketing System

Channel 1: Producer farmer-Village trader-Wholesaler-Retailer-Consumer.

Channel 2: Producer farmer-wholesaler-Retailer-Consumer

6.2 Exportability

Seed spices contribute more than 45 per cent to national spice acreage and have a lion's share of more than 24 per cent to spice export from the country in volume. In the last three decades seed spice export from India has increased more than 11 and 57 times in terms of quantity and value respectively. Cumin, coriander and fenugreek export rose to 92, 22, and 14 times in quantity and 924, 267, and 236 times in value term respectively. The higher increase in value term indicates a steep increase in prices for seed spices in the world market. Spices produced in India are largely consumed (87%) within the country due to its large population base coupled with higher per capita consumption in the country. With 13% of total production exported to the world market, India caters to half of the world's demand. Still, there is a great opportunity for Indian seeds spices in the world market but tightened SPS measure adopted by key European markets is the big challenge.

Competitiveness of production and stringent quality assurance has taken central place in the seed spices cultivation. Adoption of improved pre and post production technologies by the farmers and traders together with awareness of quality assurances will help

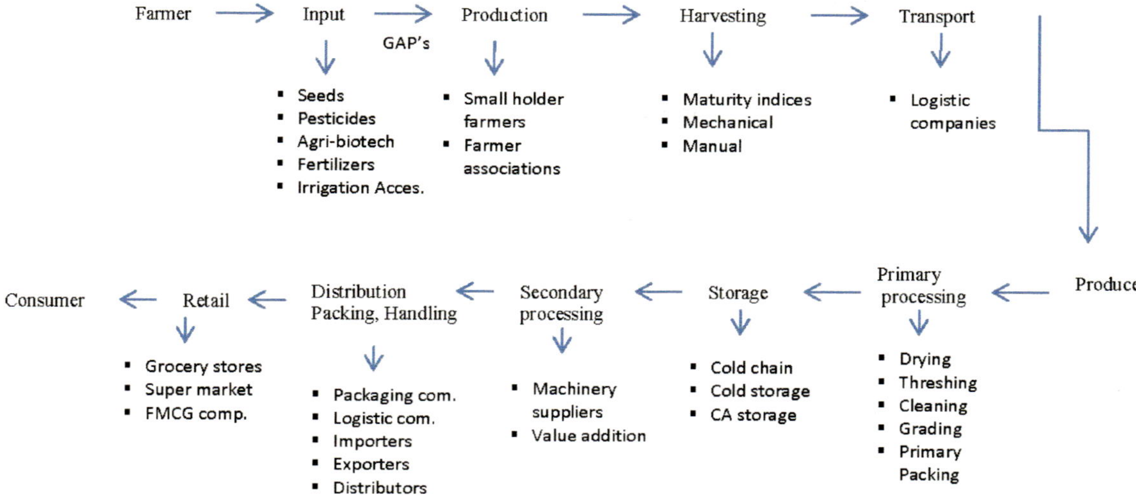

Figure 14. Complete Value Chain from Farm to Plate

increase production of exportable surplus of seed spices and foreign exchange earnings for the country. It is, therefore the complete value chain will assist in quality assurance to cater to the needs of generating exportable surplus of seed spices (Figure 14).

6.3 Export System

The export of seed spices mainly evolves indirect exports through links with agents or export management firms which will eventually lead to the handling of own exports. The other entry options are licensing, contract production (supply), management contracting, and joint ventures.

The marketing system followed for seed spices export are:

Export via independent representative (agent)

Establishment of overseas sales subsidiary by own or in collaboration with foreign party.

The Marketing Channels involved are:

Direct channel - Foreign distributor, foreign retailer, state controlled trading companies and end users.

Indirect channel - Export brokers, export agents, export management organizations and co-operative exporters.

7.0 Conclusion

Seed spices are low volume high value export oriented crops and India is a major producer and exporter of seed spices and their products. India despite being the largest producer of seed spices could not exploit their value addition potential as the majority of export is in the form of whole spices. There is a great potential for increasing the export of Indian spices in the form of value added products. To realize this potential there is a need to enhance the quality as per international standard through post harvest technology and value addition at a different level of post production system. Food losses take place at production, post-harvest and processing stages in the food supply chain. The preliminary processing is required to provide better income to the farmer, fewer losses in transport to urban areas for processing, creation of gainful employment at the rural level to enhance the processable character of farm produce and to achieve good quality. Cleaned produce fetch high prices in the market always. After threshing and cleaning the produce should be dried below 10% moisture level to avoid pathogenic infections, insect infestations, *etc.* in the storage. Spices can be graded by size, density, colour, shape, and flavour. Spices can be graded into 3-4 grades as per the quality. Packaging has played a pivotal role in increasing the price of the produce. Barrier films, vacuumed packs, MAP/CAP are effective and reliable methods of packing for these commodities to retain quality and shelf life. The extracted compound of spices serves as the alternative to whole and ground spices. They are used in various food products like sauces, confectioneries, beverages alternative, pharmaceutical, and perfume industries. Extracted oleoresins have the full aroma, flavour, and pungency of fresh or dried spices. The oleoresins and/or essential oils are encapsulated to keep the full flavour impact of spices over extended shelf life. The seed spices based beverages such as RTS and squashes have been developed and commercialized. Besides, these several other products such as mouth fresheners/chews, biscuits, toffees, dill parag, cookies, and many more products can be prepared for value addition.

8.0 Important References

Aglodia, A. 2014. Final Report NAIP Subproject "Value Chain in Major Seed Spices for Domestic and Export Promotion" Centre for Research on Seed Spices, Jagudan, Gujarat, India, pp.16.

Anonymous, 2013. Annual Report of NRCSS, Ajmer 2013-14. Pp. 34.

Anonymous, 2015. Annual Report of CISH, Lucknow 2015-16. pp. 133.

Baboo, R. and Rana, N.S. 1995. Effect of cutting management, nitrogen and phosphorus on growth and yield of coriander *(Coriandrum sativum). Indian J. Agron.* **40**: 253-55.

Babu, N., Srivastava, S.K. and Agarwal, S. 2013. Traditional storage practices of spices and condiments in Odisha. *Indian J. Tradit. Knowl.* **12**: 518–23.

Bakhshabadi, H., Mirzaei, H., Ghodsvali, A., Jafari, S.M., Ziaiifar, A.M., Farzaneh, V. 2017. The effect of microwave pretreatment on some physicochemical properties and bioactivity of black cumin. *Ind. Crop Prod.* **97**: 1–9.

Choudhary, G.R. and Gupta, O. P. 1991. Effect of weed control, sowing method and nitrogen application on growth and quality of cumin. *Haryana J. Agron.* **5**: 79-82.

Choudhary, S., Pereira, A., Basu, S. and Verma, A.K. 2017. Differential antioxidant composition and potential of some commonly used Indian spices. *J. Agric.***4**: 160-66.

DASD .2016. Directorate of Arecanut and Spices Development, 2016-17. https: //dasd.gov.in/

Douglas, M., Heyes, J. and Smallfield, B. 2005. Herbs, spices and essential oils: Post-harvest operations in developing countries. www.fao.org.

Garcia-Jimenez, N., Perez-Alonso, M.J. and Velasco-Negueruela, A. 2000. Chemical composition of fennel (*Foeniculum vulgare* Mill.) oil from Spain. *J. Essential Oil Res.* **12** : 159-62.

Kachru, R. P. 2006. Agro-processing industries in India-growth, status and prospects. Indian Council of Agricultural Research. New Delhi, India.

Kalra, A., Patra, N. K., Singh, H. P., Mengi, M., Naqvi, A. A. and Kumar, S. 1999. Evaluation of coriander (*Coriandrum sativum*) collection for essential oil. *Indian J. Agric. Sci.* **69**: 657-59.

Khan, F., Khan, T.U., Tajudin and Khan, N. 2016. Fruit processing preservation and development of value added products (squash, jam, and candy) to control wastages of fruits in Gilgit-Baltistan. *J. Agric. Biol. Sci.* **11**: 274-82.

Khare, M.N., Tiwari, S.P. and Sharma Y.K. 2014b. Disease problems in fennel (*Foeniculum vulgare* Mill) and fenugreek (*Trigonella foenum graceum* L.) cultivation and their management for production of quality pathogen free seeds. *Int. J. Seed Spices,* **4**: 11-17.

Khare, M.N., Tiwari, S.P. and Sharma, Y.K. 2014a. Disease problems in the cultivation of I. Cumin (*Cuminum cyminum* L.) II. Caraway (*Carum carvi* L.) and their management leading to the production of high quality pathogen free seed. *Int. J. Seed Spices,* **4**: 1-8.

Kinyuru, J.N., Kahenya, K., Muchui, P.M. and Mungai, H. 2011. Influence of post-harvest handling on the quality of snap bean (*Phaseolus vulgaris* L.). *J. Agri. Food Technol.* **1**: 43–46.

Lal, G. and Maheria, S.P. 2013. Prospects of mechanization in seed spice production and post harvest management. In: Compendium of short course on 'Hi-tech interventions in seed spice production for enhancing productivity, nutraceutical quality and value addition' held on 03-12 October, 2013 at National Research Centre on Seed Spices, Ajmer, Rajasthan Pp: 48-54.

Lal, G. and Mehta, R.S. 2013. Post harvest handling, processing and value addition of seed spices. In: Souvenir of Nat. Seminar on production, productivity and quality of spices. Published by Director, NRCSS, Ajmer and Director, DASD, Calicut, Kerela.

Lal, G.,Mehta, R.S., Sharma, B.K.,Bhati, A. and Barolia, B.M. 2015. Influence of storage period and packaging practices on essential oiland organoleptic quality of fennel (*Foeniculum vulgare* Mill) seed. *Int. J. Seed Spices,* **5**: 74-78.

Malhotra, S.K. 2005. Concept paper -Post harvest management problems of seed spices crops, Presented in Brainstorming Workshop of DBT, Aug, 4-5, 2005 at GBPUA and T, Pantnagar, Uttrakhand

Malhotra, S.K. 2009. Safety dimensions for exportable seed spices production, Proc. Brain Storming Session on Research Needs for Seed Spices, 21–22 March, NRCSS, Ajmer, 126–33.

Malhotra, S.K. 2010. Organic seed spices production. In: Singh, H.P. and Thomas, J. (Eds.). Organic production practices in Horticulture. Westville Publ, New Delhi, pp 83-119.

Malhotra, S.K. 2016. Research and development of seed spices for enhancement of productivity and profitability in the era of climate change. *Indian J. Arecanut, Spices and Medicinal Plants,* **18**: 3-20.

Mazaheri, L., Torbati, M., Azadmard-Damirchibi, S., Savage, G.P. 2019. Effect of roasting and microwave pre-treatments of *Nigella sativa* L. seeds on lipase activity and the quality of the oil. *Food Chem.* **274**: 480–86

Mattheis, J.P. and Fellman, J.K. 1999. Pre-harvest factors influencing flavor of fresh fruit and vegetables. *Postharvest Biol. Technol.* **15**: 227–32.

Meena, M.K., Chetti, M.B. and Naik, M.C. 2012. Impact of vacuum packaging on various seed quality and biochemical parameters of different spice crops. *Int. J. Plant Sci.* **7**: 432-41.

Meena, S. S., S. Balraj, S. Dheraj, J. K. Ranjan and R. D. Meena 2013. Pre and post harvest factors affecting yield and quality of seed spices: A review. *Int. J. Seed Spices* **3**: 1-11.

Moghimi, M., Farzaneh, V. 2018. The effect of ultrasound pre-treatment on some selected physicochemical properties of black cumin (*Nigella Sativa*). *Nutrire* **43**: 1–8

Pagala, B. and Popuuri, A.K. 2013. Preparation and encapsulation of spice oleoresins. *Int. J. Pharmaceutical Res. Devt.* **5**: 56-63.

Parfitt, J., Barthel, M. and Macnaughton, S. 2010. Food waste within food supply chains: quantification and potential for change to 2050. *Philosophical Transactions of the Royal Society B: Biological Sciences* **365**: 3065-81.

Parthasarathy, V.A. and Rajeev, P. 2006. Major spices production and processing (Eds.). Indian Institute of Spices Research, Kozhikode, India.

Philips, C. 1996. Review: modified atmosphere packaging and its effects on the microbiological quality and safety of produce. *Int. J. Food Sci. Technol.* **31**: 463-79.

Pruthi J.S. 1992. Post-harvest technology of spices: pre-treatments, curing, cleaning, grading and packing. J. Spices and Aromatic Crops, *1*, 1-29.

Pruthi. J.S. 1993. Major Spices of India, Crop Management and Post Harvest Technology, ICAR, New Delhi.

Rathore, S.S., Saxena, S.N., Singh, B. 2013. Potential health benefits of major seed spices. *Int. J. Seed Spices* **3**: 1–12.

Saxena, S.N., Sharma, Y.K., Rathore, S.S., Singh, K.K., Barnwal, P., Saxena, R., Upadhyaya, P., Anwer, M.M. 2015. Effect of cryogenic grinding on volatile oil, oleoresin content and anti-oxidant properties of coriander (*Coriandrum sativum* L.) genotypes. *J. Food Sci. Technol.* **52**: 568–573.

Sharma, L.K., Agarwal, D., Meena, S.K., Rathore, S.S. and Saxena, S.N. 2015. Effect of cryogenic grinding on oil yield, phenolics and antioxidant properties of ajwain (*Trachyspermum ammi* L.). *Int. J. Seed Spices* **5**: 82–85.

Sharma, L.K., Agarwal, D., Rathore, Y., Sharma, S.S. and Saxena, S.N. 2014. Cryogenic grinding technology enhances volatile oil, oleoresin and antioxidant activity of cumin (*Cuminum cyminum* L.) genotypes. *Int. J. Seed Spices* **4**: 68–72.

Sharma, R.N. and Israel, S. 1991. Effect of date of sowing and level of nitrogen and phosphorus on growth and seed yield of coriander (*Coriandrum sativum* 1.). *Indian J. Agron.* **36**: 180-184.

Singh, R. and Malhotra, S.K. 2005. Harvesting and maturity indices in seed spices. Chapter in Ed. Malhotra, S.K. 2005. A course manual for winter school on Advances in Seed Spices Production. NRCSS, Ajmer, p1-287.

Singh, S.D. and Rao, J.S. 1994. Yield – water -nitrogen response analysis in cumin. *Ann. Arid Zone* **33**: 29-34.

Siribel, M.K., Elemery, M.I. and Ugwu, F. M. 2009. Effect of storage duration on quality parameters in *Carum carvi. Revista Brasileira de Sementes* **34**: 63-64.

Thankamani, C.K., Chempakam, B., Jayashree, E., Agalodia, A.V., Chitra, R., Lal, G., Singh, D., Pethe, U.B., Giridhar, K and Babu, K.N. 2013. Post harvest handling and value addition in spices, In: Conference: SYMSAC VII Post-Harvest Processing of Spices and Fruit Crops, Karnataka. Pp. 44-56.

Vinod, K., Bhargev, Wanjavi, O.D. and Patil, R.T. 2011. Quality improvement in seed spices through post harvest management in proceedings of national workshop on spices and aromatic plants. Sokaki, Z.S., Panikar, G.N., Rathore, B.S. and Kunhar, S.R. (Eds.), Jodhpur, pp. 275-78.

Weston, L.A. and Barth, M.M. 1997. Pre harvest factors affecting post harvest quality of vegetables. *HortScience* **32**: 812-15.

Yildirim, B. and Gok, N. 2012. Effect of sowing date and varieties on essential oil ratio and essential oil components of coriander (*Coriandrum sativum* L.) in van ecological condition. *J. Anim. Vet. Adv.* 11: 1925-29.

Index

B

C

D

E

F

G

W

Z